Women's Empowerment Within the Tourism Industry

Gül Erkol Bayram
Sinop University, Turkey

Syed Haider Ali Shah
Bahria University, Pakistan

Muhammad Nawaz Tunio
The University of Sufism and Modern Sciences, Bhitshah, Pakistan

IGI Global
PUBLISHER of TIMELY KNOWLEDGE

A volume in the Advances in Hospitality, Tourism, and the Services Industry (AHTSI) Book Series

Published in the United States of America by
IGI Global
Business Science Reference (an imprint of IGI Global)
701 E. Chocolate Avenue
Hershey PA, USA 17033
Tel: 717-533-8845
Fax: 717-533-8661
E-mail: cust@igi-global.com
Web site: http://www.igi-global.com

Library of Congress Cataloging-in-Publication Data

Names: Erkol Bayram, Gül, 1987- editor. | Shah, Syed Haider Ali, 1985- editor. | Tunio, Muhammad Nawaz, 1984- editor.
Title: Women's empowerment within the tourism industry / edited by Gül Erkol Bayram, Syed Haider Ali Shah, Muhammad Nawaz Tunio.
Description: Hershey, PA : Business Science Reference, 2023. | Includes bibliographical references and index. | Summary: "Organized into four parts, the edited book provides modern and cutting-edge perspectives on important topics like linkages between women, travel, and tourism. As a traveler, worker, and entrepreneur in the emerging era. Part I introduces core concepts, and theories on women tourists with the current issues Part II explores the inclinations of women employees and their challenges, experiences, and behaviors with special reference to tourism. In Part III, on female entrepreneurs' tourism, the role of social media in tourism activities, marketing and management perspectives with special reference to entrepreneurs, the issues, threats, policies, and practices linked with spiritual tourism are discussed. All parts also aim to explore future challenges, and opportunities for women's participation and to propose researchbased solutions. Overall, the book will be a suitable means of getting insight into the minds of the diverse, experimental, and hard-working women in the tourism industry. This book will fill the gap of research on women's participation in tourism with different roles. As an edited book, it will add new research and knowledge base with high-quality contributions from researchers and practitioners interested in tourism management, hospitality management, business studies regional development, and destination management. The chapters bring light to perspectives from different countries, cultures, backgrounds, and religions, and utilize different methods, approaches, and styles of presentation. Women Participation in Tourism: Tourist, Employee, And Entrepreneur Perceptives Perspectives will be of interest to academics and graduate students from a range of disciplines, including tourism, leisure studies, sociology, cultural geography, anthropology, feminist and gender studies, business, economics, and management; as well as professionals working in the tourism industry, particularly those with an interest in niche markets and segmentation"-- Provided by publisher.
Identifiers: LCCN 2022062297 (print) | LCCN 2022062298 (ebook) | ISBN 9781668484173 (hardcover) | ISBN 9781668484180 (paperback) | ISBN 9781668484197 (ebook)
Subjects: LCSH: Women in tourism. | Tourism--Social aspects.
Classification: LCC G156.5.W66 W67 2023 (print) | LCC G156.5.W66 (ebook) | DDC 338.4/791082--dc23/eng20230324
LC record available at https://lccn.loc.gov/2022062297
LC ebook record available at https://lccn.loc.gov/2022062298

This book is published in the IGI Global book series Advances in Hospitality, Tourism, and the Services Industry (AHTSI) (ISSN: 2475-6547; eISSN: 2475-6555)
British Cataloguing in Publication Data
A Cataloguing in Publication record for this book is available from the British Library.

For electronic access to this publication, please contact: eresources@igi-global.com.

Advances in Hospitality, Tourism, and the Services Industry (AHTSI) Book Series

Maximiliano Korstanje
University of Palermo, Argentina

ISSN:2475-6547
EISSN:2475-6555

Mission

Globally, the hospitality, travel, tourism, and services industries generate a significant percentage of revenue and represent a large portion of the business world. Even in tough economic times, these industries thrive as individuals continue to spend on leisure and recreation activities as well as services.

The Advances in Hospitality, Tourism, and the Services Industry (AHTSI) book series offers diverse publications relating to the management, promotion, and profitability of the leisure, recreation, and services industries. Highlighting current research pertaining to various topics within the realm of hospitality, travel, tourism, and services management, the titles found within the AHTSI book series are pertinent to the research and professional needs of managers, business practitioners, researchers, and upper-level students studying in the field.

Coverage

- Health and Wellness Tourism
- Destination Marketing and Management
- Tourism and the Environment
- Food and Beverage Management
- Service Training
- International Tourism
- Service Management
- Travel Agency Management
- Customer Service Issues
- Leisure & Business Travel

IGI Global is currently accepting manuscripts for publication within this series. To submit a proposal for a volume in this series, please contact our Acquisition Editors at Acquisitions@igi-global.com or visit: http://www.igi-global.com/publish/.

The Advances in Hospitality, Tourism, and the Services Industry (AHTSI) Book Series (ISSN 2475-6547) is published by IGI Global, 701 E. Chocolate Avenue, Hershey, PA 17033-1240, USA, www.igi-global.com. This series is composed of titles available for purchase individually; each title is edited to be contextually exclusive from any other title within the series. For pricing and ordering information please visit http://www.igi-global.com/book-series/advances-hospitality-tourism-services-industry/121014. Postmaster: Send all address changes to above address.

Titles in this Series

For a list of additional titles in this series, please visit: www.igi-global.com/book-series

Prospects and Challenges of Global Pilgrimage Tourism and Hospitality
S.K. Gupta (Centre for Mountain Tourism & Hospitality Management, HNB Garhwal University, India) Lilibeth C. Aragon (College of International Tourism and Hospitality Management, Lyceum of the Philippines University, Philippines) Pankaj Kumar (Department of Tourism & Hospitality Management, Mizoram University, India) Madhurima (Department of Environmental Science, Mizoram University, India) and Rajesh Ramasamy (Department of Tourism & Hospitality Management, Mizoram University, India)
Business Science Reference • copyright 2023 • 310pp • H/C (ISBN: 9781668448175) • US $250.00 (our price)

Cases on Traveler Preferences, Attitudes, and Behaviors Impact in the Hospitality Industry
Giuseppe Catenazzo (AUS American Institute of Applied Sciences, Switzerland & ICN Business School, France)
Business Science Reference • copyright 2023 • 335pp • H/C (ISBN: 9781668469194) • US $215.00 (our price)

Handbook of Research on Innovation, Differentiation, and New Technologies in Tourism, Hotels, and Food Service
Gonçalo Poeta Fernandes (CITUR, Polytechnic Institute of Guarda, Portugal & CICS, Universidade NOVA de Lisboa, Portugal) and António Silva Melo (CiTUR, Polytechnic Institute of Porto, Portugal & CIDTFF, University of Aveiro, Portugal)
Business Science Reference • copyright 2023 • 400pp • H/C (ISBN: 9781668469859) • US $295.00 (our price)

Exploring Niche Tourism Business Models, Marketing, and Consumer Experience
Maria Antónia Rodrigues (CEOS.PP, ISCAP, Polytechnic of Porto, Portugal) and Maria Amélia Carvalho (ISCAP, Polytechnic of Porto, Portugal)
Business Science Reference • copyright 2023 • 300pp • H/C (ISBN: 9781668472422) • US $250.00 (our price)

Global Perspectives on Human Rights and the Impact of Tourism Consumption in the 21st Century
Maximiliano E Korstanje (University of Palermo, Argentina) and Vanessa G.B. Gowreesunkar (Anant National University, India)
Business Science Reference • copyright 2023 • 263pp • H/C (ISBN: 9781668487266) • US $240.00 (our price)

Dark Gastronomy in Times of Tribulation
Demet Genceli (Istanbul Kent University, Turkey)
Business Science Reference • copyright 2023 • 283pp • H/C (ISBN: 9781668465059) • US $205.00 (our price)

Inclusive Community Development Through Tourism and Hospitality Practices
Vipin Nadda (University of Sunderland in London, UK) Faithfull Gonzo (University of Sunderland in London, UK) Ravinder Batta (Shimla University, India) and Amit Sharma (Shimla University, India)

IGI Global
PUBLISHER of TIMELY KNOWLEDGE

701 East Chocolate Avenue, Hershey, PA 17033, USA
Tel: 717-533-8845 x100 • Fax: 717-533-8661
E-Mail: cust@igi-global.com • www.igi-global.com

Table of Contents

Detailed Table of Contents

Women are largely invisible in the mainstream of economic development. Therefore, the concept of sustainability to a great extent depends upon providing equal distribution of not only resources, but also the equality should be created between the various sections of society where women are vulnerable. Women's empowerment should not be associated with increasing their responsibility as a whole, but it is about respecting their role. To achieve the goals of sustainable development, the concept of women entrepreneurship gains prominence where men have always been dominating the business situations. Ventures by women not only provide them financial support but also makes them empowered. This chapter, by using qualitative research approaches of content analysis, literature review, and discourse analysis, attempts to find challenges of women entrepreneurship and suggest various policy implications to deal with these challenges. The major findings are women entrepreneurship can be a major source of women empowerment backed by a focus on entrepreneurial education.

Women's empowerment has become a crucial component of sustainable tourism development in tourist destinations. Meanwhile, achieving gender equality and empowering women constitutes the fifth goal of the United Nations Sustainable Development Agenda (UN-SDGs). Since 2007, UNWTO has been working—through its Ethics, Culture, and Social Responsibility Department—in partnership with UN Women and a range of external partners across the globe, to bring gender issues to the forefront of the tourism sector, promoting gender equality and women's empowerment and encouraging member states

Despite major advancements in recent decades, gender inequalities in the labor market still exist worldwide, and efforts to achieve gender equality appear to be in a standstill. Entrepreneurship provides important opportunities for women to stand on their feet and empower them economically and socially. Training and fostering entrepreneurial traits in people are two important variables in the growth of entrepreneurship. Aviation is one of the industries in that women are underrepresented. This study explores the results of an applied project based on entrepreneurship education to foster the empowerment of young female aviators. The project highlights female vocational high school students' entrepreneurship skills and future potential. The results reveal that female students may cultivate positive attitudes by being aware of their competencies and acquiring 21st-century abilities, including leadership, entrepreneurship, creative thinking, teamwork, communication, and persuasion.

Chapter 6
The Role of Diversity Management in Women's Empowerment in the Tourism Sector: The Role of Turkey 77
Melike Sak, Selcuk University, Turkey
Gül Erkol Bayram, Sinop University, Turkey
Ali Turan Bayram, Sinop University, Turkey

This study aimed to develop suggestions for removing the obstacles and difficulties in front of the career development of women in the tourism sector in Turkey. In addition to this situation, suggestions will be made to remove the gender discrimination faced by women in working life and the obstacles in front of their career development. In order for women to be empowered in working life, the opportunities they need should be provided. It is thought that diversity management can create an opportunity for this to happen. It is thought that diversity management can have a significant impact on the implementation of practices that can reveal the talents and potentials of women so that they have equal rights with men. Thus, many obstacles in front of women can be removed by using diversity management. Strategies have been determined to eliminate the career barriers of women and suggestions and strategies have been tried to be developed to overcome the obstacles in front of women working in the sector.

Chapter 7
Tourism as a Means of Empowering Women in Sustainable Development: The Case of Women's Cooperatives........ 91
Hatice Sari Gök, Isparta University of Applied Sciences, Turkey

In tourism, women's cooperatives support women's empowerment in the society by providing income with employment, employment and education opportunities, and their effects on eliminating economic and social injustice. With the increase in interest in natural and cultural heritage, local food and handicrafts, the development of healthy life awareness and the increase in education level, the interest in authentic and local cultures is increasing day by day. From this point of view, the similarities between sustainability, sustainable development, women's empowerment, women's cooperatives in tourism and sustainable development goals, and women's cooperatives were examined in this study. It is thought that the study will contribute to the literature and tourism sector.

Sustainable development is the practise of ensuring that current needs are met without jeopardising future generations' capacity to do the same. The goal of sustainable development requires people all over the world to work together and think carefully about how their actions will affect the planet and the next generation. One sector with the potential to advance sustainable development is the tourism industry. China's hospitality industry has grown rapidly, which could lead to sustainable development and tourism and empowers sustainable development in China's hospitality industry, according to this research chapter. Tourism can help achieve sustainability and local autonomy, two goals that have garnered attention recently. China must encourage responsible tourism practises that are kind to the local environment and culture while also boosting the economy of the host community. This research emphasises tourism's potential to empower China's sustainable development while acknowledging the need to continue efforts to distribute tourism's benefits fairly and sustainably.

This study identifies the facets of women empowerment which could be harnessed to bring paradigm shift in tourism industry and simultaneously advance gender parity in the areas with rich tourism resorts in Pakistan. The data were collected through survey questionnaire from women aged 18+ living five regions of Pakistan which are rich in cultural heritage and tourist attractions. The findings show that psychological and economic empowerment are occupying central stage in enhancing women's developmental participation in tourism development in Pakistan. The political and social aspects of women empowerment also play a crucial role in stirring tourism activities. Most importantly, empirical evidence demonstrates that cultural traditionality (CT) positively moderates the impact of women empowerment on the development of tourism in a way that the participatory role of empowered women higher in CT is significantly magnified towards the creation of sustainable tourist environment which promotes local culture and brings benefits to local communities.

Women always face issues while travelling alone or in the form of a group. However, these issues became more devastating for Muslim women due to religious, safety, personal, and familial issues. Therefore, there is a need for a thorough concentration of motivation, perception and hurdles of Muslim solo or group female tourists. In fact, there are few studies that tried to determine the motivation behind the preference for solo travelling. However, there are few studies that may highlight issues, problems, and threats that may be faced by Islamic women while travelling. Therefore, this chapter has been written purposively to highlight the obstacles that may halt females from travelling. The philosophy of research used for writing this study is epistemology, and the research strategy is archival as the chapter has been formulated by analyzing available literature.

The aim of this study is to reveal how influencer communication takes place during collaborations with travel agencies and to evaluate how they communicate with their followers in this context. The study has a descriptive nature and covers the collaborations of "Korea Tourism Organization & Meryem Can," "Visit Dubai & Meryem Can," "Visit Dubai & Gülnihal Eren (Cocobolinho)," and "Costa Cruises & Gülnihal Eren (Cocobolinho)" as case studies. Within this framework, collaborations with different travel agencies are examined through the relevant content on the official Instagram accounts of influencers. At the same time, this compilation study shows that influencers actively participate in social networks during the experiential process of travel agencies. Additionally, the preference for female influencers who collaborate in the tourism industry is considered as another feature of the study.

Travelling is essential for maintaining one's happiness and mental wellness. Tourism is a job providing industry and huge contributions to economic growth, and many studies have been undertaken to determine who is attracted to tourism. Surprisingly, women take a major stand in the tourism industry. The purpose of this study is to examine female solo travelers' risk assessment and their mitigation strategies by investigating online travel blogs. Therefore, the study aims at assessment of risk such as physical risk, destination-specific risk, and financial risk faced by solo women travellers by considering and analyzing online women travellers' blog narratives. These narratives from online travel blogs pertaining to Indian solo female travellers are considered in this study. Hence, this assessment gives an opportunity to know

as man's and woman's jobs, and services, such as cleaning, laundry, making up bed, guest greeting, and cooking similar to domestic chores, are performed by women. These challenges have a direct impact on operational efficiency and functioning, along with the status of women in tourism employment, is important for improving female employment in tourism. A holistic analysis was made in this chapter based on national and international literature on the challenges female employees face in the tourism industry.

Women tend to be disadvantaged at work due to the various problems based on gender discrimination. Women's employment tends to be segregated both vertically and horizontally and across a wide variety of sectors. Women face gender discrimination in a wide variety of sub-sectors of tourism ranging from hospitality to tour guiding. This study explores gender discrimination in the tourist guiding profession. By adopting a dyadic perspective, this study investigates the difficulties experienced by both female and male tour guides. A convenience sampling was used to collect data from 15 female and 15 male tour guides. In collecting data, semi-structured interviews were used. The data collected from the tour guides were analysed qualitatively. By comparing the difficulties experienced by the male and female tour guides, the extent and the nature of gender discrimination have been identified. Based on the outcomes of the study regarding the extent and nature of discrimination, policy guidelines are offered for the practitioners and the stakeholders.

In this study, the number and proportion of women in the boards of directors of publicly traded enterprises in the tourism sector at Borsa Istanbul in Turkey, the CSR, and sustainability performances of these enterprises were revealed by using a case study pattern from qualitative research methods. Within the framework of the document analysis, the number of female executives serving in the boards of directors of the enterprises operating in Borsa İstanbul and their ratio to the total number of board members will be recorded according to the year-end activity reports and websites announced on the public disclosure platform. The number and proportion of women on boards of directors will be the first unit of analysis to be used in the case study. MAXQDA 2020 qualitative research analysis software will be used in the study.

Career anchors are essential for the success of business life. Gender may also be a critical factor that determines the way of career orientation as well as career anchors. At this point, we assume that focusing on the career anchors of female tourist guide candidates will contribute to understanding the nature of career decisions. From this point of view, this study aims to reveal the career anchors of female tourist guide candidates studying at university. The data was gathered from female tourist guidance program students via a questionnaire. Results of the study revealed that female tourist guides perceive technical career anchors most, while the security anchor is perceived as least. Additionally, their career anchor perceptions significantly differ due to their career goals. Participants who want to be tourist guides in the future perceive higher autonomy as career anchors and service anchors, which are also inherent in the tourist guidance profession.

Tourism could be spending time away from home in search of leisure, relaxation, and pleasure while utilizing the commercial provision of services. The capacity of tourism to improve livelihoods by reducing poverty, guarantee or enhance environmental sustainability, and promote gender equality and women's empowerment are some of its strongest points. "Women's empowerment" can mean a variety of things, such as respecting women's opinion, making an effort to find them, and improving women's status through training, awareness, and education. They may have the chance to redefine gender norms and other similar roles, giving them more flexibility to pursue their objectives. The main aim of the chapter is to identify the major challenges faced by women in the tourism sector, the ways to overcome constraints, gender stereotypes, and the social stigmas while they juggle between multiple responsibilities. With this chapter, the authors aim to gain more knowledge more specifically on how women can be empowered in the tourism and hospitality sector.

Female entrepreneurship is perceived as the new beam of light for the growth of the world economy. The phenomenon is especially important for the developing world. However, female entrepreneurs are always associated with challenges and hurdles and similar is valid even for the developed world. However, due to the segregation of gender roles, female entrepreneurship is especially difficult in the Islamic world and also in Pakistan. Several studies highlighted the importance of the phenomenon and also place light on hurdles that may impair female entrepreneurs to cultivate their own businesses in the tourism industry. However, rear work has been conducted on female entrepreneurship in the tourism industry with reference to Pakistan. Thus, the importance of this study is massive for a diverse set of stakeholders. Therefore, this chapter has been written by linking the hurdles, issues, and challenges pertaining to female entrepreneurship with female entrepreneurship in the tourism industry of Pakistan.

Sustainable development along with social entrepreneurship is providing a new approach to the tourism industry. Tourism social entrepreneurship is all about how to take beneficial steps that reduce the negative effects in the tourism industry. It is the responsibility of the entrepreneurs to actively play their role in maintaining sustainability and taking advanced, innovative, and bold steps. It depends upon them how they fulfill the current needs without sacrificing future needs, as this is a key concept of sustainable development goals. Corporate socially responsible firms are sustainable firms. Those industries which are not applying these strategies are now suffering because there is a limited understanding of the fusion of SDGs and tourism social entrepreneurship. How do they apply the SDGs strategies as a part of their strategic planning industry? This chapter contributes to the emerging role of tourism social entrepreneurs and also tells how they achieve sustainability with a combination of 17 strategies and fusions of SDGs.

'Halal' travelers are a rising market niche for tourism, and they look for locations that accommodate their dietary, sartorial, and ritual needs. "Halal" implies lawful in accordance with Islamic teachings. Gambling, drinking, eating pork, and nudity are prohibited. By 2022, it is anticipated that Muslim tourists would spend $220 billion, up from 121 million in 2016, and they will number 156 million. The chapter is intended to present the concept of the role of women in halal tourism and highlight the opportunity to increase entrepreneurship in developing countries through a robust model. The majority of the developing or underdeveloped countries are Islamic, with extraordinary tourism destinations. However, the tourism business in these countries is still monopolized by males. The discussed topic promotes SDG 5 (Gender Equality) and SDG 8 (Economic Growth). Encouraging women's entrepreneurship in Halal tourism shall attract an inflow of finance into the local economy and reduce the unemployment level while maintaining the cultural limitations and securing the belief systems.

The term "voluntourism" refers to vacations in which participants give their time to charitable organizations. Humanitarian crises, either natural causes like natural disasters or human causes such as armed conflicts and civil wars, caused millions of people to be displaced. According to the United Nations Refugee Agency (UNHCR), global forced displacement has reached 103 million, with 53.2 million of those internally displaced and 32.5 million refugees. Sustainable Development Goals (SDGs) call on the global community to "leave nobody behind," which has resulted in the formation of dozens of new organizations and motivated thousands of individuals to participate in volunteer travel. In addition, non-governmental organizations have been actively looking for volunteers to aid displaced people. Women hailing from a wide range of nations have been taking on the mantle of a refugee and making the journey to a number of conflict zones in order to establish organizations and/or lend a hand in the process of resettling refugees.

Foreword

Women's Empowerment Within the Tourism Industry is not a book in the conventional sense. The book encourages us to reflect, to reconcile our actions with the spiritual and moral obligations we have as members of society. To do this, we need to change our attitudes drastically. The authors have endeavored to provide facts and figures, interspersed with real-life stories and experiences, to help us understand the need for this change in attitude, and have examined this from the perspective of the tourism sector.

The importance of education was emphasized in the case of women, as it is for any individual. Predicting how a change at the micro level can lead to results at the macro level has been able to provide actionable data. As a sector, tourism has a structure where women's care and employment is more prominent and should be more so in terms of the number of employees. In previous years, women's employment was positioned in the lower cadre, in jobs such as cleaning and kitchen, but over time these have been replaced by more senior and managerial positions. This change has also opened up the potential for women's empowerment in the sector. This potential has provided the basis for and contributed to the development of women's financial power, strengthening the link between employment and education, supporting work-life balance and eliminating discrimination between men and women in the workplace. Female entrepreneurial experience is characterized by a networking approach, where great importance is attributed to the relational dimension. It also emphasizes the tendency of entrepreneurs to integrate the professional- working dimension into private and family-based, and from this point of view, there are profound differences in the attitudes of entrepreneurs, more orientated to the separation between the various spheres of their own life. However, in women's entrepreneurship studies, the analysis focused on the network's women entrepreneurs create, their composition, the identity and characteristics of the subjects involved the role of the latter in relation to the management of the company, and their ability to influence their performance. Tourism offers more employment and career opportunities for women than other sectors in terms of the ratio of female to male employment. At the same time, the strengthening of women's employment will not only enable the financial and cultural empowerment of women to carry out quality work in the sector, but will also enable companies to develop a working model based on the employment of both men and women.

This work deals with women, who have a very important place in the existence of tourism, their current situation within the tourism system, their role within the social structure and the obstacles they face. This book, which emphasizes the importance of the concept of women in the tourism sector, which manages the most powerful economies in the world, has been prepared as a guide for students studying at associate, undergraduate and postgraduate level, relevant lecturers and representatives of the sector. This work, prepared under the guidance of academics from different countries of the world who believe in the rightful power of women in the tourism system, deals with the empowerment of women in tourism in all aspects.

I kindly invite academics, scientists, researchers, students, PhD students. academics and postdoctoral chapters that are fundamental to this important book or activity of students.

As I strongly believe that this book will provide valuable information on the one hand, their scientific background. It will provide explanations on the women empowerment issues, current situations and future directions. Finally, I would like congratulate the authors for their conscientiousness and patience at all stages of the preparation and development of the book chapters, right up to the moment of publication. It is also the perfect moment to congratulate the editors for choosing to be part of such an important and comprehensive study. Finally, I would like to express my deepest gratitude to IGI Global, for great work. It has given me the tremendous opportunity to write and analyses this book in depth.

Marco Valeri
Niccolo Cusano University, Italy

Preface

Welcome to *Women's Empowerment Within the Tourism Industry*, an edited reference book that explores the multifaceted roles of women in tourism as tourists, employees, and entrepreneurs. In this book, we aim to provide a conversant and comprehensive overview of the themes and concepts surrounding women's participation in the tourism industry, offering interdisciplinary insights from leading international researchers and academicians.

While women have been an integral part of the tourism industry for many years, their contributions and experiences have often been overlooked or understudied. This book seeks to address this gap by examining the main role of women in the tourism industry, shedding light on the nature of their work and the ways in which tourism creates tension between the attitudes and behaviors of tourists and the beliefs and conduct of local women.

Moreover, this book delves into the behavior, motivations, experiences, and needs of women as tourists and travelers, drawing from both historical and contemporary eras. Little research has explored key issues, experiences, and opportunities in the context of women's travel and employment as entrepreneurs and employees. Our edited book aims to fill this gap, exploring the discourses and discussions about all dimensions of women, travel, and tourism. This edited book provides modern and cutting-edge perspectives on important topics that examine the linkages between women, travel, and tourism in the emerging era. This book introduces core concepts and theories on women as tourists, addressing current issues and trends. It also explores the inclinations of women employees in the tourism industry, highlighting their challenges, experiences, and behaviors. We focus on female entrepreneurs in tourism, exploring the role of social media, marketing and management perspectives, and the issues, threats, policies, and practices associated with spiritual tourism. Each part also aims to explore future challenges and opportunities for women's participation, proposing research-based solutions.

Chapter 1 provides an overview of the women's empowerment framework commonly used in tourism research. Using qualitative research approaches of content analysis, literature review and discourse analysis, the chapter attempts to identify challenges to women's entrepreneurship and suggests various policy implications to address these challenges. The main findings are that women's entrepreneurship can be an important source of women's empowerment, supported by a focus on entrepreneurial education.

Chapter 2 focuses on women's empowerment in the context of sustainable tourism development. This paper also aims to present gender equality and women's empowerment in the light of the fifth goal of the United Nations Sustainable Development Agenda (UN-SDGs). Since 2007, UNWTO, through its Department of Ethics, Culture and Social Responsibility, in collaboration with UN Women and a number of external partners around the world, has been working to bring gender issues to the forefront of the tourism sector, to promote gender equality and women's empowerment, and to encourage Member States to mainstream gender issues in their respective tourism policies. The present study aims to examine the impact of women's empowerment on tourism development in India.

Chapter 3 identifies women's entrepreneurship issues on the socio-economic impact in the tourism industry. This chapter examines the changing and evolving scenario of women's entrepreneurship and identifies the challenges that have been overcome and those that continue to limit the growth of women's entrepreneurship. It attempts to identify the challenges and enablers that have changed the business dynamics for women. The chapter also analyses how women's entrepreneurship has changed over time and digs deep into the existing literature to understand whether the business environment has become more conducive for women over generations. The authors use an inductive approach to achieve their objectives.

Chapter 4 examines sustainability through women's empowerment in the tourist attractions of the Kashmir Valley. Based on the relevant issues, conclusions are drawn and solutions are offered. Government-adopted practices, policies and strategies are described and their applicability for improving and promoting sustainable tourism at destinations is explained. Most of the natural assets produced by women in the Kashmir Valley require careful monitoring and management. Strong policies, strategic frameworks, environmentally friendly tourism practices and adequate facilities for women's empowerment are recognised as essential in this context.

Chapter 5 examines the aviation industry in the context of gender inequalities. This study explores the results of an applied project based on entrepreneurship education to promote the empowerment of young female pilots. The project highlights the entrepreneurial skills and future potential of female vocational high school students. The results show that female students can cultivate positive attitudes by being aware of their competencies and acquiring 21st century skills, including leadership, entrepreneurship, creative thinking, teamwork, communication and persuasion.

The aim of Chapter 6 was to develop suggestions for eliminating the obstacles and difficulties facing the career development of women in the tourism sector in Turkey. In addition to this situation, suggestions will be made to remove the gender discrimination that women face in working life and the obstacles to their career development. In order to empower women in working life, the necessary opportunities should be provided. It is believed that diversity management can create an opportunity for this to happen. It is believed that diversity management can have a significant impact on the implementation of practices that can reveal the talents and potential of women so that they can be on an equal footing with men. In this way, many of the barriers that women face can be removed through the use of diversity management. Strategies have been identified to remove barriers to women's careers and attempts have been made to develop proposals and strategies to overcome the obstacles faced by women working in the sector.

Chapter 7 explains the role of women's cooperatives in providing income, employment and educational opportunities, and their impact on eliminating economic and social injustice. This paper also explains the importance of women in natural and cultural heritage, local food and handicrafts, the development of healthy lifestyles and raising education levels, interest in authentic and local cultures. According to this paper, there are similarities between sustainability, sustainable development, women's empowerment, women's cooperatives in tourism and sustainable development goals and women's cooperatives. It is believed that the study will contribute to literature and the tourism sector.

Chapter 8 aims to provide a summary of the response of the family tourism industry to the COVID-19 pandemic in Europe. Family tourism enterprises are usually run by a single family or a small group of families. These businesses offer exceptional, tailor-made solutions to their customers. The tourism and hospitality sector in Europe has been hit hard by the COVID-19 pandemic. As a result, family-friendly tourism businesses have had to get creative in order to survive. The importance of keeping customers safe and happy, as well as embracing digitalisation and sustainability, has never been greater. Family-friendly businesses in the tourism industry have thrived despite the pandemic due to their accommodating cancellation policies and personal approach to the industry. Although the road to recovery is still unclear, family tourism businesses in Europe are developing innovative strategies to remain competitive and build resilience.

Chapter 9 examines the impact of social media on women running family businesses in the travel industry. Facebook, Instagram and Twitter are essential resources for women business owners in the travel industry, according to research. When women use social media to promote their businesses, they gain visibility, customers and sales. The study also highlights the importance of social media-based online communities in breaking down cultural barriers and promoting women travel business owners. Social media users are more likely to be successful entrepreneurs, which has implications for women tourism entrepreneurs. It's important to recognise and support women tourism entrepreneurs who are driving growth, economic stability and social well-being. This study examines how women tourism entrepreneurs use social media and how it affects their businesses. This information can help other women entrepreneurs use social media effectively and help policy makers and industry stakeholders support women-owned tourism businesses.

According to Chapter 10, China's hospitality industry has grown rapidly, which could lead to sustainable development, and tourism can support sustainable development in China's hospitality industry, according to this research chapter. Tourism can help achieve sustainability and local autonomy, two goals that have recently received attention. China needs to promote responsible tourism practices that protect the local environment and culture, while boosting the economy of the host community. This research highlights the potential of tourism to support China's sustainable development, while recognising the need to continue efforts to distribute the benefits of tourism in a fair and sustainable way.

Chapter 11 identifies the facets of women's empowerment that could be harnessed to bring about a paradigm shift in the tourism industry while promoting gender parity in the areas with rich tourist destinations in Pakistan. The data was collected through a questionnaire from women aged 18 years and above living in five regions of Pakistan that are rich in cultural heritage and tourist attractions. The findings show that psychological and economic empowerment are central to enhancing women's participation in tourism development in Pakistan. The political and social aspects of women's empowerment also play a crucial role in stimulating tourism activities. Most importantly, empirical evidence shows that cultural traditionality (CT) positively moderates the impact of women's empowerment on tourism development in such a way that the participatory role of empowered women higher in CT is significantly magnified towards the creation of a sustainable tourism environment that promotes local culture and brings benefits to local communities.

Chapter 12 provides a detailed description of all the aspects that can help the reader understand the concept of Halal Holidays, in order to help the reader understand the relationship between the perceived value of Halal and the intention to visit the destination.

Chapter 13 highlights modern women and the role that the tourism industry has played in empowering them through the inter-role factors, the impact of digitalisation and issues for women in tourism as entrepreneurs. Sustainable tourism development in popular destinations increasingly depends on the empowerment of women. On the other hand, empowering women and achieving gender equality is one of the goals of the United Nations Sustainable Development Goals (UN-SDGs). Entrepreneurship is a key factor in promoting women's equality.

Chapter 14 discusses the widespread use of technology by businesses in the tourism and hospitality industry and how new technologies such as social media, virtual reality (VR), artificial intelligence (AI) and augmented reality (AR) technologies enhance the consumer experience of visitors, current or potential consumers. The use of technology is not only discussed within the tourism industry; its use by women entrepreneurs has also been detailed in this chapter. Finally, the author has explored how the Covid-19 pandemic has affected the tourism industry, whether there are differences between the genders and whether female entrepreneurs have been able to cope well enough with the process compared to male entrepreneurs.

Chapter 15 highlights the barriers that may prevent women from travelling as women travellers or travelling alone. The research philosophy used to write this study is epistemology and the research strategy is archival, as the chapter was formulated by analysing the available literature.

Chapter 16 aims to show how influencer communication takes place during collaborations with travel agencies and to evaluate how they communicate with their followers in this context. The study is descriptive in nature and covers the collaborations of 'Korea Tourism Organisation & Meryem Can', 'Visit Dubai & Meryem Can', 'Visit Dubai & Gülnihal Eren (Cocobolinho)' and 'Costa Cruises & Gülnihal Eren (Cocobolinho)' as case studies. Within this framework, the collaborations with different travel agencies will be examined through the relevant content on the influencers' official Instagram accounts. At the same time, this compilation study shows that influencers actively participate in social networks during the travel agency experience process. In addition, the preference for female influencers working in the tourism industry is considered as another feature of the study.

Chapter 17 examines the risk assessment of women solo travellers and their mitigation strategies by examining online travel blogs.Therefore, the study aims to assess the risks such as physical risk, destination specific risk and financial risk faced by women solo travellers by considering and analysing the narratives from online travel blogs of women travellers.These narratives from online travel blogs pertaining to Indian women solo travellers are considered in this study. Hence, this assessment provides an opportunity to know how to stay safe in uncertain situations and also mitigation strategies to stay alert. The result of the study highlights the various risks faced by women travelling alone and mitigation strategies to encourage more women travellers.

Chapter 18 presents the issues and constraints of solo women travellers to Turkey based on their travel blogs, which reflect their constraints before and during the trip and the safety tips they have adopted to stimulate their desire to travel to and within Turkish cities. The author asks, "What are the travel issues and constraints for female solo travellers?" and "How is Turkey less safe and at the same time the most preferred destination for female travellers?". The study uses netnographic research methodology to understand complex social phenomena and to help researchers develop themes from female travel bloggers' experiences of solo travel to Turkey. Using content analysis, the findings show that the constraints were mainly intrapersonal barriers to Turkey: such as personal and religious requirements, perceived risk and safety concerns. Other constraints, such as interpersonal and structural barriers, did not reflect solo travel bloggers' pre-travel and travel restrictions about Turkey. For all bloggers, safety is a key factor in the decision to travel.

Chapter 19 gives the definitions of special interest tourism and how it differs from general interest tourism. They will understand the characteristics, scope and variety of types of special interest tourism for women tourists such as adventure tourism, medical tourism, rural tourism, cultural tourism, culinary tourism, religious tourism, ecotourism, sports tourism, educational tourism, photographic tourism, dark tourism, celebrity tourism and film tourism. They will also learn about new types of trends in special interest tourism for women tourists, such as insta-tourism, virtual tourism, Pokémon Go tourism (gaming tourism) and space tourism (astro-tourism).

According to Chapter 20, all individuals can experience and relate to the nature of the job; however, these may sometimes be challenges that women face due to their biological characteristics. A holistic analysis of the challenges faced by female employees in the tourism industry has been undertaken in this chapter, based on national and international literature.

Chapter 21 examines gender discrimination in the tourist guide profession. Using a dyadic perspective, this study examines the difficulties experienced by both female and male tour guides. Convenience sampling was used to collect data from 10 female and 10 male tour guides. Semi-structured interviews were used to collect data. The data collected from the tour guides was analysed qualitatively. By comparing the difficulties experienced by male and female tour guides, the extent and nature of gender discrimination was identified. Based on the findings of the study on the extent and nature of discrimination, policy guidelines are offered to practitioners and stakeholders.

According to Chapter 22, the number and proportion of women on the boards of directors of listed companies in the tourism sector at Borsa Istanbul in Turkey, the CSR and sustainability performance of these companies was attempted to be revealed using a case study pattern from qualitative research methods. Within the framework of the document analysis, the number of female executives on the boards of the companies listed on the Borsa Istanbul and their ratio to the total number of board members will be recorded according to the annual activity reports and websites published on the public disclosure platform. The number and proportion of women on the boards will be the first unit of analysis used in the case study. The qualitative research analysis software MAXQDA 2020 will be used in the study.

Chapter 23 aims to identify the career anchors of female tourist guide candidates studying at university. The data were collected by means of a questionnaire from female students of tourist guide programmes. The results of the study showed that female tourist guides perceive technical career anchors at most, while the safety career anchor is perceived at least. In addition, their perceptions of career anchors differed significantly according to their career goals. Participants who want to become tourist guides in the future perceive higher autonomy as career anchors and service anchors, which are also inherent to the tourist guide profession.

Chapter 24 identifies the major challenges faced by women in the tourism sector and the ways in which they can overcome constraints, gender stereotypes and social stigmas while juggling multiple responsibilities. With this chapter, it is possible to gain more specific knowledge on how to empower women in the tourism and hospitality sector.

Chapter 25 identifies the barriers, issues and challenges related to female entrepreneurship with female entrepreneurship in the tourism industry of Pakistan. According to this chapter, female entrepreneurship is perceived as a new ray of light for the growth of the global economy. The phenomenon is particularly important for developing countries. However, female entrepreneurship is always associated with challenges and hurdles, and the same is true for the developed world.

Chapter 26 contributes to the emerging role of social entrepreneurs in tourism, and also tells how they achieve sustainability through a combination of 17 strategies and fusions of SDGs. According to this chapter, sustainable development together with social entrepreneurship offers a new approach to the tourism industry. Tourism social entrepreneurship is about how to take positive steps that reduce the negative impacts in the tourism industry.

Chapter 27 aims to introduce the concept of the role of women in halal tourism and to highlight the opportunity to increase entrepreneurship in developing countries through a robust model. The majority of developing or underdeveloped countries are Islamic countries with exceptional tourism destinations. However, the tourism business in these countries is still monopolised by men. The theme discussed promotes SDG 5 (gender equality) and SDG 8 (economic growth). Encouraging women's entrepreneurship in halal tourism is expected to attract an influx of funds into the local economy and reduce unemployment, while maintaining cultural boundaries and safeguarding belief systems.

Chapter 28 identifies the term 'voluntourism' refers to holidays where participants donate their time to charitable organisations. Humanitarian crises, either natural such as natural disasters or man-made such as armed conflicts and civil wars, have displaced millions of people. According to the United Nations High Commissioner for Refugees (UNHCR), the number of people forcibly displaced worldwide has reached 103 million, of which 53.2 million are internally displaced and 32.5 million are refugees. The Sustainable Development Goals (SDGs) call on the global community to "leave no one behind", which has led to the creation of dozens of new organisations and motivated thousands of individuals to participate in volunteer trips. In addition, non-governmental organisations have been actively seeking volunteers to help displaced people. Women from a wide range of nations have taken on the mantle of refugee and travelled to a number of conflict zones to set up organisations and/or assist in the resettlement of refugees.

Chapter 29 aims to identify the interest of women working in the tourism industry and their empowerment in the industry through inclusion and support for entrepreneurial activities. Further, to study the different perceptions and beliefs that women have about working in the tourism industry and by assessing their feelings towards the working environment in this industry. The main focus of the study was two main areas (i) women's interest in employment in the tourism industry; (ii) women's interest in entrepreneurial activities in the tourism industry. This study also considers whether the insecurities of the tourism industry and its inclusion practices deform women's interest in working in the tourism industry. Data was collected from 20 female respondents representing the interest of women to work in the tourism industry. The results show a positive response of 70.6% of the female respondents indicating their interest in the tourism industry, while 29.4% of the female respondents showed lack of interest in the industry.

Chapter 30 has shown that human resource management practices play a critical role in shaping the extent of discrimination that occurs within an organisation, as they influence the access that female employees have to workplace opportunities, such as training, fair performance appraisal, career advancement, equal pay and developmental assignments. Therefore, in order to change the employment conditions of women and provide them with equal opportunities in the workplace.

Chapter 31 gives a look into the tourism regarding women as it is known gender equality is a priority for national governments and international organizations, as evidenced by their adherence to a number of agreements, including the Beijing Declaration and Platform for Action, the Sustainable Development Goals, and the Convention on the Elimination of All Forms of Discrimination Against Women (CEDAW).

Chapter 32 examines the way in which the concept of entrepreneurship, which is becoming increasingly important, is discussed in the national tourism literature, in order to identify the current characteristics of the studies and to examine the subject. It also aims to contribute to the literature by compiling national studies that discuss the topic of entrepreneurship research within the tourism discipline. From this point, between 2019 and 2022, Google Academic, ULAKBIM and the National Dissertation/Thesis Centre (YOK) aim to search for terms such as women's entrepreneurship, entrepreneurship, tourism sector on these platforms. The qualitative method will be used to examine the academic papers obtained, first through document analysis and then through bibliometric analysis.

Chapter 33 explores and analyses the barriers that Muslim women experience in their professional growth, with a particular focus on barriers in the tourism industry. According to Chapter 33, although due to some adopted cultural, familial, political and social barriers, the professional growth of women is always hindered. Similarly, in the Islamic world, women face several barriers to their professional growth. However, growth in the tourism industry is a special case due to the perceptions associated with the industry and working practices.

Chapter 34 aims to explain the interrogation on the first Latin American centre fully specialised in tourism and hospitality research, CIET. Founded by Regina Schluter and her colleagues, this centre has inspired thousands of professional researchers in the region and represents a leading platform for tourism research in Spanish. CIET has been a source of advice and excellence in research in Argentina and Latin America for more than three decades.

Chapter 35 aims to find out the leadership of the high education institutions in empowering female entrepreneurship. Qualitative methodology is used in this study in which semi-structured interviews are conducted from the female candidates who are pursuing entrepreneurship and relying on the high education institutions and their leadership in initiating and maintaining support system for the females. Findings of the study show that female entrepreneurs are more determined for the entrepreneurship as they are mostly necessity driven in the developing country. Their leadership skills enable to tackle the various challenges and drive the situation in the order to carry on with their entrepreneurial motives.

Gül Erkol Bayram
Sinop University, Turkey

Syed Haider Ali Shah
Bahria University, Pakistan

Muhammad Nawaz Tunio
The University of Sufism and Modern Sciences, Bhitshah, Pakistan

Acknowledgment

All academic books require a support and powerful team, and this one is no exception. For this reason, we are grateful for the excellent support provided by IGI Global for great guidance and patience throughout the project. I am equally grateful to the members of the Editorial Advisory and Review Board. The Board reviewed several manuscripts, most of which were submitted for consideration by prospective authors.

We are also grateful to the external reviewers for the quality and timeliness of their reviews and this project, and Professor Marco Valeri for his elegant foreword to this book. book. Many thanks to the authors of the accepted papers for their professionalism and patience during the and patience during the review and publication process.

Without such previous support, this current project would never have been possible.

Finally, and most importantly, we are eternally indebted to our family for their support throughout this project. Our family are what makes life worth living.

Chapter 1
Achieving Sustainable Developmental Goals by Empowering Women Entrepreneurs:
Analyzing Challenges and Suggesting Policy Implications

Manpreet Arora
https://orcid.org/0000-0002-4939-1992
School of Commerce and Management Studies, Central University of HP, India

ABSTRACT

Women are largely invisible in the mainstream of economic development. Therefore, the concept of sustainability to a great extent depends upon providing equal distribution of not only resources, but also the equality should be created between the various sections of society where women are vulnerable. Women's empowerment should not be associated with increasing their responsibility as a whole, but it is about respecting their role. To achieve the goals of sustainable development, the concept of women entrepreneurship gains prominence where men have always been dominating the business situations. Ventures by women not only provide them financial support but also makes them empowered. This chapter, by using qualitative research approaches of content analysis, literature review, and discourse analysis, attempts to find challenges of women entrepreneurship and suggest various policy implications to deal with these challenges. The major findings are women entrepreneurship can be a major source of women empowerment backed by a focus on entrepreneurial education.

DOI: 10.4018/978-1-6684-8417-3.ch001

INTRODUCTION

Tourism has emerged as a driver of growth for many emerging economies: In the mid-1990s, low and middle-income countries' share in international tourism receipts was only 17 percent of the world's total. By 2012, that share had grown to 28 percent. This share is again increasing in the post-covid 19 period and it can be seen one of the major drivers of economic growth in various nations. The extensive demand in this industry not only stimulates service opportunities/facilities but it has the capability of stimulating certain entrepreneurial ventures like hotels, home stays, restaurants, food joints, souvenir shops, tour operators, tour guide firms etc. In the post-covid era both the developing nations as well as developed nations find entrepreneurship as a key to revival from the shocks of pandemic. The ultimate philosophy of entrepreneurship is based on reduction of unemployment, promoting economic development, bridging the gaps between haves and have nots and increase in GDP. The tourism industry world-wide has been contributing towards the economic development of nations by fostering start-ups, ventures and businesses involving transport, accommodation, catering, natural resources, entertainment, other facilities, and services, such as banks etc.

Tourism has emerged as a driver of growth for many emerging economies. In the mid-1990s, low and middle-income countries' share in international tourism receipts was only 17 percent of the world's total. By 2012, that share had grown to 28 percent. This share is again increasing in the post-covid period and it can be seen as one of the major drivers of economic growth in various nations. The extensive demand in this industry not only stimulates service opportunities/facilities but it has the capability of stimulating certain entrepreneurial ventures like hotels, home stays, restaurants, food joints, souvenir shops, tour operators, tour guide firms etc. In the post-covid era, both the developing nations as well as developed nations find entrepreneurship as a key to revival from the shocks of pandemic. The ultimate philosophy of entrepreneurship is based on reduction of unemployment, promoting economic development, bridging the gaps between haves and have nots and increase in GDP. The tourism industry world-wide has been contributing towards the economic development of nations by fostering start-ups, ventures and businesses involving transport, accommodation, catering, natural resources, entertainment, other facilities, and services, such as banks etc.

Tourism has played a critical role in sustainable development in many countries and regions around the world. In developing countries, tourism development has been used as an important strategy for increasing economic growth, alleviating poverty, creating jobs, and improving food security. Many developing countries are in the regions that are characterized by high levels of biological diversity, natural resources, and cultural heritage sites that attract international tourists whose local purchases generate income and support employment and economic development. Tourism has been associated with the principles of sustainable development because of its potential to support environmental protection and livelihoods. However, the relationship between tourism and the environment is multifaceted, as some types of tourism have been associated with negative environmental impacts, many of which are borne by host communities. Tourism has played an increasingly important role in sustainable development since the 1990s, both globally and in particular countries and regions. For decades, tourism has been promoted as a low-impact, non-extractive option for economic development, particularly for developing countries. Many developing countries have managed to increase their participation in the global economy through development of international tourism. Tourism development is increasingly viewed as an important tool in increasing economic growth, alleviating poverty, and improving food security. Tourism enables communities that are poor in material wealth, but rich in history and cultural heritage, to leverage their unique assets for economic development.

Local entrepreneurships are most of the times totally dependent on destinations especially in the areas of pilgrimage tourism and rural tourism. During and after covid-19, the excessive digitalisation and use of internet has played a great role in boosting tourism entrepreneurship. With the digitalisation and ease of use and taking advantage of the opportunities in small start-ups at various destinations, gave a levy to women to contribute in the economy along with fulfilling their responsibilities at home. Tourism seems to be the most exciting area where the possibilities for women entrepreneurs is extensive which has widened due to digitalisation of tourism industry in various spheres. Though the share of women in small and medium entrepreneurs around the world owned and manged by women is around 30 percent only, but the capability and the possibilities of increasing this share is immense. When we focus on sustainable development goals, we see the potential of tourism entrepreneurship to be quite good enough, which not only provide women possibilities of getting opportunities to decent work but it also leads the way towards equality and uplift the economy as whole. Financial access, lack of social networks and limitations relating to devoting time to entrepreneurial ventures can be the major challenges for women in developing nations. The tourism industry is driven by global trends and developments in the industry focussing on inclusivity, sustainability and innovation in tourism therefore promoting initiatives required to boost this sector can largely also depend on redefining, revisiting and highlighting the role of women in this sector. Further, amidst the gloom and despair created by Covid-19, entrepreneurs' role becomes tremendously crucial as the entrepreneurial spirit and zeal tends to fight against hardships and tide over seemingly insuperable encounters (Arora & Sharma, 2021). In these disruptions caused by Covid-19 pandemic, the labour industry has undergone serious changes. In such a scenario, the gig economy provides many solutions to the entrepreneurs to deal with the disruptions in the market (Arora & Sharma, 2022) and tourism industry can benefit a lot from it. Entrepreneurs are the significant pillars of any economy (Arora & Sharma, 2021), and the role of women entrepreneurs in tourism as well as other sectors cannot be undermined in that. More importantly, tourism offers an alternative to large-scale development projects such as construction of dams, and to extractive industries such as mining and forestry, all of which contribute to emissions of pollutants and threaten biodiversity and the cultural values of indigenous peoples.

RESEARCH METHODOLOGY

This is a conceptual qualitative paper and has used data from SCOPUS database to create theoretical perspective/ opinion. Literature has served as base of theoretical aspects of the paper. Content analysis of the top cited papers has been done. Through discourse analysis of 150 random students of PG level, the policy initiatives have been derived.

This chapter highlights the path of achieving sustainable developmental goals through qualitatively finding answers to the following research questions:

1. To analyse women entrepreneurship opportunities as a key to achieve sustainable development goals in relation to tourism sector.

2. To explore women entrepreneurship opportunities and its possibilities to act as a gateway of rural and local community development.
3. To analyse the challenges of women entrepreneurship especially in developing nations.

Discussion

Women's empowerment is much needed in order to create a fairer, more affluent, and inclusive society. It benefits not only women but has deeper positive impacts also on their families, communities, and the world as a whole. When we empower women the most important SDG i.e gender equality is achieved. Historically evidences are available where when have faced disparities and therefore steps promoting a more equitable society are needed. Discrimination and marginalization of women, limit their opportunities and rights. Further women are significant untapped potential as contributors to economic growth which is available world-wide. By empowering women and ensuring their equal participation in the workforce, societies can unlock a valuable resource and boost economic development. Various social issues can be tackled only when women are empowered. The road to empowerment of women starts from education, healthcare, and their involvement in decision-making processes. Only then they can contribute towards improving the well-being of their families, children, and communities. Women empowerment is not only needed but is a is a matter of human rights. All individuals, regardless of their gender, deserve equal rights, opportunities, and protections under the law. Empowering women helps uphold these fundamental principles and promotes a more just and inclusive society. When we empower women, it challenges gender stereotypes and traditional societal norms that not only limit women's roles and potential, but act as a stimulator of the growth of whole nation or economy. Women's empowerment challenges gender stereotypes and traditional societal norms that limit women's roles and potential. By empowering women, we can create a more inclusive and diverse society that recognizes and values the contributions of all individuals, regardless of their gender. Further, when women have equal access to education, healthcare services, career options, life decisions and reproductive rights, they can make informed decisions about their bodies, health, and well-being. Empowered women are more likely to have healthier families and communities. Though the factors ensuring the success of women entrepreneurs in emerging nations are somewhat different from the developed nations (Kothari, 2017). However, the potential of women empowerment in the tourism sector has not been fully explored (Vij et.al., 2023)

Analysing Women Entrepreneurship Opportunities as a Key to Achieve Sustainable Development Goals in Relation to the Tourism Sector

In order to achieve sustainability, women, act as key actors in the development process, they influence largely the planning as well as implementation of various policies and programs. The concept of sustainability cannot be achieved without creating ample opportunities of gender equality. One of the important factors of sustainable development is achieving economic growth, social development and environmental sustainability, which is impossible to achieve without promoting gender equality. All the policies and actions of individuals as well as governments should revolve around the sustainable development of the society which creates ample opportunities for all the genders. When we view the concept of sustainable development in a broad spectrum, then women have played an important role in the global environmental movement along with all the developmental areas of the world and they have played a pivotal role in conservation of environment, societal development of various nations. In the modern world the contribution of women in education, technology and information is commendable. More or less the fate of the world is dependent upon sustainable development by respecting and fostering equality for opportunity for every section and all genders. Another important aspect of the development of humanity which can never be neglected is the role of women in entrepreneurship. They have also played an important role in fostering entrepreneurship at various levels in various countries. Women as entrepreneurs have proven to a great extent that there is no field where women cannot excel. United Nations Development Programme has defined sustainable development as the development which not only generates economic growth, but it also distributes the benefits of sustainable growth equitably. In many societies and economies still, women are not paid for their work equally and there are so many services which they provide to society and environment which is not counted in our economic, political or social systems. Women are largely really invisible in the mainstream of economic development in various areas. Therefore, the concept of sustainability largely depends upon providing equal distribution of not only resources, but also the equality should be created between the various sections of society where women are very vulnerable. Women's empowerment should not be associated with increasing their responsibility as a whole, but it is about respecting their role in every part of life. To achieve the goals of sustainable development, the concept of women entrepreneurship also gains prominence in this modern world where men have always been dominating the business situations. In recognition of equality in every phase of life, women entrepreneurship is gaining very prominence. As women are vulnerable and lack resources, therefore their initiative in women entrepreneurship is regarded to be an important aspect of not only their individual growth but the overall growth of the nations. In order to achieve sustainable development goals, women empowerment through entrepreneurship can act as a useful tool as it brings gender parity, reduces poverty and brings social equality at many spheres (Chakraborty & Biswal,2022). Microcredit facilities for women have a great ability to boost women empowerment and further women entrepreneurship (Arora & Singh, 2018).

Entrepreneurial ventures by women not only provides them financial support, but it also makes them empowered in every field of life. Which in one way or another, contributes towards sustainable growth of every economy. Women entrepreneurship in various areas or industries is proving to be an important aspect in the growth of GDP of a nation. Entrepreneurship in itself is a challenging situation. Women being more vulnerable and weaker section of the society often struggle a lot more than males in order to become successful in recognized in the business community.

Therefore, the concept of women empowerment is an essential component of sustainable development, because equality as well as women empowerment are fundamental human rights and they are very much important for achieving the Sustainable Development Goals of each country. Further, access to education and empowering women through education is an important aspect of sustainable development. Those women who are educated, they have better economic opportunities and they get ample opportunities to participate in the various decision-making processes which ultimately upgrade their lives and also develop the communities to which they are associated. Moreover, there are various disparities which are associated with women, like health and gender inequality. When we educate women, they get access and information about health care and overall well-being of the women is improved which also leads to the Sustainable Development Goals and its better achievement.

Women have always been associated with the environment and natural resources; they can be regarded as the primary care takers of the natural resources and therefore the impacts of climate change or degradation of the natural resources have a direct impact on women. When we empower women and educate them for environmental, economic or societal decision making, then we can achieve sustainable outcomes. Entrepreneurial opportunities for women in tourism sector are plenty e.g in sustainable destinations, travel companies, pilgrimage-oriented ventures and much more.

Women entrepreneurship has a significant role to play in contributing towards the sustainable development of a country in various ways. Women entrepreneurship has the ability to foster and drive economic growth as well as development, particularly in developing nations where women are vulnerable and they face significant barriers for entering into the labour market.

When women start business and grow their business over the period of time, they create employment opportunities not only for themselves but for others also, which can contribute towards reduction of poverty of especially of the vulnerable sections of the society which can lead to sustainable economic growth. Women are very creative by nature and women entrepreneurship has the ability to lead to innovation for the development of the country. Women entrepreneurs have often unique and different perspectives by which they can address various social and environmental challenges in order to promote sustainable practices. Various women led entrepreneurship ventures have already been directed and performing well towards waste management and adopting sustainable practices, leading to sustainable development of the countries. Women by nature are regarded to be environmentally friendly, and there is a greater chance of adoption of environmentally friendly practices such as use of renewable energy and there is elimination of waste and promoting use of sustainable materials for growth and development of a country.

Women are an indispensable part of the society, and that is why they contribute a lot towards societal development when we talk about their impact on social economic systems. When women start business, they not only address their own financial problems or for other women, but they also address various social and environmental challenges which generally societies face. Women have the ability to contribute towards sustainable development goals like poverty reduction, gender equality and environmental sustainability. Economic independence of women leads to increase in their social status and also helps them getting empowered in various aspects. In order to create a more inclusive as well as equitable tourism sector, the role of women cannot be denied in the tourism sector too. In order to respect the diversity and gender equality in the tourism sector, women leadership is an indispensable action towards achieving these goals. Women in tourism sector can provide leadership in various roles like managers, executives and policymakers. Women have been showing their calibre in tourism industry as restaurant owners, proprietors of souvenir shops, tour operators, tour guides and much more. Women entrepreneurship in tourism sector can help to promote economic development and innovation to a great extent.

Various roles in tourism industry are fit for women thereby skill and talent can be nurtured and this leads to the sustainable development of this sector. It includes hospitality, food, retail as well as transportation. The financial independence of women in tourism sector has the ability to take this sector to a higher level. Another aspect of the tourism sector is community development, where women play an important and indispensable role by creating employment opportunities and contributing towards local economies. Various microfinance initiatives in the countries have been taken in order to promote small scale entrepreneurship at the grassroot level through tourism sector. From home stays to tour operating companies, women are indispensable in tourism sector also. The economic independence to women not only improves their social status but also increase their decision-making capability, making them more contributing towards sustainable growth. Sports tourism is an important developing area where women are proving their worth. Women often bear conflict and violence, and when we empower women, they help to prevent conflict and promote peace. Sports have been regarded as one of the important means of promoting peace and partnership altogether in the world. Participation of women in various sports helps them to contribute towards healthy and sustainable lives promoting overall wellbeing of women and children's Health.

In order to get an overview about what literature highlights for women in the tourism sector the author carried out a Scopus search on the key words "women" or "woman" and "tourism" by using the following search string in the articles in English in the area of Business, Management and Accounting.

```
TITLE-ABS-KEY ("women" OR "woman" AND "tourism") AND (LIMIT-TO (DOCTYPE, "ar")) AND (LIMIT-TO (SUBJAREA, "BUSI")) AND (LIMIT-TO (LANGUAGE, "English")) AND (LIMIT-TO (SRCTYPE, "j"))
```

The result of 602 documents was utilized to do the content analysis by the word cloud formation to see the most frequently occurring keywords. The most occurring keywords give us an indication of the researchers' areas of interest, dimensions or characteristics studied by them. Generally, they are important for the study. In this content analysis the important keywords found are "women analysts", "women along with tourism", "tourism management", "development", "Social", "Gender", and "sustainability". Furthermore, researchers frequently studied "Women empowerment", "cultural aspects", "travel", "employment", "destination", "social entrepreneurship", "sustainable entrepreneurship", "cultural relations", "indigenous", "ethnic", "heritage tourism", "behavioural factors", "gender", "identity", and "community development". All these words give a clear indication about the relevance of women in tourism sector as they are getting ample oppourtunities in the tourism sector in development of economy through travel and maintaining cultural relations.

Figure 1. Word cloud based on content analysis
Source: Created by author using SCOPUS data

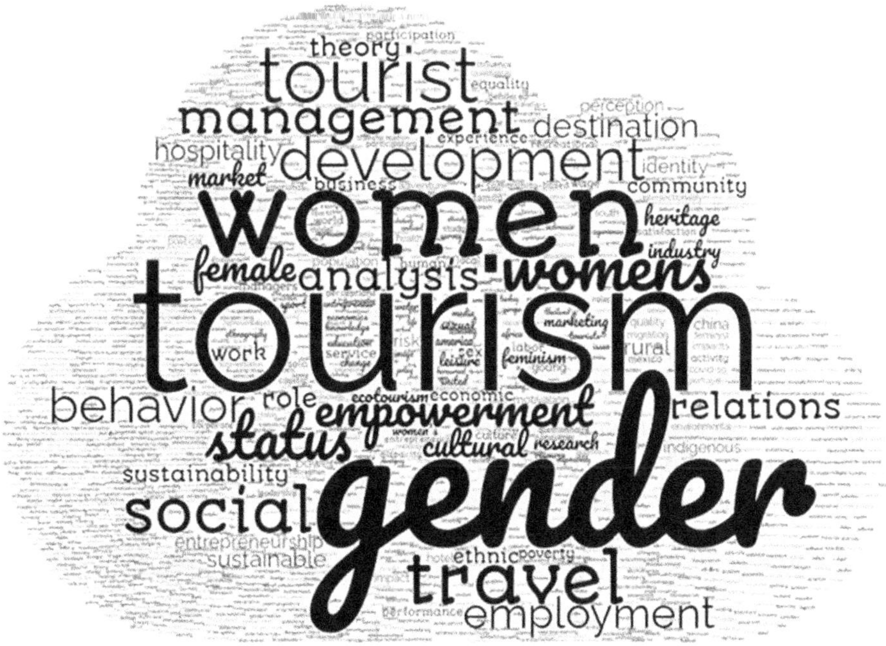

Exploring Women Entrepreneurship Opportunities and Its Possibilities to Act as a Gateway of Rural and Local Community Development

Women entrepreneurs have the ability to play a substantial role in promoting rural development by generating job opportunities, contributively providing opportunities to the local economies, and empowering other women in their communities. Through entrepreneurship, women can influence communities to enhance their skills, knowledge, and resources to start businesses that meet their needs, such as jobs, self-sustainability, agriculture, earning through local art and craft, and tourism. When women entrepreneurs start or expand their business, they can create employment opportunities for not only themselves but for others too in their communities, which can lead to increased income. Better living standard, educational opportunities and economic growth are must to follow. They can also bring/ nurture new skills, technology, and innovation to the local economy, which can lead to improved productivity, better utilization of resources, community development and increased competitiveness. In addition to contributing to economic development, women entrepreneurs can empower other women in their communities by serving as role models, influencers, achievers and mentors, providing training and skills development opportunities to many who otherwise do not come forward. They are the best examples for promoting gender equality and women's rights. This can help to address gender disparities which are prevalent in societies especially in rural areas. It can have a ripple effect in the form of increased women's participation in decision-making processes, and promoting social inclusion and cohesion. Further, women entrepreneurs can endorse sustainable and comprehensive rural development by espousing environmentally-friendly practices, supporting local suppliers and producers, and contributing to the conservation of cultural heritage and traditions. This can help to foster a sense of pride and ownership among community members and promote the sustainable use of natural resources. Women entrepreneurs exhibit multiple social identities and there is always intersectionality of these multiple identities. This in turn shifts the focus of several entrepreneurship studies beyond the traditional focus of economic growth to work life balance and at a larger level community developer. Entrepreneurial education can lay a positive as well as significant effect on the women entrepreneurial success (Maziriri et al., 2022). Therefore, efforts must be initiated in the direction of allowing entrepreneurial education of women.

Wang (2019) explored the role of gender, race, and ethnicity in the process of women entrepreneurship and how their experiences are influenced by their communities. The study involved analysing data from the 2010-2014 "American Community Survey", as well as conducting in-depth interviews and focus groups with 40 female business owners using QSR Nvivo software for data analysis. The findings indicate that women entrepreneurs face the extreme challenge of managing both their family roles and work but, they strategically negotiate their multiple roles through entrepreneurship to achieve independence, purpose, and also contribute towards social contribution. The study suggests that women entrepreneurs and their businesses are deeply embedded in local communities and highlights the significance of place in the entrepreneurial process. Digitalization and digital technologies have been playing a significant role in boosting entrepreneurship so is the case of women entrepreneurship. "Digital social innovation" (DSI) has helped overcoming the challenges in relation to various individual as well as social and structural factors. Suseno and Abbott (2021) in their research employed the individual differences theory of gender and information technology (IDTGIT) and extends it to explore how technology facilitates the social impact of women entrepreneurs' DSI ventures. The study was based on a "qualitative research design" that involved conducting interviews with 17 women entrepreneurs in Australia. The findings of the study indicate that women entrepreneurs face challenges related to their identity, individual influences, and social and structural factors, which can hinder their business ventures. However, they are using technology which ultimately helps them to overcome these challenges and create opportunities for their businesses. The impact of digital social innovations could be seen in various areas thereby facilitating the entrepreneurial opportunities in areas like tourism, healthcare, hospitality, wellness, beauty and education sector. It not only creates employment opportunities, but also is contributing in developing healthy environment and climate for social and economic growth of women, further paving way for community development. Literature supports the fact that women always experience more difficulties than men in opening their own ventures or startups (Dal Mas & Paoloni, 2020). In developing nations microfinance acts as a stimulating force for boosting entrepreneurial ventures which empowers women in many dimensions (Arora & Singh, 2018). Entrepreneurship is the key to creativity and innovation again fostering women empowerment (Arora, 2016). "Leadership and entrepreneurship, despite being discrete concepts, have to coalesce for sure success of any venture", (Sharma & Arora, 2015). Sustainable rural financing by ample credit opportunities can divert the direction of entrepreneurial ventures towards sustainable practices also (Arora & Singh, 2023).

ANALYSING THE CHALLENGES OF WOMEN ENTREPRENEURSHIP ESPECIALLY IN DEVELOPING NATIONS

For this purpose, keywords "challenges" and "women entrepreneurship" were searched on Scopus database. The search gave 256 document results that was further filtered for the subject area as "Business, Management and Accounting" and "Economics", the source type as "journal" and articles in English language only. This filtration led to limiting the results to 136 documents only. The following search string was put on the Scopus database:

```
TITLE-ABS-KEY ("challenges" AND "women entrepreneurship") AND (LIMIT TO (SUBJAREA, "BUSI") OR LIMIT-TO (SUBJAREA, "ECON")) AND (LIMIT-TO (LANGUAGE, "English")) AND (LIMIT-TO (SRCTYPE, "j"))
```

In order to understand the challenges of female entrepreneurship steps has been taken to systematically review the abstracts, keywords, titles and the full content of top 15 highly cited papers on entrepreneurship, mentioned in Table 1 below.The studies have highlighted that there are certain individuals and societal factors that lay an impact on by women.

Women have certain "women intensive constraints". Women those who have socialized in different manners in the society, face challenges of different levels as they perceive different opportunities in different ways. Men have different social networks as compared to women. So, women face a little complex environment in operating under a men operated world. A more gendered aware approach is required to boost entrepreneurship for women in order to achieve multiple SDGs. Women also take information in a different way as compared to men. Men and women also comprehend their social networks in quite a different manner. The social context of women entrepreneurship plays an important role in female entrepreneurship (Brush et.al.,2009). The research points discussed by Brush et al. (2009), addresses to major SDGs in their article which can be fulfilled by women entrepreneurship i.e., goal number 5, "gender equality" and "decent work and economic growth" (Goal number 8). Women entrepreneurship not only facilitates the gender equality but it is a wonderful way of promoting decent work and economic growth.

Table 1. Top fifteen highly cited papers as on 25th April 2023 in Scopus on "challenges of women entrepreneurship"

	Title of the Article	Authors	Year of Publication	Source Title	Citations
1	A gender-aware framework for women's entrepreneurship	Brush, C.G., de Bruin, A., Welter, F.	2009	International Journal of Gender and Entrepreneurship 1(1), pp. 8-24	516
Sustainable Development Goals mapped to this document Gender equality Goal 5 Decent work and economic growth					
2	Socio-cultural environments and emerging economy entrepreneurship women entrepreneurs in Indonesia	Anggadwita, G., Luturlean, B.S., Ramadani, V., Ratten, V.	2017	Journal of Entrepreneurship in Emerging Economies 9(1), pp. 85-96	128
	Sustainable Development Goals mapped to this document No poverty Goal 1 Gender equality Goal 5 Decent work and economic growth Goal 8				
3	Evidence of the characteristics of women entrepreneurs in the Kingdom of Saudı Arabıa: An empirical investigation	Zamberi Ahmad, S.	2011	International Journal of Gender and Entrepreneurship 3(2), pp. 123-143	125
	Sustainable Development Goals mapped to this document Quality education Goal 4 Gender equality Goal 5 Decent work and economic growth Goal 8 Partnership for the goals Goal 17				
4	Diana: A symbol of women entrepreneurs' hunt for knowledge, money, and the rewards of entrepreneurship	Gatewood, E.J., Brush, C.G., Carter, N.M., Greene, P.G., Hart, M.M.	2009	Small Business Economics 32(2), pp. 129-144	108
	Sustainable Development Goals mapped to this document Gender equality Goal 5 Decent work and economic growth				
5	Constraints faced by women entrepreneurs in developing countries: review and ranking	Panda, S.	2018	Gender in Management 33(4), pp. 315-331	100
	Sustainable Development Goals mapped to this document Gender equality Goal 5 Decent work and economic growth Goal 8 Reduced inequalities Goal 10 Partnership for the goals				

continued on following page

Table 1. Continued

	Title of the Article	Authors	Year of Publication	Source Title	Citations
6	Challenges facing women entrepreneurs in Nigeria	Halkias, D., Nwajiuba, C., Harkiolakis, N., Caracatsanis, S.M.	2011 X"	Management Research Review 34(2), pp. 221-235	94
	Sustainable Development Goals mapped to this document No poverty Goal 1 Gender equality Goal 5 Decent work and economic growth Goal 8 Industry, innovation and infrastructure Goal 9 Reduced inequalities Goal 10 Partnership for the goals Goal 17				
7	Women entrepreneurship in the Middle East and North Africa: A review of knowledge areas and research gaps	Bastian, B.L., Sidani, Y.M., El Amine, Y.	2018	Gender in Management 33(1), pp. 14-29	85
	sustainable Development Goals mapped to this document Gender equality Goal 5 Decent work and economic growth Goal 8 Partnership for the goals Goal 17				
8	Female entrepreneurship in Saudi Arabia: Opportunities and challenges	Yousuf Danish, A., Lawton Smith, H.	2012	International Journal of Gender and Entrepreneurship 4(3), pp. 216-235	85
	Gender equality Goal 5 Decent work and economic growth				
9	Saudi women entrepreneurs: A growing economic segment	Welsh, D.H.B., Memili, E., Kaciak, E., Al Sadoon, A.	2014	Journal of Business Research 67(5), pp. 758-762	78
	Sustainable Development Goals mapped to this document Gender equality Goal 5 Decent work and economic growth Goal 8				
10	Women entrepreneurs as agents of change: A comparative analysis of social entrepreneurship processes in emerging markets	Rosca, E., Agarwal, N., Brem, A.	2020	Technological Forecasting and Social Change 157,120067	77

continued on following page

Table 1. Continued

	Title of the Article	Authors	Year of Publication	Source Title	Citations
	Sustainable Development Goals mapped to this document Gender equality Goal 5 Decent work and economic growth Goal 8 Partnership for the goals Goal 17				
11	Examining successful Iranian women entrepreneurs: An exploratory study	Javadian, G., Singh, R.P.	2012	Gender in Management: An International Journal 27(3), pp. 148-164	71
	Sustainable Development Goals mapped to this document Gender equality Goal 5 Decent work and economic growth Goal 8 Partnership for the goals Goal 17				
12	Behind the veil: women-only entrepreneurship training in Pakistan	Azam Roomi, M., Harrison, P.	2010	International Journal of Gender and Entrepreneurship 2(2), pp. 150-172	64
	Sustainable Development Goals mapped to this document No poverty Goal 1 Gender equality Goal 5 Decent work and economic growth Goal 8 Partnership for the goals				
13	'And now I'm free': Women's empowerment and emancipation through entrepreneurship in Saudi Arabia and Sweden	Alkhaled, S., Berglund, K.	2018	Entrepreneurship and Regional Development 30(7-8), pp. 877-900	63
	Sustainable Development Goals mapped to this document Gender equality Goal 5 Decent work and economic growth Goal 8				
14	Exploring the false promise of entrepreneurship through a postfeminist critique of the enterprise policy discourse in Sweden and the UK	Ahl, H., Marlow, S.	2021	Human Relations 74(1), pp. 41-68	62
	Sustainable Development Goals mapped to this document Gender equality Goal 5				
15	Gender, context and entrepreneurial learning	Ettl, K., Welter, F.	2010	International Journal of Gender and Entrepreneurship 2(2), pp. 108-129	59
	Sustainable Development Goals mapped to this document Gender equality Goal 5 Decent work and economic growth				

Source: Compiled by author from SCOPUS database.

Grisna (2017) regard socio cultural factors to be one of the biggest challenges for women entrepreneurship in countries like Indonesia. Cultural diversity can lay an important impact on the behavioural intentions of women entrepreneurship. As per the study, "cultural kinship" emerged to be the most significant factor affecting the socio-cultural environment of entrepreneurship. But on the same hand, the cultural factors also act as a source of success of women entrepreneurship, as their socio-cultural values foster in them honesty, courage and readiness to face the potential risks of the society.

Ahmad (2011) have discussed the entrepreneurial intentions and performance in the Islamic society. The social control over the women in Arabic Nations is very high and there are numerous restrictions on the activities of a woman. Not only their mobility is restricted but their access to education as well as other activities is more or less controlled and governed by the social and religious norms. They are subject to stiff competition in the market, therefore the entrepreneurial intention for big successful businesses by women is very less in the societies. The major challenge is societal norms and further ease to access to credit is very less, leading to less availability of sources of finance. Moreover, the family businesses have increased their dependence on family members and limited their level of risk taking. The following are the main challenges faced by women in such cases:

- "Gender biases"
- "Family responsibilities"
- "High production costs"
- "Low networking skills"
- "Dependent on their family members"
- "Poor access to market"
- "Poor support mechanism"
- "Poor access to technology and credit"
- "Lack of business management skills"

The "Sustainable Development Goals" which are addressed in this research are relating to imparting commercial education to women and making them empowered by providing them opportunities of decent work and growth. Further, it also highlights the importance of gender equality in the development all women, entrepreneurship. Various policy initiatives regarding upliftment of Women entrepreneurs throughout the world requires partnership for their growth and development in various senses. Thereby, it becomes necessary for the policymakers to find out opportunities which can help them grow and survive. Partnership for the goals, that is Goal 17, is also evidently mapped out in this document, identifying the importance of entrepreneurship in achieving Goal 17.

Credit needs for women entrepreneurs are very peculiar and they face hurdles to raise capital. Very less opportunities are available to the women to receive venture capital investments as compared to men. Women lead industries are also in the areas which are less attractive for the investors. When there is lack of information as well as credit facilities the possibilities of growth through innovation are also very less (Gatervod et.al., 2009).

Financial constraints not only hover on the female entrepreneurial intentions especially in developing Nations, but another dimension of their challenge is the unstable" economic and political environments". Further overall orientation of women in the social systems also building their intentions. The country level constraints are different in various countries, for example; if the male dominance in a societal system is higher then the female intention towards entrepreneurship is more challenging; as they are highly dependent on the male family members. They seldom take their family decisions without consulting their male family members and more or less their business decisions are also controlled from home. Conflicts at home and work family conflicts are certain internal environmental challenges for women who are keen to be successful entrepreneurs. Whereas lack of financial opportunity, lack of favourable environment for economic and social growth of women, lack of entrepreneurship training and education, are some of the macro uncontrollable challenges for women entrepreneurship (Panda, 2018).

The growth potential of women in African countries is immense. In the African continent the major driving force of fostering entrepreneurship is microfinance. The entry barriers in some of the industries create difficulty in the access of credit, and the complex systems are some of the challenges faced by African women entrepreneurs (Halkias,2011). Cultural variables affecting entrepreneurship are very diverse and they are one of the most important challenges in Middle East in Northern or Africa. Muslim women entrepreneurs are highly impacted by their religious and cultural norms in this direction. There is uniqueness in the challenges from region to region. The complexity of social cultural and religious diversity needs to be addressed by women entrepreneurs in these Nations (Bastion et.al.,2018).

Danish & Smith (2012) in their study on Saudi women focused that though there are significant societal and institutional challenges but the trend of managing more small and medium skill enterprises is increasing. The accessibility of formal financial resources to these women is less they are more dependent on their personal resources or family wealth for starting the business. Various state as well as structural level challenges are faced by women in Arabian countries which act as an impediment in their growth. "Education" and "training programs" are very much required in order to combat these challenges. Welsh et.al. (2013) highlight that the unemployment rate of women in Saudi Arabia is very higher. Whereas women are highly qualified. Therefore, the high rate of women entrepreneurship could be seen in these nations. But the women lack management skills to run the businesses successfully. The major obstacle which was found in the survey conducted by the author was that they lack highly in financial skills. "Acquiring appropriate training", "procurement of capital" along with "gender discrimination", are some of the major challenges they face.

Rosca*et.al.* (2020) in their study have explored that "how do women engage in Social entrepreneurship process in certain BOP environments?" The concept of "Social entrepreneurship" especially for women in the bottom of pyramid is quite challenging but the concept of social entrepreneurship acts as an opportunity for them to help and support poor along with creating deep social impacts. Social entrepreneurship is also deeply associated with sustainability. Use of distribution channels for increasing their effectiveness is a big challenge for the social entrepreneurs. Lack of awareness and high uncertainty of "Bottom of Pyramid" environment are some of the major challenges for rural women who want to pursue entrepreneurial intentions. Studies support the fact especially in developing or emerging nations that self-efficacy and ability to take risks are fundamental internal qualities for women to evolve as successful entrepreneurs. Javadian & Singh (2012) have highlighted that there can be several stereo types also which act as a barrier along with cultural and traditional values acting as external barriers in their entrepreneurial journey. These stereotypes are basically the "gendered stereotypes".

In a socio-cultural setup of an emerging or developing economies along with gender stereotypes, gender related challenges have got wider dimensions and deeper perspectives. Trainings, capacity building programs, support systems and access to easy credit or some of the time-tested techniques to improve the self-efficacy and entrepreneurial skill of women derived with passion and risk-taking intentions for chasing successful entrepreneurships. Roomi & Harrison (2010) are of the view that if a focus is oriented towards enhancing the skills of women by investing in capacity building programs and trainings then it can help them to gain not only competencies but also will assist them in getting capital. They get better clarity about the market dynamics and will be able to handle the challenges posed by the external environment especially in relation to lack of appropriate skills and training. Alkhaled & Berglund (2018) are of the view that societal situations, patriarchal societies, male domination in the market place, less support from the families and overall, their "trapped lives" in the traditional family systems are the biggest challenges for women entrepreneurship. Women suffer a lot due to societal norms and entrepreneurship can act as a vehicle of change. Various entrepreneurial endeavours can help them to work on the path of freedom from the clutches of patriarchal mindsets destroying their intentions, disentangling them from where they do not want to be.

Entrepreneurship is a beautiful way of moving from employment to self-employment. But female potential in terms of women entrepreneurship is under exploited across the Nations. As men and women are different; so is their approach and that is why they require different policy measures and interventions especially in the area of entrepreneurship. Ahl & Marlow (2019) are of the view that there are many challenges associated with women entrepreneurship like;" lack of business skills", "lack of confidence" and "poor networking skills". Therefore, in order to make the more adept to the entrepreneurial ecosystem they need to adopt "appropriate attitude" and develop particular "competencies" which can further help them to increase their skills and confidence in business. Entrepreneurial learnings lighten up the path for overcoming the challenges of women entrepreneurship. Ettl & welter (2010) all of the view that macro environment lays an indirect effect on women entrepreneurship whereas macro and micro factors lay a strong effect on their personal lives as well as on the growth of their entrepreneurial ventures. Therefore, policy interventions at various levels are required to deal with the challenges of women entrepreneurship.

Foregoing discussion clearly highlights and focuses on the fact that in order to deal with the challenges of women entrepreneurship various policy implications are needed. Through discourse analysis of 150 random students of PG level the policy initiatives have been derived.

While summing up, the following are the main challenges which generally women face in women entrepreneurship.

1. There are very few sectors of business where gender equality is promoted. Men still dominate entrepreneurial ecosystem. In most of the countries, men owned businesses in the countries operate in the most profitable sectors. The male centric approach of various sector disregard women to be the powerful contenders to work in that section. That is why, many women operate in so called "women friendly sections" like education, apparel, beauty care etc. It limits their opportunities to grow and their skills as well as capabilities to a great extent.
2. Most of the women business owners, they do not get support to start their businesses from their families, peers and the ecosystem in which they operate. They lack mentorship from the business community. That is why they do not grow up in entrepreneurship at a great extent. Absence of proper entrepreneurial ecosystem which guides, promotes and nurture the talent of women in entrepreneurship is seriously lacking, which adversely impacts their level of growth and ability to take risks.
3. There are many pre notions or prejudices of the investors who invest in the businesses led by women. There is a lot of gender bias which still is prevalent in the various sections of the society. Very less people feel women as capable entrepreneurs and that is why they receive less of the funding as compared to the male counterparts. Various venture capitalist firms and Angel investors are very much hesitant and reluctant to invest in those businesses which are women led. It is always

perceived by the investors that women will be surrounded by family pressures and peer pressures. That is why they feel investing in a women led business is not a good idea. Further, in various societies, women have less assets on their names and that is why they lack collateral at the time of applying for loans.

POLICY IMPLICATIONS FOR COMBATING THE CHALLENGES OF WOMEN ENTREPRENEURSHIP IN DEVELOPING NATIONS

1. Easy access to funds and some special women-oriented programs for women are needed.
2. A community of women entrepreneurs should be built across the nations so as to discuss their ideas and problems at a common platform.
3. Along with the less rate of interest for loans for women entrepreneurship, tailor made or customised programmes should be initiated by the governments. The special need of backward communities and rural areas should be a focus too.
4. Awareness programs for start-ups especially related with local areas, local demands and local talent should be done. So that an entrepreneurial motivation amongst the women could be inculcated who otherwise do not have risk taking abilities and do not take initiative for entrepreneurial ventures.
5. Mentorship and training programs are very important to boost the confidence of women in the area of entrepreneurship. Hand holding programs, and direct support and assistance programs in the form counselling sessions are very important to increase their risk-taking abilities.
6. Organising competitions, fairs, fests, innovation fairs at higher education institutes can increase the curiosity of women for novel business ideas. It can also provide them a platform of collaboration also, where a group of women can come together and work on some business ideas.
7. Financial institutions should also impart educational and training programs to the loan borrowers for efficient handling and managing of finances from time to time. Financial skills are very important to run the business. The policy initiatives should encompass measures relating to enhancing financial management skills of women.
8. Women who want to pursue business ideas should be offered soft loans and subsidies in order to encourage them into industrial activities. Further, the financial institutions should provide more working capital assistance to women entrepreneurs.
9. Skill development is the need of the hour. It can be done in women's polytechnics and industrial training institutions. A basic understanding of skills for businesses can be imparted in local and rural areas along with educational institutions, in order to motivate women from educated or uneducated domain.
10. The success stories of women should be taught in educational institutions, right from school level to higher educational institutions. It should become part of curricula. It brings inspiration and zeal to the budding minds, where the energy can be channelised to more productive entrepreneurial ventures.

CONCLUSION

The phenomenon of women entrepreneurship is at growing stage. Across nations initiatives in various sectors can be seen being taken by women. Entrepreneurships have the ability to bring economies on track. They have the ability to solve numerous economic and social problems of the society. The findings of this chapter highlight that despite the growth in entrepreneurial ventures by women, still the access to easy credit for women is not there, procedural formalities and hindrances to gain credit needs to be dissolved by emerging nations. Men dominations in the businesses is quite challenging to face by women. Investors still favour ventures dominated by men and women mostly do businesses which are in left out areas. Cultural, societal and religious norms restrict their risk-taking ability in men dominated countries. Tourism is one such area where women have to go long way, but the opportunities are plenty. Multiple SDGs could be achieved by focussing and providing opportunities in women entrepreneurship.

REFERENCES

Gatewood, E. J., Brush, C. G., Carter, N. M., Greene, P. G., & Hart, M. M. (2009). Diana: a symbol of women entrepreneurs' hunt for knowledge, money, and the rewards of entrepreneurship. *Small Business Economics, 32*, 129-144.

Ahl, H., & Marlow, S. (2021). Exploring the false promise of entrepreneurship through a postfeminist critique of the enterprise policy discourse in Sweden and the UK. *Human Relations*, *74*(1), 41–68. doi:10.1177/0018726719848480

Alkhaled, S., & Berglund, K. (2018). 'And now I'm free': Women's empowerment and emancipation through entrepreneurship in Saudi Arabia and Sweden. *Entrepreneurship and Regional Development*, *30*(7-8), 877–900. doi:10.1080/08985626.2018.1500645

Anggadwita, G., Luturlean, B. S., Ramadani, V., &Ratten, V. (2017). Socio-cultural environments and emerging economy entrepreneurship: Women entrepreneurs in Indonesia. *Journal of Entrepreneurship in Emerging Economies*.

Arora, M. (2016). Creative dimensions of entrepreneurship: A key to business innovation. *Pacific Business Review International*, *1*(1), 255–259.

Arora, M., & Sharma, R. L. (2021). Repurposing the Role of Entrepreneurs in the Havoc of COVID-19. In Entrepreneurship and Big Data (pp. 229-250). CRC Press. doi:10.1201/9781003097945-16

Arora, M., & Sharma, R. L. (2021). Neutralizing Maleficent Effects of COVID-19 Through Entrepreneurship: Peeping Through the Lens of Communication. In Effective Strategies for Communicating Insights in Business (pp. 67-86). IGI Global.

Arora, M., & Sharma, R. L. (2022). Coalescing skills of gig players and fervor of entrepreneurial leaders to provide resilience strategies during global economic crises. In *COVID-19's Impact on the Cryptocurrency Market and the Digital Economy* (pp. 118–140). IGI Global. doi:10.4018/978-1-7998-9117-8.ch008

Arora, M., & Singh, S. (2018). Microfinance, Women Empowerment, and Transformational Leadership: A Study of Himachal Pradesh. *International Journal on Leadership*, *6*(2), 23.

Arora, M., & Singh, S. (2018). Impact Assessment of Self-Help Group Bank Linkage Programme on Women Empowerment in the State of Himachal Pradesh. Academic Press.

Arora, M., & Singh, S. (2023). Microfinance for achieving sustainable development goals: pondering over Indian experiences for the preservation of magnificent african natural resources. In *Research Anthology on Microfinance Services and Roles in Social Progress* (pp. 378–394). IGI Global.

Bastian, B. L., Sidani, Y. M., & El Amine, Y. (2018). Women entrepreneurship in the Middle East and North Africa: A review of knowledge areas and research gaps. *Gender in Management*, *33*(1), 14–29. doi:10.1108/GM-07-2016-0141

Brush, C. G., De Bruin, A., & Welter, F. (2009). A gender-aware framework for women's entrepreneurship. *International Journal of Gender and Entrepreneurship*, *1*(1), 8–24. doi:10.1108/17566260910942318

Chakraborty, U., & Biswal, S. K. (2022). Psychological empowerment of women entrepreneurs: A netnographic study on twitter. *Management Research Review*, *45*(6), 717–734.

Dal Mas, F., & Paoloni, P. (2020). A relational capital perspective on social sustainability; the case of female entrepreneurship in Italy. *Measuring Business Excellence*, *24*(1), 114–130. doi:10.1108/MBE-08-2019-0086

Danish, A. Y., & Smith, H. L. (2012). Female entrepreneurship in Saudi Arabia: Opportunities and challenges. *International Journal of Gender and Entrepreneurship*.

Ettl, K., & Welter, F. (2010). Gender, context and entrepreneurial learning. *International Journal of Gender and Entrepreneurship*, *2*(2), 108–129. doi:10.1108/17566261011050991

Halkias, D., Nwajiuba, C., Harkiolakis, N., & Caracatsanis, S. M. (2011). Challenges facing women entrepreneurs in Nigeria. *Management Research Review*, *34*(2), 221–235. doi:10.1108/01409171111102821

Javadian, G., & Singh, R. P. (2012). Examining successful Iranian women entrepreneurs: An exploratory study. *Gender in Management*, *27*(3), 148–164. doi:10.1108/17542411211221259

Kothari, T. (2017). Women entrepreneurs' path to building venture success: Lessons from India. *South Asian Journal of Business Studies*, *6*(2), 118–141. doi:10.1108/SAJBS-03-2016-0021

Maziriri, E. T., Nyagadza, B., & Chuchu, T. (2022). Innovation conviction, innovation mindset and innovation creed as precursors for the need for achievement and women's entrepreneurial success in South Africa: entrepreneurial education as a moderator. *European Journal of Innovation Management*.

Panda, S. (2018). Constraints faced by women entrepreneurs in developing countries: Review and ranking. *Gender in Management*, *33*(4), 315–331. doi:10.1108/GM-01-2017-0003

Roomi, M. A., & Harrison, P. (2010). Behind the veil: Women-only entrepreneurship training in Pakistan. *International Journal of Gender and Entrepreneurship*, *2*(2), 150–172. doi:10.1108/17566261011051017

Rosca, E., Agarwal, N., & Brem, A. (2020). Women entrepreneurs as agents of change: A comparative analysis of social entrepreneurship processes in emerging markets. *Technological Forecasting and Social Change*, *157*, 120067. doi:10.1016/j.techfore.2020.120067

Sharma, R. L., & Arora, M. (2015). Communication as the mainstay of entrepreneurial leadership: A conceptual framework. *Management Dynamics*, *15*(2), 20–28. doi:10.57198/2583-4932.1081

Suseno, Y., & Abbott, L. (2021). Women entrepreneurs' digital social innovation: Linking gender, entrepreneurship, social innovation and information systems. *Information Systems Journal*, *31*(5), 717–744. doi:10.1111/isj.12327

Vij, M., Upadhya, A., Agha, K., Begum, V., Kashmoola, B. W., & Senior, J. R. (2023). Female economic empowerment through tourism training in the United Arab Emirates. *Worldwide Hospitality and Tourism Themes*, *15*(1), 18–28. doi:10.1108/WHATT-08-2022-0099

Wang, Q. (2019). Gender, race/ethnicity, and entrepreneurship: Women entrepreneurs in a US south city. *International Journal of Entrepreneurial Behaviour & Research*, *25*(8), 1766–1785. doi:10.1108/IJEBR-05-2017-0156

Welsh, D. H., Memili, E., Kaciak, E., & Al Sadoon, A. (2014). Saudi women entrepreneurs: A growing economic segment. *Journal of Business Research*, *67*(5), 758–762. doi:10.1016/j.jbusres.2013.11.040

Zamberi Ahmad, S. (2011). Evidence of the characteristics of women entrepreneurs in the Kingdom of Saudi Arabia: An empirical investigation. *International Journal of Gender and Entrepreneurship*, *3*(2), 123–143. doi:10.1108/17566261111140206

Chapter 2
Women's Empowerment and Tourism Development: An Indian Perspective

Amrik Singh
https://orcid.org/0000-0003-3598-8787
Lovely Professional University, India

Sumitha Anantharajah
Taylor's University, Malaysia

Ruth Sabina Francis
https://orcid.org/0000-0002-1925-1155
Taylor's University, Malaysia

ABSTRACT

Women's empowerment has become a crucial component of sustainable tourism development in tourist destinations. Meanwhile, achieving gender equality and empowering women constitutes the fifth goal of the United Nations Sustainable Development Agenda (UN-SDGs). Since 2007, UNWTO has been working—through its Ethics, Culture, and Social Responsibility Department—in partnership with UN Women and a range of external partners across the globe, to bring gender issues to the forefront of the tourism sector, promoting gender equality and women's empowerment and encouraging member states to mainstream gender issues in their respective tourism policies. The present study aims to investigate the influence of women's empowerment on tourism development in India. The current study aims to investigate the perceptions of women's work in tourism and women's entrepreneurship as predictors of women's empowerment in the tourism sector. Limitations and directions for future research are also addressed.

DOI: 10.4018/978-1-6684-8417-3.ch002

INTRODUCTION AND BACKGROUND OF THE STUDY

The United Nations World Tourism Organization (UNWTO) recognizes the crucial role of empowering women in the tourism industry, aligning with the Fifth Sustainable Development Goal (UNWTO, 2020). Women constitute a valuable resource for any nation, yet they often face significant inequalities. Empowering women economically and socially is essential for achieving gender equality and inclusive development (UNWTO, 2020). The tourism sector, being a major industry worldwide, offers opportunities to promote women's empowerment and address broader societal issues such as poverty reduction, gender equality, and women's inclusion in leadership positions (Ateljevic et al., 2018).

The Indian government has acknowledged the need to prioritize women's development and empowerment within the tourism sector, shifting the focus from welfare to targeted initiatives (Gupta & Shah, 1999). As women comprise a significant proportion of the workforce in the tourism industry, their empowerment becomes crucial for sustainable development and the realization of human rights for all (UNWTO, 2020). Rural-based tourism activities in India have demonstrated positive outcomes in terms of socio-economic upliftment, decision-making within households, and increased participation in education (Slathia et al., 2015; Ambardar& Singh, 2017).

Although the concept of women's empowerment in tourism has gained attention in academic discourse, there is still a need for further research to fully understand its implications and potential (Connell &Rugendyke, 2016; Stoyanova et al., 2021). The existing literature highlights the multifaceted nature of women's empowerment and its various dimensions, including economic empowerment, access to education, and political participation (Cobanoglu et al., 2019). Additionally, the empowerment of women in the tourism industry can be seen through their increased control over natural resources and their active involvement in decision-making processes (UNWTO, 2011).

This study aims to contribute to the existing body of knowledge by showcasing positive practices and exploring the link between women's empowerment and tourism development. By examining successful initiatives, the study will shed light on how women's empowerment in the tourism sector can address broader societal issues, such as poverty reduction, gender equality, and equal opportunities (Sofield, 2017). Through an empirical analysis, this research will provide valuable insights into the impact of psychological empowerment on women's empowerment through tourism involvement.

In conclusion, women's empowerment in the tourism industry is crucial for achieving sustainable development and the United Nations' gender equality goals. The Indian context offers opportunities to explore and promote women's empowerment through tourism, addressing various societal challenges. By examining the existing literature, showcasing positive practices, and conducting empirical research, this study aims to contribute to a deeper understanding of women's empowerment in the context of tourism development.

FOREIGN TOURIST ARRIVAL IN INDIA

International tourist arrivals in India spiked to 6.19million in 2022 compared to 2021 although 43% lower than 2019. With the opening of borders, increased air connectivity, enhanced visa procurement processes in place the Indian government predicts a 100% recovery in the international tourists' arrivals in India (Phadnis, 2023). Tourism in any country is a crucial employment generator apart from being

Figure 1. Foreign tourist arrival
Source: Ministry of Tourism Report (2022)

the economy driver. The government's role evolved from being a mere regulator to a catalyst in the development of Tourism (Lok Sabha Secretariat, 2013).

DIFFERENT FORMS OF TOURISM IN INDIA

Tourism in India would contribute $512 billion to the country's GDP by 2028 and eventually would create 53million jobs by 2029 as cited by Ms.JyotiMayal, President of Travel Agents Association of India (The Hindu Bureau,2023). In tandem with this exciting prediction there is a surge in the development of tourism in India, such as Water Tourism from various states predominantly in Goa, eco-tourism sites being developed in Uttarakhand, Mokarsagar wetland in Porbandar district, Medical tourism possibilities in Amritsar, many states being declared as Heritage sites, and the recent launch of Aatithyam dashboard for real time data on domestic and international tourists in the Gujarat State (TOI, 2023). Developments in the tourism sector has further paved ways for investment opportunities and aided much in the sustenance of trade, handicrafts and the rich culture of India. In terms of foreign exchange earnings and creation of employment opportunities, tourism is the biggest sector in the Indian economy (Kumar & Singh, 2019).

UNWTO'S STANDING ON TOURISM AND WOMEN EMPOWERMENT

In itsaims to strengthen the UNSDG Goal 5 Gender Equality and Women's Empowerment various policies and action plans have been put in place by UNWTO. Since Tourism has proven to be a champion of gender equality UNWTO has established new guidelines that will help both governments and businesses exploit the Tourism sector's power to drive women empowerment (WTO,2023). The action plan further developed from the *global report on Women in Tourism* enlists 6 sectors namely, employment, Entrepreneurship, Leadership – policy -decision making, Education-Training, Community – civil society and Measurement for better policies (UNWTO,2010). In the Entrepreneurship sector opportunities for women in tourism are as listed below.

- Plans to formalize women's tourism businesses and financial development.
- To expand and diversify women's market access and fair trade for their tourism products and services.

- Support women to expand and diversify their tourism products and services.
- Measures to improve Work-life Balance for women in the tourism industry.
- Measures to expand women's access to digital technologies.

RESEARCH METHODOLOGY

Since the research scope is wide linking both theory and practice the research paradigm for this study is a post-positivist paradigm. Currently, this study is mainly based on secondary data in order to achieve the objectives of the study. The secondary data sources such as previous studies, research articles, ministry reports, industry reports, journals, magazines etc. The study would embark on a sequential mixed-method approach to thoroughly validate the propositions made.

DISCUSSION AND INTERPRETATION

Women Empowerment in Indian Tourism

Women empowerment is part of the worldwide movement and is defined as promoting women's self-worth, control over choices and right to impact social change around them.With the above being in place for the development of women in tourism sector it is also essential to know that women's empowerment in terms of economy is crucial to a nation's progress. Women entrepreneurs and women owned businesses are outpacing the overall economy as pointed out by (Sharma, Yadava, Aarif, Anandaram, Alalmai and Singh, 2022). Although there is some research about women in tourism especially mountain based tourism, ethnic tourism, impacts of tourism on village women and ecotourism for women (Scheyvens, 2000), there is still a dearth of literature in women empowerment in tourism industry in India.

With women making up half of the workforce and population globally (Shivakoti, 2022), tourism sector overall emphasizes on women empowerment. Women in the tourism and hospitality are overrepresented in the lower-level positions and they have lesser opportunities to reach the management positions according to (Clevenger & Singh, 2013; Carvalho, Costa, Lykke, Torres & Wahl, 2018; Simao& Breda, 2014). Although many hotel industries have placed women in the top management positions such as a Hilton, Marriott International, in Asia there is still room for growth opportunities expansion for women in India. The role of Government is crucial in channeling the attention towards creation of women centric incubators for entrepreneurship growths specifically in the tourism industry.

Psychological Empowerment

Psychological empowerment is a significant approach that has garnered attention in various industries. It encompasses both the structural aspect, which involves decision-making authority, and the psychological aspect, which relates to employees' motivational state within organizations (Francis & Alagas, 2020). Bandura's (1977) theory of self-efficacy supports the notion that psychological empowerment contributes to a harmonious work environment (Thomas &Velhouse, 1990).

The four key constructs of psychological empowerment—meaning, self-efficacy, self-determination, and impact—have been extensively studied in diverse industry settings. Notably, Elshaer, Moustafa,

Sobaih, Aliedan, and Azazz (2021) examined these constructs from the perspective of women and their attributes in the workplace. Meaning refers to the value that women attribute to job objectives based on their own standards (Spreitzer, 1995). Self-efficacy pertains to women's perceived competence to perform specific job tasks (Gist, 1987). Self-determination reflects women's sense of autonomy and control over their work behaviors and procedures, leading to outcomes such as flexibility, creativity, proactiveness, resilience, and self-regulation (Thomas &Velthouse, 1990; Bell &Staw, 1989). Lastly, impact refers to the extent of women's influence on strategic and administrative outcomes in the workplace (Spreitzer, 1995; Ashforth, 1989).

Understanding the psychological empowerment constructs specific to women is crucial for promoting their empowerment in the tourism industry and creating an inclusive and supportive work environment. By recognizing and fostering these constructs, organizations can enhance women's job satisfaction, motivation, and overall empowerment, contributing to their professional growth and advancement.

Tourism Involvement

Tourism is a sector that has generated many job opportunities for both men and women. The development of the tourism industry in many countries has an impact on many workers working in this industry. The women who are unemployed previouslyor just housewives could earn additional income to support their family (Irawan& Nara, 2020). Furthermore, they even support in the enhancement of women in community development. But several research state that the role of women in tourism dominates the informal role such as food and beverage services, selling hawkers, cleaning public places, making souvenirs, and working in the hospitality sector (Tristanti, Nurhaeni, Mulyanto, Sakuntalawati, 2022). These jobs of women are classified as low-skilled jobs. Women are influenced by several factors when carrying out a role in tourism such as education, women's physical condition and women's family background (Nurhaeni, IsmiDwi Astuti, et al., 2019).

Affective Events Theory (AET)

The study will be founded upon the Affective Events Theory (AET) explains how emotional events in the workplace can influence a person's emotions that in turn affect the attitudes and work behaviours (Weiss &Cropanzano, 1996). According toGreenberg, Ashton-James &Ashkanasy, (2007) AET is the seminal explanation of affective experiences at work. The relationship between an individual's experiences and herbehaviour is influenced by the role of emotion and its critical judgment (Weiss &Cropanzano, 1996). AET claims that women's affective response to workplace events or environment will extensively drive women's attitude and consequently their behaviour. Where the employee especially women employee's mood and emotions are the result of affect and the evaluative judgment is the result of attitude (Carlson, Kacmar, Zivnuska, Ferguson & Whitten, 2011). When AET was initially formulated Weiss and Cropanzano (1996) stated that they offered AET as a roadmap for future research to integrate research into studying emotions at the workplace. AET, proposes that work features play a vital role in shaping the regular events at the workplace such as emotions displayed by service employees are key to regulate discrete emotions in the hotel management research (Chen, Wang, & Lam, 2019). As implicit it is in the theory that AET includes the affective states that people experience after significant workplace events, hence emotions represent affective reactions at something (Weiss &Cropanzano, 1996).

Relationship Between Psychological Empowerment and Tourism Involvement

Tourism involvement has been highlighted as playing a significant role in many countries. These will develop tourism of local people to initiatives in developing tourism attractions in their places (Jeong, Ryu, & Brown, 2018).It is important for local people to take responsibility for their participation and engagement in tourism development (Jackson &Morpeth, 1999).Thus, the psychological empowerment of local people working related to tourism will also influence gender roles and it also may lead to the disempowerment of individuals (Aghazamani& Hunt, 2017). Furthermore ,this empowerment may result from direct economic benefits that tourism generates for local people (Scheyvens, 1999). Through this development there will be increased employment opportunities especially for women. This psychological empowerment does not pertain just for jobs but also to natural and other financial resources, political influence, government opportunities (Strzelecka, & Wicks, 2015). This is also the indication of economic empowerment which will improve local facilities and utility systems, and involve equitable distribution of economic benefits (Scheyvens, 1999).

Based on the above arguments and study samples the following proposition is developed:

PROPOSITION 1: PSYCHOLOGICAL EMPOWERMENT HAS A POSITIVE INFLUENCE ON TOURISM INVOLVEMENT

Relationship Between Tourism Involvement and Women Empowerment

Tourism involvement in woman empowerment has become an on-going issue in every country (Bayeh, 2016). Some researchers have argued that some country which recognizes only men's decision-making, and participation will not involve woman unless they are empowered(Irawan& Nara, 2020). Optimizing the role of women in tourism industry will provide a broad opportunity for them (Tristanti,et al., 2022). Woman has rights to participate in tourism activities as part of society. The involvement of women in tourism can reduce gender inequality in development. The promotion of equality and women's empowerment became one of the agendas in the UN, 2023. Scheyvens (1999), has developed an empowerment framework that can be used as the tool for analyzing the influence of tourism on the local community. This framework consists of four dimensions which are economically, socially, psychologically, and politically. In her research the women empowerment will help to increase women's self-esteem and status within their community. Furthermore, it also promotes sustainable development in their community. Even though there are many gender inequalities in the tourism sector, yet tourism has enough potential for empower women by providing them better opportunities as the employees, entrepreneurs, and leaders than other sectors of the economy (Scheyvens & Van, 2021).

Based on the above arguments and study samples the following proposition is developed:

PROPOSITION 2: TOURISM INVOLVEMENT HAS A POSITIVE INFLUENCE ON WOMEN EMPOWERMENT

Based on the Affective Events Theory it is plausible to relate women's psychological empowerment in tourism involvement and the same will have an effect on women empowerment. Through this framework

Figure 2. Conceptual framework on women empowerment through tourism involvement

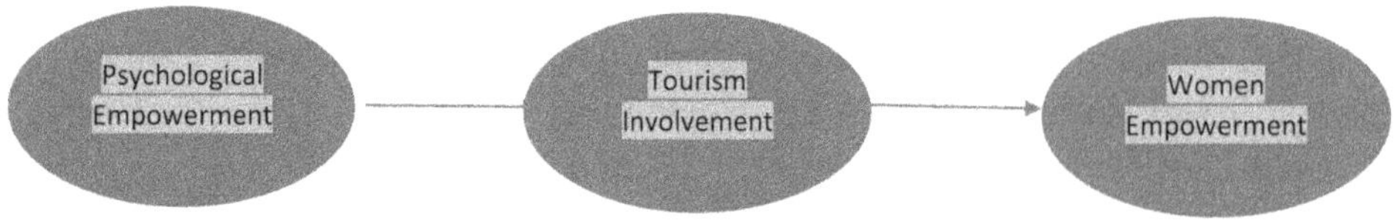

women are able to further support and develop the economy of the country. This women empowerment will be able to provide training to women in skills, language, and planning and promoting women by giving them more opportunities to participate in tourism activities.

RECOMMENDATIONS

The study strongly recommends integrating Psychological Empowerment competency into the training and development programs for women in the tourism sector. This can be achieved by implementing the following recommendations:

1. Awareness and Education: Conduct workshops and training sessions to raise awareness about the concept of Psychological Empowerment among women in the tourism industry. Educate them about the importance of self-belief, self-confidence, and resilience in overcoming challenges and achieving personal and professional growth.
2. Skill Development: Offer specific training programs that focus on enhancing Psychological Empowerment skills, such as self-efficacy, autonomy, and a sense of control over one's work environment. Provide practical tools and strategies that empower women to effectively manage their roles and responsibilities.
3. Mentoring and Coaching: Establish mentorship and coaching programs where experienced professionals can guide and support women in their career development. These programs can help women build their self-esteem, assertiveness, and leadership abilities, enabling them to navigate the tourism industry with confidence.
4. Networking Opportunities: Create networking platforms and events that facilitate connections between women in the tourism sector. Encourage collaboration, knowledge sharing, and the exchange of experiences, fostering a sense of community and empowerment among women professionals.
5. Recognition and Advancement: Recognize and reward women who demonstrate Psychological Empowerment competencies and contribute to the growth and development of the tourism industry. Provide equal opportunities for career advancement and promotions based on merit and capabilities.
6. Organizational Support: Foster an inclusive and supportive work environment that values and promotes Psychological Empowerment. This includes implementing policies that ensure equal opportunities, flexible work arrangements, and a culture that encourages open communication, respect, and diversity.

By incorporating these recommendations, the training and development programs in the tourism sector can effectively empower women, enabling them to embrace leadership roles and contribute to the growth and success of the industry.

FUTURE RESEARCH

While the current study is conceptual in nature and focuses on the tourism sector in India, it is acknowledged that empirical research is essential to validate the propositions put forward. An empirical study would provide valuable insights into the positive influence of Psychological Empowerment on Women Empowerment through Tourism Involvement within the tourism sector.

Although there is a significant proportion of empowered women in the tourism industry, it is recognized that women's empowerment still requires considerable attention. Limited knowledge and support from organizations can hinder the progress of women empowerment initiatives. Therefore, it is crucial to facilitate and promote education on women empowerment within the tourism sector. This can involve raising awareness about women's rights, gender equality, and the importance of creating inclusive and supportive work environments.

Additionally, organizations should provide necessary support for women in the community, such as mentoring programs, networking opportunities, and career development initiatives. By offering these resources, organizations can help women overcome barriers and enhance their empowerment within the tourism industry.

To comprehensively understand the dynamics of women's empowerment in the tourism sector and to develop effective strategies, further empirical research is needed. This research should focus on examining the specific challenges faced by women in the industry, identifying successful empowerment initiatives, and evaluating their impact on women's lives and careers.

CONCLUSION

Women empowerment is a crucial aspect of achieving sustainable development goals, and the tourism industry has the potential to play a significant role in fostering gender equality and empowerment. This framework aims to contribute new knowledge and insights to the existing literature on the tourism industry by emphasizing the development of women empowerment initiatives in the Indian context. By aligning with the United Nations Sustainable Development Goal 5 (UNSDG 5) on gender equality, this framework seeks to empower women through active involvement in tourism-rclated industries, hospitality sectors, and the general community. The framework emphasizes collaboration and support from policymakers, industry players, and the research fraternity to ensure its dynamic evolution with appropriate guidelines and effective communication.

1. Understanding the Current Status: The framework begins with a comprehensive assessment of the current status of women's involvement in the tourism industry in India. This includes analyzing existing data, identifying barriers and challenges faced by women, and recognizing successful initiatives that have already been implemented.

2. Capacity Building and Skill Development: To empower women, it is essential to enhance their skills and capabilities. This stage of the framework focuses on designing and implementing capacity building programs that provide training and skill development opportunities specifically tailored to the needs of women in the tourism and hospitality sectors. These programs could include vocational training, entrepreneurship development, leadership and management skills, and financial literacy.
3. Creating Supportive Policies and Regulations: Policies and regulations play a crucial role in promoting women's empowerment in the tourism industry. This stage involves advocating for gender-responsive policies that address issues such as equal pay, safe working conditions, and protection against harassment. Collaboration with policymakers and industry associations is vital to ensure that these policies are effectively implemented and enforced.
4. Encouraging Entrepreneurship and Leadership: Promoting women's entrepreneurship and leadership opportunities is a key aspect of this framework. Efforts will be made to create an enabling environment for women to start their own tourism-related businesses, such as providing access to finance, mentorship programs, and networking platforms. Recognizing and celebrating successful women leaders in the industry will also be emphasized to inspire others.
5. Fostering Community Engagement: The framework acknowledges the importance of community engagement in driving women empowerment in the tourism sector. It aims to promote awareness and change attitudes within local communities by organizing campaigns, workshops, and cultural exchange programs. Engaging community leaders, organizations, and residents will be instrumental in challenging traditional gender norms and fostering inclusivity.
6. Research, Monitoring, and Evaluation: Continuous research, monitoring, and evaluation are crucial to ensure the effectiveness of the framework and identify areas for improvement. Collaboration with the research fraternity will facilitate the collection of data, analysis of outcomes, and the identification of best practices. Findings will inform policy adjustments, program modifications, and future initiatives.

This dynamic framework for developing women empowerment through tourism involvement in India aligns with UNSDG 5 and aims to contribute to the existing literature on the tourism industry. By promoting collaboration among policymakers, industry players, and the research fraternity, this framework seeks to empower women and create a supportive environment in the tourism and hospitality sectors. Through capacity building, supportive policies, entrepreneurship promotion, community engagement, and continuous research, monitoring, and evaluation, this framework strives to create sustainable change and foster gender equality in the Indian tourism industry.

REFERENCES

Aghazamani, Y., & Hunt, C. A. (2017). Empowerment in tourism: A review of peer-reviewed literature. *Tourism Review International*, *21*(4), 333–346. doi:10.3727/154427217X15094520591321

Ambardar, M. R., & Singh, S. (2017). Empowerment and socio-economic upliftment of rural women through tourism: A case study of Bhaderwah, Jammu & Kashmir. *International Journal of Scientific Research*, *6*(5), 346–352.

Ashforth, B. E. (1989). The experience of powerlessness in organizations. *Organizational Behavior and Human Decision Processes*, *43*(2), 207–242. doi:10.1016/0749-5978(89)90051-4

Ateljevic, I., Pritchard, A., & Morgan, N. (2018). Gender and tourism: A feminist perspective. In The Routledge Handbook of Transport, Tourism and Mobilities (pp. 183-195). Routledge.

Bayeh, E. (2016). The role of empowering women and achieving gender equality to the sustainable development of Ethiopia. *Pacific Science Review B. Humanities and Social Sciences*, *2*(1), 37–42.

Bell, D. E., & Staw, B. M. (1989). People as sculptors versus sculpture: The roles of personality and personal control in organizations. In L. L. Cummings & B. M. Staw (Eds.), *Research in Organizational Behavior* (Vol. 11, pp. 1–39). JAI Press. doi:10.1017/CBO9780511625459.014

Carlson, D., Kacmar, K. M., Zivnuska, S., Ferguson, M., & Whitten, D. (2011). Work-family enrichment and job performance: A constructive replication of affective events theory. *Journal of Occupational Health Psychology*, *16*(3), 297–312. doi:10.1037/a0022880 PMID:21728437

Carvalho, I., Costa, C., Lykke, N., Torres, A., & Wahl, A. (2018). Women at the top of tourism organizations: Views from the glass roof. *Journal of Human Resources in Hospitality & Tourism*, *17*(4), 397–422. doi:10.1080/15332845.2018.1449551

Chen, B. Y., Wang, Y., Wang, D., & Lam, W. H. (2019). Understanding travel time uncertainty impacts on the equity of individual accessibility. *Transportation Research Part D, Transport and Environment*, *75*, 156–169. doi:10.1016/j.trd.2019.08.027

Clevenger, L., & Singh, N. (2013). Exploring barriers that lead to the glass ceiling effect for women in the US hospitality industry. *Journal of Human Resources in Hospitality & Tourism*, *12*(4), 376–399. doi:10.1080/15332845.2013.790258

Cobanoglu, C., Kasim, A., & Nadiri, H. (2019). The impact of women's empowerment on destination loyalty and word of mouth: The mediating role of destination satisfaction. *Journal of Travel Research*, *58*(1), 111 126.

Connell, J., & Rugendyke, B. (2016). Gender and tourism: Women 's experiences in tropical North Queensland. In Tourism and Gender (pp. 17-31). Channel View Publications.

Elshaer, I. A., Moustafa, M. M., Sobaih, A. E. E., Aliedan, A. M., & Azazz, A. M. (2021). The mediating role of psychological empowerment in the relationship between transformational leadership and women's job performance. *Journal of Organizational Change Management*, *34*(2), 541–560.

Francis, B., & Alagas, M. (2020). Structural empowerment and psychological empowerment: Effects on innovative behavior among police officers. *Journal of Police and Criminal Psychology*, *35*(4), 441–450.

Garcia-Ramon, M. D., Canoves, G., & Valdovinos, N. (1995). Farm tourism, gender and the environment in Spain. *Annals of Tourism Research*, *22*(2), 267–282. doi:10.1016/0160-7383(94)00096-4

Gist, M. E. (1987). Self-efficacy: Implications for organizational behavior and human resource management. *Academy of Management Review*, *12*(3), 472–485. doi:10.2307/258514

Greenberg, J., Ashton-James, C. E., & Ashkanasy, N. M. (2007). Social comparison processes in organizations. *Organizational Behavior and Human Decision Processes*, *102*(1), 22–41. doi:10.1016/j.obhdp.2006.09.006

Irawan, N., & Nara, V. (2020). Managing women empowerment through participation in sustainable tourism development in Kampong Phluk, Siem Reap, Cambodia. *International Journal of Economics, Business and Accounting Research, 4*(2).

Jackson, G., & Morpeth, N. (1999). Local Agenda 21 and community participation in tourism policy and planning: Future or fallacy. *Current Issues in Tourism*, *2*(1), 1–38. doi:10.1080/13683509908667841

Jeong, E., Ryu, I., & Brown, A. (2018). Moderating effect of sense of community on the relationship between psychological empowerment and tourism policy participation of local residents. *Global Business & Finance Review*, *23*(1), 36–46. doi:10.17549/gbfr.2018.23.1.36

Kumar, A., & Singh, G. (2019). Seasonal effect on tourism in India. *Journal of Financial Economics*, *7*(2), 48–51.

Lok Sabha Secretariat. (2013). *Tourism Sector in India.* Retrieved from http://164.100.47.134./internet/Tourismsectorin India.pdf

Mrema, A. A. (2015). Tourism and women empowerment in moduli district, arusha-Tanzania. *African Journal of Hospitality, Tourism and Leisure*, *4*(2), 1–14.

Nurhaeni, I. D. A., Popping, R., Sugiarti, R., Nugroho, R. A., &Pratiwi, R. D. (2019). The dynamics of gender differences on tourism governance in Indonesia. *BISNIS & BIROKRASI: JurnalIlmuAdministrasi dan Organisasi, 26*(3).

Panta, S. K., & Thapa, B. (2017). Entrepreneurship and women's empowerment in gateway communities of Bardia National Park, Nepal. *Journal of Ecotourism*, 1–23. doi:10.1080/14724049.2017.1299743

Phadnis, A. (2023, February 10). *Foreign tourists quadrupled to 6.19 mn in 2022, but 43% lower than 2019*. Business Standard News. Retrieved March 24, 2023, from https://www.business-standard.com/article/current-affairs/foreign-tourists-quadrupled-to-6-19mn-in-2022-but-43-lowerthan

Routledge.Strzelecka, M., & Wicks, B. E. (2015). Community participation and empowerment in rural post-communist societies: Lessons from the leader approach in Pomerania, Poland. *Tourism Planning & Development*, *12*(4), 381–397. doi:10.1080/21568316.2015.1013564

Russell, R., & Faulkner, B. (1999). Movers and shakers: Chaos makers in tourism development. *Tourism Management*, *20*(4), 411–423. doi:10.1016/S0261-5177(99)00014-X

Russell, R., & Faulkner, B. (2004). Entrepreneurship, chaos and the tourism area lifecycle. *Annals of Tourism Research*, *31*(3), 556–579. doi:10.1016/j.annals.2004.01.008

Scheyvens, R. (2000). Promoting women's empowerment through involvement in ecotourism: Experiences from the Third World. *Journal of Sustainable Tourism*, *8*(3), 232–249. doi:10.1080/09669580008667360

Scheyvens, R., & van der Watt, H. (2021). Tourism, empowerment and sustainable development: A new framework for analysis. *Sustainability (Basel)*, *13*(22), 12606. doi:10.3390u132212606

Sharma, N., Yadava, A. K., Aarif, M., Anandaram, H., Alalmai, A., & Singh, C. (2022). Business Opportunities And Challenges For Women In The Travel And Tourism Industry During Pandemics Covid-19. *Journal of Positive School Psychology*, 897–903.

Shivakoti, A. (2022). Women in Tourism and Hospitality-A Case Study of Bhaktapur. *Journal of Tourism & Adventure*, *5*(1), 79–96. doi:10.3126/jota.v5i1.48739

Simão, P., & Breda, Z. (2014). Gender and leadership in the tourism sector: Th e case of the restaurants industry in the city of Aveiro, Portugal. *Revista Turismo & Desenvolvimento (Aveiro)*, *1*(21/22), 341–352.

Slathia, P. S., Paul, N., & Nain, M. S. (2015). Socio-economic Empowerment of Rural Women Through Rural Tourism Projects in Jammu Region of J&K State in India. *Indian Journal of Extension Education, 51*(3-4), 40-43.

Spreitzer, G. M. (1995). Psychological empowerment in the workplace: Dimensions, measurement, and validation. *Academy of Management Journal, 38*(5), 1442-1465.

The Hindu Bureau. (2023, February 3). *The Hindu Bureau*. Retrieved March 24, 2023, from https://www.thehindu.com/business/tourism-to-contribute-512-bn-to-indias-gdp-by-2028-says-taai-chief/article66468359.ece

Thomas, K. W., & Velthouse, B. A. (1990). Cognitive elements of empowerment: An "interpretive" model of intrinsic task motivation. *Academy of Management Review*, *15*(4), 666–681.

TOI. (2023). Tourism News: Latest News on tourism - times of India. *The Times of India*. Retrieved March 24, 2023, from https://timesofindia.indiatimes.com/topic/tourism/news

Tristanti, T., Nurhaeni, I. D. A., Mulyanto, M., &Sakuntalawati, R. D. (2022). The Role of Women in Tourism: A Systematic Literature Review. *KnE Social Sciences*, 545-554.

United Nations, Global Sustainable Development Report (GSDR). (2023). https://sdgs.un.org/gsdr/gsdr2023

UNWTO Global report on women in tourism – second edition. (2010). Retrieved March 24, 2023, from https://www.e-unwto.org/doi/book/10.18111/9789284420384

Van der Cammen, S. (1997). Involving Maasai women. In L. France (Ed.), *The earthscan reader In sustainable tourism* (pp. 162–163). Earthscan.

Weiss, H. M., & Cropanzano, R. (1996). Affective events theory. *Research in Organizational Behavior, 18*(1), 74.

Wilkinson, P. F., & Pratiwi, W. (1995). Gender and tourism in an Indonesian village. *Annals of Tourism Research*, *22*(2), 283–299. doi:10.1016/0160-7383(94)00077-8

World Tourism Organization. (2023, March 23). *New Guidelines Put Women's Empowerment at Heart of Tourism's Restart*. Retrieved March 24, 2023, from https://www.unwto.org/news/new-guidelines-put-women-s-empowerment-at-heart-of-tourism-s-restart

Chapter 3
Gushing Through the Challenges:
Women Entrepreneurs Pushing Boundaries Over Generations in the Tourism Industry

Sweety Mishra
https://orcid.org/0000-0001-7851-8879
Jamia Millia Islamia, India

Nimit Chowdhary
https://orcid.org/0000-0002-7335-2379
Jamia Millia Islamia, India

ABSTRACT

Over the last two decades, women's entrepreneurship has drawn attention because of its noteworthy socioeconomic impact in every field including tourism industry. However, despite the well-documented benefits of women's involvement in entrepreneurship, they continue to confront challenges that restrict them from achieving heights. The existing literature mainly discusses the challenges women face in their entrepreneurial journey, and very few studies address the positively changing scenario. Therefore, this chapter explores the shifting and changing scenario in women's entrepreneurship and to identify what challenges have been overcome and what continues to restrict the growth of women's entrepreneurship. The study attempts to present the challenges and enablers that have changed business dynamics for women. The chapter also analyses how women's entrepreneurship has transformed over time and digs deep into the extant literature to understand if the business environments have become conducive for women over generations. The authors follow an inductive approach to achieve the objectives.

DOI: 10.4018/978-1-6684-8417-3.ch003

INTRODUCTION

Women worldwide are getting more interested in entrepreneurship, which leads to an increase in the number of women starting new businesses and becoming entrepreneurs (Wu et al., 2019; Hechavarria et al., 2019). An increasing number of women starting their businesses benefits the economy and gives them more influence (Radović-Marković, 2015). Globally, women's entrepreneurship is regarded as a significant indicator of inclusive growth (Sajjad et al., 2020), especially in developing countries where women's entrepreneurship is regarded as a significant catalyst for innovation and growth (Franzke et al., 2022). This chapter is mainly focused on women in the tourism industry as it is widely accepted that tourism plays a prominent role in the economic growth of many developing nations, particularly as a means of fostering entrepreneurship and empowering locals, especially women, as the majority of tourism workforce around the world are women (Horodnic & Williams, 2019).

Over the last two decades, women's entrepreneurship has drawn attention because of its noteworthy socioeconomic impact which helps in reducing poverty and inequality prevalent in society, specifically in developing nations where women's entrepreneurship has developed into a reliable source of income and, in some instances, the only source of income for the household (Rosca et al., 2020). However, despite the well-documented benefits of women's involvement in entrepreneurship, women entrepreneurs continue to confront challenges (economic and socio-cultural) that restrict them from achieving heights (Scott et al., 2014). Entrepreneurial behaviours, such as the willingness to take risks or chase innovations that deviate from conventions, are influenced by societal cultural values. Cultures that place high importance on conformity tend to be less open to new ideas (Mashapure et al., 2022). For example, Shane (1992) discover that individuality is associated with increased creativity and that power distance has a detrimental effect on innovation. Davidsson & Wiklund (1997) found that societies that emphasize individual freedom also tend to put a larger emphasis on the pursuit of success and that people who assess themselves highly in terms of their abilities are more likely to start ventures on their own. Aversion to risk may play a role in the creation of innovative businesses. However, if these cultural norms are restrictive, the sustainability of women's entrepreneurship could be affected in the long run (Mashapure et al., 2022). In addition, women entrepreneurs often feel the weight of expectations to manage a home and raise a family especially single moms who feel the strain of unfavourable economic growth when they must prioritise childcare (Morrison et al., 2022). Other challenges include 'gender biases' which is a hindrance for women owing to the patriarchal mentality that diminishes the sustainability of women entrepreneurs (Mandipaka, 2014; Linke et al., 2011; Winn, 2005). Tur-Porcar et al. (2017) argued that it is challenging for female entrepreneurs to have access to sufficient capital to launch their desired businesses. And they also face trust issues while working with microfinance institutions (Wasilczuk & Zieba, 2008). Although the relation between gender and repayment underlines the fact that women have a higher rate of repayment than males do (D'Espallier et al., 2011). Still, financial instability is the leading deterrent that female entrepreneurs encounter at the commencement of their firms and various studies outlined inadequate funding as the biggest rationale due to which, women entrepreneurs were forced to shut down their businesses (Boden & Nucci, 2000; Du Rietz & Henrekson, 2000; Watson, 2003). Agrawal (2018) argues that female entrepreneurs' lack of financial security and their ingenuity when bargaining with microfinance institutions act as a barrier to credit access. Most financial institutions are cautious of investing in businesses run or managed by women because they see those as a high-risk endeavours and lack faith in women's abilities in the business dynamics. It is common for women to lack creditworthiness and the inability to insure a loan with collateral (Mauchi, 2014). Patriarchal societies often discourage women business owners from accumulating such assets. So, they must rely on savings, investments, and loans from friends, relatives, partners, and co-workers (Hillman & Radel, 2023). However, things have changed over a while, and a substantial shift in appreciating and acknowledging the significance of women entrepreneurs and effectively aiding them in expanding their enterprises by providing them more assistance through funding, guidance from established professionals, and networking. Government, funding agencies, banks, and other organizations are taking various initiatives and enacting schemes and programmes to ensure that women entrepreneurs thrive. Initially, women's engagement in business was motivated by necessity rather than an opportunity, but now improved social conditions, increased access to higher education, and government support create more opportunities for women. Consequently, women entrepreneurship is increasingly becoming opportunity-based rather than necessity-based due to new possibilities (Mota et al., 2019).

In 2020, the sudden outbreak of the COVID-19 pandemic had devastating effects on the world economy, and it will still take some time to recover from it (Raman et al., 2022). The pandemic caused lockdowns all around the world due to the fear of spreading infection. This caused a sharp drop in demand for travel and tourism businesses like hotels, airlines, and tour operators, which led to a huge decline in income and jobs. This economic crisis unfairly affected female entrepreneurs too because women-led enterprises are particularly vulnerable to economic crises as they have usually lower average firm age and size and they are overrepresented in industries that are particularly hard impacted by economic crises (Popović-Pantić et al., 2020). The pandemic has had detrimental effects on business, but it has also generated new opportunities due to technological advancement. New policies are being implemented globally to help and promote women led businesses (OECD, 2017), and great strides have been made, especially during the Covid–19 pandemic, to uplift women-entrepreneurship (European Commission, 2020). The United Nations tourism organization (UNWTO) launched a project called ***'Centre Stage:** Women's empowerment during the COVID-19 recovery'*. Deutsche Gesellschaft für Internationale Zusammenarbeit (GIZ) on behalf of the Federal Ministry for Economic Cooperation and Development of Germany and UN Women is supporting the implementation of this project. As the world is recovering from the pandemic, the project aims to strengthen, coordinate, and concentrate efforts aimed at achieving gender equality in tourism governmental institutions and businesses. As per the Global Entrepreneurship Monitor (GEM) 2022 worldwide report on the status and perceptions of entrepreneurship activities, the percentage of people who perceived strong chances in establishing a new business dropped significantly in 2021 compared to 2019. Studies on the effects of covid- 19 on entrepreneurship highlighted the invisibility of women entrepreneurs in developed countries like USA, UK, and Canada due to the deep structural inequalities. Due to the lack of access to managerial and financial resources women's entrepreneurship was impacted most by the pandemic. However, the report shares some positive facts, such as nearly half of the early-stage female entrepreneurs acknowledged that the pandemic generated new business opportunities (GEM, 2022; Hughes et al., 2022).

The existing literature mainly discusses the prevailing challenges women face in their entrepreneurial journey, and very few studies address the positively changing scenario. Therefore, this chapter aims to explore the shifting and changing scenario in women's entrepreneurship and to identify what challenges have been overcome or what continues to restrict the growth of women's entrepreneurship. The chapter also analyses how women's entrepreneurship has transformed over time and digs deep into the extant literature to understand if the business environments have become conducive for women over generations.

METHODOLOGY

The chapter uses an inductive approach to achieve the objectives. Multiple cases are analysed to understand the changing scenario in women's entrepreneurship. Each case explores a distinct facet of women's entrepreneurship. The application of the case study method in practical scenarios facilitates extensive evaluations from diverse perspectives. (Crowe et al., 2011). The cases were chosen carefully to represent various locations in India. These cases highlighted examples of women entrepreneurs who overcame considerable obstacles to attain success in the tourism industry. Based on the cases, a comparative analysis is performed to analyse the transformation in women's entrepreneurship. By providing a comprehensive analysis of the research carried out in a specific discipline (Mentzer and Kahn, 1995), literature reviews support in incorporating and refining scattered knowledge (Meredith, 1993: 4). Women entrepreneurs in the tourism business are the focus of this research.

REVIEW OF LITERATURE

Women Entrepreneurs: An Evolution

Preliminary literature mainly focused on male entrepreneurs, it was only in the latter half of the 1900s that a specific sub-field of women's entrepreneurship emerged (Jennings and Brush, 2013) and in 1976 first paper on women entrepreneurship was published by Schwartz (Schaltegger & Wagner, 2011). Before the 1990s, established academic journals and major media in the United States considered women-led enterprises as single proprietorships or tiny lifestyle businesses (Baker et al. 1997), and the traditionally male-dominated business model was commonly considered as the norm (Sarfaraz et al., 2014). However, with time, the focus shifted towards women entrepreneurs and entrepreneurship, and they were defined as those who make decisions regarding the direction of a company on their own (Sutcliffe-Braithwaite & Lawrence, 2019) and then seek out, create, and use innovative approaches to business to seize profitable opportunities (Iwu et al., 2019). Their involvement in entrepreneurship is considered a way of economic empowerment and is strongly related to strengthening the status of women, families, and communities (Haugh and Talwar, 2016; Scott et al., 2012; Ardrey et al., 2006; Jamali, 2009). Characteristics of women in business suggest that they should be acknowledged as equal partners in business due to the worth of their ability, knowledge, and capability (Giantari et al., 2018; Galbreath, 2017; Kusumawardani et al., 2020).

Entrepreneurship empowers women to make their own choices and gain independence from the constraints imposed by patriarchal ideology (Batliwala, 1994). In recent years, there has been growing evidence that women's participation in entrepreneurial activity and economic growth is crucial. Globally, women entrepreneurs are significantly engaged in innovative, growth-oriented enterprises centred on international and domestic markets. Around one-third of all entrepreneurs in upper-middle-income nations create high-growth enterprises, both in terms of the number of employments generated at startup (20+) and the number of employments anticipated within five years (20+). Also, one-third of all innovators in national and worldwide marketplaces are women-owned enterprises. The innovation rates of women were highest in upper-middle-income nations, particularly at the international level, where they were on par with males. In addition, more than two-fifths of entrepreneurs in high- and upper-middle-income nations focused on foreign markets, compared to one-third of women in low-income countries (GEM, 2022). Women throughout the world are creating thriving enterprises in every sector of the economy.

Nonetheless, the mechanisms through which cultural attitudes and structural inequities perpetuate gender gaps in business creation and expansion remain poorly understood.

Changing scenario in women's entrepreneurship has been discussed through cases one by one. Here, the first case shows the initiatives taken by the government to provide skill-based training programmes to women and it is helping women to take up jobs and engage in entrepreneurial activities with the help of technology. This real-world example explored how the government's efforts to encourage women into the workforce and to take up entrepreneurial activity through training programmes have contributed to women's empowerment.

CASE 1: NISHA BEN – PROMOTING HOMESTAY TOURISM IN UTTARAKHAND, INDIA

Nisha is a 23 old young entrepreneur from Kandara village Rudraprayag district in Uttarakhand, India with a degree in Sanskrit and home science responsible for taking care of her family in the absence of her father who works in a nearby town. In her village, almost everyone is using the phone as a medium of communication between people but only a few of them are digitally literate. When Nisha got her first mobile phone in her first year, she came to know about SEWA Bharat's apprentice training programme to run homestays marketed through online platforms (Atithi). SEWA Atithi, founded in 2019 a women-run enterprise that supports women in diversifying their incomes by developing community-based tourism via training programmes. After completing three months of training programme, she was formally inducted as an *Udyami Sakhi*[1] and given charge of Athithi Homestays to manage its everyday functionality. Following the COVID-19 pandemic, the Udyami Sakhi initiative has allowed women microentrepreneurs and women-led collective social enterprises to become more resilient. The training helped Nisha learn the use of websites to make bookings, generate bills, videography and photography. She also learned about online marketing and promotion through social media (Facebook, YouTube) by posting pictures and videos of homestays. Now she manages reservations on the portal (booking.com) for the homestays and responds to queries from current as well as potential consumers. She also coordinates with guests before and during their stays about their accommodations, food preferences, and itineraries.

Nisha also manages the homestays' digital payment systems as she knows the digital payment process. As a homestay manager, one of her primary responsibilities is creating attractive videos and audio for online advertising and promotion. Nisha is a multitasker. Her day begins at 4 a.m., when she assists her mother with household chores, agricultural labour, and animal, before devoting her time to homestay responsibilities. Nisha believes that technology has made things easier and provided her with more chances with the click of a button. She has not only gained critical hard skills (such as working on Google Drive, making meeting links, and online banking), but also key soft skills, such as enhanced self-awareness, the capacity to manage time effectively, and the confidence to engage in interactions with others. She has broadened her knowledge to include effective people management and technological coordination because of being the first and, in many cases, the sole point of contact between guests and the women who run the homestays. Also, she believes that knowing how to utilise technology is essential for everyone, especially women and that the Udyami Sakhi platform is one way to achieve this. In fact, she wants to bring in more rural women to the Atithi homestay business so that the group is more acknowledged. "What brings me the most happiness is helping other women in the community," she adds proudly.

See: https://www.orfonline.org/research/women-transforming-lives-with-technology/

Technological Advancement Facilitating Entrepreneurship

With time, there has been rapid technological development which has accelerated the process of digitization across all industries (Faraj et al., 2018; Schwab, 2017). Businesses have constantly changed their working style to keep up with the ever-changing expectations of the customers brought on by digitalization (Aloulou, 2019). The changing dynamics and increased competitiveness demand businesses to incorporate technological advancement in their overall strategic plans. Digitalization has expanded the scope and reach of business, especially for small and micro-entrepreneurs by creating new opportunities through ICTs-based marketing and promotion (Nambisan et al., 2019; Sharafizad, 2016). These micro-entrepreneurs strengthen economic development at a local level and generate employment (Bernhard and Olsson, 2019). In the tourism sector, women are mainly operating as micro-entrepreneurs. Duffy and Hund (2015) and Markowska et al. (2018) acknowledge the growing significance of a distinct category of women entrepreneurs who are doing business through social media and recognized themselves as social media influencers. This is related to the idea of "digital entrepreneurship," which defines entrepreneurs who create innovative approaches to running their enterprises using digital technologies (Kraus et al., 2019). Although people were already heavily reliant on technology, it was during the covid-19 pandemic that they truly understood its value as a medium of engagement among people, a source of communication, and a catalyst for entrepreneurial activities. Tourism service interactions have been expanded beyond face-to-face contact to include pre, during, and post-service digital communication due to ICTs like social media (Olsson & Bernhard, 2020). Extensive use of the mobile phone for study, shopping, entertainment purpose, travel purpose, booking and payment, banking, etc has become prominent in the contemporary world. The government also introduced several skill-based training programmes aimed to impart digital knowledge among people and how they can use technology to run their businesses. Here, a case has been discussed that shows the initiatives taken by the government to provide skill-based training programmes to women and it is helping women to take up jobs and engage in entrepreneurial activities with the help of technology. In this study, a real-world example explored how the government's efforts to encourage women into the workforce and to take up entrepreneurial activity through training programmes have contributed to women's empowerment.

The narrative of a successful female entrepreneur emphasizes the importance of women, particularly in economically disadvantaged and resource-constrained countries, in facilitating the worldwide transition towards the goal of social, economic, and environmental sustainability. (Kimbu & Ngoasong, 2016). The next case discusses a real-world story of a first lady tour guide and her women-run travel and trekking agency in the Ladakh region of India.

CASE 2: THINLAS CHOROL AND HER LADAKHI WOMEN'S TRAVEL AGENCY

Thinlas Chorol in 2009 founded the first women-owned, women-run travel and trekking agency in Ladakh. Ladakh is a mountain range situated in the northern part of India and it is famous for its Buddhist monasteries, beautiful valleys, charming river, and mountain trekking. It is a dream destination for bikers. Thinlas Chorol was familiar with mountain trekking since childhood but her interest in becoming a tour guide emerged later when she was in college. In her college days, she starts assisting tourists as a tour guide in Ladakh itself. By realizing that many women are interested in becoming tour guides and also tourists were happy to see a female guide. By then women tour guides were generally not much involved in tourism business activities due to the lack of opportunity. and women. Through her company, Chorol initially started recruiting women and then trained them to become tour guides. Tourists were happy with the women tour guide as they mentioned that they have not seen any female tour guides while travelling to other parts of India. Though there are women who are engaged in tour guiding but they are less in number. Tourist season in Ladakh is only for five months from May to September, which is a challenge for these women because, in the off seasons, they have to find out other jobs. The covid-19 pandemic hit the business hard but parallelly she along with her employee raised the fund and did cleaning in the valley. Some of the women start working in guest houses whereas some of them were working as a labour in the Indian army. Taking care of children and doing household duties along with the job is a difficult task for the majority of women.

Thinlas Chorol is also a great supporter and promoter of ecotourism. She also operates homestays in villages that fall along the trekking routes. Tourists litter during their camping around the trek that's why Chorol promotes homestays instead of camping. When tourists lodge at homestays she helps them in the kitchen so that they can interact more and learn each other's culture apart from trekking. She also encourages tourists to buy local products and souvenirs and keep treks eco-friendly by not littering. Chorol also attempts that women on solo trips, girls' gangs' trips, and family trips hire her and her team as a guide as it makes them familiar with each other's culture well. The message that Chorol gives to every woman is "We have the same capabilities as men, it depends on us to use our skills, abilities, and knowledge."

See: https://www.hercircle.in/engage/get-inspired/achievers/how-thinlas-chorol-is-empowering-ladakhi-women-through-her-entrepreneurial-venture-2552.html

Education as a Means of Empowerment for Women

One of the best ways to empower women is to educate them. Getting an education may give women the work they deserve by equipping them with the information and skills necessary to take charge of their own lives and contribute fully to their communities. Moreover, education may aid in eliminating discriminatory attitudes towards women and fostering equality between the genders. Education empowers them to question and change the status quo of gender roles and expectations, as well as to advocate for their rights to full participation in society. Feminist voices for education advocate for a more equitable and inclusive education system around the world that addresses the unique needs and experiences of girls and women. In developing, nations women lack the required skills and education for running a business (Bandiera et al., 2020; Chinomona & Maziriri, 2015). The lack of education, skills, and business understanding prevents women from planning, organising, controlling, and operating a firm (Sumathi & Gunasundari, 2016). However, these days, practically all nations realise that without the participation of women in business, no nation can advance to its full potential. Therefore, governments now place more emphasis on women's education and entrepreneurial skill development among them (Ramadani et al., 2022). Owing to that nowadays women are more educated and experienced than previous generations and comparatively more willing to become an entrepreneur rather than to work in family businesses (Yadav & Unni, 2016). In fact, according to UNWTO (2019), more than half of the tourism workforce globally is women and there are many countries where the majority of tourism enterprises are owned and operated by women. Women's success in the tourism industry catalyzes change and encourages many other women to have confidence in themselves and engage in entrepreneurial endeavours. (Movono & Dahles, 2017).

In developing nations, the involvement of women in emerging destinations as members of the workforce and owner-managers of small or micro tourism enterprises (Kimbu & Ngoasong, 2016) makes a positive impact beyond the tourism sector (Ngoasong & Kimbu, 2016). These efforts have a multiplicative effect on the economic, environmental, and social well-being of the destination while also providing memorable and honest interactions with tourists. In this view, another real-world example from the tourism field is explored to discuss the changing scenario in the new age of women entrepreneurship.

CASE 3: SHAILZA SOOD DASGUPTA AND HER HOMESTAYS OF INDIA (HOI)

Shailza Sood Dasgupta, 42 years old is a travel writer and social entrepreneur. Years ago, Shailza with her group was on a trip to Spiti Valley, Himachal Pradesh, India, and had the idea of setting up homestays. During her trip, she realised that the homestay business is mostly dependent on walk-in guest services. In peak season, travellers have to search for accommodation as there is no vacant room left because of an upsurge in tourist arrival. She came up with the idea of creating a platform to manifest these homestays and also promote them. In April 2017, she with her partner founded Homestays of India (HOI) with five homestays in the Spiti Valley of Himachal Pradesh. Today, she operates in more than 21 states in India with 100 homestays. Shailza was previously working with Google and McKinsey which she left years ago to follow her passion for travelling and writing. Before entering into the homestays business, she founded a travel agency named Chalo Let's Go in 2010 through which she made strong ties with local communities and sourced many of them as a travel guide for her startup. Her focus also includes responsible, sustainable, and eco-friendly behaviour of travellers as well as communities. Before getting local people involved in the business they were first trained in several facets of hospitality, including guest management, basic conversational skills, health and cleanliness, homestay administration, and the adoption of environmentally friendly practices. Also, periodic problem-solving sessions are organized so that they are prepared to deal with challenges like demanding customers and unpleasant comments. The household women are also encouraged by this startup to take on responsibility and independently handle their finances. Shailza has more than 5000 clients including solo travellers, youngsters, students, senior citizens, tourists, and families. Some are happy with spending some quality time for a shorter duration whereas some of them prefer a long stay. Now, the startup intends to onboard more than 250 homestays by the end of 2020 and build sustainable homestay communities. In addition, the startup works with companies, local communities, and influencers to encourage socially responsible travel.

See: https://yourstory.com/herstory/2023/03/100-emerging-women-leaders-prianka-jhaveri-mend-packaging?utm_pageloadtype=scroll

Women as Social Entrepreneurs

Motivated by the urge to make an effect, whether by creating new products or services, utilising technology to achieve efficiency, or addressing environmental and social problems women are making a positive influence on their communities and the world (Raman et al., 2022). Modern women entrepreneurs are more inclined to prioritise social impact and sustainable practices in their businesses (Raman et al.,

2022). They believe in creating a "triple bottom line" of people, planet, and profit through a combination of business acumen with environmental and social impact. Working together and sharing available resources is key to the social enterprise business model. The development of digital networking platforms and internet-based communities possibly provided the greatest opportunity for social enterprises. Because of the proliferation of digital networking new types of financing, such as crowdfunding and peer-to-peer lending, have emerged, giving social entrepreneurs greater access to vital funds and materials (Chalermchaikit & Kozak, 2021). Overall, it can be said that women-led social enterprises may have a significant impact on achieving gender balance and sustainable development, and their success can act as an inspiration and catalyst for other women.

The takeaway from these three cases is that women in entrepreneurship have evolved with their enterprises and they are looking far beyond mere sustenance- that they are more comfortable with technology, more educated-enlightened- empowered and are now even looking forward to social causes beyond their personal desire to be by their own. Their determination and dedication are fuelling a new generation of entrepreneurs that places equal value on financial and social success. These entrepreneurs are more confident in decision-making and risk-taking; they are breaking conventional notions regarding them by following their intuition Women's participation in entrepreneurial activities, which includes empowering other women and challenging and transforming social norms, is emerging as the most vital progression. In rural areas women's involvement in small or micro-entrepreneurship, improves their living standards, leading them towards a respectful life in society with utmost dignity. The rise of modern women entrepreneurs bodes well for a more equitable and welcoming society.

DISCUSSION

The concept of women entrepreneurship is trending over the past two decades and most of the study talks about the challenges women face in their journey. Starting a business is challenging regardless of gender, but for women, it has become more difficult due to the persistent gender biases in society. Significantly the characteristics of the business and market substances are far more accurate determinants of business success than the founder's gender (Sullivan & Meek, 2012). Because of this gender prejudice, women confront several obstacles in launching and sustaining their enterprises. Women entrepreneurs encounter societal stigmas and unfavourable perceptions, due to which they face difficulties in getting loans and struggle to secure capital for their businesses (Said & Enslin, 2020). Lack of networking makes it more difficult for them to get clients, partners, and investors, also underrepresentation in positions of power makes it harder for women to be taken seriously and have equal opportunities as men. Nevertheless, an increasing number of women are engaging in business and counterattacking this prejudice. These challenges are still widespread, although not to the same extent as before. Now, women are publicly addressing these challenges and speaking out against inequality, therefore the situation is changing.

Today, women have a far easier time in the workplace and favourable circumstances for their businesses to flourish in the market. Fortunately, the unfavourable perception of most societies towards women's pursuit of independence and prosperity is being left behind as women's rights and welfare in terms of educational opportunities, political rights, and economic freedoms for women have grown and progressed in recent decades around the world (Martínez-Rodríguez et al., 2022). Similarly, the advancements in women's rights and opportunities in society have had a beneficial impact on the tourism industry, establishing a more inclusive and impartial business that benefits everyone. Thus, the diverse and dynamic nature of the tourism industry offers an abundance of opportunities for women entrepreneurs owing to low entry barriers comparatively to other sectors and supports women's empowerment and sustainable development. (In tourism, women entrepreneurs and women workers get an opportunity to interact with many people (visitors and guests), thereby broadening their mental vistas, and giving them a sense of what they could head for (a sense of direction- towards evolution and empowerment).

The cases mentioned above demonstrate how the environment of women's entrepreneurship is evolving. The first instance highlights the steps the government has made to foster an environment that allows and encourages the local communities to pick up entrepreneurial skills and gain influence through their enterprises. Also, it covers the value of technology in contemporary business as well as how social media and technology are affecting women's social position. The second case discusses the case of an empowered woman who is empowering many other women by being the first lady tour guide and running an all-women's travel and trekking agency in Ladakh, India. Additionally, her initiatives towards the responsible behaviour of tourists and local communities at tourist destinations are promoting ecotourism and sustainability in significant ways. The third case tells the story of a social entrepreneur providing an easy platform to the local communities for a convenient business of homestays. By incorporating eco-friendly and sustainable practices in her entrepreneurial principles she aims to create a positive environment and social outcome within the community. In these three instances, it is evident that instead of mourning over the lack of opportunities and facilitators for business, the women are exerting their effort and making a change in society.

CONCLUSION

The study of women entrepreneurs has grown significantly in the last few years, earning widespread acceptance among academics and, most importantly, making significant contributions to our knowledge of the factors that describe the challenges and difficulties of women in pursuing entrepreneurial careers. This study attempted to present the challenges and facilitators that have changed business dynamics, which resulted in an added transformation in women's entrepreneurship. The chapter offers a unique insight into women's entrepreneurship in the tourism sector by tracing the progress made by women by analysing the extant literature. The research will add to the advancement of knowledge and understanding in the new age of women's entrepreneurship. In addition, it will encourage many aspiring women to become entrepreneurs and offers a new perspective for budding entrepreneurs. By evaluating multiple cases, the chapter offers new insight into the changing scenario of women's entrepreneurship with time. The study suggests the importance of women's entrepreneurship has been realised worldwide, and efforts have been made to simplify women's entrepreneurship and give them equal opportunities. Based on reviewing the existing literature, the chapter attempted to present the evolution of the field and how the business environment has changed, which has provided facilities for women taking up entrepreneurship as a career choice.

REFERENCES

Agrawal, R. (2018). Constraints and challenges faced by women entrepreneurs in emerging market economy and the way forward. *JWEE*, (3-4), 1–19. doi:10.28934/jwee18.34.pp1-19

Aloulou, W. (2019). Entrepreneurship and innovation in the digitalization era. *Advances in E-Business Research*, 179–203. doi:10.4018/978-1-5225-7262-6.ch011

Ardrey, W. J. IV, Pecotich, A., & Shultz, C. J. II. (2006). Entrepreneurial women as catalysts for socio-economic development in transitioning Cambodia, Laos, and Vietnam. *Consumption Markets & Culture*, *9*(4), 277–300. doi:10.1080/10253860600921811

Baker, T. E., Aldrich, H., & Nina, L. (1997). Invisible entrepreneurs:the neglect of women business owners by mass media and scholarly journals in the USA. *Entrepreneurship and Regional Development*, *9*(3), 221–238. doi:10.1080/08985629700000013

Bandiera, O., Buehren, N., Burgess, R., Goldstein, M., Gulesci, S., Rasul, I., & Sulaiman, M. (2020). Women's empowerment in action: Evidence from a randomized control trial in Africa. *American Economic Journal. Applied Economics*, *12*(1), 210–259. doi:10.1257/app.20170416

Batliwala, S. (1994). The meaning of women's empowerment: New concept from action. Population Policies Reconsidered: Health, Empowerment and Rights, 127-138.

Bernhard, I., & Olsson, A. K. (2020). Network collaboration for local and regional development - the case of Swedish women entrepreneurs. *International Journal of Entrepreneurship and Small Business*, *41*(4), 539. doi:10.1504/IJESB.2020.111578

Boden, R. J. Jr, & Nucci, A. R. (2000). On the survival prospects of men's and women's new business ventures. *Journal of Business Venturing*, *15*(4), 347–362. doi:10.1016/S0883-9026(98)00004-4

Chalermchaikit, V., & Kozak, M. (2021). The practice of triple bottom line in the sustainability of tourism: A gender perspective. *Gender and Tourism*, 285–300. doi:10.1108/978-1-80117-322-320211016

Chawla, M., Rajan, S., & Sahni, P. (2020). *Powering the economy with her: Women entrepreneurship in India*. Bain& Company, Google.

Chinomona, E., & Maziriri, E. T. (2015). Women in action: Challenges facing women entrepreneurs in the Gauteng Province of South Africa. *The International Business & Economics Research Journal*, *14*(6), 835. doi:10.19030/iber.v14i6.9487

Crowe, S., Cresswell, K., Robertson, A., Huby, G., Avery, A., & Sheikh, A. (2011). The case study approach. *BMC Medical Research Methodology*, *11*(1), 1–9. doi:10.1186/1471-2288-11-100 PMID:21707982

D'Espallier, B., Guérin, I., & Mersland, R. (2011). Women and repayment in microfinance: A global analysis. *World Development*, *39*(5), 758–772. doi:10.1016/j.worlddev.2010.10.008

Davidsson, P., & Wiklund, J. (1997). Values, beliefs and regional variations in new firm formation rates. *Journal of Economic Psychology*, *18*(2-3), 179–199. doi:10.1016/S0167-4870(97)00004-4

Du Rietz, A., & Henrekson, M. (2000). Testing the female underperformance hypothesis. *Small Business Economics*, *14*(1), 1–10. doi:10.1023/A:1008106215480

Duffy, B. E., & Hund, E. (2015). "Having it all" on social media: Entrepreneurial femininity and self-branding among fashion bloggers. *Social Media + Society*, *1*(2). doi:10.1177/2056305115604337

European Commission. (2020). *Supporting entrepreneurship.* Internal Market, Industry, Entrepreneurship and SMEs. from https://single-marketeconomy.ec.europa.eu/smes/supporting-entrepreneurship_en

Faraj, S., Pachidi, S., & Sayegh, K. (2018). Working and organizing in the age of the learning algorithm. *Information and Organization*, *28*(1), 62–70. doi:10.1016/j.infoandorg.2018.02.005

Franzke, S., Wu, J., Froese, F. J., & Chan, Z. X. (2022). Female entrepreneurship in Asia: A critical review and future directions. *Asian Business & Management*, *21*(3), 343–372. doi:10.105741291-022-00186-2

Galbreath, J. (2017). Drivers of Green Innovations: The impact of export intensity, women leaders, and Absorptive Capacity. *Journal of Business Ethics*, *158*(1), 47–61. doi:10.100710551-017-3715-z

Giantari, I. G., Surya, I. B., Yasa, N. N., & Yasa, I. B. (2018). Development and revitalization strategies for traditional markets in Bali. *International Journal of Social Economics*, *45*(7), 1058–1070. doi:10.1108/IJSE-09-2017-0414

Global Entrepreneurship Monitor, G. E. M. (2022). *Gem Global Entrepreneurship Monitor. 2021/22 Women's Entrepreneurship Report: From Crisis to Opportunity*. GEM. https://www.gemconsortium.org/

Hashemi, S. M., & Schuler, S. R. (1993). Defining and studying empowerment of women: A research note from Bangladesh. *JSI Working Paper, 3*.

Haugh, H. M., & Talwar, A. (2014). Linking social entrepreneurship and social change: The mediating role of Empowerment. *Journal of Business Ethics*, *133*(4), 643–658. doi:10.100710551-014-2449-4

Hechavarria, D., Bullough, A., Brush, C., & Edelman, L. (2019). High growth women's entrepreneurship: Fueling social and economic development. SSRN *Electronic Journal*. doi:10.2139/ssrn.3460214

Hillman, W., & Radel, K. (2023). The social, cultural, economic and political strategies extending women's territory by encroaching on patriarchal embeddedness in tourism in Nepal. *Gender and Tourism Sustainability*, 303–324. doi:10.4324/9781003329541-19

Horodnic, I. A., & Williams, C. (2019). Tackling undeclared work in the European Union. SSRN *Electronic Journal*. doi:10.2139/ssrn.3437406

Hughes, K. D., Saunders, C., & Denier, N. (2022). Lockdowns, Pivots & triple shifts: Early challenges and opportunities of the COVID-19 pandemic for women entrepreneurs. *Journal of Small Business and Entrepreneurship*, *34*(5), 483–501. doi:10.1080/08276331.2022.2042657

Iwu, C. G., Nsengimana, S., & Robertson, T. K. (2019). The factor contributing to the low numbers of women entrepreneurs in kigali. *Acta Universitatis Danubius Economica*, *15*(6).

Jamali, D. (2009). Constraints and opportunities facing women entrepreneurs in developing countries. *Gender in Management*, *24*(4), 232–251. doi:10.1108/17542410910961532

Jennings, J. E., & Brush, C. G. (2013). Research on women entrepreneurs: Challenges to (and from) the broader entrepreneurship literature? *The Academy of Management Annals*, *7*(1), 663–715. doi:10.5465/19416520.2013.782190

Kabeer, N. (2005). Is microfinance a 'magic bullet' for women's empowerment? Analysis of findings from South Asia. *Economic and Political Weekly*, 4709–4718.

Kimbu, A. N., & Ngoasong, M. Z. (2016). Women as vectors of Social Entrepreneurship. *Annals of Tourism Research, 60*, 63–79. doi:10.1016/j.annals.2016.06.002

Kraus, S., Palmer, C., Kailer, N., Kallinger, F. L., & Spitzer, J. (2018). Digital Entrepreneurship. *International Journal of Entrepreneurial Behavior & Research*. doi:10.1108/IJEBR-06-2018-0425

Kusumawardani, K. A., Widyanto, H. A., & Deva, I. P. (2020). Understanding the entrepreneurial intention of female entrepreneurs in the Balinese tourism industry. *International Journal of Research in Business and Social Science, 9*(1), 63–79. doi:10.20525/ijrbs.v9i1.611

Linke, S., Turak, E., & Nel, J. (2010). Freshwater Conservation Planning: The case for systematic approaches. *Freshwater Biology*, *56*(1), 6–20. doi:10.1111/j.1365-2427.2010.02456.x

Mandipaka, F. (2014). An investigation of the challenges faced by women entrepreneurs in developing countries: A case of king williams' town, South Africa. *Mediterranean Journal of Social Sciences*. Advance online publication. doi:10.5901/mjss.2014.v5n27p1187

Markowska, M., Ahl, H., & Naldi, L. (2018, January 1). *Paths to entrepreneurship:life stories of Swedish Mumpreneurs*. Lancaster EPrints. Retrieved April 26, 2023, from https://eprints.lancs.ac.uk/id/eprint/127670/

Martínez-Rodríguez, I., Quintana-Rojo, C., Gento, P., & Callejas-Albiñana, F.-E. (2021). Public policy recommendations for promoting female entrepreneurship in Europe. *The International Entrepreneurship and Management Journal*, *18*(3), 1235–1262. doi:10.1007/11365-021-00751-9

Mashapure, R., Nyagadza, B., Chikazhe, L., Msipa, N., Ngorora, G. K., & Gwiza, A. (2022). Challenges hindering women entrepreneurship sustainability in rural livelihoods: Case of Manicaland Province. *Cogent Social Sciences*, *8*(1), 2132675. Advance online publication. doi:10.1080/23311886.2022.2132675

Matthew Scott, J., & Harrison, T., R., Hussain, J., & Millman, C. (2014). The role of *guanxi* networks in the performance of women-led firms in China. *International Journal of Gender and Entrepreneurship*, *6*(1), 68–82. doi:10.1108/IJGE-03-2013-0014

Mentzer, J. T., & Kahn, K. B. (1995). A framework of logistics research. *Journal of Business Logistics*, *16*(4), 231–250.

Meredith, J. (1993). Theory building through conceptual methods. *International Journal of Operations & Production Management*, *13*(5), 3–11. doi:10.1108/01443579310028120

Morrison, A., Breen, J., & Ali, S. (2003). Small business growth: Intention, ability, and opportunity. *Journal of Small Business Management*, *41*(4), 417–425. doi:10.1111/1540-627X.00092

Mota, A., Braga, V., & Ratten, V. (2019). Entrepreneurship motivation: Opportunity and necessity. *Contributions to Management Science*, 139–165. doi:10.1007/978-3-030-12342-0_8

Movono, A., & Dahles, H. (2017). Female empowerment and tourism: A focus on businesses in a Fijian village. *Asia Pacific Journal of Tourism Research*, *22*(6), 681–692. doi:10.1080/10941665.2017.1308397

Muravyev, A., Talavera, O., & Schäfer, D. (2009). Entrepreneurs' gender and financial constraints: Evidence from International Data. *Journal of Comparative Economics*, *37*(2), 270–286. doi:10.1016/j.jce.2008.12.001

Nambisan, S., Wright, M., & Feldman, M. (2019). The Digital Transformation of Innovation and Entrepreneurship: Progress, challenges and key themes. *Research Policy*, *48*(8), 103773. doi:10.1016/j.respol.2019.03.018

Narayan, D., Stern, N., Nankani, G., Page, J., & Jorgensen, S. (2002). Empowerment and poverty reduction; a source book, the world bank. *PREM.*

Ngoasong, M. Z., & Kimbu, A. N. (2016). Informal microfinance institutions and development-led tourism entrepreneurship. *Tourism Management*, *52*, 430–439. doi:10.1016/j.tourman.2015.07.012

OECD. (2017). *Policy brief on women's entrepreneurship*. Publications Office of the EU.

Olsson, A. K., & Bernhard, I. (2020). Keeping up the pace of digitalization in small businesses–women entrepreneurs' knowledge and use of social media. *International Journal of Entrepreneurial Behaviour & Research*, *27*(2), 378–396. doi:10.1108/IJEBR-10-2019-0615

Radović-Marković, M. (2015). Female entrepreneurship in transition economies: An overview. *Female Entrepreneurship in Transition Economies*, 9–30. doi:10.1057/9781137444516_2

Ramadani, V., Rahman, M. M., Salamzadeh, A., Rahaman, M. S., & Abazi-Alili, H. (2022). Entrepreneurship education and Graduates' entrepreneurial intentions: Does gender matter? A multi-group analysis using Amos. *Technological Forecasting and Social Change*, *180*, 121693. doi:10.1016/j.techfore.2022.121693

Raman, R., Subramaniam, N., Nair, V. K., Shivdas, A., Achuthan, K., & Nedungadi, P. (2022). Women Entrepreneurship and Sustainable Development: Bibliometric analysis and emerging research trends. *Sustainability (Basel)*, *14*(15), 9160. doi:10.3390u14159160

Reliance Foundation and Observer Research Foundation. (2022). *Aspirations, Access & Agency: Women transforming lives with technology*. ORF. https://www.orfonline.org/research/women-transforming-lives-with-technology/

Rosca, E., Agarwal, N., & Brem, A. (2020). Women entrepreneurs as agents of change: A comparative analysis of social entrepreneurship processes in emerging markets. *Technological Forecasting and Social Change*, *157*, 120067. doi:10.1016/j.techfore.2020.120067

Said, I., & Enslin, C. (2020). Lived experiences of females with entrepreneurship in Sudan: Networking, social expectations, and Family Support. *SAGE Open*, *10*(4). doi:10.1177/2158244020963131

Sajjad, M., Kaleem, N., Chani, M. I., & Ahmed, M. (2020). Worldwide role of Women Entrepreneurs in Economic Development. *Asia Pacific Journal of Innovation and Entrepreneurship*, *14*(2), 151–160. doi:10.1108/APJIE-06-2019-0041

Sarfaraz, L., Faghih, N., & Majd, A. (2014). The relationship between Women Entrepreneurship and Gender Equality. *Journal of Global Entrepreneurship Research*, *2*(1), 6. doi:10.1186/2251-7316-2-6

Schaltegger, S., & Wagner, M. (2011). Sustainable Entrepreneurship and sustainability innovation: Categories and interactions. *Business Strategy and the Environment*, *20*(4), 222–237. doi:10.1002/bse.682

Schwab, K. (2017). *The Fourth Industrial Revolution*. Crown Business.

Scott, L., Dolan, C., Johnstone–Louis, M., Sugden, K., & Wu, M. (2012). Enterprise and inequality: A study of avon in South Africa. *Entrepreneurship Theory and Practice*, *36*(3), 543–568. doi:10.1111/j.1540-6520.2012.00507.x

Seguino, S. (2000). Gender inequality and economic growth: A cross-country analysis. *World Development*, *28*(7), 1211–1230. doi:10.1016/S0305-750X(00)00018-8

Shane, S. A. (1992). Why do some societies invent more than others? *Journal of Business Venturing*, *7*(1), 29–46. doi:10.1016/0883-9026(92)90033-N

Sharafizad, J. (2018). Informal learning of Women Small Business Owners. *Education + Training*, *60*(1), 82–103. doi:10.1108/ET-01-2017-0006

Sullivan, D. M., & Meek, W. R. (2012). Gender and entrepreneurship: A review and process model. *Journal of Managerial Psychology*, *27*(5), 428–458. doi:10.1108/02683941211235373

Sumathi, V., & Gunasundari, K. (2016). Problems faced by the women entrepreneurs in India. *International Journal of Management Research and Reviews*, *6*(3), 348.

Suseno, Y., & Abbott, L. (2021). Women entrepreneurs' Digital Social Innovation: Linking Gender, Entrepreneurship, Social Innovation and Information Systems. *Information Systems Journal*, *31*(5), 717–744. doi:10.1111/isj.12327

Sutcliffe-Braithwaite, F., & Lawrence, J. (2019). Power and its loss in the iron lady. *Biography and History in Film*, 295–317. doi:10.1007/978-3-319-89408-9_13

Tur-Porcar, A., Mas-Tur, A., & Belso, J. A. (2016). Barriers to women entrepreneurship. different methods, different results? *Quality & Quantity*, *51*(5), 2019–2034. doi:10.100711135-016-0343-0

Vasilić, N., Popović-Pantić, S., & Semenčenko, D. (2020). Women entrepreneurship in the time of COVID-19 pandemic. *Journal of Women's Entrepreneurship & Education*, (3-4), 23–40.

Vujko, A., Tretiakova, T. N., Petrović, M. D., Radovanović, M., Gajić, T., & Vuković, D. (2019). Women's empowerment through self-employment in Tourism. *Annals of Tourism Research*, *76*, 328–330. doi:10.1016/j.annals.2018.09.004

Wasilczuk, J., & Zieba, K. (2008). Female Entrepreneurship in Transitional Economies: The case of Poland. *Journal of Small Business and Entrepreneurship*, *21*(2), 153–169. doi:10.1080/08276331.2008.10593419

Watson, J. C. (2003). *Overcoming the Challenges of Counseling College Student Athletes*. ERIC/CASS Digest.

Winn, J. (2005). Women entrepreneurs: Can we remove the barriers? *The International Entrepreneurship and Management Journal*, *1*(3), 381–397. doi:10.100711365-005-2602-8

World Tourism Organization. (2011). *Global report on women in tourism 2010*. UNWTO. https://www.unwto.org/centre-stage-project-womens-empowerment-during-the-covid19-recovery

Wu, J., Li, Y., & Zhang, D. (2019). Identifying women's entrepreneurial barriers and empowering female entrepreneurship worldwide: A fuzzy-set QCA approach. *The International Entrepreneurship and Management Journal*, *15*(3), 905–928. doi:10.100711365-019-00570-z

Yadav, V., & Unni, J. (2016). Women entrepreneurship: Research review and future directions. *Journal of Global Entrepreneurship Research*, *6*(1), 12. Advance online publication. doi:10.118640497-016-0055-x

ENDNOTE

1 Udyami Sakhi can be understood as 'women entrepreneur friend'.

Chapter 4
Sustainable Tourism Opportunities in Kashmir Valley and the Role of Women

Parvaiz Ahmad Parray
CT University, India

Puneeta Sharma
Department of Economics, CT University, India

Sammer Ahmad Wani
GHSS Zainpora, India

Khursheed Ahmad Wani
Government Degree College, Thindim Kreeri, India

ABSTRACT

This chapter looked at sustainability through women's empowerment present at Kashmir Valley tourist attractions. On the basis of the relevant issues, conclusions are reached, and solutions are offered. Government-adopted practises, concepts, and strategies are described and their applicability for improving and promoting sustainable tourism at destinations is explained. The majority of the naturally gifted goods manufactured by women in Kashmir Valley require careful monitoring and care. Strong policies, a strategic framework, eco-friendly tourist practises, and sufficient facilities for women's empowerment are recognised as essential in this context.

INTRODUCTION

Women's contributions to environmental preservation and sustainable development have gained more recognition during the past 20 years. Women's involvement in environmental conservation and management was specifically mentioned at the third world conference on women in Nairobi, Kenya, in 1985. The fourth world conference on women in Beijing, China, adopted a platform for action that included women

DOI: 10.4018/978-1-6684-8417-3.ch004

and the environment as one of its key areas of concern. According to the UNWTO and the International Ecotourism Society, women hold the majority of ecotourism jobs globally.

Humans have an insatiable need for natural wonders, and they are both the ones who degrade and defend them, always testing the limits of what is considered acceptable behaviour (Deming, 1996; Seth, 1996). Ecosystems are impacted by tourism, and we should prioritize their protection above the needs of humanity since all living things, not just humans, are valuable in their own right (Milgrath, 1989; Wearing and Neil, 1999). Strategies on tourism and sustainability were developed as part of the sustainable development model, demonstrating how ecological principles underpin the growth of economies. Sustainable tourism requires that people think about how visitors will value and utilize natural areas, how tourism can improve local communities, how tourism will have social and ecological repercussions, and how these impacts might be mitigated. A shift from practices to principles and increased effort in implementing sustainable tourism measures; being culturally sensitive when formulating a strategy for sustainable development; working within existing frameworks rather than developing new ones; involving affected stakeholders in decision making; integrating tourism with other land uses; encouraging visitor accountability toward the environment and local populations; and providing ecologically sustainable forms of tourist accommodation.

Sustainable tourism has its detractors, including the following points: tourism is not seen as an industry dependent on natural resources; it is invisible in urban areas; electoral strength is low; there is a lack of leadership; the current approach to planning and management of tourism is not sustainable; and the people in rural tourism communities rarely work together. Environmental resources and ecosystem services are just as important to any economic sector as other types of capital, which is why they are so central to the concept of sustainable development. These services provide both direct and indirect benefits to the tourist industry, but they depreciate in value over time just like any other kind of production capital. The environmental comforts that the tourist sector depends on might be harmed by the industry's operations, which in turn would harm the natural resource base. If tourism interferes with the normal functioning of ecosystem services, the quality of such services, which are crucial to economic activity, suffers, which in turn has negative indirect consequences on such capital. The industry would fail if the deterioration of necessary facilities or environmental services was not stopped. While it is generally accepted that protecting the natural resources directly utilized by the tourism industry is essential to the sector's long-term economic viability, it is often assumed that the 'free' services given by the environment do not need conservation or replenishment. As a result, they don't factor into assessments of the industry's true cost, which begs the issue of whether the industry's contribution to the national economy is in line with the use of, and consequent maintenance and/or replenishment of, equivalent reserves. Sustainable tourist development is now recognized as a top priority. Meaningful commercialization would require responsible management of resources, waste,

Throughout history, Jammu and Kashmir has been a top tourist destination. Lakes, streams, verdant grass, majestic trees, and towering mountains all drew visitors for different reasons. Jammu, Kashmir, and Ladakh are the state's three regions, and they each provide their own unique blend of natural beauty, adventure opportunities, animal viewing, religious sites, and historical sites that draw visitors from all over India and the world (Singh, 2004). Lidder valley (Pahalgam) in the Kashmir area was chosen for this investigation. Jammu and Kashmir's tourist sector has the potential to be a major contributor to the state's rising economy. There is no question that tourism has remained a tool of economic growth in the State, and that it has contributed significantly to the development of the economy, especially in the Kashmir valley and Ladakh. The aim of the present chapter is to promote Sustainable Tourism opportunities in Kashmir Valley by the involvement of women at various sectors.

MATERIAL AND METHODS

Pahalgam

This research was conducted at Pahalgam, a popular tourist destination in Anantnag District, Jammu and Kashmir, India. This hill station may be found 45 kilometres to the northeast of the district capital of Anantnag. Set at an elevation of 8989 feet (2740 meters), between latitudes 34001′N and 75011′E and longitudes 34001′N and 75019′E, it may be found on the banks of the Lidder River. Of its 5922 residents, 56% are men and 44% are women; 35% of the population is literate. The valley of Shepherds, Pahalgam, was once a popular destination for tourists and a popular filming location. Today, it is one of India's top health resorts. It is also the starting point for the well-known Amaranth ji Yatra pilgrimage. From Pahalgam, it's 16 kilometers by automobile to Chandanwari, and another 19 kilometers to the sacred Amaranth cave, but from there, the journey gets considerably higher and can only be traversed on foot or by pony. Pine trees, snow-capped mountains, a mild temperature, and expansive meadows and pastures give the valley a magnificent appearance. Pahalgam is located along the banks of the Lidder River, whose raging current contributes to the area's already stunning natural scenery.

Nothing about the valley of shepherds has been ruined by modernity. Any angler will enjoy their time here, as even a novice may pull a Rainbow trout out of the swift currents. Horseback riding, golfing, fishing, hiking, bird watching, rafting, skiing, etc. are just few of the many leisure pursuits available. The area around Pahalgam is rich in sights to see, including the gently rolling meadows of Baisaran, Hajjan, Betaab valley, Shikargah, and Aru. Famous trekking destinations include the Lidderwat, Chandanwari, and Kolhai glaciers. Within the Pahalgam tourism area are no less than eight villages, one of which, Mamal, is home to what is often regarded as Kashmir's oldest surviving Temple, a Shiva Temple dating back to the fifth century.

Climate

Climate at the Overa Aru wildlife refuge ranges from mild to cold arctic conditions (Negi, 2002). Due to the region's rocky geography, the weather is unusual. You might think of the year as having four distinct seasons: spring (March–May), summer (June–August), fall (September–November), and winter (December–February) (December-February). Strong winds from the Mediterranean depressions cause the sky to be gloomy and precipitation to range from mild to heavy during the winter months, when the average temperature is between -100°C and 150°C and the average day is just a few hours long. The summer months are very warm, with temperatures averaging between 200 and 300 degrees Celsius, with regular rains. Altitude, elevation, or height affects the air temperature, precipitation type, and coolness to varying degrees. While springtime is lovely, fall has its own splendor and charm. Extreme monsoons provide precipitation, and the average annual total is 75 centimeters. While humidity averages about 60% in most months, it peaks around 80% in December. This seasonal variation in weather gives the sanctuary a unique appearance throughout the year.

DATA COLLECTION

Secondary information was gathered from the Jammu and Kashmir Tourism Department and the Directorate of Economics and Statistics of the Jammu and Kashmir Government through in-person visits to their Srinagar offices and discussions with the relevant authorities regarding various conservation-related issues. At regular intervals during the study period, researchers travelled to Pahalgam to check in with the city's municipal office, tourism department, and public health engineering (PHE) division. Several topics, including environmental protection and unfettered tourism, were discussed with the aforementioned authorities.

Secondary data was gathered from a wide range of print sources, including academic journals, government publications, and popular media. Thesis and dissertations, both published and unpublished, as well as reference books from various libraries were also consulted.

RESULTS

Pahalgam

Pahalgam is a municipality and notified area committee in the Indian state of Jammu and Kashmir's Anantnag district. The number of people that visit Pahalgam each year attests to its status as a top tourist spot. It's classified as a tehsil (a sub district) under the Anantnag administrative division. Pahalgam, a hill resort in the Kashmir valley, is located at an elevation of 2,130 meters and roughly 95 kilometres from Srinagar. For hundreds of years, nomadic herders brought their flocks to graze in Pahalgam, also known as the "Valley of Shepherds," because of the abundance of grassland there. Beginning in the early twentieth century, its stunning natural beauty started attracting tourists, who often stayed for just a few days before heading back to Gulmarg or Srinagar. Pahalgam was also historically significant since it was the starting point for Hindu pilgrimages to the holy cave of Amarnath.

Flora and Fauna

Vegetation on land dominates the landscape in the region under study, and it is rather uniform throughout. Pinus griffithi, Abies pindrow, Cedrus deodar, Juglans regia, etc., are only few of the major tree species found there. Desmodium tiliafolium, Indofera heterantha, Artemisia vulgaris, Taxaxacum afficinale, and a number of grass species are among the shrub and plant species found there (Cynodon dactylon). Several types of fish, including Brown trout (Salmo trutta) and Rainbow trout (Oncorhynchus mykiss), and several species of birds, like the jungle crow, wagtails, Himalayan bulbul, etc., make up the bulk of the region's faunal resources. Although it is not uncommon for wild bears to wander about at night.

Tourism

Using questionnaires developed specifically for the purpose, researchers interviewed 120 visitors in Pahalgam. There were somewhat more men (58.17%) than females (40.83%) among the visitors that were surveyed. There were 40% domestic visitors, 47.50% international tourists, and 12.50% local tourists among those surveyed. For example, 41.10 percent of visitors were between the ages of 25 and 44,

30.90 percent were between the ages of 15 and 34, and the remaining 7.10 percent were between the ages of 5 and 14.10 percent, 11.9 percent were between the ages of 55 and 64, and 9.7 percent were older than 64. The money visitors spend helps support many different subsets of the local population. Tourists spend money on a wide range of products and services. According to the data collected, visitor expenditures are as follows: 35.95 percent on handicrafts, 20.60 percent on transportation, 16.95 percent on lodging, 16.42 percent on entertainment, and 10.0 eight percent on everything else (Figure 1). For 45.84% of respondents, their primary method of transportation was a heavy vehicle; for 40.83%, it was a light vehicle; and for 13.33%, it was a two-wheeled vehicle (Figure 2).

Cooked meals (45%), fast food (50.84%), and home cooking (4.16%) were the most common answers for how tourists ate while traveling. The majority of respondents (67.5%) agree that the current facilities in Pahalgam are inadequate, and 63.33 percent said that trash cans at the ecotourism site are not located in convenient areas. Seventy-six percent of those surveyed said they had hired a local guide, while 24 percent said they hadn't hired a guide because they weren't prepared to pay the going rate.

Tourists go to Pahalgam for its stunning scenery, but they also enjoy the many other activities and attractions available there. A variety of reasons were given by vacationers for being in the area. Nearly half (42.50%) of respondents claimed they were making a pilgrimage or performing religious rites when they were there. Twenty-five percent of respondents indicated they were there to have a picnic. Another 9.17% of respondents stated they came to view the local fauna. 75% of respondents favoured scenic beauty, whereas 5% favoured horseback riding, 1.6% favoured hiking, 3.3% favoured rafting, and 2.5% favoured fishing. The rest 3.33% were in favour of other activities (Figure 4).

Since 1997, Pahalgam's tourist population has been steadily growing. Approximately 80% of all tourists visit during the summer, whereas just about 20% visit during the winter. The number of yatries visiting the holy cave of Amarnathji has been steadily increasing. Government employees in Pahalgam bring in an annual salary of Rs 146 lakhs. The average daily water use is between 80 and 90 liters, rising to 120 liters per person during the summer tourist high season.

Hacker's Sabzar Park, Amusement Park, Lidder Park, Deer Park, Poshwan Park, Club Park, Nehru Park, Beetab Valley Park, Batkote Park, and others are among Pahalgam's most popular parks for leisure.

Figure 1. Percentage expenditure by tourists

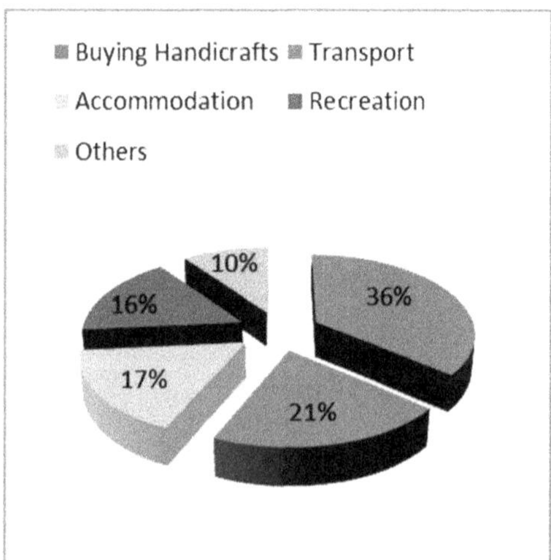

Figure 2. Mode of transportation

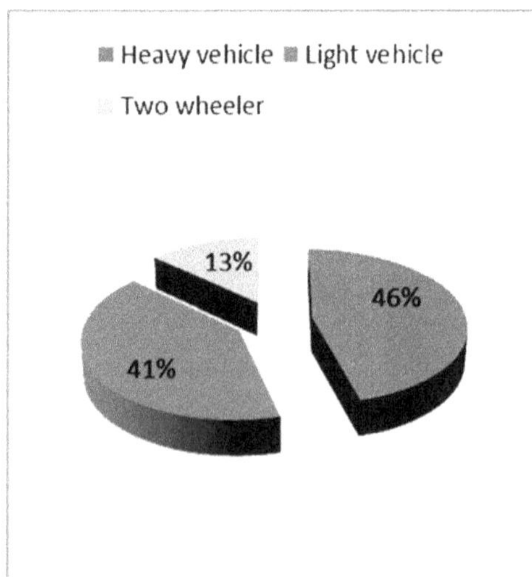

Figure 3. Involvement of people in sanctuary management
Source: Sameer et al. (2014)

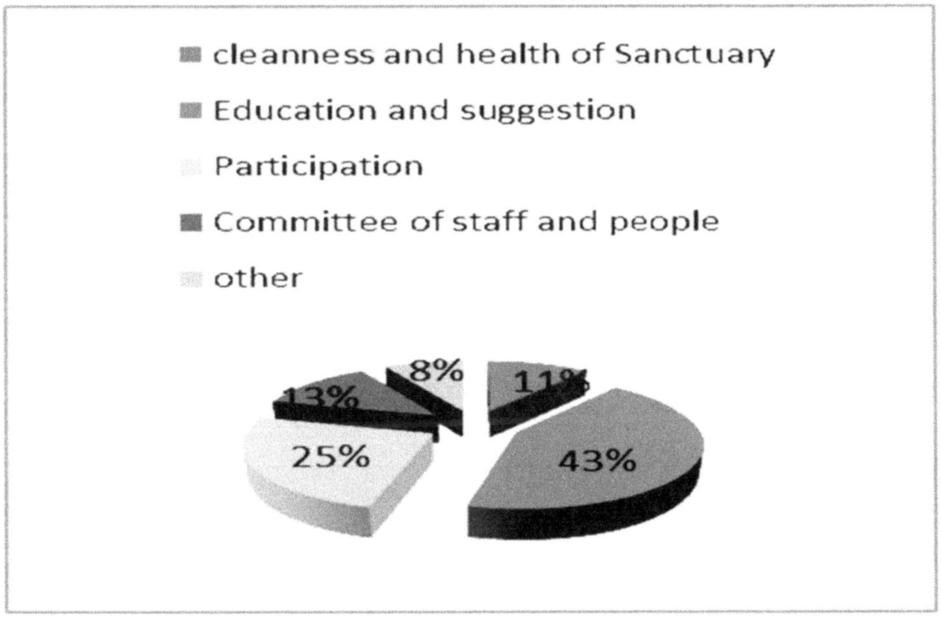

Figure 4. Types of tourist activities at Pahalgam

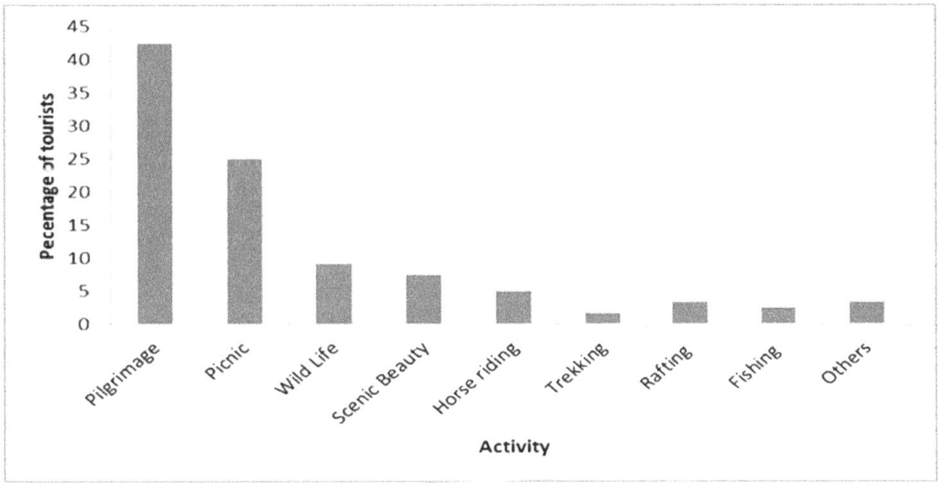

When asked about their opinions, visitors to Overa Aru Sanctuary offered a wide range of answers. The vast majority of visitors to Overa Aru refuge do not know where they are. Visitors' perceptions of numerous wildlife-related factors were measured using a questionnaire (Table 1).

Table 1. Perception of tourists visiting Overa Aru Sanctuary

S. No	Activity	Response	Tourists
1	Visit to Overa Aru Sanctuary	Yes	77(64.17)
		No	43(35.83)
2	Purpose of visit	Research	6(7.80)
		Fun	48(62.33)
		Study	23(29.87)
3	Providing food to wild animals/birds	Yes	21(27.28)
		No	56(72.72)
4	Wildlife Sanctuary a proper site to generate awareness about conservation	Yes	61(79.22)
		No	16(20.78)
5	Public should be involved in Sanctuary management	Yes	53(68.83)
		No	24(31.17)
6	Suggestion for improvement in sanctuary management	Yes	33(42.86)
		No	44(57.14)
7	knowledge of environmental laws	Yes	31(40.26)
		No	46(59.74)

Values in parentheses denote percentages

Waste Management

In the current research, data were gathered on 120 municipal personnel at the Pahalgam ecotourism location using questionnaires developed specifically for this purpose. Sixty of the employees surveyed were permanent, while the rest were seasonal workers employed by the Pahalgam municipality office for only three months during the Amarnath yatra and other busy tourist times. One hundred percent of the staff members surveyed acknowledged the presence of both solid and liquid wastes, with the former being the vast majority. Retail establishments, food service establishments, lodging establishments,

Figure 5. Waste types and composition at Pahalgam

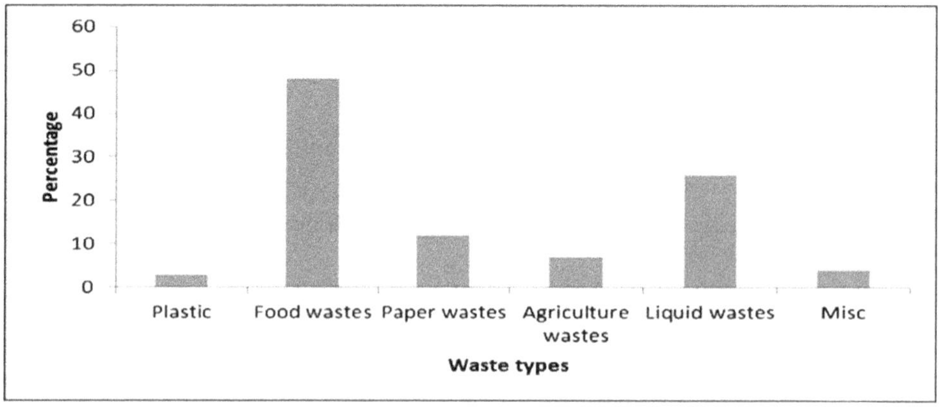

and vacationers create a wide variety of trash (Figure 5). Workers at the ecotourism site said that the Pahalgam Development Authority (PDA) and the Pahalgam municipality had donated between 600 and 700 trash cans. The ecotourism area generates 25-30 qtls of solid wastes daily during peak tourist seasons (summer), and 5-10 qtls during the off season, however according to reports from all employees, no weighing, segregation, or recycling of wastes is taking place at the site. Sixty percent of employees said they use PPE such gloves when collecting trash, while forty percent said they didn't. All responders confirmed that garbage is dumped at the Sarbal dumping site and that garbage cans are cleaned and garbage is collected every day.

Overa Aru Sanctuary

The 425-square-kilometer sanctuary may be found to the southeast of Srinagar, in the vicinity of Pahalgam. The elevation ranges from from 3,000 meters above sea level to about 5,425 meters. Numerous rare, endangered, and protected species have been bestowed upon this Wildlife Sanctuary by Mother Nature. The hangul, musk deer, serow, brown bear, Leopard, rhesus monkey, grey langur, Himalayan mouse hare, etc., are among the most notable. Vegetation is abundant in the region, with conifers making up more than 90% of the forest. The most common trees and shrubs are Cedrus deodara, Pinus griffithii, Abies pindrow, Aesculus indica, etc. Indigofera heterantha, Viburnum spp., Sorbaria tomentosa, etc. are some of the most common shrubs. Dicotyledonous plants such as Rumex patientia, Primula spp., anemone spp., etc., make up the majority of the ground cover.

Avifauna

Overa Aru Sanctuary is home to a wide variety of avifauna, including both year-round and seasonal species. From March through September, the lower portions are frequented by both resident and migrating birds. During this study, researchers counted 702 different bird species from the refuge, representing 5 different orders and 29 different families. The family Motacillidae has the most members (six), followed by the families Strigidae, Muscicapidae, Passeridae, and Columbidae (two members each). Single members of 15 different families were included (Table 2).

Table 2. Famous ecotourism places of Jammu and Kashmir

S.No	Region	Famous Ecotourism Places
1	Kashmir	Srinagar, Dal lake, Wular lake, Manasbal lake, Mughal gardens, Gulmarg, Pahalgam, Veerinag, Tulip garden, Sonamarg, Amarnath temple, Gurez, Dudhpathri, Bangus valley, Kokernag, Aharbal, Daksum and others.
2	Jammu	Vaishno Devi, Poonch, Patnitop, Sansar, Mantalai and Shivkhori, Punj old fort, Bhaderwah, Bhimgarh fort,Ramnagar fort.
3	Ladakh	Monasteries of Leh, Nubra valley, Lake Morari, Pangong lake, Zanaskar, Suru valley, Nunkun glacier, Dreass, Tiger hill

Traffic of Ecotourism in Kashmir

With so much to offer visitors, Jammu and Kashmir has become a popular vacation spot. The State's historical and natural assets, as well as its picturesque splendor, provide the foundation for a thriving tourist industry. The presence of religious sites like monasteries, temples, and subterranean temples also bodes well for tourism. Many people from all around the globe visit the State to see its many sights. The State has the ability to attract tourists interested in a wide variety of vacation activities, from relaxing to adventuring to making religious pilgrimages to participating in a variety of sports, from winter to summer. Despite fluctuations in certain years owing to unrest in the State, the overall trend for tourist visits to the different locations is upward. Information at hand strongly suggests that the number of visitors to the State is only going up.

One of the State's primary sources of income is the tourism industry. Government and public sector organizations are able to rely on the money it provides, which has grown steadily over time. Despite significant setbacks brought on by law and order issues, the sector is beginning to show signs of life. The public sector's revenue rose from INR 2081.95 lakhs in 2004-05 to INR 3101.00 lakhs in 2008-09 (Table 3).

Table 3. Revenue generated by various tourism development authorities

S.No	Dept./Organisation	Revenue Generated in Last Five Years (Rs. Lacks)				
		2004-05	2005-06	2006-07	2007-08	2008-09
01	Director Tourism Kashmir/Jammu.	11.46	14.83	35.25	62.06	54.00
02	Sheri-Kashmir International convention complex (SKICC).	46.00	49.34	52.00	64.00	71.20
03	Gulmarg Development Authority.	19.70	21.86	32.46	48.50	51.17
04	Pahalgam Development Authority.	5.12	9.16	9.45	19.00	12.73
05	Sonamarg Development Authority.	0.00	13.78	36.35	32.50	45.00
06	Patnitop Development Authority.	11.00	10.15	15.62	79.00	20.00
07	Royal Spring Golf Course	57.26	35.09	34.73	50.06	45.66
08	Cable Car Corporation	289.00	754.41	676.00	798.81	839.14
09	J&K Tourism Dev. Corporation	1642.41	1796.44	1765.39	1763.00	1963.00
	Total	2081.95	2705.06	2657.75	2916.98	3101.90

Source: Economic Survey 2009-10, Directorate of Economics and Statistics J&K

Amarnath Yatra

Himalayan pilgrimages, developed throughout time by Hindu sages, epitomize the spirit of nomadism, exploration, and spirituality, and are therefore the earliest form of organized travel. One of the holy trinity, Shiva appears as a lingam, a naturally occurring ice stalagmite. The songs of the Rig Veda, India's oldest and holiest scripture, invoke his presence. His presence has been attested to by astronomy, ritual, and mythology in the Vedas. The Himalayas are traditionally regarded as Shiva's spiritual abode. Neither he nor his future wife got a home from him. Both an ascetic and a married man, he was described as "the wild deity sports in the forest or taking his comfort on a cloud," suggesting that he was capable of living either lifestyle. It is said that in the Amarnath cave, Shiva revealed the meaning of life to Parvati. Unbeknownst to them, a mating pigeon couple overheard their chat, and now, having learnt the secret, they are reincarnated again and again, and have made the cave their forever dwelling. One day, a sadhu gave a bag of coal to a Muslim shepherd called Buta Malik, so the old narrative goes. When he got home, he opened the bag and saw that it was full with gold. Buta Malik, moved with gratitude, returned to find the sadhu and thank him; however, he instead stumbled into a cave at the site of their encounter, which subsequently became a sacred site for devotees from all over the world.

Employment Generation

Considering Pahalgam's status as a backward, rural area with few resources, the importance of tourism as an employer cannot be overstated. As a service business that relies heavily on human labor, the tourism sector is a significant employer of professionals, semi-professionals, and unskilled workers alike. Hotels, restaurants, houseboats, transportation services, tour guides, travel agencies, and similar businesses all contribute to the State's economy via direct employment. Jobs for local artisans, carpenters, jewellers, woodcarvers, and many more people who provide raw material to these producers are created through the demand for goods and services required by those with direct jobs, such as the production of handicrafts, paper mashie, cottage industries, bakeries, laundries, tea stalls, etc. There are now 412 full-time government jobs and 8063 private sector jobs available in Pahalgam thanks to the tourist industry, not including the 5300 temporary jobs that are produced every summer (Table 4).

Table 4. Employment generated due to tourism in Pahalgam

Sector	Place/Deptt.	Employees	Type
Public sector	Pahalgam Development Authority	172	Regular
	Municipality	90	
	Tourist guides	55	
	Amusement park	53	
	Other parks	42	
	Total	412	
	Hotels	2100	
	Restaurants	45	
	Shops	618	
	Total	2763	
Private sector	Pony wale	2424	Seasonal
	Tent providers	280	
	Pandy wale	156	
	Sweepers	240	
	Total	5300	
	Grand total	8475	

Revenue Generation

The tourism industry is a significant source of income for Pahalgam's residents. As an added bonus, it brings in enough money to fund public services and government programs. Successfully attracting tourists has been a boon to many states' coffers. Hotels, restaurants, taxi services, tourist attractions, theme parks, shops selling local arts and crafts, jewelry, paper mashie, etc. all play a part in the state's economy by catering to both visitors and locals. The first step in creating revenue is the spending of visitors. Every dollar spent by visitors is a dollar more for the economy as a whole. Experts in the tourism industry have calculated that it takes between 13 and 14 transactions for the money spent by visitors to vanish completely. Five or six of these deals close in the first year. The money travelers spend on things like lodging, food, transportation, entertainment, souvenirs, photography, medicine, cosmetics, and apparel all contribute to the growth of the economy and its overall worth. In the end, it sets off a domino effect in the marketplace. Pahalgam's public sector earns an annual average of 146,000,000 rupees (Table 5).

Table 5. Revenue generated in public sector due to tourism in Pahalgam

S. No.	Institution/Department	Average Revenue per Annum (Lakhs)	Source
1	Municipality	72	Vehicle entry fee, Hotel registration fee, Shop registration fee, Rent etc.
2	Amusement park	60	Park entry fee
3	Pahalgam Development Authority	10	Park entry fee
4	Floriculture Department	2	Park entry fee
5	Deer park	2	Park entry fee
	Total	146	

THREATS IN PAHALGAM AND OVERA ARU SANCTUARY

A number of threats have been observed in both Overa Aru Sanctuary and Pahalgam and their severity was taken into account by keeping in consideration the five point impact scale.

Threats to Overa Aru Sanctuary

Threats to Overa Aru Sanctuary from unrestricted tourist amenities range from negligible at site S3 to highly substantial at sites S6, S7, and S8.

Site S6, but not S2, faces a serious danger from deforestation. Sites S2 and S3 are particularly vulnerable because of their proximity to populated areas. Site S2 has a low risk of bushfires; however sites S3 and S6 have a high one. Sites S6, S8, and S1 in Overa Aru Sanctuary are very safe from bush fires.

Livestock grazing poses a very serious risk to site S3, a serious risk to sites S6 and S7, and an average risk to sites S4, S5, and S8.

Another major issue in Overa Aru Sanctuary is the gathering of firewood, which poses a serious risk to sites S3 and S4, a moderate risk to sites S6 and S7, and a low risk to sites S8 and S9 (Table 6).

Table 6. Threats observed in Overa Aru Sanctuary

S. No.	Site Code	Deforestation	Bush Fires	Grazing by Livestock	Firewood Collection	Unrestricted Tourist Facilities
1	S1	B	B	B	B	C
2	S2	A	A	B	A	C
3	S3	D	D	E	D	B
4	S4	C	C	C	D	C
5	S5	B	B	C	B	D
6	S6	E	D	D	C	E
7	S7	C	C	D	C	E
8	S8	D	B	C	B	E

A= completely no threat, B= Insignificant threat, C= Average threat, D= Significant threat, E= extremely significant

Threats to Pahalgam Ecotourism Spot

Roughly 20% of the workforce in Jammu and Kashmir State is directly or indirectly supported by the tourism industry. While the tourist industry has grown exponentially in recent years, the increase has come at the expense of the state's delicate ecosystem due to the effects of unsustainable tourism practices. These days, even Pahalgam isn't an exception. Pahalgam's once-magical beauty has been marred by an influx of tourists to the point that their effects can be seen. All of the ecotourism hotspots, including Lidder River, Beetab Valley, Nunvan base camp, Baiasaran, Amusement park and Deer park, Aru, and others, are under grave danger from unrestricted tourist amenities.

Since there aren't any effective waste management services, the neighbourhood is plagued by an offensive odor caused by the accumulation of solid and liquid garbage. The ecosystem of the Lidder River is negatively impacted by the direct entry of liquid pollutants. The hazard posed by water pollution is greatest at the Nunvan base camp and least at the Baisaran location, with the average threat level being somewhere in between. The ecotourism hotspot of Beetab valley is under danger from the high volume of automobiles that visit there every day. Unsustainable growth poses a critical risk to Beetab valley, a major risk to Lidder River, Aru, Amusement Park, and Deer Park, and a moderate risk to Baisaran (Table 7).

Table 7. Threats observed in Pahalgam tourist spot

S.No.	Site Name	Unrestricted Tourist Facilities	Wastes	Vehicular Pressure	Developmental Activities
1	Lidder River	E	C	D	D
2	Beetab valley	E	C	E	E
3	Baiasaran	E	A	B	C
4	Aru	E	C	C	D
5	Amusement park and Deer park	E	C	D	D
6	Nunvan base camp	E	D	D	D

A= completely no threat, B= Insignificant threat, C= Average threat, D= Significant threat, E= extremely significant

Threats and Conservation Issues

Pahalgam, a popular ecotourism destination in Kashmir, is set against the breathtaking backdrop of Overa Aru Sanctuary. As one of the few protected regions in J&K State that is home to Hangul, Musk deer, and other species, protecting the forest as a Sanctuary is essential to maintaining this popular tourist destination. Both regions have been exposed to a variety of serious dangers.

Rock climbers and hikers often visit several of the peaks in and near the sanctuary. Constant vigilance is necessary to avoid encroachment into the Sanctuary from the human settlements of Overa, Lardi, and Dahwattoo, whose combined population is over 7,000. Summertime brings intense grazing pressure to Overa-Aru, as local and nomad graziers use the area for their animals. Due to population growth and the need to store enough supplies of dry wood to get people through the winter, gathering firewood has become a significant challenge. Another issue is summertime bushfires, which are sometimes caused by graziers who want to get rid of dry, unappealing coarse grasses. Overgrazing had almost completely eradicated food plants edible to wild ungulates, and previous deforestation and understory clearance had caused widespread damage to Overa-Aru.

Pahalgam and the nearby Overa Aru Sanctuary are under danger from the unchecked growth of tourists. Nearly 1200 to 1300 large and small cars access the tourist destination every day, which poses a concern due to heavy traffic pressure during peak seasons. Another major issue that must be addressed with urgency and care is the harm to the environment and cultural heritage that has resulted from unsustainable development operations in both regions.

The 3MW generating power plant, horseback riding as a tourist activity, road construction, seasonal tent providers, pollution of Lidder River due to liquid wastes directly entering into River causing damage to famous trout fish species, air pollution, open drainage system, and poor water supply system are just some of the additional threats observed in the Pahalgam ecotourism spot.

DISCUSSION

There is a close connection between tourism and the environment. The natural world is considered a valuable tourist resource. There are positive economic benefits to tourism growth, but there are also a number of detrimental repercussions on the local ecology, environment, and economy (Pleumaron, 1992; Mathieson and Wall, 1982; Butler, 1990). Consequences such as biodiversity loss, deforestation, overpopulation, altered landscapes, slums, water, air, and soil pollution, siltation of water bodies, loss of wetlands, social isolation, modernization of old customs, and unsustainable development. Consistent findings between previous research and the current investigation are found. As a growing industry, wildlife tourism has the potential to negatively affect wildlife and the environments in which they live (Sinha, 2001).

The Jammu and Kashmir, India relies heavily on the tourism industry, which is also the region's most dynamic sector. During the peak tourist seasons (summer and fall), the Lidder Valley sees upwards of a thousand visitors every day. There are a lot of people trying to enjoy the State's natural attractions in the places they are. Increasing tourism in the Himalayas is putting a heavy burden on the region's wildlife and flora. Large, irreversible environmental impacts result from the development of tourism in mountain areas due to the clearing of forest cover, road construction, provision of utility services (water, sewage treatment, and power supply), accommodation services, and other tourism infrastructure, such as golf courses and recreational parks.

There can be no life without water. Water's singular features inevitably lead us to conclude that it was made with life in mind, and that no other liquid can provide an adequate substitute. Since water quality is closely related to human health, it is of paramount importance that it be protected. It's the least expensive option for dumping household trash, municipal garbage, and industrial scraps. This causes changes in the water's physical and chemical characteristics, which contributes to pollution.

Overa Aru sanctuary, a significant region in the western Himalayas, is a tourist destination in and of itself due to its rich ornithological resources. Thirty-nine different bird species were recorded throughout this investigation. In contrast to the 117 bird species reported in or around Overa sanctuary have published a list of 70 bird species in Overa Aru sanctuary. 89 species are known to breed inside its borders. Overa Aru is home to 70 different species, just one of which, the Kashmir Flycatcher, is in danger of extinction (Ficedula subrubra). Since its breeding area is so tiny in the Northwest Himalaya, Birdlife International classifies it as a restricted range species as well (2001). This species was not discovered during the current investigation.

CONCLUSION

Travel and tourism are two ways that women can be incorporated into economic and social life (Goeldner & Ritchie, 2009). Many job opportunities created by the growth of tourism have helped women find jobs and start their own businesses. Also, the opportunity to work for themselves and be financially independent that comes with tourism improves their social situation. Tourism is a significant, employment-boosting industry that is growing globally. The World Tourism Organization estimates that there are around 96.7 million people working in the tourism industry, and that number rises to over 254 million when indirect jobs are included [Reimer, J.K.; Walter; World Tourism Organization, 2017]. In many countries, this dynamic industry is the main driver of national income, job creation, and private sector expansion.

The quality of the surrounding environment is a major factor in determining the level of interaction between tourism and the environment. The general public and the conditions that govern tourist activities at any particular time each have their own unique perspectives on this. This investigation was carried out to learn how visiting Overa Aru Sanctuary and Pahalgam affected the surrounding ecology. This investigation took into account the current biodiversity state of Sanctuary as well as its water use, waste management techniques, and the Lidder River's water usage.

Using questionnaires developed specifically for the purpose, interviews with 120 visitors in Pahalgam were conducted. There were somewhat more men (58.17%) than females (40.83%) among the visitors that were surveyed. There were 40% domestic visitors, 47.50% international tourists, and 12.50% local tourists among those surveyed. Over half (45.84%) of respondents drove large cars, while 40.83%) drove small vehicles, and 13.33% rode bicycles or other two-wheeled vehicles. Respondents in Pahalgam overwhelmingly feel that the current infrastructure fails to meet their needs (67.5%).

A variety of reasons were given by vacationers for being in the area. About 42.50 percent said they were there for religious observances or pilgrimages, while 25 percent said they were going on a picnic. Responses in favor of wildlife, scenic beauty, horseback riding, hiking, rafting, and fishing were 9.17%, 7.50%, 5%, 1.67%, 3.33%, and 2.50%, respectively. The remaining 3.33 percent supported various other pursuits. The money visitors spend helps support many different subsets of the local population. Tourists spend money on a wide range of products and services. According to the data collected, vacationers shell out 35.95% of their entire budget on shopping for trinkets, followed by 20.60% on transportation, 16.95% on lodging, 16.42% on entertainment, and 10.08% on everything else.

Among the 120 visitors polled, 64.17 percent went to Overa Aru Sanctuary for reasons including research (7.80 percent), leisure (62.33 percent), and study (29.87). It was decided by 79.22% that Wildlife Sanctuary is an appropriate location to raise awareness about conservation. Seventy-two point eight three percent of visitors polled felt the general public should have a say in how the Sanctuary is run.

In addition, 120 municipal personnel at the Pahalgam ecotourism location were surveyed using questionnaires developed specifically for this research. Only 40 of the employees were permanent members of the Pahalgam Municipality, while the rest were seasonal laborers hired for the brief three-month Amarnath yatra season. One hundred percent of the staff members surveyed acknowledged the presence of both solid and liquid wastes, with the former being the vast majority. The ecotourism destination generates 25-30 qtls of solid wastes daily during peak visitor seasons (summer), and 5-10 qtls during the off season. Waste is not weighed, separated, or recycled at this ecotourism destination. Wastes are dumped at the Sarbal dumping site.

Since 1997, Pahalgam's tourist population has been steadily growing. Approximately 80% of all tourists visit during the summer, whereas just about 20% visit during the winter. The number of yatries visiting the holy cave of Amarnath Ji has been steadily increasing. Government employees in Pahalgam bring in an annual salary of Rs 146 lakhs. The average daily water use is between 80 and 90 liters, rising to 120 liters per person during the summer tourist high season. Approximately 412 full-time government jobs, 8063 private sector jobs, and 5300 seasonal jobs have been created as a result of tourism in the ecotourism location.

During the research period, a total of 702 individual birds were seen in the refuge, representing 29 species, 5 orders, and 22 families. A diverse group of villagers, including Chakur, Koklas, Monal, and Himalayan snow fowl, have found refuge in the sanctuary. The jungle crow, vultures, the Red-billed Cough, the Blue Whistling Thrush, wagtails, woodpeckers, warblers, etc. The motacillidae family has the most members at 6, followed by the sturnidae family at 4, the muscicapidae family at 2, the passeridae family at 2, and the columbidae family at 2. There was just one member of fifteen different families. Around 72% of the bird species and 65% of the people were insectivores. Over 10% of the species and 22% of the individuals were omnivores, while over 10% of the species and 8% of the people were granivores. Birds were mostly herbivores and omnivores, with a few carnivores and frugivores.

Major threats to the Overa Aru Sanctuary and the Pahalgam ecotourism destination include the collecting of firewood, bush fires, significant deforestation, grazing by animals, and unrestrained tourists. Nearly 1200 to 1300 large and small cars access the tourist destination every day, which poses a concern due to heavy traffic pressure during peak seasons. Unsustainable development activities, water and air pollution, power production plant, inadequate drainage, horseback riding as a tourist activity, road construction, and seasonal tent providers are only some of the additional concerns seen in both research regions. The status and variety of each of these regions are severely threatened by all of these dangers.

Important steps for promoting and maintaining a better environment at Pahalgam and Overa Aru Sanctuary include a massive afforestation programme, sustainable developmental measures, waterhole development, research and monitoring, promotion of wildlife tourism, promotion of eco-tourism, revival of existing tourism policy, restricted mass and other forms of tourism.

Most states have lower rates of female participation in ecotourism, and because of gender inequality, women may occasionally be excluded from participative policies altogether. Women play a significant role in the development of natural resource policies since they collect the majority of the food, fuel, and firewood used for domestic and commercial purposes in developing nations. But historically, women have not been allowed to participate in decision-making when it comes to managing these natural resources.

REFERENCES

Butler, R. W. (1990). Alternative tourism: Peons hope or Trojan Horse? *Journal of Travel Research, 28*(3), 40–45. doi:10.1177/004728759002800310

Deming, A. (1996). The edge of the civilized world: Tourism and the hunger for wild Places. *Orion, 15*, 28–35.

Goeldner, C. R., & Ritchie, J. R. B. (2009). *Tourism: Principles, Practices, Philosophies.* John Wiley & Sons.

Milgrath, L. (1989). An inquiry in to valves for a sustainable society: a personal Statement. In L. Milgrath (Ed.), *Envesconing a sustainable society*. SUNY Press.

Reimer, J. K., & Walter, P. (2013). How do you know it when you see it? Community-based ecotourism in the Cardamom Mountains of southwestern Cambodia. *Tourism Management, 34*, 122–132. doi:10.1016/j.tourman.2012.04.002

Singh, J. (2004). The economy of Jammu and Kashmir. Radha Krishna Anand and Co.

Sinha, C. C. (2001). *Tourism Geography: Issues. Challenges and the changing Nature of contemporary tourism.* Paper presented during the geography curriculum inservice conference.

Chapter 5
Women's Empowerment Through Entrepreneurship Education:
A Case From Aviation

Ferhan K. Sengur
Eskisehir Technical University, Turkey

Hilal Tuğçe Lapçın
Anadolu University, Turkey

Nazire Burçin Hamutoğlu
Eskisehir Technical University, Turkey

ABSTRACT

Despite major advancements in recent decades, gender inequalities in the labor market still exist worldwide, and efforts to achieve gender equality appear to be in a standstill. Entrepreneurship provides important opportunities for women to stand on their feet and empower them economically and socially. Training and fostering entrepreneurial traits in people are two important variables in the growth of entrepreneurship. Aviation is one of the industries in that women are underrepresented. This study explores the results of an applied project based on entrepreneurship education to foster the empowerment of young female aviators. The project highlights female vocational high school students' entrepreneurship skills and future potential. The results reveal that female students may cultivate positive attitudes by being aware of their competencies and acquiring 21st-century abilities, including leadership, entrepreneurship, creative thinking, teamwork, communication, and persuasion.

DOI: 10.4018/978-1-6684-8417-3.ch005

INTRODUCTION

Although women make up more than half of the world's population, they do not contribute nearly as much as they could to measurable economic activity, growth, and well-being, which has major macroeconomic repercussions. Women's employment is essential for a nation's growth and competitiveness (ILO, 2015). Throughout the 20th century, women's participation in the labor market significantly grew, supporting industrialization. The percentage of women in the workforce is rising globally, albeit considerably more slowly. While women in the 1970s, women made up 38% of the global workforce (ILO, 1997), the percentage of women in the labor force grew from 50.2 to 51.7 between 1980 and 2008 (Elder and Smith, 2010). Despite significant advancements during the previous 50 years, gender equality is still a work in progress. Women continue to carry most unpaid domestic and childcare responsibilities. Also, they typically labor illegally and cannot utilize their social rights when they work outside the home. Besides, any crisis has directly and disproportionately threatened women's social and economic potential, such as the 2020 financial crisis hit and COVID-19. Women are particularly vulnerable to the pandemic's consequences since they comprise most health, social service, and unpaid care employees. In addition, women continue to receive less money for doing the same job as men and experience a gender pay gap. Just around three-quarters of the legal rights granted to males are commonly possessed by women (World Bank Group, 2020).

Due to their complexity and multidimensionality, gender inequalities cannot be simplified into a single, globally accepted collection of objectives. However, women's access to education, paid work, and participation in societal governance structures undoubtedly has the potential to empower women (Kabeer, 2005). Kabeer (1999) claims that "women's empowerment is about the process by which those who have been denied the ability to make strategic life choices acquire such an ability". According to a popular definition of the United Nations Economic and Social Council, women empowerment "is the process by which women take control over their lives, acquiring the ability to make strategic choices" (United Nations Economic and Social Council, 2002). According to the European Institute for Gender Equality, the five pillars of women's empowerment are their sense of self-worth; their right to make and exercise choices; their access to opportunities and resources; their right to have the authority to manage their own lives, both inside and outside the home; and their capacity to direct social change toward the development of a more just social and economic order, both domestically and globally. In this context, key tools for empowering women and girls to assert their rights include education, training, awareness-raising, boosting self-confidence, expanding choices, increasing access to resources, and acting to transform the structures and institutions that support inequality (EIGE, 2016).

Fair treatment of men and women is not just the ethical thing to do; it is also beneficial to company performance. Women's full engagement in businesses and the greater society makes companies more competitive today and in the future. The Women's Empowerment Principles are business principles that advise empowering women in the community, the workplace, and the marketplace. They are an adaptation of the Calvert Women's Principles® and the product of a partnership between the United Nations Entity for Gender Equality and the Empowerment of Women (UN Women) and the United Nations Global Compact. A worldwide multi-stakeholder consultation process produced the Women's Empowerment Principles. It started in March 2009 and culminated in its release on International Women's Day in March 2010. Today more than 3000 companies around the world have accepted the seven principles of women's empowerment (UN Global Compact, 2022):

Principle 1: Establish high-level corporate leadership for gender equality.
Principle 2: Treat all women and men fairly at work – respect and support human rights and nondiscrimination.
Principle 3: Ensure the health, safety and well-being of all women and men workers.
Principle 4: Promote education, training, and professional development for women.
Principle 5: Implement enterprise development, supply chain and marketing practices that empower women.
Principle 6: Promote equality through community initiatives and advocacy.
Principle 7: Measure and publicly report on progress to achieve gender equality.

Women's entrepreneurship can be a powerful avenue for women's inclusion of paid, regular, and flexible work conditions that will give them a balanced work and life. Especially in recent decades, the understanding of entrepreneurship, fed by the increase in global competition, has entered the interest of women and men. Digital transformation and the rise of social media have recently increased entrepreneurship opportunities and led to new areas such as digital and social entrepreneurship. However, women's access to entrepreneurship opportunities is less easy and more challenging than men's. Although there has been an increasing trend in women's entrepreneurship in the last decade, gender equality remains a problem (Shahin et al., 2021; Bosma & Kelley, 2019; Shinnar et al., 2012). Among the sources of women's entrepreneurship, personal characteristics come to the fore, but entrepreneurship is also considered a product of the socioeconomic and cultural structure of the individual. General economic conditions in the region and the country, the family and close environment in which the individual was born, social behavior patterns, and the quality and level of the education s/he received are one of the main determinants of an individual's development of entrepreneurial spirit and entrepreneurial attitudes and behaviors.

On the other hand, social norms and stereotypes can prevent women from being included in the entrepreneurship ecosystem by creating obstacles in front of women's entrepreneurship. The fact that young women still prefer female-dominated professions maintains a substantial gender gap, especially in more science and technology-oriented work areas such as information technologies and engineering (OECD, 2017). Even women who graduate from the same fields as men make different career choices and follow different paths (Schnittker & Ettl, 2021).

One of the essential factors in the development of entrepreneurship is the dissemination of training and the development of entrepreneurial characteristics of individuals. Many researchers declared that entrepreneurship is a behavioral change gained through education and learning and a genetic characteristic. Entrepreneurship education is essential for developing the entrepreneurial skills, attitudes, and behaviors that form a country's economic growth. Entrepreneurship education has a major influence on entrepreneurial intention. In other words, attending entrepreneurship education can inspire entrepreneurial intention and increase the likelihood of realizing this goal (Zhang et al., 2014). Entrepreneurship and inquiry skills are at different levels in each individual and are known to be developed and learned. Those skills are also critical for the empowerment of individuals, and women's entrepreneurship is one of the primary instruments for women's empowerment. The factors that predict female self-employment entrance differ significantly from those that predict male self-employment entry (Bates, 1995). Women are more likely than males to start businesses in skilled service sectors. As a result, higher education and job experience—the human capital variables—are the most powerful predictors of female entrants. Building technical, digital, technological, and financial literacy skills enhance the entrepreneurship ability of women and strengthen their entrepreneurship success. For instance, among other things, having access to outside funding is a significant determinant in the growth of women's businesses Khanum et al. (2020). The rise of women empowerment via entrepreneurial skills will open new avenues for better living standards and economic prosperity, as women are motivated, enthusiastic, and possess a broader vision. It has been noted that women's entrepreneurship is not at the desired level, particularly in the air transport industry. In addition, the entrepreneurial tendencies of female students are lower than male students.

If more than 25% of occupations are women, that occupation is called non-traditional (Bagilhole, 2002). Aviation is also considered a non-traditional profession, as a usually a male-dominated industry (Clark et al., 2015; Wilson & Lohmann, 2019). Although the number and rate of female employment have been increasing in recent decades, especially female leadership and entrepreneurship in aviation are uncommon. Policymakers and practitioners are acknowledging the significance of female empowerment in society. As an international and interrelated industry, aviation is not exempt from these developments. In recent years high-level commitment and stakeholder initiatives in aviation to foster female leadership have been increasing. The UN body International Civil Aviation Organization (ICAO), IATA (International Air Transport Association) and Airport Council International (ACI). ICAO established a new programme, the ICAO's *Gender Equality Programme*, in 2018. With this programme, ICAO, governments and the industry are responsible for data, practices, and empowering women in aviation. IATA also launched the *25by25 initiative*, which seeks to raise the proportion of women in high positions in under-represented sectors by 25%, or up to a minimum of 25%, by 2025.

In parallel with high-level developments to support women in aviation, many initiatives and projects support women in aviation in the industry and academia. This study reveals the results of a project that aims to eliminate the obstacles in front of girls at an early age and to reveal their hidden potential. "Aviator Girls Step into Entrepreneurship Project" sponsored by The Scientific and Technological Research Council of Turkey (TUBITAK) 4004 Education in Nature and Science Camps/Schools Support Programme and by Eskisehir Technical University Faculty of Aeronautics and Astronautics and aviation high school female students are included in the project. The project is aimed to ensure that female students are aware of and reveal their entrepreneurial potential with entrepreneurship education. During the four-day, applied entrepreneurship education was given to 30 students from a high school in the city of Eskisehir specializing in aviation maintenance. The education was carried out by academics and teachers competent in entrepreneurship. Before and after the education, face-to-face interviews were conducted to reveal the students' views on entrepreneurship. The data were analyzed with the MAXQDA program. After the education, it was observed that the student's awareness of the entrepreneurship ecosystem increased, and their motivation to establish their initiatives was formed. Accordingly, this publication aims to bridge and emphasize the importance of women's entrepreneurial characteristics and tendencies on any digital platforms while social media usage is increasing hugely. Entrepreneurship is believed to be thoroughly related to getting empowered in the transport industry, especially in tourism. So, the current chapter proposed a curriculum for women to get empowered with the basis of entrepreneurship. Hence, the chapter also focuses on strengthening women's entrepreneurial characteristics and tendencies with the parallel of several proposed digital tools supporting empowerment.

OBJECTIVE

This study emphasized the importance of empowering women with entrepreneurship training based on the results obtained within the scope of the "Aviator Girls Step into Entrepreneurship" project supported by the TUBITAK 4004 Nature and Science Schools Program. The results of the data analysis provide evidence for the importance of the study, which is the basis for disseminating the academic and scientific outputs of the activities carried out within the scope of the project. In this context, the results obtained before and after the training carried out within the scope of the project are as follows:

- There is a significant difference between students' entrepreneurial tendencies during the pretest and posttest scores (Şengür, Lapçın, & Hamutoğlu, 2022a).
- Although there is an increase in the posttest total scores of the measurement tool, which determines the students' entrepreneurial characteristics, this increase is not significant. In addition to this;
- it was found to be the most significant in the innovation dimension.
- when the pretest and posttest scores are compared in terms of achievement sub-dimension, there is a decrease against the posttest (Şengür, Lapçın, & Hamutoğlu, 2022b).
- the results of the student's views on education show that their awareness of the entrepreneurship ecosystem and motivation to establish their initiatives has increased (Şengür, Lapçın, & Hamutoğlu, 2022c).
- Based on these findings, in this book chapter, the importance of entrepreneurship education in empowering women will be discussed with its sociological, psychological, economic, and cultural dimensions. Therefore, this study,
- is current in terms of Millennium Development Goals (MDGs) called MDG 3: Promoting gender equality and empowering women and has been the basis of Sustainable Development Goals (SDGs) and followed by SDG 5.5: Ensuring full participation in leadership and decision-making and "SDG 5.8: Supporting the empowerment of women through technology", which are the sub-target of "SDG 5: Gender Equality".
- is original in terms of the results obtained within the scope of the "Aviator Girls Step into Entrepreneurship" project supported by the TUBITAK 4004 Nature and Science Schools Program and addressing women's empowerment through entrepreneurship training with its uniqueness.
- is necessary for developing and underdeveloped countries in the world to address the importance of women's entrepreneurship training on a global scale,
- is functional in terms of considering the sociological, psychological, economic, and cultural dimensions of the steps to be taken to strengthen the visibility of women in social life by considering entrepreneurship, which has an important place in economic growth, as an intermediary concept.

WOMEN EMPOWERMENT AND ENTREPRENEURSHIP IN AVIATION

Entrepreneurship can be defined as transforming an idea into a product and service and presenting it to a market (Küçükaltan, 2009). Entrepreneurship is related to "seizing opportunities" and "developing plans and projects related to these opportunities by taking risks" (Gömleksiz & Kan, 2009). Entrepreneurs who use entrepreneurial skills to generate wealth through new ventures contribute to society by taking advantage of possibilities in a rapidly changing social, political, legal, and economic environment (Kumar & Singh, 2021). Nevertheless, there is a deviation between men and women entrepreneurs. This deviation is related to women still having fewer opportunities to make a long-life career (Costa et al. 2016). The field of women's entrepreneurship has come a long way since its emergence in the late 1970s. In 1976, Schwartz published the first academic paper on women entrepreneurship in the Journal of Contemporary Business. In the 1990s, as the number of women business owners increased, studies on women entrepreneurs also increased. Initial research listed similarities and dissimilarities between male and female business owners. Differences were reported on educational background, occupation, motivation, and business creation. Therefore, supporting women's entrepreneurship and diminishing dissimilarities is crucial (Şengür et al., 2022). Women's empowerment should be addressed to improve women, the value of their contributions to society, and advance women's entrepreneurship (Hassan et al., 2022).

Empowerment is "the ability of individuals to gain control socially, politically, economically, and psychologically through access to information, knowledge and skills, decision-making, and individual self-efficacy" (Rappaport & Zimmerman, 1988). Women's empowerment is used for the choices and productivity of individual women (Bisnath, 2001). UNIFEM (the United Nations Development Fund for Women) defines women's economic empowerment as 'having access to and control over the means to make a living on a sustainable and long-term basis and receiving the material benefits of this access and control. Facilitating and enhancing the empowerment process could be used as a vehicle for entrepreneurship (Al-Dajani & Marlow, 2013). Entrepreneurship and empowerment process to support each other, seen in Figure 1. Facilitating and enhancing the empowerment process could be used as a vehicle for entrepreneurship (Ali & Salisu, 2019; Samarakoon & Parinduri, 2015).

Women have traditionally been regarded as an "invisible" workforce, and their abilities to grow, excel, and develop are limited (Sharma & Varma, 2008). Entrepreneurship allows women to have control over resources and gives them greater self-confidence. Entrepreneurship allows women to generate income, bring flexible working hours, and make them economically independent. These all lead to women's empowerment (Sharma & Varma, 2008). There is quite an important question about whether entrepreneurship can be taught or is innate. Because entrepreneurship is an ongoing lifelong learning experience, scholars thought the best way to learn is to associate the experience with formal educational activities. Analytical thinking, accounting, finance, marketing, management information systems, and manufacturing are among the aspects of entrepreneurship that can be taught (Henry et al., 2005). Therefore, entrepreneurship education can be an important tool to empower women economically.

Figure 1. Empowerment- entrepreneurship cycle

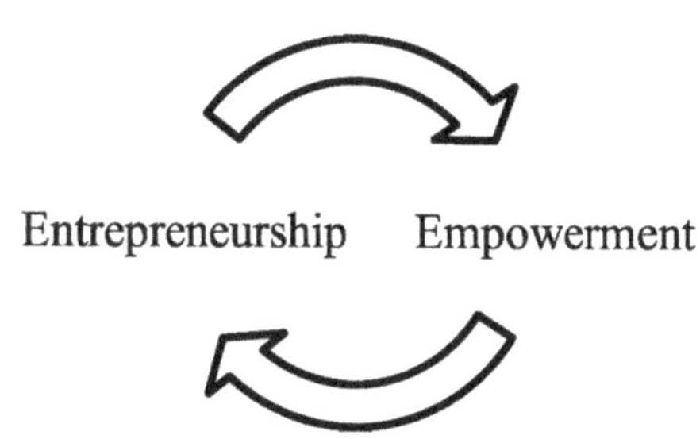

In the aviation sector, the number of women in professions such as pilot, technician, and manager is considered more "man's job". Although there is an upward trend in the number of women in aviation, this increase is limited. In the USA, in 1960, 1% of pilots or technicians were women. This number only increased to 4.3% for pilots and 2.17% for technicians in 2010. Besides, at management levels, the 'glass ceiling effect' effectively promotes women, and women are rarely ever encountered. According to International Air Transport Association-IATA, less than 5% of global airline chief executives (CEOs) are women (Şengür et al., 2022). It is seen that women do not have many opportunities in micro-entrepreneurship activities in the aviation sector. Therefore, it is very important to empower women in this sense and increase their financial literacy and awareness of entrepreneurship by informing them about the sector. The importance of entrepreneurship education is pointed out in the studies in which the views on entrepreneurship are put forward. This education must be given sufficiently, especially at a young age. Female students were determined as the target group within the scope of the study since both entrepreneurial characteristics are more common in males and females are less involved in entrepreneurial activities. It is aimed that female students will be excited about science and entrepreneurship and develop their entrepreneurial skills through applied activities (Şengür et al. 2022). Applied activities include animation and simulation, scientific trips, digital games/stories, educational games, gamification, STEAM, discussion (seminar, panel etc.), workshop, and creative drama.

PROCESS

Entrepreneurship education was carried out within the scope of the "Aviator Girls Take a Step into Entrepreneurship" project supported within the scope of the TÜBİTAK 4004 Nature and Science Schools Program. A total of 30 female students studying at Sabiha Gökçen Vocational and Technical Anatolian High School in Eskişehir/Turkey were given four-day entrepreneurship training within this project's scope. 13 academics/teachers from six different institutions in Turkey took part in the project. Among the pieces of training are "Breaking the Ice", "Entrepreneurship and Innovation", "How to start business", "Entrepreneurship with games", "Financial literacy", "I have a business idea", "My business model", "Digital entrepreneurship", "Social entrepreneurship", "3D character design", and "Green world thinking". The project aims to reveal the skills and potential of vocational high school female students in entrepreneurship. As the target group, female students, who were educated at the high school level and had not received an entrepreneurship education, were preferred. One of the reasons why vocational high school students are preferred is that they are often underestimated by society, and their self-perceptions may be settled in this way. In this sense, although students in aviation high schools can get high scores in high school entrance exams and settle in schools, their employment as high school graduates may make them think that they have no other professional alternative.

Interviews were conducted before and after the training. Descriptive analysis was used in the analysis of the data. In the descriptive analysis, direct quotations are frequently used to reflect the views of individuals. It is aimed to summarize and interpret the findings obtained in this type of analysis (Yıldırım & Şimşek, 2013). The implementation process was carried out according to the four stages of descriptive analysis.

Figure 2. Four stages of descriptive analysis

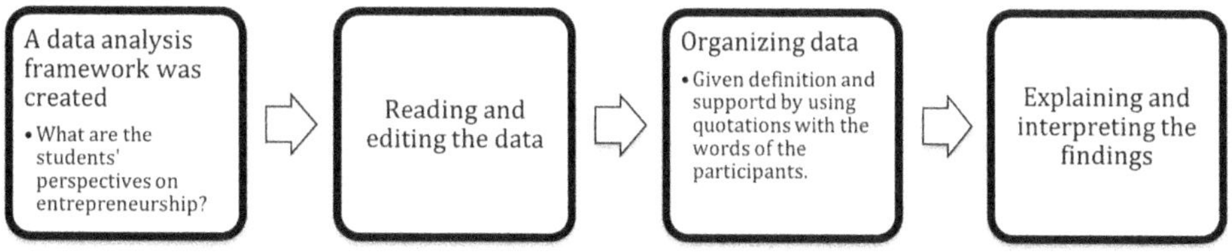

Table 1. Themes and quotations of students' views before training

Themes	Direct Quotations
"Start to own your business", "Generating business ideas."	*"...starting my own business, but a wild dream...", "...I don't know how to generate ideas...." "... generating ideas, I have an idea, but I don't know how to realize it and whether I will be successful."*

The findings of the data obtained before and after the training are included. Table 1 shows themes and quotations before the training.

However, students consider entrepreneurship a hobby and think that if they work in the aviation sector, they cannot become entrepreneurs. Students with previous entrepreneurship in their family are more moderate towards entrepreneurship.

My father was a worker in a factory. He left his job at the factory and opened a shoe store so we could live more comfortably. I'm also thinking of becoming an entrepreneur, but I don't know what to do because I don't have an idea.

It has been observed that working in the aviation sector is an obstacle to entrepreneurship. In the future, the students will either do their jobs or carry out entrepreneurial activities outside the aviation sector. However, it was observed that there was a change in students' perceptions of entrepreneurship after the training. According to students, entrepreneurship requires and is related to "teamwork", "division of work", "self-confidence", "team spirit", "creativity", "innovation", "synergy", "presenting business ideas", "communication", "business steps", "team management", "learning from mistakes", "expressing ideas", "finding a middle way in disagreements", "easy self-expression", and "harmony".

Table 2. Themes and quotations of students' views after training

Themes	Direct Quotations
"Team working" "Self-confidence" "Learning from mistakes." "Work division" "Synergy" "Women entrepreneurs"	*"...teamwork is necessary...", "... we can develop ideas by learning from mistakes...." "... creating business steps and division of work with my teammates...." "...I realized the synergy that emerged when I worked with the team... ".. entrepreneurship has nothing to do with being a woman or a man...."*

Table 3. Code, category, and themes after training

Code	Category	Themes
"Team working and work division." "Generating ideas" "Learning from mistakes."	Team working	Synergy

Table 4. Themes for "entrepreneurship" after the training

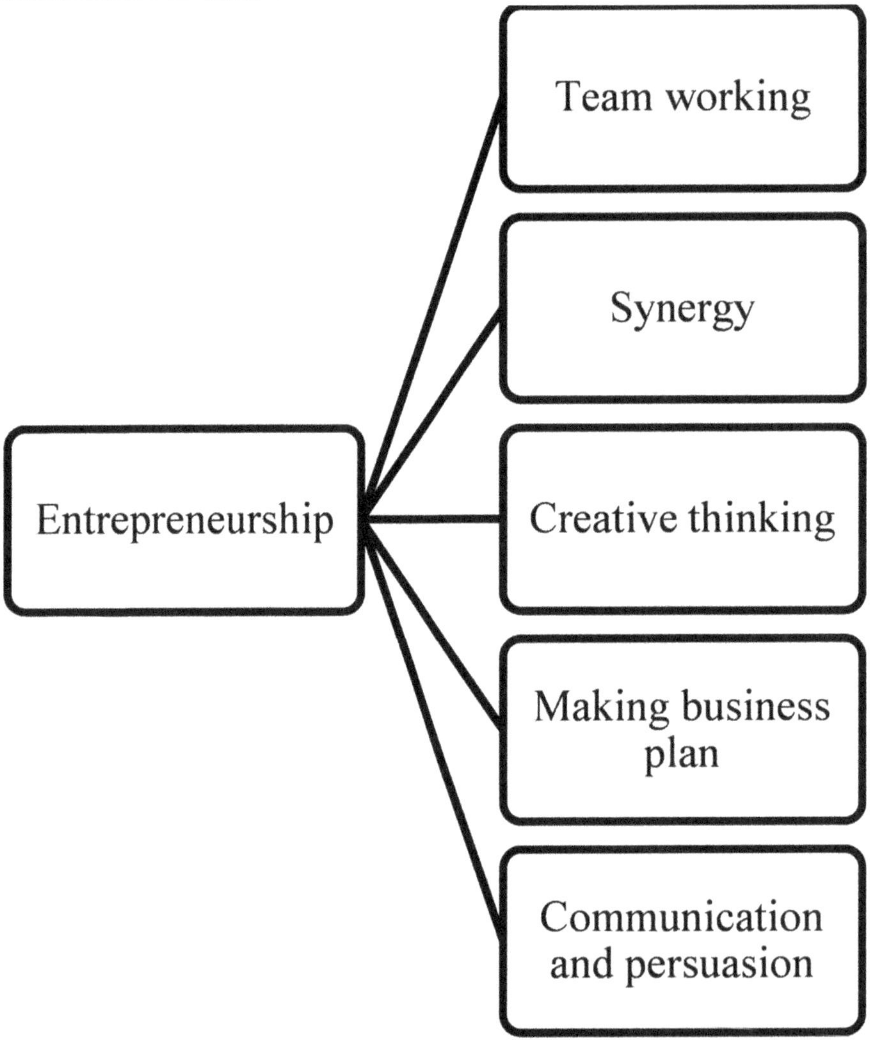

An example of code, category, and themes is given in Table 3.

As seen in Table 4, female students think entrepreneurship relates to teamwork, synergy, creative thinking, making business plans, and communication and persuasion. After the training, it has been observed that "*Entrepreneurship is gender neutral*" and "*I can do it too. I can start my own business*".

Female students develop positive attitudes by becoming aware of their competencies by developing skills such as leadership, entrepreneurship, creative thinking, team working, and communication and persuasion required by the 21st century. It is known that facilitating and enhancing the empowerment process could be used as a vehicle for entrepreneurship (Al-Dajani & Marlow, 2013). However, it is seen that family also impacts entrepreneurship skills. Having an entrepreneur in the family can be effective for other family members to be entrepreneurs.

CONCLUSION

Women's empowerment is a multidimensional economic, social and governance phenomenon. Entrepreneurship, on the other hand, is a process that has both economic and social and political benefits for the empowerment of women and girls. While it is known that entrepreneurship emerges with some innate predispositions, environmental conditions are also determinative. Entrepreneurship education has the potential to both sharpen an individual's innate competencies and increase their intrinsic motivation and courage. It can also provide women with the necessary technical, digital, and financial knowledge and skills to be included in the entrepreneurship ecosystem. Entrepreneurship education, especially at a young age, can affect individuals' tendencies and enable them to be ready when the opportunity comes or to create their opportunities.

Aviation is an industry where women's entrepreneurship is much less common than a traditional profession where the number of women is low. In this respect, this study reveals a project which aims to foster women entrepreneurship in early-age entrepreneurship education with a focus on aviation. Results of practical education suggest that entrepreneurial education can raise entrepreneurial skills, awareness and intention. The project findings show that female students may build positive attitudes by being aware of their competencies and developing skills such as leadership, entrepreneurship, creative thinking, teamwork, and communication and persuasion, which are essential skills of the twenty-first century. Entrepreneurship education at early ages can be a way of empowering women of future aviation.

PRACTICAL IMPLICATIONS AND SUGGESTIONS

In line with the results obtained in the study, it is worth to refer the advantages in terms of practical applications; First, it can be said that supporting women's digital entrepreneurship skills have a primary importance in terms of Sustainable Development Goals. Secondly, it is very valuable to consider the dynamics of the 21st century that has been lived in and the need for digitalization from the perspective of women. When this issue is combined with economic development, it is necessary to focus on women's entrepreneurship education from different perspectives. Thirdly, it can be a strong emphasis to be placed on developing entrepreneurial skills in meeting the needs of society for empowered women in almost all aspects, including economic, political, cultural, and sociological. Fourthly, when the education parameter of the Human Development Index (HDI) reports is examined, the visibility of women in developed and highly developed countries in social life and their role in economic growth are examined. Finally, it is important to establish education policies for other developing, and underdeveloped countries should be considered in terms of contributing to the global capital.

FUTURE RESEARCH DIRECTIONS

While this study also has limitations, such as the group size and focusing on one group in terms of industry and geography, the study's results can give valuable contributions to future studies. Especially longitudinal studies are necessary to reveal the impact of early entrepreneurship education on entrepreneurial intention and practices. This study is important for female entrepreneurship, which already has several social, cultural, and economic barriers comprising men's entrepreneurship. Also, studies on antecedent female entrepreneurship in aviation are needed, which does not provide a good ecosystem for women's self-employment.

Acknowledgements: "Aviator Girls Step into Entrepreneurship Project" is supported with a grant from The Scientific and Technological Research Council of Turkey (TUBITAK) 4004 Education in Nature and Science Camps/Schools Support Programme.

REFERENCES

Al-Dajani, H., & Marlow, S. (2013). Empowerment and entrepreneurship: A theoretical framework. *International Journal of Entrepreneurial Behaviour & Research*, *19*(5), 503–524. doi:10.1108/IJEBR-10-2011-0138

Ali, M. A., & Salisu, Y. (2019). Women Entrepreneurship and Empowerment Strategy for National Development. *Journal of Economics. Management and Trade*, *22*(3), 1–13. doi:10.9734/JEMT/2019/44828

Bagilhole, B. (2002). *Women in non-traditional occupations: Challenging men*. Springer. doi:10.1057/9780230501102

Bates, T. (1995). Self-employment entry across industry groups. *Journal of Business Venturing*, *10*(2), 143–156. doi:10.1016/0883-9026(94)00018-P

Bisnath, S. (2001). *Poverty in a globalizing world at different stages of women's life cycle. Içinde Globalization, Poverty and Women's Empowerment*. Sayı November.

Bosma, N., & Kelley, D. (2019). *Global entrepreneurship monitor 2018/2019 global report. Global Entrepreneurship Research Association*. GERA.

Clark, P. J., Newcomer, J. M., & Jones, A. M. (2015). Overcoming gender barriers in aircraft maintenance: women's perceptions in the United States. *The Collegiate Aviation Review International, 33*(2).

Costa, C., Breda, Z., Bakas, F. E., Durão, M., & Pinho, I. (2016). Through the gender looking-glass: Brazilian tourism entrepreneurs. *International Journal of Gender and Entrepreneurship*, *8*(3), 282–306. doi:10.1108/IJGE-07-2015-0023

EIGE. (2016). *European Institute for Gender Equality(EIGE)'s Gender Equality Glossary & Thesaurus.* https://eige.europa.eu/taxonomy/term/1102

Elder, S., & Smith, A. (2010). Women in labour markets: Measuring progress and identifying challenges. Ginebra, Suiza: International Labour Office.

Global CompactU. N. (2022). https://unglobalcompact.org/take-action/action/womens-principles

Gömleksiz, M. N., & Kan, A. Ü. (2009). Sosyal Bilgiler Dersi Öğretim Programının Eleştirel Düşünme, Yaratıcı Düşünme ve Girişimcilik Becerilerini Kazandırmadaki Etkililiğinin Belirlenmesi (Diyarbakır İli Örneği). *Doğu Anadolu Bölgesi Araştırmaları, 8*(1), 39–49.

Hassan, O. H. M., Sobaih, A. E. E., & Elshaer, I. A. (2022). The Impact of Women's Empowerment on Their Entrepreneurship Intention in the Saudi Food Industry. *Journal of Risk and Financial Management, 15*(12), 571. Advance online publication. doi:10.3390/jrfm15120571

Henry, C., Hill, F., & Leitch, C. (2005). Entrepreneurship education and training: Can entrepreneurship be taught? Part I. *Education + Training, 47*(2), 98–111. doi:10.1108/00400910510586524

ILO. (2015). *Women in business and management: gaining momentum. International Labour Organisation. International Labour Office*. ILO.

Kabeer, N. (1999). Resources, agency, achievements: Reflections on the measurement of women's empowerment. *Development and Change, 30*(3), 435–464. doi:10.1111/1467-7660.00125

Kabeer, N. (2005). Gender equality and women's empowerment: A critical analysis of the third millennium development goal 1. *Gender and Development, 13*(1), 13–24. doi:10.1080/13552070512331332273

Küçükaltan, D. (2009). Genel bir yaklaşımla girişimcilik. *Girişimcilik Gündemi, 4*(1), 21–28.

Kumar, N., & Singh, D. L. K. (2021). Status of Women-entrepreneur in Indian Startups. *International Journal of Engineering Technology and Management Sciences, 5*(2), 1–12. doi:10.46647/ijetms.2021.v05i02.001

OECD. (2017). *Report on the implementation of the OECD gender recommendations—Some progress on gender equality but much left to do*. OECD.

Rappaport, J., & Zimmerman, M. (1988). Citizen participation, perceived control, and psychological empowerment. *American Journal of Community Psychology, 16*(5), 725–750. doi:10.1007/BF00930023 PMID:3218639

Samarakoon, S., & Parinduri, R. A. (2015). Does Education Empower Women? Evidence from Indonesia. *World Development, 66*, 428–442. doi:10.1016/j.worlddev.2014.09.002

Schnittker, J., & Ettl, K. (2021). Career Paths of Women Entrepreneurs in STEM–Challenges and Opportunities. In E. Laveren, C. Ben-Hafaïedh, A. Kurczewska, & Y. D. Jiang (Eds.), *Contextual Heterogeneity in Entrepreneurship Research*. Edward Elgar Publishing. doi:10.4337/9781802200201.00012

Selanik Ay, T., & Acar, Ş. (2016). Sınıf öğretmenlerinin girişimcilik becerisi kazandırmaya yönelik görüşleri. *Elektronik Sosyal Bilimler Dergisi, 15*(58), 960–976. doi:10.17755/esosder.03138

Şengür, F. K., Hamutoğlu, N. B., & Lapçın, H. T. (2022). Aviator Girls Step into Entrepreneurship. *Aviation Turkey*. https://www.aviationturkey.com/en/content/aviator-girls-step-into-entrepreneurship-688

Şengür, F. K., Hamutoğlu, N. B., & Lapçın, H. T. (2022). Havacı Kızlar Girişimciliğe Adım Atıyor. *UTED*, 50. www.uteddergi.com

Şengür, F. K., Lapçın, H. T., & Hamutoğlu, N. B. (2022a). Determination of entrepreneurship tendencies of students after TUBİTAK 4004 "Aviator Girls Step into Entrepreneurship" Project. 8th International EMI Entrepreneurship & Social Sciences Congress, Aksaray, Turkey.

Şengür, F. K., Lapçın, H. T., & Hamutoğlu, N. B. (2022b). Improving entrepreneurship characteristics of young female aviators: The TUBITAK 4004 Project. *3rd International Congress of Women in a Global World (WGW2022).*

Şengür, F. K., Lapçın, H. T., & Hamutoğlu, N. B. (2022c). Opinions of the young aviators on entrepreneurship: TUBİTAK 4004 "Aviator Girls Step into Entrepreneurship" project. *Ege 7th International Conference on Social Sciences.*

Shahin, M., Ilic, O., Gonsalvez, C., & Whittle, J. (2021). The impact of a STEM-based entrepreneurship program on the entrepreneurial intention of secondary school female students. *The International Entrepreneurship and Management Journal*, *17*(4), 1867–1898. doi:10.100711365-020-00713-7

Sharma, P., & Varma, S. K. (2008). Empowerment of women through entrepreneurial activities of self help groups. *Indian Research Journal of Extension Education*, *8*(1), 46–51.

Shinnar, R. S., Giacomin, O., & Janssen, F. (2012). Entrepreneurial perceptions and intentions: The role of gender and culture. *Entrepreneurship Theory and Practice*, *36*(3), 465–493. doi:10.1111/j.1540-6520.2012.00509.x

United Nations Economic and Social Council. (2002). *Commission on the Status of Women: Report on the forty-sixth session (4-15 and 25 March 2002). Official Records, 2002. Supplement No. 7 (E/2002/27-E/CN.6/2002/13).* New York: United Nations.

Wilson, R., & Lohmann, G. (2019). Airline CEOs: Who are they, and what background and skill set are most commonly chosen to run the world's largest airlines? *Transportation Research Interdisciplinary Perspectives*, *2*, 100054. doi:10.1016/j.trip.2019.100054

World Bank Group. (2020). Women, business, and the law: 2020. Author.

Yıldırım, A., & Şimşek, H. (2013). *Sosyal Bilimlerde Nitel Araştırma Yöntemleri*. Seçkin Yayıncılık.

Zhang, Y., Duysters, G., & Cloodt, M. (2014). The role of entrepreneurship education as a predictor of university students' entrepreneurial intention. *The International Entrepreneurship and Management Journal*, *10*(3), 623–641. doi:10.100711365-012-0246-z

Chapter 6
The Role of Diversity Management in Women's Empowerment in the Tourism Sector:
The Role of Turkey

Melike Sak
Selcuk University, Turkey

Gül Erkol Bayram
Sinop University, Turkey

Ali Turan Bayram
Sinop University, Turkey

ABSTRACT

This study aimed to develop suggestions for removing the obstacles and difficulties in front of the career development of women in the tourism sector in Turkey. In addition to this situation, suggestions will be made to remove the gender discrimination faced by women in working life and the obstacles in front of their career development. In order for women to be empowered in working life, the opportunities they need should be provided. It is thought that diversity management can create an opportunity for this to happen. It is thought that diversity management can have a significant impact on the implementation of practices that can reveal the talents and potentials of women so that they have equal rights with men. Thus, many obstacles in front of women can be removed by using diversity management. Strategies have been determined to eliminate the career barriers of women and suggestions and strategies have been tried to be developed to overcome the obstacles in front of women working in the sector.

DOI: 10.4018/978-1-6684-8417-3.ch006

INTRODUCTION

According to the development level of a country, the employment rate and the goals that the country wants to achieve differ. There are some sectors that determine the employment rate and development levels of countries. The tourism sector is also among these sectors. The tourism sector provides a significant income effect of the countries. In this context, one of the sectors with the highest employment rate is the tourism sector.

Developed countries benefit significantly from the income effect of the tourism sector. But one of the biggest problems of developing countries is the problem of unemployment. The tourism sector has a significant impact on solving the unemployment problem. Developing countries benefit from the power of the tourism sector to increase the employment rate. The tourism sector, which is included in the service sector, is a resource that is a solution to the unemployment problem of countries (Şit, 2016).

One of the individuals that the tourism sector provides the most employment in the world is women. Women have turned to the tourism sector because it gives women economic freedom and social status. The reason for the high employment of women in the tourism sector, which is a labor-intensive sector, is that the tourism sector has more job areas for women's knowledge and skills. Despite the increase in the number of female employees in the tourism sector, many discrimination and social pressures against women still continue today. Women have faced many problems and obstacles since the day they started to take part in working life. As in every other sector, female employees face many problems and obstacles in the tourism sector. Before entering the working life, women who face social pressure are expected to deal with homework and childcare instead of working. When women enter the working life, they primarily encounter gender inequalities. Another problem faced by women is the inequality of education and opportunity. Women, who have many obstacles in their working life and career development in the face of the problems they experience, are faced with discrimination, gender inequality, not being able to be employed, and leaving the job (Harris, 2004).

Since the tourism sector is among the sectors most preferred by women in working life, the tourism sector plans opportunities for female employees. One of the countries where opportunities for female employees are realized is Turkey. Turkey's tourism sector is the country where female employees are most employed. Both the fact that the tourism sector is suitable for women's knowledge and skills and the increase in the employment rate with the development of the Turkish tourism sector is very important in the empowerment of female employees in the tourism sector in Turkey. In Turkey, where women's employment in tourism is intense, positive factors are being developed for women's career, financial and social development (Yıldırım et al., 2020).

The concept of diversity management has been developed as a new management style that can offer solutions for disadvantages by increasing the advantages that women face in working life. Diversity management is a form of management that enables individuals with many diversity such as their race, religion, language, gender, age to work together and in harmony. It is thought that diversity management can be effective as a solution to many problems arising from gender inequalities faced by female employees in their working life.

Diversity management aims to enable each individual working in the organization to use their personal skills. Women who face social inequality and career barriers in working life can use the skills they have with diversity management, so that each individual can use their characteristics instead of inequalities between men and women. In this context, social inequalities in working life will be eliminated. In the tourism sector of Turkey, where women's employment is the most intense, many obstacles faced by women, such as obstacles, difficulties, work-family conflict, will help women to overcome these obstacles and difficulties with diversity management and in this context, help women become stronger in working life.

WOMEN EMPOWERMENT IN TOURISM INDUSTRY

Since the beginning of history, the rate of women's participation in economic activities has been improving day by day in accordance with the conditions of the period. With the change of social life, women have started to take more place in society. The increase in industrialization and rural-urban migration has enabled women to take part in many sectors and participate in employment.

Despite the increase in women's employment every year, various obstacles in front of women cause women to take part in working life less than men today. Women, who started to take an active part in working life, started to encounter various obstacles and problems over time. Gender inequality has created many obstacles to women's education and career development. Women who do not have equal rights with men do not have sufficient opportunities in the fields of economic, social and cultural development. While this situation causes inequality between men and women, it puts women at a disadvantage against men in working life (Ponthieux and Meurs, 2015). The problems and obstacles faced by women under gender inequality constitute a disadvantage for women's social empowerment. Along with gender inequality, discrimination based on gender, wage inequality and being behind men in career development are among the obstacles women frequently encounter. The prejudice that prevents women from taking part in working life is another factor that causes inequality between men and women. This situation prevents the empowerment of women in working life and tourism sector.

Tegland (2008) defined the concept of empowerment as individuals' high quality of life in terms of freedom, economy, education or knowledge. From this point of view, it means that women can get the training they need to develop their personal knowledge and skills, as well as providing equal conditions and opportunities with men in order for women to be empowered in working life. Women can gain strength in sectoral competition by participating in training opportunities and career development programs in the fields they want to develop. Gender equality in women's empowerment, taking a managerial position, being involved in the decision-making process, fair working conditions, promoting professional development are the most important requirements for women to be empowered in the tourism sector.

Many countries are developing incentive plans, policies and practices for many sectors in order to prevent the problems that women face in working life, to increase women's employment and to facilitate women's participation in employment. Within the scope of the policies implemented, the employment of women in some sectors showed an intense increase. One of the sectors in which female employment has increased and has a significant impact on women's empowerment is the tourism sector (Suryawati et al., 2021).

Tourism has a high employment rate in the service sector and has very important effects that will provide advantages to countries in economic development. Tourism, which is a labor-intensive sector, plays an important role in the employment of women as it provides seasonal employment opportunities. The tourism sector contributes to gender equality and empowerment of women. Khatiwa and Silva (2015) concluded that the tourism sector provides many advantages in women's employment and the tourism sector helps to reduce gender inequality as a result of the findings they obtained in their study examining gender inequality in Namibia. In addition to this study, while the majority of tourism businesses in Malaysia, Thailand, Philippines and Indonesia are managed by women operators, more than half of tourism businesses in Panama were founded by women entrepreneurs. According to the Tourism and Women Report prepared by the World Bank in 2017, it has been stated that women's entrepreneurship and managerial activities in the tourism sector provide a higher labor force participation rate than in other sectors. Based on these examples, the features that emerge as women increase the employment rate in the tourism sector are listed as follows (Ardıç Yetiş and Çalışkan, 2020):

- **Some business lines in the tourism sector do not require experience:** The jobs in the tourism sector facilitate the employment of women as they provide various job opportunities for women who cannot benefit from educational opportunities.
- **Hospitality skills are at the forefront in the tourism sector:** Since the tourism sector is compatible with women's personal characteristics, the tourism sector is structurally more suitable for women.
- **The tourism sector offers seasonal and part-time work opportunities:** It is frequently preferred by women, as the employment rate in the tourism sector is proportional to the tourism demand and it provides part-time work opportunities, allowing women to establish a work-family balance.
- **The tourism sector does not require high financing for entrepreneurship:** A new entrepreneurial activity in the tourism sector does not require high financing. In addition to this, the implementation of support and incentive policies to support women's entrepreneurial activities in the tourism sector increases the presence of women in the tourism sector.
- **The tourism sector facilitates women's economic gain:** The tourism sector provides women with the opportunity to earn economic gain with the concept of sharing economy.

The tourism sector, which enables women to take part in working life intensively by turning the disadvantages they have due to gender inequality into an advantage, has a positive effect on women's empowerment, economic income and strengthening of their social status by increasing women's employment with the opportunities it offers to women.

Empowering women is not only to provide equal opportunities and opportunities, but also to ensure that they have the opportunity to exercise these rights and abilities. Gender equality enables women to benefit from opportunities in economic, social and political fields and to be empowered. Thanks to the policies and projects that can be carried out so that women can have equal working conditions with men and become stronger in the tourism sector, women will be able to participate in the decision-making process on business-related issues. In the case of ensuring equality of gender and opportunity, which is the most important factor in women's empowerment, women's having equal rights with men in the tourism sector will increase the rate of women in managerial positions. Achieving gender equality in empowering women is very important at this point (Vukovic et al., 2021).

Unless inequalities and sexist approaches towards women are eliminated, women's empowerment will require a slow and long process. However, in order to overcome this situation, incentive policies are implemented to empower women in the tourism sector. Incentive policies ensure that women are empowered in terms of economic independence, starting their own business, rising to managerial positions and increasing their social status. In this context, empowerment of women in the tourism sector is still an ongoing process. The tourism sector, which is one of the most important sectors that contribute to countries economically, provides many incentives for women's empowerment, as it is a sector in which women are heavily employed (Uduji et al., 2021). Legal arrangements that will protect and empower women are aimed at enabling women to better use their educational knowledge and skills and to ensure gender equality. It will be possible for women to use their qualifications in the tourism sector, to benefit from equal and fair rights, and to become stronger accordingly, by preventing social prejudices and inequalities.

WOMEN IN THE TURKEY TOURISM INDUSTRY

Tourism is among the most important sectors developing among economic activities all over the world. The tourism sector, which has a dynamic feature, provides rapid and high development to countries. Today, the tourism sector has an important place in the world economy. The tourism sector is a very important sector that provides foreign exchange income to countries. The tourism sector, which provides an important economic income for developing countries, meets the employment rate in many countries (Akın and Kamışlı, 2009).

One of the countries that generate significant income from the tourism sector is Turkey. According to the statement made by the Turkish Statistical Institute, Turkey has an annual income of 46 billion dollars from the tourism sector as of 2022. The tourism sector, which makes a significant contribution to the economic development of Turkey, also has important effects on social development. Although the tourism sector is growing day by day in Turkey, it also has a direct and positive effect on employment.

Despite the high rate of female employment in the tourism sector in Turkey, the obstacles and difficulties faced by women in other sectors are also present in the tourism sector. The tourism sector offers both advantages and disadvantages for women. Although the tourism sector, which is in the service sector, is a sector where the number of female employees is increasing, the problems and discrimination in other sectors are also found in the tourism sector. The prejudice and social inequality against women's participation in working life is also a major obstacle for women working in the tourism sector in Turkey. On the other hand, the tourism sector, which requires constant physical activity, causes women to experience various health problems. The obstacles to career development of women in the tourism sector are listed as follows (Aydın Tükeltürk and Şahin Perçin, 2008):

- Glass ceiling syndrome: The invisible and insurmountable barriers between women and management, which stand in the way of women's progress despite their success (Mızrahı and Araci, 2010),
- Showing compromise,
- Lack of career development opportunities,
- Deficiencies in legal regulations related to women's career development,
- Lack of positive organizational climate for women in the organization,
- There are programs for male employees in the organization to support each other, but there is no such support program for women,
- Traditional recruitment process,
- Work-family conflict,
- Women's workload is high (housework, childcare, etc.),
- Long working hours,
- Lack of flexible working hours for women,
- Inequality of education and opportunity between women and men,
- Social pressure and family expectations,

- Adequate child care arrangements,
- Prejudice against women's participation in working life,
- Discrimination against women's career development.

When the problems faced by women in Turkey and many other countries are examined, first of all, the inequality between women and men arising from educational opportunities creates an obstacle for women's future career opportunities. On the other hand, they are expected to take care of housework and childcare instead of taking part in working life in the roles determined for women by the society. Long working hours for women cause work-family conflict, which forces women to choose between working life and home life. Women, who have difficulty in taking an active part in working life in the face of social pressure and expectations, have difficulties in career development both in the tourism sector and in other sectors.

In Turkey, women have an important place in the tourism sector, where the majority of employment is met. Tourism sector is one of the most preferred sectors by women who take part in working life in order to increase their economic independence and social status. Women have many job opportunities in various fields in the tourism sector. Among these job opportunities, the most preferred ones are managers, office workers, travel agencies, tourist guides, many departments in accommodation businesses, food and beverage businesses and security departments. The high number of women employed in the tourism sector in Turkey is due to the fact that women can find work areas in the tourism sector for their knowledge and skills. In the "Women in Tourism Global Report" study carried out by the World Tourism Organization (UNWTO) in 2010, the reasons for women's participation in the tourism sector in Turkey are listed as follows:

- In the tourism sector, women are paid closer to men than in other sectors,
- In the tourism sector, women are more likely to start their own businesses, become managers, leaders and employers compared to other sectors,
- Women can work in the tourism sector by using less labor force by concentrating on low-skill occupations,
- There are various advantageous conditions in the working conditions of women compared to men in the tourism sector.

Women are in a very important position for Turkish tourism because half of the labor in the tourism sector in Turkey is "women's labor". Various incentives and policies are implemented in order to increase the employment rate of women, who have an important position in the tourism sector in Turkey. In this context, various non-governmental organizations, local administrations and public institutions provide women with congresses, projects, seminars, courses and trainings for tourism. Prominent among these projects are listed as follows:

- **Agro Tourism Project:** It is a project that aims to enable women to use their agricultural tourism potential in rural areas by creating job opportunities for women. Within the scope of the project, both the sustainability of agricultural tourism is ensured and women living in rural areas earn income from tourism by selling the products they produce (Çanga et al., 2018).

- **Tourism Forum from a Women's Perspective on the International Platform:** A meeting was held at the meeting held by the Antalya Business Women's Association to increase the employment rate of women in the tourism sector (Erkol Bayram et al., 2018).
- **Increasing Registered Employment and Women's Employment with Lifelong Learning, Innovative Methods:** 11 projects have been prepared in order to increase women's employment in tourism in the province of Artvin (sosyalekonomi.org).

Although there are projects carried out to increase the employment of women in the tourism sector in Turkey, these studies are not sufficient. More incentives should be implemented for women to be more involved in the tourism sector. Turkey's tourism sector is a sector where women can take part more actively and carry out entrepreneurial activities. Since the tourism sector is a labor-intensive sector, it provides women with the opportunity to carry out entrepreneurial activities with less investment. The rate of women's employment in the tourism sector in Turkey is increasing day by day. The reasons affecting this situation are listed as follows (Köse, 2014);

- Women have a desire to earn economic gain and have to work.
- The employment cycle in the tourism sector in Turkey shows seasonal intensity.
- Workload is high for women living in rural areas,
- There are studies to increase education and incentive opportunities for women,
- Increasing interest in the concept of rural tourism and creating new business lines for women living in these rural areas,
- They want to market the excess of the products produced by women living in rural areas and they can benefit from the tourism sector for this,
- Women who cannot benefit from educational opportunities benefit from rural tourism activities because they cannot find jobs in other sectors,
- Projects are initiated for women who work and want to work in the tourism sector,
- Organizations that support women's employment in the tourism sector are carried out by non-governmental organizations and local government.

Despite the many problems that women face such as social pressure and gender discrimination, the tourism sector offers various opportunities for women to take part in working life and to benefit from their entrepreneurial activities. In this context, the majority of women's employment takes place in the tourism sector compared to other sectors. Rural tourism, which has been very popular especially in recent years, has enabled women living in rural areas to be included in tourism. Women are an important part of the tourism sector in Turkey. The tourism sector is a sector that not only provides women's employment, but also provides important opportunities for women's career development. Turkey is a country that supports women in the tourism sector with projects, incentives and various career opportunity programs for female employees. Turkey's tourism sector contributes to women's economic power, employment and education opportunities. It is necessary to increase the projects and support for women. With the increase in the employment rate of women in the tourism sector, Turkish tourism will develop, so both women and Turkey's income from the tourism sector will increase.

DIVERSITY MANAGEMENT

Today, the whole world has entered the era of globalization, in which the importance of labor force is gradually increasing. This new era is realized by the cooperation of many people in organizations. With the globalization period, people have started to live in countries where employment opportunities are high. With the effect of globalization, many people have started to adapt to the language, religion, race, culture, norms and lifestyles of people from different cultures. Thus, many people from different countries and cultures have started to work in harmony because they are in constant communication. Each employee and manager in the organization is trying to understand each other's cultures and differences. On the basis of this situation, it started to form multinational business structures with the globalization period (Seymen, 2005).

The emergence of multinational enterprises has created cultural diversity. Diversity includes features such as race, religion, language, ethnicity, economic status, gender, educational status, physical characteristics, and age that distinguish a person from other people (Kreitz, 2008). Hubbard (2004) examined the diversity in two groups as "primary and secondary diversity". Primary variations include the demographic information and biological characteristics of the individual. Secondary characteristics, on the other hand, include the experiences and experiences that the individual has learned from the environment he has lived in since birth and the abilities they have.

The increase in the need for labor with globalization has led people from different countries such as women, the elderly, minority nations and immigrants to countries where employment opportunities are high. The continuity of the organization can be ensured when the diversity of people from different countries and cultures is properly managed within the organization. Along with cultural diversity, the efficiency and motivation of the organization has become important. At this point, the concept of "diversity management" has emerged (Kulualp and Demir, 2020).

When diversity management is managed positively within the organization, individuals can use their diversity in the direction of the continuity of the organization, so that both employees can be motivated and organizations can increase their success rates in the sector. However, in cases where diversity is not managed effectively within the organization, communication problems may arise between employees, other employees and managers, which can cause many problems in the organization. Diversity management has an important role in employees' motivation, organizational commitment and fulfillment of their duties (Rosado, 2006). With diversity management, each individual can show their potential and thus contribute to the organization.

Diversity management first emerged in the United States in the 1960s as a result of the reaction of a group of people against discrimination (Kulualp and Demir, 2020). The concept of diversity management is defined by Strachan, Burgess, and Sullivan (2004) as the ability of individuals to use their various characteristics for organizational purposes and for the benefit of the organization, and to achieve harmony within the organization. Diversity management contributes positively to the development of organizational performance. Each employee in the organization supports the formation of organizational culture when he uses the diversity he has effectively.

Bucher (2010) stated that the individual should accept the talents, education and experience of the individual as they are and should be able to use these features for the benefit of the organization and expressed how the individual can use the diversity of the individual for the benefit of the organization in the formation of diversity management:

- Individuals should establish their boundaries,
- Individuals should be proud of themselves,
- Individuals should preserve their culture,
- Interpersonal communication should always be active for the management of diversity.
- Individuals should be respectful to the diversity among each other,
- Individuals should be moderate towards the diversity of the person,
- Individuals should support each other,
- A solution-oriented approach should be taken against any conflict that may occur within the organization.

The management of diversity within the organization and the ability of individuals to use their diversity for the benefit of the organization will enrich the organization. The basis of this situation is the ability of individuals to form a whole with the diversity they have. Increasing the motivation and satisfaction of the employees will also increase the organizational performance and organizational commitment. In this context, organizations should develop policies to manage the diversity of individuals. Approaching the skills among employees with empathy, developing flexible thinking skills, contributing to the emotional intelligence development of individuals, increasing teamwork and enabling the individual to make self-evaluation are among the policies that the organization can develop. In cases where diversity management can be applied within the organization, the benefits that organizations can provide are listed as follows (Von Bergen et al., 2002):

- Employees will be able to gain job satisfaction and motivation,
- Individual and organization will be able to get maximum efficiency,
- Communication between employees will be strengthened,
- A possible conflict between employees will be minimized,
- The organization will be able to gain experience against the demographic and cultural differences of individuals and develop responsibility for this situation.
- Organizations will be able to solve challenging situations more easily when they provide diversity management effectively,
- With the diversity management, the profit of the organization will increase and this situation will increase the earnings of the employees.

Organizations where diversity management is successfully managed can benefit from the talents of their employees at the maximum level when they provide equal opportunities to all their employees. Diversity management is basically a change strategy. The increase in diversity within the organization prepares the ground for the organization to enter a process of change. Diversity management refers to a long-term situation. For diversity management to be successful, it must be supported by managers. Managers should proceed in a planned and gradual manner in order to manage the diversity within the organization and to improve it positively. In order for diversity management to take place, the benefits of each change that will take place within the organization to the organization and its employees should be determined in advance and acted upon in this context. Managers should explain each change that will take place within the scope of diversity management and what benefits it will provide to employees through various trainings. Organizations will be successful if the changes that will take place for the diversity management to be successful are adopted by the employees. In this context, organizations in which diversity management is managed positively and successfully will be able to increase their profit rate, gain superiority over their competitors in the sector and strengthen their businesses. In addition to this, employees will be able to protect their diversity and contribute to their personal and career development. Successful management of diversity management will have positive effects in all areas.

DISCUSSION

Before the era of globalization, women had roles such as doing housework and taking care of children. However, with the era of globalization, women have started to take part in working life. Women who want to take part in working life have faced many obstacles. Social value judgments and the roles attributed to women have created many obstacles in front of women's being able to accept their existence in working life. The fact that women are not given equal rights with men due to gender inequality has left women in the background for many years. Giving women less educational opportunities than men creates a problem for women to improve themselves, causing them to stay behind men in working life. Women who entered the working life after men were exposed to discriminatory attitudes based on gender. Many sectors that had a male-dominated structure in the past have had a hard time accepting the existence of women.

In order to overcome the difficulties and obstacles they face, female employees have turned to sectors that are suitable for them. One of these sectors is the tourism sector. The tourism sector is a sector where female employees can earn economic income, benefit from entrepreneurial activities and develop themselves by gaining social status (Bayram, 2019). The tourism sector is a sector where women's labor is intense. In addition to gaining social status by working in the tourism sector, women can also become managers by benefiting from entrepreneurial activities. In this process, the tourism sector has begun to be characterized as a sector in which women are employed more. Many opportunities have been provided for women to develop themselves in this sector, as female employees have started to take part in the tourism sector, which is suitable for their qualifications and skills.

The tourism sector has a dynamic structure that grows and develops day by day. The career opportunities offered to women through projects aimed at improving women's employment in tourism increase the importance of women in the tourism sector day by day. One of the countries where women's employment in tourism is high is Turkey. In recent years, Turkey has become one of the most important tourism countries in the world. This situation increased the number of women receiving tourism education, but also increased the rate of female employment in the tourism sector. Although Turkey creates employment opportunities for women with its many tourism enterprises, the rate of female employment in the tourism sector increases every year. The growth and development of the Turkish tourism sector every year provides new job opportunities and entrepreneurship opportunities for women. This helps women to provide new career opportunities. The increase in the number of women working in the tourism sector in Turkey helps to reduce the prejudice against women's participation in working life. On the other hand, many projects are carried out in Turkey to encourage women to take more part in the tourism sector and to support their career development. In this context, Turkey is a country that gives importance to women's employment and encourages women to work in this sector. (Istanbul Dincer et al., 2016).

Despite the fact that the Turkish tourism sector has a structure that is developing day by day and employs more and more women, today women have not yet reached the position they want. There are various inequalities between men and women from the past to the present. Education inequality is the cause of most of the problems women face in their working life. The fact that women are not given equal educational opportunities with men is a major obstacle to women's self-development. In order to prevent this situation, Turkey provides women with various educational opportunities within the tourism sector. On the other hand, men have a more advantageous position than women in the recruitment process. However, the tourism sector in Turkey does not make any distinction between men and women, as it needs women's labor more and more with its developing structure every day. There is salary inequality between men and women in many other sectors, but in order to prevent this situation, the Turkish tourism sector carries out many projects that provide economic support for women to become entrepreneurs in the tourism sector. In this context, Turkey's tourism sector is a country that gives importance and value to female employees (Tekeli, 2019).

Turkey's tourism sector is an important country for women's employment. However, women still today face gender inequalities. One of the main reasons for the obstacles and difficulties faced by women in working life is the perception of gender. There are various forms of management for this discrimination faced by female employees. It is not enough to create employment for women and enable them to achieve their economic freedom. In order to solve the problems faced by women, studies should be carried out to eliminate the stereotypical gender judgments of the society. One of the management styles that can prevent gender discrimination is diversity management. Diversity management is the way in which individuals with differences work in harmony and the way managers manage the organization by treating each individual fairly. These variations are the innate characteristics of the person such as language, religion, race, age and gender. At this point, diversity management can be applied to eliminate gender discrimination between men and women. It is thought that with diversity management, discrimination between men and women in working life will be eliminated. Gender discrimination, which started in the recruitment process, will enable women to be evaluated on equal terms with men through diversity management. The fact that women receive different wages with men will be able to change this situation with diversity management, so that women and men will be able to receive fair wages in line with their working conditions (Ataünal and Aybars, 2018). Turkey's tourism sector will be able to provide women with much better working conditions through diversity management. The similarities and differences between men and women will be managed fairly with diversity management.

With diversity management, women in the Turkish tourism sector will be able to increase their job opportunities in tourism with sustainable gender equality under better working conditions. This form of management, which will ensure gender equality in tourism, will increase the participation of women in decisions in the tourism sector. On the other hand, diversity management is a fair form of management that enables women to occupy senior positions without discriminating between men and women. It is expected that the ratio of female employees in the tourism sector in Turkey will be high and it will increase more with diversity management. Diversity management is a form of management that can eliminate the obstacles women face in the working environment. Diversity management, which is a management check that can contribute positively to the characteristics and needs of women, will be able to contribute positively to the Turkish tourism sector.

Gender equality will only be possible if women and men have equal rights and under equal conditions. Turkey's tourism sector, which is one of the countries where women's employment is most intense, will be able to provide an equal working life in economic, social and cultural fields with diversity management in order to prevent gender inequality in front of women and men and to provide equal rights. Diversity management will ensure that women and men in working life have equal rights with their similarities and differences. Thus, the Turkish tourism sector will be able to support women's career development and give women more equal conditions and rights with men. Diversity management will empower women by providing women with equal economic independence, decision-making authority and leadership in working life. It should not be forgotten that the value that a society places on women is the most important element that shows the level of development of that society.

REFERENCES

Alonso, P., Moscoso, S., & Salgado, J. F. (2017). Structured Behavioral Interview as a Legal Guarantee for Ensuring Equal Employment Opportunities for Women: A Meta-Analysis. *The European Journal of Psychology Applied to Legal Context*, *9*(1), 15–23. doi:10.1016/j.ejpal.2016.03.002

Ataünal, L., & Aybars, A. (2018). Yönetim Kurullarındaki Cinsiyet Çeşitliliği ile Çeşitliliğin Finansal Performans Etkisi: Gelişmekte Olan Piyasa Örneği. *Maliye Finans Yazıları*, (109).

Bayram, A. T. (2019). Bıblıometrıc Analysıs of Women Studıes on Turkısh Tourısm Lıterature. *Journal of Tourism And Research*, *8*(2), 55–77.

Bucher, R. D. (2000). Diversity Consciousness. Academic Press.

Çanga, A. Ç., Kutlu, T., & Çalışkan, H. (2018). Applıcatıons of Agrıcultural Tourısm n the World and Turkey. *International Journal of Tourism. Economic and Business Sciences*, *2*(2), 450–457.

Dinçer, F. İ., Akova, O., Ertuğral, S. M., & Çiftçi, M. A. (2016). Woman Labour Force in Tourism Industry in Turkey: Opportunities and Barriers. Eurasian Academy of Sciences Social Sciences Journal, 379-395.

Duffy, L. N., Kline, C. S., Mowatt, R. A., & Chancellor, H. C. (2015). Women in Tourism: Shifting Gender Ideology in the DR. *Annals of Tourism Research*, *52*, 72–86. doi:10.1016/j.annals.2015.02.017

Erkol Bayram, G., Bayram, A. T., & Sürücü, Ö. A. (2018). *Turizm ve Kadın*. Detay Yayıncılık.

Harris, H. (2004). Global careers: Work-Life Issues and the Adjustment of Women International Managers. *Journal of Management Development*, *23*(9), 818–832. doi:10.1108/02621710410558431

Hubbard, E. E. (2004). *The Manager's Pocket Guide to Diversity Management*. Human Resource Development.

Kamışlı, M., & Akın, F. (2009). The Effects Of Tourism Sector To Economic Development In Emerging And Developing Countries: An Assessment On Turkish Tourism. *Anadolu Bil Vocational School Journal*, (16), 16–29.

Khatiwada, L. K., & Silva, J. A. (2015). Mitigating Gender Inequality in Rural Regions: The Effects of Tourism Employment in Namibia. *International Journal of Tourism Research*, *17*(5), 442–450. doi:10.1002/jtr.2010

Köse, Z. (2014). *Turizmde Kadın İstihdamı ve Kadın Girişimciliği Beypazarı Örneği*. Hacettepe Üniversitesi Sosyal Bilimler Enstitüsü Antropoloji Anabilim Dalı.

Kreitz, P. A. (2008). Best Practices for Managing Organizational Diversity. *Journal of Academic Librarianship*, *34*(2), 101–120. doi:10.1016/j.acalib.2007.12.001

Mizrahi, R., & Aracı, H. (2010). A Resaerch On Women Managers And Glass Ceiling Syndrome. *Organizasyon ve Yönetim Bilimleri Dergisi*, *2*(1), 149–156.

Ponthieux, S., & Meurs, D. (2015). Gender Inequality. In Handbook of Income Distribution (Vol. 2, pp. 981-1146). Elsevier. doi:10.1016/B978-0-444-59428-0.00013-8

Rosado, C. (2006). What Do We mean By "Managing Diversity". *Workforce Diversity*, *3*(1), 1–15.

Santos, L. D., & Varejão, J. (2007). Employment, Pay and Discrimination in the Tourism Industry. *Tourism Economics*, *13*(2), 225–240. doi:10.5367/000000007780823186

Seymen Aytemiz, O. (2005). Örgütlerde Kültürel Çeşitlilik Olgusu, Boyutları ve Etkin Yönetimi Konusunda Farklı Yaklaşımlar: Yazınsal Bir Derleme. *Yönetim Dergisi: İstanbul Üniversitesi İşletme Fakültesi İşletme İktisadı Enstitüsü*, *16*(50), 3–24.

Shakeela, A., Ruhanen, L., & Breakey, N. (2010). Chapter 5 Women's participation in tourism. In Tourism in the Muslim World (Bridging Tourism Theory and Practice, Vol. 2). Emerald Group Publishing Limited. doi:10.1108/S2042-1443(2010)0000002008

Sinclair, A. (2000). Women within Diversity: Risks and Possibilities. *Women in Management Review*, *15*(5/6), 237–246. doi:10.1108/09649420010372850

Şit, M. (2016). Employment Contribution of Tourism Sector in Turkey. *Journal of Academic Approaches*, *7*(1), 101–117.

Sosyal Ekonomi. (n.d.). https://sosyalekonomi.org/kilickaya-kadin-girisimi-uretim-ve-isletme-kooperatifi/

Strachan, G., Burgess, J., & Sullivan, A. (2004). Affirmative action or managing diversity: What is the future of equal opportunity policies in organisations? *Women in Management Review*, *19*(4), 196–204. doi:10.1108/09649420410541263

Suryawati, D. A., Dewi, S. P. A. A. P., Osin, R. F., & Anggayana, I. W. A. (2021). The Role of Women in Protecting the Village and Rural Tourism in Timpag Village. *Jurnal Manajemen Pelayanan Hotel*, *5*(2), 74–79.

Tekcli, H. N. (2019). Women's Employment in Tourism Sector in Turkey, Issues Faced and the Effect of Glass Ceiling Syndrome on Women Workforce in Tourism. *Social Sciences*, *14*(2), 179–191.

Tengland, P. A. (2008). Empowerment: A Conceptual Discussion. *Health Care Analysis*, *16*(2), 77–96. doi:10.100710728-007-0067-3 PMID:17985247

Tükeltürk, Ş. A., & Perçin, N. Ş. (2008). Turizm Sektöründe Kadın Çalışanların Karşılaştıkları Kariyer Engelleri ve Cam Tavan Sendromu: Cam Tavanı Kırmaya Yönelik Stratejiler. *The Journal of American Science*, *6*(2), 113–128.

Turkiye Istatıstık Kurumu. (n.d.). https://data.tuik.gov.tr/Bulten/Index?p=Turizm-Istatistikleri-IV.Ceyrek:-Ekim-Aralik-ve-Yillik,-2022-49606#:~:text=Turizm%20geliri%202022%20y%C4%B1l%C4%B1nda%20bir,milyon%20907%20bin%20dolar%20oldu

Uduji, J. I., Okolo-Obasi, E. N., Onodugo, V. A., Nnabuko, J. O., & Adedibu, B. A. (2021). Corporate Social Responsibility and The Role of Rural Women in Strengthening Agriculture-Tourism Linkages in Nigeria's Oil Producing Communities. *Journal of Tourism and Cultural Change*, *19*(6), 754–780. doi: 10.1080/14766825.2020.1826500

Von Bergen, C. W., Soper, B., & Foster, T. (2002). Unintended Negative Effects of Diversity Management. *Public Personnel Management*, *31*(2), 239–251. doi:10.1177/009102600203100209

Vukovic, D. B., Petrovic, M., Maiti, M., & Vujko, A. (2021). Tourism Development, Entrepreneurship and Women's Empowerment–Focus on Serbian Countryside. *Journal of Tourism Futures*, 1-21.

World Bank. (2017). *Women and Tourism: Designing for Inclusion*. World Bank Group.

Yetiş, Ş. A., & Çalışkan, N. (2020). Female Employment in Tourism Sector: An Assessment of the Current Situation. *Manisa Celal Bayar University Journal of Social Sciences*, *18*(3), 105–119.

Yıldırım, G., Tarınç, A., & Kılınç, C. C. (2020). Women Entrepreneurship in Tourism: The Case of Turkey. *Journal of Tourism and Gastronomy Studies*, *8*(4), 2462–2477. doi:10.21325/jotags.2020.721

Chapter 7
Tourism as a Means of Empowering Women in Sustainable Development:
The Case of Women's Cooperatives

Hatice Sari Gök
Isparta University of Applied Sciences, Turkey

ABSTRACT

In tourism, women's cooperatives support women's empowerment in the society by providing income with employment, employment and education opportunities, and their effects on eliminating economic and social injustice. With the increase in interest in natural and cultural heritage, local food and handicrafts, the development of healthy life awareness and the increase in education level, the interest in authentic and local cultures is increasing day by day. From this point of view, the similarities between sustainability, sustainable development, women's empowerment, women's cooperatives in tourism and sustainable development goals, and women's cooperatives were examined in this study. It is thought that the study will contribute to the literature and tourism sector.

INTRODUCTION

Sustainable development, is a concept that considers ecological balance and economic growth together, ensures effective use of natural resources and attaches importance to environmental quality. Sustainable development, is believed that ecological and scientific understanding should be carried out together in order to protect the decisiveness of the world. Sustainable development was defined for the first time by the United Nations (UN) at the 2005 general assembly. The United Nations focuses on three main topics within the framework of sustainable development. These are economic sustainability, environmental sustainability and social sustainability. The Sustainable Development Goals are the goals set by the United Nations for countries to eliminate poverty, protect the planet, and live in peace and prosperity.

DOI: 10.4018/978-1-6684-8417-3.ch007

The United Nations, which aims to make the world a livable place for everyone by achieving these goals by 2030, also supports governments carrying out studies in this context (UN, 2020).

What needs to be addressed in sustainable development is how and in what way women are involved in the social development process. Gender equality and women's empowerment are seen as a more holistic strategy in creating new models and processes that are sustainable. Women's empowerment is actually both a process and an outcome, and the World Bank defines women's empowerment as the key to all programmes. One of the policies implemented to empower women is to support the establishment of women's cooperatives. Women's cooperatives in the tourism sector are a model that can be used in the socio-economic empowerment of women, reducing gender inequality and ensuring sustainable development with their economic contributions.

CONCEPTUAL FRAMEWORK

Sustainable Development

Sustainability requires making changes in thinking and lifestyle without reducing the quality of life. The basis of this change is to get rid of being a consumer society and to be in universal solidarity, environmental management, social responsibilities and economic solutions (Ar & Çelik Uğuz, 2017: 522). The concept of sustainable development developed in this direction has emerged as a result of increasing awareness of global connections, increasing environmental problems, economic and social problems, injustices among people and concerns about a healthy future (Bazyar, 2014:41). The concept of sustainable development was accepted as a *"Multidimensional Theme"* in the Brundthland Report titled *"Our Common Future"* published in 1987 by the United Nations World Commission on Environment and Development. Sustainable Development has been defined as *"a form of development that meets the needs of the present without compromising the ability of future generations to meet their own needs"* (Bodosca & Diaconescu, 2015: 230). In the Brundthland Report, the relationship between environmental protection and development is emphasized and it is mentioned that the need for change and a new global ethical understanding can be met with today's resources, technology and human resources (Ar & Çelik Uğuz, 2017: 522).

The United Nations adopted the Sustainable Development Goals in 2015 by setting the target for 2030. In determining the international goals of sustainable development, economic development, social development and environmental protection are taken as the basis. They have set 169 targets under 17 main headings for the solution of the main problems that people face all over the world. It has been developed with the aim of ensuring that no individual living in the world is left behind in terms of social, economic and environmental gains in the Sustainable Development Goals. From this point of view, in this section, the overlap of the Sustainable Development Goals and the aims of women's cooperatives has been examined. The contributions of women's cooperatives to the realization of 14 of the 17 main topics in the sustainable development goals are examined in detail below (UN, 2015);

- The first goal of the United Nations' sustainable development goals is the last goal of poverty. In this goal, it is aimed to end all kinds of poverty everywhere. With this goal, the reduction of women's poverty and the economic empowerment of women, which are among the aims of women's cooperatives in the tourism sector, overlap.

- The second goal, the last goal of hunger, includes ensuring food security, improving nutrition opportunities and supporting sustainable agriculture. Similarly, traditional and local gastronomic products and productions are at the forefront in women's cooperatives. It contributes to the sustainability of traditional productions specific to the region.
- The third goal is a healthy and quality life. It is aimed to increase the welfare level of everyone of all ages. Similarly, among the social activities of women's cooperatives, the aims of creating social difference in terms of maternal and child health and the right to life of disabled individuals overlap (KEDV & CCA, 2015: 11).
- The fourth goal is quality education. In this goal, it is aimed to provide lifelong education opportunities for everyone and to provide equitable and quality education. In women's cooperatives, sector-related knowledge and skills development and capacity building trainings are provided to their members, their families and external beneficiaries on issues related to the tourism sector.
- The fifth goal is gender equality. This goal is aimed at ensuring gender inequality and empowering all women and girls. It contributes to the improvement of gender-based unequal division of labor and power relations in women's cooperatives (KEİG, 2015: 9). It is aimed to contribute to women's overcoming difficulties such as unstable employment situation, economic dependency, lack of education and lack of access to leadership opportunities in the business world (KEDV & CCA, 2015: 21).
- The sixth goal is clean water and sanitation. Access to and sustainability of water and sanitation for all must be ensured. Many women's cooperatives cooperate with local governments and try to raise awareness for everyone's access to a healthy life and its sustainability.
- The eighth goal is decent work and economic growth. It is aimed to provide full and productive employment together with sustainable and inclusive economic development. In the tourism sector, women get the opportunity to organize their own business ideas, which they have determined under democratic conditions, by organizing through cooperatives. By producing traditional and local products, they improve their self-confidence by doing sustainable work and prevent their economic and social poverty.
- The eleventh goal is sustainable cities and communities. This goal is aimed at making cities and other settlements inclusive, safe and sustainable. Women's cooperatives help women who are unpaid agricultural and family workers to earn an income outside the home according to their skills. It also contributes to the reduction of crime rates caused by social and economic injustice in society with various studies aimed at preventing gender discrimination.
- The twelfth goal is to develop responsible production and consumption patterns. Women's cooperatives offer sustainable job and employment opportunities and a fair distribution of resources is ensured.
- The thirteenth goal is climate action. In recent years, it is aimed to take urgent action for climate change and its effects due to environmental changes. Ecological production methods are applied in many women's cooperatives. The fourteenth goal is life in the water. It is aimed to protect and sustainably use water, which is an indispensable resource for the continuation of life on earth. This goal aims to protect oceans, seas, lakes and rivers. With sustainable production methods in women's cooperatives, they prevent water pollution caused by waste and therefore reduce ecological damage in oceans, seas, lakes and rivers (Aldırmaz Akkaya, 2018: 83).
- The fifteenth goal is terrestrial life. It is aimed to protect terrestrial ecosystems, to ensure their sustainable use, to ensure the sustainable use of forests, to combat desertification, to take neces-

sary measures to prevent the loss of productivity of soils, to stop the loss of biodiversity. Women's cooperatives carry out information and awareness-raising activities on production methods and environmental waste.

- Sixteenth goal; peace, justice and strong institutions. In this goal, it is aimed to promote peaceful and inclusive societies for sustainable development and to ensure that everyone has access to justice. By strengthening the sense of socialization and participation in their members, women's cooperatives prevent fear, violence, hatred and alienation and play a major role in ensuring social peace.
- The seventeenth goal is partnerships for purposes. In this target, it is aimed to revive the global partnership for sustainable development and strengthen the implementation tools. Similarly, women's cooperatives can improve their production and marketing capacities by becoming members of local, national and international unions and higher unions (Aldırmaz Akkaya, 2018: 83).

Women's Cooperatives

Cooperatives are an organizational model with both economic and social purposes, established and managed by people who come together to achieve a common goal. According to the International Cooperative Union, a cooperative is "an autonomous structure formed by people who voluntarily come together to meet their common economic, cultural and social aspirations and needs through a jointly owned and democratically controlled enterprise" (KEIG, 2018: 7). Women's cooperatives, on the other hand, are defined as a type of cooperative established and operated by women. However, it is not different from other cooperatives in terms of legal status and obligations. Women's cooperatives are encouraged and brought to the agenda by governments with the emphasis on *"women's entrepreneurship"* and *"women's actors in the economy"*. In addition, it attracts the attention of various institutions and structures from international development organizations to the private sector. Undoubtedly, among the most important reasons for this, women and organizations such as women's cooperatives have played an important role in reducing poverty, increasing employment and ensuring development in recent years (KEİG, 2018: 8). Women's cooperatives are thought to be an important mechanism for poor women, who are considered to be the disadvantaged segment of society, especially with low income.

Cooperatives are one of the tools that are effective in closing the gender gap in social and economic life, in accessing opportunities and resources, and in empowering women. Cooperatives, as democratic organizations based on partnerships of people on the basis of equality instead of capital shares, are forms of local government that are effective in expressing women's needs, producing collective solutions and having a say in the public sphere. Cooperatives are also important models in terms of directing women's limited economic resources (capital, land, skills, credit, etc.) to effective use. Because it protects the representation of small producers and workers, their economic benefits, based on the representation of the person rather than the capital. With this aspect, cooperatives are emphasized as a sustainable and inclusive development strategy all over the world (Gönüllü Atakan & Çoban, 2022: 10).

Women mainly establish cooperatives for reasons such as accessing cheap products, benefiting from employment opportunities and accessing financial instruments. In the process of establishing and operating women's cooperatives, they face difficulties such as lack of capital, care obligations of women in the household, insufficient support for childcare and poverty (Duguid et al., 2015: 30). For this reason, women's cooperatives are supported by public institutions and the private sector in many countries.

Besides developed countries, women's cooperatives have an important place in the lives of women living in many countries such as India, Nepal, Palestine, Egypt, Guatemala, Greece, Spain, Ethiopia, Peru, Namibia, Kenya, Mali, Morocco and Bolivia. For example, neo-liberal policies followed by Liberal Democrat Party governments in Japan support people, especially women, to turn to cooperatives (Kutay, 2022: 131). In brief, cooperatives recognized as effective in empowering women in many developed and developing countries are supported and encouraged.

Empowering Women

According to the United Nations Development Fund (UNIFEM), empowerment for women is the belief that an individual has the ability to make choices, use bargaining power and gain this skill, develop self-esteem and realize all these changes. Therefore, empowerment is about making decisions, implementing the decision, showing will, choosing and of course power (Balkız & Öztürk, 2013: 8). The empowerment of women has become an agenda item that is examined and given importance not only in terms of gender equality, but also in terms of sustainable development, economic growth, politics, education, peace and security (Yıldırım & Gül, 2021: 682). From this point of view, various international institutions and organizations, especially the United Nations, have been working on issues such as women's empowerment and gender equality.

By establishing the United Nations Women's Unit in 2010, she wanted to carry out studies in the field of gender equality more actively. Within the scope of reform efforts in this unit, she established the United Nations Women's Empowerment and Gender Equality Unit. It has combined the four organs of the United Nations operating in the field of gender equality and women's empowerment. These are (UN Woman, 2010);

- Division for the Advancement of Women (DAW),
- International Research and Training Institute for the Advancement of Women (INSTRAW),
- Office of the Special Adviser on Gender Issues and Advancement of Women (OSAGI),
- United Nations Development Fund for Women (UNIFEM),

The main roles of UN Women are:

- To support inter-governmental bodies, such as the Commission on the Status of Women, in their formulation of policies, global standards and norms.
- To help Member States to implement these standards, standing ready to provide suitable technical and financial support to those countries that request it, and to forge effective partnerships with civil society.
- To hold the UN system accountable for its own commitments on gender equality, including regular monitoring of system-wide progress.

In the same year, another important development took place and the Women's Empowerment Principles (7 articles) developed by the Women's Empowerment Unit (UN Woman) and the Gender Equality Unit came into force. These principles offered suggestions to companies in order to ensure gender equality in all areas of society. These recommendations are (Yıldırım & Gül, 2021: 690);

- Demonstrating high-level leadership for gender equality in business life,
- Treating all female and male employees fairly in the workplace, respecting and supporting the principles of human rights and non-discrimination,
- Ensuring the health, safety and well-being of all individuals, regardless of gender,
- Encouraging women in education, training and professional development,
- Ensuring that practices related to entrepreneurship, business development, supply chain and marketing processes in workplaces are carried out in a way that will help women's empowerment,
- Promoting equality through social initiatives and supports,
- Measuring progress on gender equality and sharing the results with the public.

Addresses the global challenges facing the world, including poverty, climate change, gender inequality, environmental destruction, justice and peace, with the United Nations Sustainable Development Goals (UN-SDGs) (UN, 2020). The United Nations promotes the empowerment of women through the development of tourism in the fifth Sustainable Development Goals. In 2010, she prepared the global report on the situation of women in tourism on her policy agenda and officially dictated it to the World with the slogan *"Tourism opens its doors to women"*. UNWTO has identified an action plan to empower women through tourism with the input of stakeholders to promote women through international campaigns, information sharing and promoting equality in the tourism sector (UNWTO, 2011).

The first of the goals set by the United Nations for the empowerment of women is to ensure the efficient participation of women in society and to provide fair environments for them to make decisions and become leaders in economic, political and public life. The second goal is to give women fair access to economic resources as well as property rights. The ultimate goal is to implement sound policies and enforceable laws to promote gender equality and empower women at all levels (UN, 2020). The tourism sector is one of the important sectors that will help implement these goals, ensure gender equality and empower women.

According to global research, more than half of those working in the tourism sector are women. In addition, it has been determined that 23% of tourism ministers worldwide are women (WTO, 2019). According to the European Institute for Gender Equality (2016), the rate of women working in managerial and supervisory positions in the tourism and accommodation sectors is low. Based on these data, providing women with more opportunities in the tourism sector and promotion to higher levels, especially in developed countries, will contribute to ensuring and promoting gender equality in economic terms. This will certainly fulfill the objectives of the fifth UN-SDG. In addition, women's economic empowerment will provide opportunities for decision-making and equal access to resources (The World Bank, 2018).

Potentials of Cooperatives for Women's Empowerment

The empowerment of women is associated with grassroots women's organization. In this direction, women's empowerment is related to women's cooperatives and women's cooperatives have an important potential in terms of women's empowerment. The effect of women's cooperatives on women's empowerment can be more effective in cooperatives established for the purpose of an enterprise or production, including economic empowerment. In addition, one of the variables that must be taken into account in order for women's cooperatives to be effective in women's empowerment is the duration of activity. In order for women's cooperatives to contribute to women's empowerment, it is necessary to create women's employ-

Table 1. The potential of cooperatives in women's empowerment

Potentials of Cooperatives for Women's Empowerment		
Economic Empowerment	**Improvement in Working Conditions**	**Social Empowerment**
Equal access to paid work	Fair and adequate income / of equal value equal pay for work	Technical skills and management training access to learning opportunities such as
Registered sector through cooperative transition opportunities	In safe working conditions decent job	Social care and health services services and goods such as and available for purchase
Business capital, finance and market access to areas	Collective bargaining power	Developing self-confidence and respect
Skill related to economic activities and opportunities to access information	Participatory democracy in the workplace increase	
Economic development in rural area		

Source: Gönüllü Atakan and Çoban (2022)

ment in the process. Therefore, the effect of women's cooperatives on women's empowerment depends on the condition of continuing their activities for at least five years (Şahankaya Adar et al., 2023: 1187).

In Table 1, the contributions of cooperatives to the empowerment of women are discussed in three dimensions. Three potentials of cooperatives for women, economic empowerment, improvement of working conditions and social empowerment were emphasized. Cooperatives are women's equal access to paid jobs in economic empowerment, that is, equal access to the workforce where women can earn income, without gender discrimination. Despite the fact that women work in the informal sectors, they can find the opportunity to switch to the formal sectors through cooperatives. Cooperatives provide women who want to start their own businesses with access to working capital, financing and markets. Cooperatives provide women with access to knowledge and skills related to economic activities, as well as providing opportunities for economic development to women living in rural areas. he second potential of cooperatives to empower women is that they offer the opportunity to improve working conditions. It contributes to women's obtaining a fair and sufficient income, equal pay for equal work, working conditions and job opportunities in safe and decent conditions. The fact that women work in cooperatives contributes to the increase of collective bargaining power and participatory democracy in the workplace due to the structure of these institutions. The third potential of cooperatives to empower women is social empowerment. The contribution of cooperatives to women's social empowerment is that they provide access to learning opportunities for technical skills and management tendencies. Women's co-operatives, early childhood education programs and lifelong learning are essential for healthy societies. For this reason, learning services for children and adults provided by women's cooperatives are socially needed services. In addition, thanks to cooperatives, women can access or buy social care and health services. When women have a job, their self-confidence and self-esteem can also improve.

Women's Cooperatives in the Tourism Sector

Cooperatives are an important economic development tool in terms of ensuring a fair distribution of income, solving poverty and providing employment (Serinikli, 2019: 47). Cooperatives, especially women's cooperatives, contain structural features that prioritize the expectations and needs of women, such as mutual assistance and cooperation, democratic participation in management and putting people's needs first, as well as developing social relations, providing equal opportunities, working as a team. These features can also have an encouraging effect on women's entrepreneurship (Bastida et al., 2020: 4).

By putting equality and social justice at the center of business, cooperatives enable people to work together to create sustainable businesses that create jobs, jobs and well-being in the long run. Cooperatives in more than ninety countries in the world provide products, services and employment for the needs of society (Duguid et al., 2015: 28). It has been reported that there are more than one billion cooperative members in the world, that cooperatives provide employment to 280 million people and that 10% of the total working population is either directly employed by cooperatives or their main source of income (Co-operatives UK, 2022).

The potential of cooperatives for women's economic empowerment is pointed out by many other development actors as well as organizations under the ICA framework. Based on the results of 581 surveys and 14 key stakeholder interviews conducted around the world, in the 2015 study titled *"Advancing Gender Equality: The Cooperative Path"* by the ILO and ICA, it was concluded that there is a rising interest in equality in cooperatives; emphasizes that women are increasingly occupying managerial positions in cooperatives and that the number of cooperatives with predominantly female members is increasing, albeit slowly. Equality-based structure of all members of cooperatives, training, information, organization and collective bargaining and empowering structure of their members will help women to quickly close their basic deficiencies in business life (Gönüllü Atakan & Çoban, 2022: 24).

By providing women with a job field, women's cooperatives contribute both to the evaluation of the products they produce and to contribute to the family budget, and to help female partners gain the ability to manage themselves within the framework of democratic rules. In addition, the socialization of women enables them to become individuals who know their own rights, to be conscious mothers in child care and child rearing, and to be individuals who stand on their own feet (Duguid et al., 2015: 22). Women provide services in the tourism sector with the cooperatives they have formed in areas such as boarding, production and sale of local products, production and sale of local handicrafts, local restaurants, and campsite management.

Among the studies on cooperatives in the tourism sector, Berno (2011) examined the suitability of using sustainable agricultural products in tourism. Proper marketing of local food products not only contributes to tourism, but also creates a good way to sustain agriculture. Aside from the supply chain, such a strategy, in which efficient communication and cooperation is also essential, has been declared to be suitable for establishing between agriculture and tourism cooperatives. In addition, relatively new ideas where agriculture, tourism and food sectors can be processed together (such as local food cooking or crop cultivation courses on farms) are presented as suggestions.

Mohamad and Hamzah (2013) examined the community cooperative in which a community-based tourism program evolved in the village of Batu Puteh in the Kinabatangan region of Sabah state of Malaysia. In Batu Puteh, which is the research area, its contributions to creating alternative income sources and reducing poverty and unemployment through community-based ecotourism activities are mentioned. By expanding the spread of tourism income to local people, it has ensured that net tourism incomes increased four fold from 2004 to 2010. Founded in 2003 with 72 members, the cooperative increased the number of members to 260. Its members include forest restorationists, home boarders, boat tour operators, local producers, souvenir vendors, food and beverage businesses, and camping businesses.

Çınar et al. (2021), in their study, examined the empowerment of women through women's cooperatives. He concluded that women's cooperatives contribute to the economic, psychological, social and organizational empowerment of women, even under political and social constraints in Turkey. As a result of the research conducted by interviewing women working in women's cooperatives in various cities in Turkey, it was stated that women's cooperatives contributed significantly to the empowerment of women through economic independence, psychological gain and social solidarity, and opened a new way of struggle and gain on the way to a more equal society.

Agro-tourism women's cooperatives in Greece are based on the aim of developing the traditional skills of housewives in the countryside and creating a source of income for themselves. In this type of cooperatives, traditionally furnished standard rooms are provided, a breakfast based on local products is offered, traditional food is produced and guests participate in production, traditional food products are sold, recreation and cultural activities are organized, and visitors are allowed to participate in agricultural and ecological activities (Iakovidou, 2010: 49). Cooperatives established in various regions of Greece provide agro-tourism activities to the guests coming to the villages. In addition, these cooperatives manage to survive in these areas by entering niche market areas that large enterprises do not enter. Agro-tourism women's cooperatives make a significant contribution to reducing rural poverty, protecting cultural heritage, developing tourism and improving the quality of life in rural areas (Stavroulakis et al., 2013: 132).

CONCLUSION

Although women in the cooperative movement face organizational, financial and structural problems stemming from traditional gender roles, today cooperatives had significant potential in terms of empowering women in the context of the basic principles they contain. In addition, women's cooperatives make a significant contribution to improving women's incomes, skills and self-confidence. Collaboration from the grassroots in women's cooperatives has the potential to empower women through participation and applicability.

Women's cooperatives create a job and income opportunity for women and contribute to local development and economy. Cooperatives serve not only women's empowerment but also sustainable development (Bouhazzama & Guenaoui, 2020: 195). Women's cooperatives provide women with lifelong learning opportunities and contribute to the creation of a job and livelihood. In addition, it supports women's capacity to overcome difficulties and develops their self-confidence, increases solidarity among women and affects women's gender roles. In this study, sustainable development, women's empowerment and women's cooperatives in tourism are discussed. Through cooperatives, women can be provided with improvements in work and employment, as well as in working conditions.

They contribute to the provision of gender equality, which is the fifth of the women's empowerment and sustainable development goals of the cooperatives. Cooperatives also contribute to women's economic empowerment, improvement of working conditions and social empowerment. Economic empowerment should be promoted through cooperatives in terms of access to paid jobs in the formal sectors, access to work-related knowledge and skills, access to capital, finance and markets, allowing rural women to access economic opportunities.

One of the main areas where cooperatives can have a significant impact on women's empowerment is meeting the economic and social needs of women. It is known that women's cooperatives make sustainable and positive contributions to the lives of women and societies. Therefore, supporting and encouraging women's cooperatives will make positive contributions to societies. In addition, women's cooperatives that enable the development of social capital will contribute to the development of women's self-confidence, self-esteem, collective action, solidarity and common identity. It should be supported and encouraged by governments.

Women's cooperative activities in the tourism sector include rural tourism, agro-tourism, agricultural food production and sales, boarding, local/regional handicraft production and sale, gastronomy businesses, guidance services and campsite management. Women's cooperatives have a very important place especially in the development of tourism in rural areas, supporting entrepreneurship, promoting rural heritage and local production. In order to expand and sustain women's cooperatives, it is important that public incentives become functional and reach the right places.

REFERENCES

Aldırmaz Akkaya, F. (2018). *Türkiye'de Kadın Kooperatiflerinin Sürdürülebilir Kırsal Turizmdeki Önemi ve Finansal Performanslarının Analizi*. Doktora Tezi, İnönü Üniversitesi, Sosyal Bilimler Enstitüsü.

Ar, H., & Çelik Uğuz, S. (2017). Küresel Sürdürülebilir Kalkınma Hedeflerinde Turizmin Rolü: Türkiye Örneği. *Uluslararası Sosyal Araştırmalar Dergisi*, *10*(49), 521–530.

Balkız, Ö. I., & Öztürk, E. (2013). Neo-liberal gelişme anlayışı ve kadın: Mikro finans uygulamaları kadınları güçlendiriyor mu? *Mediterranean Journal of Humanities*, *3*(2), 1–21. doi:10.13114/MJH/201322451

Bastida, M., Pinto, L. H., Olveira Blanco, A., & Cancelo, M. (2020). Female entrepreneurship: Can cooperatives contribute to overcoming the gender gap? A Spanish first step to equality. *Sustainability (Basel)*, *12*(6), 1–17. doi:10.3390u12062478

Bazyar, A. (2014). Social Sustainable Development and Economic Transformation, OIDA. *International Journal of Sustainable Development*, *7*(5), 40–50.

Berno, T. (2011). Sustainability on a Plate: Linking Agriculture and Food in the Fiji Island Tourism Industry. In Tourism and Agriculture: New Geographies of Consumption, Production and Rural Restructuring (pp. 87-103). London: Routledge.

Bodosca, S., & Diaconescu, M. D. (2015). Tourism Development After the Implementation of Sustainable Strategies in Neamt County. *Procedia: Social and Behavioral Sciences*, *188*, 230–236. doi:10.1016/j.sbspro.2015.03.378

Bouhazzama, M., & Guenaoui, A. (2020). The Moroccan women's cooperative in response to lasting impacts: Social cohesion, solidarity and inclusion. *International Journal of Management*, *11*(5), 195–202. doi:10.34218/IJM.11.5.2020.01

Cinar, K., Akyuz, S., Ugur-Cinar, M., & Onculer-Yayalar, E. (2021). Faces and phases of women's empowerment: The case of women's cooperatives in Turkey. *Social Politics*, *28*(3), 778–805. doi:10.1093p/jxz032

Co-operativesU. K. (2022). https://www.uk.coop/understanding-co-ops/what-co-op/quick-facts-about-co-ops

Duguid, F., Durutaş, G., & Wodzicki, M. (2015). *The current state of women's co-operatives in Turkey*. World Bank. doi:10.1596/25977

European Institute for Gender Equality. (2016). *Gender in tourism*. Luxembourg: EIGE. Publications Office of the European Union. Retrieved from: https://eige.europa.eu/

Gönüllü Atakan, A., & Çoban, A. (2022). *Türkiye'de Kadınların Kooperatifler Yoluyla Güçlenmesi: Yasal ve Yapısal Stratejiler, Uluslararası Çalışma Örgütü*. ILO Türkiye Ofisi.

Iakovidou, O., (2010). Agro-Tourism In Greece: The Case of Women Agro-Tourism Co-operatives of Ambelakia. *Medit W, 1*(97).

KEDV & CCA. (2015). Türkiye'de Kadın Kooperatifleri: Ayakta Kalma Mücadelesinden Büyümeye. KEDV Yayınları.

KEİG. (2015). *Türkiye'de Kadın Kooperatifleşmesi: Eğilimler ve İdeal Tipler*. Kayhan Matbaa.

KEİG. (2018). *Kadın Kooperatifleri Kılavuzu*. Hazal Matbaa.

Kutay, T. (2022). Kırsal Kalkınmada Kadın Kooperatiflerinin Önemi: Türkiye Özelinde Bir Değerlendirme. *Politik Ekonomik Kuram, 6*(1), 119–150. doi:10.30586/pek.1020843

Mohamad, N. H., & Hamzah, A. (2013). Tourism cooperative for scaling up community-based tourism. *Worldwide Hospitality and Tourism Themes, 5*(4), 315–328. doi:10.1108/WHATT-03-2013-0017

Şahankaya Adar, A., Dedeoğlu, S., & Kurtuluş, G. (2023). Türkiye'de Kadın Kooperatifleri. *Mevcut Durum Analizi, Kadın Güçlenmesi ve İstihdam Yaratma Potansiyeli, Çalışma ve Toplum, 2*(77), 1171–1208.

Serinikli, N. (2019). Kırsal Alandaki Mikro Kadın Girişimcilerin Kooperatifleşmeye Karşı Tutumları: Agro-Turizm Kadın Kooperatifleri. *Girişimcilik ve Kalkınma Dergisi, 14*(1), 45–57.

Stavroulakis, D., Karagiannis, S., Mitoula, R., & Papagrigoriou, A. (2013). Female Entrepreneurship and Agritourism Cooperatives in the Greeк Periphery: An Overview. *American Journal of Economics and Business Administration, 5*(4), 129–138. doi:10.3844/ajebasp.2013.129.138

The World Bank. (2018). *Women economic empowerment study*. World Bank Group. Retrieved from http://documents1.worldbank.org/curated/en/86149155111 3547855/pdf/134846-WP-PUBLIC-march-2-WB-Women-Study-EN.pdf

UN (United Nations). (2015). *The General Assembly on 25 September 2015 - Transforming our world: the 2030 Agenda for Sustainable Development, A/RES/70/1*. United Nations.

UN (United Nations). (2020). *United Nations sustainable development goals*. Retrieved from https://www.un.org/sustainabledevelopment/gender-equality/

UNWTO. (2011). *Global report on women in tourism*. United Nation World Tourism Organization. Online available at: http://ethics.unwto.org/content/gender-and-tourism

Woman, U. N. (2010). *The world's women 2010: Trends and statistics*. Retrieved from: https://www.unwomen.org/en/docs/2010/1/worlds-women-2010

WTO (World Tourism Organization). (2019). *Global report on women in tourism* (2nd ed.). UNWTO. doi:10.18111/9789284420384

Yıldırım, F., & Gül, H. (2021). Uluslararası kalkınma politikaları ve kadınların güçlendirilmesi üzerine bir değerlendirme. *Manas Sosyal Araştırmalar Dergisi, 10*(1), 679–695. doi:10.33206/mjss.760005

Chapter 8
Family Tourism Businesses and Post Pandemic:
An Overview of the Hospitality Industry of Europe

Kamran Jamshed
Bahria University, Pakistan

Syed Haider Ali Shah
Bahria University, Pakistan

Basheer M. Al-Ghazali
King Fahd University of Petroleum and Minerals, Saudi Arabia

ABSTRACT

The purpose of this chapter is to provide a summary of the family tourism industry's reaction to the COVID-19 pandemic in Europe. Family tourism businesses are usually run by a single family or a small group of families. These companies provide their clients with exceptional, customised solutions. The tourism and hospitality sectors in Europe have been hit hard by the COVID-19 pandemic. As a result, family-friendly tourism enterprises have had to get creative to survive. The importance of ensuring the security and happiness of your customers, as well as embracing digitalization and sustainability, has never been higher. Businesses catering to families in the tourism industry have been thriving despite the pandemic because of their accommodating cancellation policies and personable approach to the industry. Though the road to recovery is still hazy, family-run tourism businesses in Europe are developing innovative strategies to maintain competitiveness and fortify their resilience.

DOI: 10.4018/978-1-6684-8417-3.ch008

INTRODUCTION

Family-owned businesses are a cornerstone of the global economy, and each nation's economic prosperity depends on their continued health and expansion. Numerous well-known brands from throughout the world are among the largest family-owned businesses. The term "family business" refers to a company that is owned and run by relatives. A long-term focus, dedication to quality, and a strong sense of family values are common threads in successful businesses like these. Across the board, family-owned businesses are crucial to the health of economies around the world. More than half of global GDP is produced by family businesses, and those businesses make up the vast majority of the world's businesses, according to a report by the Family Business Institute. Sixty percent of all businesses in Europe are owned and operated by families, and they account for more than 40 percent of all jobs in the region. Europe's hotel industry follows this pattern, with many of the best properties still being run by their original families. The Brunei Investment Agency, which is run by the country's ruling family, owns a number of high-end businesses, including the Dorchester Collection, whose properties include The Dorchester in London and Le Meurice in Paris. Family businesses are significant not just because they contribute significantly to the economy, but also because of the long-term stability they provide, the particular dedication they demonstrate to local communities, the ownership responsibility they feel, and the values they uphold. Family businesses have a tremendous impact on the health and dynamism of the real economy in Europe, where they account for between 65 and 80 percent of all firms and, on average, more than 40 to 50 percent of all jobs (European Family, 2023). The European hospitality industry is one of the biggest industries with covering the market share of above 50% (Jamshed, et.al, 2023). Services for self-isolation are inexpensive, Cheap rates for luxury accommodations for domestic travellers Collaboration with companies like Careem and Uber, as well as OYO, AirBNB, Booking.com, and Agoda booking platforms, the integration of artificial intelligence and machine learning to CSR, the encrypting of payment channels for visitors, and the best method, the sharing economy (SOEG Jobs, 2021). Family-owned hotels (FBH) improves employee digital literacy, control the sharing economy, create a unified European Tourism Data Space, revive night trains with free bus, rail, and ferry passes to reduce CO2 emissions, and create a special platform for EU destinations with travel protocols in multiple languages (Tourism Sector 2021) and be connected to various Apps (Facebook, Instagram, Twitter, YouTube etc.) Hotels should prioritize implementing cutting-edge technology, safeguarding guest health, using modern cleaning methods, and minimizing guest interaction (Phumbandit, 2021). The family-run hotels will increase their use of social media marketing (Obermayer, Leinonen & Others, 2021).

Family firms range from small businesses to large corporations with over 30 percent of the S&P 500 comprised of family firms (Zellweger, Nason and Nofrdqvist, 2012). In 2018, there were some 5.1 million family-owned businesses in the UK, representing 87.6 per cent of private sector firms. The UK family business sector employed 14.2 million people in 2018 – more than half of all employees in the private sector and almost two-fifths of the UK's total employment.

- The UK family business sector earned more than £1.9 trillion in revenue in 2018, 44 per cent of the private sector's total turnover.
- Family firms in the UK contributed £657 billion to UK gross domestic product in 2018, 31 per cent of the UK's total GDP.
- Family firms paid the UK Exchequer £196 billion in tax receipts in 2018, around one quarter of the UK Government's total revenue. According to our findings, family-owned hotels in Europe

Figure 1.
Source: https://europeanfamilybusinesses.eu/about-european-family-businesses/

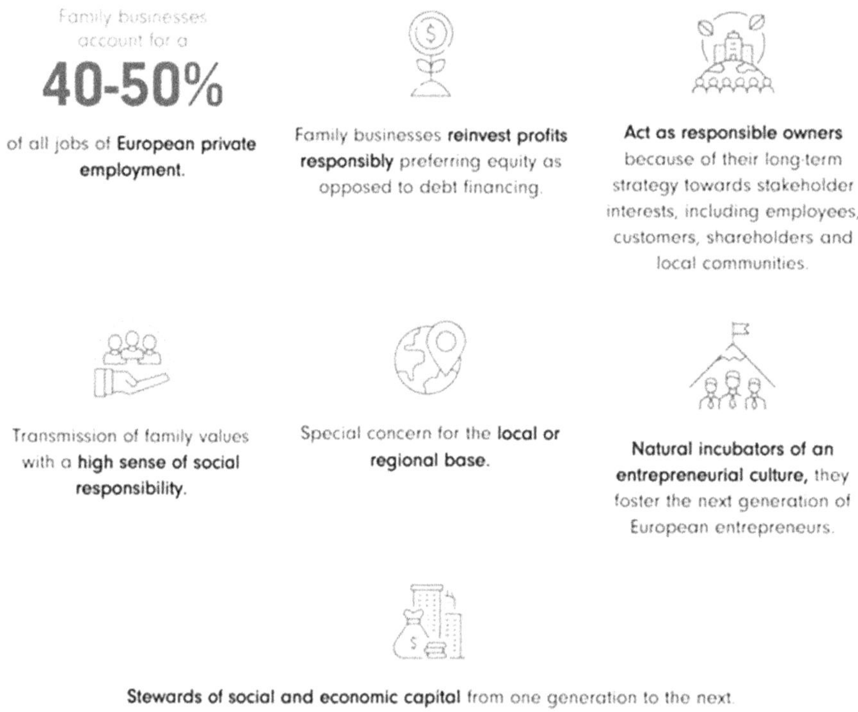

expanded their use of cleanliness procedures, employee vaccinations, EU digital COVID-19, long- and short-term domestic vacation bookings, Paperless transactions, the implementation of ecological initiatives as part of their CSR, and diverse travel packages for domestic and foreign visitors, Electronic luggage tags, mobile self-check-in, mobile booking, self-service check-in kiosks, phones as key cards, lobby media panels, bring your own content platforms, bring your own device, Voice over internet protocol phones, mobile boarding passes, hotel service optimization systems, and guest device connectivity tools.

Family-owned hotels in Europe have faced significant challenges during the COVID-19 pandemic, but their resilience, flexibility, and focus on customer service have enabled many of them to survive. As the world begins to emerge from the pandemic, these hotels will need to adapt to new trends and expectations, such as sustainable tourism, digital technology, wellness experiences, and experiential travel. Family businesses can also participate in sustainability through the adoption of environmentally friendly practises to lessen their contribution to greenhouse gas emissions, curb their excessive consumption of water and electricity, and improve the overall sustainability of their environment (Shah, et.al., 2022). By leveraging their strengths and meeting these new demands, family-owned hotels can continue to provide exceptional service and build loyal customer relationships in the post-pandemic world.

The exact percentage of family-owned businesses globally is difficult to estimate because of the differences in definitions and data availability. However, some reports suggest that around 65-70% of

Figure 2.
Source: https://www.ifb.org.uk/advocacy/about-family-business/

businesses worldwide are family-owned (Business Survey, 2016). The earliest known family business on record dates back to 15th century Italy. It was a printing press founded in Venice by the Aldine family and some of early family businesses include the Rothschild banking dynasty in Germany, which dates back to the 18th century, and the S. C. Johnson & Son Inc. manufacturing business in the United States, which was founded in 1886. There are many family-owned hotels around the world like the Ritz Hotel in London, the Adlon Kempinski Hotel in Berlin, and the Langham Hotels in both London and Chicago. The oldest known family-owned hotels are the St Moritz Hotel and Spa in Switzerland, which was founded in 1856, and Corus Hotels, which was founded in the United Kingdom in 1876. The oldest family-owned hotel in Europe is the St Moritz Hotel and Spa in Switzerland, which was founded in 1856. The St Moritz Hotel and Spa is owned by the Bronz-Lachaux family and is currently estimated to be worth around $30 million. After COVID-19 the revenue loss of hotels and restaurants had reached 85% with 60-90% reduction in bookings comparing to pre-pandemic annual revenue. Most of the family business hotels have rebuilt their businesses by different innovative steps such as the closure of spa services or kept it for families only. This chapter's goal is to evaluate how European family hotels have recovered from COVID-19 and what unique features have aided their recovery. The European hospitality sector is one of the largest in the world, accounting for more than 50% of the total market share. Following COVID-19, hotels and restaurants had a revenue loss of 85% and a decline in bookings of 60% to 90% when compared to their pre-pandemic yearly revenue. The majority of family-owned hotels have reconstructed their operations through a variety of creative measures, such as closing the spa or keeping it family-only, entry with less contact, Daytime accommodations, butler services (printing, scanning, and babysitting), strict housekeeping standards, and robot butlers.

BACKGROUND

The first research on family businesses is generally credited to the psychologist Harry Levinson and his 1957 book, The Passages of Enterprise. The book discussed various aspects of family businesses, including the psychological implications of them and their impact on the family dynamic. Levinson's book concluded that family businesses are a unique type of business structure and require careful management to succeed in the long-term. He also found that family businesses tend to be more resistant to failure than non-family businesses, and that successful family businesses can provide a long-term source of income and stability for families. . Family businesses have been around for centuries, and many of the world's oldest organisations are still run by members of the founding family. Walmart, Ford, and BMW are all family-owned businesses that have achieved global prominence. For the economy as a whole, family businesses are essential for many reasons, including job creation, technological advancement, and overall growth. Long-term thinking and investment in R&D by these companies help to create the kinds of innovative goods and services that fuel economic expansion. These companies care deeply about the impact they have on the world around them and frequently adopt eco-friendly policies. Many family-run businesses care deeply about leaving a positive impact on the world for the next generation, and this concern inspires them to practise eco-friendliness and conservation in all aspects of their business. Many businesses are started or purchased with the primary intention of catering to the requirements and preferences of their owners and their families, rather than growth, maximisation of profits, or even long-term viability. This is not always the case. That is the essence of what is meant by the term "family business," and it differentiates these businesses from others in which the owners and their families do not play a significant role in the strategy or operations of the business. Recent years have seen a surge in both interest in and research pertaining to family businesses. This attention is long overdue, given the scope of family businesses in most nations and the importance they play in those nations' economies. Emerging industries, such as nature tourism, are characterised by the preponderance of small businesses, the majority of which are run by owner-operators and families. Over the past few years, there has been a surge in research pertaining to family tourism; however, the majority of studies have concentrated on analysing family travel decision-making and the behaviour of traditional families while vacationing in conventional tourism destinations (Li et al., 2020). Organizations, particularly those in the hospitality industry, need to incorporate CSR practises, organisational identification, and co-worker pro-environmental advocacy on an individual level in order to cultivate workplace behaviours that are beneficial to the environment. This is especially true in the hospitality industry (Shah, et.al, 2021). The hospitality industry needs to be prepared for significant technological innovations as well as disruptions from other external factors such as workforce, climate change, and environment. The change will, in the broad sense, be one of the lasting effects that COVID-19 will have on the hospitality industry. Family businesses are the most common and common form of ownership and management of businesses in the majority of countries. Throughout history, the development of individual and family-run farms and businesses was followed by the formation of companies and other forms of ownership. This was primarily done to satisfy the needs of generating larger amounts of capital and dealing with legal issues such as providing protection against lawsuits and personal bankruptcy. It was the beginning of the Industrial Revolution in the 18th century that led to the replacement of smaller manufacturing businesses run by families with larger ones. More recently, the expansion into such massive proportions of the service sector created a plethora of new opportunities for family businesses. Family businesses will undoubtedly continue to thrive as the foundation of the commercial sector, despite the fact that large, publicly traded corporations typically garner the most attention, particularly in terms of share offerings, stock values, and speculation. The pandemic caused by COVID-19 resulted in a decrease of global tourism of 72% and a loss of $1 trillion in tourism receipts in 2020, having an unprecedented impact on the hospitality industry. In the year that began in January 2020 and ended in January 2021, there were 87 percent fewer international tourists than in the previous year, which equated to fewer than 10,000 travellers. In addition to this, Asia and the Pacific were the regions that were affected by travel restrictions the most severely. In point of fact, the impact of Covid-19 on international tourism (a loss of $320 billion) was "triple" that of the "2009 global financial crisis." This was due to the fact that Covid-19 occurred between March and May of 2020. Since the beginning of the hospitality industry in Europe, family-run tourism businesses have been an essential component of the field. These companies typically range in size from very small to medium-sized and are usually owned and run by a single family or a group of related families. Customers of family-owned hotels, restaurants, and other tourism-related businesses receive individualised attention and the opportunity to participate in one-of-a-kind activities. Nevertheless, the COVID-19 pandemic has had a significant impact on the hospitality industry, including family tourism businesses in Europe. The pandemic has caused widespread economic and social upheaval and has impeded global travel as a direct result. Because of lockdowns and restrictions on travel, many businesses in the hospitality industry were forced to close, either temporarily or permanently. This made the hospitality industry one of the sectors that was hit the hardest. In order to continue operating successfully in the face of the pandemic, family-owned tourism businesses in Europe have been forced to adopt new strategies and practises. Digitalization and the use of technology have taken on a greater significance than they ever have before. As a result, many companies are switching to online booking systems and contactless check-ins in an effort to reduce the amount of face-to-face interaction. Many family-owned tourism businesses now place a significant emphasis on sustainable business practises, such as minimising waste and adopting environmentally friendly procedures. Another domain in which family-run tourism businesses have needed to adapt is that of environmental responsibility. According to Cánoves-Valiente and López-Guzmán (2017), as a result of the pandemic, many businesses have prioritised the implementation of environmentally friendly practises and the reduction of waste. [Citation needed] According to Lopez-Guzmán et alresearch .'s from 2020, this has not only been beneficial for the environment, but it has also resonated with customers, who are becoming increasingly aware of the impact their travel choices have on the planet. Despite the difficulties, family-run tourism businesses in Europe have identified some opportunities that have arisen as a result of the pandemic. For instance, because of the recent trend toward domestic and regional tourism, companies now have the chance to present their one-of-a-kind products and services to a more extensive target market (Alonso & Otero-Neira, 2021). In addition, the pandemic brought to light the significance of adaptability and individualised service delivery for customers, which is a strength that family-owned businesses can capitalise on during the post-pandemic recovery period (Gruber & Mandl, 2018). Family-run tourism enterprises now place a premium on ensuring the well-being and protection of their clients. Many have implemented new health and safety protocols, such as increased cleaning and sanitation measures, social distancing, and policies that require people to wear masks at all times. Some companies have also implemented cutting-edge strategies to maintain a physical distance between themselves and their customers, such as the provision of private dining rooms and opportunities to eat outdoors. According to research, family businesses play an extremely important part in the hospitality industry. This is especially true in Europe, where family businesses account for a sizeable proportion of the sector (Gruber & Mandl, 2018). According to Canoves-Valiente and López-Guzmán (2017), the emphasis placed on personalised services and one-of-a-kind experiences is one of the most important contributors to the success of family-owned tourism businesses. However, the COVID-19 pandemic has presented difficulties for the industry that have never been seen before, and as a result, family businesses have been forced to adapt and develop new strategies in order to continue existing. In spite of the difficulties brought about by the pandemic, there have been some encouraging developments for family-run tourism businesses in Europe. For instance, many businesses have been successful in retaining customer loyalty by providing customers with personalised services and adaptable cancellation policies. In addition, there has been a renewed focus on domestic and local tourism, which has provided an opportunity for family-owned tourism businesses to showcase the unique offerings they have to offer and attract new clients. Because of the widespread COVID-19 pandemic, family-run tourism businesses in Europe have been forced to adopt new strategies and practises. Even though the path to recovery is still unclear, many businesses are coming up with innovative strategies to maintain their competitive edge and build resilience in the face of persistent uncertainty. During the pandemic, one of the most significant obstacles that family-run tourism businesses in Europe have had to overcome is the requirement to strike a balance between providing personalised services and ensuring the safety of their customers. Although the use of technology has been a key tactic in minimising contact between staff and customers, it has also been important for businesses to maintain the personal touch that differentiates them from larger corporations that are less personable (Alonso & Otero-Neira, 2021). For instance, the use of chatbots and other digital tools to communicate with customers has been effective in reducing the need for face-to-face interactions. However, many family businesses have also emphasised the importance of maintaining personal connections with customers through the use of virtual events and personalised messages. Overall, the COVID-19 pandemic has had a significant impact on family-owned tourism businesses in Europe, but these businesses have been able to demonstrate resilience and adaptability in the face of the challenges. These companies can position themselves for success during the recovery period as well as beyond it by capitalising on their strengths in personalised services, sustainability, and flexibility. According to projections made by Visa Inc. in 2016, it is anticipated that by the year 2025, more than 280 households across the globe will travel internationally on an annual basis. The total revenue generated by hotels in Europe is expected to reach approximately $110.90 billion in 2023 and $118.20 billion in 2027. (Statista,2023). These companies are making strides toward ensuring the long-term viability of their operations and the continuation of their family's legacy by adopting contemporary business practises and cutting-edge technologies. Family Business hotels in Europe have been implementing a variety of measures to adapt to the pandemic. These measures include increasing cleaning and sanitation protocols, providing contactless services, and reducing capacity to allow for social distancing (Lai et al. 2021). These are just some of the measures that have been taken. However, the importance of communication and collaboration among family-owned hotels in Europe during the pandemic is highlighted, and it was suggested that family businesses could work together to share best practises and support each other through the crisis. During the pandemic, family-owned hotels in Europe saw an increase in the number of cases of influenza A and B. (Cagnazzo et al. 2021). The pandemic forced many family-owned hotels in Europe to implement digital technologies in order to adapt to the situation. These technologies included the provision of virtual tours and online booking systems (Eisingerich et al. 2021). It was emphasised how important family-owned businesses are to the hospitality industry, and it was suggested that family-owned hotels could benefit from cultivating strong relationships with the communities in which they are located (Fichman et al). (2021). According to the findings of a study carried out by Iglesias-Sánchez and colleagues (2021), the authors hypothesised that smaller, family-owned hotels in Europe could differentiate themselves from larger hotel chains by providing more individualised services and a cosier environment. It would be beneficial for family businesses to have the ability to quickly adapt their business strategies in response to shifting market conditions (Mody et al. 2021). The authors of a study that was conducted by Soler-Guzmán et al. (2021) suggested that family-owned hotels in Europe could benefit from collaborating with other family businesses in the tourism industry to promote their destinations as safe and attractive places to visit. The research was published in the journal Tourism Management. During the pandemic, family-owned hotels in Europe used social media to communicate with their guests in a variety of ways, including providing updates on safety protocols and offering virtual experiences, as discovered by Wang et al. (2021). In a study conducted by Zhang et al. (2021), the authors suggested that family-owned hotels in Europe could benefit from highlighting their distinct heritage and history as a way to differentiate themselves from larger hotel chains. This would allow the family-owned hotels to compete more effectively with the larger hotel chains. Family businesses could reap benefits from their ability to swiftly adjust to shifting conditions and implement any necessary adjustments to their business practises (Zhou et al. 2021). Europe has been putting into action a variety of measures to adapt to the pandemic, such as improving cleaning protocols, making use of digital technologies, and working together with other businesses in the tourism industry. In order to distinguish themselves from larger hotel chains, family-owned businesses are putting an emphasis on developing robust relationships with the communities in which they operate, providing individualised services, and highlighting the distinctive history and heritage of their establishments. In general, resiliency, adaptability, and collaboration have been key factors in helping family-owned hotels in Europe to weather the challenges of the pandemic and remain competitive in the era that followed the end of the pandemic. The pandemic caused by COVID-19 has had a significant impact on family-owned hotels in Europe, which has resulted in the necessity of adaptation and innovation in order to remain profitable. Pechlaner and Innerhofer (2021), for example, found that family-owned hotels in Europe were focusing on sustainability initiatives as a way to differentiate themselves and appeal to environmentally conscious travellers. This finding was based on the findings of a study conducted by the two researchers. In a similar vein, Schegg and Stangl (2021) conducted research in which they highlighted the significance of providing guests with one-of-a-kind experiences and activities in order to attract guests in the post-pandemic era. The pre-COVID phase was very frustrating and it have worsened the mental health (Jamshaid, et.al.,2023) of everyone including the tourists and family businesses. According to the findings of a study conducted by Sigala et al., family-owned hotels have also been implementing health and safety protocols. These include frequent cleaning and disinfecting as well as the use of technology to reduce the amount of person-to-person contact (2021). According to research conducted by Topcu et al. (2021), family-owned businesses have been using social media as a tool to keep in touch with their clientele and to keep them apprised of any new developments concerning health and safety policies. According to the findings of a study that was conducted by Tussyadiah et al., one important strategy that family-owned hotels in Europe utilised during the pandemic was to work together with other companies that were involved in the tourism industry (2021). In addition, a study that was conducted by Weiermair and Fuchs (2021) highlighted the significance of ensuring the happiness and well-being of employees working in family-owned hotels, as this can ultimately have an effect on the experience that the guests have. According to the findings of a study conducted by Xiang et al. (2021), in order to ensure their continued success over the long term, family businesses have also been placing increased emphasis on financial management and cost control. In conclusion, a study conducted by Yuan et al. (2021) highlighted the significance of customer satisfaction in the post-pandemic era. The researchers suggested that family-owned hotels should place an emphasis on providing a positive experience for their guests by providing personalised services and paying close attention to detail. These family-owned hotels in Europe have been putting into action a variety of different strategies in order to adjust to the challenges that the pandemic has presented. Among these are initiatives aimed at preserving the environment, providing one-of-a-kind experiences, adhering to health and safety protocols, making use of technology, participating in social media, working in partnership with other companies, ensuring the satisfaction of employees and customers, and effectively managing financial resources. Hotels that are owned by families can ensure their continued viability and position themselves for success in the post-pandemic era by concentrating their efforts on the key areas listed above.

COVID-19 AND THE HOSPITALITY INDUSTRY OF EUROPE

Many hotels around the world have closed down or operating at reduced capacity as a result of the COVID-19 pandemic, which has led to fewer guests and fewer bookings. Although family-owned hotels in Europe were not completely safe from the pandemic's effects, many of them remained open and even thrived as a result of their quick thinking, dedication to their guests, and environmental consciousness. The ability to adapt quickly was a major benefit for family-owned hotels during the pandemic. In response to the changing needs of their guests during the pandemic, many of these hotels introduced new services or products. One trend in the hospitality industry is the "work from hotel" package, which allows guests to rent a room for the day and conduct remote work in peace and comfort. Family-run hotels were more likely to survive the pandemic because of their dedication to their guests. Many of these hotels have established a steady stream of repeat guests over the years, and they cashed in on this fact when the pandemic hit. Family-owned hotels were able to maintain a strong connection with their customers by consistently communicating with them and providing individualised service, even during economic uncertainty. The hotels that are owned by families were able to survive the pandemic because of their dedication to eco-friendliness. Because of their long-term commitment to sustainability, many of these hotels were able to cut costs and lessen their impact on the environment even as the pandemic raged. Some hotels, for instance, have installed motion sensors in guest rooms and switched to LED lighting to reduce energy consumption and operating costs. Family-run hotels had some advantages, but they still had to deal with serious problems during the pandemic. Declining occupancy rates were a major source of trouble, resulting in lost money. The European Hotel Industry Association found that hotel occupancy rates across Europe dropped from an average of 72% in 2019 to 32% in 2020. Family-owned hotels were forced to reduce services, hours of operation, and staff in order to survive these tough economic times. However, these solutions weren't always adequate, and many hotels run by families needed government aid to stay open. Family-owned hotels also faced significant operational challenges due to the pandemic. For example, hotels had to implement new health and safety protocols to prevent the spread of the virus, such as increased cleaning and disinfection, social distancing measures, and contactless check-in and check-out. These measures required significant investment and training, and many family-owned hotels had to adapt quickly to ensure that they were compliant with the new regulations. Despite these challenges, family-owned hotels in Europe were able to survive the pandemic due to their resilience, flexibility, and focus on customer service. Many of these hotels were able to adapt quickly to the changing circumstances, and they were able to maintain a strong connection with their customers, even during times of uncertainty. Looking forward, family-owned hotels in Europe will continue to face challenges as they navigate the post-pandemic landscape. However, by leveraging their strengths and focusing on innovation, sustainability, and customer service, these hotels will be well-positioned to thrive in the years ahead.

Post-Pandemic and Family Businesses Hotels

As Europe begins to emerge from the COVID-19 pandemic, family-owned hotels are facing new challenges and opportunities. While the pandemic has had a significant impact on the hotel industry, family-owned hotels have been able to weather the storm due to their resilience, flexibility, and focus on customer service. However, as the world begins to return to a sense of normalcy, these hotels will need to adapt to new trends and expectations in order to stay competitive. One of the biggest trends in the post-pandemic world is the rise of sustainable tourism. Consumers are increasingly aware of the impact that their travel has on the environment, and they are looking for hotels that prioritize sustainability. Family-owned hotels are well-positioned to meet this demand, as many of these hotels have been focused on sustainability for years. By implementing eco-friendly practices and communicating their commitment to sustainability to their customers, family-owned hotels can differentiate themselves from the competition and attract environmentally conscious travelers. Another trend that family-owned hotels will need to adapt to is the rise of digital technology. The pandemic accelerated the adoption of digital technology in the hotel industry, as hotels implemented contactless check-in and check-out, digital room keys, and virtual concierge services. Family-owned hotels will need to continue to invest in digital technology in order to meet the expectations of their customers. By offering seamless digital experiences and personalized services through digital channels, family-owned hotels can improve the guest experience and build loyalty also, family-owned hotels will also need to adapt to changing consumer preferences and behaviors. For example, many consumers are now looking for hotels that offer wellness and wellbeing experiences, such as yoga classes, spa treatments, and healthy food options. Family-owned hotels can meet this demand by partnering with wellness experts and offering specialized packages that cater to health-conscious travelers. Family-owned hotels will need to adapt to is the rise of experiential travel. Consumers are increasingly looking for unique and authentic travel experiences that allow them to connect with local culture and communities. Family-owned hotels can meet this demand by partnering with local artisans, offering culinary experiences that showcase local cuisine, and promoting local cultural events and activities. Despite these challenges, family-owned hotels in Europe are well-positioned to thrive in the post-pandemic world. These hotels have a long history of resilience and innovation, and they are known for their personalized service and attention to detail. By adapting to new trends and expectations, family-owned hotels can continue to build on their strengths and provide unique and memorable experiences for their guests. Post-pandemic world presents both challenges and opportunities for family-owned hotels in Europe. By leveraging their strengths and adapting to new trends and expectations, these hotels can continue to provide exceptional service and build loyal customer relationships. Whether it's through sustainable practices, digital innovation, wellness experiences, or experiential travel, family-owned hotels have the potential to thrive in the years ahead. In the era following the pandemic, family-owned hotels in Europe are implementing a variety of different strategies to ensure their continued success. Some businesses are placing their emphasis on environmentally friendly practises and digital innovation in order to maintain their competitive edge and fulfil the requirements set forth by their clients (Pechlaner & Innerhofer, 2021). Others are making use of technology to improve their guests' overall experiences and to make their properties more secure for everyone who visits (Schegg & Stangl, 2021; Sigala et al., 2021). According to Topcu et al. 2021 and Tussyadiah et al. 2021, social media marketing is also being used as a method to reach potential customers and engage with them. [Citation needed] When it comes to the creation of value in the tourism and hospitality industry, the employees of family businesses play an extremely important role (Weiermair & Fuchs, 2021). In addition, the pandemic's effect on the economy has had repercussions for those working in the hotel and travel industries; family-owned hotels are not exempt from this (Xiang et al.,2021). The pandemic has had a significant impact on a great number of family-owned hotels in Turkey, and these hotels are currently reevaluating their business strategies in order to remain competitive (Akgun et al., 2021). Restaurants in Cyprus that are owned and operated by families are experimenting with food tourism and adjusting to the post-pandemic era (Zopiatis et al., 2021). Family-owned hotels continue to place a high premium on providing an exceptional experience for their guests, and they are turning to service innovation to accomplish this goal (Yuan et al., 2021). In general, family-owned hotels in Europe are adopting a multi-pronged strategy in order to maintain their position as market leaders in the post-pandemic era. The pandemic has created challenges that have never been seen before, but it has also created an opportunity for family-owned hotels to innovate and adjust their services to meet the shifting requirements and expectations of their clientele. Family-owned hotels are working toward the goal of building resilience in the face of ongoing uncertainty by placing an emphasis on environmental responsibility, digital transformation, and the provision of a secure environment for their guests. There has been a significant effect that the pandemic has had on hotels in Europe that are owned by families. A great number of family businesses have been forced to make adjustments to their operations and strategies in order to remain competitive amidst the difficult market conditions brought about by the pandemic. Lockdowns and travel restrictions imposed by the government during the early stages of the pandemic caused a large number of family-owned hotels to be forced to close their doors. Because of this, many companies experienced significant drops in revenue, and their prospects for the future became less certain. However, as the pandemic has progressed and vaccines have become more widely available, a number of family-owned hotels have been able to reopen and have adjusted their business practises to account for the new reality. The following is a list of some of the changes that have been made by family businesses:

Protocol Improvements for Improved Hygiene and Safety

In order to protect both their customers and their employees, hotels run by families have been forced to adopt stringent hygiene and safety policies in recent years. This includes engaging in frequent cleaning and disinfection, taking steps to maintain a social distance, and making use of personal protective equipment. Transformation through the use of digital technologies is something that many family-owned businesses have done in order to improve their internal processes and the overall experience they provide for their customers. This includes the ability to check in and check out of a hotel room online, as well as options for contactless payments and virtual concierge services.

Diversification

In order to entice a greater number of customers, family-owned hotels have also been working to broaden the range of services and amenities they provide. This includes providing one-of-a-kind experiences such as culinary workshops and cultural tours, in addition to adding new conveniences such as fitness centres, spas, and opportunities for outdoor activities.

Collaborations

In order to promote their region as a secure and appealing destination for tourists, a few hotels that are owned by families have teamed up with other local businesses and organisations dedicated to the tourism industry. In spite of the difficulties posed by the pandemic, a significant number of hotels in Europe that are owned by families have demonstrated resiliency and adaptability in their response.

CONCLUSION

Family-owned hotels in Europe have shown remarkable resilience and adaptability during the COVID-19 pandemic, despite facing significant challenges. These hotels have been able to weather the storm due to their long history of innovation and personalized service, as well as their ability to pivot quickly in response to changing customer needs and expectations. Throughout the pandemic, family-owned hotels have been focused on keeping their guests safe and comfortable, implementing new health and safety protocols, and offering flexible cancellation policies. Many of these hotels have also shown a strong commitment to sustainability, implementing eco-friendly practices and communicating their environmental stewardship to their guests. As the world begins to recover from the pandemic, family-owned hotels will need to continue to adapt to new trends and expectations in order to remain competitive. One of the biggest trends in the post-pandemic world is the rise of sustainable tourism, as consumers become increasingly aware of the impact of their travel on the environment. Family-owned hotels are well-positioned to meet this demand, as many of these hotels have a long history of sustainability and eco-friendly practices. E-commerce operations are expanding as a direct result of digitalization, which is resulting in the creation of new business processes, methods, and products (Shah, et al., 2023) therefore family businesses must use these technologies. Another, trend that family-owned hotels will need to adapt to is the rise of digital technology, which has accelerated during the pandemic. By investing in digital technology, family-owned hotels can improve the guest experience, streamline operations, and provide personalized services through digital channels. Wellness experiences are also becoming increasingly important to consumers, with many travellers looking for hotels that offer yoga classes, spa treatments, and healthy food options. Family-owned hotels can meet this demand by partnering with wellness experts and offering specialized packages that cater to health-conscious travellers. Finally, experiential travel is becoming increasingly popular, as travellers look for unique and authentic experiences that allow them to connect with local culture and communities. Family-owned hotels can meet this demand by partnering with local artisans, offering culinary experiences that showcase local cuisine, and promoting local cultural events and activities. In order to thrive in the post-pandemic world, family-owned hotels will need to continue to leverage their strengths, including their focus on personalized service, attention to detail, and commitment to innovation. By adapting to new trends and expectations, these hotels can continue to provide exceptional service and build loyal customer relationships for years to come. In summary, family-owned hotels in Europe have demonstrated remarkable resilience and adaptability during the COVID-19 pandemic. While the pandemic has presented significant challenges, these hotels have been able to weather the storm by prioritizing the safety and comfort of their guests, implementing eco-friendly practices, and investing in digital technology. As the world begins to recover from the pandemic, family-owned hotels will need to continue to adapt to new trends and expectations, such as sustainable tourism, wellness experiences, experiential travel, and digital innovation, in order to remain competitive and provide exceptional service to their guests.

REFERENCES

About European Family Businesses - European Family. (2023). *European Family*. https://europeanfamilybusinesses.eu/about-european-family-businesses/

Akgun, A. E., Koseoglu, M. A., & Dogru, T. (2021). COVID-19 pandemic and the Turkish hospitality industry: Evaluating the impact on family-owned hotels. *Journal of Hospitality and Tourism Management*, *48*, 32–40. doi:10.1016/j.jhtm.2021.01.004

Baker, M. A., & Nelson, R. E. (2005). Creating something from nothing: Resource construction through entrepreneurial bricolage. *Administrative Science Quarterly*, *50*(3), 329–366. doi:10.2189/asqu.2005.50.3.329

Barrett, R., Jones, T., & McEvoy, D. (2016). "How do we get from there to here?": An interpretive phenomenological analysis of the lived experiences of women family business founders. *International Small Business Journal*, *34*(8), 1060–1080.

Family Business Institute. (2016). *Global Family Business Survey 2016*. Retrieved from http://fambizpwc.com/2018/03/global-family-business-survey-2016-results-announced/

Gallo, M. A. (2018). Family businesses and entrepreneurship: A literature review. *Journal of Family Business Management*, *8*(2), 141–157.

Hotels - Europe | Statista Market Forecast. (2023). *Statista*. https://www.statista.com/outlook/mmo/travel-tourism/hotels/europe

Jamshaid, S., Bahadar, N., Jamshed, K., Rashid, M., Imran Afzal, M., Tian, L., Umar, M., Feng, X., Khan, I., & Zong, M. (2023). Pre-and Post-Pandemic (COVID-19) Mental Health of International Students: Data from a Longitudinal Study. *Psychology Research and Behavior Management*, *16*, 431–446. doi:10.2147/PRBM.S395035 PMID:36814636

Li, M., Lehto, X., & Li, H. (2020). 40 Years of Family Tourism Research: Bibliometric Analysis and Remaining Issues. *Journal of China Tourism Research*, *16*(1), 1–22. doi:10.1080/19388160.2020.1733337

Li, X., Liang, X., Li, Y., & Liu, J. (2020). Family tourism research: A bibliometric review. *Current Issues in Tourism*, *23*(6), 673–693.

Pechlaner, H., & Innerhofer, E. (2021). Sustainability and digitalization in family-owned hotels in the post-COVID era. *Journal of Tourism Futures*, *7*(1), 92–101. doi:10.1108/JTF-11-2020-0092

Schegg, R., & Stangl, B. (2021). The role of technology in the recovery phase of family-owned hotels. *Current Issues in Tourism*, *24*(11), 1419–1425. doi:10.1080/13683500.2020.1851388

Shah, S. H. A., Cheema, S., Al-Ghazali, B. M., Ali, M., & Rafiq, N. (2021). Perceived corporate social responsibility and pro-environmental behaviors: The role of organizational identification and coworker pro-environmental advocacy. *Corporate Social Responsibility and Environmental Management*, *28*(1), 366–377. doi:10.1002/csr.2054

Shah, S. H. A., Jamshed, K. A., Saleem, S., Bashir, M., & Kiani, O. I. (2022). Sustainable Development in the Hospitality Industry of Europe: Role of Green HRM Practices. In Sustainable Development of Human Resources in a Globalization Period (pp. 49-67). IGI Global.

Sigala, M., Gretzel, U., & Christou, E. (2021). Technology use in the hospitality industry: Impacts on service quality and the guest experience in a post-pandemic era. *International Journal of Hospitality Management*, *94*, 102847. doi:10.1016/j.ijhm.2021.102847

Statista. (2021). *Hotel market revenue in Europe from 2016 to 2023, by country (in billion U.S. dollars).* Retrieved from https://www.statista.com/statistics/657066/hotel-market-revenue-in-europe-by-country/

Topçu, İ., Uslay, C., & Fuxman, L. (2021). Post-COVID-19 social media marketing in family-owned hotels: An empirical investigation in Turkey. *Journal of Hospitality and Tourism Technology*, *12*(3), 373–391. doi:10.1108/JHTT-10-2020-0137

Tussyadiah, I. P., Sigala, M., & Law, R. (2021). COVID-19 and the changing landscape of social media engagement with destination and hospitality brands. *Journal of Travel Research*, *60*(1), 44–56. doi:10.1177/0047287520956997

United Nations World Tourism Organization. (2021). *International tourism fell by 74% in 2020, the worst year in tourism history.* Retrieved from https://www.unwto.org/news/international-tourism-fell-by-almost-75-in-2020

Visa. (2016). *Visa Study Estimates Households Traveling Internationally Will Surpass 280 Million By 2025.* Available at https://investor.visa.com/news/news-details/2016/Visa-StudyEstimates-Households-Traveling-Internationally-Will-Surpass-280-Million-By2025/default.aspx

Weiermair, K., & Fuchs, M. (2021). Employees in family businesses: A critical resource for creating value in tourism and hospitality. *Journal of Hospitality & Tourism Research (Washington, D.C.)*, *45*(4), 518–537. doi:10.1177/1096348020966869

Xiang, Z., Du, Q., Ma, Y., & Fan, W. (2021). COVID-19 and the economy: Impacts on and implications for hotel and travel industries. *Journal of Hospitality & Tourism Research (Washington, D.C.)*, *45*(5), 704–717. doi:10.1177/1096348020969768

Yuan, S., Shen, X., Chen, W., & Zhan, W. (2021). Service innovation and customer satisfaction in family-owned hotels: The mediating role of customer experience. *Journal of Hospitality and Tourism Management*, *48*, 83–91. doi:10.1016/j.jhtm.2021.05.004

Zopiatis, A., Giannikis, S., & Theocharous, A. L. (2021). Food tourism and the post-COVID era: An exploratory study of the perceptions of Cypriot family-owned restaurants. *Journal of Culinary Science & Technology*, *19*(4), 366–387. doi:10.1080/15428052.2021.1906651

Chapter 9
Role and Impact of Social Media in Tourism for Women Entrepreneurs:
A Study on Family Business Women Entrepreneurs

Noor E. Eman
University of Agriculture Faisalabad, Pakistan

Samrah Jamshaid
Northeast Normal University, Changchun, China

Kamran Jamshed
Bahria University, Pakistan

Azzam Inayatullah Shamsi
Bahria University, Pakistan

Basheer M. Al-Ghazali
King Fahd University of Petroleum and Minerals, Saudi Arabia

Nikhar E. Kainat
University of Veterinary and Animal Sciences, Lahore, Pakistan

ABSTRACT

Recently, social media's role in tourism for women entrepreneurs has garnered attention. This study examines how social media affects the travel industry for women who run family businesses. Facebook, Instagram, and Twitter are essential resources for travel industry women business owners, according to research. When women use social media to promote their businesses, they gain visibility, customers, and sales. The study also emphasises the importance of social media-based online communities in reducing cultural barriers and promoting women travel business owners. Social media users are more likely to be successful entrepreneurs, which affects women tourism business owners. It's important to recognise and support women tourism business owners, who promote growth, economic stability, and social well-being. This study examines how women tourism entrepreneurs use social media and how it affects their businesses. This information can help other women entrepreneurs use social media effectively and policymakers and industry stakeholders support women-owned tourism businesses.

DOI: 10.4018/978-1-6684-8417-3.ch009

INTRODUCTION

There is a growing trend among businesses to use social media to advertise their products and services, and the tourism sector is not an exception to this. The operations of the tourism sector have been significantly altered as a result of the major impact that social media has had on the industry. As a result, new opportunities have emerged for women and men entrepreneurs in family companies to reach a wider audience. According to Kim, Lee, and Yoo (2015), the use of social media has changed the ways in which tourists get information, discuss their trip experiences, and make decisions on their travel plans. This article's objective is to study the role of social media in the tourist industry and its influence on female family business entrepreneurs. Women now make up around 54 percent of the workforce in the tourism and hospitality industries, as reported by the World Tourism Organization (UNWTO), which indicates that the number of women participating in tourism is growing. Also, the tourism industry is home to a sizeable number of family businesses that are run by women entrepreneurs, many of whom are involved in management roles. These female business owners face a number of challenges, such as restricted access to financial resources, restricted exposure on the global market, and insufficient attempts to promote their products and services. But, the platforms for social media have made it possible for them to sell their company, display their wares, and engage with customers who are considering making a purchase from them. According to Buhalis and Law (2008), small firms can sell their products, build their brand image, and communicate with their customers by using social media platforms because these platforms offer a low-cost platform with a high-impact potential. This study has evaluated the influence that social media has had on female business owners operating family businesses within the tourism industry. Particular attention will be paid to the marketing strategies that are utilised by tourism businesses. In addition to this, the study will analyse the challenges that are experienced by female business owners, as well as the opportunities that social media presents for addressing these challenges. The research will use a combination of methodologies, such as questionnaires and in-person interviews with women who own family businesses as part of its methodology. Social media has made it possible for women entrepreneurs in family businesses to overcome challenges and promote their companies, despite challenges such as limited access to cash and resources, cultural and societal biases, a lack of exposure to global markets, and insufficient promotional activities. It is essential to conduct additional research into the role of social media in assisting women entrepreneurs in the tourism business and their coping mechanisms in this challenging environment as we navigate the current pandemic and its influence on the tourism sector. Customers can be reached, brand awareness raised, and sales leads produced all through social media. Equally as important, it facilitates communication between customers and the women business owners who serve them. Businesses can benefit from using social media to handle concerns and complaints raised by customers in real time, which can also help them retain positive relationships with those customers and improve their reputation. This is of utmost significance in the tourist sector, since the happiness of one's clients is essential to one's financial success.

BACKGROUND

Women business owners can benefit from the growing number of tourists who are active on social media. Online communities and sharing tools like Facebook, Twitter, Instagram, and YouTube have revolutionised the travel industry. The advent of social media has fundamentally altered the ways in which people interact with one another and gather information. By promoting their businesses on social media, women business owners can do so without spending a fortune on advertising. They can easily create and distribute content to an engaged audience on social media platforms at no cost. It can help women business owners stand out from the crowd by making their products and services stand out from the competition. By leveraging the reach of social media, they can promote their products and services to a larger audience, draw in more customers, and increase sales. In addition, they can use these platforms to build lasting relationships with their customers by producing content that is interesting and relevant to their demographic. Women business owners in the travel industry now have a potent new resource at their disposal. This allows them to connect with customers, spread the word about their products and services, and ultimately grow their businesses and brands. Social media has many advantages for women business owners in the tourism sector. They can reach a wide audience in any part of the world due to these online marketplaces. To increase their business, women entrepreneurs should take advantage of the fact that social media is increasingly influencing consumers' decisions to travel. 97% of vacationers check online reviews before making reservations, and 87% say they use social media to learn more about potential destinations before booking (Tripadvisor, 2018). The significance of social media in connecting with consumers and giving them access to relevant information is highlighted. Sixty percent of Instagram users report discovering new products on the platform (Hootsuite, 2019), which is great for brand awareness, and women business owners can reach more people by using social media to promote their wares. Women make up about sixty percent of the workforce in the tourist industry worldwide, as stated by the World Tourism Organization (UNWTO,2019) in the year 2020. Since family businesses make up a sizeable portion of the overall business landscape across the globe (Fernández, Nieto, & Ramos, 2013), the contributions of female business owners working within their families' companies are essential to the success of the tourism sector. Notwithstanding the considerable contributions they make to the tourist industry, women entrepreneurs in family enterprises frequently encounter a number of obstacles that restrict the possibility for expansion of their businesses. (Garca-Sánchez, et.al., 2017) These challenges include limited access to finances and resources, cultural and social biases, a lack of exposure to global markets, and inadequate promotional activities. For women business owners in the travel industry, social media has become an invaluable resource for connecting with customers, spreading the word about their offerings, and establishing their respective brands. Family companies may leverage social media to improve their brand visibility and reputation, develop stronger relationships with their clientele, and provide superior service to their clientele (Naldi, Rasmussen, & Colombo (2018). Social media platforms such as TripAdvisor and Facebook are crucial sources of information for travellers, and when making travel decisions, tourists place a significant amount of reliance on user-generated material (Lee & Hyun, 2015). The tourism industry, being the largest employer on the planet and operating on a global scale, is home to a more diverse range of stakeholders than many other types of businesses. Probably only the growth of information and communications technologies can be compared to the industry's rapid expansion and development. In recent years, there has been a growing and mutually beneficial connection between technology and tourism, which has resulted in significant transformations in both the structure of the industry as a whole as well as our perceptions of it. The significance of exceeding the new information threshold of universal, ubiquitous communications access has propelled the tourism industry to the new heights of engagement, quickening management by wire. This has been made possible as a result of the widespread availability of the internet. The use of information and communication technologies (ICTs) is becoming an increasingly crucial factor in determining the competitiveness of tourist organisations and destinations, as well as the overall tourism sector. Increases in the carrying capacity and speed of networks have contributed to an increase in the number of passengers all over the world who utilise technologies to plan and experience their travels. Search engines have also contributed to these increases. ICTs have also significantly affected the efficacy and efficiency of tourism organisations, the way in which businesses compete in the market, and the way in which customers engage with businesses. Many new players have entered the tourism industry, there have been shifts in market share and the balance of power, there have been adjustments in political conceptions of tourism, and there has been a rising acknowledgement of the importance of tourism to an increasing number of national and regional economies (Buhalis, D., & Law, R. 2008). Moreover, the tourism industry is constantly evolving, and the impact of social media is also changing. For example, the COVID-19 pandemic has forced businesses to change the way they operate, and social media has become even more critical in this context. According to a study by Chou, Chen, and Chen (2021), social media has become an essential tool for businesses in the tourism industry during the pandemic. The study found that businesses used social media platforms such as Facebook, Instagram, and WeChat to provide real-time information to customers, promote their products and services, and maintain customer relationships. Therefore, the use of social media for women entrepreneurs in family businesses in the tourism industry is not only a way to promote their businesses but also a way to adapt to the changing landscape of the industry. While the usage of social media by family businesses is on the rise, many companies still do not have a defined plan for how to use social media (Naldi, Rasmussen, & Colombo, 2018). This is especially true for firms that are run by families, as the limits between personal and professional connections are sometimes hazy in such settings. It is essential for businesses to have a distinct social media strategy in order to accomplish their business goals and ensure that their social media activities are in line with their core values and company culture. In addition, the tourist industry is highly competitive, which means that firms operating within it need to separate themselves from their rivals in order to get new consumers and keep the ones they already have. By the use of social media, companies may differentiate themselves from their rivals by offering customers with a one-of-a-kind experience that is incomparable to that of their competitors. The usage of social media can positively influence tourist behaviour, such as the intention to visit a place, the intention to revisit a destination, and the intention to suggest a destination to others, according to research conducted by Lee and Hyun (2015). A study that was conducted by Buhalis and Law (2008) found that the use of social media can raise the chance of repeat business and improve the level of satisfaction experienced by customers. The COVID-19 epidemic has had a huge influence on the tourist industry, and as a result, firms have been forced to adjust to the shifting dynamics of the sector. Due to the fact that businesses are utilising social media during the pandemic to deliver real-time information to clients, market their products and services, and preserve customer relationships, social media has become even more crucial. According to the UNWTO (2020), the COVID-19 pandemic has had a disproportionate impact on women working in the tourism industry. In particular, the pandemic has had a significant impact on the numerous female entrepreneurs who run family companies. Women entrepreneurs who run family companies in the tourism industry may find that making use of social media platforms is helpful in overcoming the obstacles presented by the epidemic and adjusting to the shifting dynamics of the industry.

ROLE OF SOCIAL MEDIA IN TOURISM

At this time, there are 4.2 billion people using their mobile devices to explore social media sites. Users of Facebook log in to their accounts an average of more than five times each day, and 74% of marketers believe that social media platforms like Facebook, Google+, Twitter, and others are essential to the success of their businesses and product initiatives. The travel industry will be significantly impacted as a result of these effects. Travelers can share their experiences with others by using social media platforms such as Facebook, Twitter, and YouTube. People in any region of the world can use the internet, read blogs, and make use of social media to learn new things or share what they already know. These days, the terms "selling" and "seeking" have largely been supplanted in many different types of businesses by the terms "connecting" and "exchanging." For this reason, information technology (IT)-enhanced social media in the hospitality and tourism industry has become a vital tool for gaining access to a variety of tourist information sources. Therefore (Chung & Koo, 2015). A new environment that is conducive to acculturation has been made possible by the rapid development of information and communication technology (ICT), including social media. In point of fact, as social media evolves into an indispensable tool in marketing communications between companies and travellers, as well as among travellers' social networks (Niu et al., 2016), it has opened a new venue for "virtual acculturation." This refers to the process by which travellers and potential travellers are able to go through pre-transitional adaptation prior to their actual visit to a foreign country. This research makes a significant contribution by systematically explicating the functions of social media usage in the process of virtual acculturation in the context of tourism. It has been established that social media platforms play a significant part in shaping the cultural orientation and identification of their users; hence, these platforms need to be incorporated into the development of future acculturation models. In particular, the findings of our study extended the standard definition of acculturation, which had previously been defined as "continuous first-hand engagement," to encompass "indirect and/or intermittent virtual interactions," which are the outcome of the use of social media. To begin, tourists are influenced by social media in terms of learning a new culture, in terms of shaping perceptions and expectations of the surrounding cultural milieu and subsequent acculturations. This is because social media is increasingly used by tourists as a source of information to research, explore, plan, and ultimately share their travel experiences. Second, culture is learned as a result of the four sub-functions of social media. These sub-functions make it possible for pre-transitioning tourists to acquire the culturally appropriate skills necessary to thrive in the new environment. In conclusion, the social contact offered by social media platforms elevates both the inter-group and the intra-group interactions among visitors, which is positively connected with psychological and sociocultural acculturation. The current research offers conceptually more robust insights into a process framework of virtual acculturation and its connection to the utilisation of social media. In a more concrete sense, this information may be utilised to construct more successful destination marketing strategies in order to generate pleasant and beneficial tourism experiences for a variety of different tourist types. For instance, the results of our case study suggest that culturally relevant elements should be incorporated as inputs into content marketing. Visitors should be encouraged to consume content and share their experiences on social media prior to, during, and after their trips in order to facilitate a co-creational process of destination communication. This can be accomplished by destination managers and other businesses related to tourism optimising their social media marketing efforts. It is possible that a spectacular and one-of-a-kind experience of acculturation could be the outcome of providing a location where tourists can have a meaningful grasp of the social background of the area (Li, et al., 2019). The proliferation of social media platforms has fundamentally altered the way in which businesses function, and the tourism sector is not an exception to this trend. Platforms for social media have evolved into an integral component of a company's marketing strategy since they enable companies to communicate with a more diverse demographic and present their goods and services to prospective customers. In the travel and tourist industry, women entrepreneurs who run family businesses have found social media to be particularly helpful since it gives them a new platform from which to promote their companies and access a larger customer base. Women's involvement in the tourism business has grown significantly over the past several years, and as a result, their contributions to the industry's expansion have become increasingly important. The use of social media has presented women business owners in family companies with a unique opportunity to overcome these obstacles and promote their companies to a wider audience. Platforms for social media are marketing tools that have a low cost but a high impact, and they may help companies establish their brand image, engage with their customers, and boost their visibility (Buhalis & Law, 2008). It has become the most effective tool for businesses to communicate with clients and create relationships, and social media may be particularly helpful for female entrepreneurs who are involved in family enterprises. Women entrepreneurs can use social media to their advantage to boost bookings, revenue, and customer satisfaction by raising their visibility and reach, enhancing engagement, using cost-effective marketing strategies, and competing with larger companies. They can use social media to reach an exact demographic, interest, and behaviour niche and this will improve the efficiency of their advertising because they can tailor their campaigns specifically to the preferences of their target audience. It can be also used to create meaningful connections with their target audience and nurture them with relevant and interesting content. As a result, a rise in devoted customers who keep coming back and who enthusiastically spread the word about your company. Customers can easily and quickly contact the company with any inquiries, concerns, or suggestions they may have. Social media has the potential to increase brand loyalty and positive word-of-mouth recommendations, as approximately 71% of consumers who have had a positive experience with a brand on social media are likely to recommend it to others (Sprout, 2020). Several studies have been conducted to investigate the function and impact of social media platforms on the travel and hospitality sector, as well as the ways in which these platforms can be utilised to support female business owners in family enterprises. For instance, social media websites like Facebook and Twitter can be utilised to give clients with a more personalised experience and to increase customer engagement (Sigala & Gretzel, 2012). The average revenue increase for businesses owned and operated by women is 25% when they use Facebook to promote their products and services (Facebook, 2019), demonstrating the potential of social media in driving revenue growth. Social media provides businesses with a way to stay connected with customers and adapt to changing customer needs and expectations. By using social media, women business owners can provide better customer service by quickly responding to questions and resolving issues. Women business owners can use the information gleaned from these platforms to assess the value of their advertising campaigns, monitor customer reactions, and make educated business decisions. For female business owners, social media plays an important part in and influences the tourism industry. Women business owners in the tourism industry can improve their online presence, audience reach, customer loyalty, and customer satisfaction by utilising social media. Since more and more tourists are using social media, it's crucial that female business owners recognise the potential of this medium to expand their customer base and boost profits. Women business owners can differentiate themselves in a competitive market and advance their companies by developing innovative social media strategies informed by the latest trends and best practises in the field. The use of social media by female entrepreneurs operating family enterprises in the tourism industry has the potential to be advantageous, but these women also confront a number of obstacles. Women entrepreneurs in family businesses in the tourism industry face gender-related challenges, such as limited access to financial resources, a lack of recognition and support, and limited access to networks and training programmes, (Garcá-Sánchez, et.al.,2017). These challenges can be found in the tourism industry. Because of the difficulties they face, their capacity to use social media to promote their companies and compete effectively in their field may suffer as a result. Women who own family companies in the tourism industry require support in the form of training programmes and access to financial resources to help them use social media to promote their enterprises so that they can overcome the hurdles that they face. A significant amount of the effect of the touristic visual image on user attention was disclosed and native language of the users influenced how well touristic communications were received on Facebook (Shein, 2015) however, electronic word of mouth has on the decision-making process for vacation spots (Jalilvand et al. 2013).

New strategies for communicating via the internet are required as a result of increased globalisation and supply-side concentration. Internet destinations have low barriers for both financial resources and knowledge. The accessibility and cost of the internet have ushered in a new era of corporate environments and practises. Throughout the course of its brief history, the internet has evolved from a static tool for digital publishing (a tool that facilitates communication between one and many) into a fully dynamic platform for collaboration (a tool that helps promote communication between many and many) (a tool to support many-to-many communication). Many destinations rely heavily on tourism as a source of revenue; however, due to financial restraints as well as shifting patterns in the utilisation of various forms of media and technology, these locations will need to adjust their communications strategy as well as the manner in which they advertise themselves in the global market. To survive, you need to adopt new forms of media and technology. Because of modern media, marketing a place may now be done at a lower cost, and existing guests can take part in luring in new ones. The use of social media in marketing efforts for tourism is becoming an increasingly important component of those efforts. Advertising for a resort that exaggerates its advantages is no longer trusted by tourists. Visitors are looking for communication that is not just intellectually stimulating but also emotionally engaging, creatively participatory, and interactive. They want to develop tourist goods and make purchases based on relationships. According to Kiráová and Pavleka (2015), locations may maintain their competitive edge by offering content that is interesting, encouraging visitors to communicate with one another, and making use of social media.

Around 52% of travellers have altered their initial trip plans as a direct result of reading posts, 70% of travellers post to social media while they are on vacation, and 77% of travellers use social media to share their travel experiences. According to these figures, social media is becoming an increasingly important component of the travel and tourism business, particularly in terms of its capacity to motivate and persuade tourists. The increasing usage of social media among young people has both positive and negative implications for all areas of business; nevertheless, the tourist industry is the one that is most significantly impacted by this trend. Businesses in the hospitality, travel, and transportation industries, in addition to tour operators, are currently being subjected to intensive scrutiny within the tourism sector. Given that the product of the tourism industry is a service, marketing this service is of the utmost importance. Even if a particular firm provides an exceptional level of care to its customers, a single unfavourable review of any magnitude can have a devastating impact on the organisation. It is important for companies involved in tourism to make efficient use of social media by establishing official accounts and staffing these accounts with dedicated individuals. The younger generation may submit unfavourable comments along with photographs and videos on their social media accounts; nonetheless, firms should handle the discontent of their visitors and discover the most effective techniques to reduce the risk of adding other persons in this review. This condition is of utmost significance to businesses because of the lightning-fast pace at which social media operates. At this time, the vast majority of significant enterprises in the tourism industry make effective use of social media. Guests are able to rapidly share their criticism because there are many social media networks, including online booking sites. Businesses provide solutions as soon as the process begins to safeguard their brand image. In the event of a negative experience, faster refunds can be ascribed to social media, which fosters a favourable climate for travellers. Because customers generally have great things to say about businesses, using social media as a marketing tool has a significant amount of untapped potential. As a consequence of this, businesses related to tourism should be aware of this new advertising potential and make use of social media as a new communication and promotion tool in order to provide the highest possible level of service to their clients. According to the findings of the poll, organisations in the tourism industry, such as travel agencies, hotels, and transportation businesses, should make efficient use of social media (Varkaris, & Neuhofer, 2017). Hire relevant staff and instruct them to use social media platforms for social networking as a component of your overall marketing strategy. In addition, in order to shield the reputation of the organisation, irrelevant comments made through fake accounts has to be dealt with legally. According to Sahin and Sengun (2015), one of the most effective strategies for competing in the tourism industry is to take advantage of the rapidly advancing technological landscape.

ROLE OF SOCIAL MEDIA IN TOURISM FOR WOMEN ENTREPRENEURS

Women entrepreneurs running family businesses in the tourism industry could stand to benefit from social media, there are still a number of obstacles that need to be overcome. For instance, in order for businesses to utilise social media platforms in an efficient manner, they require access to dependable internet connections and digital infrastructure. In addition, businesses require the knowledge and abilities essential to make effective use of social media, and there is a demand for training and education programmes to raise the level of digital literacy among women entrepreneurs in family businesses (UNWTO, 2020). However, very little research has been done in the field of tourism management on the subject of how entrepreneurship and crises are related to one another, particularly from the point of view of gender (Castro, & Zermeo, 2020). The study of the relationship that exists between entrepreneurship and crises is referred to as "entrepreneurship and crises," and it is a relatively new subfield of research that falls under the umbrella of risk, crisis, and catastrophe studies. The findings of the study indicate that crises have significant detrimental effects on the level of entrepreneurial activity at the destination level (Ritchie, & Jiang, 2021). However, adverse conditions can serve as an incentive for business owners in the tourism industry to make investments in their companies, which ultimately results in those companies becoming more resilient (Dahles, & Susilowati, 2015). Crises can even give rise to a phenomenon known as "disaster entrepreneurship" in certain regions. This phenomenon describes the mobilisation of resources by affected communities and businesses with the goal of mitigating the negative effects of catastrophic catastrophes. The concept of "disaster entrepreneurship" refers to business ventures that are undertaken as a result of catastrophic events (Badoc-Gonzales et al., 2022). In the midst of a prolonged period in which the economy has been in a state of decline, the COVID-19 conference has been the driving force behind an abundance of research on the topic of female entrepreneurship. The research has looked into the detrimental effects that the pandemic has had on businesses that are owned and operated by women

(Grandy et al., 2020), as well as the educational experiences and coping mechanisms that women business owners have had (Afshan et al., 2021). (Hennekam & Shymko, 2020). However, there have been no studies that have specifically focused on female entrepreneurs in the tourism industry (Dayour and Adam 2022). This represents a notable lack of research when taking into consideration the negative effects that COVID-19 has had on the sector. The significance of social media platforms for owners of businesses in the tourism industry has steadily increased over the course of the past several years. Because of the proliferation of social media platforms such as Facebook, Instagram, and Twitter, it is now much simpler for female entrepreneurs in the travel and tourism industry to market their businesses and connect with a more diverse audience. In this article, we will discuss the significance of social media in the hospitality industry for female business owners, including the benefits and drawbacks of using these platforms. In addition to this, it will provide examples of successful female entrepreneurs working in the tourism industry who have made use of social media in order to grow their businesses.

Advantages of Social Media for Women Entrepreneurs in Tourism

In 2021, a poll conducted by Social Media Examiner found that 97% of marketers make advantage of social media to communicate with their target clients. For female business owners in the tourism industry, one of the most significant advantages of utilising social media is the increased audience size that can be reached. As a result of the millions of users that social media platforms have from all over the world, it is now much simpler for female business owners to market their companies to an international audience. This is evidence that social media has developed into a tool that cannot be ignored by businesses if they wish to communicate with their ideal customers. One additional benefit of using social media for women company owners in the tourism industry is that it gives them the opportunity to showcase their wares and offerings. Instagram and Pinterest are examples of visual social media networks that make it possible for companies to showcase their wares by posting high-quality images and videos to their own platforms. This might be of assistance to female-owned enterprises in extending their brand recognition and acquiring new customers. Also, women entrepreneurs in the tourist industry are able to create client relationships through the use of social media. Businesses now have a venue through which they can communicate with their customers, reply to their inquiries, and provide them with the information they require thanks to the platforms provided by social media. This might be of assistance to female entrepreneurs in the process of developing long-lasting relationships with their customers.

Social Media Obstacles for Women Businesses in Tourism

Women have distinct cultural, economic, and social hurdles while striving to engage in entrepreneurial activities and economic endeavours, as stated by Ghouse, Durrah, and McElwee (2021). These challenges can be broken down into three categories: These challenges are made even more difficult to overcome as a direct result of the cultural and economical characteristics of rural communities. Although social media affords women entrepreneurs in the tourism industry with a number of opportunities for success, it also poses a number of important challenges. One of the most significant challenges is the dynamic nature of social media, which is always changing. Because social media platforms continually update their algorithms and make other changes to their features, it can be difficult for businesses to keep up with them. Women business owners in the tourism industry really need to stay current on the latest social media trends and advances if they want to maintain their competitive edge. Another difficulty is the presence of rivalry. Given the fact that millions of companies already utilise social media to sell their goods and services, female business owners in the tourism sector need to find ways to set themselves apart from the competition. The ability to develop content that is intriguing and resonates with their audience requires inventiveness as well as an in-depth understanding of the market they are trying to reach.

Successful Examples of Women Entrepreneurs Using Social Media in Tourism

Many female entrepreneurs in the tourism industry have been successful in growing their businesses with the use of social media, despite the challenges they have faced. One such person is Lauren Bath, an Australian travel photographer who has amassed more than 500,000 followers on Instagram (Johanson, 2017). Instagram has been used as a promotional platform for Bath's photography as well as her travel-related goods and services. In addition to this, she has worked together with tourism boards and travel firms to promote the latter's wares and services on various social media platforms. Another illustration of this can be seen in Christina Tosi, who established the dessert and bakery firm Milk Bar in New York City. Tosi has been successful in selling her items and drawing attention to the unique personality of her company thanks to her use of social media. Also, she has made use of social media in order to communicate with and cultivate relationships with her clients (Kotler, 2020). Several female business owners who run family businesses have been successful in growing their companies with the use of social media, in spite of the challenges they face. One such person is Nisha Katona, who established the British restaurant chain Mowgli Street Food. She is a good example. Katona has been successful in marketing her restaurants and drawing attention to the distinctive aspects of her company by utilising social media. In addition, she has made use of social media in order to communicate with and cultivate relationships with her clients (Cohen, 2018). Another illustration of this is the American entrepreneur Jeni Britton Bauer, who established the popular ice cream brand Jeni's Splendid Ice Creams. Britton Bauer has been successful in selling her items and drawing attention to the unique personality of her company because to her use of social media. In addition, she has made use of social media in order to communicate with and cultivate relationships with her clients (Hepola, 2019).

IMPACT OF SOCIAL MEDIA IN FAMILY BUSINESSES

The Internet, which makes it possible for humans to instantly collaborate and share information with one another, has been the greatest important step forward in technical development during the past three decades. The way in which companies communicate with their customers and partners has been profoundly revolutionised by advances in technology. The collaboration tools made available by social media platforms to businesses significantly alter the traditional dynamic of working together. The development of local economies depends heavily on the success of small and medium-sized enterprises (also known as SMEs), which also play a significant part in the expansion of economies. A small or medium-sized enterprise (SME) is any company in the European Union (EU) that employs less than 250 people. The majority of SME owners are families. They make up two-thirds of the total private sector and 99 percent of all businesses in the EU. Moreover, they are responsible for 85 percent of all newly created jobs. To increase the number of small and medium-sized enterprises (SMEs) that employ environmentally responsible business practises and digital technology, the EU plan provides funding for initiatives such as the Small Business Act, the Start-up and Scale-up Initiative, and the Competitiveness for Small and Medium Businesses (COSME) Programme. The organisational structure of family-owned businesses and small to medium-sized businesses (SMEs) often makes them more adaptable to changes in technology and the requirements of their customers, and this structure also makes it easier for these businesses to make decisions quickly. Their uniqueness may be traced back to the various effects that the family has had on ownership, strategic management, governance, as well as the direction and control of day-to-day operations. But, they must contend with a number of challenges, including inefficient marketing strategies and a diminished brand value. Family businesses are looking to make marketing investments in order to compete and thrive in today's increasingly competitive and dynamic marketplace. This is because markets are becoming increasingly competitive. The development of new methods of administration and marketing, in addition to original technological advancements, absolutely need innovative thinking (Obermayer, et al., 2022). There have been substantial upheavals in the hospitality industry as a direct result of the use of social media, which has also had a huge effect on family companies. According to Kim, Lee, and Yoo (2015), social media has had an effect on the travel industry because it has changed the ways in which travellers obtain information, discuss their travel experiences, and make decisions regarding their vacations. This in turn has changed the methods by which the travel industry operates. Because of the explosion of social media platforms, women entrepreneurs who run family businesses now have additional opportunity to communicate with a bigger consumer base, build their brand image, and increase their visibility. In addition, with the help of social media, businesses may have a greater grasp of the demands, preferences, and behaviours of their customer base. This gives them the opportunity to create products and services of a higher quality, as well as supply such products and services to customers (Buhalis & Law, 2008). According to the conclusions of a research project that was carried out by Buhalis and Law (2008), social media platforms have made available to small enterprises a marketing tool that is not only inexpensive but also has a significant influence. This technology has the potential to assist these firms in increasing their visibility, engaging with their respective clients, and building their brand images. The usage of social media is currently the most efficient technique for businesses to communicate with customers and create relationships; this may be of particular help to female entrepreneurs who run family businesses. In the field of travel and tourism, social media platforms can be utilised to assist businesses in reaching a more extensive audience while simultaneously advertising certain locations, travel packages, and tourist experiences. One of the most significant advantages that social media can provide to family-run businesses in the tourism industry is the opportunity to reach a more extensive customer base with their products and services. This is especially helpful for businesses that focus on providing services to tourists. Social media platforms are a potent marketing tool that, according to Wang, Xiang, and Fesenmaier (2014), may be utilised to boost brand awareness as well as customer engagement and brand loyalty. Using the many platforms for social media can help achieve the aforementioned goals. The use of social media allows businesses that are owned by families the chance to interact with their customers in real time, respond to their questions and concerns, and deliver a more individualised shopping experience to their clients. Also, the usage of social media might be beneficial to family companies since it would assist in the development of deeper client links and a stronger feeling of community. Both of these factors have the potential to result in improved customer loyalty and repeat business (Buhalis & Law, 2008). The usage of social media makes it much simpler for family-run enterprises operating in the tourism industry to keep an eye on their competitors. The usage of social media platforms provides businesses with a plethora of information regarding the marketing techniques of their competitors, as well as client feedback and opportunities for customer participation in the firm. Family businesses have the opportunity to make use of this information in order to enhance their products and services, develop original marketing tactics, and maintain a competitive advantage over their contemporaries (Kim, Lee, & Yoo, 2015). In addition, family businesses in the tourist industry can utilise social media to communicate with their clients and gain feedback and insights from such interactions. By the use of a variety of social media channels, customers of family-owned businesses are given the opportunity to share their thoughts, experiences, and recommendations regarding the products that are sold by the companies. Family businesses have the ability to profit from this information by enhancing their products and services, determining areas that are lacking in development, and implementing novel marketing methods (Buhalis & Law, 2008). Also, by utilising social media, family-owned firms operating in the tourism industry may be able to lower the amount of money they spend on marketing. By utilising social media platforms as a low-cost marketing strategy, family businesses have the ability to reach a large audience without having to spend a significant amount of money on conventional types of advertising. In other words, they can reach a larger audience without having to spend money on traditional advertising (Buhalis & Law, 2008). Social media platforms provide companies with the ability to contact a huge number of people by publishing information, participating in conversation with customers, and building their brand image. This is made possible by the fact that social media platforms are free to use (Wang, Xiang, & Fesenmaier, 2014). But, tourism firms that are owned and controlled by families often find that social media presents a number of challenges. Family businesses, in order to continue being successful and relevant in their respective fields, are required to adapt to the ever-changing nature of the various social media platforms. This is necessary in order to preserve their efficacy (Kim, Lee, & Yoo, 2015). In addition, maintaining a presence on social media may be a time-consuming endeavour. As a consequence, family businesses need to allocate resources and people in order to properly manage their social media presence.

CONCLUSION

The tourism industry has been significantly influenced by social media, which has also provided women who run family businesses with a rare opportunity to advertise their companies to a wider audience. Social media has had a significant impact on the tourism sector. Women entrepreneurs in family businesses can now develop their brand image, communicate with customers, and increase their visibility all with the help of social media platforms, which have become a marketing tool that is both low-cost and high-impact. This is especially important as we navigate the pandemic's influence on the tourism sector. It has been determined through extensive study of the topic that social media has become an indispensable resource for women business owners in the tourism industry. Women business owners have found new avenues for expanding their customer base, generating more revenue, and promoting their companies thanks to the proliferation of social media platforms. The research centred on female entrepreneurs who were operating modest-sized tourism enterprises in various countries around the world. These businesswomen faced numerous obstacles, including a dearth of available resources, a narrow social circle, and cultural biases. However, by leveraging social media to spread the word about their businesses, they were able to triumph over the obstacles they faced. According to the research, the most popular social media sites for female business owners are Facebook, Instagram, and Twitter. They were able to promote their businesses, make connections with customers, and build a positive brand image at a low cost thanks to these channels. Also, by listening to their customers on social media, women business owners learned how to better serve their clients. The development of online networks has been one of social media's most pervasive effects on female business owners in the tourism sector. Women business owners were able to network with one another, exchange tips, pool resources, and work together on various endeavours. This online interaction reduced cultural barriers and raised the profile of female business owners in the tourism sector. Women business owners who embraced social media were found to have a higher chance of business success than those who did not. This was because they were able to expand their customer base, solidify their brand's identity, and foster personal connections with followers thanks to social media. Women business owners also benefited from the increased visibility and exposure they received thanks to the rise of social media. Finally, it's important to note the importance of social media for women business owners in the travel industry. Women business owners can now promote their companies, build relationships with customers, and establish a trustworthy reputation on social media at a low cost. It has facilitated the dissolution of cultural barriers and the development of online communities, both of which have served to raise women business owners' profiles in the travel sector. This has significant ramifications for the position of women business owners in the tourism sector, as social media users are more likely to be successful entrepreneurs. Women business owners in the tourism industry play a critical role in fostering growth, economic stability, and social well-being, so it's important to recognise and support their efforts.

REFERENCES

Afshan, G., Shahid, S., & Tunio, M. N. (2021). Learning experiences of women entrepreneurs amidst COVID-19. *International Journal of Gender and Entrepreneurship*, *13*(2), 162–186. doi:10.1108/IJGE-09-2020-0153

Badoc-Gonzales, B. P., Mandigma, M. B. S., & Tan, J. J. (2022). SME resilience as a catalyst for tourism destinations: A literature review. *Journal of Global Entrepreneurship Research*, *12*(1), 1–22. doi:10.100740497-022-00309-1

Buhalis, D., & Law, R. (2008). Progress in information technology and tourism management: 20 years on and 10 years after the Internet—The state of eTourism research. *Tourism Management*, *29*(4), 609–623. doi:10.1016/j.tourman.2008.01.005

Castro, M. P., & Zermeño, M. G. G. (2020). Being an entrepreneur post-COVID-19–resilience in times of crisis: A systematic literature review. *Journal of Entrepreneurship in Emerging Economies*, *13*(4), 721–746. doi:10.1108/JEEE-07-2020-0246

Castro, O., & Zermeño, M. G. (2020). Gender perspective in crisis and disaster management: Entrepreneurship and tourism. *Journal of Business Research*, *117*, 454–462.

Chou, C. M., Chen, Y. L., & Chen, C. J. (2021). Using Social Media to Overcome the COVID-19 Pandemic: The Influence of Social Media Use on Hotel Performance. *Journal of Travel Research*, *60*(3), 490–504.

Chung, N., & Koo, C. (2015). The use of social media in travel information search. *Telematics and Informatics*, *32*(2), 215–229. doi:10.1016/j.tele.2014.08.005

Cohen, E. (2018). *How Nisha Katona's Mowgli Street Food became a social media sensation*. Evening Standard. Retrieved from https://www.standard.co.uk/insider/fooddrink/how-nisha-katona-s-mowgli-street-food-became-a-social-media-sensation-a3824181.html

Dahles, H., & Susilowati, T. P. (2015). Business resilience in times of growth and crisis. *Annals of Tourism Research*, *51*, 34–50. doi:10.1016/j.annals.2015.01.002

Dahles, H., & Susilowati, T. P. (2015). The role of tourism in coping with poverty in rural Indonesia. *Journal of Sustainable Tourism*, *23*(4), 554–573.

Dayour, F., & Adam, A. A. (2022). Impact of Covid-19 on women entrepreneurs in tourism: The case of women tour operators in Ghana. *African Journal of Hospitality, Tourism and Leisure*, *11*(2), 1–22.

Dayour, F., & Adam, I. (2022). Entrepreneurial motivations among COVID-19 induced redundant employees in the hospitality and tourism industry. *Journal of Human Resources in Hospitality & Tourism*, *21*(1), 130–155. doi:10.1080/15332845.2022.2015239

Expedia. (2019). *2019 Expedia Group travel trends report*. https://www.expediagroup.com/travel-research/travel-trends-report-2019

Facebook. (2019). *Facebook IQ: Driving revenue for travel brands*. https://www.facebook.com/iq/insights-to-go/driving-revenue-for-travel-brands

Fernández, Z., Nieto, M. J., & Ramos, A. (2013). Family involvement and internationalization in family firms: A review of the literature. *Journal of Family Business Strategy, 4*(3), 157-166.

Fernández, Z., Nieto, M. J., & Ramos, C. (2013). Internationalization and innovation: An analysis of complementarity and substitutability. *International Business Review*, *22*(2), 371–386.

García-Sánchez, I. M., Cuadrado-Roura, J. R., & Martínez-Fernández, M. T. (2017). Gender and family business in tourism: An exploratory study of the Spanish case. *Journal of Family Business Management*, *7*(1), 50–69.

García-Sánchez, I. M., Cuadrado-Roura, J. R., & Martínez-Fernández, M. T. (2017). Determinants of women entrepreneurship in Spain: An institutional approach. *Journal of Business Research*, *76*, 1–10.

Ghouse, S. M., Durrah, O., & McElwee, G. (2021). Rural women entrepreneurs in Oman: Problems and opportunities. *International Journal of Entrepreneurial Behaviour & Research*, *27*(7), 1674–1695. doi:10.1108/IJEBR-03-2021-0209

Grandy, G., Cukier, W., & Gagnon, S. (2020). (In) visibility in the margins: COVID-19, women entrepreneurs and the need for inclusive recovery. *Gender in Management*, *35*(7/8), 667–675. doi:10.1108/GM-07-2020-0207

Grandy, G., Montagno, R. V., Campbell, J. E., & Nedic, O. (2020). Women entrepreneurs and the Covid-19 pandemic: A systematic literature review and research agenda. *Journal of Business Venturing Insights*, *14*, e00184.

Hennekam, S., & Shymko, Y. (2020). Coping strategies of female entrepreneurs in times of crisis: The impact of Covid-19. *International Journal of Entrepreneurship and Small Business*, *40*(2), 187–202.

Hennekam, S., & Shymko, Y. (2020). Coping with the COVID-19 crisis: Force majeure and gender performativity. *Gender, Work and Organization*, *27*(5), 788–803. doi:10.1111/gwao.12479 PMID:32837010

Hepola, S. (2019). *Ice cream queen Jeni Britton Bauer on building her brand through social media*. Inc. Retrieved from https://www.inc.com/sonya-hepola/jeni-britton-bauer-social-media-marketing-brand-building.html

Hootsuite. (2019). *The social media trends shaping the travel industry*. https://www.hootsuite.com/pages/the-social-media-trends-shaping-the-travel-industry

Jalilvand, M. R., Ebrahimi, A., & Samiei, N. (2013). Electronic word of mouth effects on tourists' attitudes toward Islamic destinations and travel intention: An empirical study in Iran. *Procedia: Social and Behavioral Sciences*, *81*, 484–489. doi:10.1016/j.sbspro.2013.06.465

Johanson, M. (2017). *How this woman turned her Instagram account into a $4,000-a-week job*. Business Insider. Retrieved from https://www.businessinsider.com/how-this-woman-turned-her-instagram-account-into-a-4000-a-week-job-2017-1

Kim, H., Lee, J. S., & Yoo, J. J. (2015). The impact of social media on consumers' acculturation and purchase intentions in tourism-related industries. *Journal of Travel & Tourism Marketing*, *32*(5), 564–586.

Kiráľová, A., & Pavlíčeka, A. (2015). Development of social media strategies in tourism destination. *Procedia: Social and Behavioral Sciences*, *175*, 358–366. doi:10.1016/j.sbspro.2015.01.1211

Kotler, N. (2020). *How Christina Tosi is keeping Milk Bar sweet during the pandemic*. Forbes. Retrieved from https://www.forbes.com/sites/nataliekotler/2020/05/12/how-christina-tosi-is-keeping-milk-bar-sweet-during-the-pandemic/?sh=66c5ccbc58dc

Lee, H., & Hyun, S. S. (2015). The impact of social media on tourist behavior. *Journal of Travel & Tourism Marketing*, *32*(5), 527–540.

Li, C., Guo, S., Wang, C., & Zhang, J. (2019). Veni, vidi, vici: The impact of social media on virtual acculturation in tourism context. *Technological Forecasting and Social Change*, *145*, 513–522. doi:10.1016/j.techfore.2019.01.013

Naldi, L., Rasmussen, E., & Colombo, M. G. (2018). Social media in family business: What do we know and where do we go? *Journal of Family Business Management*, *8*(2), 126–147.

Obermayer, N., Kővári, E., Leinonen, J., Bak, G., & Valeri, M. (2022). How social media practices shape family business performance: The wine industry case study. *European Management Journal*, *40*(3), 360–371. doi:10.1016/j.emj.2021.08.003

Ritchie, B. W., & Jiang, Y. (2021). Research on tourism entrepreneurship in the aftermath of the COVID-19 pandemic: Implications and directions. *Journal of Sustainable Tourism*, *29*(6), 705–720.

Ritchie, B. W., & Jiang, Y. (2021). Risk, crisis and disaster management in hospitality and tourism: A comparative review. *International Journal of Contemporary Hospitality Management*, *33*(10), 3465–3493. doi:10.1108/IJCHM-12-2020-1480

Sahin, G. G., & Sengün, G. (2015). The effects of social media on tourism marketing: A study among university students. *Management and Administrative Sciences Review*, *4*(5), 772–786.

Shein, A. (2015). *Communication language in Social Communication and its effect on the Touristic Attration-Facebook as a Model* [Unpublished Master Thesis]. Mohamed Khaidar University, Bascara, Algeria.

Sigala, M., & Gretzel, U. (2012). *Social media in travel, tourism and hospitality: Theory, practice and cases*. Ashgate Publishing, Ltd.

Social Media Examiner. (2021). *Social media marketing industry report*. Retrieved from https://www.socialmediaexaminer.com/social-media-marketing-industry-report/

Sprout Social. (2020). *Sprout Social index: Edition XVII: The social media trends shaping the future*. https://sproutsocial.com/insights/data/sprout-social-index/social-media-trends/

TripAdvisor. (2018). *The modern traveller journey*. https://www.tripadvisor.com/TripAdvisorInsights/w671

United Nations World Tourism Organization. (2020). UNWTO World Tourism Barometer.: Vol. 18. *June 2020*. World Tourism Organization.

UNWTO. (2020). *Global Report on Women in Tourism 2020*. World Tourism Organization.

Varkaris, E., & Neuhofer, B. (2017). The influence of social media on the consumers' hotel decision journey. *Journal of Hospitality and Tourism Technology*, *8*(1), 101–118. doi:10.1108/JHTT-09-2016-0058

World Tourism Organization. (2019). *UNWTO Tourism Highlights* (2019 Edition). World Tourism Organization.

World Tourism Organization (UNWTO). (2019). *Global Report on Women in Tourism 2019*. UNWTO.

Yoo, W., Choi, D. H., & Park, K. (2016). The effects of SNS communication: How expressing and receiving information predict MERS-preventive behavioral intentions in South Korea. *Computers in Human Behavior*, *62*, 34–43. doi:10.1016/j.chb.2016.03.058 PMID:32288174

Chapter 10

Tourism as an Empowerment Tool in Sustainable Development:

Evidence From the Hospitality Industry of China

Kamran Jamshed
Bahria University, Pakistan

Samrah Jamshaid
Northeast Normal University, Changchun, China

Muhammad Imran Afzal
Northeast Normal University, Changchun, China

Noor E. Eman
University of Agriculture Faisalabad, Pakistan

Basheer M. Al-Ghazali
King Fahd University of Petroleum and Minerals, Saudi Arabia

Azzam Inayatullah Shamsi
Bahria University, Pakistan

ABSTRACT

Sustainable development is the practise of ensuring that current needs are met without jeopardising future generations' capacity to do the same. The goal of sustainable development requires people all over the world to work together and think carefully about how their actions will affect the planet and the next generation. One sector with the potential to advance sustainable development is the tourism industry. China's hospitality industry has grown rapidly, which could lead to sustainable development and tourism and empowers sustainable development in China's hospitality industry, according to this research chapter. Tourism can help achieve sustainability and local autonomy, two goals that have garnered attention recently. China must encourage responsible tourism practises that are kind to the local environment and culture while also boosting the economy of the host community. This research emphasises tourism's potential to empower China's sustainable development while acknowledging the need to continue efforts to distribute tourism's benefits fairly and sustainably.

DOI: 10.4018/978-1-6684-8417-3.ch010

INTRODUCTION

As a globally significant industry, tourism has the potential to improve the lives of countless individuals and groups. It has the potential to generate economic activity, sustain cultural practises, and even generate new jobs. It can also serve to bridge gaps between cultural groups and impart lessons in tolerance and appreciation. Travel has the power to unite people and spur positive changes that will be felt for generations to come. In today's globalised economy, tourism has become a vital resource that can be used to improve the lives of individuals and communities in need all over the world. It has the potential to generate economic growth, raise living standards, protect cultural traditions, and unite people of different backgrounds. Supporting tourism helps leave a positive legacy for future generations. Sustainable development is a goal that can be advanced through tourism. It has the potential to preserve the natural and cultural resources of a destination while also promoting economic growth, creating jobs, and improving the quality of life for local communities. The creation of eco-friendly vacation spots is one way that China is putting the tourism industry to work for long-term progress. The tourism industry is one of the largest and most rapidly expanding industries in the world, making a considerable contribution to overall economic expansion, employment creation, and earnings in foreign currencies (UNWTO, 2018). Yet, the rapid rise of tourism may also have unintended consequences for the natural environment, the communities in the surrounding area, and the cultural legacy of the area. As a result of this, the idea of sustainable tourism has arisen as a solution to these issues. Sustainable tourism seeks to strike a balance between economic, social, and environmental considerations in order to guarantee that tourism can contribute to the process of sustainable development (UNWTO, 2015). China has emerged as a leading force in the travel and tourism industry around the world thanks to its massive population and extensive cultural heritage. In 2019, China's domestic tourism industry brought in 5.7 trillion yuan, which is equivalent to nearly 887 billion dollars in U.S. currency, while the country welcomed over 150 million visitors from other countries (National Bureau of Statistics of China, 2020). The growth of China's tourism business has made it possible for local communities to participate in tourism activities and benefit from the expansion of the tourism industry in a number of different ways. There is ongoing debate on the degree to which tourism has contributed to increased autonomy and environmentally responsible growth in China. The Chinese hospitality business is the focal point of this study as it investigates the role that tourism might play as an instrument of empowerment in the context of sustainable development. This essay will explicitly analyse how local communities in China have been empowered by the hotel sector in China through the creation of jobs, the generation of income, and activities geared towards community development. In addition, a critical analysis of the challenges and limitations of tourism as a tool for empowerment in China will be provided in this article. The purpose of this analysis is to establish techniques that will promote tourist development that is more inclusive and sustainable. Significant economic benefits, such as higher profits in foreign currency, increased employment opportunities, and redistribution of income, can be generated for host communities through the practise of sustainable tourism (UNWTO, 2015). Practices that are more environmentally friendly in the tourism industry can result in higher tourist spending, higher occupancy rates, and improved tourism profits, all of which can help to economic growth and development (UNEP, 2010). These locations are chosen because of their commitment to sustainable tourism and their dedication to protecting the local community and environment. To lessen its impact on the environment, the Qinghai Lake Scenic Area in western China has, among other things, restricted the number of tourists who can visit at once and encouraged the use of alternative, less carbon-intensive modes of transportation. The growth of cultural tourism is another example of how China's tourism industry is contributing to the country's sustainable development. China has a long history and a wide variety of cultural practises that can be preserved and popularised through tourism. An example of this is the ancient city of Pingyao in Shanxi province, which is now a popular cultural tourism destination thanks to its UNESCO World Heritage Site status. The city government has taken steps to preserve the city's historic buildings and encourage the preservation of local arts and culture. It promotes economic development while protecting a location's cultural and environmental treasures. Despite the difficulties presented by pollution caused by tourists, the tourism industry continues to play a vital role in China's economy and serve as a powerful tool in the fight for sustainable development. China is working to maximise tourism's positive effects while minimising its negative ones on the environment and local communities by enacting policies to encourage responsible tourism practises and developing sustainable tourism destinations. The sustainability of economic growth is dependent on the sustainability of social growth, which in turn is dependent on the sustainability of tourism industry growth. Because of the stability in growth of the country's tourism industry, the country continues to receive a large number of visitors from other countries, most of whom are business travellers. These visitors are given the opportunity to become familiar with the businesses that are operating within the country, as well as the characteristics of those businesses, such as the opportunities and risks they face, and to foster social relationships between those businesses. These social relationships come to an end when commercial transactions or contracts are completed between the travellers and the representatives of the various companies. In addition, when tourists engage in conversation with residents, they have the opportunity to share their perspectives and ways of life, which ultimately enhances the locals' standard of living (Chen, 2022). According to the United Nations World Tourism Organization (UNWTO), environmentally responsible tourism has the potential to contribute to the alleviation of poverty by generating new jobs and other sources of income, particularly in rural and remote areas where there are fewer opportunities for other types of economic activity (UNWTO, 2018). Also, the implementation of sustainable tourism practises can assist in the diversification of local economies, thereby lowering the region's reliance on a single economic sector and bolstering the economy's overall resilience (UNWTO, 2015). The promotion of cultural exchange and the maintenance of local traditions and customs are two of the considerable social benefits that can be generated by environmentally responsible tourist practises (Hall & Lew, 2009). Cultural heritage, such as traditional arts, crafts, and architecture, can be preserved and promoted through the use of sustainable tourism techniques, which in turn can assist to strengthen the cultural and social identity of the communities that are visited. In addition, sustainable tourism practises can contribute to the support of community development, which in turn can serve to empower local communities and promote social inclusion (UNWTO, 2018). This can be accomplished by the participation of local communities in the planning, development, and management of tourism, which can give such communities with chances for self-determination and control over tourist-related activities. The preservation of natural resources, the lessening of carbon emissions, and the safeguarding of ecosystems and biodiversity are only some of the environmental advantages that can result from the implementation of sustainable tourism practises (UNWTO, 2018). Tourism can have a number of detrimental effects on the environment, including the overuse of resources, increased pollution, and the destruction of natural habitats. Nevertheless, environmentally responsible tourist practises can assist to mitigate some of these effects (UNWTO, 2015). Practices that promote sustainable tourism can also encourage the use of renewable energy, cut down on trash output, and promote sustainable mobility, all of which can contribute to the reduction and adaptation of the effects of climate change (UNEP, 2010). Moreover, sustainable tourism practises can help to assist the conservation and restoration of ecosystems and biodiversity, which can increase the natural beauty and ecological value of tourism sites (UNWTO, 2018).

BACKGROUND

The tourism industry is the most important and rapidly expanding part of the world economy. It is recognised as a driver of economic progress in both industrialised nations and developing nations alike. The tourism industry helps alleviate poverty while also producing valuable foreign currency, creating jobs, and contributing to the development of essential infrastructure. Despite this, irresponsible tourism growth has catastrophic repercussions for the economy, society, and ecosystem of areas that attract tourists. When it comes to attaining sustainable development and reducing the negative consequences of tourism, sustainable tourism has emerged as the most comprehensive and effective technique. Therefore, the United Nations Environment Program (UNEP) and the World Tourism Organization (WTO) encourage all nations to develop policies and actions for the development and management of sustainable tourism that fully incorporate the principles of sustainability. These policies and actions can be found at: https://www.unep.org/wto/. "tourism development that meets the demands of present tourists and host regions while safeguarding and improving future prospects," according to the World Tourism Organization's (WTO) definition of sustainable tourism. It is anticipated that this will result in the management of all resources in a manner that achieves economic, social, and aesthetic goals while simultaneously protecting cultural integrity, essential ecological processes, biological diversity, and life support systems. It underlines the balance between environmental, social, and economic elements of tourism, the importance to implement sustainability principles in all areas of tourism, and global goals such as the eradication of poverty. Among other things, it focuses on: (Venugopalan, & Kumar, 2017). Global tourism statistics reveal the industry's significance. There were a record 1.5 billion international visitors in 2019, up 4% from the previous year, according to the World Tourism Organization. This increase in tourism has helped many nations expand their economies and provide more employment opportunities. As a matter of fact, the tourism industry is responsible for 10.4% of the global GDP and 319 million jobs. It's important to remember that tourism's negative effects on the environment and local communities, especially with regards to pollution, should not be underestimated. In 2013, tourism in China contributed to 8 percent of the country's total carbon emissions, according to a study. Given that China is one of the most popular tourist destinations in the world, with over 150 million visitors expected in 2018, this is a major issue. Despite these obstacles, tourism in China continues to play a crucial role in the country's long-term economic growth. Because of tourism's potential to boost the economy, the government has invested heavily in it. Over 28 million people in China had jobs thanks to the tourism industry in 2019. For its part, the government has enacted policies to promote eco-tourism and the use of renewable energy sources in the hospitality industry as a whole. On the one hand, the high natural tourist potential and the local residents' awareness of the significance of natural conservation for sustainable development can help to explain why the environmental component of sustainable development is the most important. On the other hand, the desire to reduce the negative effects of tourism development on the area can help to explain why the environmental component of sustainable development is the most important. The development of tourism has a positive effect on local residents' quality of life because it contributes to overall economic expansion, which in turn results in the creation of additional job openings. Rural tourism is seen as helpful for a variety of reasons, including the strengthening of the overall infrastructure and the diversification of leisure opportunities (Muresan, et al., 2016).

In addition to green tourism services, there are new opportunities for the expansion of sustainable tourism in terms of the products and services that are offered. These opportunities centre on the development of new types of tourism and distinct customer groups, such as those who are elderly or disabled. Because of globalisation, tourism is now available to a larger number of people, particularly those of the middle class; as a result, the tourism industry has grown to become one of the largest in the world. Although travel has become easier for the general public and tourism has become more accessible, certain socioeconomic groups, such as the disabled and the elderly, who form a minority with limited economic clout have typically been overlooked and ignored by the tourism industry. This is despite the fact that travel has become more accessible. Yet as a result of the declining birthrate across the globe and the growing population of elderly people, this scenario has taken on a new twist, and senior citizens have emerged as a sizable and influential demographic that utilises tourism services to a greater extent than any other demographic. Even though tourist services have effects on host regions, such as impacts on the local environment and host communities, as well as the provision of employment opportunities, the tourism industry has important consequences for the long-term sustainable development of tourism destinations. The tourism industry ought to encourage environmentally responsible patterns of consumption in order to contribute to the sustainable expansion of tourist destinations. The practise of sustainable consumption is quickly becoming standard in the tourism industry. Visitors are very interested in purchasing sustainable tourism service packages that are being offered by a variety of tourist organisations since they are currently looking for sustainable tourism services and love engaging in responsible consumption behaviours (Streimikiene, et al., 2021).

ROLE OF TOURISM IN SUSTAINABLE DEVELOPMENT

For sustainable development, it must simultaneously implement 3 objectives: Development with economic efficiency; Harmonious development of social components; improvement in living standards and living standards across demographic strata; Development with a focus on environmental sustainability. Enhance environmental environment, ensure solid long-term development for todays and future generations. It is vital to adopt the principles that assure sustainable tourist development in order to guarantee sustainable development. These principles are as follows: exploiting and using resources in an acceptable manner; preventing the excessive use of resources; and minimising waste. The expansion of tourism must go hand in hand with efforts to preserve biological diversity, and the industry must adapt its practises to the existing socioeconomic climate. benefit sharing with the local community, encouraging the participation of local communities in tourism activities, regularly exchanging and consulting with local communities and related subjects, and focusing on training and raising awareness about environmental resources are all important aspects of sustainable tourism. (Tien, et al., 2019). The significance of the tourism industry in rural development is due to the fact that by making efficient use of natural and human resources, not only does it contribute to economic growth, but it also helps to promote the agricultural sector and the production of local handicrafts, as well as to improve environmental conditions and preserve the local cultural heritage and local customs in the villages. All of these things help to ensure that the local cultural heritage and local customs are maintained. One of the most important strategies for developing rural areas is to pay attention to this type of tourism in order to remove rural environments from isolation and exclusion and to create balanced development, since by influencing on three dimensions of economic, social, and environmental development, it is possible to create a balanced development. This is especially important in villages where agricultural activities are less likely to benefit from optimal agriculture or desirable income due to climatic or other conditions. This is especially important in villages where agricultural activities are less likely to benefit from optimal agriculture or desirable income. In point of fact, the expansion of rural tourism can be seen as one of the potential responses to problems and concerns that are prevalent in rural areas. In recent years, there has been a considerable increase in the number of tourists visiting the Doroodzan region located inside the Marvdasht Township of the Fars Province. Because of its natural, cultural, and historical features, this region is visited by a significant number of people on a yearly basis. The region has the ability to draw tourists during all four seasons of the year. As a result, it has the potential to play a significant role in the economy of the region and in the empowerment of the local population (Nooripoor, et al., 2021). The cultural and natural elements of an area are extremely important to the tourism sector. It is imperative that the environment, the residents of the town, and the tourists be protected. Even though the risk is extremely high, there is still a chance. Tourism is now a prerequisite for political and economic development. It is quickly emerging as one of the key development tools at all levels of government, and at the same time, international tourism is the most lucrative export industry in the world. Tourism that focuses on culture has the potential to improve the maintenance and transmission of historical and cultural traditions. This, in turn, can frequently contribute to the maintenance of local history, the renaissance of indigenous cultures, and the revitalization of cultural arts and crafts. As a result, culture and legacy act as a motor for the growth of tourism that is truly sustainable, and we may anticipate that tourism and culture will contribute to the maintenance of the universe's delicate balance (Kumar, 2017). Since the end of World War II, there has been a significant increase in the amount of people travelling all over the world. The travel and tourism business is one of the most important parts of the world's economy. The global gross domestic product was increased by 8.9 trillion United States dollars, or 10.3%, as a result of the direct, indirect, and induced economic consequences of international travel (GDP). Almost 330 million people have occupations thanks to the globe's travel and tourism business, which accounts for one in every 10 jobs in the world. Tourism makes a large contribution to the economy, which in turn makes a significant contribution to sustainable development. Tourism has the potential to play a unique role in the generation and distribution of revenue in many developing countries, in part because of the significant multiplier effect it produces and the fact that it encourages the use of goods and services produced locally. Despite this, research conducted on the economic consequences of tourism suggests that this potential is hardly realised in its whole. Another study analysed the potential contribution of tourism to economic growth and development using panel data from 1990 to 2007 from 19 island economies throughout the world. The researchers came to the conclusion that the growth of tourism is a significant factor in explaining economic performance in the economies that were chosen for this study. The findings have policy implications for maximising the contribution of the tourism sector to overall economic growth, which can be enhanced as a result of the findings. In addition, research was conducted to evaluate the tourism industry's contribution to the economies of seven developing nations located on islands in the Pacific, Caribbean, and Indian oceans. In the vast majority of states, the transportation industry has far stronger linkages to other economic subsectors than the national average. The findings showed that taking advantage of economies of scale is beneficial for achieving maximum economic effect from tourism (Richardson, 2021).

TOURISM AS AN EMPOWERMENT TOOL

Governments and researchers alike have recently shown a growing interest in the question of how critically important it is to advance ST. For instance, the impacts that come from enormous annual tourist migrations may have contrastingly positive and bad repercussions for the cultures, economics, society, or ecosystems that they touch. A few different conceptual frameworks and analytical approaches are presented here to help ease understanding of STD. In spite of the large number of studies that have been carried out to this point, there has been a growing chorus of voices emphasising the importance of conducting additional research on the attitudes of locals as the essential building block for achieving sustainable and fruitful tourism development. The points of view were investigated and held by community stakeholders in the small town of Bridgetown, located in the Australian state of Western Australia. The research focused on two main themes: the perspectives of local stakeholders regarding the development of a ST sector and the perceived difficulties in accomplishing this goal. In the same vein, two significant contributions were made. The first thing that this article does is contribute by identifying four important types of local stakeholders and the functions that each of them plays in the community. These groups are tourism advocates, brand creators, reluctant followers, and conservative citizens. In particular, tourism advocates and brand developers identified practical methods to expand tourism in a way that is sustainable. One of these ways is through the production of events and activities that help connect tourists to the environment in which Bridgetown is located. In turn, hesitant followers and conservative citizens noted problems, particularly the predicament of striking a balance between maintaining a traditional or calm way of life while also catering to the needs of tourists. The research was enhanced by the roles played by these individuals who had distinct social situations, and as a result, numerous alignments with role theory developed. These alignments constitute the second contribution that the research brings to the table (Duarte Alonso, & Nyanjom, 2017).

One of the most important activities of modern people, tourism is responsible for enormous changes in the outward appearance of the globe, which in turn has an effect on humans' political, economic, cultural, and cultural practises as well as their way of life. The International Tourist Organization has suggested using tourism as a means to achieve the goals that have been set for the third millennium. Eighty percent of the 56 countries that have formed programmes to alleviate poverty point to tourism as a potential alternative for economic growth, employment, and poverty reduction ((Nooripoor, et al., 2021). The growth of tourism in Costa Rica is a prominent example of sustainable tourism development since the country has placed equal importance on preserving the environment, fostering social inclusion, and expanding the economy (Fennell, 2015). Significant economic benefits are generated by the tourist industry of the country, which has become a global model for sustainable tourism practises. These benefits include the protection of natural resources as well as the promotion of social and cultural diversity (Fennell, 2015). Promoting ecotourism, which places an emphasis on responsible vacationing in natural regions with the goals of preserving the natural world and providing economic advantages to the communities there, has been one of the most important components of Costa Rica's efforts to build sustainable tourism (Mowforth & Munt, 2015). The nation has created a variety of ecotourism products, such as national parks, protected areas, and community-based tourism initiatives. These developments have assisted in the promotion of environmentally responsible tourism practises and contributed to the nation's economic and social development (Fennell, 2015). Moreover, certification programmes for sustainable tourism, zoning laws for tourism development, and waste management programmes are only some of the policies and regulations that Costa Rica has put in place to encourage activities that are more environmentally friendly with regard to tourism (Mowforth & Munt, 2015). These policies and regulations have helped to guarantee that the expansion of tourism is carried out in a manner that is responsible in terms of both the environment and society, while also fostering expansion of the economy (Fennell, 2015). The positive effects that environmentally responsible tourism development in Costa Rica has had on the country's economy, society, and environment are plain to see. The tourism industry has become a significant contributor to the country's economy, generating more than 8% of its GDP and providing employment opportunities for more than 200,000 people (UNWTO, 2018). In addition, Costa Rica's ecotourism initiatives have helped to protect more than 25% of the country's land area and contributed to the conservation of the country's rich biodiversity (Fennell, 2015).

TOURISM IN CHINA

To encourage the sustainable and healthy development of tourism in China, the relationship between the economy and the environment must be handled with care, and the environment must not be harmed in order for the economy to develop. Tourism environmental protection is closely related to the sustainable development of tourism in China. While looking at the bigger picture, it is important to prioritise environmental protection in order to first safeguard tourism's long-term interests and then promote the industry's continued development in a manner that is both sustainable and beneficial to public health. Given these circumstances, there is a large amount of practical value in investigating the effects that tourist environmental protection has on the development of sustainable tourism. It is vital to strengthen environmental protection, rationally employ tourism resources, perfect the tourism operation system, standardise tourists' consumption habits, and increase tourism supervision in order to support the sustainable and healthy development of tourism. Constructing a stratified tourism environmental protection system based on the protection class of conservation areas and implementing a variety of tourism environmental protection measures are necessary in order to unify various aspects, increase the value of tourist areas, coordinate the relationship between environmental protection and economic growth, and ultimately achieve the sustainable and healthy development of tourism in China (Jia, 2018).

This Figure 1 shows the Chinese tourism market has become one of the most closely observed inbound and outbound visitor marketplaces in the world. It has the most visited place in the world.

China has an obligation to recognise the strategic importance of the tourism industry to the economies of the surrounding regions and to other stakeholders. Monitoring and control continue to be among the highest priorities throughout the entirety of the nation's twelfth Five-Year Plan. It is important to recognise that tourism in western China should not only be seen as a factor in the expansion of the economy but also as a way to broaden the base of the regional economy. More knowledge on the importance of tourism should be disseminated to each and every level of government. The governments of the provinces should make active efforts to encourage low-carbon tourism and ensure that it grows in a sustainable manner for the benefit of the local community, the environment, and the growth of the economy of the region as a whole. In order to avoid a lack of coordination and excessive development, a tourism development strategy that is well-planned and coordinated must be implemented with care. This is done to guarantee that the underlying policy and planning are conducive to sustainable growth. It is necessary, from the perspective of environmental protection, to adopt and enforce clear national guidelines in order to ensure that the environmental benefits of tourism are not limited to the eastern portion of the country. This can be accomplished by ensuring that clear national guidelines are adopted. Education

Figure 1.
Source: https://www.statista.com/topics/1210/tourism-industry-in-china/#topicOverview

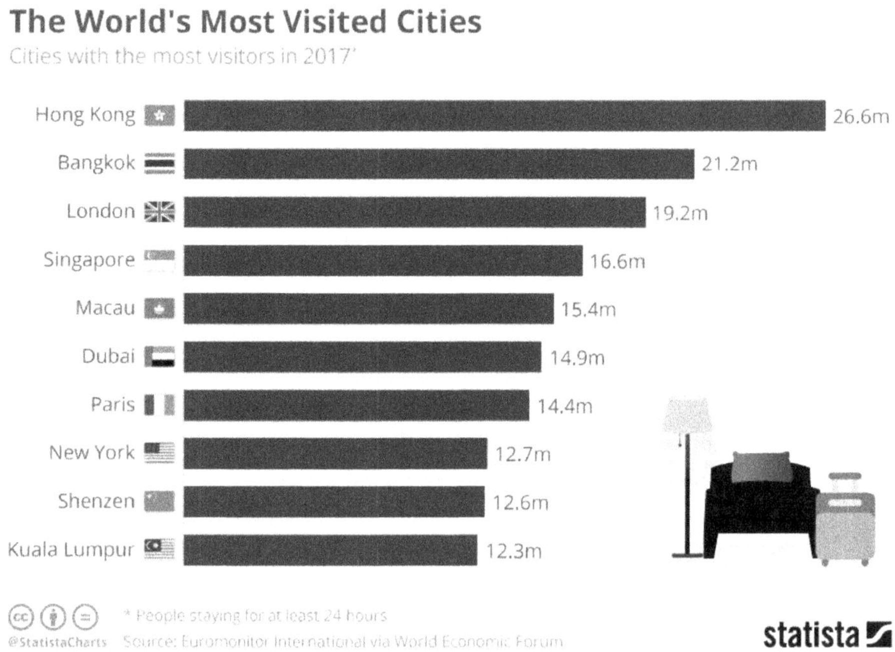

and awareness are two of the most important tools that the regional government can employ to build a tourism business that is sustainable on both an economic and an environmental level. Concurrently, there is a requirement to evaluate the existing environmental laws and policies that pertain to tourism, and if it is deemed necessary, there is also a requirement to implement new laws and regulations in order to mitigate the adverse effects of increased tourism. Marketing and promotion of low-carbon tourism at the industrial level, in particular in western China, require special attention (Zhang, & Gao, 2016). The Figure 2 illustrates the gradual increase in China's inbound tourism over the years, as depicted in a graph provided by TravelChinaGuide. According to the graph, the revenue generated by China's inbound tourism in 2019 amounted to US$127.3 billion dollars.

China has been actively engaged in the provision of tourist activities for an extended period of time and derives a variety of economic, social, and cultural benefits from these practises. Despite the fact that the tourism industry in the country has been developing over the course of the past few years, the advancement rate is still relatively modest, and there are a lot of obstacles in the way of the sustainable development of this industry (Chen, 2022).

By performing research and development on low-carbon technology and by producing and developing low-carbon tourism products, the tourism industry should expand its capacity for energy conservation and the reduction of emissions. Due to the interdependence of the primary and secondary industries, attention must also be focused on the growth of the tertiary sector during this interim period. In conclusion, understanding the economic and environmental impact of tourism is essential for Chinese governments at all levels in order to establish a sustainable tourist sector in a timely manner in light of the increasing influence of the tourism industry. This is due to the fact that the tourism industry is expected to continue

Figure 2.
Source: https://www.china-mike.com/china-travel-tips/china-tourism-statistics/

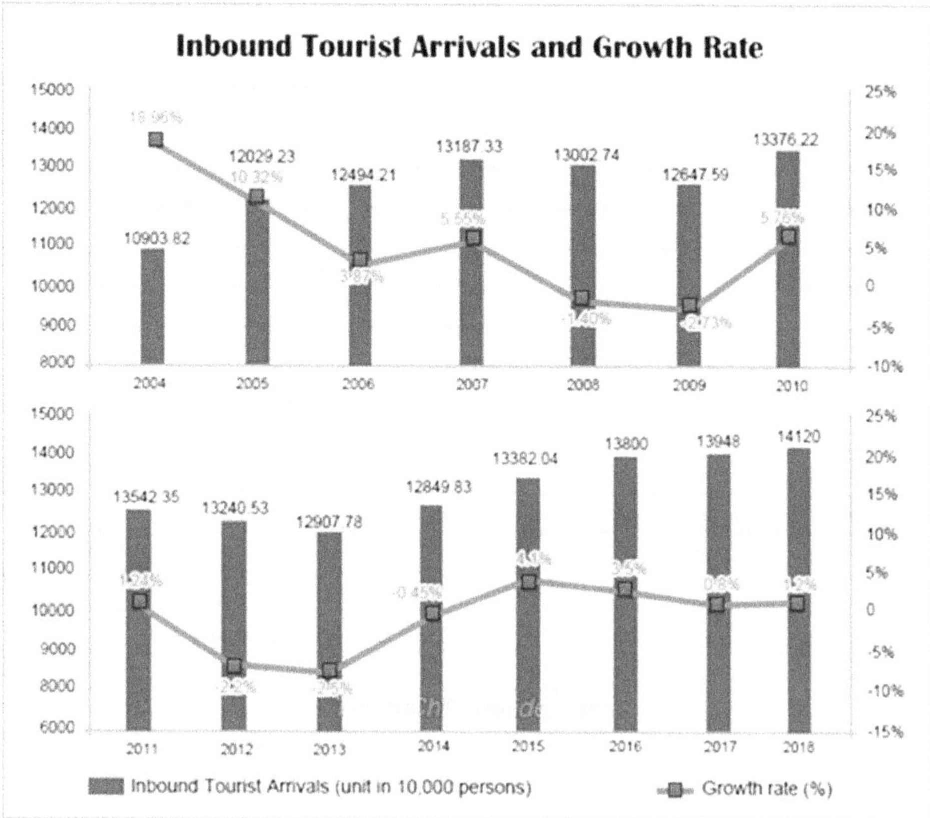

to grow in the coming years (Zhang, & Gao, 2016). It is generally agreed upon that tourism is one of the most important factors contributing to the expansion of economies all over the world. Tourism is recognised as a new tool for the expansion and development of regional economies, and it is also recognised as having a substantial influence on the socioeconomic development of a region. Tourism is one of the most important industries in the world. At the end of World War II, the travel and tourist industry saw a period of extraordinary growth. Due to the rapidity with which it has expanded, it has become one of the largest industries in the world. In the case of China, the expansion of the tourist industry has occurred rapidly, and it is widely acknowledged as one of the key factors behind the country's brisk economic expansion. At the same time, we need to look into the possibility of tourism-related destruction, in particular to the natural environment. Even if socioeconomic circumstances and factors have to be taken into account, it is still a challenging endeavour to determine whether or not tourism-related activities are related to overall economic expansion (Su, et al., 2021). This Figure 3 shows the arrivals in China for tourism and it can be seen that there is a decline in 2020 because of COVID.

Figure 3.
Source: https://tradingeconomics.com/china/international-tourism-number-of-arrivals-wb-data.html

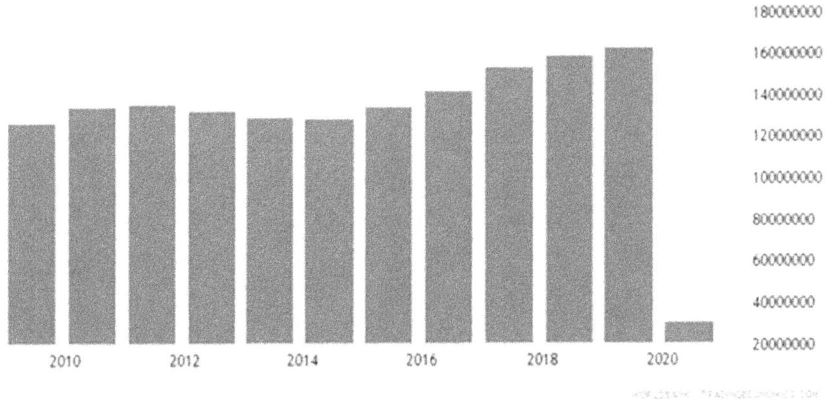

TOURISM IN CHINESE SUSTAINABLE DEVELOPMENT

During the course of the previous three decades, China's tourist industry has seen a number of significant shifts. China has risen to the position of the world's third most visited nation since 1978, when the country's reform and openness programme was first put into effect. It is common knowledge that international tourism plays a key role in the expansion of economies around the world. The most recent statistics available from the National Bureau of Statistics of China show that there were 55.98 million foreign tourists and 1.61 billion domestic tourists in the country in the year 2010, and it is anticipated that these numbers will continue to rise over the course of the next decade. Earnings from trading in foreign currency reached a record high of 45 billion US dollars in 2009, placing them fourth in the world. During this time period, domestic tourism was responsible for a combined total revenue of 777,1 billion Yuan. It is possible that the tourism business may fuel economic growth by generating positive externalities in non-tourism sectors such as the manufacturing industry, the transportation industry, and the service industry. This is due to the spillover effects that the tourism industry produces. The rise of tourism comes with a plethora of benefits, some of which include an increase in money, chances for employment, and an improvement in image. The sector's total contribution to China's GDP in 2012 was 4,783 billion Yuan, which is equivalent to 9.3% of the country's total GDP. Its direct contribution to China's GDP in 2012 was 1,361.9 billion Yuan, which is equivalent to 2.6% of the country's total GDP. In addition, the report projects that the tourist industry's direct and total contributions to GDP will increase by 8.7% and 8.9% annually to 3384.8 billion Yuan and 12,062.2 billion Yuan respectively in 2023. These figures are derived from projections made in the paper. (at constant 2012 pricing). In addition, the tourist sector was responsible for the creation of about 22.7 million direct jobs in addition to 41 million indirect jobs. In addition, investments in travel and tourism amounted to 655.2 billion Yuan in 2012, which represents 2.8% of overall investments. By the year 2023, it is expected that the amount of money invested in travel and tourism would have grown by 9.3%. It is projected that by the year 2023, the tourism industry in China would overtake that of the United States to become the largest in the world. Previous research has identified additional advantages of inbound tourism, in addition to the benefits that were mentioned above, such as the development of rural areas, the enforcement of infrastructure development, increased

Figure 4.
Source: http://www.researchinchina.com/htmls/report/2013/6771.html

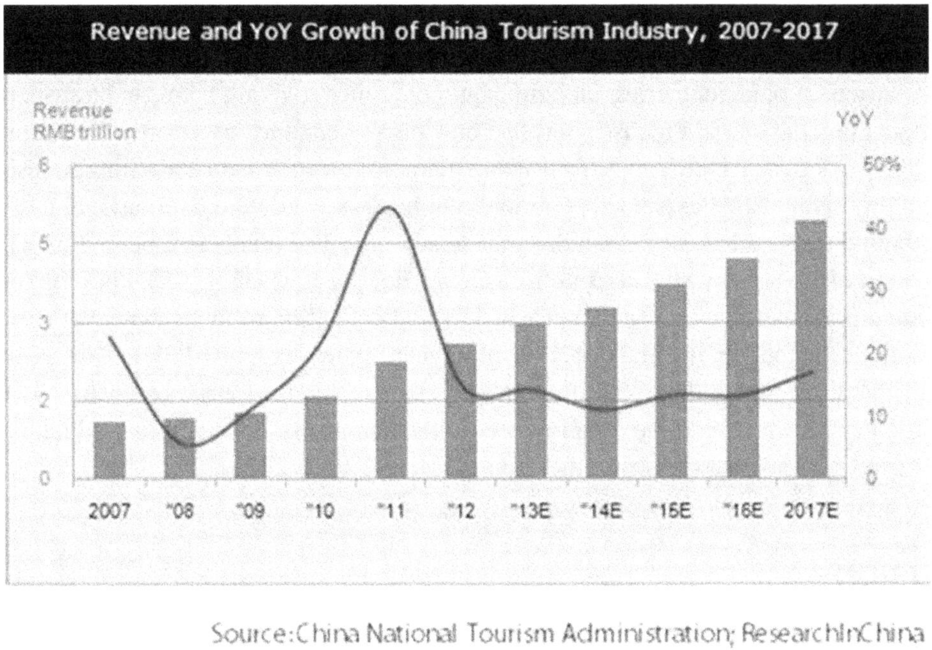

efficiency through competition, and the exploitation of the scale of local economies (Zhang, & Gao, 2016). The figure illustrates the revenue generated till 2017 through tourism.

The rise in the tourism industry in the Beijing region is the growth that has led to a good transition over the past few decades. This growth can be measured in terms of economic and social advancement. It has made it possible for the regional economy to take in foreign currency at a price that is more affordable and has supported the establishment of a great deal of connected businesses. The tourism business has spurred the rapid expansion of a number of other industries, including banking and finance, telecommunications, health care, insurance, and transportation. According to the most recent research and estimations, the tourist industry in China directly supports more than 5 million jobs and indirectly employs over 25 million people. The purpose of this study is to provide a conceptualization of the relationship between the tourism industry and economic growth by utilising an econometric model. The results of this research point to the existence of a relationship that works in both directions between the two variables (Su, et al., 2021). It is impossible to overstate the importance of a nation's tourism industry, particularly in relation to economies that are naturally endowed with abundant resources that could serve as tourist attractions or economies with less impressive economic performance from other sectors. Both of these types of economies benefit tremendously from having strong tourism industries. A nation like China, which has a long and well-preserved historical legacy consisting of rich culture and custom, remarkable antique goods, and a climate that is both natural and man-made, serves as the foundation for attracting tourists from all over the world. This is because China has a long and well-preserved historical legacy. The aftermath of World War II laid the groundwork for the development of the tourist sector, which is now rapidly rising to the position of becoming the largest industry in the world. This has resulted in and will continue to result in outstanding outcomes for the Chinese economy in terms of the economic and social development of the country. The open-door policies that were implemented in the 1980s, the rise of an affluent middle class in China, and the loosening of mobility limitations for both domestic and international tourists have all contributed to a boom in travel, which has, in turn, helped to enhance the tourism industry. Tourism is an interdependent industry that has a considerable impact on a variety of other industries, including but not limited to retail, hotel, and transportation. The massive impact that the tourism industry has had on China's economy, as seen by the country's efforts is to build and modernise its tourism infrastructure, demonstrates the industry's immense significance. The domestic tourism market expanded at a rapid rate in 2018, while the international tourism market and the outbound tourism industry expanded at a steady rate. The number of people travelling within their own country increased by 10.8 percent from the previous year, reaching 553.9 million (representing 411.9 million urban residents with an increase of 12.0%, 142 million rural residents with an increase of 7.3%, and a domestic tourism income of 5.13 trillion yuan, an increase of 12.3%). This results in a total of 282.6 million direct jobs and 516.5 million indirect jobs within the tourism industry. As a result, the combined total of direct and indirect employment in the country is 799.1 million, which is equivalent to 10.29% of the total number of jobs in the country (Haibo, et al., 2020).

CONCLUSION

In conclusion, tourism makes a contribution to sustainable development since it encourages economic growth, the creation of new jobs, and the decrease of poverty. Nonetheless, it has a negative impact on the local inhabitants, the environment, and the cultural legacy of the area. As a direct consequence of these challenges, an industry known as sustainable tourism has developed in an effort to find a happy medium between issues pertaining to the economy, society, and the environment. China has emerged as a key player in the global tourist sector, providing local communities with numerous chances to participate in tourism-related activities and to profit from the expansion of the tourism business. Nonetheless, the extent to which tourism has helped empower Chinese people and contribute to the country's long-term sustainable development is still up for debate. Therefore, additional research is required to investigate the perspectives of residents about the growth of tourism that is both sustainable and effective, the part that tourism plays in the protection of the environment, and the essential steps to take to achieve sustainable tourist development. It is vital to accept sustainable tourism development concepts and approaches, such as the proper extraction of resources, the conservation of biodiversity, the benefit sharing with local populations, and training and awareness-raising about natural resources. By putting these principles into practise, tourism has the potential to become a powerful tool for empowering people and fostering sustainable development not only in China but also everywhere else in the world.

REFERENCES

Duarte Alonso, A., & Nyanjom, J. (2017). Local stakeholders, role and tourism development. *Current Issues in Tourism*, *20*(5), 480–496. doi:10.1080/13683500.2015.1078782

Fennell, D. (2015). Ecotourism in Costa Rica: Environmental Impacts and Management. *Journal of Sustainable Tourism*, *23*(2), 169–189.

Haibo, C., Ayamba, E. C., Udimal, T. B., Agyemang, A. O., & Ruth, A. (2020). Tourism and sustainable development in China: A review. *Environmental Science and Pollution Research International*, *27*(31), 39077–39093. doi:10.100711356-020-10016-7 PMID:32638313

Hall, C. M., & Lew, A. A. (2009). *Understanding and Managing Tourism Impacts: An Integrated Approach*. Routledge. doi:10.4324/9780203875872

Jia, X. (2018, May). Consideration of the role of tourism environmental protection on the sustainable development of Tourism. In *2018 7th International Conference on Energy, Environment and Sustainable Development (ICEESD 2018)* (pp. 1642-1646). Atlantis Press. 10.2991/iceesd-18.2018.296

Kumar, A. (2017). Cultural and heritage tourism: A tool for sustainable development. *Global Journal of Commerce & Management Perspective*, *6*(6), 56–59. doi:10.24105/gjcmp.6.6.1709

Luo, F., Moyle, B. D., Moyle, C. L. J., Zhong, Y., & Shi, S. (2020). Drivers of carbon emissions in China's tourism industry. *Journal of Sustainable Tourism*, *28*(5), 747–770. doi:10.1080/09669582.2019.1705315

Mowforth, M., & Munt, I. (2015). *Tourism and Sustainability: Development, Globalisation and New Tourism in the Third World*. Routledge. doi:10.4324/9781315795348

Muresan, I. C., Oroian, C. F., Harun, R., Arion, F. H., Porutiu, A., Chiciudean, G. O., Todea, A., & Lile, R. (2016). Local residents' attitude toward sustainable rural tourism development. *Sustainability (Basel)*, *8*(1), 100. doi:10.3390u8010100

National Bureau of Statistics of China. (2020). *China Statistical Yearbook 2020*. China Statistics Press.

Nooripoor, M., Khosrowjerdi, M., Rastegari, H., Sharifi, Z., & Bijani, M. (2021). The role of tourism in rural development: Evidence from Iran. *GeoJournal*, *86*(4), 1705–1719. doi:10.100710708-020-10153-z

Richardson, R. B. (2021). The role of tourism in sustainable development. In Oxford Research Encyclopedia of Environmental Science. doi:10.1093/acrefore/9780199389414.013.387

Streimikiene, D., Svagzdiene, B., Jasinskas, E., & Simanavicius, A. (2021). Sustainable tourism development and competitiveness: The systematic literature review. *Sustainable Development (Bradford)*, *29*(1), 259–271. doi:10.1002d.2133

Su, Y., Cherian, J., Sial, M. S., Badulescu, A., Thu, P. A., Badulescu, D., & Samad, S. (2021). Does tourism affect economic growth of China? A panel granger causality approach. *Sustainability (Basel)*, *13*(3), 1349. doi:10.3390u13031349

Tien, N. H., Thai, T. M., Hau, T. H., Vinh, P. T., & Long, N. V. T. (2019). Solutions for Tuyen Quang and Binh Phuoc tourism industry sustainable development. Comparative analysis. *International Journal of Research in Marketing Management and Sales*, *2*(1), 101–107. doi:10.33545/26633329.2020.v2.i1b.31

UNEP. (2010). *Tourism and the Green Economy: Background Report*. United Nations Environment Programme.

UNWTO. (2015). *Tourism and the Sustainable Development Goals – Journey to 2030*. United Nations World Tourism Organization.

UNWTO. (2018). *Tourism Highlights* (2018 Edition). United Nations World Tourism Organization.

Venugopalan, T., & Kumar, D. (2017). Sustainable development through sustainable tourism in India-A case study of Kerala tourism. *Asian Journal of Research in Business Economics and Management*, *7*(12), 10–27. doi:10.5958/2249-7307.2017.00189.X

World Tourism Organization. (2020). *UNWTO World Tourism Barometer and Statistical Annex, November 2020*. https://www.e-unwto.org/doi/pdf/10.18111/wtobarometereng.2020.18.1.2

Zhang, L., & Gao, J. (2016). Exploring the effects of international tourism on China's economic growth, energy consumption and environmental pollution: Evidence from a regional panel analysis. *Renewable & Sustainable Energy Reviews*, *53*, 225–234. doi:10.1016/j.rser.2015.08.040

Chapter 11
Women's Empowerment and Sustainable Tourism Development:
A Moderating Role of Pakistan's Cultural Traditionality

Imamuddin Khoso
University of Sindh, Jamshoro, Pakistan

Maria Shaikh
University of Sindh, Jamshoro, Pakistan

Erum Shaikh
Shaheed Benazir Bhutto University, Sanghar, Pakistan

ABSTRACT

This study identifies the facets of women empowerment which could be harnessed to bring paradigm shift in tourism industry and simultaneously advance gender parity in the areas with rich tourism resorts in Pakistan. The data were collected through survey questionnaire from women aged 18+ living five regions of Pakistan which are rich in cultural heritage and tourist attractions. The findings show that psychological and economic empowerment are occupying central stage in enhancing women's developmental participation in tourism development in Pakistan. The political and social aspects of women empowerment also play a crucial role in stirring tourism activities. Most importantly, empirical evidence demonstrates that cultural traditionality (CT) positively moderates the impact of women empowerment on the development of tourism in a way that the participatory role of empowered women higher in CT is significantly magnified towards the creation of sustainable tourist environment which promotes local culture and brings benefits to local communities.

DOI: 10.4018/978-1-6684-8417-3.ch011

1. INTRODUCTION

Great strides have been made in empowering women across the globe, but still women are far lagging behind in getting level playing field in socioeconomic spheres. In comparison to men, women earn less for doing the same jobs, less likely to own land & start their ventures, have fewer legal & property rights, less upward social mobility, and spend disproportionately more time on household chores. It is well documented that women are not afforded the equal basic human rights and privileges, consequently, they trail much behind men in accessing almost all decent & high-paying economic opportunities (Hutchings et al., 2022). These inequalities are more pronounced in developing countries. Empowering women and attaining gender equality is a 5th goal of Sustainable Development Goals (SDGs) set out by United Nations. Giving women empowerment falls under an essential human right category and is viewed as a critical factor to reduce poverty and achieve inclusive growth. Under the goal of gender equity, United Nations targeted to enhance participation of women in communities and provide equitable opportunities for efficient engagement in economic, political, and social life. In this regard, United Nation stresses countries to design and implement the laws and policies to ensure women's ownership rights to economic resources and legal participation at all levels of decisions. Removing gender inequality and improving women empowerment are critical to reducing poverty and ensuring inclusive socio-economic development of communities. To this end, Tourism is one of the most attractive and impactful sectors that has a potential to attain gender parity and reduce poverty by engaging women in economic activities.

Women constitute nearly 49 percent of Pakistan's population, most of them stay at home doing household chores or do informal work in agriculture, which is not counted as formal employment. The woman labor participation is 22 percent, which is the lowest in South Asia, causing substantial losses in productivity. Recent, World Bank Group (2022) pointed out that Pakistan has failed to tap into the talent of human capital, especially women. The report further estimated that closing the gender gap in employment by putting women to productive uses, Pakistan can add up a whopping 23 percent gain to GDP. Pakistan's each province and region is unique with respect to cultural heritage, food, sports, and tourist attractions such as coastline lines, hills, plateaus, forests, and deserts. The economic power of tourism in Pakistan, still largely untapped, can substantially improve socioeconomic landscape of local communities, which are presently rich in tourist attractions but are in state of disempowerment, poverty and destitute. The boom in tourism activities can create versatile and flexible labor market, catering the job needs of local people as per their level of skills and seasonal availability. For instance, building of tourist resorts and other small ventures form a supply chain creating a host of jobs including tour guides, derivers, cooks, cleaners, accountants, computer operators, managers, among other service industry jobs. Many of such jobs are flexible in nature and could be taken on by local women while in tandem doing household jobs or working on agricultural forms. Additionally, to cater the needs of national and foreign visitors, tourism contributes to building of crucial infrastructure such as roads, flyovers, sports complex, quality health care centers and restaurants, benefiting the local communities. Tourism is rapidly increasing in the global economy and has become 4th largest industrial sector. In the developing economies, promotion of tourism can help achieve economic stability by exploiting the unique comparative advantage of each region to create income-generating opportunities, offer economic diversification, protect historical cultural heritage, and foster cross-cultural understanding and engagement. Such situation would ultimately transform local communities, currently marred in conflict and poverty, into the culture of peace and economic prosperity. Successful practices of countries flourishing in tourism include the removal of the socio-political barriers that limit local women's control over infrastructure development and a conducive legal structure was put in place for safeguarding the needs and interests of local communities especially women by providing them adequate representation on all decision-making boards. The benefits of empowering women go beyond positive contribution towards tourism development, but spreads throughout broader economic development and growth. Consequently, in a bid to enhance women empowerment, government across the globe launched initiatives to provide women with quality education, training, and entrepreneurial skills targeted to hone their skills, cultural values, and self-confidence which in turn will empower them in all dimensions.

2. RESEARCH PROBLEM

There is an economic rationale that no industrial sector can progress and reach to its full potential without an active and empowered role of women. It holds more relevance when it comes to build a sustainable development tourism industry which is primarily grounded in the rural regions of developing countries, thus requiring the participatory role of local community members, especially women for inclusive and equitable growth and development. Previous research studies have examined the role of women in tourism, using different set of empowerments with respect to psychological, political, sociological, cultural, cognitive, and educational spheres (Chatterjee et al. 2018; Elshaer et al. 2021). Nevertheless, the dimensions through which women can materially influence the development of tourism are vaguely explained. In this regard, Valeri et al. (2021) claimed that the issue of developmental role of gender in tourism is still unclear due to its interdisciplinary nature and can be better understood through contextual studies. The economic, social and political aspects of women empowered have been extensively investigated in the previous literature, however, the psychological dimension of empowerment and moderating role of cultural traditionality have largely remained unexplored in the context of tourism development. These variables have been included with the assumptions that Pakistan is seemingly enriched in tourist attractions but have not been succeeded in transforming tourist industry due to cultural rigidity and women's low self-esteem, preventing participation of local women which is viewed as critically essential in the development of tourism. It is assumed that women who have respect for cultural values and customs tend to contribute more towards the development of regional tourism sector. Based on these arguments, the main objective of this study is to evaluate the extent to which four aspects of women empowerment (i.e., psychological, economic, social and political), with cultural traditionality as a moderating variable, influence the development of tourism sector in Pakistan.

3. LITERATURE REVIEW AND DEVELOPMENT OF HYPOTHESES

The conceptual model of this study is based on the theoretical assumptions of social identity, social exchange and stakeholder theories. Social identity theory (SIT) postulates that an individual can be intrinsically driven to make contribution to the common good if the development process is collaborative based on local community-group and its outcomes bring social recognition and prestige to the group members a per the level of their input efforts (Scheepers and Ellemers 2019). SIT has been used to explain the role of empowered individuals in providing leadership skills for the promotion community development (Arshad et al., 2022; Hogg et al., 2012). Social exchange theory (SET) posits that individuals are motivated to put in their best efforts if they gain monetary or non-monetary benefits from an exchange transaction

(McGehee & Andereck, 2004). The assumptions of SET have been applied to determine the level of local women involvement into triggering tourism in developing economies (Abou et al., 2021). Similarly, Stakeholder theory (ST) stresses on the involvement of all the relevant stakeholders at managerial and operational decisions, which will lead to efficient use of resources and ensure effective attainment of desired outcomes. Local women are one of the most important stakeholders in planning and executing projects aimed at tourism development and thus, previous research studies have investigated the impact of empowered women through the lens of stakeholder theory (Vij et al., 2023; Vukovic et al., 2021).

Women empowerment is multifaceted term referring to a woman's control and power over economic resources, and freedom to make social and political decisions. Empowerment fundamentally points to an individual, institute, or society's ability to exert control and establish command in addressing the issues that matter to them. Scheyvens (1999) identified four facets of women empowerment (i.e., psychological, political, social and economic) which are having impact on the development of tourism sector. Though in subsequent research studies, scholars have added additional dimensions such as environmental (Ramos & Prideaux, 2014), and entrepreneurial facets of empowerment (Movono & Dahles, 2017), but this study focuses on four-dimensional framework of women empowerment used by a number of research studies in the context of developing tourism sector (Abou et al., 2021; Scheyvens, 1999).

3.1 Psychological Empowerment

The women who are self-reliant and have pride in their cultural traditions and heritage, are viewed as psychologically empowered, a condition necessary for promotion of tourist ventures (Boley et al., 2016). In the developing countries regions with tourist attractions are likely unindustrialized and require preservation of cultural heritage in order to boost self-esteem of local people. Ecotourism that is cognizant of cultural norms and values lead to psychological empowerment of local people especially women, who in turn, proudly share the products and traditional customs with visitors (Arroyo et al., 2019). The women who are aware of their cultural customs and values and take pride into them, tend to develop positive attitude and deliver sting support to tourist activities, which in turn provides women a sense of pride, self-independence and self-esteem, ultimately strengthening their psychological empowerment (Boley & McGehee, 2014). Hence, it is hypothesized that:

Hypothesis#1: Psychological empowerment of women significantly supports the tourism development.

3.2 Economic Empowerment

It refers to the two points. First, the degree to which local women have access to economically productive resources and their capacity to utilize them in the benefit of local community and tourist business. Second, how the benefits generating from tourist activities are distributed. Economic empowerment entails that the earnings should be controlled by local communities and profit should be used to improve the living standard of local people, especially women should reap the benefits as per their contribution. The equitable disruption of profit would ensure high commitment of women towards promoting tourist products and services. The expansion of tourist activities provides numerous business and employment opportunities, leading to higher earnings for local people, especially for women who then can support their families and getting a sense of economic independence (Strzelecka et al., 2017). Likewise, Scheyvens

(2000) evidenced that the economic dimension of women empowerment is the most important manifestation of local support to the development of tourism trades and thus, it is hypothesized that:

Hypothesis#2: Economic empowerment of women is a significant positively influencing the tourism development.

3.3 Political Empowerment

Under political empowerment, women get active participation in development of tourism eco-environment within respective regions in terms that their voices are heard, and concerns are addressed in designing and executing ecotourism projects. Scheyvens (1999) pointed that the pace of tourist development can be greatly enhanced through engaging youth and women in project decision making. The politically charged local women carry huge capacity to set tourism on the path of sustainable development. In this regard, (Arroyo et al., 2019) argued that higher representation of women in leadership positions is associated with higher tourism growth. Political empowerment enables women to control resources and use them in a manner that brings benefits to local community, inspiring local people to promote cultural items as tourist products ((Strzelecka et al., 2017), leading to the following hypothesis.

Hypothesis#3: Political empowerment of women significantly drives up the tourism development.

3.4 Social Empowerment

A community is said to be socially empowered if development of tourism activities fosters social integrity and cohesion, which is manifested in creation of strong local community organizations and platforms working in social development such as improving education, health, and recreational facilities in the local area. A community is disempowered if rise in tourist entities does not improve social conditions and instead, lead to loss of lands from local people, higher crime, begging, and prostitution. Socially empowered women actively interact with visitors and develop strong attachment with local community and its historical resources. The tourist sector can prosper and thrive if it has working force who are socially active and have potential to gather community members for preservation and promotion of cultural heritage and other tourist attractions (Strzelecka et al., 2017). Boley et al. (2016) found social empowerment as a critical factor in deriving local people's positive attitudes to the development of tourism at regional level.

Hypothesis#4: Social empowerment of women is a positive determinant to the tourism development.

UN World Tourism Organization Report (2019) showed that women empowerment in its all dimensions leads to sustainable tourism development at regional level. Tinker (2006) evidenced that women who are actively engaged in tourism ventures, tend to have higher economic status and claimed that if women are given more jobs and promoted to higher managerial positions in the tourism industry, they would enhance its competitiveness. It is well acknowledged that role of empowered women is pivotal to enhancing development and growth of tourism sector, resulting in improved livelihood and resilience of local people (Bagheri, et al., 2022). Gannon et al. (2020) argued that empowering women is single most important developmental strategy aimed at developing regional communities with respect to enhanced education and health services through tourism process. Empowered local people can make significant contribution to the attainment of sustainable goals linked to tourism businesses. Likewise, European Institute for Gender Equality (2016) showed that empowering women result in a plethora of innovative businesses (eg., handicrafts, traditional & heritage units) developed and owned by local women that create a great attraction for tourists, thereby benefiting women owners as well as supporting tourism at national level. Furthermore, Boley et al. (2016) found that the best strategy to engage local people into the sustainable tourism development process is to empower women and seek their input. Duffy et al. (2015) reported that social and economic independence of a woman is correlated with her job performance in tourism. According to Aghazamani et al. (2020) tourism industry is developed and flourished on the back of local people's support and engagement.

The availability of empowered women is viewed as a key deriver to the development of tourist ventures such as guest houses, homemade crafts, food stalls and other small businesses (Mahmoud et al., 2019). In the same line, Carvalho et al. (2019) argued that women with psychological, economic, political and social empowerment greatly enable tourism sector in unleashing its potential, thereby economically benefiting local communities and making big contribution to national economic growth.

3.5 Cultural Traditionality

Cultural traditionality (CT) refers to an individual's commitment, openness, respect and acceptance towards cultural values, norms, and customs. CT shows a person's evaluative, inspirational, attitudinal, and behavioral traits towards a given culture (Yang, 2003). People having higher CT are more motivated to share all aspects of their culture with others, creating a commercial market. The belief in cultural values and norms sufficiently influence a woman's attitude towards working in tourism business. Women who believe working in the tourist sector is not in congruence with cultural values and is linked with moral failure, choose to stay away from joining tourist organization despite being empowered (Hutchings et al., 2020). The women are less likely to promote cultural heritage in cases where women's working in tourism is perceived as negative, leading only man to run the tourist resorts. Culture also prevents women from joining formal workplace on the notion that women can not perform well a job task due to lack of skills and household responsibilities (Alrwajfah et al., 2020). Conversely, local women who have high pride in their cultural customs and heritage, can take the tourist ventures to next higher level with their passionate approach to introduce tourist destinations to the visitors (Aghazamani et al., 2020). Cultural traditionality has been dominantly used in the field of human resource practices (Wan et al., 2021), however, recently Shen et al. (2022) explored the role of CT in the development of tourism and showed empirical evidence that people with higher cultural traditionality demonstrate strong pro-tourism attitudes which positively influence tourism in the region. Thus, the following hypotheses are proposed:

Women's cultural traditionality significantly moderates the relationship between psychological (H5), economic (H6), political (H7) and social (H8) empowerment of women and tourism development.

Based on the review of literature, the following conceptual model is developed.

Figure 1. Proposed conceptual model

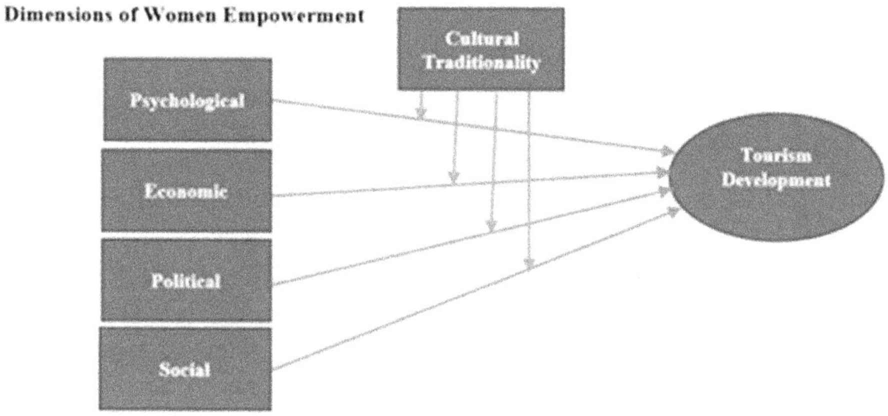

4. RESEARCH METHODOLOGY

The quantitative method is used in data collection and analysis. The five regions containing mountains and exquisite valleys (i.e., Chitral, Swat and Kaghan Valleys situated in Khyber Pakhtunkhwa; Leepa Valley in Azad Jammu & Kashmir and Hunza Valley in Gilgit Baltistan) were identified as potential tourist destinations and women aged 18+ living in those regions constituted the population. The convenience and snowball sampling techniques were used to reach out to the targeted women respondents. The non-probability sampling approach is used because it is cost effective, consumes less time to find and access women respondents who are knowledgeable about the concept of empowerment and finally, it enhances response rate as most of women living in those targeted regions are generally reluctant and unwilling to participate in research surveys for a number of socio-cultural reasons. The survey questionnaire was designed to collect the primary data. The measurement scale of psychosocial empowerment was adapted from (Samad & Alharthi, 2022) and measurement items of economic, political and social empowerment of women and state of tourism development were taken from (Abou et al., 2021). The scale of perceived cultural traditionality was adapted from (Wang & Zhang, 2020). The minimum required sample size was determined using GPower software and at 0.15 effect size, 0.05 alpha error probability, and 95% power level, the recommended threshold of sample size is 74 and above. In total 500 women were approached, 452 showed willingness and participated in this study.

5. DATA ANALYSIS AND DISCUSSION

Structural Equation Model (SEM) technique is used to analyze data as SEM considerably facilitates the estimation and interpretation of conceptual model with moderating paths. In SEM, the measurement model (MD) is examined for validity before analysis of structural model (SM) is undertaken. MD defines relationship between latent constructs and their observed indicators. The observed items are of importance because they represent variables which are abstractly conceptualized in SEM. The accuracy at which observed items define their latent construct is analyzed using reliability and validity.

5.1 Convergent and Discriminant Validity

Convergent validity entails that all the observed items developed to measure an unobservable construct, in fact, measure the same trait, which is evidenced if items of the same construct are correlated to one another. All the latent variables used in this study are reflectively measured and hence, Cronbach's Alpha, Rho_A and Composite Reliability tests are applied to determine the extent to which items are internally consistent in representing their respective construct. The results of these three tests in table-1 show that all latent variables are having values above the threshold level of 0.7, indicating that measurement model is empirically reliable. Further, the value of Average Variance Extracted (AVE) for latent variables fall in the range of 0.6 to 0.8, which exceed the cut-off value of 0.5 (Hair et al., 2016), showing appropriate convergent validity is established. The discriminant validity implies that latent constructs used in SEM are theoretically different and are operationalized with a unique set of indicators, in fact, are empirically distinctive at indicator and construct levels. The Fronell-Larcker Criterion (FLC) and Heterotrait-Monotrait ratio (HTMT) tests have been used to determine the discriminant validity. FLC calculates the square root of average variance extracted (AVE) of a variable and compares it to shared correlation with other variables used in the structural model. FLC assumes that average correlation among items of the same construct tend to be higher than that of items representing other constructs. The results in table-2 show that square root of AVE is greater than that the corresponding correlation with other constructs, establishing distinctiveness of each construct. Likewise, HTMT evaluated discriminant validity by measuring heterotrait-heteromethod correlations and compare them with the monotrait-heteromethod correlations among constructs of a single structural model. The results of HTMT are lower than the suggested value of 0.9 (Henseler et al., 2015), implying that constructs are empirically unique. Furthermore, the results of Variance Inflation Factor (VIF) are much lower than the suggested value of 05 (Craney & Surles, 2002), demonstrating that there is no serious issue of multicollinearity among explanatory variables.

Table 1. Convergent validity

	Cronbach's Alpha	Rho_A	Composite Reliability	AVE
PE	**0.763**	**0.809**	**0.849**	**0.590**
SE	**0.731**	**0.742**	**0.846**	**0.648**
POE	**0.874**	**0.910**	**0.922**	**0.797**
EE	**0.785**	**0.810**	**0.857**	**0.658**
TD	**0.869**	**0.892**	**0.905**	**0.765**

Table 2. Discriminant validity (Fronell-Larcker criterion)

	EE	PE	POE	SE	TD
EE	**0.818**				
PE	0.263	**0.869**			
POE	0.332	0.278	**0.885**		
SE	0.336	0.244	0.537	**0.897**	
TD	0.568	0.052	0.307	0.308	**0.858**

Table 3. Discriminant validity (Heterotrait-Monotrait ratio)

	EE	PE	POE	SE	TD
EE					
PE	0.406				
POE	0.420	0.292			
SE	0.469	0.279	0.569		
TD	0.558	0.081	0.334	0.339	

Table 4. Multicollinearity: Variance inflation factor (VIF)

	EE	PE	POE	SE	TD
EE					1.205
PE					1.131
POE					1.490
SE					1.473
TD					

5.2 Structural Model: Path Coefficients

The evaluation of structural model begins with determining its explanatory and predictive power and the value of R-square adjusted in table-5 is 0.602, implying that 60.2% of variance in tourism development is jointly explained by exogenous variables of the model, showing strong explanatory power. Similarly, the value of SRMR is 0.082, which is much lower than 0.1 cut-off, echoing sufficient fitness between implied and observed model. With respect to predictive power of structural model, Q-square test is administered, and its value is 0.427, indicating that model's predictive capacity falls in the category of being strong.

The results of direct path coefficients show that all of four dimensions of women empowerment positively contributing to the development of tourism with psychological empowerment ($\beta 1 = 0.624$ and $p < 0.01$), social empowerment ($\beta 1 = 0.399$ and $p < 0.01$), political empowerment ($\beta 1 = 0.233$ and $p < 0.05$), and economic empowerment ($\beta 1 = 0.559$ and $p < 0.01$). As per magnitude of beta coefficient, a woman's psychological empowerment is found to be the most important factor deriving up tourism activities, followed by economic empowerment. These findings underline the significant fact that empowerment of women should, in the first place, begin from within, under which women need to have self-confidence, self-esteem and strong belief that they are equally able to make contribution towards the betterment of tourism industry, leading to their own betterment in social, political, and economic terms. The previous research studies have highlighted that women who psychological empowered tend to feel pride in their cultural heritage and are excited in sharing it with the visitors (Elshaer et al. 2021). Psychological empowerment is generally manifested through a women's feeling of independence and pride which emanates from a sense that a woman' achievement is appreciated and provide socioeconomic status within a community. Such psychologically empowered women put forth their best efforts in supporting and reshaping the tourism in a way that effectively attains its development goals and also maintains the alignment with their traditional cultural customs and values (Boudrias et al. 2004). Further, it is inferred from the significance of economic empowerment that for achieving growth in tourism activities at regional level, it is vital to enhance the participatory role of women who are economically empowered and thus, would be incentivized to double up their contributions in unique capacities towards taking tourism to next higher level with hope of reaping personal economic benefits. Recently, Vij et al. (2023) evidenced that economically empowered woman is significant stakeholder and holds entrepreneurial, leadership and destination knowledge, spurring tourism activities by a set of new and innovative ways.

Path coefficients related to social and politically empowerment are positive and statistically significant, implying that the availability of socially and politically empowered women bodes well for tourism. The communities characterized by power equilibrium, cooperation, and participatory approach to community decision-making are likely to empower women socially and politically, enabling women to create community-driven tourism rather than tourism simply occurring in the community (Scheyvens & Hughes, 2019). Dolezal and Novelli (2022) echoed the same argument that with social and political empowerment, a woman can participate as the main "subject" instead of being "object" in tourism. Women especially in developing countries are increasingly using social and political dimensions of empowerment to disrupt patriarchal community structure and break the vicious circle of poverty in which they have been trapped for decades in an attempt to improve their lives and address the needs of their families, underscoring the future role of women in providing vision in the development of industrial sectors including tourism. These findings fully support hypotheses 1 to 4 and point out that empowering women psychologically, socially, politically and economically should be central to government's policy efforts to transforming tourism sector in Pakistan. Based on the positive and statistically significant direct path coefficients of variables (i.e., psychological, economic, political and social empowerment of women), the hypotheses 1-4 are fully supported.

With respect to the moderating role of cultural traditionality (CT), the results show that CT significantly moderates the relationship between three dimensions of women empowerment (i.e., psychological, economic and social empowerment) and tourism development, such that women having higher element of cultural traditionality play more positive role in boosting up the tourism sector. This is because the attachment with values and customs of one's culture is a critical determinant of an individual's developmental behavior (Wang & Zhang, 2020). In this context, it is assumed that women who are aware of cultural history, heritage and practical customs and subsequently internalize them, are more likely act in a way that promotes their cultural traditions and tend to have high commitment towards development of local tourisms primarily due to their high passion to introducing cultural products and services to visitors. The intervening role of cultural traditionality is fundamentally rooted in the cultural self-representation theory, in which an individual's preferential behaviors are guided by his or her belief and value in cultural norms (Nazariadli et al., 2019; Wang & Zhang, 2020). Recently, Huang et al. (2023) used the assumptions of cultural self-representation theory to establish that a person's cultural values towards green environment cultivates a sense of green personal identity, leading to favorable or unfavorable green behavior, implying that supportive behaviors are largely function of cultural beliefs that a person adopts in a real time. Thus, it is inferred that the women who are low in cultural traditionality tend to become comparatively insensitive towards actions that trigger the promotion of their social and cultural activities. Contrary, women high in their respective cultural traditionality, put their individual passion into action, causing an activation of collective passion among local people which ultimately leads a big spike in tourism activities.

Based on statistically significant moderating coefficients, the hypotheses H5, H6 and H8 are confirmed. Whereas, H7 is rejected, inferring that cultural traditionality is not moderating relationship

Table 5. Structural path coefficients

	Original Sample (O)	Sample Mean (M)	Standard Deviation (STDEV)	T Statistics (|O/STDEV|)	P Values
PE -> TD	0.624	0.624	0.072	8.694	**0.000**
SE -> TD	0.402	0.399	0.069	5.806	**0.000**
CT -> TD	0.263	0.229	0.139	1.891	**0.059**
POE -> TD	0.273	0.233	0.114	2.384	**0.017**
EE -> TD	0.554	0.559	0.094	5.868	**0.000**
Moderating Effect 1 (CT x PE -> TD)	0.349	0.353	0.113	3.093	**0.002**
Moderating Effect 2 (CT x SE -> TD)	0.495	0.499	0.104	4.751	**0.000**
Moderating Effect 3 (CT x POE -> TD)	0.239	0.192	0.141	1.694	**0.090**
Moderating Effect 4 (CT x EE -> TD)	0.296	0.306	0.088	3.353	**0.001**
	R-square Adjusted				
TD	0.602				
	Q-square (1-SSE/SSO)				
TD	0.427				
	SRMR (Estimated Model)				
TD	0.082				

between political empowerment of women and tourism development. One plausible explanation for the rejection of this hypothesis is that in Pakistan tourism attractions exist in rural areas in which women are relatively less active in political activities and consequently women participation in political events is constrained. In such situation of diminished political role of women, the moderating impact of cultural traditionality is not perceptible.

6. CONCLUSION

Tourism industry holds tremendous potential to offer job opportunities to women who have ambitious to grow their career and earn a living for family. Currently, women are underrepresented in tourism, particularly at supervisory and managerial positions. The leadership role of local women is critical for Pakistan to fast-track its tourism industry and to bring inclusive development, leading to shared prosperity. The gender stereotype still tilts the socioeconomic balance in favor of men in tourism, marginalizing women from the developmental process and causing poverty primarily driven by gender inequalities. Women empowerment is viewed as a significant enabler to a number of Sustainable Development Goals (SDGs), which have a potential to generate high socio-economic externalities and impact. Pakistan is blessed with a plethora of tourism attractions for sightseers, mountaineers, hikers, and skiers including one of the world's highest mountain peaks & the oldest cultural civilizations, unique handicrafts & distinctive folk arts, and diverse natural scenic beauty spread across the regions. This study identifies the facets of women empowerment which could be harnessed to bring paradigm shift in tourism industry and simultaneously advance gender parity in the areas with rich tourism resorts in Pakistan. The findings show that psychological and economic empowerment are occupying central stage in enhancing women's developmental participation in tourism development in Pakistan. The political and social aspects of women empowerment also play a crucial role in stirring tourism activities. Most importantly, empirical evidence demonstrates that cultural traditionality (CT) positively moderates the impact of women empowerment on the development of tourism in a way that the participatory role of empowered women higher in CT is significantly magnified towards the creation of sustainable tourist environment which promotes local culture and brings benefits to local communities. If women are provided with level playing field by creating equal opportunity environment and are empowered in all dimensions, then they would be inspired to use their full potential in uplifting the tourism sector, while simultaneously progressing themselves and their families. The following recommendations are made to boost women's empowerment and their subsequent participation in tourism activities:

Firstly, Women empowerment should be mainstreamed in national and provincial policies aimed at developing and promoting tourism. This initiative would institutionalize the attainment of women empowerment and subsequently attract Non-Governmental Organizations (NGOs) and private sector investment which are essential for creating a vibrant women entrepreneurship in tourism. Secondly, women entrepreneurship in tourism can be leapfrogged through creating women-supportive legal environment and thus, gender-specific legal framework be devised to ensure women's rights are protected in all spheres. Thirdly, more women be recruited in tourism at managerial positions with equity in pay and promotional opportunities. Fourthly, the representation of women and other local members can be ensured by decentralizing ecotourism developmental projects from the national level to the local level. Finally, targeted training programs be designed based on training need analysis and tasked to achieve a broader set of goals including enhancing a bundle of women's soft skills, building women network, and improving their access and use to relevant digital technology in tourism which would make women more innovative and empowered.

REFERENCES

Abou-Shouk, M. A., Mannaa, M. T., & Elbaz, A. M. (2021). Women's empowerment and tourism development: A cross-country study. *Tourism Management Perspectives*, *37*, 100782. doi:10.1016/j.tmp.2020.100782

Aghazamani, Y., Kerstetter, D., & Allison, P. (2020). Women's perceptions of empowerment in Ramsar, a tourism destination in northern Iran. *Women's Studies International Forum*, *79*, 102340. doi:10.1016/j.wsif.2020.102340

Alrwajfah, M. M., Almeida-García, F., & Cortés-Macías, R. (2020). Females' perspectives on tourism's impact and their employment in the sector: The case of Petra, Jordan. *Tourism Management*, *78*, 104069. doi:10.1016/j.tourman.2019.104069

Arroyo, C., Barbieri, C., Sotomayor, S., & Knollenberg, W. (2019). Cultivating women's empowerment through agritourism: Evidence from Andean communities. *Sustainability*, *11*(11), 1–14.

Arshad, M., Qasim, N., Farooq, O., & Rice, J. (2022). Empowering leadership and employees' work engagement: A social identity theory perspective. *Management Decision*, *60*(5), 1218–1236. doi:10.1108/MD-11-2020-1485

Bagheri, F., Ghaderi, Z., Abdi, N., & Hall, C. M. (2022). Female entrepreneurship, creating shared value, and empowerment in tourism; the neutralizing effect of gender-based discrimination. *Current Issues in Tourism*, 1–18. doi:10.1080/13683500.2022.2126749

Boley, B. B., Ayscue, E., Maruyama, N., & Woosnam, K. M. (2016). Gender and empowerment: Assessing discrepancies using the resident empowerment through tourism scale. *Journal of Sustainable Tourism*, *25*(1), 113–129. doi:10.1080/09669582.2016.1177065

Boley, B. B., & Gard McGehee, N. G. (2014). Measuring empowerment: Developing and validating the resident empowerment through tourism scale (RETS). *Tourism Management*, *45*, 85–94. doi:10.1016/j.tourman.2014.04.003

Boudrias, J. S., Gaudreau, P., & Laschinger, H. K. S. (2004). Testing the structure of psychological empowerment: Does gender make a difference? *Educational and Psychological Measurement*, *64*(5), 861–877. doi:10.1177/0013164404264840

Carvalho, I., Costa, C., Lykke, N., & Torres, A. (2019). Beyond the glass ceiling: Gendering tourism management. *Annals of Tourism Research*, *75*, 79–91. doi:10.1016/j.annals.2018.12.022

Chatterjee, S., Dutta Gupta, S., & Upadhyay, P. (2018). Empowering women and stimulating development at bottom of pyramid through micro-entrepreneurship. *Management Decision*, *56*(1), 160–174. doi:10.1108/MD-04-2017-0296

Craney, T. A., & Surles, J. G. (2002). Model-dependent variance inflation factor cutoff values. *Quality Engineering*, *14*(3), 391–403. doi:10.1081/QEN-120001878

Dolezal, C., & Novelli, M. (2022). Power in community-based tourism: Empowerment and partnership in Bali. *Journal of Sustainable Tourism*, *30*(10), 2352–2370. doi:10.1080/09669582.2020.1838527

Duffy, L. N., Kline, C. S., Mowatt, R. A., & Chancellor, H. C. (2015). Women in tourism: Shifting gender ideology in the DR. *Annals of Tourism Research*, *52*, 72–86. doi:10.1016/j.annals.2015.02.017

Elshaer, I., Moustafa, M., Sobaih, A. E., Aliedan, M., & Azazz, A. M. (2021). The impact of women's empowerment on sustainable tourism development: Mediating role of tourism involvement. *Tourism Management Perspectives*, *38*, 100815. doi:10.1016/j.tmp.2021.100815

European Institute for Gender Equality (EIGE). (2016). *Gender Equality Index*. Author.

Gannon, M., Rasoolimanesh, S. M., & Taheri, B. (2021). Assessing the mediating role of residents' perceptions toward tourism development. *Journal of Travel Research*, *60*(1), 149–171. doi:10.1177/0047287519890926

Hair, J. F. Jr, Hult, G. T. M., Ringle, C., & Sarstedt, M. (2016). *A primer on partial least squares structural equation modeling (PLS-SEM)*. Sage Publications.

Henseler, J., Ringle, C. M., & Sarstedt, M. (2015). A new criterion for assessing discriminant validity in variance-based structural equation modeling. *Journal of the Academy of Marketing Science*, *43*(1), 115–135. doi:10.100711747-014-0403-8

Hogg, M. A., van Knippenberg, D., & Rast, D. E. III. (2012). The social identity theory of leadership: Theoretical origins, research findings, and conceptual developments. *European Review of Social Psychology*, *23*(1), 258–304. doi:10.1080/10463283.2012.741134

Huang, L., Guo, Z., Deng, B., & Wang, B. (2023). Unlocking the relationship between environmentally specific transformational leadership and employees' green behaviour: A cultural self-representation perspective. *Journal of Cleaner Production*, *382*, 134857. doi:10.1016/j.jclepro.2022.134857

Hutchings, K., Moyle, C. L., Chai, A., Garofano, N., & Moore, S. (2020). Segregation of women in tourism employment in the APEC region. *Tourism Management Perspectives*, *34*, 100655. doi:10.1016/j.tmp.2020.100655

Mahmoud, M., Abou-Shouk, M., & Fawzy, N. (2019). The role of governmental authorities in supporting entrepreneurship in tourism industry. International Journal of Heritage. *Tourism and Hospitality*, *13*(2), 107–123.

Movono, A., & Dahles, H. (2017). Female empowerment and tourism: A focus on businesses in a Fijian village. *Asia Pacific Journal of Tourism Research*, *22*(6), 681–692. doi:10.1080/10941665.2017.1308397

Nazariadli, S., Morais, D. B., Bunds, K., Baran, P., & Supak, S. (2019). Rural tourism microentrepreneurs' self-representation through photography: A counter-hegemonic approach. *Rural Society*, *28*(1), 29–51. doi:10.1080/10371656.2019.1576294

Ramos, A. M., & Prideaux, B. (2014). Indigenous ecotourism in the Mayan rainforest of Palenque: Empowerment issues in sustainable development. *Journal of Sustainable Tourism*, *22*(3), 461–479. doi:10.1080/09669582.2013.828730

Samad, S., & Alharthi, A. (2022). Untangling Factors Influencing Women Entrepreneurs' Involvement in Tourism and Its Impact on Sustainable Tourism Development. *Administrative Sciences*, *12*(2), 52. doi:10.3390/admsci12020052

Scheepers, D., & Ellemers, N. (2019). Social identity theory. *Social psychology in action: Evidence-based interventions from theory to practice*, 129-143.

Scheyvens, R. (1999). Ecotourism and the empowerment of local communities. *Tourism Management*, *20*(2), 245–249. doi:10.1016/S0261-5177(98)00069-7

Scheyvens, R. (2000). Promoting women's empowerment through involvement in ecotourism: Experiences from the Third World. *Journal of Sustainable Tourism*, *8*(3), 232–249. doi:10.1080/09669580008667360

Scheyvens, R., & Hughes, E. (2019). Can tourism help to "end poverty in all its forms everywhere"? The challenge of tourism addressing SDG1. *Journal of Sustainable Tourism*, *27*(7), 1061–1079. doi:10.1080/09669582.2018.1551404

Shen, K., Yang, J., & Geng, C. (2022). How residents' attitudes to tourists and tourism affect their pro-tourism behaviours: The moderating role of Chinese traditionality. *Frontiers in Psychology*, *12*, 792324. doi:10.3389/fpsyg.2021.792324 PMID:35126243

Strzelecka, M., Boley, B., & Strzelecka, C. (2017). Empowerment and resident support for tourism in rural central and Eastern Europe (CEE): The case of Pomerania, Poland. *Journal of Sustainable Tourism*, *25*(4), 554–572. doi:10.1080/09669582.2016.1224891

Tinker, I. (2006). Empowerment just happened: the unexpected expansion of women's organizations. *Women and gender equity in development theory and practice: Institutions, resources and mobilization*, 268-303.

UN World Tourism Organization (UNWTO). (2019). *Global report on women in tourism.* UNWTO.

Valeri, M., & Katsoni, V. (2021). Female entrepreneurship in tourism. In *Gender and Tourism* (pp. 1–9). Emerald Publishing Limited. doi:10.1108/978-1-80117-322-320211001

Van Nguyen, S., & Habók, A. (2021). Designing and validating the learner autonomy perception questionnaire. *Heliyon*, *7*(4), e06831. doi:10.1016/j.heliyon.2021.e06831 PMID:33981892

Vij, M., Upadhya, A., Agha, K., Begum, V., Kashmoola, B. W., & Senior, J. R. (2023). Female economic empowerment through tourism training in the United Arab Emirates. *Worldwide Hospitality and Tourism Themes*, *15*(1), 18–28. doi:10.1108/WHATT-08-2022-0099

Vukovic, D. B., Petrovic, M., Maiti, M., & Vujko, A. (2021). Tourism development, entrepreneurship and women's empowerment–Focus on Serbian countryside. *Journal of Tourism Futures.*

Wan, M. M., Zhang, Y., & Li, M. (2021). Do narcissistic employees remain silent? Examining the moderating roles of supervisor narcissism and traditionality in China. *Asian Business & Management*, 1–25.

Wang, X., & Zhang, C. (2020). Contingent effects of social norms on tourists' pro-environmental behaviours: The role of Chinese traditionality. *Journal of Sustainable Tourism*, *28*(10), 1646–1664. doi: 10.1080/09669582.2020.1746795

World Bank Group. (2022). *From Swimming in Sand to High and Sustainable Growth: A Roadmap to Reduce Distortions in the Allocation of Resources and Talent in the Pakistani Economy*. World Bank.

Yang, K. (2003). Methodological and theoretical issues on psychological traditionality and modernity research in an Asian society: In response to Kwang-Kuo Hwang and beyond. *Asian Journal of Social Psychology*, *6*(3), 263–285. doi:10.1046/j.1467-839X.2003.00126.x

Chapter 12
Halal Holidays:
A New Concept Associated With Halal – Tourism With Special Concern Towards Gender-Segregated Facilities

Muhammad Faisal Sultan
Khadim Ali Shah Bukhari Institute of Technology, Pakistan

Muhammad Asim
University of Karachi, Pakistan

Imamuddin Mahar
Institute of Business Management, Karachi, Pakistan

Muhammad Nawaz Tunio
University of Sufism and Modern Sciences, Pakistan

ABSTRACT

Halal tourism is getting more popular day after day due to the increase in Muslim tourists all around the globe. In fact, non-Muslim countries are also striving towards providing Muslim-friendly facilities to attract Muslim tourists. However, Halal holidays is a bit different from Halal tourism as these are the holidays that provide complete leverage to Muslim tourists. In fact, Halal holidays are not associated with any religious act like a pilgrim, etc., but all the aspects associated with leisure and fun will be according to Sharia principles and will never obstruct any of the religious act of Muslims. However, the concept has not been discussed yet, especially with reference to Muslim women. Therefore, this study has been written purposively to describe all the aspects that may make readers understand the concept of Halal holidays in a detailed manner in order to make readers understand the relationship of Halal's perceived value with the intention to visit the destination.

DOI: 10.4018/978-1-6684-8417-3.ch012

INTRODUCTION

Halal is the term perceived as the way of slaughtering animal for meat and non-Muslims mainly associate it with food (Wingett & Turnbull, 2017). Although in reality, Halal is a term that encompasses all the activities related to eating, drinking, and use of cosmetics, etc (Boğan, 2020). Thus, the term "*Halal*", is not restricted to any one domain but it is a way of living life (Wingett & Turnbull, 2017). On the other side, Muslims are contributing around 30% to the overall world's population & the share of Halal tourism was around 12% till 2019. However, it was expected to raise phenomenally in the future (Rodrigo & Turnbull, 2019). Thus, over the past two decades, research is observing a significant surge in the use of terminologies associated with Islamic tourism for e.g., Halal Tourism, Muslim-friendly Tourism, and Islamic Tourism. However, the most common types are Halal Tourism and Islamic Tourism (Boğan, 2020).

Thus, in a nutshell, Islamic or Halal Tourism is a form of tourism directed toward Muslims to provide them tourism destination that is non-offensive and also follows Sharia rules. Halal Tourism is relatively a new concept. However, the raise in concern towards Halal leisure holidays is one of the fastest-growing domains in the industry. Thus receives phenomenal attention & gains high-level of interest from entrepreneurs, intrapreneurs & media. Halal Holidays are unique as these are not associated with any religious activities like Hajj or Umrah etc (Wingett & Turnbull, 2017). In fact, Halal tourism is composed of activities undertaken by Muslims to fulfill recreational, leisure & social purposes. Hence the demand and its growth rate have also been examined by different segments like air companies, hotels, traveling agencies & other segments associated with hospitality providers. Halal Tourism needs understanding that goes beyond food and clothing & may cover aspects like facilities for prayers, gender of hotel representatives, sexual activities & showing affection in public places (Rodrigo & Turnbull, 2019). Previous statements are in line with Wingett and Turnbull (2017) that Islam does not make travel or tourism prohibited & this can be done for spiritual, health, social & physical benefits. Although it may also fulfill spiritual purposes and make Muslims observe all the gifts of God (ALLAH), through keen observation of history and wonders that are indicating towards the greatness of God.

However, in Islam purpose is always required to encourage tourism as any sort of tourism is prohibited if it is for the sake of anyone's responsibilities or daily duties. In fact, tourism packages that are fully aligned with Islamic rules, regulations, and practices are termed "*Halal Holidays*" (Battour & Ismail, 2016). Halal Holidays are the sub-division of religious tourism (Wingett & Turnbull, 2017). In fact, one of the recent studies defines Halal Holidays as the form of holidays that provide Muslims the liberty of fulfilling all of their religious needs in addition to leisure & fun to provide the real joy of tourism to Muslims (Rodrigo & Turnbull, 2019). In fact, Halal-Holidays include Halal-Hotels and Halal-Resorts with Halal foods and drinks along with all the other activities in line with Sharia principles. These may include separate swimming pools, separate spa and leisure activities, separate open beach facilities & family oriented facilities like prayer facilities & Islamic swimming dress code, etc. (Battour & Ismail, 2016).

STATEMENT OF PROBLEM AND GRAY AREA

The motivation behind religious pilgrimages like Hajj etc is understandable. However, there is a lack of understanding regarding Halal Holidays as these Holidays are not associated with performing any religious pilgrimage. In fact, the purpose of Halal holidays is leisure & fun. However, regardless of differences in the definition given by various scholars and researchers, there is still much to be grasped regarding Halal Holidays. Especially important are those points that may uncover factors that may shape holiday choices among Muslim tourists (Wingett & Turnbull, 2017).

The need for research became more vibrant with a continuous and significant increase in Muslim tourists. Hence there is potent a need to understand the requirements of Muslim-friendly destinations. Thus, studies from Muslim, as well as Non-Muslim areas, must be increased to understand the criteria of Muslim-preferred tourism locations. However, initial studies might be done through qualitative methods followed by quantitative methods (Battour & Ismail, 2016). Hence the investigation may be based upon the consumption value theory given by Sheth et al (1991) in order to reflect the preference of female Muslim tourists as indicated by the future need for research by Rodrigo and Turnbull (2019).

On the other side, there is also a lacking of studies that reflect the impact of halal attributes on the perception of Muslim customers. However, Halal attributes of location have the ability to enhance the perceived value of the destination. Thus, may also attract Muslim tourists to that location. Therefore S-O-R Model may also be used to reflect the impact of stimulus created by Halal aspects of the location on perceived value and might be observed as the response by Muslim tourists (Sodawan & Hsu, 2022). Although for relating the discussion with the perception of Muslim Female tourists this study uses the reference of the S-O-R model for making a better understanding. Halal attributes can produce a positive impact on Muslim perception as well as the perception regarding the location. In fact, Muslims may not travel to those locations that are unable to provide them with Halal facilities. Thus, legitimate to indicate that Halal's perceived value is the outcome of Halal attributes of the location that may foster customers' willingness to visit the location (Sodawan & Hsu, 2022). However, rear research work has been based on the linkage among Halal attributes, Halal Perceived Value, and intention to visit, especially with reference to Non-Muslim countries (Sodawan & Hsu, 2022). Therefore, legitimate to relate the study with the S-O-R model that has been presented by Mehrabian and Russel (1974) as quoted by Sodawan and Hsu (2022) to reflect the impact of halal tourism.

THE S-O-R MODEL AND HALAL HOLIDAYS

The linkage will assist researchers and scholars in understanding the impact of Halal Attributes (stimulus), on Halal perceived Value (internal state of Muslim, i.e., (organism) that will cause consumer behavior (response), against external responses. Thus, the study bifurcated Attributes (S), with Islamic Physical Attributes and Islamic Non-Physical Attributes (Sodawan & Hsu, 2022).

Stimulus

Physical Attributes are tangible dimensions e.g., personnel, area, dresses, and material used for communication, etc. Islamic Non-Physical Attributes include the response of Muslims toward the environment which may include the conservative way of dressing, Islamic-friendly decoration, Muslim-friendly ways of entertainment, and a Ban on gambling, etc. (Sodawan & Hsu, 2022).

Organism

The study further explains that organism (O) is an internal state that includes emotional and cognitive behaviors triggered by internal and external environments. In fact, some of the studies mentioned that stimulus generated by external and internal environments is of three types. These types are classified as physiology, cognition, and emotion. The physiological component is composed of autonomic, somatic, and humoral signals. Cognitive elements are required to interpret the physiological condition of the environment & emotion is required to evaluate and express feelings. Thus, previous studies evaluated organisms on the bases of cultural memory, emotions, and satisfaction. In fact, the evaluation of organisms as the part of S-O-R model revealed that Muslim perceived values were affected positively by the presence of Halal Attributes of the location (Sodawan & Hsu, 2022).

Response

Last but not least Sodawan and Hsu (2022) discussed response as the individual internal response that might be received in two different ways. These ways are classified as approach behavior and avoidance behavior Avoidance behavior is reciprocal to approach behavior as it is grounded on sadness and anxiety etc. On the other side approach behavior has been grounded upon willingness and desire to stay etc. A study by Rodrigo and Turnbull (2019) also indicated that Islamic values are a potent predictor of customer expectations associated with perceived consumption of tourists' behavior. Thus, the study also supported Eid and El-Gohary (2015), that the incorporation of religious (Islamic) attributes is essential to understand the behavior of Muslim tourists it is essential to incorporate (Sodawan & Hsu, 2022).

HALAL TANGIBLE AND INTANGIBLE ATTRIBUTES IN ASSOCIATION WITH MUSLIM WOMEN

Research inventory that may reflect Halal Tangible and Halal Intangible aspects as a stimulus is rear. However, some of the available and recent inventory of research includes Battour (2017); Battour and Ismail (2016); Sodawan and Hsu (2022), and Wingett and Turnbull (2017), etc.

However, Muslims have to follow Shari'ah compliance on daily bases even during traveling. Thus, there are three levels o needs that may provide a pleasant impact on Muslim travelers (Mannaa, 2020). These levels are categorized as Basic, Preferable, and Nice to have. The first level needs to encompass Halal foods and Prayers facilities, the second level includes toilets with bidet showers & services associated with Holy Months and the third level is based on THE availability of conservative entertainment and a separate gym and spa (Mannaa, 2020). Hence through following these studies, this chapter will provide a deeper explanation for all the elements (S) that may reflect positively upon Halal Perceived value (O) to make Muslims prefer any location for tourism (R) for Muslim women. The postulate is true as previously Muslim women face a lot of struggles and hindrances in the process of tourism. However, recently, the percentage of Muslim women travelers has increased significantly, and in 2018 it reaches up to 45% of the overall Muslim tourists. Thus, legitimate to declare Muslim women as the sub-sector of Muslim tourism that are involved in tourism activities for business, leisure, and fun either alone or with their spouses, families, and friends (Tavakoli & Mura, 2021).

Halal or Muslim-Friendly Hotels

The researchers observed a significant surge in the numbers of Halal or Sharia Compliant hotels in Muslim as well as Non-Muslim Countries. In fact, countries (locations) are also marketing these hotels for influencing the Muslim segment of travelers. In fact, these places offer Halal food, non-alcoholic drinks, Directions of Qibla (Kaaba) as well as prayers rooms & calls for prayers etc. Some examples of Halal or Muslim-friendly hotels include Al-Meroz Hotel in Thailand. On the other side, Kyoto City in Japan is also putting high attention to cater to Muslim tourists by proving a list of Muslim-friendly hotels on its website. Moreover, hotels in Kyoto city also place a higher focus on Muslim needs (Battour, 2017).

In fact, it is right that due to the increase in Islamic tourism destinations and hotels are continuously trying to incorporate Islamic values into their activities. However, most of the hotels are still unable to incorporate Islamic values into all of their activities. In fact, these hotels were also criticized for opening their pools for common use despite being promoted as Halal (Duman, 2012).

Halal Foods and Beverages

Halal foods do not only mean slaughtering of animals but it also means the way of providing top-quality products to address ever-increasing customer requirements. Hence Halal foods are not restricted to any particular region or area but it is the demand of Muslims & Non-Muslim from all over the globe. In fact, Non-Muslim countries are top-ranked Halal meat exporters (Rahman et al., 2020). On the other side, it is mandatory for locations to provide Halal foods and drinks in order to attract Muslim tourists. In recent times there is a significant increase in Muslim tourists in countries like France, the United Kingdom, Germany, etc. In fact, Muslim tourists always requested Halal foods and beverages whenever they visit any Non-Muslim location. The availability of Halal foods makes Muslim tourists their favorite food in full faith that eating will not harm their religion and religious beliefs. Thus, will motivate these tourists to prefer the locations that provide Halal foods and beverages. On the other side increase of Muslim travelers to Europe there is a substantial increase in sales of Halal foods and beverages especially in consumer foodservice outlets (Battour, 2017).

Hence it has also been predicted that an increase in investment will also be observed in other Non-Muslim destinations. Studies claimed that the reason behind the increase in investment in this particular area is based upon continuous increases in the percentage of Muslim or Halal tourism. Therefore several places, markets, and hotels from Non-Muslim markets are also getting popular for serving Halal Foods e.g., Ryokans the traditional Japanese hostels are serving Halal foods. Similarly Manhattan fish market is one of the well-known Malaysia has also opened its outlet in Tokyo to serve Halal Food to attract Muslim tourists from all over the globe (Battour, 2017). Similarly has been highlighted by Tavakoli and Mura (2021), that Muslim women always adhere to Halal foods and most of the time will never prefer a location that does not offer Halal foods. Thus, a high level of Halal food assurance is required to motivate Muslim female tourists to visit the location.

Water Friendly Washrooms

Water Friendly washrooms are required by Muslims in order to perform wudu to make them prepare for offering prayers.

Therefore destination must provide washrooms with showers and hoses that must be situated next to the toilet to assure Muslim purity as well as to attract Muslim tourists. However, the destination most of the time does not have an ample supply of water-friendly washrooms which may create hurdles for Muslim females. In fact, difficulties also exist in Malaysia which is ranked one of the premier and friendliest destinations for Muslim women by the Global Muslim Tourism Index (Tavakoli & Mura, 2021)

Muslim Swimming Suit

Availability of Burqini, i.e., Muslim swimming suits for females may also increase the preference of Muslim tourists towards the place. This dress is made up of neoprene that is tailored made to fulfill Muslim women's needs through compels with religious beliefs and preferred way of modesty for Muslim women while swimming. The dress is light enough to assist females in swimming but covers the entire body except for the face, hands, and feet (Battour, 2017).

Muslim Friendly Smart Phone Application

One of the recent fashions of Halal-Tourism is to supplement Muslim tourists with Smart-Phone Applications that leverage tourists in making their tour more desirable and memorable. The purpose of the application is to trace hotels that provide halal foods and shopping malls with praying areas etc (Battour, 2017). The application is getting popular in several Non-Muslim countries like Thailand etc where the application is used to enhance the tourism experience by making users access Muslim-friendly hospitality services. Thus, the application is also available for different forms of Android applications through Google Inc's Android & for the operating system of Apple through Apple Inc's iOS. Other than this tourism authorities are also trying to update this application to other languages like Arabic, Bahasia, and Indonesian languages (Battour, 2017). On the other side Muslim females are rarely been discussed as the host and therefore there is a significant lacking of studies that may make one understand the behavior of tourist Muslim women. However, technology may provide liberty to Muslim women and may decrease resistance that Muslim women may face during tourism (Feldbauer & Jaffrey, 2021).

Halal Cruise

One of the other recent up gradations in Halal Tourism is Muslim friendly cruise that is initially launched in Turkey and sails through Greece. The cruise is not only free from haram foods, beverages but also provides gender-segregated washrooms, a spa, and prayer areas, and sports centers. The cruise also does not accommodate any form of gambling and therefore it is a form of family-friendly Halal cruise for Muslims (Battour, 2017). In fact, these provisions will remove a few of the largest hurdles in front of Muslim women like separate swimming, spa, and sports facilities as mentioned by Tavakoli and Mura (2021).

Muslim-Friendly Airport

The increase in Muslim tourists, especially from South-Asian countries makes Non-Muslim destination to take steps that may facilitate Muslim tourists. One of the prominent is the initiative taken by Kansai International Airport which allocated three separate praying rooms for Muslim tourists. Praying rooms are also segregated on the bases of gender and also at different floors to accommodate Muslim tourists effectively. Moreover, separate washing areas that may serve the purpose of ablution for praying purposes are also available 24/7 at airports. Similarly, Narita Airport of Tokyo also separated a praying room for Muslim tourists back in 2014 to accommodate and satisfy Muslim travelers and tourists (Battour, 2017)

METHODOLOGY

The compilation of this chapter is based on qualitative inquiry through previously published scholarly articles. Hence it is based upon a mono-qualitative methodological choice & archival research strategy (Saunders et al., 2015). Hence legitimate to declare that study collected details from previous scholarly articles and therefore the purpose of the research work is descriptive and the time horizon is cross-sectional (Sekaran & Bougie, 2016). However, the concept of Halal-Holidays is still in the infancy stage and still needs to understand about the concept. However, this study tries to make a thin line of differentiation between Halal-Tourism and Halal Holidays and therefore the inquiry is based upon epistemology as the philosophy of research (Saunders et al., 2007).

REFERENCES

Battour, M. (2017). *Muslim Travel Behavior in Halal Tourism, Mobilities, Tourism and Travel Behavior-Contexts and Boundaries*. Intech Open. Available online: https://www. intechopen. com/books/mobilities-tourism-and-travel-behavior-contexts-and-boundaries/muslimtravel-behavior-in-halal-tourism

Battour, M., & Ismail, M. N. (2016). Halal tourism: Concepts, practises, challenges and future. *Tourism Management Perspectives*, *19*, 150–154. doi:10.1016/j.tmp.2015.12.008

Boğan, E. (2020). Halal tourism: The practices of halal hotels in Alanya, Turkey. *Journal of Tourism and Gastronomy Studies*, *8*(1), 29–42. doi:10.21325/jotags.2020.534

Creswell, J. W., & Creswell, J. D. (2017). *Research design: Qualitative, quantitative, and mixed methods approaches*. Sage Publication.

Duman, T. (2012). The value of Islamic tourism: Perspectives from the Turkish experience. *ICR Journal*, *3*(4), 718–739. doi:10.52282/icr.v3i4.513

Eid, R., & El-Gohary, H. (2015). The role of Islamic religiosity on the relationship between perceived value and tourist satisfaction. *Tourism Management*, *46*, 477–488. doi:10.1016/j.tourman.2014.08.003

Feldbauer, I. G., & Jeffrey, H. L. (2021). Disempowered Hosts? A Literature Review of Muslim Women and Tourism. *Women in Tourism in Asian Muslim Countries*, 13-23.

Grönroos, C. (2007). *Service management and marketing: customer management in service competition* (3rd ed.). John Wiley.

Mannaa, M. T. (2020). Halal food in the tourist destination and its importance for Muslim travellers. *Current Issues in Tourism*, *23*(17), 2195–2206. doi:10.1080/13683500.2019.1616678

Mehrabian, A., & Russell, J. A. (1974). *An approach to environmental psychology*. The MIT Press.

Rahman, M., Moghavvemi, S., Thirumoorthi, T., & Rahman, M. K. (2020). The impact of tourists' perceptions on halal tourism destination: A structural model analysis. *Tourism Review*, *75*(3), 575–594. doi:10.1108/TR-05-2019-0182

Rodrigo, P., & Turnbull, S. (2019). Halal holidays: How is value perceived by Muslim tourists? *International Journal of Tourism Research*, *21*(5), 675–692. doi:10.1002/jtr.2290

Saunders, M., Lewis, P., & Thornhill, A. (2007). Research methods. Business Students, 6(3), 1-268

Saunders, M. N. K., Lewis, P., Thornhill, A., & Bristow, A. (2015). *Understanding research philosophies and approaches: Research methods for business students*. Pearson.

Sekaran, U., & Bougie, R. (2016). *Research methods for business: A skill building approach*. John Wiley & sons.

Sheth, J. N., Newman, B. I., & Gross, B. L. (1991). Why we buy what we buy: A theory of consumption values. *Journal of Business Research*, *22*(2), 159–170. doi:10.1016/0148-2963(91)90050-8

Sodawan, A., & Hsu, R. L. W. (2022). Halal-Friendly Attributes and Muslims' Visit Intention: Exploring the Roles of Perceived Value and Destination Trust. *Sustainability (Basel)*, *14*(19), 1200. doi:10.3390u141912002

Tavakoli, R., & Mura, P. (2021). Muslim women travellers' constraints: A critical review. *Women in Tourism in Asian Muslim Countries*, 25-40.

Wingett, F., & Turnbull, S. (2017). Halal holidays: Exploring expectations of Muslim-friendly holidays. *Journal of Islamic Marketing*, *8*(4), 642–655. doi:10.1108/JIMA-01-2016-0002

Chapter 13
Tourism as a Women's Empowerment Tool:
The Interrelated Factors and Issues

Eman Zameer Rahman
https://orcid.org/0000-0003-2420-625X
Bahria University, Pakistan

Haider Ali Shah
Bahria University, Pakistan

ABSTRACT

There is an increasing rate of women travelers and entrepreneurs globally, generally associated with the movement of 'women empowerment'. The tourism industry has increased the opportunity for businesses, while women are playing a full role in overtaking these opportunities. The chapter highlights modern women and the role that the tourism industry has played to empower them through the inter-role factors, the impact of digitization, and issues to women in the tourism industry as an entrepreneur. Sustainable tourism development in popular destinations increasingly depends on women's empowerment. On the other hand, empowering women and achieving gender equality is one of the goals of the United Nations' sustainable development agenda (UN-SDGs). Entrepreneurship is a key factor in advancing women's equality.

INTRODUCTION

A major global movement emphasizing the advancement of women and the protection of their rights has emerged in recent years and is making strides toward its goals. Women empowerment is itself a fundamental human right that's necessary to make the world a better place. The term "women's empowerment" refers to efforts to increase women's confidence in their abilities, freedom to make their own decisions, and access to political and social power. Women empowerment is the process by which women both individually and collectively develop into knowledgeable and purposeful actors who either lead or back

DOI: 10.4018/978-1-6684-8417-3.ch013

efforts to eliminate gender inequality (Cole, 2018). Women's empowerment can be broken down into four distinct areas: psychological, political, social, and economic (Abou-Shouk, Mannaa, & Elbaz, 2021).

One of the modern world's most important industries, tourism has grown into a huge worldwide economic force in recent decades. Edgell (2016) urges for "empowerment" to be included as a benefit to local communities from tourism development as part of a sustainable strategy. Women can gain economic independence through entrepreneurship. Improvements in women's and families' economic standings as well as the general prosperity of their communities can be traced directly to women's increased involvement in business (Scott, Dolan, Johnstone-Louis, Sugden, & Wu, 2012; Haugh & Talwar, 2016).

The United Nations' sustainable development agenda is to promote gender equality (UN-SDGs). Hence, the ability to start a business is a crucial part of women's empowerment. The extent to which women can participate in society is heavily influenced by cultural norms. Women's empowerment in the tourist industry can be predicted in large part by how people view women's roles in the industry and the success of women entrepreneurs (Abou-Shouk, Mannaa, & Elbaz, 2021).

While the world has achieved progress towards gender equality and women's empowerment under the Convention on the Elimination of all forms of Discrimination against Women, women and girls continue to suffer discrimination and violence in every part of the world. (UNWTO, 2023)

Traveling might boost women's creativity and provide them with a sense of freedom which is strategically translated into businesses by many women today. The tourism industry can support women by enhancing their experiences to explore new cultures, trends, and ideas and providing them opportunities to empower themselves. Traveling might boost their creativity and provide them with a sense of freedom which is strategically translated into businesses by many women today. Yet, women have limited time to travel (Women's Entrepreneurship: Issues and Policies, 2004).

The tourism industry has also increased the opportunity for businesses, while women are playing a full role in overtaking these opportunities which can be assessed by Figure 1 which shows the involvement of women in different parts of the tourism industry. The percentage in the figure represents the

Figure 1. Employment characteristics in tourism
Source: EU28 (2014).

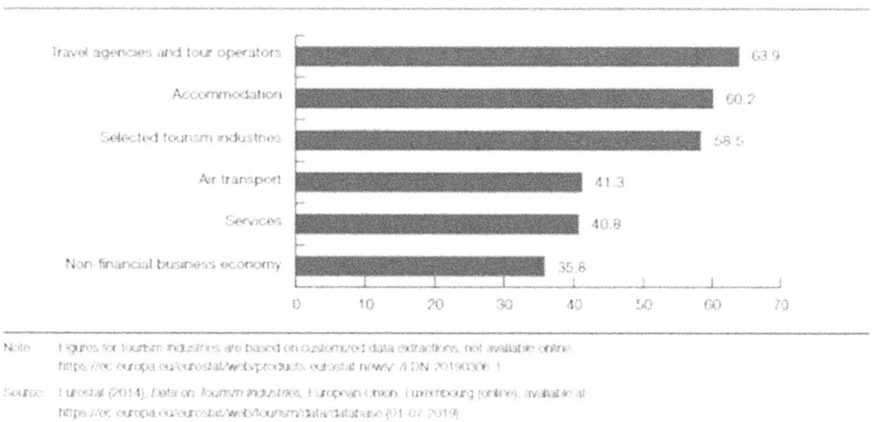

share of employed women. Hence, the chapter focuses on the importance of women entrepreneurship, the issues women entrepreneurs face and the factors and opportunities such as digitization that lead to empowering women in the tourism sector.

UNDERSTANDING THE ROLE OF WOMEN'S EMPOWERMENT

Empowering women is becoming an essential part of developing sustainable tourism in popular places. Women's entrepreneurship is a complex phenomenon (De Bruin, Brush, & Welter, 2006) and it can be studied from a variety of angles, including as a form of activism when women invest in other women and inspire them to take action (Calas, Smircich, & Bourne, 2009) as an economic development when viewed through the lens of business investment (Hania, Rachmaniaa, Setyaningsih, & Putri, 2012) as a tool of social policy when material security influences the encouragement of self-employment (Kimbu & Ngoasong, 2016); and as sustainable development when women's entrepreneurship is viewed as a means of balancing economic growth with environmental and social growth via environmentally and socially conscious enterprise (Shams, 2017).

There is an understanding that women can gain economic empowerment through entrepreneurship. Improvements in women's and families' economic positions as well as the general prosperity of their communities can be traced directly to women's increased involvement in business (Peeters & Ateljevic, 2009). The tourist industry has the potential to greatly narrow the gap between the economies of industrialized and developing nations by penetrating their very foundations (Kulkarni & Harshitha.Rao., 2022).

Tourism is seen as a powerful tool for attaining development due to its rapid and sustained expansion. Women business owners that focus on tourism as a means of economic growth can make significant contributions to their communities and nations. For societies around the world still in the process of modernizing and advancing, women empowerment is a crucial issue. Women's self-employment has the potential to challenge and transform the existing power structure because it improves women's social position both individually and collectively (Soroushmehr, Kalantari, Fami, & Sarani, 2012). Cultural interchange and international cooperation are bolstered by tourism, and it also serves to promote social and political understanding (Kulkarni & Harshitha.Rao., 2022). The level of women's empowerment has a considerable impact on the expansion of the tourism industry in Arab nations (Abou-Shouk, Mannaa, & Elbaz, 2021).

Furthermore, women empowerment to prosper in the tourism industry, especially in undeveloped regions, must establish and maintain productive relationships with local stakeholders (Vukovic, Petrovic, Maiti, & Vujko, 2021). Self-actualization leads to women's empowerment which can be attained through local tourism stakeholder support as reported in the regions of Malaysia (Kunjurman & Hussin, 2016), Iran (Soroushmehr, Kalantari, Fami, & Sarani, 2012), Tanzania (Mrema, 2015) and even in European regions (Nassani et al., 2019). Moreover, women in many rural areas have formed self-help groups (SHGs), which improve communication with multiple parties.(Vukovic, Petrovic, Maiti, & Vujko, 2021). Group membership can give women more recognition, and exposure and equip them with the resources they need to make a real change (Soroushmehr, Kalantari, Fami, & Sarani, 2012).

WOMEN'S EMPOWERMENT TOOL AND THE INTERRELATED FACTORS

Empowerment of women is a dynamic process that has been variously defined, characterized, and quantified (Mahmud, Shah, & Becker, 2012). There are various dimensions and determinants of women's entrepreneurship. The incorporation of women entrepreneurs in the tourism industry is considerably and favorably influenced by psychological empowerment, perceptions of women's work, self-efficacy, psychological empowerment, and empowering leadership. Moreover, women entrepreneurs' involvement in tourism has a major effect on sustainable tourism development (Samad & Alharthi, 2022).

Women's empowerment is influenced by a variety of social, economic, demographic, family, and information-exposure factors (Abbas, Isaac, Zia, Zakar, & Fischer, 2021). Women in the hospitality industry are substantially influenced by their level of risk-taking, empowerment, subjective norm, and self-efficacy. The most crucial aspect is one's perspective on taking risks, as taking risks indicates more aspirations to be an entrepreneur (Yoopetch, 2021). Empowerment manifests itself along four dimensions: more confidence, involvement in decision-making, control of material security, self-efficacy, and increased independence in terms of mobility; while the determinants are information exposure, social status, economic status, and demographic status of the women as with these the level and factors that influence empowerment of women varies (Mahmud, Shah, & Becker, 2012; Abbas, Isaac, Zia, Zakar, & Fischer, 2021). There are two crucial facets of women's empowerment: the ability to make decisions and the right to possess property as illustrated in Figure 2. Issues of healthcare, economics, and mobility all factor into the decision-making process.

Social entrepreneurship and conventional entrepreneurship are distinct from one another. Conventional entrepreneurship emphasizes bold market initiatives like product development, market penetration, new ventures, business model and innovation. In contrast, social entrepreneurship is characterized by a greater commitment to environmental and social responsibility, a focus on measurable outcomes, and

Figure 2. Determinants and dimensions of women's empowerment
Source: Abbas, Isuuc, Zia, Zakar, and Fischer (2021)

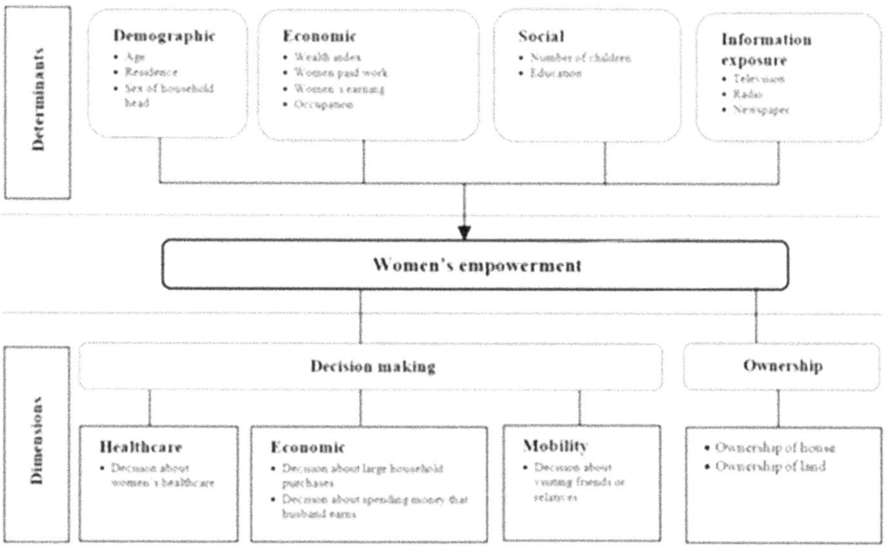

a clear plan for effecting positive social change (Branzei, 2012). Women empowerment, social change and entrepreneurship are related factors (Haugh and Talwar, 2016) hence, social entrepreneurship and women entrepreneurs are strongly correlated.

Consequently, we must consider the green economy's intrinsic potential for advancing women's rights and the possibility of empowering women through tourism. Green entrepreneurs in the tourism sector, including areas of digitalization and green growth, from a female perspective. The positive effects of this field on women have received insufficient research attention so far (Radovi-Markovi & Ivanovi, 2019). Women entrepreneurs can greatly benefit from embracing digital technologies and marketing strategies or else known as digital entrepreneurship (Zainol, Esa, Muhamad, & Mohamad, 2022).

Digitization has led to digital entrepreneurship that has generated various opportunities for all but has proven to be a barrier breaker for women. Promoting niche tourism among younger generations via OTP and app-driven services, the startup and entrepreneurship process in the travel industry creates jobs in this dynamic and expanding sector. Commercial tourism's adoption of electronic commerce (E-commerce) was hastened by the advent of travel automation, which in turn boosted the prominence of websites, OTPs, and app-driven M-commerce. As a result of the wide variety of niches available for tourists to explore, the travel industry is ripe with opportunities for new businesses and entrepreneurs to capitalize on the need for intermediate services (Krishnamurthy, Rajasekar, & Jebasingh, 2019).

Digitization has led to more empowerment of women and involvement in the tourism sector surpassing most of the pre-defined determinants and dimensions that were pre-defined. However, certain factors can be concluded to be absolute such as risk-taking, self-efficacy, decision-making, exposure, and resources.

WOMEN'S EMPOWERMENT TOOL AND THE ISSUES

A UNTWO report investigated the factors that contribute to gender equality in the tourism industry, to identify the barriers to achieving gender equality in this field, and propose solutions that will allow us to overcome these barriers and use the tourism industry to its fullest potential in empowering women and girls around the world. The report highlighted that decent work for women in the tourism industry can be promoted by targeted interventions by public, private, and civil society actors, such as encouraging equal pay, combating sexual harassment, and recruiting women into high-level jobs.

The UN World Tourism Organization studied women's roles in the tourism industry and found the following. One striking fact about the tourism industry is the prevalence of female workers. Women's earnings can be up to 15% lower than men's. At the parent companies, women do most of the unpaid work (UWTO Report Women in Tourism, 2010). Since there is a large gap between the number of men and women working in the industry, gender inequality is rampant. Also, unskilled or low-skilled women are disproportionately represented in the lowest-paying, most precarious occupations. Separate and unequal educational and career opportunities for women contribute to the fact that they earn significantly less than their male counterparts for equivalent work. Migrant workers and women from marginalized socioeconomic groups are more likely to suffer these inequities, hence it is important to consider problems of race, ethnicity, and migration when examining gender inequality.

In light of the promise made by the 2030 Plan for Sustainable Development to "leave no one behind," it is essential that these intersecting inequities be addressed. The International Labor Organization (ILO) suggests establishing tourist policies that promote full and productive tourism employment, sustainable

Figure 3. Goals for gender equality in tourism

tourism enterprises, and decent employment for women and youth, as a means of generating full and productive employment and decent work for all (A global report on women in tourism, 2019).

There are steps suggested by the global report on women in tourism, 2019:

- Put into effect international labor standards and ensure their enforcement
- Ensure freedom of association and the right to collective bargaining
- Reduce the prevalence of nonstandard forms of employment
- Encourage workers to become formally employed
- Provide social security and maternity protection
- Invest in the training and education of workers
- Promote equality and nondiscrimination
- Eliminate forced and child labor
- Improve working condition

To ensure gender equality in the sector of tourism a few goals and objectives are required as depicted in Figure 3, such as:

- ***Employment*:** To provide good and decent opportunities in the tourism sector for women.

- ***Entrepreneurship*:** The involvement of women in tourism businesses leads to economic empowerment in the economy.
- ***Education and training*:** Providing education and training in the education sector promotes the empowerment of women in the tourism sector.
- ***Leadership and decision-making*:** Women have to be a part of the major roles to make decision-making at all tourism levels.
- ***Community and civil society:*** Tourism supports women to address issues related to gender inequalities.

Moreover, the financing needs of green tourism entrepreneurs have been shown to differ significantly from the financial system's willingness to assist women entrepreneurs with funding. Commercial banks do not provide enough suitable loan products. In the tourism industry, women are underrepresented in startup ventures because of a lack of combined financial support and appropriate financial tools (Radovi-Markovi & Ivanovi, 2019).

Public and commercial sectors have taken and should continue to take significant steps towards a gender-responsive recovery and in support of women business owners in the tourism industry. They include the maintenance of programs for training and education, the opening of new markets, the provision of wage subsidies, the elimination or reduction of some taxes, the broadening of access to credit, the lowering of interest rates, and the decrease of social security payments. These are some measures that can be taken by the commercial sector and government to support women entrepreneurs (UN, 2022).

Furthermore, female mindsets can be influenced and their interest in becoming entrepreneurs can be piqued by the sharing and promotion of the success stories of female entrepreneurs (Yoopetch, 2021).

CONCLUSION

The chapter highlights the increasing rate of women travelers and entrepreneurs globally, and how the tourism industry has provided opportunities for women to play a full role in overtaking these opportunities. It also explores the interrelated factors and issues related to women entrepreneurship and the attainment of the gender equality SDG.

The methodology of the chapter is secondary analysis of the existing data available on the topic, which suggests that the chapter is based on a thorough review of relevant literature and data. Additionally, the chapter notes that sustainable tourism development in popular destinations increasingly depends on women's empowerment, and that empowering women and achieving gender equality is one of the goals of the United Nations' sustainable development agenda (UN-SDGs). Entrepreneurship is also identified as a key factor in advancing women's equality.

Women's economic empowerment in the tourism industry is enhanced by well-implemented gender-sensitive national legislation and macroeconomic policies. More results for gender equality can be achieved by investing in skills training for women, including training in soft skills and raising awareness of available training options. In order to truly empower women, gender equality measures in the tourism industry need to be supported by both institutions and financial resources.

As connections are formed between the tourism industry and local communities and non-governmental groups, women are able to take a more active role in civic life and the political sphere. Connecting women entrepreneurs with women's initiatives in the tourist industry show how far we have come in recognizing the importance of women entrepreneurs in driving both traditional business and the kind of social innovation and change that arises from that. Women empowerment factors include risk-taking, self-efficacy, decision-making, exposure, and resources. Most importantly digitization is playing a huge role in women empowerment.

The solution of women empowerment is in the digitalization of tourism as it can create exciting new prospects for women's innovation and empowerment if women have access to appropriate technology and receive gender-sensitive training. With more gender-disaggregated tourist data available, organizations may more precisely address gender inequality and boost women's economic participation.

Overall, the chapter effectively addresses its agenda by providing a comprehensive overview of the role that the tourism industry plays in empowering women, as well as the factors and issues related to women entrepreneurship and the attainment of the gender equality SDG.

SOLUTIONS AND RECOMMENDATIONS

The chapter highlights the significance of women entrepreneurship in the tourism industry and the importance of women's empowerment. It emphasizes that women's involvement in business can contribute to the economic growth of their communities and nations, as well as narrow the gap between the economies of industrialized and developing countries. The chapter also discusses the factors and opportunities such as digitization that lead to empowering women in the tourism sector.

To enhance women's empowerment in the tourism industry, it is recommended to establish and maintain productive relationships with local stakeholders and promote the formation of self-help groups. The tourism industry can support women by enhancing their experiences to explore new cultures, trends, and ideas and providing them opportunities to empower themselves. Women entrepreneurs' involvement in tourism has a major effect on sustainable tourism development.

To further empower women in the tourism industry, it is suggested to focus on psychological empowerment, perceptions of women's work, self-efficacy, psychological empowerment, and empowering leadership. The chapter highlights that women's empowerment is influenced by various social, economic, demographic, family, and information-exposure factors, which needs to be focused on when considering women empowerment.

FUTURE RESEARCH DIRECTIONS

In the future, it is essential to address the issues that women entrepreneurs face in the tourism industry and provide them with the necessary support to overcome these challenges. Additionally, there is a need to focus on bridging the gender gap and promoting gender equality in the tourism industry. Furthermore, it is recommended to promote entrepreneurship education and training programs for women to enhance their skills and knowledge, which will lead to more successful and sustainable businesses in the tourism sector.

REFERENCES

A global report on women in tourism. (2019). UNWTO. Retrieved from https://www.unwto.org/publication/global-report-women-tourism-2-edition

Abbas, S., Isaac, N., Zia, M., Zakar, R., & Fischer, F. (2021). Determinants of women's empowerment in Pakistan: Evidence from Demographic and Health Surveys, 2012–13 and 2017–18. *BMC Public Health*, *21*(1), 1328. doi:10.118612889-021-11376-6 PMID:34229626

Abou-Shouk, M. A., Mannaa, M. T., & Elbaz, A. M. (2021). Women's empowerment and tourism development: A cross-country study. *Tourism Management Perspectives*, *37*, 100782. Advance online publication. doi:10.1016/j.tmp.2020.100782

Branzei, O. (2012). *Social change agency under adversity: how relational processes (re)produce hope in hopeless settings*. Routledge.

Calas, M., Smircich, L., & Bourne, K. (2009). Extending the boundaries: Reframing 'entrepreneurship as social change' through feminist perspectives. *Academy of Management Review*, *34*(3), 552–569. doi:10.5465/amr.2009.40633597

Cole, S. (2018). *Gender Equality and Tourism: Beyond Empowerment*. CABI. doi:10.1079/9781786394422.0000

De Bruin, A., Brush, C., & Welter, F. (2006). Introduction to the Special Issue: Towards building cumulative knowledge on women's entrepreneurship. *Entrepreneurship Theory and Practice*, *30*(5), 585–593. doi:10.1111/j.1540-6520.2006.00137.x

Edgell, D. L. (2016). *Managing Sustainable Tourism: A Legacy for the Future*. Routledge. doi:10.4324/9781315688343

Hania, U., Rachmaniaa, I., Setyaningsih, S., & Putri, R. (2012). Patterns of Indonesian women entrepreneurship. *Procedia Economics and Finance*, 4.

Haugh, H., & Talwar, A. (2016). Linking social entrepreneurship and social change: The mediating role of empowerment. *Journal of Business Ethics*, *133*(4), 643–658. doi:10.100710551-014-2449-4

Kimbu, A., & Ngoasong, M. (2016). Women as vectors of social entrepreneurship. *Annals of Tourism Research*, *60*, 63–79. doi:10.1016/j.annals.2016.06.002

Krishnamurthy, L., Rajasekar, A., & Jebasingh, D. (2019). An Empirical Study on Role of Travel Start-Ups and Entrepreneurship in E-Tourism. *Journal of Service Science and Management*, *12*(5), 620–627. doi:10.4236/jssm.2019.125042

Kulkarni, R., & Harshitha.Rao, R. (2022). A Study on Challenges and Opportunities of Women Entrepreneurs in Tourism Sector. *Journal of Positive School Psychology*, *6*(5), 6732–6743.

Kunjurman, V., & Hussin, R. (2016). Women participation in ecotourism development: Are they empowered. *World Applied Sciences Journal*, *34*(12), 1652–1658.

Mahmud, S., Shah, N. M., & Becker, S. (2012). Measurement of Women's Empowerment in Rural Bangladesh. *World Development*, *40*(3), 610–619. doi:10.1016/j.worlddev.2011.08.003 PMID:23637468

Mrema, A. (2015). Tourism and women empowerment in monduli district, arusha-Tanzania. *African Journal of Hospitality, Tourism and Leisure*, *4*(2), 1–14.

Nassani, A., Aldakhil, A., Abro, M., Islam, T., & Zaman, K. (2019). The impact of tourism and finance on women empowerment. *Journal of Policy Modeling*, *41*(2), 234–254. doi:10.1016/j.jpolmod.2018.12.001

Peeters, L., & Ateljevic, I. (2009). Women empowerment entrepreneurship nexus in tourism: processes of social innovation. In J. Ateljevic & S. J. Page (Eds.), *Tourism and Entrepreneurship*. Routledge.

Radović-Marković, M., & Živanović, B. (2019). Fostering Green Entrepreneurship and Women's Empowerment through Education and Banks' Investments in Tourism: Evidence from Serbia. *Sustainability (Basel)*, *11*(23), 6826. Advance online publication. doi:10.3390u11236826

Samad, S., & Alharthi, A. (2022). Untangling Factors Influencing Women Entrepreneurs' Involvement in Tourism and Its Impact on Sustainable Tourism Development. *Administrative Sciences*, *12*(2), 52. Advance online publication. doi:10.3390/admsci12020052

Scott, L., Dolan, C., Johnstone-Louis, M., Sugden, K., & Wu, M. (2012). Enterprise and inequality: A study of Avon in South Africa. *Entrepreneurship Theory and Practice*, *36*(3), 543–568. doi:10.1111/j.1540-6520.2012.00507.x

Shams, S. (2017). Transnational education and total quality management: A stakeholder-centred model. *Journal of Management Development*, *36*(3), 376–389. doi:10.1108/JMD-10-2015-0147

Soroushmehr, H., Kalantari, K., Fami, H., & Sarani, V. (2012). Investigation of self-help groups (SHGs) effect on rural women empowerment (Hamedan county, Iran). *Journal of Agricultural Science*, *4*, 1–11.

UN. (2022). *Advancing Women's Economic Empowerment in the Tourism Sector in COVID-19 Response and Recovery*. UN. https://www.unwomen.org/sites/default/files/2022-05/Advancing-womens-economic-empowerment-in-the-tourism-sector-in-COVID-19-en.pdf

UNWTO. (2023a). *Global Report on Women in Tourism: Second Edition*. UNWTO. Retrieved from https://www.e-unwto.org/doi/pdf/10.18111/9789284420384

UNWTO. (2023b). *Women's empowerment and tourism*. Retrieved from UNWTO: https://www.unwto.org/gender-and-tourism

UWTO Report Women in Tourism. (2010). Retrieved from UWTO: UWTO Report Women in Tourism. 2010. Available online: https://cf.cdn.unwto.org/sites/all/files/pdf/global_report_on_women_in_tourism_2010.pdf

Vukovic, D. B., Petrovic, M., Maiti, M., & Vujko, A. (2021). Tourism development, entrepreneurship and women's empowerment – Focus on Serbian countryside. *Journal of Tourism Futures*, 1-21.

Women's Entrepreneurship, Issues, and Policies. (2004). *2nd OECD conference of ministers responsible for small and medium-sized enterprises (SMEs) promoting entrepreneurship and innovative SMEs in a global economy: towards a more responsible and inclusive globalisation*. Istanbul, Turkey: OECD.

Yoopetch, C. (2021). Women empowerment, attitude toward risk-taking and entrepreneurial intention in the hospitality industry. *International Journal of Culture, Tourism and Hospitality Research*, *15*(1), 59–76. doi:10.1108/IJCTHR-01-2020-0016

Zainol, N., Esa, M. M., Muhamad, N., & Mohamad, S. S. (2022). Factors Affecting The Success of Digital Entrepreneurship among Suburban Women: Malacca. *International Journal of Academic Research in Business & Social Sciences*, *12*(12), 153–163. doi:10.6007/IJARBSS/v12-i12/15426

Chapter 14
The Role and Impact of Technology, Digitalisation, and Social Media on Consumer Experience in the Tourism and Hospitality Industry:
From the View of Women Entrepreneurs

Aysegul Eda Kop
Sinop University, Turkey

ABSTRACT

This chapter discusses the widespread use of technology in the tourism and hospitality industry by companies and how new technologies such as social media, virtual reality (VR), artificial intelligence (AI), and augmented reality (AR) technologies enhance consumer experience among visitors, current consumers, or potential consumers. Technology use is discussed within the tourism industry; its use by women entrepreneurs has also been detailed in this chapter. Lastly, the author has explored how the COVID-19 pandemic influenced the tourism industry, if there are any differences among different genders and if women entrepreneurs could manage the process well enough compared to men entrepreneurs.

1. INTRODUCTION

The tourism and hospitality industry offer a variety of services such as sleep, transportation, food, tours, attractions, entertainment and shopping that are provided in many different locations such as hotels, restaurants, transports, airlines, tourism destinations, heritage tourism sites, museums, art galleries, protected areas, science festivals, theme parks that provide the experience of hospitality (Quan and Wang, 2004). The main services that consumers expect to find in these places and be satisfied can be listed as “social atmosphere”, “location”, “staff”, “cleanliness”, “security” and “facilities & equipment”. These

DOI: 10.4018/978-1-6684-8417-3.ch014

criteria are found to be important factors when consumers assess a touristic place. For instance, visitors review the tourism destinations, accommodations, and hotels based on these criteria to rank the places. In another study, sustainability is identified as another factor that increases the quality of consumer experience and satisfaction along with development of touristic places such as archaeological heritage sites, as well as other types of heritage and tourism attractions (Daengbuppha et al., 2006).

The experience in tourism can vary between visual and non-visual experience. These both types of experience are important for customer satisfaction. These visual and non-visual experiences can be either the main or the supporting performance of the consumer experience (Quan and Wang, 2004). Satisfied customers are found as eager to share their experience with others by providing offline and online reviews rather than giving feedback or complaining to the company (Sotiriadis, 2017). Thus, it is possible to say that variety of factors such as consumer satisfaction, and service quality influence consumer experience positively and increase customer loyalty and positive word-of-mouth (WOM) (Sotiriadis, 2017). Although, satisfied consumers are more eager to share their experience in the tourism and hospitality industry, both positive and negative reviews as well as positive and negative Word–of–Mouth (WOM) can be seen on both online and offline platforms. Therefore, this chapter discusses mainly the consumer experience and how to increase consumer experience to obtain more satisfied and loyal customers. Increasing consumer experience to obtain more satisfied and loyal customers is essential because the consumers' positive or negative online comments will always follow the company and its products' future.

This chapter initially discusses the consumer experience in the tourism and hospitality industry and how technology influences the consumer experience. In this respect, Technology use is evaluated from 2 perspectives. Firstly, the use of the Web and social media usage is explained for the tourism and hospitality industry. Secondly, the use of other interactive Technologies such as Virtual Reality (VR) and Artificial Intelligence (AI) Technologies is detailed by explaining how touristic places can benefit from those technologies and how these technologies impact consumer experience in the tourism and hospitality industry. Last but not least, women entrepreneurs and their activities during covid-19 pandemic is discussed by focusing on consumer experience in tourism and how women entrepreneurs in the tourism and hospitality industry benefit from new technologies during this period.

2. CONSUMER EXPERIENCE IN TOURISM AND HOSPITALITY INDUSTRY

Consumer experience in the tourism and hospitality industry has aspects of service approach than physical product (Otto and Ritchie, 1996). However, consumer satisfaction is not only about service quality (Otto and Ritchie, 1996). In addition to quality, entertainment, keeping safe, and the physical environment are important criteria for consumer satisfaction (Otto and Ritchie, 1996, Walls, 2013). The factors that influence consumer satisfaction in different tourism and hospitality locations are researched by Rojas et al. (2020, p. 223) who summarised the important factors as follows:

Concerning hotels, customer needs were classified into seven categories: (1) room/bathroom aspects, (2) location, (3) food aspects, (4) environment, (5) staff aspects, (6) security facilities, and (7) others. Related to restaurants, customer needs were classified also into seven categories: (1) food quality, (2) food variety, (3) environment, (4) location, (5) staff aspects, (6) service aspects, and (7) others. Regarding airlines, customer needs were classified also into four categories: (1) staff aspects, (2) cabin, (3) service aspects, (4) ground, and (5) others.

Knowing these criteria is of cardinal importance for the companies to understand their consumers and satisfy their needs in order to generate a positive customer experience in tourism and hospitality industry (Rojas et al., 2020). Cultural differences along with emotional and social values influence consumer needs and satisfaction levels (Teng, 2011). Since the satisfied consumer experience is essential for the tourism and hospitality industry, companies have to consider the consumers' expectations as well (Chen, 2021). One of the recent expectations of the visitors is environmental protection. Therefore, 'green hotel' concept has become a criterion for choosing the hotel by consumers. For these reasons, companies in the tourism and hospitality industry need to consider a wide range of aspects for the satisfaction of their consumers.

Additionally, the employees' body language, nonverbal behaviour engagement with visitors, co-creation, insight exchange, and relationship building are indicators of visitors' satisfaction and positive relations with visitors in the tourism and hospitality industry (Islam and Kirillova, 2021). Consumer experience is significantly important to create bonds between the consumers and the companies and create loyal customers (Mitchell and Mitchell, 2001). The interaction between consumers and the company staff helps consumer satisfaction and value creation in the tourism industry (Barnes et al., 2020). Thus, it is essential for companies to build this bond between customers and the companies. Building the host-guest connection and encouragement of consumers to get the hotel experience eagerly by themselves are among the key points of success in the tourism and hospitality industry (Alcántara-Alcovera et al., 2013).

In Gupta et al.'s (2022) recent study on the luxury service experience in the context of tourism and hospitality, it is highlighted that the tourism and hospitality industries are industries that offer emotional and hedonic values and create positive customer reactions. These values and positive consumer experiences improve the connection with the brand and increase brand loyalty as well as brand love. The personalized and memorable consumer experiences offered by luxury brands help to achieve this customer value in the eyes of the consumers. 'Store atmosphere, online website experience, social media management, or personalized brand communication' are identified as factors that create customer value (Gupta et al., 2022, p. 1). The consumer experience from the women entrepreneurs' point of view will be explained in the next section.

3. UNDERSTANDING CONSUMER EXPERIENCE IN TOURISM AND HOSPITALITY INDUSTRY BY WOMEN ENTREPRENEURS

Although the consumer experience is essential for all industries, how women entrepreneurs can make a difference to obtain consumer experience in the tourism and hospitality industry may vary. As mentioned in the previous section, environmental issues have become a highly important factor among the modern responsible consumers along with their choices and preferences. Therefore, such concerns are set to become central topics that entrepreneurs evaluate in the tourism and hospitality industry. One of the examples of how women entrepreneurs can make a difference in that area is researched by Ertac and Tanova (2020) whose study is about developing a sustainable tourism model on a small island in the Mediterranean Sea, Cyprus. Women in Cyprus can help improve ecotourism. The results of this study reveal that ecotourism can make a difference in women's lives in a constructive way. Empowerment through ecotourism activities encourages especially women entrepreneurs to have a wellfare and growth mindset (Ertac and Tanova, 2020). Change can start with women in the society and influence the whole society by increasing women's employment rate and, thus, helping not only women entrepreneurs but also women in society to acquire the necessary knowledge, skills and abilities.. This approach will also have a positive impact on the overall society by increasing the sensitivity to social and environmental issues and improving environmental policies. Women have considerable influence on changing the society. Taking the society, their needs, expectations together with sustainability or green consumption into consideration can make noticeable difference in consumers' minds with respect to consumer experience. Furthermore, women entrepreneurs can have an influence on economic growth and development (Kunene et al., 2022; Alonso-Almeida, 2013).

However, women entrepreneurs come across some challenges in the tourism industry and other industries. The challenges that women entrepreneurs commonly face are mainly due to gender inequality (Ertac and Tanova, 2020; Alonso-Almeida, 2013; Vujko et al., 2019). Some of the challenges that women entrepreneurs mostly come across are summarised by (Kunene et al., 2022, p. 19060):

Access to credit facilities and finances; lack of awareness of business support services; lack of mentors; poor access to business networks; gender discrimination; lack of confidence; lack of entrepreneurial training; no access to profitable markets; lack of information; balancing family role and business; various societal norms and expectations; and lack of business management skills.

While finance and access to credit are identified as the main challenges for women entrepreneurs by these authors, Kunene et al. (2022) summarised these challenges into 3 main categories: (i) starting a business venture, (ii) operating their business, (iii) growing their business venture.

Likewise, Alonso-Almeida (2013) emphasizes the financial challenge that women entrepreneurs face more than men entrepreneurs. This challenge is identified as starting the business with less capital and receiving less external financial support compared to men entrepreneurs. On the contrary, considering corporate growth, Alonso-Almeida (2013) observed that businesses owned by women entrepreneurs outperform businesses owned by men entrepreneurs when given the same financial opportunities in the tourism and hospitality industry.

The challenges that women entrepreneurs –particularly tourism entrepreneurs- face is summarised by Meera (2014) who studied especially Tourism-Specific issues such as (i) unreliable income, (ii) seasonality in terms of work and income, (iii) profit earning opportunity within a short span of time, (iv) Lack of professional Education, (v) lack of innovations, (vi) Lack of Infrastructure, (viii) strong competition.

Since women entrepreneurs face these challenges, their performance in gaining profits, market share, contribution, and sustainability gets low (Meera, 2014). Alonso-Almeida (2013) disagrees with women entrepreneurs' low impact on sustainability by highlighting that women entrepreneurs have a positive role in fighting against water scarcity. By managing water resources well and following practices that help to protect the environment, the wealth of society increases. Similarly, Vujko et al.'s (2019) study in the Serbian business environment shows that women entrepreneurs face an unsupportive environment which has a negative impact on the development of women's entrepreneurship. However, it is stated that in terms of the need for training and support, women need less than men (Figueroa-Domecq et al., 2020). Thus, it is possible to say that if women and men entrepreneurs were given equal opportunities, the success of women would not be subordinate to men entrepreneurs. Although women entrepreneurs may face such kind of challenges because of gender, Meera (2014) argues that the tourism and hospitality industry is considered as a suitable and good area for women entrepreneurs due to women's entrepreneurial abilities.

Being aware of all these challenges will help society to mitigate gender inequities among organizational and community-basedbased entrepreneurs who can contribute to social change (Ozkazanc-Pan et al., 2018). In line with this, it was found that establishing strong ties with the government agencies and other businesses may help women entrepreneurs access more resources (Ribeiro et al., 2021). Considering the increasing number of women entrepreneurs, government support, education for women, and encouragement for economic independence are needed to overcome these challenges (Meera, 2014).

In terms of competitive service industry especially under this gender inequality, creating a unique and memorable consumer experience in addition to the services and products is needed for providing a priviledged service for customers compared to rivals (Walls, 2013). Therefore, technology and social media can be used in different ways by companies in the tourism and hospitality industry to provide a priviledged service to consumers as will be detailed in the following section.

4. THE ROLE AND IMPACT OF TECHNOLOGY, DIGITALISATION, AND SOCIAL MEDIA ON CONSUMER EXPERIENCE IN THE TOURISM AND HOSPITALITY INDUSTRY

To begin with, engagement and better use of the Internet, Web and social networking services have various positive outcomes on consumer experience, such as (i) better online value creation by museums, (ii) increases in Internet ranking and the number of followers of museums in social media, (iii) enhanced information exchange before and after the visit, (iv) motivation of the public and encouragement of visitors to return to the museum (Padilla-Melendez and del Aguila-Obra, 2013). Accordingly, social media, one of these internet technologies, provides enhanced brand interaction and engagement opportunity to consumers (Tuten and Solomon, 2014). The consumer contributions such as consumer experiences, user feedback, and product reviews, via social mediaare cited as the main sources of information from consumers for the companies (Duan et al., 2016).

Social media is widely used in the tourism and hospitality industry (Carvalho and Carvalho, 2020). Various social media platforms are used for information exchange, recommendations and to understand the visitors' quality expectations and so on. This helps other potential visitors to make their decisions. For instance, consumers can check the reviews on the websites such as Google reviews, trip advisor, booking.com before they plan a visit to any destination. In addition, the services such as hotels and airlines benefit from social media which provides control over the travel shopping experience (Chatzigeorgiou and Christou, 2020). These Web 2.0 technologies allow consumers to add comments, reviews and feedback about their experience of their expenditures on different platforms. That is useful for both sides: for other consumers and the company itself. While the company can observe and analyse what people like, dislike, whether they are satisfied or dissatisfied with their experience. Thus, it is a good opportunity for those companies to improve their products and services. They will take advantage of observing their online reputation, and developing their management strategies to increase the consumer value they created. Online platforms help for managing these reviews, react quickly and turn negative reviews into positive vibes by managing good communication via online platforms.

Social media platforms can help for interacting with consumers emotionally (Manthiou et al., 2016). When consumers share their experience online, give feedback, and write reviews about their experience, this two-way communication model allows the company to interact with consumers more efficiently. In line with this, Lia et al. (2020) argue that consumers who have strong dissatisfaction or satisfaction with the services such as restaurants are eager to post their online reviews earlier than averagely satisfied consumers. For the other consumers, they would know the positive and negative sides of the experience before they visit. This feedback allows the consumers to keep their expectations lower or higher before they experience or analyse how much they value the similar positive and negative sides of the company. From the company side, the companies can use AI Technologies to help for influencing potential customers' reactions when they would like to analyse online data (Chuh and Yu, 2021). For instance, expressions of surprise and happiness about machine learning tools unveil a positive influence on consumers. Furthermore, the discussions, contributions, collaborations on these online platforms, and sharing with brands and with other consumers make consumers more involved and influence each other (Tuten and Solomon, 2014). Using online platforms to engage with consumers is key to provide more options for consumers, but also gives the opportunity to the consumers for information exchange to learn more about the consumer experience, interact with each other, and socialize (Wei et al., 2016). These online platforms that allow exchanging information help co-creation between businesses and consumers to create a memorable experience as well (Wei et al., 2016). One example of this memorable experience that these technologies provide is online gamification. Gamification in tourism has been found as bringing a higher satisfaction rate among tourists and increase in brand awareness (Dahl, 2018).

New technologies can provide more than consumer experience and information exchange. Pantano and Priporas (2016, p. 553) highlight that mobile technology helps enhance the consumer experience by "saving them time, saving them money, supporting their lifestyle, offering security in transactions, and offering quality collection services". Such technologies encourage consumers to benefit from them by providing advantages such as less time waiting in queues, satisfaction, convenience, easy and fast consumption experience, access to new information and offers together with personalised offers and deals.

From the tourism industry perspective, the use of technology influences the tourism experience significantly (Liberato et al., 2018). Especially in the context of a smart tourism destination, internet access in the destination, airports and hotels improve tourism experience and consumer satisfaction considering the intensive use of smartphones and laptops. Providing personalized online tourist services via these technologies increases consumer satisfaction (Liberato et al., 2018). For instance, research conducted in Uganda shows that the use of smartphones is found to improve business and bring more benefit for women entrepreneurs for various reasons such as (i) fewer costs, (ii) minimum risks, (iii) improved business networks and communication, (iv) reaching out to better markets and prices, (v) access to more information and knowledge, (vi) money and time-saving, (vii) dealing with emergencies, (viii) in consequent more productivity and more chance in competition (Komunte, 2015).

The second way of technology use in the tourism and hospitality industry is interactive technologies. The use of interactive technologies is beneficial in providing visitors with a sensory experience through various means such as sight, hearing, touch and smell, creating a better consumer experience that appeals to consumers' visual senses (Manthiou et al., 2016). Mixed Reality (MR) is becoming a new trend in enhancing museum/cultural sites engagement and the visiting experience when it is incorporated within tours. Engagement is defined in different manners such as attachment, interaction or emotional connection. to illustrate, engagement in museums enhances the visitors' consumption, provides real-time experience, visualised experience, and enhances visitors' imagination. Normally, successful engagement is measured in cultural sites and museums by the average time spent in the space. Therefore, engagement for visitors is better represented in spending time with the interpretation techniques and creative presentations. Some of these New Technologies such as Virtual Reality (VR) and Augmented Reality (AR) in tourism have become an important part of tourism and destination marketing (Griffin et al., 2017; Yung and Khoo-Lattimore, 2019; Alcañiz, Bigné and Guixeres, 2019). Virtual reality is used to create a real-like virtual environment to visualise visitors' imagination and the touristic experience and to enable interaction via computer-based technologies (Griffin et al., 2017; Varinlioğlu, 2020). By using

VR Technologies, the feeling of being in an environment through virtual platforms arouses that feeling of real-like experience in the user's mind (Varinlioğlu, 2020). Virtual stores that are created with these new technologies help interaction between the consumers, the products and sellers in virtual platforms (Alcañiz et al., 2019). Virtual stores allow consumers to have at-home consumer experiences, without the time consumption of travel, and without collaboration with friends to make purchases. These virtual stores compete with physical stores which allow the consumers with physical sensations.

The improvement of technology such as virtual reality (VR) technology helps enhance consumer experience and satisfaction in the tourism and hospitality industry (Wei et al., 2019). Virtual reality (VR) technology creates a virtual environment that presents the experience virtually and immerses all aspects in virtual platforms which creates a real-like feeling. Thus, Virtual reality (VR) technology provides a full experience such as participation, enjoyment and etc. (Wei et al., 2019). For instance, in theme parks, Virtual reality (VR) technology has increased the consumer experience and encourages and increases the intention of visitors to revisit and recommend that experience to their friends and families in various ways such as revolutionizing traditional roller coaster rides, using virtual reality headset (Wei et al., 2019). The reason for the satisfaction level is that VR technology enhances the feeling and sense of joyfulness of presence. This experience can get more visitors interested in the attraction and provide more innovative and immersing experience for their visitors. Virtual reality (VR) technology is not necessarily limited to theme parks, but is also used for other tourism and hospitality areas such as museums, art galleries, and other tourism destination attractions (Wei et al., 2019).

VR Technologies can be used for various purposes such as designing places and experiences, increasing the accessibility of remote places, educating and guiding visitors, enhancing interaction between visitors globally (Griffin et al., 2017; Yung and Khoo-Lattimore, 2019), increasing the ability to communicate with a distant place or feel experience, help users to have more informed decisions with more realistic expectations (Griffin et al., 2017). With the help of VR, people have more positive emotions towards the destination because of feeling like they are in the destination and participating in the activities (Griffin et al., 2017). VR experience in tourism provides different vicarious experiences to consumers: visual, audio, and spatial aspects of the destination with advantages in usability and costs (Yung and Khoo-Lattimore, 2019). Similarly, the use of visual imagery in tourism while promoting the destinations has positive influence on consumers that encourage better communication through experiences and emotions (Griffin et al., 2017). Such technologies influence customer and tourist experience positively and helps promoting touristic places. Similarly, virtual tourism opportunities can also provide benefits to tourism consumers who have restrictions on travel such as physical disabilities, financial difficulties and so on (Yung and Khoo-Lattimore, 2019). Thus, these new technologies can be the source of sustainability in tourism (Yung and Khoo-Lattimore, 2019).

Likewise, new technologies such as 3D virtual world can be used actively in tourism by offering an enhanced consumer experience and giving the consumer the opportunity to explore in depth (Huang et al., 2016). However, only consumers who have the technology access among the potential consumers can efficiently use this technology. Thus, it may have limitations to access to all levels of consumers. Such 3D virtual tours enhance the consumer experience and provide them entertainment with an enjoyable visit, but also market them to plan a trip to the destination by encouraging them to intend to travel or arrange a trip (Huang et al., 2016; Griffin et al., 2017). Thus, it is possible to say 3D virtual tours are useful for online and real-world visitors. These tours can be planned based on potential consumers'/ visitors' needs and what they prefer to experience in the touristic destination. Especially with the Covid-19 pandemic, such technologies are used intensively by touristic places. For instance, museums and touristic sites opened up their doors virtually or enhanced their virtual tours if they had already started to use this technology before Covid-19.

Sensors, interconnected devices, use of the Internet of Things (IoT) are other new digital Technologies that are used in the tourism industry (Pencarelli, 2020). These technologies are the concepts that are used by the tourism companies in order to improve service quality, enhance consumer experience and obtain a competitive advantage compared to rivals (Pencarelli, 2020). The use of such technologies create a collaborative ecosystem with all related parties to co-create an enriched consumer experience in both the online and offline World of the tourism industry (Pencarelli, 2020). For instance, Airbnb is one of the examples of the use of new technology in the tourism and hospitality industry that brings a differentiated approach to the hosting industry (Pencarelli, 2020). Besides, self-service technologies and kiosks offer more personalised services, and convenience along with saving time (Wei et al., 2017). Thus these technologies influence consumer experience and satisfaction positively by increasing the interaction among consumers and creating a more enjoyable and memorable customer experience (Wei et al., 2017). The use of new technologies by women entrepreneurs will be detailed next.

5. THE USE OF TECHNOLOGY, DIGITALISATION, AND SOCIAL MEDIA IN THE TOURISM INDUSTRY BY WOMEN ENTREPRENEURS TO ENHANCE CONSUMER EXPERIENCE

Although there is a considerable amount of literature on technology use, digitalization, social media and women entrepreneurs, the studies based on technology use, digitalization, social media and women entrepreneurs in the tourism and hospitality industry remained relatively fragmented. Therefore, this area in the tourism and hospitality industry is still underresearched.

Research on women entrepreneurs is inconclusive on how small businesses are encountering digitalization and learning to use social media for their businesses. One study explains that the lack of digitalisation among women entrepreneurs is the challenge that women entrepreneurs have to deal with. As explained by Ozkazanc-Pan et al. (2018), one of the challenges that are associated with women entrepreneurs is that women are disadvantaged in the technology industry. The main reasons for that can be summarised as limited access to resources and gender norms in society. In order for women entrepreneurs to gain digital skills to adaptdigitalization tools to their business, they need to get through informal learning, practice, and recruit digitally skilled employees (Olsson and Bernhard, 2021). The training and new skills required are acquiring new digital skills, benefitting online sources, as well as scanning and maintenance of social media platforms (Olsson and Bernhard, 2021). Their participants emphasized that they do not experience any significant gender stereotyping based on the technology contrary to the previous studies. In the era of digitalization, women entrepreneurs have to apply digital tools into their businesses and use innovative ways (Olsson and Bernhard, 2021). How women entrepreneurs can benefit from digitalization and social media at work and adapt digitalization to their businesses are the areas that need further research. To fill this gap, Olsson and Bernhard (2021) analysed digitalization and women entrepreneurs in the tourism industry. Their findings revealed important and novel results for women entrepreneurs in the tourism industry. Firstly, women entrepreneurs in the tourism industry need to learn new digital skills to keep up with the competition. Secondly, they will have to deal with the disadvantages of digitalisation and social media. For instance, although there are lots of benefits of digitalization to businesses, digitalisation will need extra resources, and add extra stress to deal with the social media presence, maintenance of digital tools, social media listening, and negative comments on social media. This extra stress can influence women entrepreneurs' emotions negatively. Thirdly, women entrepreneurs can prefer to stay in offline business after evaluating the positive and negative sides of digitalization. Fourthly, the respondents have not mentioned any significant gender inequality in online platforms. Therefore, gender differences or being women does not cause any extra stress for them in online platforms. As a result, learning and adapting digitalization and social media to their businesses is an essential step for women entrepreneurs in the tourism and hospitality industry, however, this learning process is accepted as a usual part of learning new skills for their business. This learning process of digitalization and social media is needed for all small businesses in order to adapt to novelty, engage in competition, and generate local and regional growth.

Likewise, another study that focuses on women entrepreneurs argues that they need different types of support which includes support in technology, digitalisation, and social media, in order to compete in the tourism industry (Samad and Alharthi, 2022). The findings of the study in Saudi Arabia suggest that women entrepreneurs expect to get more support from the government, such as training, financial support, and family and society support (Samad and Alharthi, 2022). One of the trainings that women need most is identified as innovation and innovation skills. Women entrepreneurs in Saudi Arabia are more interested in the tourism industry. If such kind of support and training that women entrepreneurs are needed were given, women would be more confident with technology use and be able to apply technological and innovative solutions to their entrepreneurs.

6. HOW THE COVID-19 PANDEMIC INFLUENCED THE USE OF TECHNOLOGY IN THE TOURISM AND HOSPITALITY INDUSTRY

The covid-19 pandemic has changed many daily routines all over the world including tourism. The digitalisation process of the tourism industry became especially important during the COVID-19 pandemic, not only for tourism but for all industries (Olsson and Bernhard, 2021). Touristic places had applied various novelties such as new sanitary measures and applications, the limited number of visitors, bookings, new mobile applications and more virtual visits by using various technological tools, more outdoor activities etc. (Lebrun, Su and Bouchet, 2021). However, the digitalization process was not a smooth transition for all organisations such as small businesses or women entrepreneurs in the tourism and hospitality industry. The small businesses needed financial support to employ staff to deal with digital platforms to support and develop the digitalization of small businesses and to focus on local and regional development (Olsson and Bernhard, 2021).

Although new Technologies bring many opportunities to the tourism industry, women entrepreneurs are not without any challenges. Initially, women entrepreneurs have disadvantages in the technology world compared to male colleagues because of gender inequality in technology entrepreneurship (Ozkazanc-Pan and Muntean, 2016). As Komunte (2015) mentioned, the use of technology- mobile phones in this study- by women entrepreneurs is still limited in developing countries, in rural and urban areas, although they found that technology and mobile phone use brings more profit to their business, secure better markets and prices, save time and money, negotiate for suitable prices and communicate business timely. Women entrepreneurs may come across other difficulties as well. For instance, cultural traditions, results of covid, limited resources and financial capital, lack of empowerment and confidence, gender-centric factors and lack of internal and external support and cooperation, low-level risk-taking attitudes, and the male-female competition are some of the challenges that women entrepreneurs confront all around the world (Siddiqui, 2012; Meera, 2014; Tajeddini et al., 2017; Suseno and Abbott, 2021; Filimonau et al., 2022). All these challenges they come across made their survival during the Covid-19 pandemic more challenging. However, as Suseno and Abbott (2021) found, women entrepreneurs are more capable of using technology to pursue social innovation contrary to the arguments that were discussed in the previous studies.

Although many challenges are identified for women entrepreneurs that may limit them from benefitting new technologies, taking precautions against the Covid-19 pandemic, and maintaining their businesses; women entrepreneurs are capable of finding solutions. Firstly, Suseno and Abbott (2021) argue that technology provides opportunities for women entrepreneurs to overcome some of these challenges. Likewise, appropriate training, incentives, encouragement and motivation, social recognition of their entrepreneurial abilities, and family moral support help women entrepreneurs to overcome such challenges (Siddiqui, 2012). Lastly, while Covid-19 affected the tourism industry negatively as it did to many industries, women entrepreneurs could manage to find solutions. Women entrepreneurs that were examined by Filimonau et al. (2022) highlighted that they used alternative ways to generate more income and overcome the challenge of Covid-19 and stay in the tourism industry. For instance, they focused on food delivery and farming and had the chance to invest in self-education. Secondly, they used new Technologies such as Augmented reality (AR). Augmented reality (AR) can be efficiently used in marketing in various ways such as the company's interaction with the consumers, value creation with consumers, and enhancing the consumer experience, especially in the retailing, tourism and advertising in both online and offline platforms (Rejeb et al, 2021; Du et al., 2022). Augmented reality (AR) technologies make the experience more interactive, enjoyable and personalized which increases customer satisfaction. Similarly, the use of emerging technologies such as VR, AR and social media helps achieving high levels of customer engagement and luxury brand experience (Gupta et al., 2022). Luxury brands use emerging technologies, such as VR, AR devices efficiently by creating virtual shopping and visiting experience which increase consumer experience and satisfaction. Such technologies are used by many entrepreneurs that work in the tourism and hospitality industry.

Especially with the Covid-19 pandemic, the online experience has been substituted for the offline experience mostly. Thus, luxury brands use new online platforms, and new technologies to design better consumer experience (Gupta et al., 2022). The luxury experience covers many areas such as 'hospitality, cruises, fine dining, private jets, cars, yachts, fine wines, and spirits, as well as gourmet food, compared to luxury fashion' (Gupta et al., 2022, p. 2).

Lastly, the covid-19 pandemic also brought quarantine and isolation experience for the entrepreneurs and customers in the tourism and hospitality industry (Chia and Xiong, 2023). The services offered during quarantine are analysed under 6 main headings such as 'staff, room, hygiene and disinfection, food, healthcare and location'. The visitors' reviews during the quarantine helped hotels improve their services, visitors' experience, perceived value, satisfaction, and loyalty. Especially hygiene was the most essential need during covid period. During the visitors' quarantine period, the staff's support of empathy and understanding is found important by visitors to feel relieved from quarantine psychology such as anxiety and loneliness. As seen, covid-19 pandemic has influenced entrepreneurs in the tourism and hospitality industry as heavy as other industries. However, especially women entrepreneurs were able to find solutions to overcome most of the challenges they come across because of covid-19 pandemic and gender inequality.

7. CONCLUSION

To conclude, this chapter has covered a variety of topics based on consumer experience in the tourism and hospitality industry. The use of technology (VR and AI technologies, mobile technologies, IoT etc.) and social media platforms have been discussed to understand how these technologies influence consumer experience in the tourism industry. Additionally, this chapter has not only included technology use in the tourism and hospitality industry, but also the role of women entrepreneurs has been covered in this chapter. Finally, the influence of the Covid-19 pandemic and how this period has influenced women entrepreneurs, their businesses, and their use of new technologies by women entrepreneurs have been discussed from the women entrepreneurs' perspective.

REFERENCES

Alcañiz, M., Bigné, E., & Guixeres, J. (2019). Virtual Reality in Marketing: A Framework, Review, and Research Agenda. *Frontiers in Psychology*, *10*, 1530. doi:10.3389/fpsyg.2019.01530 PMID:31333548

Alcántara-Alcover, E., Artacho-Ramírez, M. A., Martínez-Guillamóna, N., & Campos-Sorianoa, N. (2013). Purpose of stay and willingness to stay as dimensions to identify and evaluate hotel experiences. *International Journal of Hospitality Management*, *33*, 357–365. doi:10.1016/j.ijhm.2012.10.006

Alonso-Almeida, M. del M. (2013). Influence of gender and financing on tourist company growth. *Journal of Business Research*, *66*(5), 621–631. doi:10.1016/j.jbusres.2012.09.025

Barnes, S. J., Rutter, R., Mattsson, J., & Sørensen, F. (2020). Patron Sentiment of Employee–Customer Interaction: Exploring Hotel Customer Reviews through Machine Learning: An Abstract. In S. Wu, F. Pantoja, & N. Krey (Eds.), *Marketing Opportunities and Challenges in a Changing Global Marketplace. AMSAC 2019. Developments in Marketing Science: Proceedings of the Academy of Marketing Science*. Springer. doi:10.1007/978-3-030-39165-2_161

Carvalho, S., & Carvalho, J. V. (2020). The Implications of Digital Marketing and e-Commerce in the Tourism Sector Growth. In Á. Rocha, A. Abreu, J. de Carvalho, D. Liberato, E. González, & P. Liberato (Eds.), *Advances in Tourism, Technology and Smart Systems. Smart Innovation, Systems and Technologies* (Vol. 171). Springer. doi:10.1007/978-981-15-2024-2_15

Chatzigeorgiou, C. & Christou, E. (2020). Adoption of social media as distribution channels in tourism marketing: A qualitative analysis of consumers' experiences. *Journal of Tourism, Heritage & Services Marketing*, *6*(1), 25-32. doi:10.5281/zenodo.3603355

Chen, X. (2021). Research on the Influencing Factors of Hotel "Green and Low Carbon" Behavior on Consumer Experience. *IOP Conference Series. Earth and Environmental Science*, *692*(3), 032083. doi:10.1088/1755-1315/692/3/032083

Chia, K. W., & And Xiong, J. (2023). Once upon a time in quarantine: Exploring the memorable quarantine hotel experiences of Chinese student returnees during the COVID-19 pandemic. *Tourism and Hospitality Research*, *23*(1), 72–87. doi:10.1177/14673584221085209

Chuah, S. H.-W., & Yu, J. (2021). The future of service: The power of emotion in human-robot interaction. *Journal of Retailing and Consumer Services*, *61*(102551), 1–8. doi:10.1016/j.jretconser.2021.102551

Daengbuppha, J., Hemmington, N., & Wilkes, K. (2006). Using grounded theory to model visitor experiences at heritage sites Methodological and practical issues. *Qualitative Market Research*, *9*(4), 367–388. doi:10.1108/13522750610689096

Dahl, S. (2018). *Social Media Marketing Theories & Applications* (2nd ed.). SAGE Publications Ltd.

Du, Z., Liu, J., & Wang, T. (2022). Augmented Reality Marketing: A Systematic Literature Review and an Agenda for Future Inquiry. *Frontiers in Psychology*, *13*, 925963. doi:10.3389/fpsyg.2022.925963 PMID:35783783

Duan, W., Yu, Y., Cao, Q., & Levy, S. (2016). Exploring the Impact of Social Media on Hotel Service Performance: A Sentimental Analysis Approach. *Cornell Hospitality Quarterly*, *57*(3), 282–296. doi:10.1177/1938965515620483

Ertac, M., & Tanova, C. (2020). Flourishing Women through Sustainable Tourism Entrepreneurship. *Sustainability (Basel)*, *12*(5643), 1–17. doi:10.3390u12145643

Figueroa-Domecq, C., Anna de Jong, A., & Williams, A. M. (2020). Gender, tourism & entrepreneurship: A critical review. *Annals of Tourism Research*, *84*(102980), 1–13. doi:10.1016/j.annals.2020.102980

Filimonau, V., Matyakubov, U., Matniyozov, M., Shaken, A., & Mika, M. (2022). Women entrepreneurs in tourism in a time of a life event crisis. *Journal of Sustainable Tourism*, 1–23. Advance online publication. doi:10.1080/09669582.2022.2091142

Griffin, T., Giberson, J., Lee, S. H., Guttentag, D., Kandaurova, M., Sergueeva, K., & Dimanche, F. (2017). Virtual Reality and Implications for Destination Marketing. *Travel and Tourism Research Association: Advancing Tourism Research Globally*, 29. https://scholarworks.umass.edu/ttra/2017/Academic_Papers_Oral/29

Gupta, D. G., Shin, H., & Jain, V. (2022). Luxury experience and consumer behavior: A literature review. Marketing Intelligence & Planning. doi:10.1108/MIP-12-2021-0438

Huang, Y. C., Backman, K. F., Backman, S. J., & Chang, L. L. (2016). Exploring the Implications of Virtual Reality Technology in Tourism Marketing: An Integrated Research Framework. *International Journal of Tourism Research*, *18*(2), 116–128. doi:10.1002/jtr.2038

Islam, M. S., & Kirillova, K. (2021). Nonverbal communication in hotels as a medium of experience co-creation. *Tourism Management*, *87*, 104363. doi:10.1016/j.tourman.2021.104363

Komunte, M. (2015). Usage of Mobile Technology in Women Entrepreneurs: A case Study of Uganda. *The African Journal of Information Systems*, *7*(3), 3. https://digitalcommons.kennesaw.edu/ajis/vol7/iss3/3

Kunene, B., Mashau, P., & And Nyawo, J. (2022). Ascertaining The Challenges Faced By Women Entrepreneurs in Ethekwini Municipality, Kwazulu-Natal Province. *Gender & Behaviour*, *20*(1), 19060–19083.

Lebrun, A.-M., Su, C.-J., & Bouchet, P. (2021). A More Sustainable Management of Domestic Tourists in Protected Natural Parks: A New Trend in Sport Tourism after the Covid-19 Pandemic? *Sustainability (Basel)*, *13*(14), 7750. doi:10.3390u13147750

Li, H., Xie, K. L., & Zhang, Z. (2020). The effects of consumer experience and disconfirmation on the timing of online review: Field evidence from the restaurant business. *International Journal of Hospitality Management*, *84*(102344), 1–11. doi:10.1016/j.ijhm.2019.102344

Liberato, P., Alen, E., & Liberato, D. (2018). Smart tourism destination triggers consumer experience: The case of Porto. *European Journal of Management and Business Economics*, *27*(1), 6–25. doi:10.1108/EJMBE-11-2017-0051

Manthiou, A., Kang, J., Sumarjan, N., & Tang, L. R. (2016). The Incorporation of Consumer Experience into the Branding Process: An Investigation of Name-Brand Hotels. *International Journal of Tourism Research*, *18*(2), 105–115. doi:10.1002/jtr.2037

Meera, S. (2014, March). Challenges of Women Entrepreneurs: With Specific Reference to Tourism Industry, TMIMT. *International Journal (Toronto, Ont.)*, *1*(1), www.tmimtjournal.org

Mitchell, M. A., & Mitchell, S. J. (2001). Consumer Experience Tourism in the Nonprofit and Public Sectors. *Journal of Nonprofit & Public Sector Marketing*, *9*(3), 21–34. doi:10.1300/J054v09n03_02

Olsson, A. K., & Bernhard, I. (2021). Keeping up the pace of digitalization in small businesses–Women entrepreneurs' knowledge and use of social media. *International Journal of Entrepreneurial Behaviour & Research*, *27*(2), 378–396. doi:10.1108/IJEBR-10-2019-0615

Otto, J. E., & Ritchie, J. R. B. (1996). The service experience in tourism. *Tourism Management*, *17*(3), 165–174. doi:10.1016/0261-5177(96)00003-9

Ozkazanc-Pan, B., & Susan Clark Muntean, S. C. (2018). Networking towards (in)equality: Women entrepreneurs in technology. *Gender, Work and Organization*, *25*(4), 379–400. doi:10.1111/gwao.12225

Padilla-Meléndez, A., & del Águila-Obra, A. R. (2013). Web and social media usage by museums: Online value creation. *International Journal of Information Management*, *33*(5), 892–898. doi:10.1016/j.ijinfomgt.2013.07.004

Pantano, E., & Priporas, C.-V. (2016). The effect of mobile retailing on consumers' purchasing experiences: A dynamic perspective. *Computers in Human Behavior*, *61*, 548–555. doi:10.1016/j.chb.2016.03.071

Pencarelli, T. (2020). The digital revolution in the travel and tourism industry. *Information Technology & Tourism*, *22*(3), 455–476. doi:10.100740558-019-00160-3

Quan, S., & Wang, N. (2004). Towards a structural model of the tourist experience: An illustration from food experiences in tourism. *Tourism Management*, *25*(3), 297–305. doi:10.1016/S0261-5177(03)00130-4

Rejeb, A., Rejeb, K., & Treiblmaier, H. (2023, April 03). And Treiblmaier, H. (2021): How augmented reality impacts retail marketing: a state-of-the-art review from a consumer perspective. *Journal of Strategic Marketing*, *31*(3), 718–748. Advance online publication. doi:10.1080/0965254X.2021.1972439

Ribeiro, M. A., Adam, I., Kimbu, A. N., Afenyo-Agbe, E., Adeola, O., Figueroa-Domecq, C., & de Jong, A. (2021). Women entrepreneurship orientation, networks and firm performance in the tourism industry in resource-scarce contexts. *Tourism Management, 86*(104343), 1–15. doi:10.1016/j.tourman.2021.104343

Rogers, D. L. (2016). *The digital transformation playbook: rethink your business for the digital age.* Columbia University Press. doi:10.7312/roge17544

Rojas, L., Quiñones, D., & Rusu, C. (2020). Understanding User Needs and Customer eXperience in Tourism Area. HCII. , LNCS 12195. doi:10.1007/978-3-030-49576-3_16

Samad, S., & Alharthi, A. (2022). Untangling Factors Influencing Women Entrepreneurs' Involvement in Tourism and Its Impact on Sustainable Tourism Development. *Administrative Sciences, 12*(2), 52. doi:10.3390/admsci12020052

Siddiqui, A.B. (2012). Problems Encountered by Women Entrepreneurs in India. *International Journal of Applied Research & Studies, 1*(2), 1-11.

Sotiriadis, M. (2017). Experiential Dimensions And Their Influence On Behavioral Intentions Within The Context Of Nature-Based Tourism. *Tourism and Hospitality Management, 23*(1), 35–50. doi:10.20867/thm.23.1.7

Suseno, Y. and Abbott, L. (2021). Women entrepreneurs' digital social innovation: Linking gender, entrepreneurship, social innovation and information systems. Information Systems Journal, 31(5), 714-744.

Tajeddini, K., Ratten, V., & Denisa, M. (2017). Female tourism entrepreneurs in Bali, Indonesia. *Journal of Hospitality and Tourism Management, 31*, 52–58. doi:10.1016/j.jhtm.2016.10.004

Teng, C.-C. (2011). Commercial hospitality in restaurants and tourist accommodation: Perspectives from international consumer experience in Scotland. *International Journal of Hospitality Management, 30*(4), 866–874. doi:10.1016/j.ijhm.2011.01.007

Tuten, T. L. and Solomon, M. R. (2014). Social Media Marketing. Pearson.

Varinlioğlu, F. (2020). Understanding Virtual Reality Applications in Digital Heritage Through Teos. *MEGARON, 15*(1), 161–170. doi:10.14744/megaron.2019.85619

Vujko, A., Tretiakova, T. N., Petrović, M. D., Radovanović, M., Gajić, T., & Vuković, D. (2019). Women's empowerment through self-employment in tourism. *Annals of Tourism Research, 76*, 328–330. doi:10.1016/j.annals.2018.09.004

Walls, A. R. (2013). A cross-sectional examination of hotel consumer experience and relative effects on consumer values. *International Journal of Hospitality Management, 32*, 179–192. doi:10.1016/j.ijhm.2012.04.009

Wei, W., Qi, R., & Zhang, L. (2019). Effects of virtual reality on theme park visitors' experience and behaviors: A presence perspective. *Tourism Management, 71*, 282–293. doi:10.1016/j.tourman.2018.10.024

Wei, W., Torres, E., & And Hua, N. (2016). Improving consumer commitment through the integration of self-service technologies: A transcendent consumer experience perspective. *International Journal of Hospitality Management, 59*, 105–115. doi:10.1016/j.ijhm.2016.09.004

Wei, W., Torres, E. N., & And Hua, N. (2017). The power of self-service technologies in creating transcendent service experiences, The paradox of extrinsic attributes. *International Journal of Contemporary Hospitality Management*, *29*(6), 1599–1618. doi:10.1108/IJCHM-01-2016-0029

Yung, R., & Khoo-Lattimore, C. (2019). New realities: A systematic literature review on virtual reality and augmented reality in tourism research. *Current Issues in Tourism*, *22*(17), 2056–2081. doi:10.1080/13683500.2017.1417359

ADDITIONAL READING

Ert, E., Fleischer, A., & Magen, N. (2016). Trust and reputation in the sharing economy: The role of personal photos in Airbnb. *Tourism Management*, *55*, 62–73. doi:10.1016/j.tourman.2016.01.013

Huang, X., Zhang, Y., & Li, C. (2022). Assessing the Compensatory Potentiality of Hot Spring Tourism in the COVID-19 Post-Pandemic Environment. *Sustainability (Basel)*, *14*(14), 8579. doi:10.3390u14148579

Lugalla, I. M., Jacobs, J. P. A. M., & Westerman, W. (2023). What Drives Women Entrepreneurs in Tourism in Tanzania? *Journal of African Business*, 1–19. Advance online publication. doi:10.1080/15228916.2023.2168970

KEY TERMS AND DEFINITIONS

Consumer Experience: The interaction between the consumer and the company during the consumers' journey, not only for shopping or purchasing but a relationship between a consumer and a company.

Destination Marketing: A part of tourism marketing that involves marketing activities to promote especially specific destinations (a country, a city, a town, a hotel etc.) to increase the number of visitors.

Digital Marketing: Use of digital and mobile technologies that are connected to the internet for marketing purposes which contain email, social media, and web-based applications.

Digitalization: Transformation process from analogue to digital forms.

Smart Tourism: The way of tourism that is supported by technological tools such as mobile applications, augmented reality (AR), and virtual reality (VR) to provide tourism experience to the consumers.

Social Media: The technologies that enhance the interaction between two parties by allowing 2-way communication.

Social Networks: Online platforms where consumers and organisations can connect, interact, communicate, and exchange information.

Tourism Marketing: All types of marketing activities that focus on the tourism industry to promote touristic products and services including a variety of destinations such as hotels, airlines, restaurants, travelling etc.

Women Entrepreneurship: Business and enterprise owned by women who take the risk, find the capital, and gather human and monetary resources together to start up their own businesses.

Chapter 15
Stigma, Double-Stigma, and Associated Hurdles:
Discussion and Understanding for Muslim Female Tourists

Muhammad Faisal Sultan
Khadim Ali Shah Bukhari Institute of Technology, Pakistan

Abdul Kabeer Kazi
Baqai University, Karachi, Pakistan

Muhammad Furquan Saleem
Iqra University, Karachi, Pakistan

Sadia Khurram
Binazir Bhutto Shahid University, Pakistan

ABSTRACT

Women always face issues while travelling alone or in the form of a group. However, these issues became more devastating for Muslim women due to religious, safety, personal, and familial issues. Therefore, there is a need for a thorough concentration of motivation, perception and hurdles of Muslim solo or group female tourists. In fact, there are few studies that tried to determine the motivation behind the preference for solo travelling. However, there are few studies that may highlight issues, problems, and threats that may be faced by Islamic women while travelling. Therefore, this chapter has been written purposively to highlight the obstacles that may halt females from travelling. The philosophy of research used for writing this study is epistemology, and the research strategy is archival as the chapter has been formulated by analyzing available literature.

DOI: 10.4018/978-1-6684-8417-3.ch015

INTRODUCTION

Religion is a significant predictor of choice behavior. In terms of tourism religion plays a significant role in selection of destination as well as foods and beverages (Battour et al., 2011). Social Distance Theory by Thyne et al (2006) claimed that people are found to be comfortable with those who possess social and cultural values. Thus, the ability of the location to be selected may be increased in the minds of desired set of targeted tourist (Battour et al., 2011). Muslims are perceived as the one of the largest global markets. Similar is found to be true for tourism as Muslims were perceived to contribute $300 billion till 2020. Thus several non-Muslim destinations are trying to establish their image in a way that it may increase likelihood of Muslim tourists. Although, due to lack of understanding of Islamic principles, notions associated with Islamic hospitality and Muslim cultures several studies reported dissatisfaction of Muslim tourists from non-Muslim destinations (Han et al., 2021).

However, in recent times increase in female solo as well as group tourism is getting popular day after another. Studies further highlighted that solo female travellers are highly motived due to sense of self-dependency, empowerment and self-identity. Similarly female group tourism is also gaining mass attention. However, solo-travelling is not the matter of personal rights and to prove the point that women have equal liberty as of men (Hamid et al., 2021).

On the other side some of the studies also declare travelling alone or in a group is perceived as one of the largest barriers in front of female tourists. However, the hurdle is much higher for Muslim females. In fact, Muslim females have to take permission from their guardians or spouse officially to initiate any plan like tourism. However, the restrictions may found to be varying from location to location on the bases of implementation of religious norms within different socio-cultural contexts. Hence research classified these barriers in the form of two major types i.e., micro-level and macro-level. The micro level encompasses individual and family level challenges and the macro level includes national & International level challenges (Tavakoli & Mura, 2021).

GRAY AREA FOR RESEARCH

There are few studies that may relate tourism and religion. Especially there is a severe lacking of studies found in relating tourism with Islam. However, there are studies that may relate tourism with the needs of tourists but lacking of literature is evident in relating tourism with the needs of Muslim tourists (Battour et al., 2011). Moreover, research is also evident that there is severe lacking of literature pertaining to vacations of females. Still than literature indicate that vacations of female especially of solo-females are full of challenges. Furthermore, it has also been reflected that the available literature is mainly based upon vacations and tourist's behavior of western females that is much different from Muslim women (Moufakkir *et al.,* 2020).

STIGMA AND ITS ASSOCIATION WITH BARRIERS ASSOCIATED WITH MUSLIM FEMALE TOURISTS

Study of Moufakkir et al (2020) highlighted stigma as the main element that creates meta-stereotyping. In fact, stigma is the used to criticize negatively to any of the person of out group. In fact, study of

Moufakkir et al (2015) relates stigma theory with tourism behavior of tourists from developing countries and indicated that these people are perceived as disadvantaged tourists due to the stigma of association. However, Muslim women are suffered from double-stigma, i.e., religious stigma and immigrant stigma. Thus, upcoming part will discuss about different forms of stigmas that may create hurdles for Muslim women in the process of travelling (Moufakkir et al., 2020).

Immigrant Stigma

Immigrant Stigma is also known as Othering Stigma. However, it is not described as othering stigma due to the difference between tourist and host but because of the confusion they bear due to their appearance like immigrant. Studies also highlighted that in some of the cultures immigrants' remains handicap due to the stereotyped attitude towards their belongingness and attributes (Moufakkir et al., 2020).

On the other side tourists are not only perceived as incompetent but also un-trustworthy. Thus, host may have variety of threats from immigrants like economic, social & cultural threats etc. Hence 38% of the inhabitants from 15 European countries believe that number of immigrants to these countries must be reduced. Similar has been the belief of 42% Dutch are also found agree with the statement (Moufakkir et al., 2020). In fact, the stigma may also make one feel alone even in crowded place. Similar has been highlighted by Iranian female solo travellers that they feel alone even in crowded cafe. Similarly Muslim female tourists are also seems to be worried about the perception of solo female tourists as it may be perceived as prostitute. Similarly girl and boy travelling together are mostly perceived as the couple having sexual relationship. Moreover, Muslim societies are perceived to be patriarchal society where men are praised more than the women and women who travelled alone are perceived to be a part of open minded family (Nikjoo et al., 2021).

Religious Stigma

This form of stigma may address issues faced by Muslim women more adequately as according to Moufakkir et al (2020), many of the European countries holds unfavorable views for Muslims. These issues include fear of criminality, economic instability, reluctance to integration & high probability of terrorism and extremism due to continuous increase in refugees from Iraq, Syria and other Muslim countries. European Inhabitants are found have mixed perception about the coherence of Islam and their values. However, 44% of the Dutch believes that Islam is incompatible with their values and culture. Similar has been found by other studies who claimed that Muslim countries all over the globe are not only perceived as economically backward but also as conservative and terrorism supportive. Thus, it can be easily determined that Islamophobia has positive significant relationship with psychological distress and discrimination. Therefore, understanding or perception of Arabs or Muslims regarding the social or religious discriminations must not be overlooked. Especially important are the scenarios, countries or locations where Islam phobias, blatant and subtle form of religious discrimination affect tourism.

BARRIERS ASSOCIATED WITH TOURISM OF MUSLIM FEMALES'

Micro-Level Barriers

Tavakoli and Mura (2021) highlighted individual and family level challenges as the major micro-level hindrances for female Muslim tourists. There are as under:

Individual-Level Challenges

Women are not perceived safe as of numbers of crimes that are committed against women. Therefore, Muslim female tourists may found hindrance in planning alone or group tourism. Moreover, due to following religious norms of dressing like Hijab may make Muslim female distinguished in comparison to others and therefore increase concerns towards safety may also cause decrease in fun and enjoyment. On the other side leaving family for the purpose of tourism may make female guilty of her act. Therefore Muslim female prefer matriarchal duties in comparison to fun and tourism. Similar has been true for singles as Muslim female are need to be a good daughter or sister through effective fulfillment of family objectives and relationships. Last, but not the least lesser access to jobs also prevent Muslim females in becoming financially independent and therefore Muslim women is restricted to opt for alone or in a group of females. In short Muslim women are surrounded by socio-cultural and financial issues that may resulted in four groups. The first group is the one that does not have socio-cultural and financial issues and has the ability to overcome these barriers. The second group has the power and financial liberty but facing socio-cultural barriers. Third one is the group that has social freedom but does not have financial liberty and the fourth one is the group that does not have financial as well as social liberty to opt for tourism. However, researcher, scholars and academicians are focusing on the first group of Muslim females that has financial as well as socio-cultural liberty to access tourism.

Family-Level Challenges

Approval from family members, like father, mother, husband and brothers are found to be essential for women to travel alone or in groups of females. In fact, Muslim women have to provide sufficient explanation that travelling will not produce any harm and will also not produce negative affect over their familial and work responsibilities. Thus legitimate to declare for Muslim Women leisure and self-fulfillment activities were never preferred over socio-cultural responsibilities. In fact, guardians have the power to strict females from travelling abroad & to leave country for leisure purposes. In fact, guardians may not allow females to travel abroad only due to religious & issues but may also because of previous traditions of family.

On the other side influencing members like husband and father have believed that they have right to ban or influence travelling of their wives and daughters. In Iran Muslim women have to take formal permission every time from their spouse before travelling and denial from spouse will create ban on travelling and participation in any International event. Similar was the case of Iranian soccer player who could not participate in International women soccer competition due to the restriction from her husband.

Macro-Level Barriers

Tavakoli and Mura (2021) also highlighted some macro level challenges that may create hindrance in the travelling of Muslim women. The major hindrances are classified as national and international and are discussed as under:

National Level

Several Socio-Religious norms may restrict Muslim women to opt tourism alone or in a group of females. Like Iran does not permit women to use swimming facilities openly, also females cannot opt for hotel reservation without booking made in consent from official authorities or from proper travelling agency. Moreover, women are also prohibited to appear publically without wearing hijab and travelling abroad with boy-friends are also not allowed by the Iranian government and it is punishable for females and men.

International Level

Some of the international restrictions also hinder in the way of travelling activities of Muslim women as they may fail to convince officials that they will return to their parent country after the tour. The condition is special applicable on individual or group of Muslim females. On the other side political relationships between host and other country may also affect the tourism as previously USA declined visa request for several Muslim countries. On the other side countries like USA also imposed travelling bans over the inhabitants of Muslim countries due to the threat of terrorism.

ISSUES FACED BY MUSLIM WOMEN DURING INTERNATIONAL TOURISM

After highlighting micro and macro level barriers that lies in the way of individual and in-group female tourism the study of Tavakoli and Mura (2021) also shed light over issues that are faced by Muslim females. The list is massive however major issues includes search of Halal food, Halal drinks, ways to mosque and areas to offer prayers and recreation privacy.

Threat of Being Attacked Due to Islamophobia

Mature and devoted Muslim women wear scarf to cover their body as per the religious norms and Practices. However, in some of the western countries the wearing of scarf is perceived as the sign of Islamophobia especially after the incident of 9/11. This may put Muslim women under severe threat of being attacked physically of verbally (Tavakoli & Mura, 2021). Thus, legitimate to declare that issues and concern over safety halts Muslim women from travelling. In fact, safety is always a point of concerned for decisions associated with travelling type, travelling destinations as well as travelling itineraries. Thus, threats of personal attacks and sexual harassments are not only halting Muslim females but also do not make them enjoy their trips (Ratthinan & Salamat, 2019).

Dressing

Muslim women who are on tour are also required to dress adequately as per the rules and norms of their parent country. Several athletes, actresses and models lose their job due to the violation of dress codes applicable on Muslim women (Tavakoli & Mura, 2021).

Halal Foods

Halal foods are perceived as one of the major requirements of the Muslim tourists. However, these requirements must also adhere with the Muslim requirements through matching with Islamic faith. In fact, Halal tourism encompasses high assurance of Halal foods and beverages in order to attract more Muslim tourists (Tavakoli & Mura, 2021). Religious requirements are also important elements for Muslim women especially availability of halal foods.. Hence, Muslim women always preferred to travel to only those locations that will provide Muslim friendly facilities (Ratthinan & Salamat, 2019).

Mosque and Prayer Rooms

With the growth in Muslim tourists market there is also a significant raise in the requirement of places where Muslim can practice their faith. In fact, host destinations are required to have more mosques and prayers rooms in order to address needs of Muslim tourist. However, most of the locations do not have required level of facilities available to attract Muslim tourists (Tavakoli & Mura, 2021). Less availability of mosques and praying rooms are also an obstacle in tourism of Muslim women. In fact, in western world most of the mosques are for men only and therefore it is quite difficult for Muslim women to travel to Europe or to other parts of the world (Ratthinan & Salamat, 2019).

Water-Friendly Washroom

Muslim have to perform Wudu before every prayers and therefore there is a need of water friendly washroom with ample supply of water supplemented with showers and hoses to ensure Muslim purity. However, research observed lacking of these facilities in most of the host location. Even these facilities are not abundant in Muslim country like Malaysia which is not only perceived as friendliest destination for Muslim women (Tavakoli & Mura, 2021). Thus, lacking of Muslim friendly washrooms may hinder in Muslim purity and ablution and therefore creates resistance in the minds of Muslim women (Ratthinan & Salamat, 2019).

REFERENCES

Battour, M., Ismail, M. N., & Battor, M. (2011). The impact of destination attributes on Muslim tourist's choice. *International Journal of Tourism Research*, *13*(6), 527–540. doi:10.1002/jtr.824

Hamid, S., Ali, R., Azhar, M., & Khan, S. (2021). Solo travel and well-being amongst women: An exploratory study. *Indonesian Journal of Tourism and Leisure*, *2*(1), 1–13. doi:10.36256/ijtl.v2i1.125

Han, H., Lee, S., Ariza-Montes, A., Al-Ansi, A., Tariq, B., Vega-Muñoz, A., & Park, S. H. (2021). Muslim travelers' inconvenient tourism experience and self-rated mental health at a non-Islamic country: Exploring gender and age differences. *International Journal of Environmental Research and Public Health*, *18*(2), 758. doi:10.3390/ijerph18020758 PMID:33477400

Moufakkir, O. (2015). The stigmatized tourist. *Annals of Tourism Research*, *53*, 17–30. doi:10.1016/j.annals.2015.04.004

Moufakkir, O. (2020). Experience of Arab/Muslim women visiting relatives in the West and the management of stigma by association. *Tourism Management*, *78*, 104073. doi:10.1016/j.tourman.2019.104073

Ratthinan, S. P., & Selamat, N. H. (2019). Negotiating travel constraints via technology: A study of Malay Muslim women through a hierarchical constraint model perspective. *Asian Journal of Business Research*, *9*(2), 55. doi:10.14707/ajbr.190060

Tavakoli, R., & Mura, P. (2021). Muslim women travelers' constraints: A critical review. *Women in tourism in Asian Muslim countries*, 25-40.

Thyne, M., Lawson, R., & Todd, S. (2006). The use of conjoint analysis to assess the impact of the cross-cultural exchange between hosts and guests. *Tourism Management*, *27*(2), 201–213. doi:10.1016/j.tourman.2004.09.003

Chapter 16
Influencers and Travel Agencies in the Focus of Influencer Communication:
An Evaluation of Brand Collaborations

Zeynep Alkan
Near East University, Cyprus

ABSTRACT

The aim of this study is to reveal how influencer communication takes place during collaborations with travel agencies and to evaluate how they communicate with their followers in this context. The study has a descriptive nature and covers the collaborations of "Korea Tourism Organization & Meryem Can," "Visit Dubai & Meryem Can," "Visit Dubai & Gülnihal Eren (Cocobolinho)," and "Costa Cruises & Gülnihal Eren (Cocobolinho)" as case studies. Within this framework, collaborations with different travel agencies are examined through the relevant content on the official Instagram accounts of influencers. At the same time, this compilation study shows that influencers actively participate in social networks during the experiential process of travel agencies. Additionally, the preference for female influencers who collaborate in the tourism industry is considered as another feature of the study.

INTRODUCTION

Today, it is a fact that we are able to promote and advertise anything through social media applications as well as to reach more audiences. The changes and transformations experienced with social media platforms in every sector also can also be seen in the field of tourism. It is seen that tourism companies are sharing posts on the social media platforms in order to reach their target audience. In this way, tourism companies aim to announce their desired travel routes, campaigns, discounts, and draws through social media platforms. In this context, we can say that tourism companies have started to focus on social media platforms as well as the traditional promotional types.

DOI: 10.4018/978-1-6684-8417-3.ch016

In this direction, it is possible to talk about influencers who are included in our lives as a result of social media applications and environments. Influencers are people who produce content in line with a specific topic and share this content with their followers through social media platforms. Influencers are considered as a new type of third party that uses social media platforms to change and transform their followers' thoughts (Freberg et al., 2010). It is stated that the influencers have a guiding power over their followers.

Therefore, it is seen that influencers share their experiences, opinions and thoughts about a brand or product with a sincere language. Producing contents consistently on social media platforms enables influencers to establish a virtual bond with their followers. In this context, one of the factors that creates a virtual connection between influencers and their followers is that they are able to appeal to their common tastes and likes. As a result the followers identifies with the influencers and they see them as one of their own friends. In this regard, influencers are considered as people who appeal to large audiences and are in a trusted position (Lou & Yuan, 2018).

Followers are more likely to interact with influencers than other celebrities. The posts of influencers on social media accounts for brand collaborations are usually in the form stories, posts, videos, reels and visuals. Brands that realize the interaction and virtual bond between influencers and their followers are entering the process of cooperation with influencers through social media platforms. Brands carry out the influencer collaboration process in order to reach more people from their target audience, direct them to buy their products, to remind their brand and to create a bond. In this context, influencers are considered as a communication tool by brands. In this way, the process of communicating with the target audience proceeds more actively. In addition, as a result of the cooperation, the followers are able to gain information about the brand / product in a faster and more practical way. Influencer followers are able to discover more than one brand as they are trying to gain information about a new product in this process. Therefore, the followers are aware of the brand and its product variety or the products of the trending brands. Influencers who go to the brand collaboration process are aware of which content and which products their followers might like. Therefore, influencers pay attention to this process when choosing the products of the brands that they will produce content and promote. Otherwise, the influencers are aware that their cooperation made with the brand or product may not be liked and besides, their followers trust in them will be damaged. As a result, they tend to carry out a meticulous and regular process during the brand cooperation process. In addition, influencers today use concepts such as "collaboration", "paid partnership", "advertisement" in their contents they share. In this way, it can be said that they are displaying a more transparent approach towards their followers.

Influencers are experiencing many different brands, products and environments from categories such as clothing-fashion, beauty and make-up, lifestyle, parent-child, fitness, sports, food, do-it-yourself and travel. In this context, it is possible to say that many influencer categories exist today. When travelling is observed among these categories, we are able to observe influencers who cooperate with tourism agencies. Travel agencies on social media platforms are also aware of influencers' ability to direct and influence their followers. So as a result, nowadays it can be seen that travel agencies and influencer cooperation processes are taking place more frequently.

The compatibility of travel agencies and the influencers is also an important factor within the cooperation process. Therefore, influencers who are involved in the cooperation process with travel agencies should be people who love to travel and can actively produce and share content while they are experiencing a new environment. With the cooperation process that has taken place, the recognition of travel agencies, as well as the recognition of the introduced country in terms of tourism, cultural heritage, restaurants etc. and the recognition of the settlements increases. Therefore, the influencer and the cooperated travel agency may come to the minds of the followers before they plan their trips. In other words, the travel planning processes of the followers can be realized through the agency introduced by the influencer. In this context, a change in the holiday planning practices of their followers can be observed. Therefore, the followers want to draw the travel route preferred by the influencer in order to live the same experience which is a result of trusting the influencers. In short words, every content shared by the influencer becomes important by its followers.

The study aims to reveal how influencer communication takes place during the cooperation process of influencer/influencers with travel agencies and in this context, to evaluate how they communicate with their followers. Within the scope of the descriptive study, the collaborations of "Korean Tourism Organization & Meryem Can", "Visit Dubai & Meryem Can" and "Visit Dubai & Gülnihal Eren (Cocobolinho) " "Costa Cruises & Gülnihal Eren (Cocobolinho)" are presented as case studies. In this context, collaborations made by different travel agencies and related content on the official Instagram accounts of these influencers will be examined. The study is limited to "Meryem Can" and "Gülnihal Eren (Cocobolinho)" who cooperated with different travel agencies in 2022. The study, which is a review, is important in terms of revealing the communication process established by influencers through collaborations and referring to the realization process of this situation through travel agencies. The preference of female influencers who cooperate in the field of tourism in both collaborations is also considered as another feature of the study.

INFLUENCER COMMUNICATION AND THE CONCEPT OF INFLUENCERS

The concept of an influencer, in its simplest form, is a person or group that has influence over a specific audience (Hennessy, 2022). In this context, an influencer is defined as a person who communicates with their followers through social media platforms and shares their own emotions, thoughts, and experiences related to a brand or purchased product (Alkan & Ulas, 2023). In other words, influencers are individuals who express their opinions online (Walter & Brüggemann, 2020). Influencers who create a wide network of followers in line with the ideas they have stated; They follow trends as well as lead new trends (Geyser, 2023; De Veirman et al., 2016). Influencers have the characteristic of being both famous and ordinary people due to their positions (Jin et al., 2019). In addition, it is now seen that being an influencer has become a profession and a new career field. Therefore, it should be evaluated as an area beyond just a lifestyle (Markides et al., 2020; Szczurski, 2017).

It can be seen that differences arise in individuals' daily life practices with the presence of influencers. Therefore, it can be said that influencers play an effective role in individuals' decision-making processes (Alkan & Ulas, 2022; Hennessy, 2022). In this process, influencers displaying an honest, sincere, and open approach towards their followers enable them to become a trustworthy and effective person (Grafström, Jakobson, & Wiede, 2018). In addition, it can be said that influencers have become a new source of recommendation through social networks (Vrontis et al., 2020). In the process of sharing their recommendations with their followers, influencers' reliance on a story helps create a virtual bond (Messiaen, 2017). In this regard, influencers are putting more time and effort into social media platforms to build a loyal following and gain more followers (Sukardani, 2010).

The content shared by influencers should be interesting, exciting, and intriguing to the target audience. The originality that provides value for followers to follow the influencers should be present in the

Table 1. Steps for influencers to consider in the content creation process (Cevher, 2020)

Steps	Descriptions
1. Creating a Brand Identity	The content shared by an influencer should be unique to themselves.
2. Setting up a good work schedule.	A specific working schedule should be created, and it should be ensured that followers can be active during those hours.
3.Consistency in Content	Influencers need to be consistent and organized in the content they produce.
4.Collaborating with People	Entering into a brand collaboration process can help you reach different target audiences.
5.Use of Hashtags (#) in Content	The use of hashtags in content shared on social media platforms is considered an effective method to reach different target audiences.

content on their social media profiles (Bor & Erten, 2019). On the other hand, influencers are individuals who prioritize personal persuasion. In this context, they actively exhibit their persuasion methods on social networks along with communication frequency (Non Business Advisor, 2015). Based on this information, when looking at the personality traits of influencers (Williams, 2017), it is seen that they prioritize the following characteristics:

1. Reliability
2. Expertise in a subject matter
3. Ability to set a schedule
4. Ability to reach a wide target audience
5. Persuasion skills

Individuals who possess these characteristics can potentially succeed on social media platforms as influencers (Williams, 2017).

In order for influencers to maintain or continue their position, they need to pay attention to their content creation processes. In this context, necessary steps include (Cevher, 2020) (See table 1).

The interaction power of influencers is usually measured by parameters such as the number of followers, likes, views, and comments (Li et al., 2011). In this regard, influencers need to be actively involved in the content production process to maintain their presence on social media platforms (Ecmercer, 2015). In this context, terminologies such as "follow", "subscribe", and "invite" have become popular with influencers today (Nurhandayani et al., 2019). Therefore, influencers guide their own social media platforms and brand's social media platforms with these expressions. Influencers are classified into nano, micro, macro, and mega categories (Geyser, 2022) (See table 2).

In light of all this, influencers are individuals who leave a big mark in online environments. In this context, influencers that are noticed by brands are seen as an important communication tool to reach target audiences (Frederick, 2023). Reasons for influencer and brand collaboration (Bor & Erten, 2019):

1. Through influencers, your brand, service or product can become a topic of conversation on social media platforms.
2. Through influencers, purchase actions can be taken for your brand, service or product.
3. Influencers are effective in crisis communication that may occur at any time related to your brand, service or product.

Table 2. Classification of influencers (Geyser, 2022)

Nano Influencer	These are individuals with less than 10,000 followers on social media platforms. They are considered as the newest influencers and are evaluated as daily social media users (Geyser, 2022).
Micro Influencer	Micro influencers are individuals with 10,000-50,000 followers on social media platforms. They are individuals with a high level of interaction power. They are seen as more natural and sincere by their followers (Bor & Erten, 2019; Geyser, 2022).
Macro Influencer	They are influencers with 100,000 to 1 million followers on social media platforms. They are the influencers with the highest local interest. (CreatorDen, 2018).
Mega Influencer	They are individuals with one million or more followers on social media platforms. They can reach many audiences at once (Elli, 2017).

4. Comment processes related to your brand or product through influencers are guiding the brand's new planning.
5. Through influencers, a personalized communication process related to your brand or product is realized on social media platforms (Bor & Erten, 2019).

In the light of all these, as brand ambassadors, they are in direct interaction with potential buyers (Nurhandayani et al., 2019). Influencers are perceived as role models by their followers, and personal communication with them gains importance in the eyes of their followers (Djafarova & Rushworth, 2017). Influencers who have a good relationship with their followers can increase the reputation of a brand or product and can help plan new strategies for the brand based on feedback from their followers (Booth & Matik, 2011). Influencers state the hashtags #sponsored, #ad, #paidpartnership, and #collaboration in the description section of the content they share in brand collaborations or advertising processes (Audrezet et al., 2018). In this context, there are many influencers who engage in collaboration processes in various categories. For example, it is possible to mention categories such as fashion, makeup and personal care, travel, food, sports, gaming, entertainment, parenting, and children. Accordingly, each brand selects influencers according to its own product and service category and includes them in the communication process with the target audience.

COLLABORATION PROCESS BETWEEN INFLUENCERS AND TRAVEL AGENCIES

Digital transformation and change have been effective in every field, as well as in the tourism sector. In recent years, the effectiveness of social media in brand communication is an undeniable fact. In this context, the effects of social media are seen in almost every individual and corporate context (Ulas, 2020). In this sense, there are differences in the campaign and advertisement processes carried out by travel agencies. While social media platforms bring a new breath to the tourism sector and enable it to reach many audiences, influencers also appear as an undeniable reality in this process.

Influencers and influencer agencies provide processes that highlight content produced in brand collaborations, reach different target audiences, and enable direct messaging. In this context, these elements facilitate campaigns to be more efficiently conducted and increase sales and engagement (influencermarketinghub.com, 2022). In other words, influencers are individuals who have the potential to influence their followers and direct them towards purchasing through social networks. They are high-potential individuals in terms of influencing their followers (More & Lingam, 2017). In this regard, influencers who share their travel experiences encourage their followers to take action in the tourism industry (Gudani, 2020).

According to the literature, followers consider influencers as a reliable source of information and place importance on the experiential aspect of travel (Gretzel, 2018). Travel-related content shared by influencers has a positive impact on followers, providing inspiration, cultural information, and recommendations for new places to explore. However, the personal interests of influencers also play a role in this process (Yılmaz et al., 2020; Ay et al., 2019). In the tourism industry, influencers are used as a communication tool to promote destinations and engage in collaboration (Xu & Pratt, 2018). As travel is an experiential process, influencer communication is used to convey services related to travel, tourism, luxury cruise trips, and accommodation (redpointmarketingpr.com, 2023; Agenic.io, 2022). Influencers are used to create awareness, drive traffic to websites, and target messages to specific audiences for travel agencies (Duve, 2022). Through traveling influencers, many followers are exposed to new places, cultures, and routes, which can have a significant impact on their travel choices. Brands, agencies, and hotel groups recognize this and engage in influencer collaborations to promote their products and services (Influence4You, 2022). In other words, travel companies increase their brand awareness through influencer collaborations, sharing their special travel programs and reservations on social media platforms (Rosenfeld, 2022). In this process, potential consumers attach importance to the interpretation of the service they intend to purchase by those who have experienced it (Bor & Erten, 2019).

Collaboration between travel influencers and travel agencies contributes to your advertising campaigns in five steps. In this regard (Nizami, 2023):

1. New audiences: Influencers sharing their travel-related content on social media platforms helps you connect with new audiences.
2. Awareness: Influencer and brand collaborations have become one of the most effective ways for companies to create awareness on social media platforms.
3. Organic reach: Collaborating with travel influencers helps you reach an organic audience.
4. Increase sales: Campaigns related to tourism can be organized, promotions can be announced through influencer collaboration, and potential customers can be gained for agencies. In this regard, purchasing intent can also be revealed.
5. Trust: The promotion content of an influencer, who is followed more willingly by followers than traditional advertisements, is considered more reliable (Nizami, 2023).

When it comes to influencer collaborations in the tourism industry, people who share common interests and appeal to their target audience tend to drive visitors towards the products or services they promote, resulting in purchase intent (Avcı & Bilgi, 2020; Sabuncuoğlu & Gülay, 2014). Looking at the benefits of influencers in the tourism industry, who create engaging content for their target audience, 10 advantages can be listed (Modernspeak, 2022) (See table 3).

Travel agencies that collaborate with influencers can gain trust at some point. The followers who make this process possible see influencers as friends and trust them (Baran, 2023). Influencers provide a new third-party support for individuals who love to travel (Rosenfeld, 2022). In this respect, followers choose their vacation destinations based on the location recommendations of the influencers they follow, rather than searching for a budget-friendly holiday route on travel websites. In this process, followers

Table 3. Benefits of influencers in the tourism industry (Modernspeak, 2022)

1. Increase Brand Awareness	A collaboration in travel or any other field provides a wide reach on social media platforms. This naturally means that brand awareness reaches larger audiences.
2. Increase Engagement	It is seen that the content shared by influencers in the travel industry greatly increases engagement on social media platforms.
3. Develop Brand Sensitivity	Brand awareness is important for the longevity of a brand. Influencers who travel around the world in the tourism industry are seen as more trustworthy individuals. In this regard, they are more likely to trust the travel agencies and activities promoted by the influencer,
4. Create Authenticity and Trust	The most important element in gaining a positive brand awareness is authenticity. In this regard, businesses prefer individuals who create original content for their establishments.
5. Create Organic, Visually Striking Content	High-resolution photos and videos, brightness, color schemes, and visual aesthetics of influencer content lead to higher engagement.
6. Produce Highly Shareable Content	Popular travel content appeals to and inspires anyone who loves to travel.
7. Opportunity for Inter-Brand Collaboration	In the digital world, the process of collaboration between brands is seen as an effective way to enter a large market by entering common interests and areas.
8. Attract New and Existing Customers' Interest	People who are not in a constant state of travel still engage with the brand for a long time, even if they are not traveling.
9. Get More Reservations	Influencer marketing campaigns can lead to an increase in reservations for your business.
10. Encourage Positive Reviews	Generally, it is believed that influencers, whether it is a collaboration or a personal choice, behave truthfully and honestly about the product or service they promote.

become knowledgeable about all aspects of travel planning, from purchasing flight tickets for countries they have never seen before to making arrangements for accommodation (Bor & Erten, 2019).

The process of influencer communication in the content arising from the collaboration between travel agencies and influencers is as follows (Company.shopltk.com, 2023):

1. Provides insights about new locations and experiences,
2. Triggers followers' desire to travel and escape from their current environment,
3. Shares travel details,
4. Brings a new perspective to existing services,
5. Creates demand for travel activities,
6. Includes tags and mentions related to the trip and the company,
7. More effective than professionally shot images and advertisements by the company,
8. Calms potential customers with safety concerns,
9. Enables the creation of natural and authentic content.

Influencers share all their travel adventures through social media, from hotels, restaurants, museums, and historical sites they visit to the most beautiful and unique locations of the city, as well as transportation. Additionally, they provide tips, tricks, and recommendations related to their travels in their content (Modernspeak, 2022). It has been found that female influencers are more trustworthy and attractive compared to male influencers. Moreover, female followers think that female influencers' expertise is more prominent (Yıldız, 2022). In this respect, it can be stated that influencers have a different experience during their travels compared to other people. Capturing the right lighting, the right shot, or adjusting their outfit for a photo takes a lot of time. They plan their trips step by step and immortalize every moment with photos and videos to share with their followers. (Bor & Erten, 2019).

Travelers who want to plan their vacations use the internet to their advantage, and it can be said that they benefit from influencers' experiences. Influencers who recommend a destination have been shown to be effective in influencing individuals' intentions to visit the destination through the elements of credibility, innovation, and proximity (Avcı & Bilgili, 2020). According to a survey conducted by Rakuten Marketing in 2019 with five countries and 3,600 consumers, it was found that 88% of consumers choose their destinations based on content presented by an influencer and are inspired by them (Degreef, 2019). By inviting influencers to their event activities, brands and companies can not only increase brand awareness but also increase knowledge about their products and brands. With the content produced by influencers, brands can successfully reach consumers (Evel & Şen, 2020). In this context, influencers are seen as individuals who have a positive impact on the image of travel, events, and destinations (Pop et al., 2021).

In this context, collaborations between Meryem Can and Gülnihal Eren (Cocobolinho) are being examined.

1. Collaboration Between Korea Tourism Organization and Meryem Can

In this context, the collaboration between "Korea Tourism Organization & Meryem Can" is examined. Meryem Can, who is included in the mega influencer classification, has 1.7 million followers on her Instagram account. Meryem Can, who produces many extraordinary contents, especially entertainment content, expresses her special interest in Korea on her own social media platforms at every opportunity. In this regard, it can be seen that the Korea Tourism Organization has attracted the attention of the company and realized a collaboration process in this direction. It can be said that she dresses in a style suitable for the location she is in, uses cultural outfits, and makes food choices according to the location she is in, as seen in the visuals below. In this way, she directs and informs her followers. It is seen that she also tags the travel agency in every shared story. In addition, it can be seen that she also tags the institution or location where she participated in different events. In this way, she provides references to her followers in every moment and every share. In this regard, she directs her followers to the institution and its social media accounts. In addition, she shares with her followers that she has collaborated with the institution during her travels by tagging the institution. In this way, she gains the trust of her followers and maintains the existing trust (See Figure 1).

Meryem Can's trip to Korea, which was a collaboration with Korea Tourism Organization, has been announced on the company's website in Turkey. The travel agency directs consumers to Meryem Can's social media accounts to learn the details of the trip. Additionally, it is mentioned that Meryem Can visited historical palaces and places in Korea while wearing the traditional Hanbok costume, took dance lessons at the same studios as famous K-Pop artists, got her makeup done like K-Pop artists, and explored interesting places in Seoul and Busan regions during her trip (ktroturkiye.com, 2022) (See Figure 2).

Meryem Can, who shares her travel story on the Instagram platform, posts every detail on Instagram as posts, stories, videos, and also as content on YouTube. It is also observed that there is a redirect to the influencer's social networks on the brand's website. In this context, it is seen that the influencer is evaluated as a communication tool within the scope of the collaboration carried out by the Korea Tourism Organization Turkey. In addition, the influencer has pinned her stories under the name "Korea" on

Figure 1. Meryem Can and Korea Tourism Organization collaboration images
Source: @meeryemcan (2022)

Figure 2. Meryem Can and Korea Tourism Organization collaboration images
Source: ktoturkiye.com (2022)

her Instagram profile. In this way, her followers can watch Meryem Can's Korea travel experience at any time. Therefore, with the collaboration made, access and interaction continue at all times.

2. Visit Dubai and Meryem Can Collaboration

In this context, the "Visit Dubai & Meryem Can" collaboration is being examined first. Meryem Can, who is classified in the mega influencer category, has 1.7 million followers on her Instagram account. It is seen that Meryem Can, who produces many extraordinary content, especially entertainment content, attracted the attention of Visit Dubai company and a collaboration process was carried out in this direction. It can be said that Meryem Can actively shared her travel adventure with her followers, from the hotel she stayed in to the luxury restaurants she visited, from the helicopter tour to the luxury hotels, the entertainment nights specific to Arabs, and of course the Dubai desert. In every experience she had, it is observed that she tagged Visit Dubai, the institution she collaborated with. Therefore, she not only directed her followers but also announced the collaboration. It is noticeable in her posts where she tries to convey all the details of Dubai's flashy world. Meryem Can's notification of every location she visited creates a guiding effect on her followers during her travel. In this context, it is also seen that she

Figure 3. Meryem Can and Korea Tourism Organization collaboration visuals
Source: @meryeemcan (2022)

expressed that collaboration with Visit Dubai institution can be provided at every location she visited with the help of visuals (See Figure 3).

Meryem Can, who tells the story of her travels on the Instagram platform, shares every detail on Instagram posts, stories, videos, and on YouTube as well. In this context, it can be seen that she was evaluated as a communication tool for the partnership she carried out with Visit Dubai. In addition, the influencer has pinned her stories under the name "Dubai" on her Instagram profile. In this way, her followers can watch Meryem Can's Dubai travel experience at any time. Therefore, the partnership made will sustain access and interaction.

3. Costa Cruises and Gülnihal Eren (Cocobolinho) Collaboration

The partnership between Gülnihal Eren (Cocobolinho) and Costa Cruises is being examined. Gülnihal Eren, who falls under the classification of macro-influencers, has 881,000 followers on her Instagram account. With the username Cocobolinho on Instagram, she creates content in many different categories, including fashion, hair care, parent-child relationships, home decor, and travel. It is seen that she actively shares the moments of her luxury cruise trip organized by Costa Cruises with her family on her Instagram account. In this context, she shares detailed visuals and written content about the facilities, ports to be visited, room prices, and activities for families with children provided during the cruise organized by Costa Cruises travel agency. It is observed that she uses the Costa Cruise tag in her stories and directs her followers to social media platforms with links and profile tags. In addition, the #advertisement tag is also included in every story (See Figure 4).

In addition, Cocobolinho shares content introducing the "Costa App" created by the Costa Cruises agency. In this context, she informs her followers that they can stay informed about all on-board activities such as dining, entertainment, sports, etc. as soon as they download the application. Furthermore, she leaves a question box in her Instagram stories to answer the questions that her followers are curious about. This way, she can interact with her followers in real-time about her travel experience and express all the details. At the same time, she shares the other routes the ship goes to with her followers and makes direct referrals by adding links. It is also seen that she gives her followers a 10% discount code for the trips they want to make with the ship during this process (See Figure 5).

Figure 4. Gulnihal Eren (Cocobolinho) and Costa Cruises collaboration visuals
Source: @cocobolinho (2022)

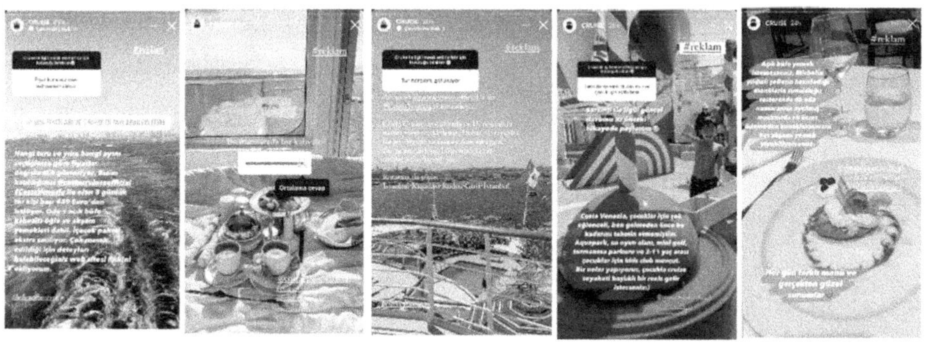

Figure 5. Gülnihal Eren (Cocobolinho) and Costa Cruises collaboration visuals
Source: @cocobolinho (2022)

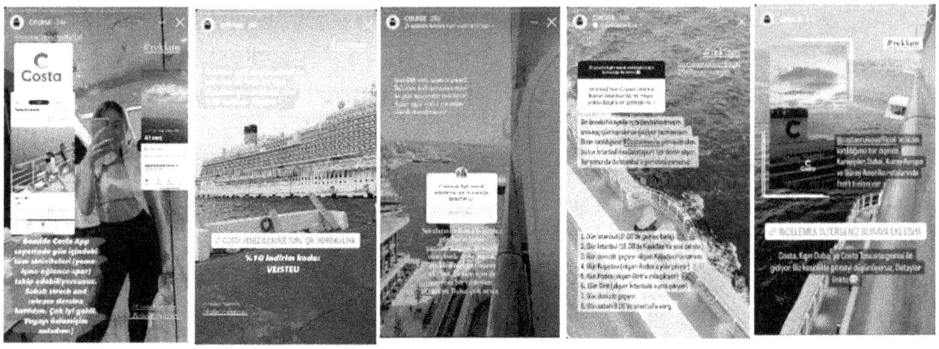

Providing detailed explanations and sharing her family in her stories can make Cocobolinho appear more sincere and trustworthy to her followers. In light of all this, it can be seen that Cocobolinho is being considered as a communication tool for the brand at all times. In this context, the influencer actively creates content and involves her followers in the experience. In addition, the influencer has pinned her stories under the title "Cruises" on her Instagram profile. Therefore, with this collaboration, access and interaction will become sustainable.

4.Visit Dubai and Gülnihal Eren (Cocobolinho) Collaboration

The collaboration between Gülnihal Eren (Cocobolinho) and Costa Cruises is being examined. Gülnihal Eren, classified as a macro influencer, has 881 thousand followers on her Instagram account. With the Instagram name Cocobolinho, she creates content in many different categories such as fashion, hair care, parenting, home decor, and travel. It can be observed that she actively shares moments from her trip to Dubai with her family, which was organized by Visit Dubai, with her followers (See Figure 6).

Figure 6. Gülnihal Eren (Cocobolinho) and Visit Dubai collaboration images
Source: @cocobolinho (2022)

Figure 7. Gülnihal Eren (Cocobolinho) and Visit Dubai collaboration visuals
Source: @cocobolinho (2022)

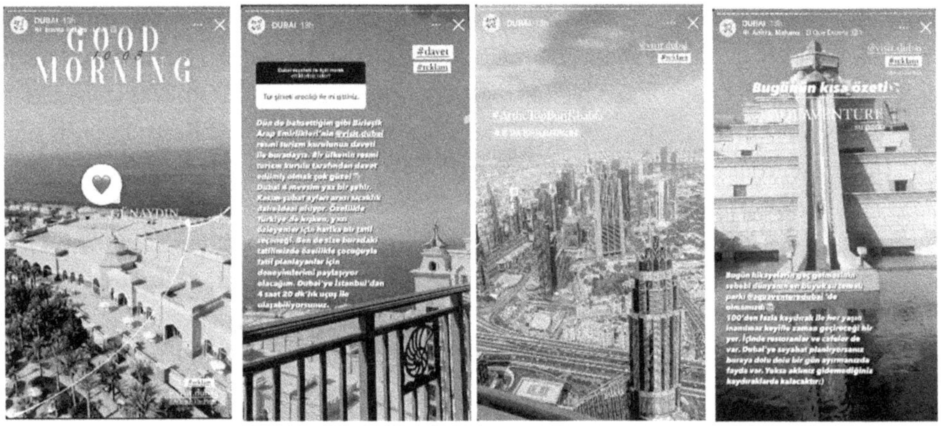

Gülnihal Eren (Cocobolinho) is sharing on her Instagram stories that she and her family have been invited by Visit Dubai and they have traveled to Dubai as part of this collaboration. She tags the hotel they are staying at and provides details about their trip. Cocobolinho frequently uses the phrase "a warm holiday for your child in the middle of winter" in her stories. She also shares a question box in her stories to answer her followers' questions about their Dubai trip. In this way, Cocobolinho provides an opportunity for her followers to ask questions about Visit Dubai. She answers all the questions with an open and honest approach, tagging the places she visits and providing short descriptions of them (See Figure 7).

It is important that Gülnihal Eren highlights the Dubai trip as a suitable vacation program for children, as it is expressed by someone who has experienced this process with their own children. Throughout the collaboration process, it is seen that Cocobolinho is constantly evaluated as a communication tool for the brand. In this context, the influencer actively creates content and involves their followers in the experience. Additionally, the influencer has pinned their Dubai stories to their Instagram profile. Therefore, the collaboration will lead to sustainable accessibility and interaction.

CONCLUSION

With the development of internet technology, social media applications have emerged. In this context, with the inclusion of social media platforms in our daily life practices, differences are also observed in the process and forms of interpersonal interaction and communication. Social media platforms provide users with more accessibility (Arora et al., 2019). It is possible to talk about influencers who use social networks effectively in this process. As expressed by Geyser, influencers are individuals who gain reputation with their expertise and knowledge in a certain field. Accordingly, they create regular content on the social media platforms they prefer. In this way, they reach a wide audience who share common interests (Geyser, 2023). This leads to the emergence of a social connection. Influencers' ability to express their own experiences in a sincere manner makes them appear as a friend and fosters a sense of trust. Therefore, influencers are in a position to reveal the intention to purchase a product or brand. As a result of brands discovering this directing power, brand collaborations are becoming more and more popular.

It can be said that every sector is affected by social media and the influencers defined as its benefit. In this context, the tourism industry is also among these sectors. Nowadays, reaching multiple audiences, promoting campaigns and being in constant interaction is made possible through influencers. The tourism industry, which cannot remain indifferent to this situation, has started to participate in the collaboration process. Influencers actively sharing their own experiential process on social media platforms also reveal potential consumers for travel agencies and businesses. Especially in the context of the examined literature, female influencers pay more attention to details such as dressing appropriately for the location, makeup, travel planning, and local clothing during their travels. In addition, in vacations with children, the vacation planning chosen by female influencers stands out for their attention to detail. The importance of storytelling cannot be ignored in the experiential process. Influencers must convince their followers of the reliability of the content they transmit during the experiential process. This persuasion can eventually turn into a potential purchase and positively impact the brand image. Therefore, influencers need to explain the organization of the travel agency in detail as well as show it through photos and videos. Influencers' travels are like an adventure for their followers, who want to hear and see every detail of that adventure from start to finish.

This study, which is evaluated as a descriptive study, examines "Korea Tourism Organization & Meryem Can", "Visit Dubai & Meryem Can" and "Visit Dubai & Gülnihal Eren (Cocobolinho)" and "Costa Cruises & Gülnihal Eren (Cocobolinho)" as case studies. In this context, although there are common travel agencies in both studies, there are differences in the way they are expressed. In this context, it is seen that Gülnihal Eren emphasizes family and children, while Meryem Can shares posts about fun and different activities. It is observed that Gülnihal Eren actively engages with her followers by opening a "question box" and being in communication on every trip she takes. Both influencers tag travel agencies and direct their followers in line with the collaboration they have made. In addition, it is seen that Gülnihal Eren received a discount code as part of her collaboration. In the brand collaboration process, it is necessary to announce "discount codes, campaigns, and events" and reach the influencer's audience. In this way, an incentive is created to motivate followers to take action. In this context, it can be said that both influencers act as communication tools in the brand collaboration process. At this point, it can be expressed that each influencer has a unique way of producing content and expression. It is seen that in the focus of the influencers, an advertising process takes place and all details are fixed on their Instagram profiles. In this context, followers can look at influencers' travel experiences anytime, anywhere and have the opportunity to be familiar with all the details again.

REFERENCES

Advisor, N. B. (2015). Focus social media outreach on cultivating 'influencers' among your followers. *Nonprofit Business Advisor Report, 6–8*. Advance online publication. doi:10.1002/nba

Agenic.io. (2022). *5 Influencer Marketing Campaign Examples In Travel Industry*. https://agenic.io/5-influencer-marketing-campaign-examples-in-travel-industry/

Alkan, Z., & Ulas, S. (2022). Practices of social media and influencer: A research based on influencer's followers. *MEDIAJ*, *5*(2), 187–205.

Alkan, Z., & Ulas, S. (2023). Trust in social media influencers and purchase intention: An empirical analysis. *Online Journal of Communication and Media Technologies*, *13*(1), e202301. doi:10.30935/ojcmt/12783

Arora, A., Bansal, S., Kandpal, C., Aswani, R., & Dwivedi, Y. (2019). Measuring social media influencer index- insights from facebook, Twitter and Instagram. *Journal of Retailing and Consumer Services*, *49*, 86–101. doi:10.1016/j.jretconser.2019.03.012

Audrezet, A., de Kerviler, G., & Moulard, J. G. (2018). Authenticity under threat: When social media influencers need to go beyond self-presentation. *Journal of Business Research*, *18*(July). Advance online publication. doi:10.1016/j.jbusres.2018.07.008

Avcı, E. & Bilgili, B. (2020). Sosyal Medya Fenomen Özelliklerinin Takipçilerin Destinasyonu Ziyaret Etme Niyeti Üzerindeki Etkisi. *Tourism and Recreation, 2*(1), 83-92.

Ay, E., İpek, K., Özdağ, N. B., Özekici, E., & Alvarez, M. D. (2019). Travel bloggers as influencers: What compels them to blog. In N. Kozak & M. Kozak (Eds.), *Tourist destination management* (pp. 159–175). Springer International Publishing. doi:10.1007/978-3-030-16981-7_9

Baran, M. (2023). *Under the influence*. https://www.travelweekly.com/Travel-News/Travel-Technology/social-media-influencers

Booth, N., & Matic, J. A. (2011). Mapping and leveraging influencers in social media to shape corporate brand perceptions. *Corporate Communications*, *16*(3), 184–191. doi:10.1108/13563281111156853

Bor, H. M., & Erten, A. (2019). Dijital Çağın Mesleği-Nasıl Influencer Olunur? İstanbul: Hürriyet Kitap.

Can, M. (2022). *Meryem Can Instagram Hesabı*. https://www.instagram.com/meeryemcan/?hl=tr

Cevher, F. M. (2020). *Influencer marketing ve sosyal medya fenomenleri*. Nobel Bilimsel Eserler.

Cocobolinho. (2022). *Cocobolinho Insgram Hesabı*. https://www.instagram.com/cocobolinho/?hl=tr

Company.shopltk.com. (2023). *The ultimate guide to travel influencer marketing*. https://company.shopltk.com/ultimate-guide-influencer-marketing-travel-brands

CreatorDen. (2018). *Tüm Pazarlamacıların Bilmesi Gereken 3 Influencer Türü*. https://creatorden.com/tr/tum-pazarlamacilarin-bilmesi-gereken-3-influencer-turu/

De Veirman, M., Cauberghe, V., & Hudders, L. (2016). Marketing through Instagram influencers: Impact of number of followers and product divergence on brand attitude. *International Journal of Advertising*, *36*(5), 798–828. doi:10.1080/02650487.2017.1348035

Degreef, A. (2019). *Rakuten Marketing 2019 Influencer Marketing Global Survey*. https://blog.rakuten-advertising.com/insights/rakuten-marketing-2019-influencer-marketing-global-survey-report/

Djafarova, E., & Rushworth, C. (2017). Exploring the credibility of online celebrities' Instagram profiles in influencing the purchase decisions of young female users. *Computers in Human Behavior*, *68*, 1–7. doi:10.1016/j.chb.2016.11.009

Duve.com. (2022). *The ultimate guide to working with travel influencers*. https://duve.com/guide-to-travel-influencers/

Elli, D. M. (2017). *The phenomenon and rise of Influencer Marketing and how it affects customer opinion and helps or damages brands* [Master Thesis]. International Hellenic University.

Eyel, C. Ş. & Şen, B. (2020). Influencer pazarlama ve tüketicilerin influencer'lara yönelik tutumlarinin satin alma niyeti üzerindeki etkisi. *Atlas Sosyal Bilimler Dergisi, Atlas Sosyal Bilimler Dergisi,* 7-29.

Freberg, K., Graham, K., McGaughey, K., & Freberg, L. A. (2010). Who are the social media influencers? A study of public perceptions of personality. *Public Relations Review*, *37*(1), 90–92. doi:10.1016/j.pubrev.2010.11.001

Frederick, B. (2023). *The Top 100 Social Media Influencers Worldwide*. https://www.searchenginejournal.com/top-social-media-influencers/475776/#close

Geyser, W. (2023). *What is an influencer? – Social Media Influencer Defined*. https://influencermarketinghub.com/what-is-an-influencer/#toc-0

Gretzel, U. (2018). Influencer marketing in travel and tourism. In M. Sigala & U. Gretzel (Eds.), *Advances in Social Media For Travel, Tourism And Hospitality: New Perspectives* (pp. 147–156). Routledge.

Gretzel, U. (2018). Tourism And Social Media Sage Handbook Of. *Tourism Management*, 1–23.

Gudani, A. M. I. (2020). *Travel influencer*. https://www.academia.edu/43198447/Travel_Influencer?email_work_card=reading-history

Hennessy, B. (2022). *Influencer Sosyal Medya Çağında Marka Yaratmak*. Nova Kitap: Kadiköy-İstanbul.

Influence4You. (2022). *Top 10 Travel Influencers in 2022*. https://blogen.influence4you.com/the-top-10-travel-influencers/

Influencermarketinghub.com. (2022). *Must-Have Features To Look for In An Influencer Marketing Platform*. https://influencermarketinghub.com/must-have-features-to-look-for-in-an-influencer-marketing-platform/#:~:text=Many%20influencer%20marketing%20platforms%20offer,your%20sales%20and%20engagement%20rates

Jin, S. V., Muqaddam, A., & Ryu, E. (2019). Instafamous and social media influencer marketing. *Marketing Intelligence & Planning*, *37*(5), 567–579. Advance online publication. doi:10.1108/MIP-09-2018-0375

Ktoturkiye.com. (2022). *Meryem Can's Korea Trip*. https://ktoturkiye.com/meryem-can-ile-kore-seyahati

Lou, C., & Yuan, S. (2018). Influencer marketing: How message value and credibility affect consumer trust of branded content on social media. *Journal of Interactive Advertising*, 1–45.

Messiaen, J. (2017). *Influencer Marketing How The Popularity Threshold Of Istara Influencers Impacts Consumer Behavıour: The Moderating Role Of Purchase Involvement* [Master's Thesis]. Universiteit Gent.

Modernspeak. (2022). *How travel influencer marketing can impact branding*. https://www.modernspeak.co/stories/how-travel-influencer-marketing-can-positively-impact-your-branding

More, J. S., & Lingam, C. (2017). *A SI model for social media influencer maximization.* Applied Computing and Informatics. doi:10.1016/j.aci.2017.11.001

Nizami, M. (2023). *13 Top Travel Influencers.* https://smallbiztrends.com/2022/05/top-travel-influencers.html?utm_content=cmp-true

Nurhandayani, A., Syarief, R.. & Najib, M. (2019). The impact of social media influencer a brand images to purchase intention. *Journal of Applied Management, 4*(17), 650-661.

Pop, R.-A., Săplăcan, Z., Dabija, D.-C., & Alt, M.-A. (2021). The impact of social media influencers on travel decisions: The role of trust in consumer decision journey. *Current Issues in Tourism.* Advance online publication. doi:10.1080/13683500.2021.1895729

Redpointmarketingpr.com. (2023). *Social influencer marketing services.* https://redpointmarketingpr.com/services/social-influencer-marketing/

Rosenfeld, K. (2022). *Travel Advisors Should Be Working With Influencers — Here's How to Start.* https://www.travelagewest.com/Industry-Insight/Business-Features/working-with-influencers

Sabuncuoğlu, A., & Gülay, G. (2014). Sosyal Medyadaki Yeni Kanaat Önderlerinin Birer Reklam Aracı Olarak Kullanımı: Twitter Fenomenleri Üzerine Bir Araştırma. *İletişim Kuram ve Araştırma Dergisi,* (38).

Sukardani, P. S. (2010). *Brand communication melalui digital influencer.* https://www.academia.edu/33955942/BRAND_COMMUNICATION_MELALUI_DIGITAL_INFLUENCER?email_work_card=reading-history

Szczurski, M. M. (2017). *Social media influencer- A Lifestyle or a profession of the XXIst century?* https://depot.ceon.pl/bitstream/handle/123456789/15166/Social%20media%20influencer.pdf?sequence=1&isAllowed=y

Ulas, S. (2020). International Journal of Marketing, Communication and New Media. *Social Media Marketing.*

Vrontis, D., Makrides, A., Christofi, M., & Thrassou, A. (2021). Social media influencer marketing: A systematic review, integrative framework and future research agenda. *International Journal of Consumer Studies, 45*(4), 617–644. doi:10.1111/ijcs.12647

Walter, S., & Brüggemann, M. (2020). Opportunity makes opinion leaders: Analyzing the role of first-hand information in opinion leadership in social media networks. *Information Communication and Society, 23*(2), 267–287. doi:10.1080/1369118X.2018.1500622

Williams, S. (2017). *Characteristics of Social Media Key Influencers.* https://www.the24hoursecretary.com/characteristics-of-social-media-key-influencers/

Xu, X., & Pratt, S. (2018). Social media influencers as endorsers to promote travel destinations: An application of self-congruence theory to the Chinese Generation Y. *Journal of Travel & Tourism Marketing, 35*(7), 958–972. doi:10.1080/10548408.2018.1468851

Yıldız, S. Y. (2022). Instagram Influencerlarının Cinsiyetine Göre Tüketici Profillerinin Belirlenmesi. *Cumhuriyet Üniversitesi İktisadi ve İdari Bilimler Dergisi, 23*(4), 943–953. doi:10.37880/cumuiibf.1130303

Yılmaz, M., Sezerel, H., & Uzuner, Y. (2020). Sharing experiences and interpretation of experiences: A phenomenological research on Instagram influencers. *Current Issues in Tourism*, *23*(24), 3034–3041. Advance online publication. doi:10.1080/13683500.2020.1763270

KEY TERMS AND DEFINITIONS

Advertisement: Used for promoting a brand, product, or service.

Influencer: Individuals who share their opinions and thoughts about a brand or product with their followers through social media platforms.

Influencer and Brand Collaboration: An agreement between an influencer and a brand for promoting a particular product.

Influencer and Tourism: The tourism industry prefers to use influencers for promotional activities.

Influencer and Travel Agency: Travel agencies collaborate with influencers to reach their potential customers on social media platforms.

Influencer Communication: The communication process where influencers interact with their followers directly through social networks.

Instagram: One of the most popular social media platforms. It is often preferred in influencer collaborations.

Chapter 17
Women's Empowerment Within the Tourism Industry:
Risk Assessment and Mitigation Strategies for Solo Women Travellers

R. Ranjitha
BNM Institute of Technology, India

Sonia Mathew
CHRIST University (Deemed), India

ABSTRACT

Travelling is essential for maintaining one's happiness and mental wellness. Tourism is a job providing industry and huge contributions to economic growth, and many studies have been undertaken to determine who is attracted to tourism. Surprisingly, women take a major stand in the tourism industry. The purpose of this study is to examine female solo travelers' risk assessment and their mitigation strategies by investigating online travel blogs. Therefore, the study aims at assessment of risk such as physical risk, destination-specific risk, and financial risk faced by solo women travellers by considering and analyzing online women travellers' blog narratives. These narratives from online travel blogs pertaining to Indian solo female travellers are considered in this study. Hence, this assessment gives an opportunity to know how to stay safe in uncertain situations and also mitigation strategies to stay alert. The result of the study indicates the various risks faced by solo female travel and mitigation strategies for promotion of more female travellers.

INTRODUCTION

Travelling is essential for maintaining one's happiness and mental wellness. Studies have demonstrated the benefits of travel for mental health. In a society where people are continuously under pressure from their jobs, relationships, and raising a family, it's critical that they take some time to unwind in order to retain their ability to function properly. Tourism is job providing industry and huge contributions to

DOI: 10.4018/978-1-6684-8417-3.ch017

economic growth, many research have been undertaken to determine who is attracted to tourism and travel surprisingly, most review of literature suggest that women travel the most, despite the fact that males are generally thought to travel more. To compare female to male travellers in US, 63% are female and 37% are male, this is actually trend with worldwide statistics. Despite the lack of safety and the pervasive belief that women are more vulnerable alone, 40% of all overnight travels with just one other person were made by women in India. In fact, rural India had a greater percentage (41%) than urban areas, which had a little lower rate (37 per cent).In all of India, Punjab has the highest percentage of single female travellers (66% of single member journeys originated there), followed by Telangana (approximately 60% of single member travels there).The purpose of this study is to examine female solo travellers risk assessment and their mitigation strategies by investigating online travel blogs.Therefore, the study aims at assessment of risk such as physical risk, Destination specific risk and financial risk faced by solo women traveller by considering and analysing online women traveller blogs such as solo female travellers, maproute travel and the solo globetrotter blog's narratives.These narratives from online travel blogs pertaining to Indians solo female travellers are considered in this study. Hence fore, this assessment gives an opportunity to know on how to stay safe in uncertain situations and also mitigation strategies to stay alert. The result of the study indicates the various risk faced by solo female travel and mitigation strategies for promotion of more female travellers.

WORLDWIDE OVERVIEW ON SOLO WOMEN TRAVEL

In 2018 and 2017, Google searches for "solo women travel" increased by 59% and 32%, respectively. The number of searches has increased by 56% in 2019. This increased search is significant for tour operators since it indicates a shift in interest in the travel industry (Kow, 2018).

Also, the number of female pinners of articles regarding "solo female travel" on Pinterest has increased by 35%." Solo travel is growing in popularity and accounts for around 18% of all bookings made worldwide.

Over 6.7 million post in Instagram are tagged with solo women travellers.

As more and more women in India opt to travel alone, hospitality technology platform like Oyo has announed strategic partnership with adventure women to support solo female travel by offering best suitable hotels preference, attractive offers and also decided to promote the trend of women travelling alone by contributing in some little way.

Traveling alone by women is increasingly prevalent in today's culture, but it also carries high risk, which is an inherent component of every travel experience. Hazards have variously been characterised as elements that make it difficult for a person to engage in favourite activities. However, travelling alone for women is not trouble-free because there are several risks associated.

THEORETICAL BACKGROUND

The issue of risk assessment and mitigation strategies for solo women travelers has become increasingly important in recent years due to a growing number of women who are traveling alone for business or leisure purposes. This trend reflects a shift in social norms and values, as well as advancements in technology and transportation that have made solo travel more accessible.

However, women who travel solo face unique risks related to their gender, such as sexual harassment, assault, and violence. These risks can be compounded by cultural norms and expectations, unfamiliarity with local customs, and language barriers. As a result, women travelers must take precautions to ensure their safety, including conducting thorough research on their destination, choosing safe accommodations, being aware of their surroundings, and dressing appropriately.Several organizations and initiatives have emerged to address the safety concerns of women travelers. For example, the Women's Travel Safety Association (WTSA) is a non-profit organization that provides resources and training programs to empower women travelers to travel safely and confidently. The WTSA also offers a certification program for hotels and accommodations that meet their safety standards.

Additionally, some travel companies and organizations have started offering women-only tours and travel experiences to provide a sense of community and safety for solo women travelers. These initiatives reflect a growing recognition.

RISK ASSESSMENT

It's critical to consider the dangers and risks that may be involved while travelling solo.This should include evaluating the overall safety of the area, as well as specific risks such as crime, terrorism, natural disasters, and disease outbreaks. Information on the current safety situation can be obtained from the local authorities, travel advisories, and news sources.

It is also important to consider the cultural and social norms of the destination, as these can impact the safety and security of solo female travelers. For example, in some cultures, women may be expected to dress modestly and adhere to strict gender roles, which could make them stand out as foreigners and potentially increase their vulnerability.

MITIGATION STRATEGIES

There are several strategies that solo female travelers can use to mitigate risks and stay safe while on their trip. These include:

1. Planning ahead: Before traveling, it is important to research the destination, obtain travel insurance, and make arrangements for emergency situations. This includes having access to a local doctor or hospital, as well as a local police or embassy.
2. Being aware of surroundings: When traveling solo, it is important to remain aware of one's surroundings at all times, including the people and environment. This includes avoiding dark or deserted areas, being mindful of what is being carried, and not displaying valuables.
3. Staying in touch: Solo female travelers should inform a trusted friend or family member of their itinerary, and keep in touch regularly. This includes using a travel tracker or GPS device, and having a backup plan in case of emergency.
4. Being cautious with personal information: Solo female travelers should be careful about who they share their personal information with, and avoid discussing their travel plans with strangers. This includes avoiding posting personal information on social media, and using a private email account.

5. Seeking local advice: Solo female travelers should seek local advice on safety and security, and be mindful of the advice they receive. This includes using local guides, talking to hotel staff, and seeking out local resources, such as a women's center or a local travel agency.

PURPOSE STATEMENT

The purpose of this study is to assess the risk assessment of female solo travellers and their mitigation strategies by examining online travel blogs such as maproute travel, solo female travellers and the solo globetrotter blog's narratives written by female solo travellers. Therefore, the study aims at assessment of risk such as physical risk, Destination specific risk and financial risk faced by solo women traveller by considering and analysing online women traveller blogs narratives. These narratives from online travel blogs pertaining to Indians solo female travellers are considered in this study. Hence fore, this assessment gives an opportunity to know on how to stay safe in uncertain situations and also mitigation strategies to stay alert. The result of the study indicates the various risk faced by solo female travel and mitigation strategies for promotion of more female travellers.

LITERATURE REVIEW

Risks associated with travel have a major influence on travellers' choices for locations and travel decisions (Roehl & Fesenmaier, 1992; Sönmez & Graefe, 1998b; Leep & Gibson, 2003; Fuchs & Reichel, 2006), risk perception in tourism has received extensive research consideration (Yang et al., 2018a). According to Bauer (1960), perceived risk is a hazard that consumers individually interpret when faced with a decision about a brand, retailer, or method of purchase. Travelers' perceived risk is described by Tsaur et al. (1997) as a hypothetical fictitious bad experience at the vacation destination As they cannot experience or evaluate actual level of safety before reaching the destination, travellers are forced to evaluate the hazards of travel destinations based on the information readily accessible. Hence, perceived risk is a situation that results from uncertainty as well as its consequences. (1994; Dowling & Staelin)

The numerous forms and structures of risks examined were perceived by travellers. For instance, in their fundamental work, Roehl and Fesenmaier (1992), thoroughly analysed perceived risk by taking into account all of the hazards, including those relating to money, health, happiness, and time. Their study's conclusions revealed three types of perceived danger: risk related to physical equipment, risk related to travel, and risk related to destination. Sönmez and Graefe (1998b) expanded on this research by examining three more risks identified, including those related to health, terrorism, and political instability.Their research revealed the perceived risk associated with a destination is strongly predicted by its safety and security characteristics, which in turn will cause travellers to stay well clear of that spot. Moreover, Maser and Weiermair (1998) examined at Australian citizens and risk factors connected to diseases that can be contracted while travelling, criminality, natural disasters, hygiene concerns, transportation, cultural and Language obstacles and regionally distinct legal systems. In contrast, Fuchs and Reichel (2006) classified risks into six categories: human-induced risk, financial risk, service quality risk, socio-psychological risk, risk from natural disasters and automobile accidents, risk from problems with food safety, and risk from weather.

While Sönmez and Graefe (1998) stated that perceptions of travel hazard do not appear to differ by gender, some researches disagree. Park & Reisinger (n.d.) and D. Karagz et al. (2010) found substantial variances. Yang et al. (2017) revealed that female vulnerability to physical assault was widely documented in 70% of the papers they examined in their comprehensive assessment of the danger literature in tourism. Female tourists, who were less risk-taking than male ones, assessed travel-related risk as being higher and/or took less risk overall. In their research of Norwegian students, Moreover, Heimtun and Abelsen (2014) found that being a solo traveller had a significant gender effect. The majority of female participants said travelling alone was unpleasant, in contrast to the male participants.

According to the research currently available, the gendered risk, which includes personal safety, violence, sexual harassment, and assault, is the main danger for single female travellers (Wilson & Little, 2005; Wilson & Little, 2008; Jordan & Aitchison, 2008; Yang et al., 2018a; Khoo-Lattimore & Gibson, 2018).

The most typical gender danger for female travellers travelling alone is sexual harassment (Seow & Brown, 2018; Toh et al., 2017; Valaja, 2018; Su & Wu, 2020). For instance, Jordan and Aitchison (2008) looked at how surveillance affected solo female travellers' experiences. The results indicated that as single females are frequently They find it challenging to avoid the possibility of becoming the target of sexual harassment and gazing since they are portrayed as sexualized objects of the gaze.

"Social-psychological risk" is another risk factor for female travellers travelling alone and is associated to emotions of exclusion or loneliness (Adam, 2015). One of the greatest worries for solitary female travellers, according to Wantono(2020), is the social risk of being perceived as someone who doesn't fit in. Due to the conservative beliefs and social conventions of Asian cultures, for instance, the notion of a solo female traveller continues to be viewed as somewhat alien (Teo & Leong, 2006), which makes Asian single female travellers perceive greater dangers. Gang et al. (2018) also discovered that Asian women travelling solo perceive certain risks related to prejudice and social ostracisation.Their research showed that women have travel anxieties related to how other individuals perceive them, are vulnerable, feel like they have limited access, and feel conspicuous. Traveling liberates females from the ethical expectations and expectations of their home society, yet this space is nevertheless subject to societal restraints because of the gender norms and power structures of the host society (Bui, Wilkins, & Lee, 2014; Heimtun, 2011). Women who travel are at risk of not being recognised by their families, friends, and residents in the host country.

Overall,Female travellers who travel alone could be concerned about geographical dangers. Destination specific risk, as defined by Rasoolimanesh et al. (2020); Maser & Weiermair (1998); Roehl & Fesenmaier (1992), is the perception of risk related to travelling to a particular location, and it frequently includes factors related to the environment, such as food safety, hygiene, natural disasters, disease, culture, and language barriers.

Physical risks are a common concern for solo female travelers. A study by Wilson and colleagues (2015) found that sexual harassment, assault, and theft were the most significant physical risks reported by female travelers. To mitigate these risks, they recommended that female travelers should avoid walking alone at night, be aware of their surroundings, and carry pepper spray or other self-defense tools.

Destination-specific risks are another important consideration for solo female travelers. A study by Hyslop and Roberts (2019) found that female travelers visiting countries with high levels of political instability or natural disasters reported increased levels of anxiety and stress. To mitigate these risks, they recommended that female travelers should research their destination before traveling, purchase travel insurance, and register with their embassy.

Financial risks are also a concern for solo female travelers. A study by Zaman and colleagues (2017) found that theft and loss of money were the most common financial risks reported by female travelers. To mitigate these risks, they recommended that female travelers should carry only the necessary amount of cash, keep credit cards safe, and monitor their bank accounts for any suspicious activity.

Several studies have also explored the role of technology in risk assessment and mitigation for solo female travelers. A study by Xu and colleagues (2019) found that female travelers use mobile apps to research their destination, book accommodations, and communicate with locals. They recommended that female travelers should use apps like Google Maps or Uber to navigate their destination safely.

A study by Haider and O'Connor (2020) on "Women Solo Travellers' Perception of Risk and Safety: A Qualitative Study" found that women who travel solo often face risks related to their gender and lack of familiarity with the local culture and customs. The study recommended that women travelers should conduct thorough research on their destination, stay in safe accommodations, and stay aware of their surroundings to mitigate risks.

A study by McCabe and Johnson (2013) on "Safety and Security in Tourism: Relationships, Management, and Marketing" found that women travelers face unique risks related to their gender, such as sexual harassment and assault. The study suggested that women travelers should take precautions such as dressing modestly, avoiding walking alone at night, and carrying a personal safety alarm.

In a study by Joerin et al. (2018) on "Empowering Women Travelers: A Literature Review on Risk Perception, Travel Constraints, and Travel Risk Management," the authors found that women travelers often face constraints related to cultural norms and expectations, and they may need to navigate complex social hierarchies. The study recommended that women travelers should engage in risk management strategies such as creating a network of trusted contacts, taking self-defense classes, and carrying a first aid kit.

A study by Adeyemo and Adebayo (2017) on "Empowering Women to Overcome Constraints in Solo Travel" suggested that women travelers should build confidence and self-efficacy through social support networks, educational programs, and cultural immersion experiences. The study recommended that women travelers should also develop risk assessment and mitigation strategies such as carrying a travel insurance policy and staying informed about local safety issues.

PROBLEM STATEMENT

Women are encouraged to travel alone for a variety of reasons, such as experience, escape, relaxation, social, self-esteem, self-discovery, self-empowerment, and a sense of community (Chiang & Jogaratnam, 2006). (Wilson & Harris, 2006). In other words, women likely anticipate that by leaving the "ideal" society and entering a completely new setting on their own, they would advance their personal growth, which will enhance their true selves and therefore, their sense of self (Michael, 2017). (Wood and others, 2008) However prior study has found a number of obstacles that stop women from travelling alone,including social pressure, a lack of confidence, and a fear of being dangerous (Wilson, 2004; Wilson & Little, 2005, 2008).

In this study, following research questions are formulated:

RQ1: What are the types of risk faced by women travelers during solo trips?
RQ2: What is the probability of specific risk managed ?

RQ3: What are the risk mitigation strategies which women solo travelers engage in order to minimize the impact?

It is anticipated that by addressing these issues, the research would considerably enhance the existing research on risk and tourism and produce a vast amount of helpful data that solo female travellers and other industry participants may use. The purpose of this research is to provide relevant and valuable information regarding risk assessment and risk mitigation approaches for female solo travellers, which may then be applied by different solitary female travellers for their own gain.

RESEARCH GAP

The goal of the conducted literature review was to provide an summary of earlier studies on solo travel, female travellers, and risk in the tourist industry. According to the literature review, the most of the research on solo travel have concentrated on examining the individual characteristics of solo travellers, their reasons for travelling alone, and the issues they face. This study aims at filling gap by analysing the different types of risk and their mitigation strategy of solo women travellers . The results from earlier research have been examined and contrasted using data from the assessment phase of the project.The data similarities and difference can reveal how people assess the type of risk and suggest mitigation strategy.

RISK ASSESSMENT AND MITIGATION MATRIX

Physical risk includes the potential for harm or violence, especially in unfamiliar environments. Female travelers may be more vulnerable to sexual harassment, assault, or theft. Therefore, it's essential to take precautions like avoiding isolated areas, not walking alone at night, and keeping important documents and valuables safe.

Destination-specific risks are those associated with specific countries or regions. For example, political instability, natural disasters, or health risks like disease outbreaks. It's essential to research potential risks before traveling and plan accordingly. This may involve taking vaccinations, purchasing travel insurance, and being aware of local customs and laws.

Financial risk is another important factor to consider. This includes issues like theft or loss of money and identity theft. It's important to carry only the necessary amount of cash, keep credit cards safe, and regularly monitor bank accounts for any suspicious activity.

RISK ASSESSMENT MATRIX

Table 1. Risk 1

Risk 1: Physical Risk (Kidnap / Sexual Harassment / Assault)	
Probability	Low
Severity	Very High
Avoidance	Verify information, evaluate intentions, and foresee potential pitfalls. Do not trust blindly strangers, but trusting random strangers beyond limit assess the risk involved. Pre-travel research is important to determine how secure a location is, which locations to avoid, and what frequent tactics fraudsters use to trap their victims. Keep others well informed about your whereabouts. Consider the suggestions from other solo women travellers who have experienced.
Minimizing Impact	Self-defence skills Women safety accessory tools

Table 2. Risk 2

Risk 2: Destination Specific Risk (Terror Attack, Natural Calamities, Health Issues, Crime)	
Probability	Very low
Severity	High
Avoidance	Avoid extremely crowded areas – these are primary targets Be aware of others and take note of any potential escape routes, hiding places, or barriers that could provide some degree of defence against gunshots or sharp objects. Have adequate travel insurance to cover potential hospitalization or injury while travelling. Search online for any potential health risks. Follow traffic rules and regulations clearly. Stay on top of news to know if there's anything predicted for your location.
Minimizing Impact	Keep an eye out for potential escape routes, hiding places, and obstacles that might provide some protection from crowded areas. Stay away from going if a certain location is known for terror attacks or has experienced a number of incidents recently.

HYPOTHESES DEVELOPMENT

Table 3. Risk 3

Risk 3: Financial Risk (Get Scammed, Losing Valuables / Robbed)	
Probability	Very low
Severity	High
Avoidance	Securing valuable items in lockers, not carrying liquid cash, not leaving valuable items unattended and always staying aware of surroundings. Use designed products and safety measures like Pacsafe or a travel bra.
Minimizing Impact	The finest risk-reduction tactic against financial troubles is travel insurance. Emergency cash and carrying several bank cards are advised to minimize the risk impact.

Ho: There is no significant difference in the risk assessment of solo female travelers across different regions and countries.

H1: Destination-specific risks for solo female travelers vary significantly by region and country, and mitigation strategies need to be tailored to the specific destination.

INTERPRETATION OF HYPOTHESES

Ho: This hypothesis assumes that there is no significant difference in the risk assessment of solo female travelers across different regions and countries. In other words, it suggests that risk assessment is not influenced by the destination. If this hypothesis is supported by research, it would indicate that the same risk assessment strategies could be applied to solo female travelers regardless of the destination.

H1: This hypothesis assumes that destination-specific risks for solo female travelers vary significantly by region and country, and mitigation strategies need to be tailored to the specific destination. If this hypothesis is supported by research, it would suggest that destination is an important factor to consider when conducting risk assessments for solo female travelers.

LIMITATIONS AND RECOMMENDATIONS FOR FUTURE RESEARCH

There are several limitations to the existing research on risk assessment for solo female travelers that should be addressed in future studies.

Firstly, there is a lack of quantitative data on the prevalence of physical risk, destination-specific risk, and financial risk for solo female travelers. Most of the existing research has relied on anecdotal evidence, self-report data, or case studies. Future research should aim to collect more objective and reliable data to inform risk assessment and mitigation strategies.

Secondly, most of the existing research has focused on Western travelers visiting non-Western countries. More research is needed to understand the risks and challenges faced by solo female travelers from non-Western countries or those traveling within their own countries.

Thirdly, the existing research has largely ignored the experiences of solo female travelers with disabilities, LGBTQ+ identities, and those from marginalized communities. Future studies should aim to explore the unique risks and challenges faced by these travelers and develop strategies to address their needs.

In conclusion, future research on risk assessment for solo female travelers should aim to address these limitations and provide more comprehensive and evidence-based strategies for enhancing the safety and security of solo female travelers.

OUTCOME

Findings and analysis of the performed study have shown the responses to formulated research questions, giving insights into the concept of risk existence in travel blogs, types of risk among solo female travellers, and the risk-reduction tactics used by these travellers proposed findings of this study are current and valuable, taking into account both society and the research in the field of risk and tourism, particularly when considered in view of the study's presented background and, in particular, the arguments regarding the rise in solo female travel and reduction strategies as well as the transformation of modern society into a risk society.

IMPLICATION OF STUDY TOWARDS SOCIETY AND FUTURE RESEARCH

Research focused on solo female travelling should continue, not only because of the rising influence of solo travel on the tourism industry, but also because meaningful gender equality cannot be realised as long as women cannot travel under the same conditions as men. Gender equality in tourism industry must be empowered till women is responsible for their own security, financial independence, freedom and more safe. In addition to that, further research can be enhanced to know satisfaction level of solo women travellers and increase in the number of it.

CONCLUSION

Solo travel can be rewarding and empowering experience for women, but it is important to assess the potential risks and implement mitigation strategies to ensure safety and security. In this research paper it clearly says the physical risk, destination specific risk, financial risk and its minimizing impact to make trip more secured.More study is needed to determine whether restrictions on women persist throughout their lives and how they could effect solo or group travel and the tourist sector as a whole.

REFERENCES

Baratta, R., Vigolo, V., & Simeoni, F. (2021). Perceived risks and future travel intentions: The experience of solo female travelers. *Tourman 2021 Book of*… https://aisberg.unibg.it/retrieve/c04b9d53-fa1b-4e1c-804e-9860dfef2296/Baratta%20et%20al.%202021_EA_Tourman.pdf

Baratta, R., Vigolo, V., & Simeoni, F. (2021). Perceived risks and future travel intentions: The experience of solo female travelers. *…: Restarting tourism* … https://aisberg.unibg.it/retrieve/c04b9d53-fa1b-4e1c-804e-9860dfef2296/Baratta%20et%20al.%202021_EA_Tourman.pdf

Berdychevsky, L., & Gibson, H. J. (2015). Sex and risk in young women's tourist experiences: Context, likelihood, and consequences. *Tourism Management*. https://www.sciencedirect.com/science/article/pii/S0261517715000850

Bernard, S., Rahman, I., & McGehee, N. G. (2022). Breaking barriers for Bangladeshi female solo travelers. *Tourism Management Perspectives*. https://www.sciencedirect.com/science/article/pii/S2211973621001458

Bernard, S., Rahman, I., & McGehee, N. G. (2022). Breaking barriers for Bangladeshi female solo travelers. *Tourism Management Perspectives*. https://www.sciencedirect.com/science/article/pii/S2211973621001458

Breda, Z., Santos, A., Kliček, T., & Dinis, G. (2020). Profile, motivations, and experiences from Portuguese solo female travelers. ... *management approaches to* ... https://www.igi-global.com/chapter/profile-motivations-and-experiences-from-portuguese-solo-female-travelers/251672

Das, A. (2020). *Impact of travel journalism on female solo travelers' experience*. https://www.academia.edu/download/63370019/AnantaDas_NtccDissertation20200520-123320-1q2l0ro.pdf

Hassan, S. B., & Damir, A. K. (2022). *Sustainable Development and Gender Equality: Empowerment Through Solo Female Travel Experiences. In Planning and Managing Sustainability in Tourism*. Springer. doi:10.1007/978-3-030-92208-5_12

Heimtun, B., & Abelsen, B. (2013). Singles and solo travel: Gender and type of holiday. *Tourism Culture & Communication*. https://www.ingentaconnect.com/content/cog/tcc/2013/00000013/00000003/art00002

Jordan, F., & Gibson, H. (2005). "We're not stupid... But we'll not stay home either": Experiences of solo women travelers. *Tourism Review International*. https://www.ingentaconnect.com/content/cog/tri/2005/00000009/00000002/art00006

Lepp, A., & Gibson, H. (2003). Tourist roles, perceived risk and international tourism. *Annals of tourism research*. https://www.sciencedirect.com/science/article/pii/S0160738303000240

McNamara, K. E., & Prideaux, B. (2010). *A typology of solo independent women travellers. International Journal of Tourism*. doi:10.1002/jtr.751

Nguyen, D. T. A., & Hsu, L. (2022). Exploring visit intention to India among Southeast Asian solo female travelers. *Journal of Destination Marketing & Management*. https://www.sciencedirect.com/science/article/pii/S2212571X22000373

Nguyen, D. T. A., & Hsu, L. (2022). Exploring visit intention to India among Southeast Asian solo female travelers. *Journal of Destination Marketing & Management*. https://www.sciencedirect.com/science/article/pii/S2212571X22000373

Nguyen, H. N. (2018). *An Investigation into Female Solo Travelers: Solo Travel Exposure and Perceived Constraints, Personal Development, Authentic Personality and Self-Esteem*. https://search.proquest.com/openview/5f27aff116766f5334371c4bc4baba90/1?pq-origsite=gscholar&cbl=18750&diss=y

Nguyen, D.T.A., & Hsu, L. (n.d.). Travel experience of Southeast Asian solo female travelers. *Journal of Indonesian Tourism, Hospitality and* ... https://ejournal.upi.edu/index.php/Jithor/article/view/51021

Ngwira, C., Tse, S., & Vongvisitsin, T. (2020). Negotiation strategies and constraints for solo female travelers in Africa. *Tourism Culture &* ... https://www.ingentaconnect.com/content/cog/tcc/2020/00000020/00000001/art00004

Nikjoo, A., Markwell, K., & Nikbin, M. (2021). The flag-bearers of change in a patriarchal Muslim society: Narratives of Iranian solo female travelers on Instagram. *Tourism Management*. https://www.sciencedirect.com/science/article/pii/S2211973621000301

Nisha, F., & Cheung, C. (2022). Locating Muslimah in the travel and tourism research. *Tourism Management Perspectives*. https://www.sciencedirect.com/science/article/pii/S2211973622000058

Nunkoo, R., Thelwall, M., Ladsawut, J., & Goolaup, S. (2020). Three decades of tourism scholarship: Gender, collaboration and research methods. *Tourism management*. https://www.sciencedirect.com/science/article/pii/S0261517719302547

Osman, H., Brown, L., & Phung, T. M. T. (2020). The travel motivations and experiences of female Vietnamese solo travellers. *Tourist Studies*. doi:10.1177/1468797619878307

Otegui-Carles, A., Araújo-Vila, N., & Fraiz-Brea, J. A. (2022). Solo Travel Research and Its Gender Perspective: A Critical Bibliometric Review. *Tourism and Hospitality*. https://www.mdpi.com/2673-5768/3/3/45

Pereira, A., & Silva, C. (2018). Women solo travellers: motivations and experiences. *Millenium*. https://repositorio.ipv.pt/handle/10400.19/5257

Perkov, D., Primorac, D., & Perkov, M. (2015). Position of female managers in Croatian tourism. *Economic and Social*. https://www.esd-conference.com/upload/book_of_proceedings/Book%20of%20Proceedings_Belgrade%202015.pdf#page=107

Ragsdale, K., & Difranceisco, W. (2006). *Where the boys are: sexual expectations and behaviour among young women on holiday. In Culture, health*. Taylor &Francis. doi:10.1080/13691050600569570

Schwab, K. (2019). *Because I Can, So I'm Going To: Solo Female Travelers and the Absence of Constraints*. scholarworks.umass.edu, https://scholarworks.umass.edu/ttra/2019/qual_research_papers/6/

Schwab, K. (2019). *"Seriously, shit has to change." Spatial Constraint Negotiation among Solo Female Travelers*. scholarworks.umass.edu, https://scholarworks.umass.edu/ttra/2019/research_papers/96/

Sebova, L., Pompurova, K., & Marcekova, R. (2021). *Solo female travelers as a new trend in tourism destinations. In Culture and Tourism in a*. Springer. doi:10.1007/978-3-030-72469-6_20

Seow, D., & Brown, L. (2018). *The solo female Asian tourist. In Current Issues in Tourism*. Taylor &Francis. doi:10.1080/13683500.2017.1423283

Su, C.P., & Wu, T.C. (2022). The dark side of solo female travel: Negative encounters with male strangers. *Innovation and Impact of Sex as Leisure in* … doi:10.4324/9781003288206-8

Thomas, T. K., & Mura, P. (2019). *The 'normality of unsafety'-foreign solo female travellers in India. In Tourism Recreation Research*. Taylor &Francis. doi:10.1080/02508281.2018.1494872

Uatay, G., Reid, E. L., & Lee, H. Y. (2019). The impact of female travelers' travel constraints on attitude toward solo travel and travel intention. *Culinary Science &Hospitality Research*. https://www.dbpia.co.kr/Journal/articleDetail?nodeId=NODE09216304

Valaja, E. (2018). *Solo female travellers' risk perceptions and risk reduction strategies-as expressed in online travel blog narratives*. https://lup.lub.lu.se/student-papers/record/8947243/file/8947244.pdf

Wilson, E., & Little, D. E. (2005). A "relative escape"? The impact of constraints on women who travel solo. *Tourism Review International*. https://www.ingentaconnect.com/content/cog/tri/2005/00000009/00000002/art00004

Wilson, E., & Little, D. E. (2008). *The solo female travel experience: Exploring the 'geography of women's fear'. In Current Issues in Tourism*. Taylor &Francis., doi:10.2167/cit342.0

Yang, E. C. L., & Khoo-Lattimore, C. (2018). Constructing space and self through risk taking: A case of Asian solo female travelers. *Journal of Travel …* doi:10.1177/0047287517692447

Yang, J., Zhang, D., Liu, X., Li, Z., & Liang, Y. (2022). Reflecting the convergence or divergence of Chinese outbound solo travellers based on the stimulus-organism-response model: A gender comparison perspective. *Tourism Management Perspectives*. https://www.sciencedirect.com/science/article/pii/S2211973622000472

Yang, E.C.L. (2016). Risk perception of Asian solo female travelers: An autoethnographic approach. *Women and travel: Trends, journeys and experiences …* doi:10.1201/9781315365879-21

Chapter 18
Women's Travel Issues and Constraints:
Travel Issues and Constraints of Solo Female Travelers in Travelling to Turkey

Esra Anış
https://orcid.org/0000-0002-6970-3180
Adnan Menderes University, Turkey

ABSTRACT

In this chapter, the author presents ıssues and constraints of solo female travelers to Turkey based on their travel blogs reflecting their pre-travel and during-travel constraints and safety tips they adopted to energize their desire to travel to and within Turkey's cities. The author asks, "What are the travel issues and constraints for solo women travelers?" and "How is Turkey less safe and at the same time the most preferred destination for women travelers?" The study employs netnographic research methodology to understand complex social phenomena and assist researchers in developing themes from female travel bloggers' experiences of solo travels to Turkey. Using content analysis, the findings show that the constraints were mainly intrapersonal barriers about Turkey: like personal, religious requirements, perceived risk, and safety constraints. Other constraints such as interpersonal barriers and structural barriers have not reflecting solo travel bloggers' pre-travel and during-travel restrictions about Turkey. For all bloggers, safety is a key factor in the travel decision.

INTRODUCTION

People travel, particularly to countries or societies new or unfamiliar to them. Negotiating their movements through unfamiliar spaces and places can be a daunting task for many tourists, particularly those choosing to travel independently or alone (Wilson & Little, 2005; 2008). In particular, studies claim that women today are increasingly choosing to travel abroad alone (or for solo). Research suggests that women's motivations for independent travel relate to the desire to learn self-development, challenge

DOI: 10.4018/978-1-6684-8417-3.ch018

themselves, find a sense of identity and autonomy, meet new people and experience new life and adventure moments. These motivations influences the solo women travel experiences turning it into a memorable tourism experiences (Pereira and Silva, 2018). Despite these benefits, studies of women's leisure have consistently revealed that females tend to be more highly and intensely constrained than men in seeking access to leisure and tourism activities. Moreover, constraints appear to be heightened for women when they take part in activities out of the home or in the outdoors, due to their apparent fears regarding harassment or male violence (Wilson & Little, 2005; 2008, p. 168).

Previous research conceptualized travel constraints as a three-dimensional concept including intrapersonal constraints, interpersonal constraints, and structural constraints. Intrapersonal constraints include psychological or cognitive constraints (e.g., stress, anxiety, depression, etc.) perceived by an individual. Interpersonal constraints are essentially social constraints (such as finding a travel partner; the unavailability of friends or acquaintances to accompany the individual in travel). Lastly, structural constraints include various types of physical or operational constraints (such as limited financial resources, lack of time, season, climate, work schedule, or family commitment, and insufficient places to visit) (Crawford et al., 1991; Nyaupane et al., 2004; Chen et al., 2013; Chen and Petrick, 2016; Shin et al, 2022).

Despite the distinct challenges and travel patterns of solo women travelers in Muslim countries, particularly in Europe and Asia, travel ıssues and constraint research in countries such as Turkey has not been explored. In this context, the author focuses on the subject of solo women's travel issues and constraints (What are the travel issues and constraints for solo women travelers?), and focuses on the topic of safer destinations and safety tips for solo women travelers (How is Turkey less safe and at the same time the most preferred destination for women travelers?). It is aimed to be evaluated with the postmodern tourism paradigm and to contribute to the gap in the literature. Moreover, it gives theoretical and practical contribution for students, academy and for those who concerned.

1. CONCEPTUAL FRAMEWORK

1.1. Solo Female Travelers' Motivations, Experiences, Travel Issues, and Constraints

Tourists generally seek experiences that provide escape, freedom, and pleasure (Wilson & Little, 2005). The relationship between travel and experience is the key to why women choose to travel alone. The goal is participation and qualifications acquired, not journey. Solo female travelers seek adventure, social interaction, education, and self-understanding (Bond, 1997). The main motivations dimensions of solo female travelers and research studies are challenge and overcoming (Bond, 1997; Jordan & Gibson, 2005; Chiang & Jongaratnam, 2006; Wilson & Little, 2005, 2008; McNamara & Prideaux, 2010), self-recognition (Wilson & Harris, 2006), contact with other travelers (Jordan & Gibson, 2005; Wilson & Little, 2005; Wilson & Harris, 2006), new life prespectives (Jordan & Gibson, 2005; Chiang & Jongaratnam, 2006; Wilson & Little, 2005, 2008; McNamara & Prideaux, 2010), escape to routine, learning and new experiences (Crompton, 1979; Bond 1997) autonomy (Bond; 1997; Jordan & Gibson, 2005; Chiang e Jongaratnam, 2006; Wilson and Harris, 2006; Wilson & Little, 2005, 2008; McNamara & Prideaux, 2010), adventure and leisure (Bond, 1997; Pereira and Silva, 2018, pp. 101-102) and so on.

The tourist is a consumer who is motivated towards a tourism experience desire (Quan & Wang, 2004). According Oh et al. (2007), a tourist experience can be defined as something "pleasant, engaging, and memorable" (p.120), allowing each tourist to build his or her own travel experiences so that these satisfy a wide range of personal needs, from pleasure to a search for meaning (Pereira and Silva, 2018, p.102). Schmitt (1999) conceptualized experience as individual and shared experiences. The model comprises sensing, feeling and thinking in individual experiences and acting and relating are considered as shared experiences (Loureiro, 2014). Finally, according Pereira and Silva (2018), the women's motivations for independent travel and tourism experiences are related to the desire to learn, self-development, challenge themselves, find a sense of identity and autonomy, meet new people and experience new life and adventure moments. These motivations influences the solo women travel experiences turning it into a memorable tourism experiences (p.105).

Crawford, Jackson, and Godbey (1991) proposed a hierarchical model of family leisure constraints that suggests potential tourists should encounter intrapersonal constraints first, followed by interpersonal constraints, and finally structural constraints (Chen and Petrick, 2016). They argued these constraints occur hierarchically. This proposed model was empirically tested by Raymore, Godbey, Crawford, and von Eye (1993), and again later by Hawkins, Peng, Hsieh, and Eklund (1999). Both studies "verified that the constraint categories can be replicated and extended with subtle distinctions" (Chen et al., 2001). The most commonly used conceptual constraints are "internal" (attributes of the individual) and "external" (characteristics of the physical and social environments) (Jackson, 1988; Chen et al., 2013).

1.1.1. Intrapersonal Barriers: Personal, Religious Requirements, Perceived Risk, and Safety Constraints

In 1991 Crawford, Jackson, and Godbey introduced the hierarchical constraints model (HCM), the primary conceptual framework guiding the study of leisure constraints. They claim that leisure restrictions exist at three levels intrapersonal, interpersonal and structural. Intrapersonal leisure constraints include factors such as perceived self-skill, sense of entitlement and subjective evaluations of the appropriateness of a certain activity. For example, a woman who finds cycling inappropriate in her community, or a man who finds rock climbing with friends intimidating and unattainable, might be examples of an intrapersonal leisure constraint (Crawford and Godbey, 1987; Shahvali et al., 2016, p. 16).

The women's expressions of fear during the trip ranged from concerns about actual violence to expressions of self-doubt instilled through socially constructed messages and stories about what had happened to other female travelers. Moving between gender-based tourism environments and venues, these solo female travelers had a hard time adjusting and taking the "objective traveler" stance. Instead, many women described a sense of restraint regarding the safety and appropriateness of their movements. In turn, they had developed cognitive maps of which countries would be safe and secure, whether culturally correct or not (Caesar, 1999; Valentine, 1989). In addition, research shows that female travelers traveling alone feel the need to justify and explain their actions when they are sexually harassed. According to Koskela (1997), the effect of social fear production is so great that even when women successfully negotiate risky situations, they feel that they are doing something wrong and that they should be more afraid. Such thoughts were shown not only in their self-deprecation for being in the "wrong place" but also in their desire to avoid stories about other women's travel problems. While they hold themselves or other solo female travelers responsible for the abuse, the responsibility never belongs to the perpetrators (Valentine, 1989).

Accordingly, several studies of female tourists in western societies affirmed that they became more confident and independent after engaging in adventure travel (Wilson & Harris, 2006; Doran, 2016).

Bernard et al. (2022), claim that Asian Muslim women's travel habits were sorely under-researched. In response to various calls for research in this area, their study utilized Hofstede's five cultural dimensions to determine how Bangladeshi cultural values inhibit and/or enhance travel constraints for solo Muslim female travelers and the subsequent effects on solo travel behavior. They proposed solo travel as a strategic tourism development tool to achieve mobility rights and gender equality particularly for destinations that are highly populated with more women than men. They reveal that these 307 women-only English-speaking Facebook Bangladeshi travel group was constrained by a unique combination of intrapersonal, interpersonal, and structural factors. Power-distance, masculinity, and uncertainty-avoidance were also play key roles. Yang et al. (2017) reviewed of research on Asian travelers found only one study that discussed the constraints of Muslim female travelers from Southeast Asia (Asbollah et al., 2013). This work found that women's travel behavior was constrained by a 'culture bubble' characterized by domesticated gender roles and religious rituals that required women to comply with Islamic dress codes, choose specific destinations to access halal food, and seek prayer room spaces (Bernard et al., 2022). Furthermore, Kaba and Emekli (2018) suggest that security is the most important factor in destination choice. Gender-based risks are the main determining factor on how women travelers perceive and decide solo travelling. Women are highly satisfied with their solo travel experiences and have strong intention to continue their travels. The level of satisfaction is strongly related with the motives such as learning and experience, meeting new people and cultures, expansion of vision, self-confidence and empowerment.

In tourism, risk was defined as what is perceived and experienced by the tourists during the process of purchasing and consuming traveling services and at the destination (Tsaur et al., 1997). Risk has been identified as a major concern for international travelers (Yavas 1990). Perception of risk in tourism varies depending on tourists' characteristics. Roehl and Fesenmaier (1992) distinguished among three groups of tourists based on their perceptions of risk: risk neutral, functional risk, and place risk. The risk-neutral group of tourists does not perceive its vacations or traveling to destinations as risky. The functional-risk group considers the possibility of mechanical, equipment, and organizational risks. The place-risk group perceives tourism and traveling as risky (Reisinger & Mavondo, 2005). Maser and Weiermair (1998) examined travel risks associated with diseases, crime, natural disasters, problems with hygiene, transportation, culture/language barriers, and uncertainty related to destination-specific laws and regulations. Moreover, also terrorism risk and concern about safety became major factors when choosing a vacation destination (Reisinger & Mavondo, 2005, p. 213).

Schintler et al. (2000)'s study examined women's travel patterns, needs and psychology, as well as the impact of these patterns and behaviors on efforts to promote sustainable development. They found that women's travel patterns differed from men's in important ways. Genders differences are particularly evident in (a) distance traveled, (b) mode of travel, and (c) complexity and purpose of travelling. The specific psychology of women contributed to these patterns. Compared to men, women are prone to ambivalent emotions, but analyzing these emotions can lead to leaps in thought and creative solutions to problems. Also, women are more risk averse than men. Risk aversion can influence women's travel decisions, for example when it comes to safety. Changes in the economy also place new demands and constraints on women, their lives, and the way they travel.

To create a "favorable environment" for tourism development, it is important to understand how potential tourists experience their environment in terms of safety (Reisinger & Mavondo, 2005, p. 213). Safety-related risk perception can impede travel patterns as well as the intention to travel independently. Constrained travel behavior can be expressed in a variety of ways; for example, women may avoid certain trips and travel modes and might change their destinations and routes. Those behavioral adaptations can be dependent on the time of day; more specifically, they often are based on lighting conditions, not only daylight but also artificial light in buildings. To cope with fear and to increase perceived safety, women sometimes employ a strategy of "arming measures," which includes carrying items for self-defense, such as pepper spray, pocket alarms, or knives; holding or using a mobile phone while walking alone; or attending a self-defense course (Stark & Meschık, 2018).

Previous research showing that women view security and fear as important determinants in their travel activities (Jordan & Gibson, 2005; Westwood et al., 2000; Wilson & Little, 2005; 2008). Prior to travel, women showed a fear of anticipation stemming from their own inner perception of the safety of going alone and concerns about others' solo travel plans. Friends, family members, and acquaintances expressed concern and bewilderment at women's willingness to travel alone, reflecting and imposing the idea that such activities are unsafe for women (Valentine, 1989; Wilson & Little, 2005; 2008). Such social stigma underlines the perceived connection between women in isolated areas, the potential for incidents of sexual violence (Hanmer & Saunders, 1984; Sheffield, 1995; Valentine, 1989), and socio-cultural messages that fear should be a feminine quality. (Greer, 1999; Koskela, 1997; Wilson & Little, 2005; 2008). Wilson & Little (2008) claim that for the solo women travelers, the geography of fear had manifested itself in a number of overt and covert ways through the women's personal fear, others' fear, and the sexually harassing actions of others. By extension to Valentine's thesis, then, they argued that there was 'geography of women's travel fear'. This geography of women's travel fear worked to perpetuate the notion that solo travel was somehow unsafe and, at certain times and in certain places, inappropriate. Sexualized, the objects of male gazes and different from the norm, the women had experienced fears that were based predominantly on an anticipation of male violence and harassment. As a result, these women perceived, experienced and used their tourist landscapes and spatial environments differently from how they had thought male travelers would. Through producing and enforcing the geography of fear, a patriarchal system of control dictated women's appropriate use of tourist spaces and places. Women solo travelers accommodated by doing what was safe, right and sensible and by keeping themselves out of harm's way.

1.1.2. Interpersonal Barriers: Roles, Responsibilities, and Family Constraints

Crawford et al. (1991) claim that the second level, interpersonal constraints, are experienced when an individual is unable to find a suitable partner with whom to engage in a particular activity; for example, not having someone with whom to play tennis.

According to a common understanding, Gender refers to the roles and responsibilities of men and women created in our families, societies and cultures. It includes expectations about the characteristics, abilities and possible behavior of both women and men, where conforming to one's gender role is understood as 'gendering', which, ironically, leads to differences between men and women. For example, personality traits such as independence, assertiveness, logic, rationality, competitiveness, and focus on individual goals are considered traits associated with masculinity. On the other hand, being understanding, caring, nurturing, responsible, thoughtful, sensitive, intuitive, passionate, and having the ability to focus on social goals are the characteristics associated with femininity (Khan, 2011, pp. 106-107).

According to Khan's (2011, p. 113) research results, women are more restricted than men due to family responsibilities. For the majority of women, family responsibilities and obligations are perceived as a constraint. This is so because the family is often the axis of women's lives. Women give primary importance to family responsibilities, which put their individual independence in the background, in a way. While men can travel whenever they want and do not have to compromise on the wishes of other family members, women cannot do this as the daily functioning of the house is disrupted in their absence. Thus, it appears that family commitments continue to dominate women's lives and is an important factor constraining their leisure travel decisions. On the other hand, the changing nature of gender roles in contemporary times, where the trend in families today is more towards an equal sharing of responsibilities. Men, too, have come to take on many caring responsibilities in the home and in modern families, and both sexes participate in family duties jointly, giving each other equal freedom.

1.1.3. Structural Barriers: Time, Money, and Lack of Information Constraints

Crawford et al. (1991) claim that the last level, structural constraints are represented by factors such as lack of money, time or transportation. Jackson and Henderson (1995) described various difficulties perceived by women in leisure participation, such as finding others to join, lack of time due to too much dedication to family, lack of information, lack of transportation, feeling uncomfortable in social situations, and being physically unable to participate.

In the past, women's social role and gendered 'female' position had tremendously and intensely constrained their access to leisure participation (Jackson & Henderson, 1995). For example, social expectations for women were strongly linked to stereotypical images of marriage and children, emphasizing women's core responsibilities as family keepers and taking care of their husband and children (Cambronero-Saiz, 2013; Wilson & Little, 2005). Because of the concept of care ethic, women often prioritize meeting the needs of others over their own leisure pursuits (Henderson & Allen, 1991). In fact, many women even believed they had no right to leisure, thus placing a severe limitation on their construction of time for leisure and their attitudes toward it (Henderson & Dialeschki, 1991).

Restrictions on travel decisions are largely related to gender, life cycle and cultural characteristics (Jackson, 1991; Shaw, 1994). It is stated that women experience more limitations than men. Factors such as work, family, domestic obligations determine the amount and quality of time devoted to leisure and tourism, and in some cases act as constraints for certain individuals. One's gender is combined with personal and socio-economic factors to exacerbate restrictions. Although limiting factors are effective on both genders, the degree to which they affect both genders is different. However, it appears that the restrictions affect more women than men. For example, when relating economic constraints to gender, women are often described as being more restricted in pursuing leisure travel due to their low economic status (Khan, 2011, p. 109).

Men hold on strongly to their dominant and superior status in society. It is usually men who have more power in the family. Some theorists suggest that the power difference between men and women in the family can be attributed entirely to the higher income of men. In terms of power and income relationship, the power of a man or woman in the household also determines the individual economic situation. Higher individual earnings tend to have a more equal say in family decisions, while having a lower economic status acts as a constraint for decisions about spending discretionary income (supports trust and increases dependency). (Khan, 2011, p. 114).

Proponents of resource theory argue that when women make a significant economic contribution to the household, power and control in decision-making shifts in their favor (Blood and Wolfe, 1960; Green and Cunningham, 1976; Blumstein and Schwartz, 1983). Nowadays, women who earn more in the family can have more power. Independent working women have more authority and freedom to make leisure travel decisions and can even influence their spouse or family to their advantage (Khan, 2011, p. 115).

1.2. Safety Tips for Solo Female Travelers

"As a woman, being able to move about freely all over the world without having to worry about personal safety, is my definition of safety in the context of future travel."

Unfortunately, the pandemic has negatively affected women's safety, going far beyond secure accommodations and personal safety to include medical facilities, transportation and health and hygiene. It has also made us realize that safety isn't just a travel issue; it affects women at home too. Therefore the pandemic made women's safety even more important. According to JourneyWomen (2022) recommends twenty safety tips for solo women travelers as follows;

1. Maintain awareness
2. Avoid certain areas
3. Trust tour intuition
4. Walk with confidence
5. Dress appropriately fort the culture
6. Pack lightly
7. Learn self-defense skills
8. Manage alcohol responsibly (or avoid it altogether!)
9. Make friends with locals
10. Learn a few words of the local language
11. Have a local contact before you travel
12. Spend longer in a place
13. Have your phone ready
14. Travel with a friend
15. Ask for two room keys
16. Leave jewelry at home
17. Beware of inappropriate flattery
18. Never leave a drink unattended
19. Stay in at night
20. Never say you are travelling alone

On April 2021 Women's' Safety survey showed that 97% of women do not feel they have the physical skills to defend themselves. Moreover, women feel most at risk on a street at night (%52), on a street in the daytime (%30), while in transit (i.e. bus, plane, taxi, train; %28), at a hotel (%10) and in a restaurant or a bar (%3) (Ray, 2021). Self Defense and Personal Safety Training Companies, Safe4 Life CEO recommends four keys to personal safety to keep women protected and safe, regardless of where they are in the world, of how they are showing up and what they environment looks like (JourneyWomen, 2022):

1- Be aware: Put the phone away, make eye contact, briefly, trust your instincts, walk confidently.
2- Establish and protect your boundaries: Step back. Walk around and go to the back of the line-up.
3- Confidence and Skills: Learn self-defense skills and practice them.
4- Have a Safety plan: It should include fighting, shouting, running to safety, and breathing.

Table 1. Women's safe destination experience

"Safer" Destinations	"Less Safe" Destinations
1- United Kingdom + Ireland	1- India
2- Canada	2- Mexico
3- New Zealand	3- Morocco
4- Iceland	4- Egypt
5- The Netherlands	5- Turkey

Source: JorneyWomen (2022)

1.3. Why Study Solo Female Travelers to Turkey?

The Author study solo female travelers to turkey, due to the statistics which made by JorneyWomen and Forbes Advisor. She wants to found out the ıssues and constraints of solo female travelers by choosing Turkey, as a less safe and one of the world's best destinations for solo travelers. According to JorneyWomen (2022) safety is not specific to a destination, but certain places require extra planning and preparation for solo women. In addition, safety is a subjective for people who they found more or less safe, based on their own personal experiences. According to women's safe destination survey, the following countries are listed more or less 'safe' based on women's personal experience.

Table 1 shows women's safe destination though their experiences. The safer destinations are UK and Ireland and the less safe destination found India. As we saw Turkey is the less safe in this list with ranking fifth. However, Forbes Advisor looked at the most popular destinations for solo travelers to create a list of the best cities around the world, ranked for people looking to go it alone in 2022. They employed the help of three independent solo travel experts (Adventurous Kate, Vicky Flip Flop Travels and Global Grasshopper) to find the essential aspects of a successful solo trip. Then they used this insight to decide on 12 ranking metrics, including the number of hostels, local crime rate and average cost of a meal. The analysis ranked 40 cities around the globe. Each city was awarded from 1 to 40 points for each of the 12 metrics, depending on its position in each ranking, with a maximum possible score of 480. Each city was then judged on these factors, findings show Ho Chi Minh City in Vietnam emerging as the world's best destination for solo travelers. Its follows then Marrakech second and Istanbul third respectively (Howard, 2022). Finally, with these findings, it is desired to clarify how Turkey is a less safe and at the same time a most preferred destination.

Why travel solo to Turkey? A luxury solo group tour in Turkey unites classical architecture and an azure coastline with the country's famously warm hospitality. From sailing along the Mediterranean coast and eating fresh fish, to visiting the UNESCO World Heritage Site of Ephesus and soaking up the dark hues of Cappadocia's fairy chimneys, Turkey is a country that packs a punch. Its cities are also its own - a stroll through the atmospheric streets of Istanbul will present you with minarets, mosaics and mosques, not to mention the dizzying smells of the spice bazaar. And of course, no trip to Turkey would be complete without soaking up the country's rich food heritage, from gözleme (flatbread) pit stops to kebab feasts meze washed down with a bottle of delicious fruit wine. Top places to visit in Turkey (Flashpack, 2023); Fethiye, Selcuk, Cappadocia and Istanbul.

2. METHOD

The aim of the research is to present the issues and constraints of solo female travelers to Turkey, based on travel blogs reflecting their pre-travel and during-travel restrictions and the safety tips they have adopted to stimulate their desire to travel to and within the cities of Turkey. The author asks, "What are the travel issues and constraints for solo women travelers?" and "How is Turkey less safe and at the same time the most preferred destination for women travelers?" is asking questions. Research is limited by time and financial means and to five solo female bloggers. Blogs provide relevant, honest, truthful, and more self-representing data at a lower cost than traditional diary writing or other qualitative data collection methods (Hookway, 2008; Schmallegger & Carson, 2008). The study employs netnographic research methodology to understand complex social phenomena and assist researchers in developing themes from female travel bloggers' experiences of solo travels to Turkey by using content analysis.

Ethnographic research, one of the qualitative research methods, was used in this study. Ethnography is included in qualitative research studies. The researcher can use different methods such as observation, survey, interview by entering and not disturbing the culture and cultural behaviors (ceremony, tale, ritual, epic, work, anecdote, religion, language, lifestyle, etc.) of a certain group of people, an institution, a region, a hospital worker, a fan group, a social media user or a blue-collar worker group through their eyes, by entering their natural environment and not disturbing them. It is to describe, analyze and interpret culture in depth by taking advantage of it, looking holistically and objectively, understanding all aspects of culture. On the other hand, this study uses netnographic research methodology, as it is thought to be effective in understanding complex social phenomena and assists researchers in developing themes from the respondent's perspective (Kozinets, 2002). It is a type of ethnographic research, and it is the analysis and interpretation of the culture experienced by certain groups or individuals in the virtual environment by describing them in depth, just like in ethnographic research.

The study group of the research consists of solo female travelers traveling to Turkey. Its sample is the experiences of five independent female solo travel experts traveling to Turkey. Data were collected from five female travel experts who traveled alone to Turkey from ForbesAdvisor, JorneyWomen and CreditDonkey. Therefore, since bloggers are from all over the world, this study chose blogs as data sources for researchers to collect data on the topic (Enoch & Grossman, 2010). The data focus is on bloggers thoughts about Turkey and its destinations, thoughts about ıssues and constraints of solo female travelers to Turkey, thoughts about safe and preferred destinations of Turkey and safety tips for female travelers on solo travel experience.

3. FINDINGS

In this section, the travel experiences of five travel experts who traveled alone to Turkey are presented. It also seeks answers to research questions "What are the travel issues and constraints for solo women travelers?" and "How is Turkey less safe and at the same time the most preferred destination for women travelers?" The five bloggers demographic information are presented in Table 2.

3.1. Bloggers Thoughts about Turkey and Its Destinations

What am I doing here? Why Turkey or its other destinations?

Table 2. Demographics of 5 solo woman traveler bloggers

Blogger	Place of Origin	Gender	Past Occupation	Current Occupation	First Trip to Turkey	Year Blog Created	Turkey Visited Destination
1.	American, USA	Female	Online marketing	Full-time travel blogger	In 2008	2010	Istanbul, Cappadocia, Pamukkale, Kaş, Antakya, Lycian Way, Ephesus, The Kaçkars, Ani, Diyarbakir or Şanlıurfa or Gaziantep.
2.	British, United Kingdom	Female	Writer and photographer	Full-time travel blogger, travel writer, adventurer and photographer	In 2020	-	Bodrum, Istanbul, Cappadocia, Fethiye, Oludeniz, Pamukkale, Butterfly Valley, Kas, Patara, Dilek Penninsular-Buyuk Menderes Delta National Park, Iztuzu Beach, Dalyan, Ayvalik, Side, Icmeler, Koprulu Canyon, Antalya, Ankara, Alanya, Kure Mountains National Park, Temple of Hadrian and Ephesus.
3.	Afro-Arab	Female	Travel blogger	Full-time foodie travel blogger, travel writer, adventurer Solo digital nomad	In 2015	2016	Istanbul, Cappadocia, Konya, Antalya, Kas, Fethiye, Pamukkale, Selcuk, Canukkale,
4.	Indian, India	Female	Market research consultant	Freelance writer, travel photographer, compulsive blogger	In 2008	2003	Istanbul, Ephesus, Sirince, Izmir, Kusadasi
5.	American, USA	Female	Solo digital nomad	Full-time travel blogger, travel writer, adventurer Solo digital nomad	In 2020	14 + years	Istanbul, Antalya, Cappadocia, Kas, Patara Beach, Fetiye, Kabak Bay, Marmaris, Bodrum. Mugla, Aydin, Selcuk, Ephesus, Izmir and Ankara

Blogger 1

I love Turkey. The jagged landscapes, the steam-filled hamams, the warm hospitality, the blue Mediterranean, the endless tulip-shaped cups of tea. I wish more people realized this! Turkey is a destination that gets a lot of bad publicity, particularly for women on their own — but if you have at least a little bit of travel experience, Turkey is a great destination for solo female travelers.

Blogger 2

There is far more to Turkey than the packed holiday resorts. Explore the sites away from the crowds to find ancient architecture and some extraordinary landscapes. From snow-like natural wonders to hippy beach retreats, here is my choice for 20 of the best and most beautiful places to visit in Turkey…

Blogger 3

Hot air balloons rising with the sun, the call to prayer echoing across the sky and the glimmering blue seas: welcome to Turkey, where enchantment and beauty are a part of everyday life. Discover the kaleidoscope of colors that are Cappadocia's hot air balloons, explore the natural pools of Pamukkale and soak up the unique East-meets-West culture in Istanbul – the only transcontinental capital in the world. Epic landscapes, mouthwatering cuisine, and delights at every turn: Turkey is guaranteed to captivate you.

Intrepid's Turkey Encompassed tour was my third trip with Intrepid and it surpassed all expectations. It's no secret that I love Intrepid but I also love Turkey – it's a unique and fascinating country with friendly people, spectacular landscapes and delicious cuisine (if you're on the fence, this is me telling you to go ASAP).

Blogger 4

Turkey in November. A few days in Istanbul, with perfect weather and then a couple of days among the ruins of Ephesus, based in the pretty hill village of Sirince, tasting the local fruit wine and watching the sun set in the valley deep below.

Blogger 5

Turkey's natural places are absolutely stunning. And many of them are unique to Turkey. For example, in central Turkey, Cappadocia is filled with spectacular canyons, each with a different color of stone cliff walls. There are tall, bizarre rock pinnacles, cave homes carved into cliffs, carved nooks for pigeons and ancient underground cities.

Turkey's Mediterranean coast has vibrant turquoise-colored water, soaring cliffs, rock pinnacles, long coastlines backed by rough jagged mountains, dozens of beautiful beaches, narrow fjord-like bays and other dramatic geological land features. And these are just two examples of Turkey's amazing geography. There are heaps more around the vast nation.

3.2. Blogger Thoughts about Issues and Constraints of Solo Female Travelers to Turkey

What are the travel issues and constraints for solo women travelers? The five bloggers intrapersonal issues and constraints are presented in Table 3.

What are the travel issues and constraints for solo women travelers? The five bloggers interpersonal issues and constraints are presented in Table 4.

What are the travel issues and constraints for solo women travelers? The five bloggers Structural issues and constraints are presented in Table 5.

Table 3. Intrapersonal issues and constraints of solo female travelers to Turkey

INTRAPERSONAL BARRIERS				
Bloggers	**Personal Constraints (Fear, Stress, Anxiety Depression)**	**Religious Requirements (Gender Roles and Religious Rituals-Islamic Dress Codes)**	**Perceived Risk (Diseases, Crime, Natural Disasters, Problems with Hygiene, Transportation, Accommodation, Culture/Language Barriers, Uncertainty Related To Destination-Specific Laws and Regulations, Terrorism Risk, Incident Risk, Financial Risk,)**	**Safety Constraints (The Time of Day, Lighting Conditions (Daylight and Artificial Light In Buildings), Arming Measures (Pepper Spray, Pocket Alarms, or Knives; Holding or Using a Mobile Phone) Attending a Self-Defense Course, Isolated Areas, Safe and Secure Destinations, Male Violence and Harassment)**
1	Fear, Stress, Anxiety	Gender Roles and Religious Rituals-Islamic Dress Code, I have a light cotton scarf that fits in my bag, and I carry that if I'm going to visit a mosque.	Culture/Language Barriers, The main complication in Turkey is that, once you leave the heavily touristic parts of Istanbul (like Sultanahmet), fewer people speak English. Terrorism Risk, Geographically, Turkey is sort of in the middle of everything, with Russian submarines slipping through the Bosphorus Strait, war continuing to roil the bordering Syria to the southeast, and refugees transiting on their way to Europe. Problems with Hygiene, The water isn't safe to drink in most parts of Turkey Diseases, Crime, Incident Risk, Financial Risk, It's absolutely vital to have travel insurance before traveling to Turkey.	Male Violence and Harassment, The street harassment in Turkey varies wildly, and is mostly concentrated in the most touristic areas, like Sultanahmet or the Grand Bazaar. Safe and Secure Destinations, People won't touch you in Turkey, generally. All the harassment that takes place is usually verbal. And it is usually from merchants trying to sell you things. Get out of the touristy parts of destinations and it happens much less frequently.
2	Fear, Stress, Anxiety	-	Incident Risk, Check The Crime Rates or Reports	Reliable Wi-Fi, Safe and Secure Destinations
3	Fear, Stress, Anxiety	Gender Roles and Religious Rituals-Islamic Dress Code, Although Turkey is a Muslim-majority country, you don't need to dress particularly conservatively there,...Ladies, I recommend bringing a scarf and cardigan for any potential visits to mosques.	Culture/Language Barriers, It can be a tad difficult to navigate on your own, especially if you don't speak the language. Terrorism Risk, You must avoid cities near the Syrian border. Incident Risk, Istanbul is a big, cosmopolitan city. I wouldn't even consider going to Turkey without travel insurance anyway. You honestly never know what could happen. Technologic Risk Unprotected Wi-Fi network	Safe and Secure Destinations, Travel Insurance, Purchased a Local SIM, Reliable Wi-Fi,
4	Fear, Stress, Anxiety	Gender Roles and Religious Rituals-Islamic Dress Codes, I remove my shoes near the entrance, though I am allowed to carry them inside the mosque in plastic covers. I am a little intimidated, both by its size and grandeur and by the fact that this is the first time I am stepping into a mosque and I do not know what to expect.	Crime, Istanbul has some of the most happening night clubs in Europe as well as a high rate of honour killings. Hang on to your patience, sense of humour and your wallet (watch out for pickpockets) and you will have a great experience.	Safe and Secure Destinations, Male Violence and Harassment, Turkey is a patriarchal society and according to some reports, the country ranks second in Europe, just after Russia, in abuse of women.
5	Fear, Stress, Anxiety, I've also been preparing myself mentally and physically to travel in this new Corona Virus world.		Diseases, Corona Virus Risk, Give me a fairly safe (corona virus wise) place to stay during the summer travel season in Turkey and Europe. Problems with Hygiene, Transportation, Accommodation, Every time I take a bus or stay in a new hotel / private room, it's a risk. I'll make sure I stay a safe distance from people, wear a mask, sanitize bus seats & handles, sanitize my rooms upon entry, carry disinfectant and hand wipes with me, among other measures. Uncertainty Related To Destination-Specific Laws and Regulations, When Corona Virus cases started escalating in Turkey, the president began several safety/ preventative/lock-down measures around the country. Many things going on here are very unclear to me, even inexplicable. So to a large extent, I simply don't know what's going on. Culture/Language Barriers	Isolated Areas, Safe and Secure Destinations, I do believe this coastal region of Turkey through to Bodrum is relatively safe, especially compared to Istanbul and even other European countries. I think that is just too risky in a Corona Virus world. So I have to find cheap local hotels, private Airbnb rooms or camping places. If the place is open-air and is empty or near-empty, and I can sit in a space well away from staff moving around, then I might choose to eat there. Keeping 1.5 M away from other passengers and people I carry a mask with me all the time. Carrying wet-one wipes with disinfectant to use in stores, etc. and Not touching any door handles, grocery carts, etc. with my bare hands

Table 4. Interpersonal issues and constraints of solo female travelers to Turkey

INTERPERSONAL BARRIERS			
Bloggers	**Roles ('female' position being sister, being a wife for marriage and being a mother for children)**	**Responsibilities (family keepers and taking care of their husband and children, work)**	**Family/Friends Constraints (Finding a travel partner; the unavailability of friends or acquaintances to accompany the individual in travel, Thoughts about traveling solo)**
1	-	-	If you're not quite sure if you're ready to travel completely solo, another option is joining a group tour.
2	-	Check Internet Speed Before Arriving, If my Wi-Fi keeps dropping out, is extra slow or I can't upload images it hugely impacts my business.	-
3	-	-	If it's your first time visiting to a destination, travel with a friend or booking a tour.
4	-	-	-
5	-	A coronavirus personal safety plan for travels.	-

3.3. Bloggers Thoughts about Safe and Preferred Destinations of Turkey

How is Turkey less safe and at the same time the most preferred destination for women travelers?

Blogger 1 says:

The hangover from the Very Bad Year of 2016 lingers when people think of traveling to Turkey. That was the year we had terror attacks... Since 2016, the country has been quiet and safe, with no major terror attacks and no coups. Even in the thick of the Very Bad Days, life in Istanbul seemed to hum along... We live in an age of terror, unfortunately, and I don't believe that Turkey is any less safe than anywhere else in the world...Turkey touches the Balkans, the Caucasus, the Mediterranean, and the Middle East; it's one of the reasons there is such a rich cultural and culinary history here, and why Turkey always seems to come up in the news...

Turkey is quite safe and I find that pickpocketing and stealing are much less common here than you would expect, both in Istanbul for a city of its size and in throughout the country... It's easy to get around Turkey as a solo female traveler. There are flights, good roads, some trains, and a comprehensive bus system that covers the whole country in depth....

If you're looking to meet people in Turkey while traveling solo, you're in luck. It's easy to connect with travelers while doing organized activities, and locals are friendly, inquisitive, and love a good time... So come to Turkey, see what it's really like to travel as a solo woman here. Swim in the Mediterranean or stroll the backstreets of ancient Byzantium or trek through Cappadocia's fairy chimneys. Turkey has a way of wiggling into your heart. There's nowhere quite like it.

Table 5. Structural issues and constraints of solo female travelers to Turkey

STRUCTURAL BARRIERS				
Bloggers	**Lack of Time**	**Lack of Money**	**Lack of Information**	**Other Constraints (Season, climate, work schedule or tour program, family commitment, insufficient places to visit)**
1	-	-	-	Season, climate, Turkey is a big country and the weather can vary wildly between regions. Istanbul is generally best around April-May and September-October, when the temperature is perfect.
2	-	-	-	-
3	It's doable, but you definitely can't cover all these places at the pace of a tour if you're on your own.	-	If it's your first time visiting and you'd prefer not to worry about all these tips and the hassle of transporting yourself from place to place, I highly recommend booking a tour. You should have a few friends living in that destination.	Season, climate, Turkey is a year-round destination, busy from March – October and peaking in the summer. You can visit Turkey in the winter for fewer tourists and low prices, as long as you don't mind the cold. The flow of the tour program
4	-	-	-	-
5	I am currently over-staying my 30-day visa due to the corona virus lock-downs. However, once borders open, we are supposed to depart Turkey in a fairly quick manner. I might just have to skedaddle in early July. So I'd better have my coastal trip completed by then.	As usual, I'm on a fairly tight budget. That's how I live and travel normally. At this point, my travel work will run out fairly soon, so I must be extra careful about my expenses. Unfortunately, once I start traveling along the coast, my room expenses will go up a lot. And this will be a major factor in which towns and cities I can actually stay in!		Season, climate, If I had traveled along the coast in April and May, as originally planned, I would have had many accommodation options within my budget. The temperatures in July and August soar up to 40-45C / 105-115F, and sometimes even higher! July and August are the peak tourist season here. Both Turkish and international travelers flock to all these coastal cities, beaches and resorts for their annual vacations (and in the case of Turks, also for weekend getaways).

Blogger 2 says:

Turkey offers much more than just cheap package holidays. It's a treasure of a country overflowing with history, great food, and rich culture, all of these being displayed over amazing backgrounds. Spread above two continents, Turkey is a colorful mix of philosophies and traditions. With 5000 miles of coastline laid along three different seas, the country connects Europe with Asia, borrowing their traditions and mixing them together.

One of the most beautiful places in Turkey is the Cappadocia region, where the moonscape hills seem to be unreal like a wonder of nature. Istanbul is a memorable city where the East meets the West. It's famous for its amazing mosques, the Grand Bazaar, rich culture, and stunning architecture. In such a diverse country there is something for everyone, from beautiful beaches to the archeological sites of Anatolia, from the lively holiday favorite Bodrum to Ephesus which was once the trade center of the ancient world. When at Ephesus make sure you see the remains of The Temple of Artemis which was one of the Seven Wonders of the Ancient World.

Blogger 3 says:

First of all, let me just say that yes, Turkey is safe and can be traveled to independently. I visited once before with a friend and we traveled to both Istanbul and Cappadocia. However, Turkey can be a tad difficult to navigate on your own, especially if you don't speak the language. It's doable, but you definitely can't cover all these places at the pace of a tour if you're on your own. If it's your first time visiting and you'd prefer not to worry about all these tips and the hassle of transporting yourself from place to place, I highly recommend booking a tour…

I've visited Turkey twice: once in the winter of 2015 and again on this trip in the summer of 2018, shortly after elections. I also have several friends who have lived in Turkey over the past few years and I can assure you that Istanbul is safe. So are other parts of Turkey – I say this with confidence since I've now visited 11 cities across the country. Of course, use common sense in Istanbul – it is a big, cosmopolitan city after all – and avoid cities near the Syrian border (duh). It's safe for solo female travelers as well, in case you're reading this and wondering.

Although I felt safe the entire time I was there (both with and without the group), there's something to be said about having all the details of your trip already planned and having an awesome guide that is knowledgeable both about the history and the customs. English is not widely spoken (especially outside of Istanbul and Cappadocia), making it difficult to get around for non-Turkish speakers. After visiting Turkey twice and spending a total of three weeks there, my resounding conclusion is this: it's time for you to get to Turkey. With a welcoming and fascinating culture and plenty of beautiful places to see, Turkey will surely shatter any misconceptions you have about visiting.

Blogger 4 says:

Turkey is a patriarchal society and according to some reports, the country ranks second in Europe, just after Russia, in abuse of women… Like many other things in Turkey, the contradictions here too are stark and startling, especially in the urban areas. You see women in veils and women in short skirts. Istanbul has some of the most happening night clubs in Europe as well as a high rate of honour killings.

I am not a terrorist or a thief; I am only trying to help you". This is a new one. Walking through the Sultanahment area on an autumn morning, I am stopped (is the word accosted?) by many men offering their ahem, services – from proposing to act as unauthorized guide inside the Topkapi Palace to taking me to the cheapest and best place for carpets. Most of them go away when ignored; I don't think they really expect to be taken seriously. It is just one of the things Turkish men seem to do to pass time during the long off-season days.

Blogger 5 says:

After one month of exploring fascinating Istanbul, I set out to travel around the country. It wasn't long before Turkey became one of my favorite countries. A couple months later I got 'stuck' at Turkey's gorgeous Mediterranean coast in Antalya city, due to Covid-19 lock-downs. I ended up 'stuck' there for several months. That was perfectly fine with me. In fact, I'll wager I enjoyed one of the most wonderful corona-pandemic-world lives possible.

In total, I ended up spending more than seven months in Turkey (waaayyy beyond my allowed 90-day visa) because of the pandemic. I eventually set out in mid-September, 2020 to nearby Croatia and Bosnia. This year, when it turned out that I basically needed to wait out the pandemic yet again, I quickly opted to return to Antalya. And here I am! All told, there must be dozens and dozens of reasons I love Turkey so much. But following are the primary reasons I think Turkey is just divine…
1. Stunning & Diverse Nature
2. Intriguing & Diverse Historic Architecture
3. Inexpensive Living
4. Street Cats
5. Warm, friendly, hospitable polite people
6. Safe and theft-free
7. Lokantas
8. Turkish Delights
9. Nut Shops
10. Fresh Produce
11. Delicious & inexpensive bakeries
12. No Loud Music

The first week of June things began re-opening in Turkey. On June 1st the restaurants, cafes, parks, beaches, museums, national parks & cultural sites were given an official government go-ahead to re-open while adhering to Covid-safe practices like mask-wearing indoors and social distancing.

By traveling in regions of Turkey with relatively few virus cases to date, by adhering to social distancing, avoiding crowded places, wearing a mask, disinfecting new areas, not touching any public surfaces Math my bare hands and other measures, I hope I am minimizing the risks and can prevent myself from contracting the virus.

I feel fairly safe traveling along the coast for the time being. At least during June, when the Turkish people are just tentatively starting to begin traveling and international visitors are still banned from entering the country…

3.4. Bloggers Safety Tips for Female Travelers on Solo Travel Experience

What are my top tips for a safe and enjoyable solo experience?

Blogger 1 says:

I always recommend that solo travelers budget extra money for safety issues. That could be taking a cab home at night instead of public transportation, or rebooking accommodation in a better location or neighborhoods, or taking a scuba course with the company with the better reputation rather than the cheapest one.

I believe that nearly every destination on the planet can be travelled safely by a woman on her own — as long as she does the proper research and preparation. Research is so important, and every country and city is different. Learn about the best areas, cultural basics, neighborhoods, and the activities you want to do beforehand.

I also recommend keeping your valuables on you in transit — that means in a bag that stays with you, not locked away in a luggage hold somewhere.

My biggest tip is to consider the source when people give you travel advice. A well-meaning relative might tell you 'Don't go to Mexico City! It's dangerous!' Well, the questions you should ask yourself are: does this person travel? Does this person travel in my style of travelling? Has this person been to this destination? Has this person been to this destination in the last 5-10 years?

If the answer to all four questions is yes, you've got an accurate source and you should listen to them; otherwise, seek out information from people who actually know what they're talking about.

Blogger 2 says:

As a digital nomad I also always check internet speed before I head out to a destination. If my Wi-Fi keeps dropping out, is extra slow or I can't upload images it hugely impacts my business.

A reliable Wi-Fi is so important for me that I even go as far as conducting a speed test before arriving. I do this by contacting my chosen accommodation beforehand, sending them a speed test link, asking them to run the speed test and then share the results with me.

As a female solo traveler I always check the crime rate of the destination before I leave for that destination. Even though there's still every chance you could travel around the world and arrive back without any incidents, sadly it's still safer to travel solo if you're a man.

There have also been many recent high profile events in the press and I myself have had a few close scrapes in the past. Now I'm extra careful and always check crime reports and if there are any areas in a country I would be wise to avoid.

Blogger 3 says:

I'll be the first to say that overall, the world is a pretty safe place to be. I'm a firm believer that the media shouldn't scare you from traveling because most of the time, places aren't even half as scary as the news makes them out to be. That being said, as a solo female traveler, there are certain countries that I've visited that I know would be more enjoyable as part of the group. Whether it's a cultural tendency toward sexual harassment or a pickpocketing problem I'm worried about, traveling in a group always makes me feel a little more relaxed about my surroundings. And there's no doubt that feeling safe makes a trip better...here are some tips to help you not only survive your first solo trip, but to have an awesome time while you're at it!"

Research a lot. Do yourself a favor, and do some research about where you're going: tourist tricks to be wary of, unsafe areas, local customs, where to eat, and the works. If you're going to several cities, make sure that there is actual transportation between each place. In a foreign country, learn how to say a few key phrases: "hello", "goodbye", "thank you", and "do you speak English?" are key...

Book accommodation. You know what always gives me peace of mind when I'm alone? Knowing that I have a place to stay for the night. When you're armed with the knowledge that you won't be lugging your backpack around at midnight looking for a bed, anything seems possible. I like to use Booking or Hostelworld...Don't forget Airbnb either!

Make some friends. Sometimes, when you're traveling alone, you actually want to BE alone. As in, you want to be by yourself with your thoughts and reflections. Other times, you're more than ready to get out of the weird wonderland that is your thoughts, and socialize with some actual people.

Share Your Itinerary. Tell your family/loved ones where you are so you don't have to cut off your arm.

Make good choices. Please, please don't wander through an unknown city late at night alone. It's almost always a bad idea. If you're dying to go out, find some friends (see point #2), but stay alert and trust your gut! And don't be flashy with your stuff, especially in a developing country. Know when it's time to whip out your shiny iPhone for some photos, and when to tap into your photographic memory instead. Check out additional tips on how to stay safe while traveling solo.

Have a plan B. Keep some emergency money stashed outside of your wallet. Keep copies of your passport handy. Buy some Skype credit before you leave. Have easy access to emergency numbers. I know it's no fun to think of the bad things that could happen, but it never hurts to be prepared.

Embrace the unknown. You get one life, and you should live it adventurously, even if you're scared. When you travel alone, you have the opportunity do whatever you want, according to your own agenda. There's something to be said for being in a place where nobody knows who you are. So get out of your comfort zone and do it! If you hate it, then at least you gave it a shot, right?

Blogger 4 says:

Finally, the question of safety. Some European countries are notorious for pickpockets and petty thieves and everyone we know has a story of how someone they know lost their bag containing their passport and all their money while in Spain or Italy. In most places, it is not necessary for you to carry your passport around, as long as you have a copy of it with you and some valid form of identification.

Keep your money, credit cards and important documents (if you happen to be carrying them around) close to your person and don't let that bag out of sight even if you are seated at a cafe or a seemingly safe place...Stay close to the main areas after dark and always keep someone informed about your plans.

In Turkey...Hang on to your patience, sense of humour and your wallet (watch out for pickpockets) and you will have a great experience.

Blogger 5 says:

I believe your safety mainly boils down to yourself: attitude, education, preparation, and wits (awareness).

Many philosophies believe that 'You get what you expect.' That goes for your health, fitness, finances, successes, appearance, relationships, safety, experiences of safety, injuries, accidents, tragedies.

What do you expect to happen to you? What do you expect to achieve, to find, to experience in your life? Your attitudes will directly affect your actions, your decisions, your demeanor, and your 'vibe' that you give off to others. That, in turn, will produce the results you experience. In effect, you will produce/create your experiences.

I expect to be safe in this world. Therefore, I exude an air of confidence, strength, in-vulnerability. I appear to be strong, not someone to tinker with. People pick up on that vibe, and they leave me alone. I do not appear an easy target or helpless victim. I also make decisions that ensure my safety. I go to safe places, stay in safe hotels and hostels. If I get in a situation that seems potentially dangerous, I get out.

But what if you're actually not so confident? What if you're not sure whether or not you'll be safe? It's quite possible that you don't know how safe a particular country or region is. Or maybe you feel confident about certain aspects of travel, but not others, such as traffic, pick-pockets, illnesses or food safety?

Perhaps the biggest concerns would-be travelers have are for their health due to illnesses or accidents, aggression by other humans, be it various forms of theft, trickery, or physical assaults, getting lost, and not being able to communicate.

Get a clear realization of what you're afraid of and then: If you're unsure or fear something, the first step to alleviating your fears and doubts is to get educated on that topic. For example, if you'd like to go to Pakistan but don't know if it's safe, then go get educated about safety in Pakistan. Where? How? You can:

- *Read guidebooks on Pakistan.*
- *Read blogs of travelers who have gone to Pakistan.*
- *Talk to people who have gone to Pakistan.*
- *Read the US State Dept. 'Travel Warnings' and 'Travel Alerts'*
- *You can find information online, at libraries, at bookstores, at local travel clubs.*

I never leave my valuables unattended. Those items are always (I mean always) with me, and usually on my body. They are with me when I'm walking around, when I'm sleeping, when I take a shower or go to the toilet, and even when I'm swimming or snorkeling! I never leave them unattended. Ever.

I keep my passport, cash and ATM cards in a simple money belt, which is always concealed and strapped on my body. If I'm in public, (i.e. outside my room) my valuables are strapped on my body. I wear my money belt in one of two positions:

1. Clipped around my waist, under my clothes, not showing at all. Not outside my clothes!
2. In a concealed pocket in my special hip bag, which is clipped around my hips/waist? (i.e. It's tied on my body.)

I never display my valuables in public…leave my valuables in a room alone… Using hotel safes… Outdoor activities… I have my valuables in my money belt, clipped on my body, and concealed under my clothes or strapped on my hip bag. (unless I have them in a hotel safe) and I never leave my valuables with other people.

4. CONCLUSION AND RECOMMENDATIONS

The aim of the research is to present the issues and constraints of solo female travelers to Turkey, based on travel blogs reflecting their pre-travel and during-travel restrictions. It also aims to present safety tips that bloggers have adopted to stimulate their desire to travel to and within the cities of Turkey. The findings show that the constraints were mainly intrapersonal barriers about Turkey: like personal, religious requirements, perceived risk and safety constraints. Other constraints such as interpersonal barriers: roles, responsibilities and family constraints and structural barriers: time, money and lack of information constraints, have not reflecting solo travel bloggers' pre-travel and during-travel restrictions about Turkey. Because they are full time female solo travel professionals looking for an independent and adventurous lifestyle.

Safety is an important determinant in the travel decision-making process, including women's mode of transportation, type of accommodation, and especially destination selection (Uysal, McGehee, & Loker-Murphy, 1996). So, for all bloggers the focus is often the safety of solo female travelers. Is it because they are woman? They don't think so. It seems to be because women are seen as more vulnerable than men. Traveling solo safely is pretty well the same for women and men but the reasons for concern, the dangers, are different, so this is why they offer their travel safety tips for solo female travel (Waugh, 2017). According to Khan's (2011, p. 112) research, the male gender is advantageous, privileged, independent and superior to the female gender. Being born male is an advantage. More than half of female respondents think their gender is a strong limiting factor. In many societies, being born a woman automatically connotes subordination, gender stereotypes and taboos. Developing societies, in particular, attribute a lower status to women than men. Society sees women as the 'weaker sex', vulnerable, in need of protection and ideally staying within the confines of the home. Because of societal pressure, most women fear breaking down gender stereotypes and empower men in many areas of life, choosing to conform to "gender roles" by accepting submission to men and limiting their lives to specific "do's and don'ts."

Today, rapid development changes and social progress have begun to change the balance between the sexes, to raise the status of women gradually, and to act according to equal respect and equality with men. Equal treatment of the sexes and the gradual abandonment of gender stereotypes strengthen women's position in societies, and especially educated and empowered women no longer see their gender as a constraint. Therefore, tourism and travel statistics now reveal a large proportion of women who take part in leisure travel (Khan, 2011, p. 112). As it is seen, the power is in the hands of women, according to Ray (March 2, 2022), women make 80 to 85 percent of travel decisions and make up two-thirds of all travelers. Key influencers go far beyond their role as travel consumers. Whether traveling alone or in a group, there is much you can do to support and empower women in local communities. In the fact that women are the largest growing niche market in travel, the age ranges women travel to be young women aged 18 to 25 and older women aged 40 to 65 lead the way.

In conclusion, the author recommends minimizing pre-travel and during-travel issues and constraints of solo female travelers to Turkey. Encourage solo female travelers to discover themselves as they discover the world, to do what they want to do when they want to do it without being judged, to build confidence, to try new things, to have a last hurrah before they get married or have children and to refresh and return to caregiving. Moreover, continue for solo female travelers to reflect the safety tips they adopted to energize their desire to travel to and within Turkeys' cities. One of the blogger suggests getting information about to know if it is safety destination. For example, if you want to go to Turkey but don't know if it is safe, then you should get information about safety in Turkey. Read guidebooks, blogs of travelers, the US State Dept. "Travel Alerts" on Turkey (see Travel.State.Gov. 2023) and talk to people who have gone to Turkey. You can get information online, in libraries, bookstores, and local travel clubs.

Tourism is known as one of the most important economic sectors in the world. It employs one out of every 10 people in the world and provides a livelihood for hundreds of millions more. Most of them are women. Tourism also contributes to UN's Sustainable Development Goal No. 5: "Achieving gender equality and empowering all women and girls." Therefore, tourism is an opportunity to redefine the gender balance. By lifting one woman up, an entire household, and ultimately an entire community can be lift one step at a time. According to Ray (March 2, 2022), there are 12 ways to empower women who travel to advance the UN Sustainability Goals to support women, equality and women-owned businesses, both in our own countries and in countries that rely on tourism. These; choose a woman-owned business and hire a woman, seek ındigenous experiences when you travel, choose a female guide, choose a woman-owned accommodation, support women writers, authors, and storytellers, volunteer with women when you travel, donate when you travel or when you return home, bring menstrual supplies when you travel, embrace a curious mind, celebrate women's stories and histories, show gratitude and be kind to those you meet on your travels that are doing their best to give you a memorable experience.

REFERENCES

Asbollah, A. Z. B., Lade, C., & Michael, E. (2013). The tourist's gaze: From the perspective of a Muslim woman. *Tourism Analysis*, *18*(6), 677–690. doi:10.3727/108354213X13824558188703

Bernard, S., Rahman, I., & McGehee, N. G. (2022). Breaking barriers for Bangladeshi female solo travelers. *Tourism Management Perspectives*, *41*, 100932. doi:10.1016/j.tmp.2021.100932

Blood, R. O., & Wolfe, D. M. (1960). *Husbands and Wives: The Dynamics of Married Living*. The Free Press.

Blumstein, P. W., & Schwartz, P. (1983). *American Couples: Money, Work, Sex*. William Morrow.

Bond, M. (1997). *Women Travellers: A New Growth Market*. Pacific Asia Travel Association.

Cambronero-Saiz, B. (2013). Gender policies and advertising and marketing practices that affect women's health. *Global Health Action*, *6*(1), 1–7. doi:10.3402/gha.v6i0.20372 PMID:23806207

Chen, C. C., & Petrick, J. F. (2016). The Roles of Perceived Travel Benefits, Importance, and Constraints in Predicting Travel Behavior. *Journal of Travel Research*, *55*(4), 509–522. doi:10.1177/0047287514563986

Chen, H.-J., Chen, P.-J., & Okumus, F. (2013). The relationship between travel constraints and destination image: A case study of Brunei. *Tourism Management*, *35*, 198–208. doi:10.1016/j.tourman.2012.07.004

Chiang, C. Y., & Jogaratnam, G. (2006). Why do women travel solo for purposes of leisure? *Journal of Vacation Marketing*, *12*(1), 59–70. doi:10.1177/1356766706059041

Crawford, D. W., & Godbey, G. (1987). Reconceptualizing barriers to family leisure. *Leisure Sciences*, *9*(2), 119–127. doi:10.1080/01490408709512151

Crawford, D. W., Jackson, E. L., & Godbey, G. (1991). A hierarchical model of leisure constraints. *Leisure Sciences*, *13*(4), 309–320. doi:10.1080/01490409109513147

Crompton, J. (1979). Motivation for pleasure vacation. *Annals of Tourism Research*, *6*(4), 408–424. doi:10.1016/0160-7383(79)90004-5

Enoch, Y., & Grossman, R. (2010). Blogs of Israel and Danish backpackers to India. *Annals of Tourism Research*, *37*(2), 520–536. doi:10.1016/j.annals.2009.11.004

Flashpack. (2023). *Group Solo Travel in Turkey*. Retrieved from https://www.flashpack.com/destinations/turkey/

Green, R. T., & Cunningham, I. C. M. (1976). Empowerment Status, Feminine Role Perception and Family Purchasing Decisions. *Journal of Business Research*, *4*(4), 313–324. doi:10.1016/0148-2963(76)90041-2

Greer, G. (1999). *The Whole Woman*. Anchor.

Hanmer, J., & Saunders, S. (1984). *Well-founded Fear: A Community Study of Violence to Women*. Hutchinson.

Henderson, K. A., & Allen, K. R. (1991). The ethic of care: Leisure possibilities and constraints for women. *Society and Leisure*, *14*(1), 97–113. doi:10.1080/07053436.1991.10715374

Henderson, K. A., & Dialeschki, M. D. (1991). A sense of entitlement to leisure as constraint and empowerment for women. *Leisure Sciences*, *13*(1), 51–65. doi:10.1080/01490409109513124

Hookway, N. (2008). Entering the blogosphere': Some strategies for using blogs in social research. *Qualitative Research*, *8*(1), 91–113. doi:10.1177/1468794107085298

Howard, L. (2022). *Best International Destinations for Solo Travelers in 2022*. Retrieved from https://www.forbes.com/uk/advisor/travel-insurance/best-international-destinations-for-solo-travellers-in-2022/

Itchfeet. (2023). Retrieved from http://charukesi.com/itchyfeet/category/europe/turkey/

Jackson, E. L. (1991). Leisure Constraints/Constrained Leisure: Special Issue Introduction. *Journal of Leisure Research*, *23*(4), 279–285. doi:10.1080/00222216.1991.11969860

Jackson, E. L., & Henderson, K. A. (1995). Gender-based analysis of leisure constraints. *Leisure Sciences*, *17*(1), 31–51. doi:10.1080/01490409509513241

Jordan, F., & Gibson, H. (2005). 'We're not stupid . . . but we'll not stay home either': Experiences of solo women travellers. *Tourism Review International*, *9*(2), 195–212. doi:10.3727/154427205774791663

JourneyWomen. (2022). *Safety Resources for Women Travelers*. Retrieved from https://journeywoman.com/solo-travel-safety/

Kaba, B., & Emekli, G. (2018). A rising Trend in Tourism: Solo Women Travelers (The Case of Turkey). *Aegean Geographical Journal*, *27*(2), 111–126.

Kaiser, M. (2016). *Best Female Travel Blogs: Top Influential Views*. Retrieved from https://www.creditdonkey.com/female-travel-blogs.html

Khan, S. (2011). Gendered Leısure: Are Women More Constraıned in Travel for Leısure? *TOURISMOS: An Internatıonal Multıdıscıplınary Journal of Tourısm*, *6*(1), 105–121.

Koskela, H. (1997). 'Bold walk and breakings': Women's spatial confidence versus fear of violence. *Gender, Place and Culture*, *4*(3), 301–319. doi:10.1080/09663699725369

Kozinets, R. V. (2002). The field behind the screen: Using netnography for marketing research in online communities. *JMR, Journal of Marketing Research*, *39*(1), 61–72. doi:10.1509/jmkr.39.1.61.18935

Loureiro, S. (2014). The role of rural tourism experience economy in place attachment and behavioral intentions. *International Journal of Hospitality Management*, *40*, 1–9. doi:10.1016/j.ijhm.2014.02.010

Maser, B., & Weiermair, K. (1998). Travel Decision-Making: From the Vintage Point of Perceived Risk and Information Preferences. *Journal of Travel & Tourism Marketing*, *7*(4), 107–121. doi:10.1300/J073v07n04_06

McCulley, K. (2022). *Solo Female Travel in Turkey — Is Turkey Safe?* Retrieved from https://www.adventurouskate.com/solo-female-travel-turkey-safe/

McNamara, K. E., & Prideaux, B. (2010). A typology of solo independent women travelers. *International Journal of Tourism Research*, *12*(3), 253–264. doi:10.1002/jtr.751

NadwornyK. (2013). *Turkey Posts*. Retrieved from https://katrinkaabroad.com/turkey/

Nguyen, H. N. (2018). *An investigation into female solo travelers: Solo travel exposure and perceived constraints, personal development, authentic personality and self-esteem* (Order No. 10981186). Available from ProQuest Dissertations & Theses Global. (2269043425). Retrieved from https://www.proquest.com/dissertations-theses/investigation-into-female-solo-travelers-travel/docview/2269043425/se-2

Ngwira, C., Tse, S., & Vongvisitsin, T. B. (2020). Negotiation Strategies and Constraints for Solo Female Travelers in Africa. *Tourism, Culture & Communication*, *20*(1), 35–47. doi:10.3727/109830420X15859302993081

Nyaupane, G. P., Morais, D. B., & Graefe, A. R. (2004). Nature Tourism Constraints: A Cross-Activity Comparison. *Annals of Tourism Research*, *31*(3), 540–555. doi:10.1016/j.annals.2004.01.006

Oh, H., Fiore, A. M., & Jeoung, M. (2007). Measuring experience economy concepts: Tourism applications. *Journal of Travel Research*, *46*(2), 119–132. doi:10.1177/0047287507304039

Passportandplates. (2021a). Retrieved from https://passportandplates.com/destination-guides/intrepid-travel-turkey-review/

Passportandplates. (2021b). Retrieved from https://passportandplates.com/turkey-travel-guide/

Passportandplates. (2023). Retrieved from https://passportandplates.com/aboutme/

Pereira, A., & Silva, C. (2018). Women solo travellers: Motivations and experiences. *Millenium*, *6*(6), 99–106. doi:10.29352/mill0206.09.00165

Quan, S., & Wang, N. (2004). Towards a structural model of the tourist experience: An illustration from food experiences in tourism. *Tourism Management*, *25*(3), 297–305. doi:10.1016/S0261-5177(03)00130-4

Ray, C. (2021). *JourneyWoman and Safe4Life Host Special Self-Defence Class for Young Women*. Retrieved from https://journeywoman.com/feature-articles/journeywoman-and-safe4life-host-special-self-defence-class-for-young-women/

Ray, C. (2022a). *Expert Solo Travel Safety Tips for Young Women*. Retrieved from https://journeywoman.com/travel-tips/solo-travel-safety-tips-for-young-women/

Ray, C. (2022b). *12+ Ways to Empower Women When You Travel*. Retrieved from https://journeywoman.com/feature-articles/12-things-you-can-do-to-empower-women-when-you-travel/

Reisinger, Y., & Mavondo, F. (2005). Travel Anxiety and Intentions to Travel Internationally: Implications of Travel Risk Perception. *Journal of Travel Research*, *43*(3), 212–225. doi:10.1177/0047287504272017

Roehl, W., & Fesenmaier, D. (1992). Risk Perceptions and Pleasure Travel: An Exploratory Analysis. *Journal of Travel Research*, *30*(4), 17–26. doi:10.1177/004728759203000403

Schintler, L., Root, A., & Button, K. (2000). Women's Travel Patterns and the Environment an Agenda for Research. *Transportation Research Record: Journal of the Transportation Research Board*, *1726*(1), 33–40. doi:10.3141/1726-05

Schmallegger, D., & Carson, D. (2008). Blogs in tourism: Changing approaches to information exchange. *Journal of Vacation Marketing*, *14*(2), 99–110. doi:10.1177/1356766707087519

Schmitt, B. (1999). Experiential marketing. *Journal of Marketing Management*, *15*(1), 53–67. doi:10.1362/026725799784870496

Shahvali, M., Shahvali, R., & Kerstetter, D. (2016). Women's Travel Constraints in a Unique Context. In M. Kozak & N. Kozak (Eds.), *Tourist Behaviour: An international perspective* (pp. 16–25). CABI. doi:10.1079/9781780648125.0016

Shaw, S. M. (1994). Gender, Leisure and Constraint: Toward a Framework for Analysis of Women's Leisure. *Journal of Leisure Research*, *26*(1), 8–22. doi:10.1080/00222216.1994.11969941

Sheffield, C. J. (1995). Sexual terrorism. In J. Freeman (Ed.), *Women: A Feminist Perspective* (5th ed., pp. 1–21). Mayfield.

Shin, H., Nicolau, J. L., Kang, J., Sharma, A., & Lee, H. (2022). Travel decision determinants during and after COVID-19: The role of tourist trust, travel constraints, and attitudinal factors. *Tourism Management*, *88*, 1–14. doi:10.1016/j.tourman.2021.104428 PMID:34539019

Stark, J., & Meschık, M. (2018). Women's everyday mobility: Frightening situations and their impacts on travel behaviour. *Transportation Research Part F: Traffic Psychology and Behaviour*, *54*, 311–323. doi:10.1016/j.trf.2018.02.017

Travel.State.Gov. (2023). Retrieved from https://travel.state.gov/content/travel/en/international-travel/International-Travel-Country-Information-Pages/Turkey.html

Tsaur, S.-H., Tzeng, G.-H., & Wang, K.-C. (1997). Evaluating Tourist Risks from Fuzzy Perspectives. *Annals of Tourism Research*, *24*(4), 796–812. doi:10.1016/S0160-7383(97)00059-5

Valentine, G. (1989). The geography of women's fear. *Area*, *21*, 385–390.

Waugh, J. (2017). *Solo Female Travel Tips: Why, Where, How and Safety*. Retrieved from https://solotravelerworld.com/solo-female-travel-tips/

Westwood, S., Pritchard, A., & Morgan, N. J. (2000). Gender-blind marketing: Businesswomen's perceptions of airline services. *Tourism Management*, *21*(4), 353–362. doi:10.1016/S0261-5177(99)00069-2

Wilson, E., & Harris, C. (2006). Meaningful travel: Women, independent travel and the search forself and meaning. *Tourism (Zagreb)*, *54*(2), 161–172.

Wilson, E., & Little, D. E. (2005). A 'relative escape'? The impact of constraints on women who travel solo. *Tourism Review International*, *9*(2), 155–174. doi:10.3727/154427205774791672

Wilson, E., & Little, D. E. (2008). The solo female travel experience: Exploring the 'geography of women's fear'. *Current Issues in Tourism*, *11*(2), 167–186. doi:10.2167/cit342.0

Yang, E. C. L., Khoo-Lattimore, C., & Arcodia, C. (2017). A narrative review of Asian female travellers: Looking into the future through the past. *Current Issues in Tourism*, *20*(10), 1008–1027. doi:10.1080/13683500.2016.1208741

Yavas, U. (1990). Correlates of Vacation Travel: Some Empirical Evidence. *Journal of Professional Services Marketing*, *5*(2), 3–18. doi:10.1300/J090v05n02_02

Chapter 19
Special Interest Tourism and Women

Ülker Çolakoğlu
Adnan Menderes University, Turkey

ABSTRACT

The growth of special interest tourism is seen as a reflection of the growing diversity of leisure interests in the leisure society of the early 21st century. Post-modern tourism is slowly moving from traditional tourism attractions to being part of a general lifestyle that corresponds to people's (women's) daily lives and activities. As a conceptual study, in this chapter, readers will understand the definition of special interest tourism and the difference between general interest tourism. They will understand the characteristics, scope, and wide variety types of special interest tourism for women tourist such as adventure tourism, medical tourism, rural tourism, cultural tourism, culinary tourism, religious tourism, ecotourism, sports tourism, educational tourism, photographic tourism, dark tourism, celebrity tourism, and film tourism. Moreover, they will learn new types of trends in special interest tourism for women tourists like Insta-tourism, virtual tourism, Pokémon-go tourism (gaming tourism), and space tourism (Astro-tourism).

INTRODUCTION

A widely known and accepted definition of tourism is travel for leisure, business or recreational purposes (Kruja and Gjyrezi, 2011, p.79). Special interest tourism occurs in many forms of tourism and is defined by certain activities, geographies or groups of tourists with similar interests (Trauer, 2006). Special interest tourists enjoy activities in contact with nature, beauty of a conserved natural area, the pleasure of watching the fauna in its natural habitat, of exploring, discovering and learning, overcoming obstacles and feeling the pleasure of overcoming them. Furthermore they like to discover how easy it is to read a map, navigate a river and learn climbing techniques, diving, horse riding, canoeing, driving cattle, packing and carrying a rucksack. For that tourist the physical effort involved is a small price to pay for enjoying the view of a mountain, camping next to a lake of still clear water or watching an eagle soaring in the wind. Above all there is the pleasure of sharing experiences with friends and relatives (Kruja and Gjyrezi, 2011, p.80).

DOI: 10.4018/978-1-6684-8417-3.ch019

Special interest tourism has found a place in the modern age of throat-cutting competition; this has emerged as a new concept that gives travel agents and tour operators relaxing results for their business. Now, travel agents and tour operators have much more options for their business, which was previously a traditional business, focused on certain forms of tourism. With the advent of this special interest tourism, the companies are now playing with new business strategies, new ideas and new concept which are mutually beneficial to both the parties in terms of female tourist satisfaction, monitory benefit and exploration of new destinations and over all economics of the destination. This activity forced the government to think about things and as a result made many plans and proposals to develop isolated places and areas of special interest. Female tourists benefiting from special interest tourism activities expect to learn more, enrich their awareness of the place they intend to visit, and express themselves about their perceptions of the place or service. Such female tourists expect high-quality services and personalized touches. Many want to travel soundly and have genuine social, cultural and environmental experiences in an authentic environment, with a positive attachment with the host population.

As it is known, tourism is a discipline with a wide range of products and services to be offered to female tourists. The female tourists come with a wide variety of personal interests and motivations. For example, according to Mitten (1992) woman benefits from participating in all female outdoor adventures (such as flatwater and whitewater canoeing, bicycle touring, trekking, backpacking, sea kayaking whitewater rafting, snorkeling, scuba diving, rock climbing, mountaineering, horse packing, and skinning). Women discover how enjoyable and powerful it feels to be part of an all-woman group. They sense that emotional sharing is easier so they absorb in a potentially stressful environment. Knowing that within the group they, feel safe, independent, gain self-esteem (support, love and acceptance; work and accomplishment; play and fun; and time for reflection and introspection), discover their own power and expand their self-images. Moreover, millions of woman patients around the world, travel other countries that provide highly specialized medical care for various reasons, such as cosmetic surgery, dental work, to orthopedic procedures (Stephano, 2022), fertility (reproductive), wellness at affordable rates. According to Skwirczyfiska et al., (2022) plastic surgery, apart from its aesthetic values, has a positive effect on the health of female patients in the form of increasing well-being in its various forms and increased self-esteem. Likewise, special interest women tourists' motivations for astro tourism are associated with learning, imagining and discovering while observing the deep sky that arouses intellectual and competence mastery motivations (Cater, 2010; Soleimani et al., 2019). Travel influencers in social media platform such as Insta tourism can motivate their women followers to participate in touristic activity by sharing their experiences in an aesthetic way (Asan, 2022, p.980), and influence their decision making and behavior intention to travel the destination safe. As a result, we have a wide range of special interest tourism for women tourist. From this point of view, a conceptual study was made in this chapter based on the literature on special interest tourism and various types of special interest tourism for female tourist.

1. CONCEPT OF SPECIAL INTEREST TOURISM

1.1. Definition and Concept of Special Interest Tourism

Special interest tourism, which developed especially in the 1980s, is not a new phenomenon, but is a new concept. Special interest tourism is considered to be one of the oldest types of tourism since people have traveled for special purposes such as interest, curiosity and adventure since the early ages. Sometimes both active tourism and alternative concepts are used as synonyms for special interest tourism (Akinci and Kasalak, 2016, p.178). To set special interest tourisms in a broader overall tourism context, Brotherton and Himmetoglu (1997) suggest a ''Tourism Interest Continuum''. Based on Culligan's framework, they propose that through increasing travel experience, confidence and affluence, a maturation or tourist life cycle transition from ''safe to more adventurous kinds of travel and holidays'' occurs, with the tourist ''trading up'' and purchasing social prestige and ego-enhancement (Trauer, 2006, p. 187).

Brotherton and Himmetoglu (1997) put forward the notion that the questions a tourist would ask in the decision-making process are (see Figure 1),

- General Interest Tourism (GIT): Where would I like to go?
- Mixed Interest Tourism (MIT): Where do I want to go and what activities can I pursue there?
- Special interest tourisms (SIT): What interest/activity do I want to pursue, and where can I do it?

General interest tourism (GIT) when, in accordance with the characteristics described above, the destinations tourists go to offer an extensive complementary offer composed of several special interest tourisms (SIT). Many contributions have been made today to tourist typologies that fall under the GIT category (see Figure 2) (Fernández et al., 2019, p. 2).

As an example, activities such as hunting or risky sports are tourist market niches: they pre-exist and are produced in the same way whether they become tourist attractions or not. In contrast, gastronomic routes, routes dedicated to some cultural aspect that motivates the trip and/or the adaptation of activities to make them accessible to tourism (agritourism, language tourism, safari tourism, astro-tourism, etc.) generate SIT around particular needs (Soleimani et al., 2018; Fernández et al., 2019, p. 2). Moreover, according to Fernández et al., (2019, p. 2-3) olive oil tourism can generally be described as special interest tourism based on three major GITs (rural tourism, nature-based tourism and cultural tourism) that, shares some common features with certain SIT typologies (gastronomic, industrial, ethnographic, landscape, agrotourism, creative, flora observation tourism, wildlife observation, business and health tourism).

Furthermore, alternative tourism types are part of special interest tourism. They take natural life as a starting point, which is directly linked to the protection of the natural environment, relations with local people, customs and traditions and the preservation of cultural heritage; furthermore, such tourists encourage the establishment and development of businesses owned and controlled by local residents (Karasoula, 2010). Alternative and special interest tourism provides an option for mitigation as it helps to create alternatives to mainstream tourism. Alternative tourism does not have a single definition, but rather is seen as the opposite of mass tourism and "targets a smaller number of tourists who seek direct engagement with the local communities in the host country and to develop intercultural understanding" (Isaac and Eid, 2019), and where "tourists are the ones that often choose a way of organizing and shaping the journey to which they are headed" (Belias et al., 2018). Alternative tourism is an umbrella term for all kinds of responses to mass tourism, of which special interest tourism refers to "customized leisure and recreational experiences driven by the specific expressed interests of individuals and groups" (Soleimani et al., 2019; Garanti, 2022, p. 452).

The advent of SIT has been an acknowledgment of the need in a postmodern consumer culture for recognition and provision of increasingly higher levels of customization of tourism experiences. SIT may be defined as the provision of customized leisure and recreational experiences driven by the specific expressed interests of individuals and groups (Derrett, 2001), and to this Hall and Weiler (1992) have added a geographical component, so that a particular interest can be pursued in a particular region or at

Figure 1. Leisure–tourism interest cycle
Source: Brotherton and Himmetoğlu (1997), Prosser (2001), Ruys and Wei (2001), Schofield (2001), and (Akinci and Kasalak, 2016, p. 185)

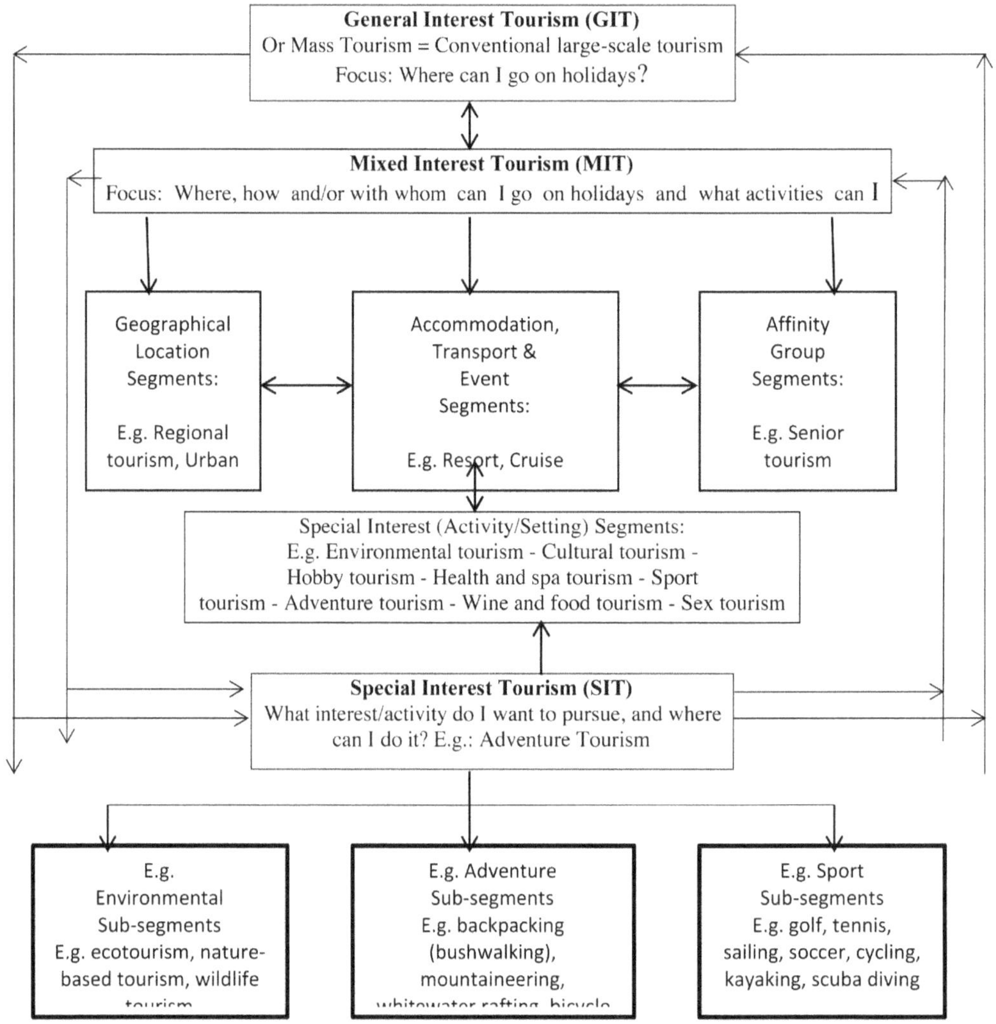

Figure 2. General interest tourism
Source: Fernández et al. (2019, p. 2)

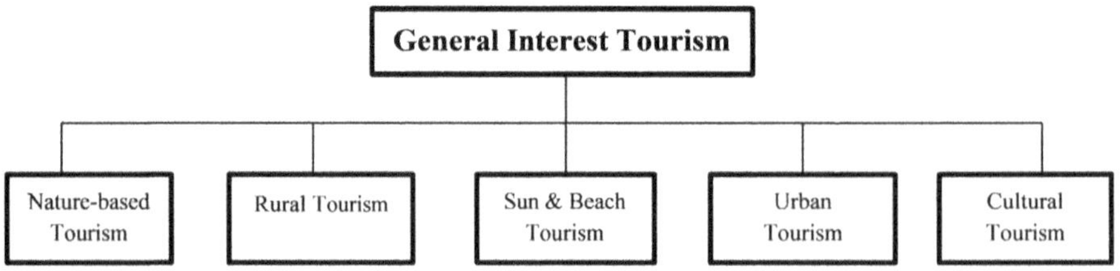

a particular destination. (Soleimani et al., 2019, p. 2300). Trauer (2006) identified SIT in first, certain activities like sport tourism, adventure tourism, cultural tourism and many forms of tourism, second, geographies such as Antarctica, Europe and Africa and finally, groups of tourists with similar interests such as grey tourism to identify groups of seniors who travel for similar reasons, pink tourism to recognize gay and bisexual visitors, or girly/female tourism. Based on this classification, Trauer (2006) defines SIT as the opposite of mass tourism. Similarly, Robinson and Novelli (2007) characterize culture-based tourism and special interest tourism as a part of niche tourism as opposed to mass tourism in this context.

Douglas et al. (2001) defined SIT as "a kind of customized recreational or tourism experience with specific interests presented by an individual or group." So-called "specific interests" included the preference for the activities, places, scenery, and even the emotional attachment to the companions (Hall and Weiler, 1992; Trauer and Ryan, 2005). In regard to the meaning of "special interest tours,", it primarily refers to a type of travel product designed in accordance with the unique interests of travelers subsequent to market segmentation (Yung, 1993) (e.g. tours arranged primarily for the purposes of mountain climbing, photography, fashion clothing, culinary arts, or golf; or tours with a focus on overseas study, family education or even great castles visits). In general, any theme that consumers are interested in can be used to design a type of special tour. This kind of product has also brought to the travel market a new area of business operations (Sheng et al., 2007, p. 273).

SIT is a derivation of the generic marketing term for niche markets, which consist of small market segments that other firms in an industry overlook or ignore (Kotler, Armstrong, Brown, & Adam, 1998, p. 725). When properly measured and diagnosed, SIT can serve as a significant contributor to a destination's visitation for multiproduct destinations and a core activity for single-product destinations (McKercher & Chan, 2005). This creates opportunities for niche firms to carefully target customers on the basis of their demographic, psychographic, or geodemographic profiles with highly personalized product offerings (Schiffman et al., 2001, p. 235). Niche implies that there is a place in the market for the product, and that there is a consumer for the product. It is based on the premise that the market should not be seen as a simplistic, homogenous whole with general needs, but rather as sets of individuals with specific needs relating to the qualities and features of particular products. In a tourism context, niche tourism can be divided into macro-niches (such as cultural tourism or rural tourism), and these can be further sub-divided into micro-niches such as gastronomy, sport, volunteer, and adventure (Robinson & Novelli, 2005), and in the instance of the present study, astro-tourism. (Soleimani et al., 2019, p. 2300-2301).

1.2. Characteristics of Special Interest Tourism

SIT has evolved from the search for new ways that are likely to add value and attract traditional tourism. It provides a unique experience to tourists who are of special interest. The tourist goes on an unusual journey. Tourists travel to a particular destination in order to realize a particular interest that can only be realized in that destination. It offers many alternative forms as it has been developed considering the preference of tourists. Also, there is the possibility of offering various experiences in the same place or destination. These are the characteristics that make SIT unique.

1.3. Scope of Special Interest Tourism

The scope of SIT is very broad because it offers countless opportunities. When it comes to the scope of SIT, this emerging form of tourism has great scope for every stakeholder i.e. tourist, destination, travel agencies and tour operators, host, government and indirectly beneficiaries. Tourists get acquainted with some new attractions that come into their interest with some unique and unrivaled features.

SIT includes different forms of special interest travel, known as niche markets and gaining popularity among the actors of the tourism industry over the years. While there is no limit to these niche markets, some have become more sustainable and are considered SIT categories. These main categories are, with a brief description for some of them (Kruja and Gjyrezi, 2011, p.79-80):

a) Adventure tourism,
b) Medical tourism,
c) Rural Tourism,
d) Cultural tourism,
e) Culinary tourism,
f) Religious tourism,
g) Ecotourism,
h) Wildlife tourism,
i) Heritage tourism and so on.

Moreover, special interest tourists are generally adults or third age persons, adolescents and children (members of family or school groups), with the following characteristics (Kruja and Gjyrezi, 2011, p. 80):

- Spirit of adventure,
- Curiosity and
- Sharing experiences.

If we exclude some markets, this type of tourism is relatively expensive. Although tourist forms such as students, backpackers and low income earners can afford some types of tourism, such as ecotourism, cultural tourism or adventure tourism, the requirements for organizing such trips need to be organized independently. Another feature of special interest tourism is that they are more middle-aged and travel with or without children (exceptions may be included in some forms of nature-based or wildlife tourism), because many forms of special interest tourism are not preferred or attractive to children. It is also common for tourists to travel alone.

1.4. Types of Special Interest Tourism for Women Tourists

There is no such classification of special interest tourism, but it depends entirely on the nature and motivation of the tourist choosing a particular tourism product. Thus, the definition never focuses on any particular tourist destination, but emphasizes the purpose of satisfying a particular interest that can be satisfied in that particular destination.

1.4.1. Adventure Tourism

Adventure tourism includes a broad range of activities, often provided by a commercial operator, which involves close interaction with the natural environment away from the participant's home and contains elements of perceived and real risk (Doran, 2016).

Adventure tourism is a type of tourism which usually takes place in destinations with specific geographic features and landscape and tends to be associated with a physical activity, cultural exchange, interaction and engagement with nature. This experience may involve some kind of real or perceived risk and may require significant physical and/or mental effort (UNWTO, 2019). Moreover, it is seen as a fantastic and energetic place with adventure activities, like rock climbing, mountain climbing, mountain hikes, white water rafting, paragliding, snorkeling, skiing, skydiving, bungee jumping, zip-lining, etc. The holidaymakers will go to places with the most fun things that the customers can do and experience, which includes destinations like Australia, Africa, Thailand, and many more (Stainton, 2022).

Adventure tourism may involve 'soft' activities, such as backpacking, hiking, cycling or flat-water canoeing, or 'hard' activities, such as climbing and mountaineering, white-water rafting, wilderness backpacking and sky-diving. While some definitions also include safaris, camping, birdwatching, ecotourism and volunteer tourism as 'soft' forms of adventure tourism (Doran, 2016).

- ***Mountain tourism*** is a type of "tourism activity which takes place in a defined and limited geographical space such as hills or mountains with distinctive characteristics and attributes that are inherent to a specific landscape, topography, climate, biodiversity (flora and fauna) and local community. It encompasses a broad range of outdoor leisure and sports activities" (UNWTO, 2023). Such activities are climbing, hiking, mountain biking or skiing. Popular mountain ranges hosting tourism activities include the Alps, the Himalayas, and the Andes (Stainton, 2022).
- ***Extreme tourism***, also commonly referred to as danger tourism or shock tourism, is the act of traveling to a destination to engage in extreme activities. As an extension of adventure tourism, this type of travel is often considered dangerous. Extreme tourist activities include BASE jumping off the cliff in Norway, volcanic bungee jumping in Chile, and climbing Mount Hua in China (Stainton, 2022).

Many females love going on outdoor adventures. They join outdoor experiences for variety of reasons and drive a variety of personal benefits like feel empowered, relax, have fun, gain sense of renewal, network and/ or for spiritual healing. Women choose all outdoor adventures without any hesitate; they cope with hardship or difficult situations that arise from weather, trail conditions, choices that participants make, and other predicaments that occur in the wilderness. However some women join all female adventures to work with their fears and safety issues. Moreover they choose all female trips to have freedom to

Table 1. Types of benefits for women in adventure tourism (Doran, 2016)

Feelings of Empowerment				
Sense of Freedom	*Self-Development*	*Social Encounters*	*Heightened Bodily Experience*	*Female Company*
• Time for yourself • Independence • Escape from everyday life	• Education/pursuing a hobby • Self-understanding • Fitness, exercise & risk taking • Construct adventurous identity	• With people of different cultures • In new contexts • Emerging friendships • Relationship with guide • Sense of belonging	• Bodily movement • Sensual experience • Heightened & different emotions	• Freedom to be yourself • Supportive & non-competitive • Less intimidating • Escape gender roles • Increase self-understanding • Shared experiences

express their feelings and get support when they needed it. It is important to note that women perceive that they can meet certain emotional needs, such as unconditional support, attention, acceptance for who they are, and personal time more readily in all female environment (Mitten, 1992).

Women's participation in adventure tourism is rapidly growing, Doran's (2016) research seeks to extend our understanding of female adventure tourists by examining the empowering journey women can take through constraint negotiation to enjoy the benefits of adventure tourism. Their research model showed that constraints, negotiations and benefits could be experienced simultaneously, at different points in a woman's adventure tourism journey and used as a vehicle for empowerment. Women will also reevaluate these categories before, during and after their adventure tourism experienced. Therefore, the categories were not fixed and evolve each time a woman participates in adventure tourism throughout her life. According to Doran (2016) five comprehensive categories with sub-categories are proposed based on the key benefits expressed by women within the literature (see Table 1).

1.4.2. Medical Tourism (Health Tourism)

Millions of patients' travel each year from countries due to their lacking healthcare infrastructure or less advanced in a particular area of medical care to countries that provide highly-specialized medical care. This has birthed a robust global medical tourism market that was worth over $37 billion in 2019. Patients book flight trips to countries for various medical procedures ranging from cosmetic surgery, dental work, to orthopedic procedures at affordable rates. For the health tourism destinations and healthcare providers, the competition is fierce, requiring interplay of factors to drive medical travel and improve their brand in the medical tourism market. According to the Medical Tourism Index, which assesses the attractiveness of countries for medical travel, a country's economy and public image, healthcare costs, and quality of care are the major factors that drive medical tourism growth in a destination. The top 10 medical tourism destinations are Canada, Singapore, Japan, Spain, United Kingdom, Dubai, Costa Rica, Israel, Abu Dhabi and India (Stephano, 2022).

Medical tourism, also known as health tourism, refers to the act of traveling to another destination for medical treatment (i.e. India, Turkey and Panama). Medical tourists' motivations may include lower costs for treatments or higher quality of service. Medical tourists may seek life-saving treatments, cosmetic surgery or dental procedures that cannot reach them at home, among a host of other medical needs. In addition, health tourism allows clients to go to a destination for purposes such as medical treatments, visiting spas, wellness centers and physical activities (yoga, swimming, fitness) - to "cleanse" and focus on their minds and bodies; While most destinations seem like a relaxing and calm place to go, some destinations can be seen as expensive, exotic places to travel; popular destinations include Asia, the Middle East, the Indian Ocean and many more (Stainton, 2022).

Medical tourism is a type of tourism activity which involves the use of evidence-based medical healing resources and services (both invasive and non-invasive). This may include diagnosis, treatment, cure, prevention and rehabilitation. Health tourism is the umbrella term for the subtypes wellness tourism and medical tourism (UNWTO, 2019).

The health and wellness tourism sector is a steadily growing global industry, driven by factors such as human stress and lifestyle, demand for personalized services and an aging population among others (Kazakov and Oyner, 2020). Andreou (2017) proposed health tourism as a pillar for economic development, suggesting that health tourism development can lead to a better quality of health services and economic gains.

- ***Wellness tourism*** is a type of tourism activity which aims to improve and balance all of the main domains of human life including physical, mental, emotional, occupational, intellectual and spiritual. The primary motivation for the wellness tourist is to engage in preventive, proactive, lifestyle-enhancing activities such as fitness, healthy eating, relaxation, pampering and healing treatments (UNWTO, 2019). Popular wellness tourism destinations include Iceland for spa holidays, India for Ayurveda treatments, and Bali for yoga retreats (Stainton, 2022).
- ***Esthetic surgery*** *tourism* has defined by the American Medical Association as the transformation of normal body structures in order to improve the patient's appearance and self-esteem (Skwirczyfiska et al., 2022, p.2).
- ***Dental tourism***, also known as a dental vacations or dental holidays, is the act of traveling to a destination for dental treatment. It is a subset of the medical tourism industry. Tourists will often visit places where treatment is available at a lower cost than in their home country. Popular dental tourism destinations include India, Thailand, Bulgaria, Turkey and Vietnam. Dental tourism is another type of tourism that has grown considerably in recent years (Stainton, 2022).

Female patients deciding to undergo a procedure in the field of aesthetic medicine (or other medical procedures) are most often guided by criteria related to the functional improvement of the body parts that cause them discomfort, whereas aesthetic reasons are another motivation. Regardless of their motivation, female patients treat the surgeon as a vehicle of hope for improving not only their image, but also their own self-esteem and satisfaction with life (Skwirczyfiska et al., 2022, p.2).

1.4.3. Rural Tourism

Rural tourists can travel anywhere in the countryside; means that tourists can experience rural life, culture, beliefs and religion and participate in activities that take place in rural life, such as farming, picking vegetables by hand, riding a tractor to mow the land; It means that the tourist will not stay in expensive hotels, tourists will spend their nights in a small house in the village with local tourists or sleep in tents and camps, destinations include Poland, the Middle East, the Philippines and more (Stainton, 2022).

According to the World Tourism Organization (UNWTO, 2023), rural Tourism as "a type of tourism activity in which the visitor's experience is related to a wide range of products generally linked to nature-based activities, agriculture, rural lifestyle / culture, angling and sightseeing. Rural tourism activities take place in non-urban (rural) areas with the following characteristics: i) low population density, ii) landscape and land-use dominated by agriculture and forestry and iii) traditional social structure and lifestyle". According to Eurostat (1998) rural tourism "regards the activities of a person travelling and staying in rural areas (without mass tourism) other than those of their usual environment for less than one consecutive year for leisure, business and other purposes (excluding the exercise of an activity remunerated from within the placed visited) " (Kruja and Gjyrezi, 2011, p.80).

- ***Agritourism,*** also referred to as agricultural tourism, argotourism or farm tourism is a subset of the rural tourism industry. It focuses on agricultural activities and includes tourist activities in or around farms. This includes activities such as wine tours, horseback riding, clay bird shooting, animal petting, and historical agricultural exhibits. Agritourism is more common than many people think and is one of the types of tourism that is often overlooked (Stainton, 2022).

1.4.4. Cultural Tourism

"Cultural tourism is a type of special interest tourism based on seeking and participating in new and profound cultural experiences, whether aesthetic, intellectual, emotional or psychological" (Stebbins, 1996). This definition includes various cultural forms including history, religion(s), art, architecture and other elements that contribute to the formation of the unique culture of the visited region (Kruja and Gjyrezi, 2011, p.80).

Cultural tourism is a type of tourism activity in which the visitor's essential motivation is to learn, discover, experience and consume the tangible and intangible cultural attractions/products in a tourism destination. These attractions/products relate to a set of distinctive material, intellectual, spiritual and emotional features of a society that encompasses arts and architecture, historical and cultural heritage, culinary heritage, literature, music, creative industries and the living cultures with their lifestyles, value systems, beliefs and traditions (UNWTO, 2019). For example, Bali has been declared as cultural tourism destination, and together with other destinations all over the world it brings out cultural tourism grows faster than the general tourism industry in average and emerges as one of the potential segments in the tourism industry (Abdillah et al., 2022, p.).

- ***Ethnic tourism*** is tourism that focuses on learning and experiencing about a particular ethnicity. Ethnic tourism is a type of cultural tourism. Ethnic tourism can involve a deep cultural experience, for example, through volunteer work or homestay. It may also include a lighter cultural experience, for example, by watching games or visiting museums (Stainton, 2022). Harron and Weiler (1992) have made a significant effort to define this form of special interest tourism as "travel motivated by the pursuit of first-hand, authentic and sometimes intimate contact with people whose ethnic and/or cultural backgrounds differ from those of tourists". For example, the hill tribes in Thailand, remote villagers in Nepal, islanders in the Pacific, the nomadic tribes of the African desert regions, the Native peoples of the American (USA) west, the Aboriginal peoples of Australia's outback, the Old Order Amish and Mennonite communities in Pennsylvania and Ontario (Fagence, 2000, pp.77-78).

1.4.5. Culinary Tourism (Food Tourism, Gastronomy Tourism)

Culinary tourism, also known as food tourism, is the act of pursuing unique and memorable eating and drinking experiences. Culinary tourism, which is seen as a sub-sector of cultural tourism, allows tourists to taste local authentic flavors and participate in traditional eating and drinking activities. Some people go to another destination (going famous restaurants, trying street food, going to the food festivals, visiting the local food market, etc.) to learn how to cook the country's famous dishes from the locals and try dishes from all over the world. For example, traveling to Russia to have vodka shots with meals, Northern Thailand, France, Italy, Korea, Middle East, USA and more for taking a cooking class (Stainton, 2022).

According to the World Tourism Organization (UNWTO, 2012, p.31),culinary tourism is a special interest for the tourist whose interest in food is primary reason influencing his travel behavior and falls on the upper end of the food tourism interest continuum. The culinary tourist is also a cultural tourist. Thus, the obvious overlap of food as a special interest component as well as a cultural component makes the culinary tourist possibly both a special tourist and a cultural tourist.

- ***Wine tourism (vinitourism, enotourism or oenotourismn)*** as a sub-type of Gastronomy Tourism, refers to tourism whose purpose is visiting vineyards, wineries, tasting, consuming and/or purchasing wine, often at or near the source. World's top ten wine producer destinations - Argentina, Chile, France, Georgia, Germany, Italy, Portugal, South Africa, Spain and United States of America (UNWTO, 2023).
- ***Olive oil tourism (oleotourism, OOT)*** is a novel type of tourism, which takes place mainly within rural areas where olives are the principal crop grown, as the repercussion of the cultivation of the olive tree is undeniable on a landscape, cultural, territorial, and economic level (Agüera et al., 2017; Aguilar and García, 2019). Although on occasion, owing to the diversity of complementary activities, the term OOT could also be used to refer to oil tastings in restaurants, visits to museums linked to olive growing, or the purchase of Extra Virgin Olive Oil (EVOO) in specialized shops. (Fernández et al., 2022, p. 2). Popular oleotourism destinations include mainly in Mediterranean basin countries: Spain, Italy, Greece, Morocco, Turkey, Portugal, although cases of oleotourism have also been found in countries such as Argentina, Australia, Chile, Japan, the USA or Lebanon (Campon-Cerro et al., 2017; Millan et al., 2015; Fernández et al., 2019).
- ***Vegan tourism*** is tourism designed to meet the needs of vegans. A number of tourism providers have begun to emerge in popularity specifically designed to suit the needs of vegan tourists. This includes hotels, tour operators and event organizers (Stainton, 2022).

Food is an essential part of any travel, but in recent years, travelers have become more adventurous, wanting to experience local cuisine and experiment with local dishes and food culture, making food a significant part of destinations' image (Du Rand and Heath, 2006).

1.4.6. Spiritual/Religious Tourism (Faith Tourism)

Spiritual/ Religious Tourism is that form of tourism where groups of people or individuals travel for pilgrimage, leisure of missionary purposes (Kruja and Gjyrezi, 2011, p.80). Religious Tourism, so-called "Faith Tourism", is a form of tourism, whereby people of faith travel individually or in groups for reasons related to religion or spirituality in their quest for meaning. It could be under pilgrimage, missionary, or leisure purposes. Many of today's most popular tourist destinations are related to ancient places of worship or to the site of apparent miracles (Jongmeewasin, 2016).

Spiritual/religious tourists can go abroad and connect with the world (i.e. god and religions) and even themselves; It means that spiritual tourism doesn't always connect with specific religions, it can also connect people's bodies, minds and soul; While faith tourism is asking for blessings from Allah based on religious beliefs and beliefs; such destinations include: Middle East, Spain, India, Israel, Mecca, UK and more. While most of the tourists visit pilgrims, churches, holy places, others visit places where there are quiet and calm areas like Greenland (Stainton, 2022).

- ***Christian tourism*** is a sub-sector of faith tourism. It is the largest segment of the faith tourism industry, focusing on tourism that includes religious practices or pilgrimages. Christian tourism activities include visiting destinations of importance to Christian beliefs, such as Bethlehem or Jerusalem. It may also include visits to monasteries, stays in Christian camps, fellowship vacations, missionary travels, crusades, rallies and retreats (Stainton, 2022).

- ***Halal tourism*** is tourism designed according to the needs of Islam. Halal resorts, hotels and attractions popular with strict Muslims will only serve meat that has been suitably handled, will not serve alcohol, and have separate swimming and spa facilities for men and women. Halal tourism is popular in Indonesia, Turkey and Croatia and is growing in many destinations around the world. This is one of the few types of tourism designed around a particular religion (Stainton, 2022).
- ***Kosher tourism*** is tourism designed to meet the needs of Orthodox Jews. Meals are designed according to religious requirements and the accommodation is within walking distance of the Jewish Synagogues (Stainton, 2022).
- ***Pilgrimage tourism*** is a branch of religious tourism. Pilgrimage, as a part of religious tourism, is the act of moving from one place to another, often traveling through foreign lands; an ordered march of a group of people, usually with religious connotation. It is a ritual journey with a hallowed purpose; every step along the way has meaning; the pilgrim knows that the journey will be difficult and that life-giving challenges will emerge. Basically, sacred destinations and places of pilgrimage associate with the mainstream faiths: Christianity, Islam, Hinduism, Buddhism, Judaism, and Sikhism (Jongmeewasin, 2016).

The pilgrimage functions as an intermediary between the natural world and the cultural world, and between the natural world and the supernatural world. This is not a holiday, but a journey of transformation where important changes are experienced; new perceptions are given, perceptions are deepened, old and new places are visited in the hearts, blessings are received, healings are given, and life is viewed from different eyes on the return of pilgrimage and nothing will be the same as before. While the ultimate spiritual purpose of pilgrimage is to seek eternal truth and become one with God, it is a journey for religious reasons, externally to a holy place and internationally for spiritual purposes and inner understanding. Home to relics, the sanctuary encourages ritual practices and often attracts religious travelers or pilgrims who return home with memories, marking time and expanding the space of the journey (Jongmeewasin, 2016).

1.4.7. Ecotourism

Ecotourism is a type of nature-based special interest tourism (Hassan, 2012, p.2). According to The International Ecotourism Society (1990) ecotourism is "responsible travel to natural areas that conserves the environment and improves the wellbeing of local people" (Kruja and Gjyrezi, 2011, p.80). Ecotourism is a sub-branch of tourism centered on the tourist's desire for engagement in a relatively nature-based environment where the tourist and the supporting facilities have lower impact on the environment (Rymer, 1992). In general, ecotourism as special interest tourism has some specific features, like: services or products based on nature, dynamic management for reducing environmental blows, a component of environment education and contribution towards the conservation of natural environment through closer liaison with the host communities (Buckley, 2003; Hassan, 2012, p.2).

Ecotourists are clustered in another five different groups according to their characteristics. These are 'nature exclusive', 'social and nature oriented', 'autonomous and nature oriented', enthusiastic generalists' and 'less passionate' (Wallace and Smith, 1997).

Special interest tourism has exploded and a significant market share has attracted high-income people who prefer long and expensive tours characterized by diversity and difficulty (Alpine, 1986). The links between special interest tourism and the environment developed in various stages in the mid-twentieth century, so that typologies such as ecotourism, nature-based tourism or wildlife tourism developed under one umbrella (Hudman, 1991).

Fredman and Tyrväinen (2010) proposed one of the comprehensive definitions of nature-based tourism that is human activities occurring when visiting nature areas outside the person's ordinary neighborhood. Along with their definition, they confirmed that nature-based tourism is becoming more specialized and diversified, which highlights the importance of further investigating the process of this specialization in different contexts (Soleimani et al., 2019, p. 2301).

1.4.8. Sports Tourism

Tourism and sports are interrelated and complementary. Sports – as a professional, amateur or leisure activity – involves a considerable amount of traveling to play and compete in different destinations and countries. Major sporting events, such as the Olympic Games, football and rugby championships have become powerful tourism attractions in themselves – making a very positive contribution to the tourism image of the host destination (UNWTO, 2023). Sport tourism, which is defined as sport-based travel away from the home and where sport based and non-sport activities are integrated (Ito and Higham, 2020).

Sports tourism occurs when tourists travel aboard to watch or take part in sports or sports events. Sporting events include things like the Olympics, Formula One, and Football World Cup. Sporting activities may include playing football, golf, volleyball, taking part in car races, yoga, surfing, watch tennis match, etc. Popular sport tourism destinations include Germany, Italy, Middle East, Spain, Belgium and more (Stainton, 2022).

1.4.9. Educational Tourism

Education tourism covers those types of tourism which have as a primary motivation the tourist's engagement and experience in learning, self-improvement, intellectual growth and skills development. Education Tourism represents a broad range of products and services related to academic studies, skill enhancement holidays, school trips, sports training, career development courses and language courses, among others (UNWTO, 2019). According to Ritchie (2003) educational tourism is defined by the desire to learn. Education tourism can take a variety of directions and serve a diversity of visitor interests, "such as satisfying curiosity about other people and their language and culture; stimulating interest in art, music, architecture or folklore; inspiring concerns for natural environments, landscapes, flora and fauna; or, deepening the fascination of cultural heritage and historic places" (Kalinowski & Weiler 1992, p 17). Educational tourism goes beyond a curiosity, interest or fascination for a particular topic but includes an element of organized learning (Kalinowski & Weiler 1992).

- ***Linguistic tourism*** is tourism which involves learning a language. This could be part of a formal course or it could be part of wider cultural experience. Popular destinations for language learning include much of Spain and South America to learn Spanish, France for French and Italy for Italian (Stainton, 2022).
- ***TEFL tourism*** is the act of travelling for the purposes of teaching English as a foreign language as part of a wider tourism experience. As defined in my PhD thesis a TEFL tourist can be defined as 'a person who travels outside of their usual environment to teach English as a foreign language, whose role shifts between tourist, educator and educatee at various points in their trip'. Popular

TEFL Tourism destinations include China, Thailand, Costa Rica and Mexico. TEFL tourism is one of the types of tourism that many people are unfamiliar with, but it is definitely a sizeable industry! (Stainton, 2022).

1.4.10. Photographic Tourism

Photographic tourism is the act of visiting a particular destination with the intention of capturing it on camera. This branch of special interest tourism is often associated with picturesque destinations that tourists want to photograph for their unique appearance, unusual imagery or personal interest. The scope of photography ranges from landscapes, artworks, cultural imagery, wildlife, food and architectures. Photographic tourism is one of the most fun types of tourism to participate in and helps capture all those wonderful memories (Stainton, 2022).

1.4.11. Dark Tourism (Black Tourism, Thana Tourism, or Grief Tourism)

Dark tourists will travel to places where certain (bad) things happen. Dark tourism usually means tourists visiting places where a high number of deaths occur; some people come to these places to commemorate and honor their loved ones who were involved, while others go there to learn more about what happened and to see the experience. They can visit places such as battlefields, prisons, cemeteries, Mount of Olives in Jerusalem, Recoleta Cemetery in Buenos Aires and Arlington National Cemetery in Washington (grave tourism, cemetery tourism or tombstone tourism), castles, Chernobyl, Nevada test site and Hiroshima (nuclear tourism), Kathmandu earthquake or New Orleans hurricane (disaster tourism), Jewish concentration camps, such as Auschwitz (holocaust tourism), Pablo Escobar (Narcho tourism) and the 9/11 memorial site (Stainton, 2022).

1.4.12. Celebrity Tourism

Celebrity tourism is tourism where celebrities are the main attraction. Tourists may try to visit a famous tourism destination or attraction because a celebrity is currently or has been there before. Many destination management organizations (DMOs) use celebrity tourism to promote a destination or attraction. Tourists may also try to visit places centered around a celebrity, even if the celebrity has never been there. Popular celebrity tourism destinations/activities include Hollywood, the Cannes film festival, Harry Potter studio and Madam Tussauds (Stainton, 2022).

1.4.13. Film Tourism

Film tourism, also referred to as screen tourism is a sub-sector of the cultural tourism industry. It focuses on the concept of filmmaking and production, in which tourists try to visit places featured in films or film recordings. Popular film-driven tourism destinations include The Beach in Thailand, Dubrovnik and Northern Island as featured in the Game of Thrones and Petra which is famously known for its use in Indiana Jones and the Last Crusade. Film set examples include the Harry Potter Studios in Hertfordshire, Universal Studios in California and Pinewood Studios in London (Stainton, 2022).

2. TRENDS IN SPECIAL INTEREST TOURISM FOR WOMEN TOURISTS

Trend in tourism demands have clearly been changing over years. They have become increasingly demanding and focused on specific interest, derived from very specific and personal motivations to adapt these motivation and interests to tourism products and services for women. For example, Insta tourism, virtual tourism, Pokemon-Go tourism (gaming tourism), space tourism (astro tourism) and so on.

2.1. Insta Tourism

Backaler (2018, p.1) defines social media influencers as "individuals who can influence the actions/decisions of a loyal group of local online followers according to their area of expertise" (Asan, 2022, p.978). The concept of influencer includes other alternative terms such as "creator", "talent", "YouTuber", "blogger", "vlogger", "Instagrammer" and "phenomenon" (Brown & Hayes, 2008; Veirman et al., 2017; Backaler, 2018; Asan, 2022, p.978). In tourism the term of influencer in the social media platform Instagram is "Insta tourism" which is alternative term.

Insta tourism is a new form of tourism that emerged in response to the use of the social media platform Instagram. More than a billion people use Instagram every month and share photos with their followers from all over the world. Some Instagram users predominantly share content with their friends and family, while others are charged as "Influencers" with thousands of loyal followers. Users are often inspired by the photos they have been exposed to through the social network, which has seen a surge in tourism to the places featured in said images, especially those shared by large-scale Influencers (Stainton, 2022).

Marketers have developed the "influencer marketing" approach by collaborating with influencers who have the power to influence consumer behavior (Brown & Hayes, 2008; Gretzel & Yoo, 2013; Veirman et al., 2017; Ge & Gretzel, 2018; Asan, 2022, p.978).

- ***Travel influencers***, inform their followers about unknown destinations, geographies, cultures, and travel formalities such as passport, visa, transportation services, foreign exchange transactions, and issues not included in formal tourist information sources. Thus, as a reliable source of information, it is accepted that followers are influential in their decisions (Asan, 2022, p.980). According to Asan (2022, pp. 978-994)'s research, which aims to explain "the impacts of travel influencers on their followers in the context of bicycle touring as a sustainable special interest tourism type", travel influencers in Turkey on their followers can be explained into four factors: informative impacts, motivating impacts, impacts as a role model, and communal impacts. Accordingly, travel influencers inform their followers about travel procedures and destinations with their expertise and trustworthiness. Also, travel influencers motivate their followers to participate in touristic mobility by sharing their experiences in an aesthetic way. In addition, travel influencers' challenging, extraordinary, and unique experiences broaden their followers' horizons. Finally, travel influencers establish social bonds not only with their followers but also travel communities, and they support touristic socialites.

2.2. Virtual Tourism

Virtual tourism is the act of using technology to simulate a travel destination and its features. Although virtual tourism is unlikely to replace physical visits to a destination, it has been used as a powerful marketing tool. It is also used to improve user experience in various attractions. Virtual tourism most commonly consists of film and image sequences, although it can also include 3D and sensory experiences (Stainton, 2022).

Virtual reality technology is increasingly used in hospitality and tourism guest and marketing strategies, such as operational booking processes, virtual tours, translation and other tools that expand market participation, help customers plan and fulfill reservations, and maximize overall satisfaction (Nayyar, 2018). Moreover, training organizations have successfully used online platforms in cooperation with Taiwanese businesses' factory tours in cultural tourism education (Chiao et al., 2018), and virtual field trips (Patiar et al., 2017) without any physical travel. Finally, the use of virtual reality has become even more important, with changes in the hospitality and tourism industry due to COVID-19, subsequent changes in industry training mechanisms, and the potential impact of work-integrated learning internship placements for the next generation of students, graduates and industry professionals (Bilsland, 2020).

2.3. Pokemon-Go Tourism (Gaming Tourism)

The world famous gaming example is Pokémon GO tourism; an open, mobile multi-player augmented reality (AR) based game since its release in July 2016. AR is an integration of the real world and the virtual world, with the aim of providing additional information about something in the real world with information displayed in the virtual world. Pokémon GO is an app where players can collect points by walking around in the real world, while holding up their smartphone, in a quest to find, catch and collect in their phone camera view, virtual characters, called "pocket monsters", which are located near and linked to specific geographical locations. As of April 2017, it already had over 65 million players and continues to grow rapidly. With these numbers, it became the most successful game of all time. It is praised internationally by healthcare professionals for getting people out of their homes and motivating them to get outside and connect with others (Nayyar, 2018, pp. 156-157).

Gastronomy providers were already taking advantage of the Pokémon GO craze, and there were more ways Pokémon GO can benefit the tourism industry. The game makes use of "PokéStops", strategically placed at sightseeing locations, such as monuments and other public areas of interest; while in the game, these locations are displayed with a photo and a brief description. Tour operators can design new tours according to the location of PokéStops or the known locations of rare collectable Pokémons. During the tour, players can expand their knowledge of the world around them in a fun and spontaneous way as they encounter PokéStops of interesting historical, artistic or cultural value. Pokémon GO has made AR technology popular among consumers, and many variations on this theme can now be easily envisioned and introduced. This new technology can bring new opportunities to the Hospitality and Tourism industry (Nayyar, 2018, p. 157).

2.4. Space Tourism (Astro Tourism)

Astro tourism is a type of space that involves an interest in space and also refers as space tourism. Trip to space is one of the most innovative types of tourism (Virgin Galactic and SpaceX) and space tourism can also include visiting space focused museums, watching rocket launches or travelling to destinations popular for stargazing (Stainton, 2022). Astro tourism is a form of tourism based on the special interest of tourists in sky-related activities such as sky observation and astrophotography, and is also classified as a subcategory of nature-based tourism (Cater, 2010; Weaver, 2011). Scholars assert that astronomy and space have a long-standing association with the field of science, and the intellectual stimulation that results in creating unique experiences, as well as engaging in mental stimuli such as 'learning, exploring, discovering, creating or imagining' (Beard & Ragheb, 1983; Cater, 2010; Soleimani et al., 2019, p. 2299-2300).

According to Soleimani et al., (2019, p. 2302-2303) it is important to understand the preferences of astro tourists due to success of this market. The previous literature indicated that astro tourists can be classified in special-interest and nature-based tourists segments, depending on their primary motivation and interest in astronomical attractions (Fayos-Sola, Marin, & Jafari, 2014). Special interest tourists range from amateur (D'Auria, 2012) and professional astronomers (Krco, 2005) to astrophotographers (Thommes, 2012) where they travel to different places to see astronomical phenomena (Fayos-Sola et al., 2014). Bill Kramer, an eclipse chaser, has listed 138 such enthusiasts (Kramer, 2012) under the name of Eclipse Chaser Log Summary. These and many other people travel to different places to chase eclipses as well as to gather at dark-sky sites for a few nights under stars. For example, 90% of amateur astronomers were highly interested in observing objects in the night sky. The astro tourism market often comprises socially and environmentally aware, highly educated tourists and visitors who travel both to learn and to achieve personal and social goals (Pine & Gilmore, 1999, Storksdieck, 2002).

By contrast, nature-based tourists are those who are generally interested in natural settings and environment. Especially, astronomy is an outdoor nature hobby for tourists (MacRobert, 2012). Staying in contact with the natural environment makes a tourist go for rural tourism (Burns & Novelli, 2008). Natural phenomena like solar eclipses gather millions of nature-based tourists in many destinations that are appropriate for observation (Editors of Sky & Telescope, 2006). These observations provide novel experiences for the tourists. Therefore, in developing a nature-based tourism product, destination marketers can propose new ideas for organizing new tours such as aurora observing (Soleimani et al., 2019, p. 2303).

CONCLUSION AND SUGGESTIONS

Special interest tourism is defined as travelers visiting a particular destination with the motivation to fulfill a particular interest that can be pursued only at that visited place. Moreover, SIT is about active personal growth and gaining new experiences, rather than escaping and relaxing from reality. While activities can be fun and therapeutic, they often have educational or personal growth dimensions, such as developing a fitter, healthier body or a more creative or vibrant mind. Special interest tourism products, therefore, offer a wide range of options that can be preferred by woman tourists.

As a conceptual study, this chapter contributes to readers (e.g. students, lecturers, professionals, and others) and international literature to understand the definition of special interest tourism and the difference between general interest tourism. Especially, it introduces the characteristics, scope, and wide variety of types of special interest tourism for women tourists such as adventure tourism, medical tourism, rural tourism, cultural tourism, culinary tourism, religious tourism, ecotourism, sports tourism, educational tourism, photographic tourism, dark tourism, celebrity tourism, and film tourism. Moreover, it makes them learn new types of trends in special interest tourism for women tourists like insta-tourism, virtual tourism, Pokémon-go tourism (gaming tourism), and space tourism (Astro-tourism). Further studies may find different types of SIT that have feelings of empowerment and benefits for women.

REFERENCES

Abdillah, Y., Supriono, S., & Supriyono, B. (2022). Change and innovation in the development of Balinese dance in the garb of special interest tourism. *Cogent Social Sciences*, *8*(1), 2076962. doi:10.1080/23311886.2022.2076962

Agüera, F. O., Cuadra, S. M., López-Guzmán, T., & Morales, P. C. (2017). Estudio de la demanda existente en torno al oleoturismo. El caso de Andalucía. *Cuadernos de Turismo*, (39), 437–453. doi:10.6018/turismo.39.290641

Aguilar, M. J. C., & García, E. M. (2019). Olive oil tourism and rural development: Advances and challenges in the case of the province of jaén (Andalusia, Spain). *Journal of Tourism and Development*, (32), 255–264.

Akinci, Z., & Kasalak, M. A. (2016). Management of Special Interest Tourism in Terms of Sustainable Tourism. In Global Issues and Trends in Tourism. St. Kliment Ohridski University Press.

Asan, K. (2022). Measuring the impacts of travel influencers on bicycle travelers. *Current Issues in Tourism*, *25*(6), 978–994. doi:10.1080/13683500.2021.1914004

Backaler, J. (2018). *Digital influence: Unleash the power of influencer marketing to accelerate your global business*. Palgrave Macmillan. doi:10.1007/978-3-319-78396-3

Beard, J., & Ragheb, M. (1983). Measuring leisure motivation. *Journal of Leisure Research*, *15*(3), 219–228. doi:10.1080/00222216.1983.11969557

Belias, D., Velissariou, E., Kyriakou, D., Varsanis, K., Vasiliadis, L., Mantas, C., Sdrolias, L., & Koustelios, A. (2018). Tourism consumer behavior and alternative tourism: the case of agrotourism in Greece. In *Innovative Approaches to Tourism and Leisure* (pp. 465–478). Springer. doi:10.1007/978-3-319-67603-6_35

Bilsland, C., Nagy, H., & Smith, P. (2020). Virtual internships and work-integrated learning in hospitality and tourism in a post-COVID-19 world. *International Journal of Work-Integrated Learning*, *21*(4), 425–437.

Brotherton, B., & Himmetoğlu, B. (1997). Beyond Destinations—Special Interest Tourism. *Anatolia*, *8*(3), 11–30. doi:10.1080/13032917.1997.9687118

Brotherton, B., & Himmetoglu, B. (1997). Beyond destinations— special interest tourism. *Anatolia: An International Journal of Tourism and Hospitality Research, 8*(3), 11–30.

Brown, D., & Hayes, N. (2008). *Influencer marketing: Who really influences your customers?* Routledge. doi:10.4324/9780080557700

Burns, P. M., & Novelli, M. (2008). *Tourism development: Growth, myths, and inequalities*. CABI Publishing. doi:10.1079/9781845934255.0000

Campón-Cerro, A. M., Folgado-Fernández, J. A., & Hernández-Mogollón, J. M. (2017). Rural destination development based on olive oil tourism: The impact of residents' community attachment and quality of life on their support for tourism development. *Sustainability (Basel)*, *9*(9), 1624. doi:10.3390u9091624

Cater, C. I. (2010). Steps to space; opportunities for astro-tourism. *Tourism Management, 31*(6), 838–845. doi:10.1016/j.tourman.2009.09.001

Chiao, H.-M., Yu-Li, C., & Wei-Hsin, H. (2018). Examining the usability of an online virtual tour-guiding platform for cultural tourism education. Journal of Hospitality, Leisure, Sports and Tourism Education, 23, 29–38.

D'Auria, T. (2012). *2013 winter star party*. Southern Cross Astronomical Society (SCAS) Inc. Retrieved from http://www.scas. org/wsp.html

Doran, A. (2016). Empowerment and women in adventure tourism: A negotiated journey. *Journal of Sport & Tourism, 20*(1), 1–24. doi:10.1080/14775085.2016.1176594

Douglas, N., Douglas, N., & Derret, R. (Eds.). (2001). *Special Interest Tourism*. Wiley.

Du Rand, G. E., & Heath, E. (2006). Towards a framework for food tourism as an element of destination marketing. *Current Issues in Tourism, 9*(3), 206–234. doi:10.2167/cit/226.0

Eurostat. (1998). *Community Methodology on Tourism Statistics*. European Commission.

Fagence, M. (2000). Ethnic Tourism in Developed Countries: Special Interest or Specialized Mass Tourism? *Tourism Recreation Research, 25*(2), 77–87. doi:10.1080/02508281.2000.11014913

Fayos-Sola, E., Marin, C., & Jafari, J. (2014). Astro-tourism: No requiem for meaningful travel. *Pasos (El Sauzal), 12*(4), 663–671. doi:10.25145/j.pasos.2014.12.048

Fernández, J. I. P., Montilia, J. C., & Hidalgo, I. C. (2019). Introducing olive-oil tourism as a special interest tourism. *Heliyon, 5*, 12. PMID:31872132

Fernández, J. I. P., Montilla, J. C., Hidalgo, I. C., & Fernández, M. C. P. (2022). Evaluating olive oil tourism experiences based on the segmentation of demand. *International Journal of Gastronomy and Food Science*, 27.

Fredman, P., & Tyrväinen, L. (2010). Frontiers in nature-based tourism. *Scandinavian Journal of Hospitality and Tourism, 10*(3), 177–189. doi:10.1080/15022250.2010.502365

Garanti, Z. (2022). Alternative and special interest tourism to mitigate the effects of tourism seasonality: The debate from Cyprus. *Worldwide Hospitality and Tourism Themes, 14*(5), 451–460. doi:10.1108/WHATT-07-2022-0083

Ge, J., & Gretzel, U. (2018). A taxonomy of value co-creation on Weibo – A communication perspective. *International Journal of Contemporary Hospitality Management, 30*(4), 2075–2092. doi:10.1108/IJCHM-09-2016-0557

Gretzel, U., & Yoo, K. H. (2013). Premises and promises of social media marketing in tourism. In S. McCabe (Ed.), *The Routledge handbook of tourism marketing* (pp. 491–504). Routledge.

Hall, C. M., & Weiler, B. (Eds.). (1992). *Special Interest Tourism*. Bellhaven Press.

Harron, S., & Weiler, B. (1992). Ethnic tourism. In B. Weiler & C. M. Hall (Eds.), *Special Interest Tourism*. Belhaven Press.

Hassan, A. (2012). Developing Country Studies 'Package Eco-tour' as Special Interest Tourism Product-Bangladesh Perspective. *Developing Country Studies*, *2*(1), 1–9.

Isaac, R. K., & Eid, T. A. (2019). Tourists' destination image: An exploratory study of alternative tourism in Palestine. *Current Issues in Tourism*, *22*(12), 1499–1522. doi:10.1080/13683500.2018.1534806

Ito, E., & Higham, J. (2020). Supplemental tourism activities: A conceptual framework to maximise sport tourism benefits and opportunities. *Journal of Sport & Tourism*, *24*(4), 269–284. doi:10.1080/14775085.2020.1850322

Jongmeewasin, S. (2016). *Religious Tourism, Pilgrimage, and Cultural Tourism*. Silpakorn University. Retrieved from https://www.academia.edu/12252237/Religious_Tourism_Pilgrimage_and_Cultural_Tourism

Kalinowski, K. M., & Weiler, B. (1992). Review. Educational Travel. In C. M. Hall & B. Weiler (Eds.), *Special Interest Tourism*. John Wiley & Sons, Inc.

Karasoula, K. (2010). *The competitiveness of Greece in the touristic sector. The case of prefecture Achaea and the development of the specific and alternative forms of tourism* [MBA thesis]. University of Patras.

Kramer, B. (2012). *Eclipse chasers log. Eclipse chasers log summary*. Retrieved from http://www.eclipsechasers.com/ tseChaserLogSums.php

Krco, M. (2005). *Curious about astronomy? Ask an Astronomer*. Retrieved from http://curious.astro.cornell.edu/question. php?number=657

Kruja, D., & Gjyrezi, A. (2011). The Special Interest Tourism Development and the Small Regions. *Turizam*, *15*(2), 77–89. doi:10.5937/Turizam1102077K

MacRobert, A. (2012). How to start right in astronomy. *Sky and Telescope Magazine*. Retrieved from http://www. skyandtelescope.com/howto/basics/How_to_Start_Right_in_Astronomy.html

Millan, M. G., Amador, L., & Arjona, J. M. (2015). El oleoturismo: una alternativa para preservar los paisajes del olivar y promover el desarrollo rural y regional de Andalucía (Espana). *Rev. Geogr. Norte Gd., 60*, 195–214.

Mitten, D. (1992). Empowering Girls and Women in the Outdoors. *Journal of Physical Education, Recreation & Dance*, *63*(2), 56–60. doi:10.1080/07303084.1992.10604117

Nayyar, A., Mahapatra, B., Le, D., & Suseendran, G. (2018). Virtual Reality (VR) & Augmented Reality (AR) technologies for tourism and hospitality industry. *International Journal of Engineering & Technology*, *7*(2.21), 156-160.

Patiar, A., Ma, E., Kensbock, S., & Cox, R. (2017). Students' perceptions of quality and satisfaction with virtual field trips of hotels. *Journal of Hospitality and Tourism Management*, *31*, 134–141. doi:10.1016/j.jhtm.2016.11.003

Pine, B. J., & Gilmore, J. H. (1999). *The experience economy, work is theatre and every business a stage*. Harvard Business School Press.

Ritchie, B. W. (2003). *Managing Educational Tourism.* Channel View Publications. doi:10.21832/9781873150528

Robinson, M., & Novelli, M. (2007). Niche tourism: An introduction. In M. Novelli (Ed.), *Niche tourism: Contemporary issues, trends and cases* (pp. 1–11). Elsevier.

Rymer, T. M. (1992). Growth of U.S. Ecotourism and Its Future in the 1990s. *FIU Hospitality Review, 10*(1), 1–10.

Sheng, C.-W., Shen, M.-J., & Chen, M.-C. (2008). An exploratory study of types of special interest tour preferences and preference demographic variables analysis. *International Journal of Culture, Tourism and Hospitality Research, 2*(3), 271–284. doi:10.1108/17506180810891627

Skwirczy ´nska, E., Piotrowiak, M., Ostrowski, M., Wróblewski, O., Tejchman, K., Kwiatkowski, S., & Cymbaluk-Płoska, A. (2022). Welfare and Self-Assessment in Patients after Aesthetic and Reconstructive Treatments. *International Journal of Environmental Research and Public Health, 19*(18), 11238. doi:10.3390/ijerph191811238 PMID:36141511

Soleimani, S., Bruwer, J., Gross, M. J., & Lee, R. (2019). Astro-tourism conceptualisation as specialinterest tourism (SIT) field: A phenomonological approach. *Current Issues in Tourism, 22*(18), 2299–2314. doi:10.1080/13683500.2018.1444021

Stainton, H. (2022). *150 types of tourism! The ultimate tourism glossary.* Retrieved from https://tourismteacher.com/types-of-tourism-glossary/

Stephano,R.M.(2022).*Top10MedicalTourismDestinationsintheWorld.*GlobalHealthcareResources&Medical Tourism Association. Retrieved from https://www.magazine.medicaltourism.com/article/top-10-medical-tourism-destinations-world

Thommes, J. (2012). *Jim's astrophotography. Astrophotography by Jim Thommes.* Retrieved from http://www.jthommes. com/Astro

Trauer, B. (2006). Conceptualizing special interest tourism—Frameworks for analysis. *Tourism Management, 27*(2), 183–200. doi:10.1016/j.tourman.2004.10.004

Trauer, B., & Ryan, C. (2005). Destination image, romance and place experience – an application of intimacy theory in tourism. *Tourism Management, 26*(4), 481–491. doi:10.1016/j.tourman.2004.02.014

Veirman, M., Cauberghe, V., & Hudders, L. (2017). Marketing through Instagram influencers: The impact of number of followers and product divergence on brand attitude. *International Journal of Advertising, 36*(5), 798–828. doi:10.1080/02650487.2017.1348035

Wallace, G. N., & Smith, M. D. (1997). A comparison of motivations, preferred management actions, and setting preferences among Costa Rican, North American and European visitors to five protected areas in Costa Rica. *Journal of Park and Recreation Administration, 15*, 59–82.

Weaver, D. (2011). Celestial ecotourism: New horizons in nature-based tourism. *Journal of Ecotourism, 10*(1), 38–45. doi:10.1080/14724040903576116

Wen, J., & Wu, M. Y. (2020). How special is special interest tourism – and how special are special interest tourists? A perspective article in a Chinese context. *Current Issues in Tourism*, *23*(16), 1968–1972. doi:10.1080/13683500.2020.1750575

World Tourism Organization (2012). *Affiliate Members Global Report, Volume 4 - Global Report on Food Tourism*. UNWTo. doi:10.18111/9789284414819

World Tourism Organization. (2019). *UNWTO Tourism Definitions*. UNWTO. doi:10.18111/9789284420858

World Tourism Organization. (n.d.-a). *Gastronomy and Wine Tourism*. Retrieved from https://www.unwto.org/gastronomy-wine-tourism

World Tourism Organization. (n.d.-b). *Mountain tourism*. Retrieved from https://www.unwto.org/mountain-tourism

World Tourism Organization. (n.d.-c). *Rural tourism*. Retrieved from https://www.unwto.org/rural-tourism

World Tourism Organization. (n.d.-d). *Sports tourism*. Retrieved from https://www.unwto.org/sport-tourism

Yang, L. X. (1993). *An analysis of R.O.C. nationals tourism consumer behavior – the case of the residents of Taipei* [Unpublished Master's thesis]. Department of Business Administration, National Sun Yat-Sen University, Kaohsiung.

Chapter 20
Challenges and Issues of Woman Employees in Tourism

Erdem Korkmaz
Anadolu University, Turkey

Seher Gülenç
Anadolu University, Turkey

Dönüş Çiçek
Anadolu University, Turkey

ABSTRACT

Today, it is well-known that all individuals may experience and related to the nature of the job; however, these may sometimes be challenges that woman face due to their biological characteristics. Mostly dominated by female employees, the tourism industry has longer and more tiresome working hours, has seasonal employment relationships, requires less educated and qualified labor compared to the other industries, and has lower unionization and labor contract rates. The jobs in the tourism industry are defined as man's and woman's jobs, and services, such as cleaning, laundry, making up bed, guest greeting, and cooking similar to domestic chores, are performed by women. These challenges have a direct impact on operational efficiency and functioning, along with the status of women in tourism employment, is important for improving female employment in tourism. A holistic analysis was made in this chapter based on national and international literature on the challenges female employees face in the tourism industry.

INTRODUCTION

The extent of the tourism industry has been continuously expanding. Accordingly, the industry offers employment for 10 out of every 100 individuals. According to the 2019 International Labor Organization (ILO) data, the male labor force participation rate was 74.2%, while the female labor force participation rate was only 47.2%. Considering the tourism industry, the ratio of women workers is 54% of the whole

DOI: 10.4018/978-1-6684-8417-3.ch020

employees (Nalçacı, 2020). Besides, 27.230.000 individuals were employed in Türkiye in 2019, and the female labor force participation rate was 31.9% (Ardıç Yetiş and Çalışkan, 2020). According to the information gathered by Turizmdatabank from the Turkish Statistical Institute (TUIK) and Social Security Institution (SGK), more than 400.000 women are employed in the tourism industry. Accordingly, 95.000 women work in the hospitality industry, 256.000 women work in the food & beverage industry, 17.000 women work in travel agencies, and 12.400 women work in the aviation industry (Karabulut, 2022).

Tourism is one of the industries that boost development, and it also plays a significant role in empowering women (Pritchard et al., 2007: 9). The tourism industry, which increases the visibility of women in society by providing benefits in terms of employment, income, education, and social status (Chant, 1997), also empowers women by supporting their involvement in the labor force, entrepreneurship, and leadership (Kodagoda and Jayawardhana, 2022). Within this context, it may well be concluded that tourism is crucial for women employees' socioeconomic independence. In fact, some international organizations (The United Nations, International Labor Organization, Organization for Economic Cooperation and Development, etc.) consider the tourism industry as a field of opportunity for facilitating women's participation in the labor force, and thus, minimizing unemployment rate for women (Ardıç Yetiş and Çalışkan, 2020).

Today, it is well-known that women face numerous challenges in work life. These may sometimes be challenges that all individuals may experience and related to the nature of the job; however, these may sometimes be challenges that woman face due to their biological characteristics. Mostly dominated by female employees (Koç and Uşaklı, 2022), the tourism industry has longer and more tiresome working hours, has seasonal employment relationships, requires less educated and qualified labor compared to the other industries, and has lower unionization and labor contract rates. Tourism involves activities such as tour guiding, accommodation, food catering, etc., which are mostly fulfilled by women on a daily basis. This leads to the formation of a mostly gender-based employment structure in tourism and determines the role and position of women who will be employed in the tourism industry (Ardıç Yetiş and Çalışkan, 2020). In other words, the jobs in the tourism industry are defined as man's and woman's jobs, and services, such as cleaning, laundry, making up bed, guest greeting, and cooking similar to domestic chores, are performed by women (Akoğlan, 1996; Aykaç, 2009). Therefore, the specific work conditions of the tourism industry have a negative impact on the female labor force, and female employees face various challenges, from job seeking to promotion, because of the nature and characteristics of the industry. Knowing these challenges, which have a direct impact on operational efficiency and functioning, along with the status of women in tourism employment (Demirdağ, 2022), is important for improving female employment in tourism. From this point of view, a holistic analysis was made in this chapter based on national and international literature on the challenges female employees face in the tourism industry.

THE CHALLENGES FACED BY WOMEN WORKING IN THE TOURISM INDUSTRY

Numerous studies in tourism literature focus on the challenges female employees face in the tourism industry. While some of these studies focused on a single challenge, such as mobbing (Kensbock, 2013; Akbolat, Yılmazer and Tutar, 2014; Koyuncu and Sevin, 2017), glass ceiling (Kattara, 2005; Aydın Tükeltürk and Şahin Perçin, 2008; Anafarta, Sarvan and Yapıcı, 2008; Ciğer and Uyar, 2016; Akdağ and Üçyıldız, 2021), sexual harassment (Reguera and García-Izquierdo, 2021; Worke et al., 2021), and wage discrimination (Munoz-Bullon, 2009; Campos-Soria et al., 2015; Çelik Uğuz and Topbaş, 2016; Kaya and

Topbaş, 2019), others discussed these problems with a holistic approach (Table 1). Performed in different periods with different samples, these studies reported that the women working in the tourism industry face various and similar challenges worldwide. For example, Demirkol, Fidan and Pelit (2004) listed the challenges faced by female employees as excessive workload and work shifts, lack of authorization, lack of union rights, inadequate on-the-job training and rotation opportunities, inadequate employment security, and harassment. Yanardağ and Avcı (2012) found that female employees experienced similar problems, such as wage inadequacy, irregular work schedules, lack of social security, etc. Birdir, Derinalp and Çanakçı (2013) reported that low wages, extended work hours, and unequal wage distribution were the primary issues faced by the kitchen staff in hotel facilities. Çelik and Akar Şahingöz (2018) stated that female cooks are exposed to discrimination in employment, and they reported gender-based division of labor, stereotyped and biased behaviors, and that they are forced to work in lower-level positions regardless of their skills. In another study, these challenges were listed as gender discrimination, division of labor as men's and women's jobs, promotional challenges, unbreakable glass ceilings, wage discrimination, and having no rights to take legal action (Balcı and Aslan, 2022).

Similar challenges were also found in the international literature. In a study they conducted in Sukkur, Pakistan, Avais et al. (2014) found that sexual harassment and being unable to spend enough time with the family were the most significant challenges. Gender discrimination, workload, stress, lower wages, promotional challenges, and working hours were found to be the most significant challenges faced by women working at fast-food restaurants in Egypt (Faten Ali, 2014). Orido (2017) reported that the challenges faced by female chefs in Kenya were patriarchal social structure, gender discrimination, sexual harassment, unsanitary work environment, workload, and work-life balance. Garrigos, Haddaji, Segovia, and Signes (2019) stated that the female chefs were having difficulties with work-life balance, they were forced to postpone getting married and starting a family, and they had to sacrifice their personal lives and travels.

Categorizing the challenges women face in the tourism industry is crucial in indicating the necessary changes and regulations for an efficient solution. Within this context, İstanbullu Dinçer et al. (2016) stated that the most significant challenges women face in the tourism industry are discrimination, glass ceiling syndrome, and sexual harassment. Karaset and Tanrısever Yiğit (2017) investigated these issues through working hours and wage system, profession and position, sexual and emotional harassment, training and orientation, promotion and glass ceiling syndrome, and stereotyped behaviors. While Güçer et al. (2018) categorized these as psychological, physical, and personal factors, Eryılmaz (2022) categorized these as discrimination, familial, work, and time issues. Albayrak (2019) categorized the challenges faced by women working at food & beverage facilities as discrimination (promotion, wage, social rights), mobbing and emotional harassment, preferences and perceptions forced upon women, organizational culture and policies, failure to participate in the informal and external communication network, and sexual harassment. This chapter discusses the challenges women face in the tourism industry as discrimination, career obstacles, workplace violence, long/rotatory working hours, and social and familial challenges.

Discrimination

Discrimination of individuals based on their distinct economic, social, cultural, and demographic characteristics is one of the most significant issues in social life (Alparslan, Çetinkaya Bozkurt and Özgöz, 2015). Discrimination is defined as being paid lower or being treated differently because of one's race, gender, religious views, or any other personal traits that do not have any impact on work productivity, despite having equal or similar skills, educational background, experience, and productivity (Loureiro et al., 2004). Discrimination can be experienced in recruitment, promotion, dismissal, benefiting from on-the-job training opportunities, retirement, and salary payments (Tokgöz, 2011: 237). It may well be argued that the most common discrimination issues at hotel facilities are gender discrimination, wage discrimination, and discrimination during recruitment.

Gender discrimination involves evaluations based on gender rather than personal skills or job performance. Despite all legal regulations made against gender discrimination in various industries and rapidly evolving social structures at an international level, gender discrimination still exists in work life. Social gender roles have a great impact on gender discrimination. Gender discrimination arises because of the duties expected to be fulfilled by women based on their social roles and how these duties affect their social hierarchies and economic conditions (Tunca, 2021). Gender-based discrimination is among the most important reasons for the challenges women face in work life regarding access to employment opportunities, wages, work conditions, social security, etc. It also stands in the way of sustainable development (Aydoğdu, 2021).

Discrimination takes place at two levels: (a) horizontal discrimination refers to not recruiting female employees or recruiting female employees only for lower-level positions with no career opportunities, and (b) vertical discrimination refers to lower rates of female employees in managerial positions (Hutchings et al., 2020: 2). In horizontal discrimination, the female employees may have different duties and responsibilities from their male counterparts despite having the same status (Birdir, Derinalp and Çanakçı, 2013). For example, while women are mainly employed for jobs such as housekeeping and reception, men are usually employed as bellboys, bartenders, maintenance personnel, etc. (Campos et al., 2011). Vertical discrimination occurs when women work in lower positions, have fewer responsibilities than men, and have fewer opportunities for promotion (Birdir, Derinalp and Çanakçı, 2013). The most important sign of gender discrimination in the workplace is that the jobs are divided as women's jobs (lower status, lower wages, no continuity, less social security benefits, etc.) or men's jobs (higher paid and stable jobs that require more skills and responsibilities, more social security benefits) (Kinnaird and Hall 1996; Giddens, 2012:805; Karaset and Tanrısever Yiğit, 2017). Within this context, while cleaning and housekeeping services employ female labor, men are employed for more technical jobs, such as landscaping and pool maintenance, and they are also employed for more prestigious positions, such as management (Biswas and Cassel 1996; Aykaç, 2009; Masadeh, 2013). On the other hand, while there are no doubts about women's kitchen skills, creativity, and qualifications, the common belief that men belong to professional kitchens and women belong to their kitchens at home is one of the indicators of this discrimination (Druckman, 2010 cited by: Kabil 2022: 74). This traditional way of thinking pushes women into the background and stands in their way in work life (Karaset and Tanrısever Yiğit, 2017), and women have to do jobs with lower status and lower wages with no security because of these gender stereotypes and male domination (Mohanty, Kumar Swain, Besra, 2018). This is not because women have unskilled labor; it is because they are believed to have no skills or qualities (Toksöz, 2011: 148).

Another significant form of discrimination is ***wage discrimination***. In general terms, a wage is a tangible or intangible income that employees receive as a result of their participation in an economic activity. It may also be defined as the core income required for employees to live on and look after themselves and their dependents. Wage discrimination occurs when women turn towards different jobs and institutions (underpaying institutions and jobs), and this discrimination may be explained in three different ways. The first one is wage discrimination resulting from distribution discrimination, and it starts with the recruitment process and continues in the promotion process. The second one is wage discrimination resulting from the nature of the job, and it refers to the fact that women are paid lower even though they do the same job within the same organization as men. The third one is wage discrimination resulting from value discrimination. This refers to the fact that women are paid lower in male-dominated jobs, traditionally known as men's jobs, even though they have the qualities and skills required for that job (Özkan and Özkan, 2010). It may be argued that women earn less and men earn more because they are employed for higher-status jobs, and wage discrimination can be regarded as a form of gender discrimination (Çelik Uğuz and Topbaş, 2014).

Many studies explained the reasons for wage discrimination through various theories. Within this context, the human capital theory, which is explained as self-investment, suggests that women are paid less and work in lower positions, and jobs with lower status result from insufficient human capital. The statistical discrimination theory developed by Phelps explains why employers discriminate against female employees as having to make a decision under imperfect competition conditions with incomplete knowledge and ambiguity. On the other hand, Becker's emotional discrimination theory suggests that employers do not behave objectively since they act based on their personal feelings or selections and offer lower wages to female employees. According to overcrowding theory, women are paid less since they do not compete with men under the same conditions, even though they have the same human capital and skills as their male counterparts. The distinctive feature of this theory is that it abandons perfectly-competitive labor market analysis and focuses on occupational discrimination. The institutional approach suggests that there are primary jobs (higher-paying jobs with more promotional opportunities and less labor turnover) and secondary jobs (lower-paying jobs with no promotional opportunities and higher labor turnover rates). It is also suggested that women are usually employed in secondary jobs, while men are employed in higher-paying primary jobs. According to feminist economists, women may be exposed to discrimination both at home and in the workplace because of the male-dominated system and traditional standards of judgment (Palaz, 2003; Orhan, 2009; Yalçın et al., 2019).

Munoz-Bullon (2009) investigated the wage discrimination between female and male employees in the tourism industry in Spain. They found that male employees in the tourism industry earn 6.7% more monthly wages on average than their socially-comparable female counterparts. They stated that this difference resulted from the type of agreements, the skills required by the job, and a certain sub-industry of employment. Another study found that male tourism employees in Norway earn 20% more than their female counterparts (Thrane, 2008). The results of Çelik Uğuz and Topbaş's (2016) study performed based on the 2011 Household Labor Survey prepared by the Turkish Statistical Institute (TUIK) also supported this idea. Accordingly, male tourism employees earn approximately 18% more than their female counterparts, and 6% of this wage difference results from gender discrimination. According to the 2020 Global Gender Equality report published by World Economic Forum, the women-men ratio was 0.628 in terms of wages. On the other hand, it is suggested that it will take at least 100 years for women to have equal rights with men and 257 years to get paid equally. Men are represented more in better-paying jobs in the tourism industry, and women are employed for lower wages for the same jobs and are not usually employed for jobs that require an audit mandate, and this is a clear indication of discrimination (Casado-Díaz et al., 2020). Within this context, wage discrimination has a negative impact on women's personal growth and social integration, and it also stands out as one of the significant issues that prevent economic growth (Yalçın et al., 2019).

Hiring discrimination means enforcing some boundaries during recruitment even though they are not related to the qualifications for the job and imposing these boundaries in the required qualifications for that position even though they are not actually necessary (Demir, 2011). It may also be defined as not being able to benefit from the existing opportunities because of any qualification the employee has or does not have. The statistical discrimination theory, which was developed by institutionalist economists, is an approach based on the idea of assessing potential employees based on the group that the individuals belong to rather than assessing the candidates individually because education, experience, age, and test results are the only certain information on the employer's side, and sometimes they may be insufficient to guess the candidates' potential productivities (Duruoğlu, 2007). For example, male and female candidates are evaluated through different criteria in job interviews. Women are also forced to talk about their thoughts on marriage and having kids in the future (Mooney and Ryan, 2009). This shows that women's future decisions about their private lives are scrutinized more than men's (Dalkıranoğlu, 2006). Çelik and Akar Şahingöz (2018) found that the majority of female cooks experience similar discriminatory behaviors and work at lower positions compared to their skills. Similarly, Cano (2019) found that women are only some significant symbols for the kitchen, and they are secondary choices when it comes to getting hired in industrial kitchens, they usually work in lower positions, and men are more privileged in industrial kitchens when compared to women. Dönmez and Dağaşan (2017) stated that female employees are not usually preferred for the kitchens in five-star hotels in the Mediterranean Region.

When the job postings at the recruitment stage were analyzed, similar discriminative traces could also be seen. Ulu and Kutanis (2016) found that some job postings expect applications only from male candidates, and women are usually fired because of pregnancy or parental leave. Demir (2011) found that male candidates are usually recruited as waiters, bussers, bartenders, tour guides, transfer personnel, cooks, and dishwashers, and women are usually recruited as housekeepers, cashiers, airport ground service officials, customer relations, and sales representatives. Moreover, this discrimination against women at the recruitment stage can only be based on evaluations and practices that are merely made based on the applicant's gender (Masadeh, 2013). Therefore, individual skills or job performances are overlooked. This is also supported by other studies. For example, in another study, female cooks stated that they were discriminated against at the recruitment stage, and they worked at lower positions compared to their skills (Çelik and Akar Şahingöz, 2018). All of these issues may lead women not to have the jobs and positions they deserve in the tourism industry, and they may also lead to decreased productivity and motivation in their social lives.

Even though female employment rates are high in the tourism industry, it is known that women are represented less at the senior managerial position. One of these results of this situation could be career barriers in the industry. Career is another field of discrimination, and it was discussed in the following chapter.

Career Barriers

Career is defined as progressing and advancing in any field and gaining skills and experience throughout one's work life. Career barrier is a term used to express the factors that prevent individuals from raising their expectations and obtaining results, such as power, prestige, and payment (Anafarta, Sarvan and Yapıcı, 2008). Career barriers, which shape the career that an individual would choose and advance and lead them to go out of their career plans, may come from the individuals themselves, the organization, or the social structure (Bozkurt, 2019). It is well-known that female tourism employees face various career barriers. These barriers may generally be listed as tokenism, glass ceiling, glass elevator/glass escalator, glass cliff, and queen bee.

The concept of ***tokenism (so-called participation)*** was first introduced by R. M. Kanter. In theory, it was suggested, based on gender discrimination, that tokenism leads to lower chances of promotion since women are left with jobs with career dead ends and neglected for senior positions in a male-dominated workplace. The groups with less than 15% representation in workplaces or at managerial positions are called "tokens"; the dominating groups with more than 85% representation are called the "tokenists" (Kanter, 2003). According to the tokenism theory, the tokens face some obstacles that must be overcome to advance in their careers against the dominating group. These obstacles emerge as alienation from the formal organization networks (failure to join in social networks, to make friends, to gain social capital), lack of culture (failure to learn about the corporate culture), and lack of mentorship (failure to be taken under the wings of a protector and mentor). This puts them under pressure and obliges them to work harder. In addition, tokenist approaches towards women prevent them from gaining experience in decision-making processes and weaken their self-confidence (Demirel, 2019). The fact that women work in the background or do jobs that require fewer skills in tourism facilities turns them into potential "tokens." For example, the fact that women cannot endure jobs that require physical force is shown among the reasons for the underrepresentation of women in professional kitchens (Kabil, 2022). However, today's technological advancements facilitate things in industrial kitchens and help create an environment where men and women can work under equal conditions (Doğan, 2017). However, the ongoing male dominance in industrial kitchens can be interpreted as an indicator of tokenism. In order to prevent women from being involved in this negative process, Boley et al. (2017) stated that women should be involved in all stages of tourism development and all decision-making processes to prevent them from becoming so-called participants (tokens). Moreover, well known chefs and associates as pioneers of gastronomy community created "Platform of Inspiring Women of Gastronomy" in order to encourage awareness and women employment for this industry (Yılmaz and Akay, 2020).

The glass ceiling is another career barrier for women. Researchers, institutions, and individuals frequently address the glass ceiling of invisible barriers that prevent women from advancing in their careers even though they have equal skills, experiences, and qualifications with their male counterparts. The glass ceiling concept was first introduced in 1984 in an interview with a magazine editor, who stated that "women can advance up to a certain point in their careers, and then they get stuck at that point." This concept was later introduced to the literature through an article published in the Wall Street Journal (Utma, 2019). The glass ceiling syndrome is defined as an invisible and abstract perception that prevents women from developing and advancing in their careers, regardless of their adequacy, skills, and whether they deserve to rise and advance in their careers (Öktem, Kubat and Kızıltan, 2018). The obstacles that affect the glass ceiling stem from three factors: individual (assuming multiple roles, personal preferences and perceptions of women), social (occupational segregation, gender-related stereotyped biases), and organizational (organizational culture and policies, lack of mentorship, and failure to join in informal communication networks) (Dreher, 2003). Even though many jobs at hotel facilities are considered women's jobs, it is observed that women face various obstacles in terms of career development. In their study, Anafarta, Sarva and Yapıcı (2008) found that the main career obstacles that affect the glass ceiling for female hotel managers were stereotypes, discrimination, role conflict, and lack of mentorship. Mooney and Ryan (2009) found informal communication networks and occupational segregation factors as the most important obstacles affecting the glass ceiling for young female managers.

Glass elevator/glass escalator is another phenomenon similar to the glass ceiling. This concept describes situations where men get promoted faster than their female counterparts in female-dominated institutions/positions. In other words, the men working at female-dominated jobs benefit from the glass elevators based on a series of invisible factors that facilitate their career advancements (Hutchings et al., 2020). The glass elevator or glass escalator concept refers to the trend that men escalate/rise faster than their female counterparts in female-dominated work environments (Williams, 2013). Therefore, it can be argued that female managers, who cannot advance further than low- and mid-level management, get a lower chance of promotion than their male counterparts (Alhas, 2020). In this respect, it is observed that the glass ceiling syndrome, which puts stone walls against women's career development, can turn into a glass elevator/glass escalator for men. When considered within the context of the tourism industry, which mostly employs female labor, the women mostly work at lower-level jobs. Still, the men are mostly represented in upper-level positions, which can be considered an indicator of the glass elevator/glass escalator phenomenon in the tourism industry.

Glass cliff means promoting female employees to dangerous leadership positions or when there is a high risk of failure for the company. The concept of glass cliff refers to the *"[tricky] positions that women are promoted during negative financial performance periods in order to move them away from their career goals, put them under a lot of stress, and make them responsible for the company's failure, not because of their skills or because they are believed to bring the company to success*" (Yıldız et al., 2016: 1132). Male managers underestimate the glass cliff concept and prefer to base the emergence of this situation on suitability to tough leadership tasks, the need for making strategic decisions, and corporate factors unrelated to gender. On the other hand, female managers acknowledge the glass cliff concept and state that they perceive it as an injustice, a hazard, sexism, inequality of opportunity, and an obstacle (Ryan et al., 2007). The glass cliff concept is associated with fewer female managers in upper-level positions. However, this leads to the assignment of women, rather than men, for upper-level positions that involve more risks and hazards (Alhas, 2020). This has a negative impact on the job performances of these women, who are promoted to upper-level positions and put at risk (Yıldız, 2017). However, considering there are different pursuits for solutions in companies as a strategy for avoidance from financial crises, the glass cliff phenomenon may also be regarded as an area for women to prove themselves as managers.

The queen bee is defined as women's changing behaviors in the workplace, starting to act like their male counterparts and treating other female employees like male managers do (Zel, 2002: 41- 42). The queen bee behavior is usually seen in work environments where there is a female manager at the organization or the department, and female employees are working under this female manager. The female leaders and composers of this syndrome would have an irresistible desire to control everything and everyone at all times. While doing this, they may become relentless and unmerciful (Bune, 2008). It is suggested that this particular behavior develops in male-dominated organizations, where women most likely become subjects of gender discrimination and have to fight against negative expectations about their leadership. In addition, the three ways female managers usually follow to become the queen bee are presenting themselves as men (pretending), keeping their physical and psychological distance from other women, and approving and legitimizing the existing gender hierarchy (male-dominated) (Derks, 2016). In fact, women using these male behaviors against other women, trying to eliminate other women from the competition, and ignoring all approaches regarding discrimination may be counted among the characteristics of this syndrome (Zel, 2002: 42). This is actually an obstacle women impose on other women. They prefer to hinder other women because they try to protect their own areas and see other women as rivals. In their study, Çelen and Tuna (2021) found that female employees find male managers more successful, that it is easier for them to work with male managers, female managers put obstacles before female employees, and that they do not want to have more than one female manager. Since there are not too many female managers in upper-level positions in the tourism industry, it is clear that the queen bee syndrome is not the sole reason for this situation (Avcı, 2018).

Workplace Violence

The word violence (şiddet in Turkish) comes from the Arabic word "şedd." It may be defined as roughness, rough and hard behaviors, and brute force. The World Health Organization (2002) defines violence as follows: "violence is the intentional use of physical force or power, threatened or actual, against oneself, another person, or against a group or community, that either result in or has a high likelihood of resulting in injury, death, psychological harm, maldevelopment, or deprivation." When the theories developed regarding violence, an interdisciplinary field of work, were analyzed, it can be seen that various approaches have been suggested. Laying the groundwork of the theory of instinct, Freud believed that violence is inherent in human beings, while Dollard's frustration-aggression theory suggests the positive violence inherent in all humans exists to protect the desire to achieve goals; however when it is prevented for various reasons, it may turn into behavior that leads to negative violence. Bandura's approach, which developed the social learning theory, suggests that violence is something learned (Çayır and Çetin, 2011).

The violence phenomenon, which can be seen in many areas of everyday life and should definitely be fought against, also exists in work life. Violence or aggression in the workplace is considered one of the most significant problems in occupational health and safety. The European Commission states that workplace violence covers all kinds of explicit or implicit (hidden) actions threatening the health and safety of the employees, such as abuse, threats, or assault (Aytaç and Dursun, 2013). The Violence and Harassment Convention-C190 by the International Labor Organization (ILO) defines violence as "a range of unacceptable behaviors and practices, or threats thereof, whether a single occurrence or repeated, that aim at, result in, or are likely to result in physical, psychological, sexual or economic harm, and includes gender-based violence and harassment."

The primary source of violence is the individual, and we should get to the root of human behavior in order to categorize the types of violence because the employees facing physical or psychological violence may suffer from problems with job satisfaction, organizational commitment, reduced communication, job dissatisfaction, alienation, desperation, etc. All of these issues have a negative impact first on the individual's and then the organization's productivity. Therefore, it would be better to categorize and explain the issues of workplace violence against women in the tourism industry, which involves a considerable amount of human relations and offers larger areas of employment for women compared to other industries. Even though there are various categorizations of violence in the literature, it is not easy to categorize violence since it is a multidimensional and complex phenomenon. However, it is possible to categorize workplace violence as physical and psychological (Yıldız, Kaya and Bilir, 2011).

Physical violence may occur in various ways, such as hitting, injuring, punching, fighting, etc. Physical violence aims to take an individual under control through pain and fear (Alkan, Yılmaz and Abar, 2020). Physical violence against women is one of the most common forms of domestic violence. The rate of domestic violence against women is 35% in Türkiye. *Assault* is another concept that defines physical violence, and it is a deliberate behavior that physically harms another person (Yıldız, Kaya and Bilir, 2011). Today, physical violence in the workplace decreases gradually and is replaced with psychological violence.

Psychological (emotional) violence may occur in the form of humiliation, insulting, mocking, preventing personal development, restricting one's relationship with their surroundings, threatening, bullying, or intimidation. Mostly regarded as communication problems, psychological violence materially and morally affects women in work life. It also poses an obstacle to the active participation of women in social, economic, and communal life (Çötok, 2020: 10).

Various concepts, such as threat, abuse, harassment (sexual-racial), and mobbing/bullying, are used to describe psychological violence in the workplace. *Threat* refers to verbal assault that leads to a fear of suffering from psychological or sexual assault. *Abuse* means using physical or psychological force in an offending and humiliating way. *Harassment* is an undesired behavior that affects the respectability of individuals because of the reasons such as age, gender, race, political and religious beliefs, different national and social origins, belonging to a minority group, etc. It involves two types of behaviors as sexual and racial harassment. *Mobbing* is a type of harassment that involves harmful behaviors, and these behaviors force the targeted individuals to quit their jobs and leave the workplace (Yıldız, Kaya and Bilir, 2011). Within the frame of these definitions, it may be argued that the temporary and rarely-observed conflicts are not included within the scope of psychological violence because an individual should be subject to negative behaviors at least once a week and for at least six months to consider these behaviors within this context (Leymann, 1996: 169).

When considered in terms of tourism facilities, it may be argued that the studies on workplace violence mostly focus on mobbing (Kensbock et al., 2013; Akbolat, Yılmazer and Tutar, 2014; Koyuncu and Sevin, 2017) and sexual harassment (Ram, 2018; Öz, Dönmez and Akdağ, 2020; Reguera and García-Izquierdo, 2021; Worke et al., 2021; Brobbey et. al., 2022). Mobbing and sexual harassment appear as the cases that cause women working in the tourism industry to remain silent or quit their jobs (Balcı and Aslan, 2022). Supporting this situation, Koç and Uşaklı (2022) reported that particularly the egoist manager behaviors, rudeness in communication, oppressive behaviors that increase the fear of making a mistake, and physical-psychological harassment increase women's tendency to quit their jobs in the tourism industry. Tunca (2021) found that female cooks working in kitchen departments face physical and psychological violence, such as sexual harassment, physical assault, vulgarism, and mobbing. In addition, women offering services in various areas of the tourism industry may also be exposed to inappropriate jokes, verbal abuse, and even sexually explicit behaviors by customers (La Lopa and Gong, 2020: 90). Reguera and García-Izquierdo (2021) reviewed 61 court decisions taken between 2000 and 2016 regarding sexual harassment in the hospitality industry. The researchers found that the number of sexual harassment cases has increased over the years. About half of these lawsuits resulted in material compensation, and the labor contracts of the victims were terminated in many of these cases. The cases of psychological violence against female employees significantly decrease overall living standards and work life standards (Kara, Kim and Uysal, 2018). Therefore, it would be the right decision to adopt policies that will increase positive organizational behaviors for all employees regardless of gender (Demirdağ, 2022).

Long/Rotatory Working Hours

Working hours are defined as the period that an employee spends at work, according to the Regulation on Working Hours regarding the Labor Law (https://www.mevzuat.gov.tr) Tourism is regarded as an industry with long and rotatory working hours (Cheong, 2009: 70, cited by: Kaya and Atçı, 2015). The employees work for longer periods in the tourism industry, particularly in hotels that offer 24/7 service, and this is one of the main issues for the employees in the tourism industry (Kaya, 2017:8). For example,

Yenipınar (2005:165) reported that 37% of the hotel employees work more than 16 hours a day, 29% work between 13 to 15 hours a day, 18% work between 9 to 12 hours a day, and 1% work 8 hours a day. In a study conducted with 6291 wage workers in the tourism industry, Kaya and Atçı (2015) reported long working hours in all groups involved. It was also found that the working hours were within more acceptable limits in medium- and large-scale facilities.

Rotatory working hours bring more challenges to women since they have more domestic responsibilities than their male counterparts, and women working in shifts experience higher levels of work-family conflicts. Long working hours and rotatory working systems have a negative impact on the job performance of the employees. The increasing number of occupational accidents, worsened health conditions, and damaged family life and social relations may be listed as examples. These adverse conditions lead women to avoid their professional lives (İstanbullu Dinçer et al., 2016).

Social and Familial Challenges

Another problem regarding female employment in the tourism industry may be related to the social structure and the general perspective on female employment (Kocacık and Gökkaya, 2005). However, the roles of which shifting women and men could be interchangeable among societies, ordinarily division of labor arising from gender differences makes genders different in terms of dominance, reputability and wealth (Giddens, 2008).

The social standards of judgment, the common view that women should stay at home, take care of the kids and domestic chores, and the fact that a mother is the only person that can take care of her children lead women not to participate in or take breaks from their professional careers (Çakır, 2008). The International Labor Organization explains the reasons for limited female employment as the fact that women usually focus on domestic chores and childcare. Thus, they cannot seek paid and permanent jobs (Tüzünkan, 2019). Focusing on first becoming a good wife, mother, and housewife, the women push working into the background with these thoughts in mind (Yüksel, 2005). Moreover, some patriarchal societies believe that jobs other than domestic chores are against the nature of women (Erikli, 2020;). The research findings of Alrwajfah et al. (2020) indicate that the roles of women are shaped within the axis of cultural norms, particularly in rural areas (getting married and taking care of the family and the kids, etc.), female employment in the tourism industry, which does not go well with the traditional values, is not appreciated, women from these regions are oppressed by their families and/or the society to change their professions, even though they studied tourism and hospitality, and thus, they are alienated from the industry. Therefore, the factors that prevent women from taking part in work life may be related to cultural and social norms rather than the business itself or other legal obstacles (Li and Leung, 2001).

Women's entry into work life also brought some issues, such as the formation of new sexist roles and identities, dual role conflicts, and disputes regarding domestic chores (Duffy, Kline, Mowatt and Chancellor, 2015). If the working woman is qualified and has a high salary, she can hire someone for childcare or domestic chores and reduce her responsibilities. Otherwise, women cannot participate in or have to leave the labor market to stay home and care for their kids. This is supported by the fact that the labor force participation rate of child-free women is higher than that of women with children. This dual role brings work-family conflicts and is considered one of the primary reasons for women to quit their jobs (Nalçacı İkiz, 2020). In other words, besides the external factors that prevent women from participating in the labor force, their own mental states and roles are also among these problems (Li and Leung, 2001: 191).

The viewpoints of society and the family are among the significant factors that prevent female employment in the tourism industry. Although the overall attitude towards female employment is positive in Türkiye, women can sometimes be restrained or not supported by their spouses or family members (Kuzgun and Sevim, 2004: 17). The most important reason for this is the belief that women should choose a profession that is suitable within the social gender roles (Trauth, 2002: 101). Skills and qualifications are actually the most important points to consider when choosing a profession. However, both society's and family's attitudes towards female employment may sometimes force them to choose a job without considering their own desires and skills. For example, nursing is considered a woman's job, and driving is considered a man's job (Tunca, 2021). The studies show that more than half of women do not receive any support from their families and have difficulty establishing a work-life balance (Li and Leung, 2001).

CONCLUSION AND SUGGESTIONS

Despite the increasing rates of female employment in the tourism industry, it can be clearly seen that women are in an unfavorable position considering the nature of the job and the current work conditions both in international and national organizations. Even though female employment has been increasing in recent years in the tourism industry, it is observed in the literature that they are mostly employed in semi-skilled or low-skilled, seasonal, part-time, low-paid jobs with heavy working conditions (Santos and Varejão, 2007; Gebelek, 2008; Cave and Kılıç, 2010; Masadeh, 2013; Kodagoda, 2022). As in other work areas in the service industry (nurses, secretaries, teachers, etc.), since a considerable number of jobs in the tourism industry are considered "woman's jobs," the female labor force is usually preferred (Akoğlan, 1996). This gender-based structure of tourism employment (Pritchard and Morgan, 2016) brings various issues for women, from job selection to wages and promotion. These challenges prevent women from using their capacities to make greater economic, political, and social contributions (Baum, 2013). There is evidence proving that women working in the tourism industry have increasing tendencies towards quitting their jobs because of the challenges they face in the workplace (Pelit, Güçer and Demirdağ, 2016; Koç and Uşaklı, 2022). Moreover, women have a hard time establishing a work-life balance; they postpone their marriage and familial plans and have to give up on their personal lives (Garrigos, Haddaji, Segovia and Signes, 2019). The traditional stereotyped beliefs based on the idea that women are unsuitable for some jobs should be abandoned.

Giving active roles to women in the tourism industry by eliminating all factors that have a negative impact on women is a powerful instrument for sustainable development. Particularly while emphasizing the contribution of women and men equalization for sustainable tourism, unfavorable conditions of women should be prevented, more righteous, reliable and convenient work environments should be ensured and efficient policies should be adopted to support women in establishing a family-work balance (Erkol, 2022).

Female-friendly policies regarding eliminating the social and sectoral challenges women face are clearly required to increase female employment in the tourism industry and ensure women keep their jobs. As stated by İstanbullu Dinçer et al. (2016), women can only be fairly and accurately positioned within society and the tourism industry by eliminating the challenges they face in the workplace and by seizing the opportunities. Accordingly, first, the challenges to female employment in the tourism industry should be determined by relevant institutions and organizations, and solutions should be provided and realized (Ardıç Yetiş and Çalışkan, 2020). Within this context, legal regulations regarding gender equality and female empowerment may be implemented in all tourism facilities, awareness may be raised regarding gender and female empowerment, and businesses may work on improving lower wages paid to female employees. In addition, necessary measures may be taken to eliminate discrimination, career barriers, and all sorts of violence (Mohanty, Swain and Besra, 2018).

Because the core purpose of tourism facilities is to satisfy their customers and offer the best service to increase their revenues and profitability, performance-based evaluations of the employees become crucial for the future of these businesses. The male-female job divisions in the tourism industry should be eliminated, and women and men should have the same opportunities and conditions. On the other hand, tourism facilities embody various jobs and offer a holistic service. Therefore, performing each job carefully and aesthetically becomes crucial for customer satisfaction and service quality because the customer evaluates the facility as a whole. Therefore, each employee should be treated fairly.

REFERENCES

Akbolat, M., Yılmazer, A., & Tutar, H. (2014). Konaklama işletmeleri çalışanlarının algıladıkları mobbingin iş tatmini ve işten ayrılma niyetine etkisi. *Manas Sosyal Araştırmalar Dergisi*, *3*(3), 1–17.

Akdağ, G. & Üçyıldız, U. H. (2021). Seyahat acentası çalışanlarının cam tavan sendromu algılamaları: Ege bölgesi seyahat acentalarında bir uygulama. *İşletme Araştırmaları Dergisi, 12*(3), 2540–2654.

Akoğlan, M. (1996). Konaklama endüstrisinde kadının konumu. *Anatolia Turizm Araştırmaları Dergisi*, *7*(3-4), 16–23.

Al-Ismail, S., Carmichel, F., & Duberley, J. (2019). Female employment in hotels in Saudi Arabia and UAE. *Gender in Management*, *34*(7), 554–576. doi:10.1108/GM-11-2017-0151

Albayrak, A. (2019). *Yiyecek içecek endüstrisinde kadının yeri ve sorunu. In Turizmde Kadın Olmak*. Değişim Yayınları.

Alhas, F. (2020). Kadınların üst düzey yönetici olmasına engel olan nedenler cam uçurum, cam yürüyen merdiven ve cam asansör arasındaki ilişkinin incelenmesi. *Ağrı İbrahim Çeçen Üniversitesi Sosyal Bilimler Enstitüsü Dergisi*, *6*(2), 101–113.

Alkan, Ö., Yılmaz, F. & Abar, H. (2020). Türkiye'de kadına yönelik aile içi şiddeti etkileyen faktörlerin sıralı probit regresyon modeliyle belirlenmesi. *İnsan ve Toplum Bilimleri Araştırmaları Dergisi, 9*(5), 3338-3360.

Alparslan, A. M., Çetinkaya, B. Ö., & Özgöz, A. (2015). İşletmelerde cinsiyet ayrımcılığı ve kadın çalışanların sorunları. *MAKÜ İktisadi ve İdari Bilimler Fakültesi Dergisi*, *2*(3), 66–81.

Alrwajfah, M. M., Almeida-García, F., & Cortés-Macías, R. (2020). Females' perspectives on tourism's impact and their employment in the sector: The case of Petra, Jordan. *Tourism Management*, *78*, 104069. doi:10.1016/j.tourman.2019.104069

Anafarta, N., Sarvan, F., & Yapıcı, N. (2008). Konaklama işletmelerinde kadın yöneticilerin cam tavan algısı: Antalya ilinde bir araştırma. *Akdeniz İ.İ.B.F. Dergisi*, *15*, 111–137.

Ardıç Yetiş, Ş., & Çalışkan, N. (2020). Turizm sektöründe kadın istihdamı: Mevcut duruma ilişkin bir değerlendirme. *Manisa Celal Bayar Üniversitesi Sosyal Bilimler Dergisi*, *18*(3), 105–119.

Arslan, A., & Şimşek, G. (2018). Kadın turist rehberlerinin yaşadıkları mesleki sorunlar: Aydın turist rehberleri örneği. *Journal of Travel and Tourism Research*, *13*, 23–49.

Avais, M. A., Wassan, A., & Shah, S. (2014). A case study on problems of working women in city Sukkur. *Academic Research International*, *5*(2), 325–333.

Avcı, N. (2018). *Turizm sektöründe çalışan kadınların karşılaştıkları engeller ve bu engellere yönelik algıları* [Unpublished Master Thesis]. Muğla Sıtkı Koçman Üniversitesi, Muğla.

Aydın Tükeltürk, Ş., & Şahin Perçin, N. (2008). Turizm sektöründe kadın çalışanların karşılaştıkları kariyer engelleri ve cam tavan sendromu: Cam tavanı kırmaya yönelik stratejiler. *Yönetim Bilimleri Dergisi*, *6*(2), 113–128.

Aydoğdu, G. (2021, March 11). *Toplumsal Cinsiyet ve Turizm*. https://kalkinmaguncesi.izka.org.tr/index.php/2021/03/11/toplumsal-cinsiyet-ve-turizm/

Aykaç, A. (2009). *Yeni işler, yeni işçiler*. İletişim Yayınları.

Aytaç, S., & Dursun, S. (2013). Çalışma hayatında kadına yönelik şiddet: Sağlık sektöründe bir uygulama. *Journal of Istanbul University Law Faculty*, *71*(1), 67–78.

Balcı, E., & Aslan, E. B. (2022). Turizmde çalışan kadın sorunları: Antalya'da 5 yıldızlı otellerin yiyecek-içecek departmanında çalışan kadınlar. *Journal of Tourism and Gastronomy Studies*, *10*(3), 1674–1698.

Baum, T. (2013). *International perspectives on women and work in hotels, catering and tourism*. International Labour Organisation (No. 289). Gender Working Paper 1/2013, Sector Working Paper.

Birdir, K., Derinalp, S., & Çanakçı, T. (2013). Otel mutfaklarında çalışan iş görenlerin sorunları, 14. In Ulusal Turizm Kongresi (pp. 980-997). Kayseri.

Biswas, R., & Cassel, C. (1996). Strategic HRM and gendered division of labour in the hotel industry: A case study. *Personnel Review*, *25*(2), 19–34. doi:10.1108/00483489610147870

Boley, B. B., Ayscue, E., Maruyama, N., & Woosnam, K. M. (2017). Gender and empowerment: Assessing discrepancies using the resident empowerment through tourism scale. *Journal of Sustainable Tourism*, *25*(1), 113–129. doi:10.1080/09669582.2016.1177065

Bozkurt, Ö. (2019). Kariyer engelleri ve kariyer sorunları. In Kariyer, Temel Kavramlar, Yönetimi, Güncel Konular (pp. 63-75). Konya: Eğitim Yayınevi.

Brobbey, V. A., Asiedu, R., Mensah, P. O., Antwi, F. A., Sie, E. F., & Kumah, A. (2022). Employees' Perception of Sexual Harassment and Its Impacts on Employee Performance in Ghana. *Social Values and Society*, *4*(2), 41–45. doi:10.26480vs.02.2022.41.45

Bune, K. (2008, June 4). *Queen Bee Syndrome Produces Stinging Victimisation*. https://www.lawofficer.com/queen-bee-syndrome-produces-stinging-victimization/

Çakır, Ö. (2008). Türkiye'de kadının çalışma yaşamından dışlanması. *Erciyes Üniversitesi İktisadi İdari Bilimler Fakültesi*, *31*, 25–47.

Çakır, P. G., Barakazı, M., & Barakazı, E. (2017). Turizm Sektöründe Çalışan Kadınların Karşılaştığı Sorunları Değerlendirmeye Yönelik Bir Araştırma. *Journal of Academic Social Science Studies*, *61*, 461–474.

Campos-Soria, J. A., García-Pozo, A., & Sánchez-Ollero, J. L. (2015). Gender wage inequality and labour mobility in the hospitality sector. *International Journal of Hospitality Management*, *49*, 73–82. doi:10.1016/j.ijhm.2015.05.009

Campos-Soria, J. A., Marchante-Mera, A., & Ropero-García, M. A. (2011). Patterns of occupational segregation by gender in the hospitality industry. *International Journal of Hospitality Management*, *30*(1), 91–102. doi:10.1016/j.ijhm.2010.07.001

Cano, M. (2019). *Masculinity in the Kitchen: Gender Performance in the Culinary Arts Industry* [Master's Thesis]. University of Texas, El Paso.

Casado-Díaz, J. M., Driha, O., & Simón, H. (2020). The gender wage gap in hospitality: New evidence from Spain. *Cornell Hospitality Quarterly*, *63*(3), 399–417. doi:10.1177/1938965520971273

Cave, P., & Kılıç, S. (2010). The role of women in tourism employment with special reference to Antalya, Turkey. *Journal of Hospitality Marketing & Management*, *19*(3), 280–292. doi:10.1080/19368621003591400

Çayır, C., & Çetin, Ö. (2011). Din ve şiddet üzerine psikolojik bir yaklaşım. *Dicle Üniversitesi İlahiyat Fakültesi Dergisi*, *13*(1), 1–34.

Çelen, O., & Tuna, M. (2021). Konaklama işletmelerindeki kadın çalışanların kraliçe arı sendromu algılarının ölçülmesine yönelik bir ölçek geliştirme çalışması. *Türk Turizm Araştırmaları Dergisi*, *5*(3), 2127–2148.

Çelik, M.B. & Akar Şahingöz, S. (2018). İş yaşamında cinsiyet ayrımcılığı: Kadın aşçılar örneği. *Journal of Tourism and Gastronomy Studies, 6*(3), 370-383.

Çelik Uğuz, S., & Topbaş, F. (2016). Turizmde kadın istihdamı ve ücret ayrımcılığı: Karşılaştırmalı bir analiz. *Anatolia: Turizm Araştırmaları Dergisi*, *27*(1), 62–78.

Çelik Uğuz, S., & Topbaş, T. (2014). Toplumsal cinsiyet ve kadın emeği: Turizm sektörüne betimsel bir yaklaşım. *Akademik Sosyal Araştırmalar Dergisi*, *2*(7), 487–509.

Chant, S. (1997). Gender and tourism employment. In Gender, Work and Tourism (pp. 120-180). London: Routledge.

Çiçen, C., Boyacıoğlu, E. Z. & Oğuzhan, A. (2020). Turizmde kadın işgücünün cinsiyet ayrımcılığı algısı: Edirne ilinde bir araştırma. *Tourism and Recreation, 2*(1), 54-63.

Ciğer, A., & Uyar, G. F. (2016). Konaklama işletmelerinde kadın muhasebecilerin cam tavan algısına yönelik bir araştırma: Antalya ili örneği. *MANAS Sosyal Araştırmalar Dergisi*, *5*(2), 43–58.

Çötok, N. A. (2020). *Şiddet, çalışma yaşamı ve medya bağlamında kadın*. Pegem Akademi.

Dalkıranoğlu, T. (2006). *Çalışma yaşamında kadın işgücü ve cinsiyet ayrımcılığı: konaklama işletmelerinde bir uygulama* [Unpublished Master Thesis]. Anadolu University.

Demir, M. (2011). İş yaşamında ayrımcılık: Turizm sektörü örneği. *Uluslararası İnsan Bilimleri Dergisi*, *8*(1), 760–784.

Demirdağ, Ş. A. (2022). Otel işletmelerinde karşılaşılan fiziksel sorunların örgütsel bağlılık ve iş tatmini üzerindeki etkisi: Kadın çalışanlar üzerine bir araştırma. *KADEM Kadın Araştırmaları Dergisi*, *8*(1), 95–121.

Demirel, E. T. (2019). Tokenizm teorisinin Türkiye bağlamında işlerliğine yönelik bir ölçek geliştirme çalışması. *Fırat Üniversitesi Uluslararası İktisadi ve İdari Bilimler Dergisi*, *3*(2), 1–34.

Demirkol, Ş., Fidan, F., & Pelit, E. (2004). Turizm sektöründeki bayan işgörenlerin karşılaştıkları sorunlar ve otel işletmelerinde bir uygulama. *Abant İzzet Baysal Üniversitesi Sosyal Bilimler Enstitüsü Dergisi*, *1*(8), 71–88. doi:10.11616/AbantSbe.128

Derks, B., Van Laar, C., & Ellemers, N. (2016). The Queen Bee Phenomenon: Why Women Leaders Distance Themselves From Junior Women. *The Leadership Quarterly*, *27*(3), 456–469. doi:10.1016/j.leaqua.2015.12.007

Doğan, M. (2017). The role of cookery vocation in the tourism sector on women employment. *The Online Journal of Science and Technology*, *7*(1), 127–132.

Dönmez, B., & Dağaşan, G. (2017). Kadın aşçıların dikkatine! Profesyonel mutfakta yemek yapmak evde iki kap yemek yapmaya benzemezmiş. In Uluslararası Turizmin Geleceği Kongresi (pp. 1053-1064). Mersin.

Dreher, G. (2003). Breaking the glass ceiling: The effects of sex ratios and work–life programs on female leadership at the top. *Human Relations*, *56*(5), 541–562. doi:10.1177/0018726703056005002

Duffy, L. N., Kline, C. S., Mowatt, R. A., & Chancellor, H. C. (2015). Women in tourism: Shifting gender ideology in the DR. *Annals of Tourism Research*, *52*, 72–86. doi:10.1016/j.annals.2015.02.017

Duruoğlu, T. (2007). Emek piyasasında cinsiyetçi ücret ayrımı: Bursa Organize Sanayi bölgesinde bir araştırma. *İletişim Kuram ve Araştırma Dergisi*, *24*, 61-76.

Erdoğmuş, N. (2003). *Kariyer geliştirme: kuram ve uygulama*. Nobel Yayınları.

Erikli, S. (2020). Çalışma yaşamında toplumsal cinsiyet ayrımcılığının görünümü. *Hacettepe Üniversitesi Sosyal Bilimler Dergisi*, *2*(1), 39–60.

Erkol, G. B. (2022). Kadınların turizm sektöründe karşılaşmış olduğu engellerin demografik değişkenler kapsamında incelenmesi: Kadın rehberler üzerine bir araştırma, *Yönetim. Ekonomi ve Pazarlama Araştırmaları Dergisi*, *6*(3), 144–161.

Eryılmaz, G. (2022). Restoran işletmelerinde istihdam edilen kadın çalışanların karşılaştıkları sorunlar ve çözüm önerileri: Örnek olay çalışması, *Journal of Social. Humanities and Administrative Sciences*, *8*(56), 1056–1071. doi:10.29228/JOSHAS.64314

Faten Ali, M. H. (2014). Challenges facing female employees in quick service restaurants: Egypt as a case study. *Journal of Association of Arab Universities for Tourism and Hospitality*, *11*(3), 59–74. doi:10.21608/jaauth.2014.57689

Garrigos, J. A., Haddaji, M., Segovia, P. G., & Signes, A. P. (2019). Gender Differences in the Evolution of Haute Cuisine Chef's Carerr. *Journal of Culinary Science & Technology*. Advance online publication. doi:10.1080/15428052.2019.1640156

Gebelek, G. (2008). *The new international migration from a gender perspective: A case study of post-Soviet servants of tourism in Antalya* [Unpublished Master Thesis]. Koç University.

Giddens, A. (2012). *Sosyoloji*. Kırmızı Yayınları.

Güçer, E., Keleş, Y., Demirdağ, S. A., & Çelikkanat, N. (2018). Kadın isgörenlerin karşılaştığı sorunların örgütsel sessizlikleri üzerindeki etkisi: Bodrum'daki otel işletmelerinde bir araştırma. *Afyon Kocatepe Üniversitesi Sosyal Bilimler Dergisi*, *20*(2), 1–18. doi:10.32709/akusosbil.414693

Hutchings, K., Moyle, C. L., Chai, A., Garofano, N., & Moore, S. (2020). Segregation of women in tourism employment in the APEC region. *Tourism Management Perspectives*, *34*, 100655. doi:10.1016/j.tmp.2020.100655

ILO Şiddet ve Taciz Sözleşmesi (C190). (2021). *Çalışma yaşamında şiddet ve taciz*. Friedrich Ebert Stiftung.

İstanbullu Dinçer, F. İ., Akova, A., Muğan Ertuğral, S., & Aydoğan Çifçi, M. (2016). Türkiye'de turizm sektöründe kadın istihdamı: Imkanlar ve engeller. *Eurasian Academy of Sciences Social Sciences Journal*, *1*, 379–395.

Kabil, E. N. (2022). Toplumsal cinsiyet eşitsizliklerinin kadınların motivasyon ve verimliliği üzerindeki rolü: Mutfak departmanı çalışanları üzerine bir araştırma. *Journal of New Tourism Trends*, *3*(1), 70–85.

Kanter, R. M. (2003). *Men and women of the corporation*. Basic Books.

Kara, D., Kim, H., & Uysal, M. (2018). The effect of manager mobbing behaviour on female employees' quality of life. *Current Issues in Tourism*, *21*(13), 1453–1467. doi:10.1080/13683500.2015.1078298

Karabulut, E. (2022, April 19). *Turizmde çalışan kadın Sayısı*. https://www.turizmdatabank.com/turizm-istatistikleri/turizmde-kadin-istihdami/

Karaset, Ş.B., & Yiğit, T. (2017). Turizm sektöründe kadın istihdamı ve sorunları. III. In *Uluslararası Girişimcilik, İstihdam ve Kariyer Kongresi* (pp. 1114-1130). Muğla.

Kattara, H. (2005). Career challenges for female managers in egyptian hotels. *International Journal of Contemporary Hospitality Management*, *17*(3), 238–251. doi:10.1108/09596110510591927

Kaya, İ., & Atçı, D. (2015). Türk turizm sektöründe çalışma saatlerinin analizi ve değerlendirilmesi. *Cag University Journal of Social Sciences*, *12*(2), 23–47.

Kaya, İ., & Topbaş, F. (2019). Turizm sektöründe cinsiyete dayalı ücret ayrımcılığı. *Izmir Democracy University Social Sciences Journal*, *2*(1), 74–87.

Kaya, Ş. (2017). Turizmde Kadın Emeği. *Karatahta İş Yazıları Dergisi*, *9*, 1–22.

Kensbock, S., Jennings, G., Bailey, J., & Patiar, A. (2013). The lowest rung': Women room attendants' perceptions of five star hotels' operational hierarchies. *International Journal of Hospitality Management*, *35*, 360–368. doi:10.1016/j.ijhm.2013.07.010

Kinnaird, V., & Hall, D. (1996). Understanding Tourism Processes: A Gender-Aware Frame Work. *Tourism Management*, *17*(2), 95–102. doi:10.1016/0261-5177(95)00112-3

Koç, B., & Uşaklı, A. (2022). Kadın çalışanların turizm sektöründen ayrılma nedenleri: Keşifsel bir araştırma. *Journal of Tourism and Gastronomy Studies*, *10*(1), 589–609.

Kocacık, F. & Gökkaya, B.V. (2005). Türkiye'de çalışan kadınlar ve sorunları. *C.Ü. İktisadi ve İdari Bilimler Dergisi, 6*(1), 195-219.

Kodagoda, T., & Jayawardhana, S. (2022). Forced or voluntary reluctance or voluntary preference to work? Women in hotel industry: Evidence from Sri Lanka. *Gender in Management*, *37*(4), 524–534. doi:10.1108/GM-01-2021-0011

Koyuncu, M. B., & Sevin, H. D. (2017). Konaklama işletmelerinde çalışanların psikolojik şiddet (mobbing) algısının iş tatminine etkisi. *Yönetim. Ekonomi ve Pazarlama Araştırmaları Dergisi*, *1*(3), 33–52.

Kuzgun, Y., & Sevim, S. A. (2004). Kadınların çalışmasına karşı tutum ve dini yönelim arasındaki ilişki. *Ankara Üniversitesi Eğitim Bilimleri Fakültesi Dergisi*, *37*(1), 14–27.

La Lopa, J. M., & Gong, Z. (2020). Sexual harassment of hospitality interns. *Journal of Hospitality & Tourism Education*, *32*(2), 88–101. doi:10.1080/10963758.2020.1726767

Leymann, H. (1996). The content and development of mobbing at work. *European Journal of Work and Organizational Psychology*, *5*(2), 165–184. doi:10.1080/13594329608414853

Li, L., & Leung, R. W. (2001). Female managers in Asian hotels: Profile and career challenges. *International Journal of Contemporary Hospitality Management*, *13*(4), 189–196. doi:10.1108/09596110110389511

Loureiro, P. R. A., Carneiro, F. C., & Sachsida, A. (2004). Race and gender discrimination in the labor market: An urban and rural sector analysis for Brazil. *Journal of Economic Studies (Glasgow, Scotland)*, *31*(2), 129–143. doi:10.1108/01443580410527114

Masadeh, M. (2013). Women in the hotel industry: What's missing from this picture? *International Journal of Academic Research in Business & Social Sciences*, *3*(1), 573–580.

Mohanty, P., Kumar Swain, S., & Besra, S. (2018). Women at work: Exploring the issues and challenges of women employees in travel and tourism. *Indian Journal of Economics and Development*, *6*(1), 1–5.

Mooney, S., & Ryan, I. (2009). A woman's place in hotel management: Upstairs or downstairs? *Gender in Management*, *24*(3), 195–210. doi:10.1108/17542410910950877

Munoz-Bullon, F. (2009). The gap between male and female pay in the Spanish tourism industry. *Tourism Management*, *30*(5), 638–649. doi:10.1016/j.tourman.2008.11.007

Nalçacı İkiz, A. (2020). Turizm işletmelerinde cinsiyet ayrımcılığı ve kadın çalışanlar: Yöneticilerinin konuya ilişkin tutum ve uygulamaları. *Van Yüzüncü Yıl Üniversitesi Sosyal Bilimler Enstitüsü Dergisi*, *50*, 373–396.

Öktem, Ş., Kubat, G., & Kızıltan, B. (2018). Otel İşletmelerinde Kadın İşgören Davranışlarını Etkileyen Algılara İlişkin Bir Araştırma. *Anatolia: Turizm Araştırmaları Dergisi*, *29*(2), 197–208.

Okumuş, F., Sarıışık, M., & Naipaul, S. (2010). Understanding why women work in five-star hotels in a developing country and their work-related problems. *International Journal of Hospitality & Tourism Administration*, *11*(1), 76–105. doi:10.1080/15256480903539693

Orhan, G. (2009). *Ücret ayrımcılığının önlenmesinde iş değerlemenin rolü* [Unpublished Master Thesis]. Marmara University.

Orido, C. O. (2017). *Challenges faced by female chefs in the kenyan hospitality industry: a study through an african oral tradition of storytelling* [Doctoral dissertation]. Auckland University of Technology.

Öz, H., Dönmez, B., & Akdağ, G. (2020). İş yerinde cinsel taciz ile iş doyumu ve işten ayrılma niyeti ilişkileri: Kadın mutfak çalışanları örneği. *Türk Turizm Araştırmaları Dergisi*, *4*(4), 3842–3854. doi:10.26677/TR1010.2020.594

Özkan, G. S. & Özkan, B. (2010). Kadın çalışanlara yönelik ücret ayrımcılığı ve kadın ücretlerinin belirleyicilerine yönelik bir araştırma. *Çalışma ve Toplum, 1*(24), 91-104.

Palaz, S. (2003). Türkiye'de cinsiyet ayırımcılığı analizinde neo klasik yaklaşıma karşı kurumcu yaklaşım: Eşitliği sağlayıcı politika önerileri. *Balıkesir Üniversitesi Sosyal Bilimler Enstitüsü Dergisi*, *6*(9), 87–109.

Pelit, E., Güçer, E., & Demirdağ, Ş. A. (2016). Kadın işgörenlerin karşılaştığı sorunların iş bırakma eğilimlerine etkisi: Otel işletmelerinde bir araştırma. *Mehmet Akif Ersoy Üniversitesi Sosyal Bilimler Enstitüsü Dergisi*, *8*(15), 43–65. doi:10.20875b.16565

Pritchard, A., & Morgan, N. (2016). Encyclopedia of Tourism. Springer Cham.

Pritchard, A., Morgan, N., Ateljevic, I., & Harris, C. (Eds.). (2007). *Tourism and gender: Embodiment, sensuality and experience*. Cabi. doi:10.1079/9781845932718.0000

Ram, Y. (2018). Hostility or hospitality? A review on violence, bullying and sexual harassment in the tourism and hospitality industry. *Current Issues in Tourism*, *21*(7), 760–774. doi:10.1080/13683500.2015.1064364

Reguera, C., & García-Izquierdo, A. L. (2021). Women as victims of court rulings: Consequences of workplace harassment in the hospitality industry in spain (2000–2016). *Sustainability (Basel)*, *13*(14), 7530. doi:10.3390u13147530

Ryan, M. K., Haslam, S. A., Hersby, M. D., Kulich, C., & Atkins, C. (2007). Opting out or pushed off the edge? The glass cliff and the precariousness of women's leadership positions. *Social and Personality Psychology Compass*, *1*(1), 266–279. doi:10.1111/j.1751-9004.2007.00007.x

Santos, L. D., & Varejão, J. (2007). Employment, pay and discrimination in the tourism industry. *Tourism Economics*, *13*(2), 225–240. doi:10.5367/000000007780823186

Thrane, C. (2008). Earnings differentiation in the tourism industry: Gender, human capital and socio-demographic effects. *Tourism Management*, *29*(3), 514–524. doi:10.1016/j.tourman.2007.05.017

Tokgöz, G. (2011). *Kalkınmada kadın emeği*. Varlık Yayınları.

Trauth, E. M. (2002). Odd girl out: An individual differences perspective on women in the IT profession. *Information Technology & People*, *15*(2), 98–118. doi:10.1108/09593840210430552

Tunca, A. (2021). *Otel mutfaklarında cinsiyet engelleri: Kadın aşçılar üzerinde bir araştırma* [Unpublished Master Thesis]. Necmettin Erbakan Üniversitesi.

Tüzünkan, D. (2019). Turizmde kadın istihdamı. In A. Vatan (Ed.), Turizmde kadın olmak içinde (pp. 67-84). Adapazarı: Değişim Yayınları.

Ulu, S., & Özen Kutanis, R. (2016). İşgücü piyasalarında ayrımcılığın kaynakları. *Selçuk Üniversitesi Sosyal Bilimler Enstitüsü Dergisi*, (35), 359–372.

Utma, S. (2019). Kadına yönelik cinsiyet ayrımcılığı ve cam tavan sendromu. *Sosyal ve Beşerî Bilimler Dergisi*, *11*(1), 44–58.

Williams, C. L. (2013). The glass escalator, revisited: Gender inequality in neoliberal times, SWS feminist lecturer. *Gender & Society*, *27*(5), 609–629. doi:10.1177/0891243213490232

Worke, M. D., Koricha, Z. B., & Debelew, G. T. (2021). Perception and experiences of sexual harassment among women working in hospitality workplaces of Bahir Dar city, Northwest Ethiopia: A qualitative study. *BMC Public Health*, *21*(1), 1–18. doi:10.118612889-021-11173-1 PMID:34116672

Yalçın, E. C., Bayram Arlı, N., Aytaç, S., Başol, O., & Aydemir, M. (2019). Türkiye'de emek piyasalarında cinsiyete dayalı ücret ayrımcılığı: IBBS düzey 1'e göre farklılıklar. *Kocaeli Üniversitesi Sosyal Bilimler Dergisi*, *2*(38), 293–314. doi:10.35343/kosbed.524255

Yanardağ, Ö., & Avcı, M. (2012). Turizm sektöründe istihdam sorunları: Marmaris, Fethiye, Bodrum ilçeleri üzerine ampirik bir inceleme. *Ege Stratejik Araştırmalar Dergisi*, *3*(2), 39–62. doi:10.18354/esam.81733

Yenipınar, U. (2005). Otel işletmelerinde ücretlendirme: İzmir ili 4-5 yıldızlı otel işletmelerinde bir analiz. *Dokuz Eylül Üniversitesi Sosyal Bilimler Enstitüsü Dergisi*, *7*(3), 148–176.

Yıldız, A. N., Kaya, M., & Bilir, N. (2011). *İşyerinde şiddet*. Türkiye İşçi Sendikaları Konfederasyonu.

Yıldız, S. (2017). Toplumsal cinsiyetin şirketlere yansımasında ortaya çıkan ayrımcılık kavramlarının ardılları üzerine bir model önerisi. *ESOGÜ Sosyal Bilimler Dergisi*, *18*(1), 121–138. doi:10.17494/ogusbd.330753

Yıldız, S., Alhas, F., Sakal, Ö., & Yıldız, H. (2016). Cam uçurum: Kadın yöneticiler cam tavanı ne zaman aşar? *Ankara Üniversitesi SBF Dergisi*, *71*(4), 1119–1146. doi:10.1501/SBFder_0000002425

Yılmaz, İ., & Akay, E. (2020). Gastronomi sektöründe kadınların durumunun değerlendirilmesi. *International Journal of Social and Economic Sciences*, *10*(2), 81–86.

Yüksel, İ. (2005). İşsizlik olgusunun psikolojik boyutu: Görgül bir araştırma. *Ankara Üniversitesi SBF Dergisi*, *60*(3), 255–274.

Zel, U. (2002). İş arenasında kadın yöneticilerin algılanması ve kraliçe arı sendromu. *Amme Idaresi Dergisi*, *35*(2), 39–48.

Chapter 21
Gender Discrimination in Tour Guiding:
A Comparative Study of the Problems Experienced by Female and Male Guides

Evrim Bilgi
Aydın Adnan Menderes University, Turkey

Ahu Yazici Yazıcı Ayyıldız
Aydin Adnan Menderes University, Turkey

ABSTRACT

Women tend to be disadvantaged at work due to the various problems based on gender discrimination. Women's employment tends to be segregated both vertically and horizontally and across a wide variety of sectors. Women face gender discrimination in a wide variety of sub-sectors of tourism ranging from hospitality to tour guiding. This study explores gender discrimination in the tourist guiding profession. By adopting a dyadic perspective, this study investigates the difficulties experienced by both female and male tour guides. A convenience sampling was used to collect data from 15 female and 15 male tour guides. In collecting data, semi-structured interviews were used. The data collected from the tour guides were analysed qualitatively. By comparing the difficulties experienced by the male and female tour guides, the extent and the nature of gender discrimination have been identified. Based on the outcomes of the study regarding the extent and nature of discrimination, policy guidelines are offered for the practitioners and the stakeholders.

INTRODUCTION

Gender, unlike biological sex, is a concept that is not innate, but is learned later, that determines the roles of women and men and their positions in society. Gender stereotypes are formed by observing gender role models starting from the family and through social interaction. How gender and gender differences are viewed may vary according to the culture of each society (Koc and Yazici-Ayyildiz, 2021), and it

DOI: 10.4018/978-1-6684-8417-3.ch021

can also change over time. The roles that society imposes on women and men can also result in gender discrimination according to the society's perception of gender roles. Compared with men, women tend to be more negatively influenced by gender discrimination, both in their education life and in their working and family life. women; Factors such as women's marital status, whether they have children or not, and their responsibilities within the family may have a direct impact on their career development. On the other hand, these conditions may result in women being exposed to discrimination in working life due to factors arising from gender roles. (Tekin, 2017). While women contribute significant amount of labor to tourism businesses and the implementation of community development initiatives, gender inequalities still persist. Gender problems at national, regional and global levels cause women not to use their capacity to contribute more economically, politically and socially (Akgül 2020).

While the main reasons for women's social discrimination may be the patriarchal structure in societies, it can also be through the continuation of the extension of the social structure in social practices. With the accepted view that women's place is naturally in the home, the woman's adopting this task assigned to her and the relations established with the woman as an individual through this adoption gain a direction in this context, often forming the basis and legitimacy of the distinction (Avcı, 2021).

Since job opportunities in the tourism sector are similar to the jobs which women do at home, women are intensely employed in tourism. However, the excessive involvement of women in the tourism sector does not prevent the problems, inequalities and discriminations they experience in other sectors from being seen in the tourism sector as well (İkiz, 2020).

The tourism sector is a sector in which more female employees are employed compared to other sectors and offers great opportunities for women. Equality between men and women in the sector is a phenomenon that is generally accepted but not fully realized in practice. In addition, there are some problems experienced by female employees, especially since it requires an intense workforce (Ateş, 2022). In the research, not only the problems of the female guides were discussed, but also the comments of the male guides were discussed, revealing whether there was gender discrimination in terms of both male and female tourist guides, and a comparison was made in terms of female and male guides. It is aimed to reveal the problems of the guides, to produce solutions to them and to make suggestions by determining the problems caused by gender discrimination in the profession.

THE PROFESSION OF TOURISM GUIDANCE

Promotion and marketing activities are particularly important for growing industries such as tourism. Only public institutions and tourism enterprises are not responsible for the promotion and marketing of tourism destinations. In addition to these, tourist guides are the people who transfer information about the destination in many fields such as history, culture, social life, education, health, economy, industry to the tourists visiting the tourism destination, introduce the destination in every aspect, communicate with the tourists and are the promotional face of the destination (Albuz, 2008). 2018).

Tourist guiding, which is one of the basic elements of tourism, emerges as a very comprehensive and versatile professional activity. Tourist guides, one of the main stakeholders of the tourism sector, are tourism employees who have undertaken the task of mediating between the local people who are the hosts and the tourists who are the guests. Tourist guides, as volunteer tourism ambassadors of countries, interact with tourists one-by-one, spend more time with tourists than other employees in the tourism sector, accompany tourists as representatives of countries / destinations and travel businesses they serve, and provide positive experiences. They play an important role in this interaction and communication process as sector representatives who ensure that they live and leave the country in a happy and satisfied way (Çetinkaya and Öter, 2021).

According to the definition of the World Federation of Tourist Guide Associations, a tourist guide is a person with an area-specific qualification granted and/or recognized by the relevant authority, who guides visitors in the language of their choice and interprets the cultural and natural heritage of an area (WFTGA, 2022).

Ap and Wong (2001) tour guides are one of the most important players in the tourism industry. They are people who, with their knowledge and interpretation of a destination's attractions and culture, and their communication and service skills, have the ability to transform a tourist's visit from a tour into an experience.

Tourist guiding has an important and versatile role in contemporary tourism. A safe and quality experience with a guided tour includes the guide, tourists, and the environment, and occurs when all three components interact at the same point in space and time (Rabotic, 2010).

Huang and Chan (2010) define a tourist guide as someone who works at the forefront of the tourism industry who plays an important role in shaping the experience of tourists at a destination. Tour guide service is the core component of various tour services offered by tour operators. Whether tour guides can provide quality service to tourists is critical not only for the business success of the tour operators they are affiliated with, but also for the overall image of the destination they represent.

The image that tourists have about the country is one of the key points in choosing that country in terms of tourism. Creating a positive image for the country in terms of marketing in tourism and transferring this image to people correctly are issues that need attention. Tourist guides come to the fore in transferring the image of the country to tourists. Their performance can create recurring and new business and can also affect the image of tour companies and even destinations (Ayyıldız et al.; Zhang and Chow, 2004).

Tourist guides have some responsibility as providers of tourism experiences and sometimes have competing roles to meet the needs of visitors, employers and host communities. Visitors expect their safety and health to be a high priority, while at the same time expecting an enjoyable and rewarding travel experience. Some have special needs and expectations associated with their cultural background, physical and intellectual abilities, and passions and interests in certain subjects. To meet these expectations, employers expect the guide to provide high-quality service to visitors, as well as manage the group, itinerary, and other logistical aspects of the experience in a way that maximizes not only visitor satisfaction but also profit margins (Weiler and Ham, 2002).

One of the most systematic researches on tourist guides was made by Cohen (Çetinkaya and Öter, 2021). According to Cohen (1985), tourist guides are evaluated under two main headings as guiding guides and guiding guides. Considering these two basic roles of tourist guides, the role of guiding is defined as leadership and the role of guiding is defined as mediation.

GENDER DISCRIMINATION IN TOURISM GUIDANCE

Gender inequality is one of the most important problems of today's world. It is a global problem where gender-based prejudice and discrimination against women exist in both developed and developing countries (Cheema and Jamal,2022). Although there have been some important changes in recent years, gender inequality continues to be an important economic problem. Gender discrimination in the field of tourism is not only limited to women employed in the tourism sector, but is also reflected in the scientific studies of academicians (Genç, 2018).

Gender discrimination in business life is mostly perceived as barriers to women. Sociological studies, which reveal that there are women disproportionately in some occupational groups and in lower-paid jobs, are significantly questioning about discrimination (Blau and Tatum, 2000). Perception of gender discrimination for women is related to perceived prestige at work and specific outcomes such as low power, high work conflicts, low self-esteem, increased depression and anxiety. The fact that women have more family responsibilities and the role of a wife and mother can create a disadvantage among men in business life (Demir,2011).

There is also gender-based discrimination in the tourism guiding profession within the tourism sector. Although the number of female tourist guides is increasing, discrimination, which is a problem in the society in general, is also noticeable here. Especially female guides are faced with wages, working hours, working conditions, sexual harassment and mobbing (Zengin et al., 2014). In their study, Zengin et al, (2014) found that female guides experienced some problems related to their gender while performing their profession. Almost all of the participants in the study experienced concrete distress related to their gender. It has been found that travel agency employees have rude behavior towards female guides and do not want to take orders due to being a woman. The conclusion that being a woman is not an obstacle for the guidance profession, but requires a little more struggle, was obtained from the research data. Most of the participants stated that they would not be able to do this job in the future, and they stated social and familial reasons as reasons. According to the data obtained, the most common problems are sexual harassment, not being given tours because they are women, rude behavior and words of the people they work with, family and social pressure. Sexual abuse is one of the main problems caused by the fact that the guides are women. Both domestic and foreign tourists participating in the tour and the personnel working with the guide themselves commit this abuse.

Ateş et al., (2022) stated in their research that female tourist guides faced positive discrimination in some cases compared to their male friends during their work. However, it was also determined that most of the female tourist guides who participated in the interview were exposed to negative discrimination in their professional lives. It was stated that female tourist guides were subjected to verbal/sexual harassment by drivers, tourists and colleagues during the tour.

Arslan and Şimşek (2018) examined the professional problems of female tourist guides, in order of importance, the problems of female tourist guides, social and family problems, problems with colleagues and colleagues, problems with travel agencies, problems arising from the characteristics of the sector and the guidance profession, problems experienced with customers. These problems are classified in six categories as control and education problems.

Barcia and Quijano (2017), the general perception is that access to jobs and leadership positions are determined by talent and knowledge, not gender. However, even when the women in question perceive that recruitment and distribution of duties depend on the skills of the candidates, for example, the fact that women are not addressed as their male peers by their academic titles shows that there is an obvious gender discrimination in business life, which leads to adherence to conscious or unconscious stereotypes.

In his research, Koc (2020) revealed that women can possess quite a few of the qualities necessary to fulfill a wide variety of service duties efficiently and effectively. The findings of the study are not intended to show that women are superior; rather, it means that they are not inferior to men. However, one must be aware that an employee's education, personality traits, abilities, and other skills can play an important role in determining his level of performance on the job and therefore job performance. It is emphasized that family and community approval is required for guides to practice the profession, and the roles within the family constitute the biggest obstacle to continuing the profession, and accordingly, travel agencies mostly prefer male guides.

The personal characteristics of women are more suitable for practicing the profession than men; It is thought that they can be successful in the profession despite the perception of gender. Although it is accepted that men are more advantageous in terms of physical characteristics in the tourist guide profession, their empathy skills and communication skills take women one step further in the profession. The responsibilities of family life and the social environment are seen as the biggest obstacle for female tourist guides to continue their profession (Yumuk and İl, 2022).

In the research of Yumuk et al., (2020), it was determined that female tourist guides are exposed to both positive discrimination and harassment and violence. In order to deal with discrimination, female tour guides behave in their own way. They adopt strategies such as being persistent in their profession, keeping up with difficult conditions, improving themselves by constantly reading and experiencing new places, being shy against travel agencies, drivers and tourists, demanding and defending their rights.

Köroğlu et al., (2020) One of the important results obtained from their research is that the main problems faced by female tourist guides in the profession are conflicts with drivers and transportation issues after the tour that ends at a late hour. Female tourist guides are constantly interfered with the thought that they will not succeed during the Tour. Being seen as defenseless and being harassed are among the problems.

Zafarghandi (2016) conducted a study on the perceptions of Iranian tourist guides. According to the results of the research, the destructive perceptions of agency managers towards married female guides cause economic distress and reduce their income. They believe that married women cannot handle long tours or heavy workloads. Male tour guides have very positive opinions about their colleagues. These views have been influential in their personalities. From the men's point of view, patient personality, high social contact and considerable language skills are common characteristics of female guides.

The latest paradigma in tourism development policies should be reconsidered in order to maximize the tourism potential, so that women are empowered and further steps should be taken to promote gender equality (Genç, 2018).

Unless we understand the gendered complexity of tourism and the power relations they involve, we cannot recognize the strengthening and construction of new power relations that emerge from tourism processes. All parts of the tourism experience, from the values and activities of the transnational tourist operator to the diverse experiences of individuals participating as hosts or guests, are influenced by our collective understanding of the social construction of gender (Ferguson, 2013).

PURPOSE AND IMPORTANCE OF THE RESEARCH

Women are disadvantaged in the workplace due to various problems based on gender discrimination. Gender discrimination is common in tourism, which is a labor-intensive industry. Women are subject to gender discrimination in a wide variety of sub-sectors of tourism, from hotel management to tour guides. This study investigates gender discrimination in the tourist guiding profession. This study examines the difficulties experienced by both male and female tour guides by adopting a dual perspective. Studies on discrimination in tourism guidance are limited. The study is important in terms of determining the size and nature of gender discrimination by comparing the difficulties experienced by both male and female tour guides. Based on the results of the study on the scope and nature of discrimination, it creates a strategic plan for practitioners and stakeholders.

METHODOLOGY

Women are exposed to gender-based discrimination not only in the tourism sector, but also in many different areas where they work. Especially in the tourism sector, it is known that they experience discrimination because they are women and they encounter undesirable situations in the sector. In this study, the problems experienced by female guides will be discussed and necessary solutions will be revealed. In the study, a comparison will be made in terms of male and female guides, taking into account not only the opinions of female guides but also the opinions of male guides about discrimination in tourism guidance. With the results of the research, it will be emphasized what should be done about the discrimination experienced by women guides in the sector, and suggestions will be presented so that women can hold on to this sector and carry out their profession with peace of mind without being exposed to discrimination.

Qualitative research method was preferred in the study. Qualitative research; It is a type of research in which the data are analyzed using content analysis, one of the qualitative data analysis techniques such as observation, interview, focus group interview, and document analysis, and a process is followed to reveal the events in a realistic and holistic way by the researcher in the natural environment (Sığrı, 2022). Within the scope of the research, convenient sampling was used to collect data from 15 female and 15 male tourist guides. Interviews with the guides were provided to collect the data. The guides were asked 10 questions about discrimination in tourism guidance. In the preparation of these questions directed to the guides, the studies of Ateş et al., (2022), Köroğlu et al., (2020) and Zafarghandi (2016) were used. The universe of the study was determined as active female and male tourist guides. Purposive sampling method was used in the selection of the sampling method. Purposeful sampling is a non-probability sampling method and is widely used in qualitative research. It is based on the principle of determining and selecting people who are competent and knowledgeable in the researched subject, with the personal opinion of the researcher. Since the researcher acts with his own judgments and evaluations, he should have an idea about the universe (Ural & Kılıç, 2018; Altunışık vd. 2022).

In order to collect the data, the questions directed to both male and female guides are given below;

- Do you think there is positive discrimination based on gender in your profession? If you think this way, what are the circumstances that cause it?
- Do you think there is negative discrimination in your profession? If you think this way, what are the circumstances that cause it?
- When you compare it with the past, do the developments you experience in the first years of your career and afterward go in a positive or negative direction? What kind of developments made you happy? What kind of developments made you sad ?
- Do you encounter the situation of being preferred/not preferred by the agency on tours depending on your gender? If so, do you use any method to solve this problem?
- Can male and female guides benefit equally from job opportunities in the profession?
- How does the counseling profession affect your family life? If you are married, how does it affect your marriage? What are the problems? What are you doing to overcome them?

If you are single, will you continue your career? How do you continue? How do you plan to overcome the problems you may encounter? Do you intend to continue your profession?

- Can you tell us the opinions of both male and female tourists about your profession? How do they behave towards you? What kind of problems and obstacles did you encounter during the tour due to your gender? With whom/with whom did you have more problems? What are your suggestions for solving the problems?
- Are you being harassed due to your gender? Do you think that female guides are subjected to verbal abuse/sexual harassment during the tours?
- Do you see men or women more advantageous in terms of income? Why?
- Despite the problems you have experienced in your profession, have you considered continuing to work in this profession or leaving this profession? How do you deal with problems in the profession?

FINDINGS

Within the scope of the study, 30 people were interviewed, including 15 female guides and 15 male tourist guides. Each participant is P-coded and numbered. The numbering of the participants consists of female participants from P1 to P15 and male participants from P16 to P30. Demographic characteristics of female participants are given in Table 1, and demographic characteristics of male participants are given in Table 2.

Table 1. Demographic characteristics of female guides

Participant	Age	Marriage Status	Education	Professional Experience	Chamber of Tourist Guide
P1	49	Single	Undergraduate	25 years	IRO
P2	35	Married	Undergraduate	13 years	ATRO
P3	52	Married	Lisansüstü	3 years	ATRO
P4	51	Single	Undergraduate	26 years	IRO
P5	36	Married	Postgraduate	13 years	ÇARO
P6	46	Single	Postgraduate	20 years	ATRO
P7	60	Single	Undergraduate	39 years	IZRO
P8	53	Married	Undergraduate	34 years	IRO
P9	50	Married	High School	30 years	IZRO
P10	56	Single	Undergraduate	35 years	IZRO
P11	30	Single	Undergraduate	3 years	IZRO
P12	51	Single	Undergraduate	27 years	ANRO
P13	27	Single	Undergraduate	5 years	IRO
P14	42	Married	Undergraduate	20 years	ATRO
P15	38	Married	Undergraduate	17 years	BURO

Table 2. Demographic characteristics of male guides

Participant	Age	Marriage Status	Education	Professional Status	Chamber of Tourist Guide
P16	57	Married	Undergraduate	34 years	IZRO
P17	56	Single	Undergraduate	32 years	IZRO
P18	29	Single	Postgraduate	6 years	IZRO
P19	45	Single	Undergraduate	21 years	MUTRO
P20	32	Married	Undergraduate	6 years	IRO
P21	33	Married	Undergraduate	10 years	MUTRO
P22	27	Single	Undergraduate	3,5 years	GARO
P23	34	Single	Undergraduate	15 years	IRO
P24	27	Single	Undergraduate	2 years	ÇARO
P25	36	Married	Undergraduate	13,5 years	MUTRO
P26	33	Single	Postgraduate	7 years	ATRO
P27	27	Single	Undergraduate	5 years	IRO
P28	31	Single	Undergraduate	7 years	IRO
P29	32	Married	Postgraduate	9 years	ARO
P30	28	Single	Postgraduate	4 years	ARO

When the demographic findings of the female guides in Table 1 are examined, it has been determined that the ages are between 27-60, 7 people are married, 8 people are single, 11 people have the Bachelor's degree, 3 people have a Master's degree and one person is high school graduated. When we look at their professional experience, it varies between 3 and 39 years. 4 people are members of IRO, 4 people from ATRO, 4 people are from IZRO, 1 person is from ÇARO, 1 person is from BURO and 1 person is a Guide of ANRO's.

When we look at the findings of the male guides regarding Table 2, the ages ranged between 27 and 57 years. It has been determined that 10 of the male guides are single and 5 are married, 4 have a Master's degree and 11 have a Bachelor's degree. Among the guides, 3 people are a member of IZRO, 2 people are from MUTRO, 4 people are from IRO, 2 people are from ATRO, 1 person is from GARO, 2 people are members of ARO and 1 person is an affiliated ÇARO.

Table 3. Gender-based positive discrimination in the profession of tourism guidance

	Sub-Theme	F
Main Theme Positive Discrimination	Female groups prefer female guides	5
	Preferring a male guide for challenging tours	6
	The preference of religious and conservative groups	7
	The presentation and appearance of female guides	2
	Seeing female guides more friendly	2
	Preferring a male guide in terms of flexible working hours	5

In the study, the comments of male and female guides about whether there is discrimination in tourism guidance were taken by asking some questions to the guides. First of all, "Do you think there is positive discrimination based on gender in your profession? If you think that way, what are the circumstances that cause it?" question was posed (Table 3). There are guides who both think and do not think that there is positive discrimination. The answer given by one of the female participants is as follows:

In some tours, there is more demand for female guides. This is sometimes done as by groups of women, or sometimes to guide their protocol or protocol mate. I mostly received protocol/ peers' requests. **(P1)**

Response of the participants **(P2)** is:

I think in some situations. I usually guide private tours. I have received feedback from my clients many times that they are very happy to see a female guide.

I believe and think that there is positive discrimination. Because I heard from the Travel agencies in the area that the female guides are more friendly and warmer to the ship customers, and recently many agencies have started an initiative to recruit women guides. Agencies always say that they used to work with men, but they said that they are considering changing the staff now that the female guides seem better and more friendly. **(P3)**

When I first started my profession, I thought there was positive discrimination. I thought that the profession was more suitable for men and that male guides could do it easily. Of course, it is difficult to be accepted as a woman in many sectors, including the tourism sector, to adapt to living and working conditions, and to carry out all tasks at the same time. At first, I had such feelings. But later on, I can say that this situation has disappeared. If such positive discrimination is in question, it is actually positive discrimination in the structure of society. Roles assigned to women and men, responsibilities undertaken by women in business or family life. We know that women's responsibilities exist in our social structure, mostly at home and more in family-related processes. Accepting a woman in business life, especially in the guidance profession, and working under these conditions is challenging at first. **(P5)**

I think there is positive discrimination against female guides. I think that female guides are preferred in terms of first impression, communication and problem-solving skills, especially among guides serving upper segment customers. **(P6)**

Positive gender discrimination is preferred when there are women and when there is a religious group. There are religious groups coming to Spain from Turkey. The group of women prefers a female guide. **(P8)**

I think that there is positive discrimination based on gender in my profession. While looking for a guide to go on a tour, I have witnessed many times that a nice woman is preferred over her male counterpart, and I have experienced it myself. **(P10)**

Yes, I think. In general, I see that women guides are more preferred by agencies and in closed groups. I think the biggest reason for this is that women are more visually presentable than men. I think that the soft tone of voice, style of dressing and appearance are important factors. **(P11)**

I think there is positive discrimination. I can exemplify the thought that it would not be right for me to get out of the vehicle during any security check, as I am a female guide, and that I would be more tired because I am both a mother and a guide. **(P12)**

There is positive discrimination for men. It is still the case that people do not attribute travel-requiring occupations to women based on traditional norms. **(P13)**

On some tours, especially on tours with female guests, Arab, Indian etc. I know that women guides are specifically requested in the tours of families from other nationalities. In some groups, female guides are preferred more. Women have more advantages. Sometimes, if closed groups coincide, they may ask for a female guide. **(P14)**

The comments of male guides on positive discrimination were examined. The thoughts of **P16,** one of the participants, are as follows:

I heard that some religious groups ask for female guides. When they go to Konya to visit Mevlana, they wanted a female guide, but it did not happen to me.

Sometimes there is discrimination. For example, if a closed group requested service, a female guide may be preferred in line with the request of that group. **(P18)**

I can clearly say that more men are preferred in scheduled tours which have a difficult and long road and also not a specific day and night. **(P19)**

*I think there is positive discrimination against men. Except for some special requests, I have witnessed that some agencies do not want to work with women in particular. (***P21)**

In general I don't think so, but some conservative groups request female guides. I don't support this, it's not a nice behavior. **(P22)**

No, I do not think so. In our profession, there can only be choices. This can be shaped according to the wishes of the groups and agencies you will work with. **(P24)**

*As a man, I experience firsthand that men often experience positive discrimination in the profession. This is due to the fact that men object less to long and flexible working hours due to the fact that the family and domestic burden (child care, housework, etc.) is mostly on women, and the trust in men is more likely due to the widespread belief that men are more physically and psychologically resilient. For example, men are often car drivers and so on. This is due to reasons such as reducing the cost of hotel reservations because they do not have a problem sharing the same room with other people. However, from a labor perspective, it is a strange situation that men are more open to exploitation in terms of physical conditions, creating employment for them. In general, female guides need to prove themselves more than men in order to earn the same average wages as men. It is also among my personal observations that tours that are more qualified in terms of both financial and professional satisfaction are mostly given to male guides easily. (***P25)**

Table 4. Negative discrimination based on gender in the profession of tourism guidance

	Sub-Theme	F
Main Theme Negative Discrimination	The thought that women cannot do the guidance profession	5
	Requesting a male guide for tours with seller meetings	3
	Re-preferring the agency according to the sales made by the guides	3
	Preference of elderly guides	1
	Preference of young guides	1
	Inequality between tours given to husband and wife	1

I think that there is positive discrimination against men in my profession. The reason for this is undoubtedly the effects of the patriarchal social structure, but when it comes to flexible working hours in the sector, it is in the interest of travel agencies that men are more physically durable. **(P26)**

Sometimes female guests especially ask for female guides. Likewise, the opposite can also happen. There is also a situation like that, since we are a patriarchal society, the number of female guides is also low, as it is a sector where men work more. **(P28)**

I don't think there is any positive discrimination based on gender. It is a fact that sometimes some groups especially demand female guides. **(P29)**

There is no discrimination except for the reasons like the need of a female doctor and seeking a female guide.İt is the same in different ways. Since Pakistani groups do not see many women in their lives, they cannot be spoiled when a female guide is in front of them, so a female guide is sought. **(P30)**

"Do you think there is negative discrimination in your profession? If you think that way, what are the circumstances that cause it?" question has been asked (Table 4). The most common view is that women cannot do the guidance profession. The answers are as follows:

In groups that are mostly men, male guides have a higher employment rate, especially in incoming and outgoing tours with dealer meetings. This can be an advantage when a female guide does not want to meet. **(P1)**

Some agencies give shopping tours to Couple guides, one of the spouses, while the other is given the crew, working staff, etc. I've heard of tours that don't even pay a tip. It didn't happen to me. **(P2)**

I think there is negative discrimination. There is a group that thinks that women cannot do the counseling profession. **(P4)**

In sport groups, for example, because the atmosphere will be very tense, there is a tendency of the agencies not to let the women into such environments for protection purposes. **(P8)**

I think there is negative discrimination in my profession. I have experienced that the guide is preferred in later groups, especially in sales-oriented or purely sales-based tours, according to the results of the marketing and sales made by the guide, and groups with stronger financial means and more people are given to them. **(P10)**

I encountered negative discrimination in two different ways, mostly in matters related to age. First of all, older guides are more preferred than younger guides. This is because older guides are seen as more knowledgeable. However, I see that young guides are preferred most of the time. The reason for this is that young people or newly graduated guides are more dynamic and enthusiastic about the profession. Fresh knowledge and visually appealing to younger people more to the group. **(P11)**

I think there is more negative discrimination. Thinking that I will be forced to take the route just because I am a woman. Because I am a female guide, it interferes with my habits how I sit or stand up, paying more attention. **(P12)**

Negative discrimination is more intense. I think it should be perceived as gender discrimination as well. As like in any job, men are thought to be more resourceful. When women make mistakes, she is already a woman, she will definitely do it. Let her stay at home and take care of his child, why is she doing this job? But when men make mistakes, I've seen that it will be covered up because it's considered a mistake. We are in a profession where mostly men work. Since the first years, I have been aware that the Drivers who use the bus are male, especially since we travel by vehicles, and that the drivers behave very differently, especially towards female guides. If they realize that you are a novice guide, they come at you more as a woman. I have experienced these one-by-one. **(P14)**

I think negative discrimination is against women. The excuse of those who did this was that they could not find a reliable female guide. **(P21)**

As a man, although it was not possible for me to experience this situation in person, I can easily say that there is gender discrimination as far as I have observed what my female colleagues have experienced. I often observe that Women have more difficulty in collecting their receivables on time than men. **(P25)**

I do not think that I have been subjected to negative discrimination. The fact that only conservative female tourist groups do not prefer male guides can be considered as negative discrimination for these tours, which are few in number. **(P26)**

I don't think there is negative discrimination. However, in some cases, men may be preferred for challenging and intense programs instead of female guides. **(P29)**

Especially in the tourism sector, women are always seen as very spoiled and unsuitable for the job due to their low education and intellectual level among employers and employees. Since the qualifications of the job are not adjusted in a way that a person can handle, the idea that a woman will always be delicate and the understanding that men can keep up with long-term and aggressive working conditions causes women to be less preferred. Few women who can keep up with these conditions are still able to get some jobs by standing out from other women. **(P30)**

Table 5. Evaluation of tourism guidance in terms of years

Main Theme	Sub-Theme	F
Positive Aspects	Improving yourself	2
	Gaining experience	3
	Recognition	2
	Emphasis on personal and professional development	1
	Increasing specialization training	1
	Awareness of guides	1
Negative Aspects	Political crises, decrease in business due to epidemic	5
	Lack of female guides in the industry	3
	Negative news in the foreign press	1
	Negative impact of the all-inclusive system	1
	Insufficiency of the base fee for Tourist guides	3
	The troubles which female guides have during their profession	2
	A decline in tourist quality	3
	Negative working conditions of guides	2
	Considering the cheapness of the tours rather than the quality	1
	Transforming museums into places of worship	2

"Comparing to the past, do the developments you experience in the first years of your career and afterwards does it go in a positive or negative way? What kind of developments made you happy? What kind of developments made you sad?" question has been asked (Table 5). Most of the guides stated that the guidance has tended towards the negative over the years. Among the participants, **P1:**

There were those who went negative as well as positive. There has been progress in self-development, gaining experience, recognition, reputation, gaining self-confidence and confidence, these are the positive aspects. On the negative side, the decrease in business due to reasons such as terrorism and epidemics in recent years, the decline in the quality of incoming tourists in recent years, affects this income to a great extent.

It is a pleasing development to see that there are female guides who have made a name in the very successful sector and are especially sought after. It is undoubtedly a sad development to see that there are far fewer female guide candidates than in the past. **(P4)**

P5 has been positive in many ways.

The distinction between men and women is not only in tourist guidance, it exists in many different areas. This is actually the roles and responsibilities undertaken within the order and system that life has brought. In fact, these duties are attributed to us by society. When we thought like this, there were things that I had difficulty in the first place. But I can say that it turned out to be more positive. In this, the profession and the industry accept me. Despite all the difficult conditions, acceptance of the profession, my better feeling in a positive sense, giving importance to personal and professional development took me from positive to more positive.

In the first years of my career, the quality of service that came to Turkey was higher. Although we are better than all our competitors in terms of service quality, Turkey's being seen as a cheap destination which has some disadvantages. The negative news and bad reputation in the foreign press about the shopping experienced by tour groups in Turkey created a very bad image for the guide profession and guides. **(P6)**

It is going negative. The all-inclusive system has no return for the country. Opening the places that are used as museums and receive many visitors (Hagia Sophia, Chora) to the public make the visits extremely difficult. The unfinished and long restorations reduce the interest here. With the renewal of the guidance base only once a year, the wage, which is the equivalent of such a labor-intensive work, cannot be received. The only positive development is the increase in the number of people who want to explore and visit the country due to Turkish TV series. **(P10)**

In the first years of my career, I had to deal with more things than male guides. I felt like I had more problems. Over the years, it's gotten better with experience. As the years progressed, I was able to cope more easily with the negative discrimination against me. I believe I can even respond better. We are the first profession to be affected by the economic or health crises experienced in the country. Because of this, our jobs are cut off and we may be in financial difficulties. **(P14)**

I think it's going negative. The tourism sector is one of the sectors that suffer the most from external factors. Due to the pandemic, there was a high decrease in both the domestic market and the number of foreign tourists. Compared to the past, there is an incredible decrease in the quality of tourists. Turkey is preferred mostly by countries with low purchasing power due to economic concerns. **(P11)**

I started at 2016. Since then, we've had a loss of income versus minimum wage. We still earn more than the minimum wage earner, but with our monthly income, we used to be considered middle class, but now we are close to the lower class or minimum wage in monthly income. In addition, as a country policy, we switched to mass tourism. There was a situation such as Quantity over quality. **(P20)**

It is possible to say that Turkish tourism, especially the cultural tourism sector, is getting worse day by day in terms of quality. Factors such as the ever-increasing price competition, tour operators and entrepreneurs unaware of the tourism culture, the cost of all these by sacrificing quality, and the tourist profile that does not worry about this poor quality, worsen the working conditions of tourist guides. **(P25)**

When I compare it with the past, I see that it is getting harder for us to do our job day by day. Agencies have started to see guides as a cost and a burden. The focus is now on making the tours cheaper and more salable rather than their quality. It's getting harder and harder to get a base wage. Even having a guide has become a luxury for some tours. **(P29)**

It is going negative every day. Guidance will disappear for political reasons in Turkey, which has completely changed its shape in the rest of the world due to digitalization and political, economic reasons.

Table 6. Whether or not they are preferred in the tours based on gender, problems experienced due to gender during the tour and the verbal abuse/sexual harassment of the guides

Main Theme	Sub-Theme	F
Problems Faced by the Profession	The fact that Agencies do not prefer female guides	4
	Giving better tours to male guides	3
	Inability of male tourists to distance themselves from female guides	2
	Problems with tour drivers	7
	The fact that female tourists become jealous of female guides	2
	The believe of Tourists that female guides cannot do the tour	2
	The idea of male guides to establish close relationships with female tourists	1
	Sexual harassment by female guides	18
	Sexual harassment by male guides	2

The services offered by the profession will also disappear completely as the demand tourist typology changes. **(P30)**

"Whether or not they are preferred in the tours based on gender, problems experienced due to gender during the tour and the verbal abuse/sexual harassment of the guides?" questions were posed. (Table 6). According to the results, it was determined that most of the female counselors were exposed to verbal/ sexual harassment. In addition, the problems that the guides have with the tour drivers are the most common problems. Zengin et al., (2014), Ateş et al., (2022) reached similar results in their studies.

It happened to us when we were affiliated with an agency in the past. But there are no agencies with such intense potential anymore, so I am not exposed to such a distinction. **(P1)**

Since I am a woman in the agencies I work for, I do not have a problem of not being preferred. But I know I have colleagues who have this problem. From time to time, I observe that there are companies that discriminate by giving fewer tours or giving tours that are thought to be better to male guides. **(P4)**

I've never had a situation where I didn't get a job because of my gender. But I have friends who have this problem. The solution is education. **(P12)**

Some groups may prefer women in particular. These are almost all women, or in private family groups, in family groups from Middle Eastern nationalities, it is preferred to be a woman rather than a foreign man. This is also seen in the advertisements. I have not seen it particularly preferred as a male. More activity-oriented tours such as camping and trekking. Agencies give priority to men. Male guides are preferred for tours that need to be managed more based on physical strength. **(P14)**

I haven't come across any tour advertisements that specifically sought a male guide, so I've never gotten a job like this. On the other hand, I did not encounter any situation where being a man was disadvantageous. **(P25)**

As a man, I think it is one of the rare professions that I can do in Turkey. But if I had been born a woman, I would have left Turkey no matter what. As a man I can't stand it, as a woman I wouldn't spend two seconds and I would never choose this profession. **(P30)**

I have had many American tourists commenting that female guides are more detailed and interested. This interpretation is actually a veiled reference to a maternal structure, and it is still a sexist interpretation in a way. Although male tourists behave respectfully and distantly, there are also those who try to overcome this distance because I am a female guide. Of course, the biggest problem in performing the profession is that we live with the Drivers. They don't like to take orders from a woman. They think that female guides can't be good. When problems arise in hotels, restaurants etc. the difference between a male guide and a female guide becomes clear. Because the other side behaves accordingly. It is very important for the female guide to adjust her tone and words while solving the problem. Another problem I have because of my gender is that female tourists can sometimes be jealous of us, and accordingly, they can act irritably, humiliatingly and sarcastically. **(P4)**

Women guests often wonder that there will be no family togetherness and that it is difficult to have such a free profession as a woman. I didn't experience anything very remarkable among the male guests. But in one of my tours there was a Gentleman thinking since I was a woman, he thought it would be difficult for me to do that tour. He had been a little rude. Due to my patience and conscious, uncompromising respectful stance, he thought differently at the end of the tour. **(P12)**

Tourists in particular wonder how we conduct our family life, whether our earnings are worth making such sacrifices. The problems I have due to gender are mainly the tour drivers' difficulties in obeying a female guide, their pressure on the guide to increase their own income, and their requests for your help because of their interest in female tourists in the group. It is imperative that tourism drivers go through a definite training and do this job. **(P10)**

There were those who implicitly implied that a male guide could have a more intimate relationship with a foreign guest. In fact, there are many tourism professionals who claim that male guides are lucky in this regard and that they have constant close contact with female guests. **(P29)**

Most tourists do not think tourist guiding as a full-time job. It is often thought of as if we are doing it for part time or hobby purposes. The more I talk about the profession and our professional law, the more the respect to our profession increases after the quality service they receive from us. **(P20)**

The opinions of women about my profession were mostly at the level of curiosity. When they learned that I was married, They asked me if my being away all the time had a negative effect on my marriage, and if I would encounter situations such as jealousy. Because of my gender, I have never had a problem with male and female tourists. But I often had a hard time communicating well with gay guests. The prejudices and concerns they had before coming to Turkey, combined with my being a man, caused them to have some prejudices against me that were hard to break. **(P25)**

Are you being harassed due to your gender? Do you think that female guides are subjected to verbal abuse/sexual harassment during the tours?

Table 7. Benefits and opportunities

Main Theme	Sub-Theme	F
The evaluation in terms of job opportunities	The fact that male guides are at the forefront of job opportunities	8
	Unable to benefit from job opportunities for female guides who are married and have children	5
	Take equal advantage of job opportunities	4

I think that female guides are subjected to verbal abuse/sexual harassment during the tours. Being harassed about my gender is not something that happens to me all the time in this profession, but I have been subjected to serious harassment attempts in the past, especially in hotels. Other colleagues have had similar experiences. **(P4)**

I think that the female guides, if not sexually, are subjected to verbal or behavioral harassment. **(P10)**

There is verbal abuse. I have not experienced sexual harassment, but there are those who do. **(P12)**

I know that female guides are often subjected to unwanted attention on tours, and this sometimes escalates to harassment. The source of this is sometimes tourists, and sometimes the people we work with, such as car drivers, hotel and restaurant employees. Due to the widespread belief that men will not be harassed, male guides mostly prefer to remain silent and hide in cases of harassment etc. Otherwise, situations such as being mocked, being accused of lying and slander are encountered. In the past, I have been subjected to persistent attention and sexual harassment from my female guests on many of my tours. I have encountered situations such as verbal contact harassment, taking various inappropriate photos without my knowledge and permission, trying to enter my room by force. When I stated that I would make a complaint to the other party, I was exposed to the threat that the other party would complain that "I harassed them and slandered them when I could not get a response." **(P25)**

Women are completely pushed out of the profession, especially by employers and drivers. When I suggested a female guide, I got the answer from many agencies that we do not work with women. **(P30)**

I have not been harassed myself. However, I think that female guides are verbally or physically abused. I think this is a general problem of our country, not just for the guides. **(P29)**

"Do male and female mentors benefit equally from job opportunities in the profession?" question was posed (Table 7). In response to the question **P14**:

It may vary according to the situation. The situation across the country, we can see that women are more in the background in terms of job opportunities. There is also the woman's holding back about her family life. I worked as a single guide for many years. I have been a married guide for the last 5 years. I took a tour of Anatolia when I was single. But now, as a married guide with children, I have to choose to go on daily tours because I cannot leave my child. Sometimes, a male guide is preferred in situations that require physical strength. The fact that women in all occupational groups have fewer job opportunities.

Table 8. The impact of tourism guidance on family life

Main Theme	Sub-Theme	F
The evaluation of family life in guidance	Inability of guides to see their children often	4
	The negative effect on marriage	6
	The responsibilities of single guides to their families	1
	Not affecting family life	4

They do not enjoy equal employment opportunities. For example, we have guides who want to add their second or third language. Male guides, they can eliminate this responsibility very quickly and gain speed on this issue very quickly. Going to Spain for Spanish, going to China for Chinese. There are guides who can go to different countries and stay there for a long time to learn a different language. They are comfortable in this regard, but unfortunately, female guides cannot make decisions so quickly and are not fast with the responsibilities that life brings. When most of them are married and have children, they can abandon this decision. We have guides who have become new mothers and they miss the job opportunity. This is not the case in newly fathered male guides. It is related to the duties and responsibilities attributed to the society. **(P5)**

I do not think that women and men in the profession benefit equally from job opportunities. **(P25)**

I don't think they benefit equally. **(P30)**

How does the counseling profession affect your family life? If you are married, how does it affect your marriage? What are the problems? What are you doing to overcome them? If you are single, will you continue your career? How do you continue? How do you plan to overcome the problems you may encounter? Do you intend to continue your profession? Question was posed. (Table 8). According to the answers given by the guides, marriage life is negatively affected and for this reason, many guides are considering leaving the profession.

I am not married, but I am responsible for my family. A process of acceptance by members of the family. I do not face any problem and I plan to continue my profession. These problems also depend on the profession of the other party. **(P13)**

I was not married when I was starting tourıst guiding. I haven't had a problem. I am currently married, but now that I do student tours, they are tours that start in the morning and end early. It doesn't pose a problem. **(P5)**

I am a guide for 26 years, I am unmarried. Not being able to see the children of married guides as often as they want is the biggest problem I've heard from them. It is a profession that requires understanding from both parties. Being away from home for a long time and sometimes being home for a long time can affect the relationship dynamic. **(P12)**

Table 9. Income status of male and female guides in the guidance profession

Main Theme	Sub-Theme	F
The evaluation of income status in guidance	The fact that male guides earn more from tours	9
	The difficulty of accessing to well-paid tours by female guides	3
	Status of equal income	6

I'm single and that's not a problem for me. If I decide to marry in the future, I will continue to do my job. In order for this not to be a problem, I am thinking of combining my life with a lady who accepts the benefits and costs of my profession. **(P26)**

I stopped actively guiding shortly after I got married. What I experienced during the engagement process showed what kind of problems I might encounter during the marriage process. It is a profession that requires a lot of sacrifice in the name of marriage. Being away from home for a long time will not be the right choice for most couples. I think this situation will be more difficult, especially after having children. I do not think that a father or mother who is far from his family can be beneficial to his family. I took a break from the guidance because I thought that this situation would negatively affect my marriage. **(P29)**

Do you see men or women as more advantageous in terms of income? Why? (Table 9). According to the answers given, it was determined that male guides earned more income and women had problems in accessing well-paid tours.

Male guides can earn better when agencies give more or better tours to male guides. But there are other items that affect earnings, of course, where a female guide can be very successful, earning more than her male colleagues. **(P4)**

I don't think there is much difference in our profession, as there is a basic wage. **(P12)**

I think men are more advantageous in terms of income. Men have less difficulty than women in accessing better-paying jobs. **(P25)**

Because men have a higher verge behaviorally, they have a greater flow of work and money. **(P30)**

In terms of income, I think men are more advantageous than women. I think the reason for this is the man's perspective on working life. A male guide wants to return home from the tour with as much profit as possible. Therefore, he tries to earn as much income as possible from non-tour activities and sales. The female guide, on the other hand, sometimes ends the tour by acting conscientiously or by earning less than the male guide because she does not focus on sales and earnings as much as a male guide. **(P29)**

"Despite the problems you have experienced in your profession, have you considered continuing to work in this profession or leaving this profession? How do you deal with problems in the profession? question was posed (Table 10). When the answers given are examined, most of the guides think about continuing their profession despite all the negativities.

Table 10. The condition of progressing to work in the profession of tourism guidance

Main Theme	Sub-Theme	F
The persistence of tourism guidance	The desire to do and continue the profession with love	8
	Quitting work due to country problems	3
	Quitting work due to marriage	1
	Thinking about leaving the profession	2

Keep going as far as I can. I don't plan on leaving. I try to struggle by minimizing the costs in line with the conditions of work, incoming tourist profile, living and cost of living. **(P1)**

Maybe not because of the problems I experienced in the profession, but because of the increasing problems in the current country, I left the profession. I don't think I will be guiding again, with the disappearance of the features that will promote my country as enthusiastically as before. **(P10)**

No matter how many problems I had, I love my job and I will continue to do it. **(P12)**

From time to time, I was so depressed that I thought of changing professions. But since my desire to continue always dominates, I still continue my profession and I will continue for many years. Even though I have established my own agency, I still prefer to go on tours rather than just working in the office. **(P23)**

I've seen a lot of trouble and hardship for 32 years, but I never thought of changing my job. I chose to keep money aside for hard times. **(P17)**

During my marriage, as I thought that this profession would harm my marriage, I searched for different ways and stopped actively guiding. Yet I continue to describe myself as a tourist guide. **(P29)**

When the comments of the participants are examined, there are problems experienced by both female and male guides. According to the comments, the profession of guidance is seen as a profession that can be done mostly by men in the society. Such comments are received by both male and female tourist guides. Among the reasons for this, the effects of the mentality of our society, men are thought to be more suitable for this job, both physically and psychologically. But we have very good female guides and they love to do this job. Our female guides should be given the opportunity to stay in the sector by getting rid of this thought. There are guides expressing the problems they experience in this sector by being exposed to gender discrimination. By investigating the causes of these problems, eliminating this problem will increase the commitment to the sector.

CONCLUSION AND RECOMMENDATIONS

In this study, it was investigated whether there is discrimination in tourism guidance. In the study, the difficulties experienced by both male and female tour guides were examined by taking comments from both male and female tourist guides. Within the scope of the study, interviews were conducted with 15 female and 15 male tourist guides. The study is a qualitative research and was created on the basis of purposeful sampling technique. Based on the results of the study on the scope and nature of discrimination, recommendations are offered for practitioners and stakeholders.

We aim is to reveal the gender discrimination by giving a comparison of the tourist guide profession in the tourism sector in terms of male and female guides. One of the most important problems preventing women's empowerment in the tourism sector is gender discrimination. Determining the level of gender discrimination in the sector will be beneficial for solving the problems. It is important to eliminate discrimination in the tourism sector so that women feel safe and empowered in the sector. There should be substantial opportunities for women to work.

Studies on discrimination in tourism guidance are limited and this study adds multidimensionality to the study by revealing the extent of gender discrimination over male and female tourist guides.

When the findings of the study are examined, both female guides and male guides think that there is gender-based positive discrimination in tourism guidance. It can be said that this is the case for both men and women. There are both female and male guides who say that women are generally preferred when there are groups of women or when a closed group comes. They say that some agencies especially want female guides when they advertise. This means that women are thought to be more visually presentable than men. From the point of view of men, there are guides who say positive discrimination is made to male guides. It is thought that the reason for this is that we are a patriarchal society and more men adapt to flexible working hours in the sector, and that male guides are more preferred in matters that require physical strength. Considering that women have both a house and children among the responsibilities they have, it is among the answers given that agencies do not prefer female guides and male guides are advantageous in this regard.

Among the answers given, it was revealed that positive discrimination was experienced mostly in terms of female guides. When evaluated in terms of negative discrimination, it is clearly stated that there are people who think that women cannot do this profession. But this profession is not just for men. Being a woman should not be an obstacle for the guidance profession. Women have very good characteristics and are talented in performing the profession of guidance. Women should be supported and given the opportunity to make a breakthrough in the sector. Aries (2020); Rich et al. (2014) obtained similar results in his study. It is the results of the research that being a woman is not a problem for the guidance profession, it just requires a little more struggle. Within the results of the research, it was determined that the female guides were exposed to verbal abuse / sexual harassment. There are guides who say this happens by tourists, by drivers, by hotel and restaurant staff. While this problem is mostly faced by our female guides, it has been determined as a result of the research that male guides also have problems in this regard. There is a belief that men cannot be harassed, and this is why it is not brought up. However, as a male guide, there is also a male guide who is subjected to harassment and has problems on his tour. Both female and male guides agree on this. In their research, Yıldırım and Özbek (2019) found that in professional tourist guiding, tourist guides see inappropriate behaviors involving sexual harassment from tourists and their colleagues, there is gender discrimination in this profession, and there is a perception that male guides provide better quality services in this profession. However, according to the results obtained, it is the female guides who have the most problems. One of the problems especially experienced by women is the problem they have with their professional tour drivers. One of the problems expressed is that the drivers do not want to take orders from a female guide and that the guides will fail to manage the tour.

Koroglu et al. (2020); Arslan and Şimşek (2018); Fire et al. (2022) also obtained the same results in their research. It has been concluded that female tourist guides have problems with drivers and colleagues. In this regard, there should be a system that requires drivers to undergo a certain training to do this job. Not everyone should be hired as a driver. Because this has been identified as one of the important problems that our female guides have shared.

One of the problems mentioned by the guides is that they should consider leaving the profession, considering that the profession affects marriage. The fact that women have more family responsibilities and a more effective mother role is a disadvantage in business life (Demir, 2011). According to the results obtained in the study, there are some among both female and male guides who left the guidance profession after they got married, thinking that it had a bad effect on their marriage. This is especially the case among female guides. The presence of physically difficult conditions in tourism guidance negatively affects the guidance profession (Lee and Kang, 1998). Especially after female guides have children, they cannot adapt to flexible working hours and their responsibilities increase. However, the same is not the case for a guide who is a father. For this reason, agencies want to hire more male guides and male guides have more job opportunities. In this case, it is thought that men have more job opportunities and are more advantageous in terms of income . Zafarghandi (2016) touched upon a similar issue in his research. The fact that agency managers think that they cannot handle long-term tours or heavy workloads for married women guides causes economic distress for women guides and reduces their income. Yumuk and Ili (2022) confirm this in their study. The responsibilities of family life are the biggest obstacle for female tourist guides to continue their profession. Our female guides should also be given a job opportunity and it should be ensured that the agencies have such a tendency, knowing that they will perform this job well, just like men.

The study is limited only to the profession of tourist guides. Different tourism sectors can also be included in the study. A comparison can be made from two different fields in the tourism sector by comparing the male and female employees working in the accommodation sector with those working in the tourism guidance profession. Thus, it will add more dimension to the work. Gender discrimination in tourist guidance is limited only to male and female tourist guides in Turkey. By including a different country in the study, the extent of gender discrimination can be evaluated in terms of different countries.

REFERENCES

Akgül, S. Ö. (2020). Turizmde Toplumsal Cinsiyet Eşitliği ve Kadınların Güçlendirilmesi ile Mutluluk Endeksi Arasındaki İlişki [The Relationship Between Gender Equality and Women's Empowerment in Tourism and Happiness Index.] *Turar Turizm ve Araştırma Dergisi*, *9*(2), 160–178.

Albuz, N. (2018). *Turist Rehberliğine Giriş, Profesyonel Turist Rehberlerinin Turizm Sektörü içindeki Yeri ve Önemi*. Detay Yayıncılık.

Altunışık, R., Boz, H., Gegez, E., Koç, E., Sigri, U., Yildiz, E., & Yuksel, A. (2022). *Sosyal Bilimlerde Araştırma Yöntemleri: Yeni Perspektifler*. Seçkin Yayıncılık.

Ap, J., & Wong, K. K. (2001). Case Study on Tour Guiding: Professionalism, Issues and Problems. *Tourism Management*, *22*(5), 551–563. doi:10.1016/S0261-5177(01)00013-9

Arslan, A., & Şimşek, G. (2018). Kadın Turist Rehberlerinin Yaşadıkları Mesleki Sorunlar: Aydın Turist Rehberleri Örneği. *Journal of Travel and Tourism Research*, (13), 23–49.

Ateş, M., Kazan, İ., & Soybalı, H. H. (2022). Turist Rehberliğinde Ayrımcılık: Kadın Turist Rehberlerine Yönelik Bir Uygulama. *Turist Rehberliği Dergisi*, *5*(1), 1–16. doi:10.34090/tured.1064528

Avcı, C. (2021). Turizm İşletmelerinde Cinsiyet Ayrımcılığı: Kavramsal Bir Değerlendirme. *Turizm ve İşletme Bilimleri Dergisi*, *1*(2), 1–14.

Ayyıldız, T., Ayyıldız, A. Y., & Turan, O. K. U. L. (2019). Turizm Rehberliği Öğrencilerinde Türkiye İmajı Algılamaları: Aydın Adnan Menderes Üniversitesi Turizm Fakültesi Örneği. *Journal of Travel and Tourism Research*, (14), 93–100.

Basurto-Barcia, J., & Ricaurte-Quijano, C. (2017). Women in Tourism: Gender (in) Equalities in University Teaching and Research. *Anatolia*, *28*(4), 567–581. doi:10.1080/13032917.2017.1370781

Blau, G., & Tatum, D. (2000). Correlates of perceived gender discrimination for female versus male medical technologists. *Sex Roles*, *43*(1/2), 105–118. doi:10.1023/A:1007043813257

Çetinkaya, M.Y., & Öter, Z. (2021). Turizm Toplum Kültür, Turist Rehberlerinin Turizm, Toplum ve Kültür Etkileşimi Üzerine Yansımaları. *Nobel Yayıncılık.*

Cheema, S. N., & Jamal, W. N. (2022). An Empirical Study on Gender Based Discrimination at Pakistani Workplaces: Determination of the Causes of Gender Based Discrimination in Pakistan's Private Service Sector Workplaces. *Sustainable Business and Society in Emerging Economies*, *4*(2), 327–338. doi:10.26710bsee.v4i2.2272

Cohen, E. (1985). The Tourist Guide: The Origins, Structure and Dynamics of a Role. *Annals of Tourism Research*, *12*(1), 5–29. doi:10.1016/0160-7383(85)90037-4

Demir, M. (2011). İş yaşamında ayrımcılık: Turizm sektörü örneği. *Uluslararası Insan Bilimleri Dergisi*, *8*(1), 760–784.

Ferguson, L. (2013). Promoting Gender Equality and Empowering Women? Tourism and the Third Millennium Development Goal. In Tourism and the Millennium Development Goals (pp. 73-87). Routledge.

Genç, R. (2018). The Impact of Tourism in The Reduction of Gender Inequality. *Trakya Üniversitesi İktisadi ve İdari Bilimler Fakültesi Dergisi*, *7*(1), 20–35.

Huang, S., Hsu, C. H., & Chan, A. (2010). Tour Guide Performance and Tourist Satisfaction: A Study of the Package Tours in Shanghai. *Journal of Hospitality & Tourism Research (Washington, D.C.)*, *34*(1), 3–33. doi:10.1177/1096348009349815

İkiz, A. N. (2020). Turizm İşletmelerinde Cinsiyet Ayrımcılığı ve Kadın Çalışanlar: Yöneticilerinin Konuya İlişkin Tutum ve Uygulamaları. *Yüzüncü Yıl Üniversitesi Sosyal Bilimler Enstitüsü Dergisi*, (50), 373–396.

Koc, E. (2020). Do Women Make Better in Tourism and Hospitality? A Conceptual Review from a Customer Satisfaction and Service Quality Perspective. *Journal of Quality Assurance in Hospitality & Tourism*, *21*(4), 402–429. doi:10.1080/1528008X.2019.1672234

Koc, E., & Ayyildiz, A. Y. (2021). Culture's Influence on the Design and Delivery of the Marketing Mix Elements in Tourism and Hospitality. *Sustainability (Basel)*, *13*(21), 11630. doi:10.3390u132111630

Köroğlu, Ö., Yıldırım, H. U., & Kılıç, A. (2020). Kadın Turist Rehberlerinin Mesleki Sorunlarının Belirlenmesi. *Mehmet Akif Ersoy Üniversitesi Sosyal Bilimler Enstitüsü Dergisi*, (31), 26–40. doi:10.20875/makusobed.586054

Lee, C. K., & Kang, S. (1998). Measuring earnings inequality and median earnings in the tourism industry. *Tourism Management*, *19*(4), 341–348. doi:10.1016/S0261-5177(98)00033-8

Mahdavi Zafarghandi, M. (2016). *Iranian Female Tour Guides' Perceptions of Working in the Tourism Industry*. Academic Press.

Rabotić, B. (2010, March). Tourist Guides in Contemporary Tourism. In *International Conference on Tourism and Environment* (pp. 353-364). Academic Press.

Sığrı, Ü. (2022). *Sosyal Bilimlerde Araştırma Yöntemleri: Yeni Perspektifler*. Nitel Araştırma Yöntemleri, Seçkin Yayıncılık.

Tekin, Ö. A. (2017). Turizm Sektöründe Toplumsal Cinsiyet Algısı: Beş Yıldızlı Otel Çalışanları Üzerine Bir Araştırma. *Avrasya Sosyal ve Ekonomi Araştırmaları Dergisi*, *4*(12), 669–684.

Ural, A., & Kılıç, İ. (2018). *Bilimsel Araştırma Süreci ve SPSS ile Veri Analizi*. Detay Yayıncılık.

Weiler, B., & Ham, S. H. (2002). Tour Guide Training: A Model for Sustainable Capacity Building in Developing Countries. *Journal of Sustainable Tourism*, *10*(1), 52–69. doi:10.1080/09669580208667152

Yıldırım, G., & Özbek, Ö. (2019). Profesyonel Turist Rehberlerine Yönelik Cinsel Taciz ile ilgili Bir Araştırma. *Turist Rehberliği Dergisi*, *2*(2), 72–87. doi:10.34090/tured.591586

Yumuk, Y., & İli, N. D. (2022). Turist Rehberliğinde Mesleki Yatkınlık ve Kariyer Sürdürme İlişkisinin Cinsiyet Değişkeni Açısından Değerlendirilmesi. *Nevşehir Hacı Bektaş Veli Üniversitesi SBE Dergisi*, *12*(3), 1314–1333. doi:10.30783/nevsosbilen.1106689

Zengin, B., Erkol, G., & Eker, N. (2014). Rehberlik Mesleğinde Bir Engel Olarak Cinsiyet Ayrımcılığı: Bayan Turist Rehberleri Üzerine Bir Araştırma, 15. *Ulusal Turizm Kongresi*, *1144*, 1159.

Zhang, H. Q., & Chow, I. (2004). Application of Importance-Performance Model in Tour Guides' Performance: Evidence from Mainland Chinese Outbound Visitors in Hong Kong. *Tourism Management*, *25*(1), 81–91. doi:10.1016/S0261-5177(03)00064-5

Chapter 22
Women Members on Boards of Directors:
Corporate Social Responsibility and Sustainability in the Tourism Sector

Serdar Yener
Sinop University, Turkey

Helena Belchior Rocha
Instituto Superior de Ciências do Trabalho e da Empresa, Portugal

ABSTRACT

In this study, the number and proportion of women in the boards of directors of publicly traded enterprises in the tourism sector at Borsa Istanbul in Turkey, the CSR, and sustainability performances of these enterprises were revealed by using a case study pattern from qualitative research methods. Within the framework of the document analysis, the number of female executives serving in the boards of directors of the enterprises operating in Borsa İstanbul and their ratio to the total number of board members will be recorded according to the year-end activity reports and websites announced on the public disclosure platform. The number and proportion of women on boards of directors will be the first unit of analysis to be used in the case study. MAXQDA 2020 qualitative research analysis software will be used in the study.

INTRODUCTION

Corporate Social Responsibility (CSR) and sustainability are increasingly important concepts in the tourism sector. As tourism often has significant environmental, social, and economic impacts on destinations, it is crucial for companies in the industry to adopt responsible and sustainable practices (Mayer & Voght, 2016).[1]

The tourism sector, which has a significant share in the field of service, is perhaps the type of service most affected by the trends of consumer (Yachin, 2018). In addition, international relations, macro and microeconomics parameters, and sociocultural parameters affect the tourism sector (Mihalic, 2016).

DOI: 10.4018/978-1-6684-8417-3.ch022

Businesses that want to succeed in the tourism sector need to follow trends and changes in all areas of daily life (Camilleri, 2020). From the point of view of strategic management, enterprises operating in the tourism sector should use all their resources, especially human resources, effectively and efficiently within the framework of a resource-based strategy approach (Garay and Font, 2012). In addition, having different resources to be least affected by close and general environmental factors and evaluating opportunities gives flexibility to the business. To give an example of the effects of close and general environmental factors, regional conflicts in the recent past, migration, in particular COVID-19 pandemic, and polarizations at the global level adversely affect the tourism strategies of countries and therefore the strategies of tourism enterprises creating uncertainty in the sector. Despite all these uncertainties, it is thought that the sustainability and corporate social responsibility approaches of the enterprises may vary according to the structure of the resources of the enterprises (Lee and Jin-Li (2018).

Corporate social responsibility (CSR) can be defined as the human-centered human resources and community management process of enterprises (Kotler and Lee, 2002). Enterprises with CSR try to establish supply chain management that fairly combines the expectations of all stakeholders, especially employees, with the expectations of the enterprise.

Sustainability, in this sense, can be defined as the creation and development of long-term future-oriented environmental, economic, and social value by enterprises. The three pillars of sustainability are the society, economy and environment, (people, profit and planet). Social Sustainability refers to the continuation of the activities of the enterprises by preserving the lower and upper social structures and socio-cultural structures of the enterprise without deteriorating the internal environment, the close environment (sectoral environment), and the general environment (political, economic, social, technological) (Nugraheni, et al., 2019). In this context, business strategies that will not jeopardize the transfer of a healthy society structure to future generations are defined as socially sustainable strategies. Economic sustainability stands for the economic activities of the enterprise without damaging social and environmental factors and without wasting the resources will need in the future (Qiu et al.,2019). Environmental sustainability links the continuation of the activities of the enterprise without damaging the ecological environment (Heesup, 2021), in which it operates and without losing the resources to be transferred to future generation.

Engaging and collaborating with various stakeholders, including local communities, government bodies, non-profit organizations, and other businesses, is essential for effective CSR and sustainability in the tourism sector. Also, can help in understanding and addressing the concerns and expectations of different groups, building partnerships, and fostering sustainable tourism development.

However, CSR and social sustainability are very confused. CSR can be defined as the level of social accountability of the enterprise. Social sustainability defines the protection and development of the social structure around the enterprise. CSR focuses on the impact of all activities of the enterprise on the social structure (Madanaguli et al., 2021). Within the framework of CSR, it is questioned what kind of contribution or damage the enterprise makes to the social structure in its activities. For example, an activity of the enterprise aimed at the education of girls can be evaluated within the framework of CSR. Social sustainability defines the preservation of the existing sociocultural structure and its transmission to future generations without harm. For example, employees with intensive and balanced over time programs and working mothers who have negative working conditions that will prevent them from doing their family responsibilities are evaluated within the framework of social sustainability. CSR is the basis for the development of businesses in many ways, especially social sustainability. Because CSR activities can primarily ensure that the enterprise maintains its social structure, that employees are satisfied and therefore performs high and do not harm environmental resources. The role of CSR activities in the success of businesses has been questioned since the 1970s (Brammer, & Millington, 2008). Especially in industries in developed countries, the fact that the work lives of employees are difficult and problematic in a way that adversely affects their family and social lives has started to adversely affect employee satisfaction. The behaviors of unhappy employees have begun to lead to negative consequences in all areas of life, especially in business performance. Anticipating that these changes could, directly and indirectly, lead to bigger problems in the future, researchers have begun to examine the CSR approaches of enterprises. The continuation of the economic activities of the enterprise without disrupting the social structure later began to be defined as social sustainability.

Sustainable and corporate social responsibility approaches are attracting more attention from stakeholders of businesses, especially consumers (Chang, 2017). Although an index has not been created for this purpose in Turkey, enterprises in the international arena are evaluated accordingly. For example, the Governance Accountability Institute (G&A) evaluates the Corporate Social Responsibility (CSR) and Sustainability performance of publicly traded businesses in the United States. This assessment made by G&A is attracting more and more stakeholders. Stakeholders of enterprises are also questioning the impact of the enterprises they invest in or receive services from on society and other areas, especially the environment.

From the point of view of the tourism sector, it is observed that domestic and foreign tourists are more conscious and questioning than in the past. The determinants of the expectations of tourists, who are the target audience of the tourism sector, have started to diversify. When choosing a destination or business, tourists not only prefer a choice that meets personal expectations but also consider their behavior of fulfilling their responsibilities towards the social and natural environment in which they are located. With the creation of indexes for this purpose in Europe and America, enterprises in Turkey have started to show interest in CSR and sustainability issues.

The role of women employees in the management of enterprises is the subject of many studies both at national and international levels. In 2012, the European Union started to work to increase the representation of women on the boards of directors of enterprises operating in the member states to 40% (European Commission, 2012). In many academic studies, the participation of women in business management has been explained within the framework of "resource dependency theory", non-discrimination approach, diversity management approach, social gender theory, and stakeholder theories (Aluchna and Krejner-Nowecka, 2016) Within the framework of resource dependency theory, it is claimed that enterprises that adopt a resource-based competition strategy affect the efficiency and effectiveness of their human resources, including female employees. The non-discrimination approach proposes that the proportion of women in society as a whole should be taken into account when determining the boards of directors of enterprises. Within the framework of the diversity management approach, it is emphasized that the effective and efficient use of gender differences in human resources such as gender, religion, race, etc. can affect the performance of the enterprise. Within the framework of social gender theory, it is emphasized that prejudices against women and men exist in societies and that women should be given more roles and positions to break these prejudices. Within the framework of Stakeholder theory, it is emphasized that women will operate by taking more into account the expectations of business stakeholders.

In studies where these theories are used, positive and negative results have been revealed about the effects of women on the performance of enterprises (Roberts, Reynolds and Dolasinski, 2022, Gruszczynski, 2020). The fact that CSR and Sustainability parameters are more decisive in business performances has led to the questioning of women's positions in enterprises. When the studies conducted in the literature are examined, it is observed that the social culture in which the enterprises operate affects the role of women (Halliday et al., 2020). In addition, the role of women in business management can be explained in terms of organizational psychology theory.

In this context, in this study, the number and proportion of women on the boards of directors of publicly traded enterprises in the tourism sector at Borsa Istanbul in Turkey, the CSR and sustainability performances of these enterprises were tried to be revealed by using a case study pattern from qualitative research methods. Within the framework of the document analysis, the number of female executives serving on the boards of directors of the enterprises operating in Borsa İstanbul (Istanbul Stock Exchange Market) and their ratio to the total number of board members will be recorded according to the year-end activity reports and websites announced on the public disclosure platform. The number and proportion of women on boards of directors will be the first unit of analysis to be used in the case study. For the data treatment of this study, was used MAXQDA 2020 qualitative research analysis software.

LITERATURE REVIEW

Women's Empowerment in Tourism Sector

The concept of empowerment is used in different research in many disciplines (Tengland, 2007). Generally, increasing the talents, skills, and equipment of individuals can be explained by definitions such as obtaining more talent in their own living spaces. Empowerment can be associated with the increase in the level of the agency along with the skills, equipment, and abilities possessed within this framework. The concept of empowerment can be evaluated in a direct relationship with participation in decision-making processes in agency organizations (Ibrahim and Alkire, 2007). Another factor that is effective at the agency level is the corporate structure. Formal and informal practices and tools that are effective in the structure of organizations can affect the level of representation. As the level of the agency increases, the positive effect of the corporate structure on the agency may also increase. As the positive impact of the corporate structure increases, the agency level may also increase. Therefore, from an empowerment point of view, there is an interactive relationship between the agency level of the target audience and the corporate structure. Apart from these, the resources in the organization can determine the impact of both the agency and the institutional structure on empowerment. Resources can be considered as physical and intellectual resources. Physical resources can be financial resources for strengthening, resources in work areas, level of access to organizational resources, etc. Intellectual resources can be evaluated as consciousness level, psychological safety level, and social relations in terms of empowerment and these resources affect the organizational structure and agency level. For example, increasing the representation level of women without sufficient awareness may pose problems in terms of employee readiness even if corporate practices are appropriate. Therefore, empowerment in an organization can be demonstrated by evaluating the above-mentioned agency, organizational structure, and resource components together.

Women's empowerment describes the work done to ensure that women have more choices in their own lives within the framework of power relations in society (Eederwijk et al. 2017). In terms of women, the concept of empowerment may also vary according to concepts such as religion, race, ethnicity, and sexual orientation. For example, religions can bring different interpretations according to the concept of women's empowerment. Again, the age variable changes the perception of women's empowerment. The empowerment of women at different ages within the framework of their representation and role can be evaluated differently. In collectivist societies, relationships can be observed between the growth of women's age and the level of authority. Women also affect the development of gender role perception between generations due to their active role in raising children and their undeniable role in the vital activities of the family. From childhood, boys and girls gain the perception of social gender in a process of upbringing in which their mothers are active. This may have an impact on the representation of men and women in society and other areas of empowerment in the future (Huis, et al., 2017).

Women's empowerment can generally be defined in society within the framework of power relations. Power relations, which are considered a criterion in the individual, collectivist, horizontal, and vertical classification of societies, are the main factor in women's empowerment. The distance of power in society can be mutually evaluated. The measure of the intimacy of the relationships of individuals with each other affects the distance of power and therefore the level of empowerment (Abdalla, 2022). Since sincerity will be high in societies with low power distance, the level of self-expression of individuals will also increase. Self-expression or realization of wishes is associated with the level of choice. The ability of individuals to make choices also shows their level of empowerment. As the level of determination in women's own lives, having a say, and having autonomy increases in the society they are in, their empowerment may also increase (Eyben, 2011). Women can make choices about their own lives and authority and express their opinions through a process of development;

- First of all, the participation and representation of women in all institutions of society should be fair. In all institutions in society, starting from within the family, participation in these representation and decision-making processes should be determined according to the number and proportion of women.
- In societies and social institutions where women are not fairly represented, there should be public practices that will positively support representation. Training, vocational training, etc. will increase the talents and skills of women and develop their leadership and other skills. Women can be given responsibility. In this way, their development can be supported. Management and leadership positions can be shared fairly according to the population.

Power is the main factor determining the statue and position of women in society. Power can be defined as the degree to which an individual can do or get anything done. Power can also be exploited by the initiative over the resources in society. Power relations can also be indicative of the positions of the parties involved. The position of a party that is not involved in power relations cannot be said to be strong either. Empowerment can be achieved through the transformation of power relations. Gender is one of the main variables that determine the relationships of individuals in society. Gender influences perceptions of social privileges, property rights, leadership, and authority. The role of gender in social statuses, policies, concepts, sanctions, and rules is undeniable (ILO, 2023). For women, the process of empowerment requires a change or transformation of power relations. Women's increased representation and fair positioning can be achieved through a change in power relations. Changing power relations can also positively impact women's empowerment by enabling the change of institutional structures and other resources. Changing power relations in terms of agency, institutional structures, and resources can also change the perception of empowerment. From the point of view of the agency, the statue of women in participatory decision-making processes, the place of women in decision-making processes, and the role of women in management are questioned. The role of women in these three areas is an indicator of empowerment. In the female empowerment literature (Pansard and Bindi, 2021; Dutta and Bhakta, 2017), several dimensions can be found, at power-to, power-within, and power-with levels. Power-to refers to the process of making decisions in a woman's own life. Power-within expresses the woman's perception of empowerment towards herself, consciousness, and self-esteem. The concept of power-with refers to the process of making decisions in a fair way in the society in which women are involved. In society, there may be a perception of power at the levels of formal practices and laws, values, norms, etc, and hidden (secret agendas towards women). The corporate structure is often identified with hidden power. Corporate policies and practices are identified with visible power, but social norms are identified with invisible power. Therefore, the change or transformation of power requires the transformation of many variables, starting from social norms and rules.

Representation refers to the ability of the individual to express himself in an institution and to act with his intrinsic motivations. The three versions of power expressed above (visible, invisible, hidden) affect representation. The realization of representation can be measured by the participation of individuals in decision-making processes, participatory decision-making, and leadership levels. Leadership enhances decision-making while at the same time supporting collective behavior or decision-making. Decision-making processes are related to their level of ability to take initiative on the resources in the organization. As participation in decision-making processes increases, levels of initiative may also increase. The participation of women in decision-making processes in institutions in society is also an indicator of their representation. Another indicator of participation in decision-making processes is autonomy. Autonomy shows the level of behavior of women according to their value perceptions and thoughts. Autonomy, which is defined as the level of women's ability to make their own decisions on issues such as having children, marriage, and choosing a profession, can vary according to the characteristics of societies.

When the distribution of employees in the tourism sector is examined in terms of gender, it is a sector where more than half of them are women. Since there is a customer expectation based on complete satisfaction, both the role and the effort of women are felt more. In all sub-branches such as hospitality, accommodation, food, beverage, and transportation, women have vast amount of roles. Despite this, they do not receive the fair output they deserve in terms of representation and other aspects.

Empowerment of women board members involves providing them with equal opportunities for leadership roles and career advancement. This includes ensuring fair and unbiased promotion processes, mentorship programs, and professional development opportunities that support the growth and progression of women in the organization and requires breaking gender stereotypes and challenging traditional perceptions of leadership (ILO, 2023). Companies should create inclusive and supportive work environments that value diverse perspectives and skills, allowing women board members to contribute effectively and be recognized for their abilities.

Corporation Social Responsibility

Corporate Social Responsibility is a concept used to describe people-oriented management practices of enterprises (Chilufya et al., 2019). All of the human-oriented management practices of the enterprises both in the internal business environment and in the external business environment can be evaluated within the framework of CSR and compared with other enterprises. Businesses have an impact on the lives of their employees and the lives of people in their environment. Especially for businesses operating in the service sector, this is a prominent indicator. The service sector is a sector where human labor is visible and human motivation can directly affect the quality of service. In this context, the fair, respectful, and ethical behaviors of the enterprises operating in the service sector toward the individuals of all the stakeholders they interact with, especially their human resources, are evaluated within the framework of CSR (Hadj, 2020). However, human-oriented management practices also include indirect issues. When the natural resources or other resources of the society in which they live are damaged, the individuals in the society are also harmed. In this respect, this creates a harmful effect in terms of human rights. Therefore, CSR includes applications and activities in a very wide area. CSR requires taking into account the effects of all direct or indirect stakeholders that affect people in any society. Human-centered behaviors and practices have economic, social, and natural dimensions (Jenkins, 2004). The tourism sector covers many different business lines such as accommodation, food and beverage, transportation, conference tourism, and health tourism. In this respect, the tourism sector has an interdisciplinary structure. In addition, intermediary services are increasingly taking hold in the tourism sector (Lund-Durlacher, 2015). With the development of the Internet and technology, CSR approaches are also changing in the tourism sector, where different actors have started to take more share.

In terms of women's roles in business life and empowerment, corporate social responsibility is considered a useful tool (Malik, 2018). Women are involved in almost all positions in the tourism sector. In terms of the service sector, the impact of women can affect the sustainable competitive advantage of enterprises in the sector. Especially in areas such as accommodation, food and beverage, and health tourism, the role of women is undeniably high. Despite this, almost all over the world, the role of women does not receive the value it deserves. For example, in the tourism sector, the share of female employees is reported as 55% all over the world (Malik, 2018). However, in terms of representation, there is a very low representation rate. The scope of this problem is enormous. In fact, "Achieve gender equality and empower all women and girls" has been identified as the fifth sustainable development goal of the United Nations. In line with this UN goal, there is a greater emphasis on women's empowerment in activities carried out within the framework of both the UN World Tourism Organisation and the UN Development Programme. Women have certain expectations such as obtaining representation rates appropriate to their shares in all positions in the tourism sector, fair practices in personal rights, and acting following human rights. From the point of view of CSR, human-oriented practices and behaviors are valid for women. In the tourism sector, where men and women with different characteristics operate, there are expectations of fair behavior physically, psychologically, and cognitively. Most fundamentally, women who continue their quest for rights with the slogan "equal pay for equal work" still do not get what they deserve in terms of representation and other aspects. Today, the fact that sustainability and CSR approaches find more parties may be an opportunity to empower women in business life, especially in the tourism sector.

Sustainability

Approaches that question the profit-oriented perspectives of businesses are increasing, especially in Western societies. While making a profit by using the resources of the environment in which the enterprises are located, it becomes an expectation to strive for the security and maintenance of these resources. Growing controversies and concerns have given rise to the sustainability approach. It has become an expectation for businesses to direct their profit-oriented perspective on the environment and society in a balanced way (Camilleri, 2020). The physical resources of the environment in which they operate, human resources, and environmental dynamics changed by enterprises that carry out production and commercial activities by using natural resources threaten the present and the future (Hadj, 2020; Mihalic, 2016). Sustainability is becoming an increasing expectation in the field of tourism (Ristic et al., 2019). Concepts such as ecotourism, green tourism, and sustainable tourism are often used to emphasize sustainability in the field of tourism (Dileep, 2018). The tourism sector, also called a chimneyless factory, is an important source of income for many countries. The sustainability of this important source of income has become important for countries. Apart from this, since the activities in the field of tourism are generally based on regional natural and social resources, the maintenance of resources is another concern. In this respect, sustainability discussions in the tourism sector attract a lot of attention. It is a fact encountered in many studies, observations, and daily life that women are more compassionate and more interested in their environment than men in physical, psychological, and cognitive terms. The empowerment of women in the tourism sector, where the share of women is higher than men both quantitatively and qualitatively, can also positively affect the sustainability of the sector. Women can use more initiative to ensure that the natural environment, economic environment, and social environment elements interacting with the tourism sector are not damaged and transferred to future generations. Especially when it is considered that the natural and social environment is the capital of the tourism sector, the importance of responsible behavior increases even more. In the tourism sector, it is conceivable that women can act more responsibly to meet today's needs without spending or damaging the resources of the future. In addition, the fact that women are more than men in the tourism sector is a good opportunity for this. However, when it relates to empowerment, the fact that women are underrepresented is an obstacle to this. Creating fair and equitable conditions for numerically more women to reach the positions they deserve can positively affect sustainability discussions in the tourism sector.

Aim of the Study

Thinking about the society we live in, the role of women in the performance of enterprises can be explained in many ways. However, in this study, the aim will be tried to explain, in the context of Turkey, the effect of women's representation in business management on corporate social responsibility (CSR) and sustainability behaviors of enterprises.

Method

For this study will be used a case study method with a qualitative approach. All accessible documents such as web pages, press releases, annual activity reports, and sustainability reports of the enterprises determined as case studies will be classified separately. Within these documents, the objectives of the research will be tried to be achieved through the codes determined through content analysis. In the study, MAXQDA 2020 software will be used for the analysis of the contents.

In order to obtain objective results, were used year-end activity reports and websites of tourism enterprises operating in the Istanbul stock exchange. In these activity reports published on the Official Noticement Platform (KAP, Kamu Aydınlatma Platformu), were observed some issues in tourism enterprises;

- Proportions of women on boards of directors and emphasis on women,
- Corporate social responsibility performance of Enterprises
- The social, economic, and environmental sustainability performances of the enterprises were tried to be revealed.

These enterprises are;

1. AVTUR, EURASIA PETROLEUM, AND TUR.
2. AYCES, ALTINYUNUS CESME
3. DOCO,
4. ETİLER
5. MAALT, MARMARIS ALTINYUNUS
6. MARTI, MARTI HOTEL
7. MERIT, MERIT TOURISM
8. PKENT, PETROKENT TOURISM
9. TEKTU, TEK-ART TOURISM
10. ULASLAR

Since the number of enterprises was not large enough for a significant quantitative data analysis, the corporate social responsibility and sustainability data of these enterprises were evaluated through document analysis. In the research process where the qualitative research method was followed, the case study was selected as the research design. The case study method is used to reveal the relationships between the event and the content analytically. The case study method is used in cases where the minimum data needed for regression analysis cannot be obtained, requiring in-depth analysis, and in cases where the relationships between variables are not systematic (Sığrı, 2018). Leavy (2014) provides case study method types;

- Historical-Observational
- Purpose-Oriented Instrumental
- It is classified as single-multiple.

In a single case study pattern, an analysis unit is investigated in a case study. In the Multi-Case Study method, it is possible to examine more than one analysis unit in more than one case study. In this study, analysis categories of enterprises operating in the tourism sector are presented below;

- Proportion of women on boards of directors,
- Corporate social responsibility performances of enterprises,
- Multiple case studies were selected because of the social, economic, and environmental sustainability performances of the enterprises were questioned.

The three above-mentioned analysis units were examined in 10 tourism enterprises (10 case studies) operating in Borsa İstanbul. The document analysis method was used as the data collection method. In the document analysis method, the analysis units determined for this research;

- The proportion of women on boards of directors
- Corporate social responsibility performance of Enterprises
- The social, economic, and environmental sustainability performances of the enterprises have been determined.

The themes and codes to be used in the document analysis to be made in the annual activity report of the enterprises are stated below.

Table 1. Theme used in content analysis: Code list

Theme	Code	Explanation
Corporate Social Responsibility	corporate	Within the framework of the literature, the word "institutional" in the documents was evaluated as an indicator.
Corporate Social Responsibility	social	Within the framework of the literature, the word "social" in the documents was evaluated as an indicator.
Corporate Social Responsibility	Sustainable	Within the framework of the literature, the word "sustainable" in the documents was evaluated as an indicator.
Women Empowerment	woman	Within the framework of the literature, the word "woman" in the documents was evaluated as an indicator.
Women Empowerment	gender	Within the framework of the literature, the word "gender" in the documents was evaluated as an indicator.
Women Empowerment	family	Within the framework of the literature, the word "family" in the documents was evaluated as an indicator.

Results

In the document analysis using the content analysis method, information was tried to be collected on 3 topics, specified in the method section.

- 2 members of the 5-person board of directors of AVTUR are women.
- 3 members of the 7-person board of directors of Çeşme Altınyunus Enterprise are women.
- 2 of the 11-member board of directors of Merit Turizm are female members. At the last board meeting, it was announced that the board members had set a target for 25% to be women in the first stage.
- There are no women on the board of directors of Petrokent. However, no statements were made to empower women employees or women.
- There are no women on the two-person board of directors of the Do-Co business.
- There are no female members in the 4-member board of directors of Etiler.
- Marmaris Altınyunus has 1 female member in the 6-person board of directors.
- Martı Hotel has 2 female members on the board of directors of 7 people.
- 1 member of the 6-person board of directors of Tek-Art is a woman.
- There are no female members in the 8-member board of directors of Ulaşlar Tourism.

Considering the number of members and female member ratios of enterprises, it can be said that the number and proportion of female management members are insufficient in a sector where more than 55% of the workforce is female all over the world. It is considered necessary to have a female perspective in management to take constructive and supportive measures for so many women's workforce, to create business policies, and to carry out social responsibility activities for women in the environment where the enterprise operates. It is not easy to understand the problems experienced by women through the eyes of men, to support and empower women. It is also the responsibility of the board of directors to create career plans, especially for women and to create policies for them in enterprises. Without being a woman, without being a mother, and without taking responsibility for a single-parent family as a woman, it is difficult to create activities aimed at women's empowerment. When the statistics given in Table 2 above

Table 2. Number and proportion of women members in the boards of directors of designated enterprises

Business Name	Total Number of Board Members	Number of Women Members on the Board of Directors	Female Member Ratio
AVTUR	5	2	% 40
ÇEŞME AY	7	3	% 43
MERIT	11	2	% 18
PETROKENT	5	0	% 0
DO-CO"	2	0	% 0
ETILER	4	0	% 0
MARMARİS AY	6	1	% 17
MARTI	7	2	% 29
TEK-ART	6	1	% 17
ULAŞLAR	8	0	% 0

are examined, it is observed that the number and ratio of female board members of AVTUR operation and Çeşme Altınyunus enterprise is high. However, when the policy documents, activity reports, sustainability, and social responsibility documents of both enterprises were examined, it was observed that there was no policy statement or special application for women. Although the number of female board members of MERİT enterprise is 2 and the rate is 18%, it is observed that they set a target of 25%. It has been observed that there are no women members on the boards of directors of PETROKENT, DO-CO, ETİLER, and ULAŞLAR enterprises and there is no policy declaration or implementation for women's empowerment. However, since DO-CO is an international enterprise operating in other countries, it has been observed that it has published documents on some issues such as sustainability and social responsibility. The reason for this may be that especially in Western countries, corporate social responsibility and sustainability indicators are seen as important indicators in commercial life. In enterprises with a high number of female board members, it is also common for female board members to be relatives of business owners. When we look at the tables above, this issue is not very clear. Board members who use the same surname may be relatives. This may create an inability to produce the expected effect.

Descriptive statistics of the codes used in the study are presented in the table below. The means, standard deviations, variances, medians, and intervals of the codes used in the content analysis within the framework of descriptive statistics are presented in Table 3.

The frequencies of the codes determined as a result of the content analysis in the determined cases are presented in Table 4 below.

The total frequencies of the specified codes are presented in Table 4 above. When we look at the results, the code "corporate" has the highest frequency. Based on this result, it is thought that the word "corporate" increases awareness of some concepts such as sustainability and corporate social responsibility. When the documents were examined, it was observed that unfortunately, this result did not occur. The code "social" was observed to be the code with the highest frequency after the code "corporate". However, the "woman" code, which was determined as an indicator of the themes of "woman" and "woman empowerment", which is the focus of the study, was observed in only 41 frequencies in 10 documents. The code "sustainable", which is determined as an indicator of corporate social responsibility and sustainability themes, has been used only 33 times. The "gender" code, which was determined as

Table 3. Descriptive statistics of codes

	Family	Gender	Woman	Corporate	Social	Sustainable
Average	3,30	2,20	7,30	90,60	14,70	31,30
Standard deviation (example)	7,590	3,156	9,435	59,149	14,615	69,596
Variance (example)	57,610	9,960	89,010	3.498,640	213,610	4.843,610
Minimum (min.)	0,00	0,00	0,00	0,00	0,00	0,00
Q1	0,25	0,00	1,25	39,75	7,00	0,25
Median	1,00	1,50	5,00	107,00	11,50	4,00
Q3	1,00	2,75	8,00	127,50	15,50	19,75
Max (max.)	26,00	11,00	34,00	192,00	55,00	238,00
December	26,00	11,00	34,00	192,00	55,00	238,00

Table 4. Frequencies of specified codes

Codes	Word length	Frequency	%	Degree	Codes	Codes %
Corporate	8	866	77,88	1	9	150,00
Social	6	147	13,22	2	9	150,00
Woman	5	41	3,69	3	8	133,33
Sustainable	14	33	2,97	4	6	100,00
Gender	8	18	1,62	5	4	66,67
Family	4	7	0,63	6	4	66,67

an indicator of the theme of female empowerment, was used 18 times and the "family" code was used 7 times. Based on these results, it has been observed that policies and practices such as women, women empowerment, social responsibility, and corporate social responsibility are inadequate in tourism enterprises operating in Turkey. As mentioned above, in a sector where more than 55% of the workforce is female, this situation does not seem hopeful.

As a result of the content analysis made in the documents, the frequencies and other statistics of the codes in each enterprise are presented in Table 5 below.

Since the study focuses on enterprises operating in the tourism sector in general, business-based evaluations will shed light on researchers in addition to general evaluations. In Table 5, the distribution of the codes determined within the framework of the study according to the enterprises is presented. When businesses are evaluated in general, it is observed that they have different characteristics. Businesses operating entirely in accommodation-based and related tourism business lines as well as enterprises operating in food and beverage or unrelated business lines but classified as tourism enterprises were also observed. Based on cases, the AVTUR (Avrasya Petrol ve Turistik Tesisleri Yatırım Anonim Şirketi) is currently seen as an enterprise related to the Metro Bus. Business lines have been expanded based on the sectors related to the transportation sector. The frequencies of the codes determined in the content analysis made in the documents of the AVTUR enterprise are presented above. It was observed that the code "female" was used 8 times, and the codes "family" and "gender" were not used at all. It

Table 5. Distribution of codes by business

Codes	FRQ	%	DocS %	AVTUR	ÇEŞME AY	ETILER	MARMARİS AY	MARTI	MERIT	PETROKENT	TEK-ART	ULAŞLAR
Corporate	866	77,88	90,00	130	191	17	108	14	150	90	58	108
Social	147	13,22	90,00	16	55	7	11	7	21	12	4	14
Woman	41	3,69	80,00	8	14	1	5	0	7	2	1	3
Sustainable	33	2,97	60,00	4	22	0	1	1	4	0	0	1
Gender	18	1,62	40,00	0	11	0	2	0	3	0	0	2
Family	7	0,63	40,00	0	4	0	0	1	0	0	1	1

was observed that the "corporate" code was used 130 times, the "social" code 16 times, and the "sustainable" code 4 times. From this point of view, it can be said that the awareness of women's empowerment regarding these frequencies is almost non-existent and the awareness of sustainability and corporate social responsibility is insufficient.

The frequencies of the codes determined in the content analysis made in the documents of Çeşme Altınyunus enterprise are presented above. It was observed that the "female" code was used 14 times, the "family" code 4 times, and the "gender" code 11 times. It was observed that the "corporate" code was used 191 times, the "social" code 55 times, and the "sustainable" code 22 times. From this point of view, it can be said that the awareness of women's empowerment is almost non-existent and the awareness of sustainability and corporate social responsibility is insufficient. Despite this, Çeşme Altınyunus business has a higher frequency in the themes mentioned than other enterprises.

The documents of the DO-CO enterprise have generally been prepared by taking into account the operating center abroad. Their overseas notifications are very comprehensive, detailed policy statements. It has been observed that the documents they have published in Turkey are insufficient.

The frequencies of the codes determined in the content analysis made in the documents of the Etiler (Etiler Gıda Ve Ticaret Yatırımlar Sanayi Ve Ticaret A.Ş.) are presented above. It has been observed that the code "female" is used 1 time, and the codes "family", "gender" and "sustainable" are not used at all. It was observed that the "institutional" code was used 17 times and the "social" code was used 7 times. From this point of view, it can be said that women's empowerment, sustainability, and corporate social responsibility awareness are almost non-existent.

The frequencies of the codes determined in the content analysis made in the documents of Marmaris Altınyunus enterprise are presented above. It was observed that the code "female" was used 5 times, the code "gender" was used 2 times, and the code "family" was not used at all. It was observed that the "corporate" code was used 108 times, the "social" code 11 times, and the "sustainable" code 1 time. From this point of view, it can be said that women's empowerment and sustainability awareness is almost non-existent, and corporate social responsibility awareness is insufficient.

The frequencies of the codes determined in the content analysis made in the documents of Martı Hotel are presented above. It was observed that the "female" code and the "gender" code were never used, and the "family" code was used 1 time. It was observed that the "corporate" code was used 14 times, the "social" code was used 7 times, and the "sustainable" code was used 1 time. From this point of view, it can be said that women's empowerment and sustainability awareness is almost non-existent, and corporate social responsibility awareness is insufficient.

The frequencies of the codes determined in the content analysis made in the documents of Merit enterprise are presented above. It was observed that the "female" code was used 7 times, the "gender" code 3 times, and the "family" code was not used at all. It was observed that the "corporate" code was used 150 times, the "social" code was used 21 times, and the "sustainable" code was used 4 times. From this point of view, it can be said that women's empowerment and sustainability awareness is almost non-existent, and corporate social responsibility awareness is insufficient.

The frequencies of the codes determined in the content analysis made in the documents of the Petrokent enterprise are presented above. It was observed that the code "female" was used 2 times and the code "gender" and "family" were not used at all. It was observed that the "corporate" code was used 90 times, the "social" code was used 12 times, and the "sustainable" code was not used at all. From this point of view, it can be said that women's empowerment and sustainability awareness is almost non-existent, and corporate social responsibility awareness is insufficient.

The frequencies of the codes determined in the content analysis made in the documents of Tek-Art are presented above. It was observed that the code "Woman" and "Family" were used 1 time and the code "gender" was not used at all. It was observed that the "corporate" code was used 58 times, the "social" code was used 4 times, and the "sustainable" code was used 1 time. From this point of view, it can be said that women's empowerment and sustainability awareness is almost non-existent, and corporate social responsibility awareness is insufficient.

The frequencies of the codes determined in the content analysis made in the documents of the Ulaşlar enterprise are presented above. It was observed that the "female" code was used 3 times, the "Family" code 1 time, and the "gender" code 2 times. It was observed that the "corporate" code was used 108 times, the "social" code was used 14 times, and the "sustainable" code was used 1 time. From this point of view, it can be said that women's empowerment and sustainability awareness is almost non-existent, and corporate social responsibility awareness is insufficient.

Result

Documents were analyzed using the content analysis method to reveal awareness of the themes of women's empowerment and corporate social responsibility or sustainability in tourism enterprises operating in Turkey. As documents, 10 enterprises operating in Borsa Istanbul (BIST) and classified as tourism enterprises were selected. Content analysis was applied by collecting the annual activity reports, web pages, and other publications published by the enterprises. In the analyzes made using MAXQDA 2020 software, results that will not meet expectations have been revealed. It was observed that none of the 10 identified enterprises had an effective policy and practice on women's empowerment or corporate social responsibility. Even more negatively, the level of awareness of these enterprises on these issues is almost non-existent. The number and proportion of women on boards of directors are very low, in some enterprises there are none. Considering that these enterprises operate only in Turkey, it is obvious that they need a serious change of perspective on this issue.

When the research model considered, no similar study model was found in the literature. The relationship between female board members of enterprises operating in the tourism sector and CSR-sustainability performances of institutions has come to the fore as a very unique study topic. While there have been several studies that have looked at the impact of female board members on the aggressive strategies of businesses, such a broad pattern has not been observed. Therefore, it is difficult to compare the results of the research with previous studies. Applying the same model in a western society where CSR and sustainability culture is more prevalent could have produced more significant results. In societies such as Turkey, where the culture of CSR and sustainability has just begun to develop, such studies cannot go beyond raising awareness.

Based on the data observed in the study, it is seen that it is impossible to establish a relationship between female board members and the sustainability of their enterprises. The number of female managers of enterprises is insufficient, and corporate social responsibility and sustainability activities are almost non-existent or even absent in some enterprises. At the beginning of the research process, the targeted results could not be achieved. The only positive result of the research is that it reveals the current situation. After that, strategies can be determined with grounded theory studies that will reveal the perception of women in the tourism sector.

Although the impact of female board members on the CSR and sustainability performance of enterprises could not be observed, it was disappointing that the codes determined for qualitative study were not used satisfactorily in the written texts of the enterprises or in other social media posts. When it is considered for Turkey, there is a need for serious mentoring activities on CSR and sustainability in all sectors where women are concentrated, especially in the tourism sector.

Similar studies have not been observed to compare the results of the research, there are studies that reveal that the proportion of women in the boards of directors can have an impact on the performance of the enterprises in different areas. Ahmadi et al. (2018) report that gender diversity approaches in boards of directors enable the decision-making processes of enterprises. Simineosci et al. (2021) examined relationship between the gender diversity and operational performance of the boards of directors in Standart&Poors 500 Information Technology Enterprises. In the study, it was observed that the gender diversity approach in the boards of directors of the enterprises in the sample increased the performance of the enterprises. In the studies of Martinez and Rambaud (2019), it was observed that the ratio of women in the boards of directors had a positive effect on the competitive strategies and performance of the enterprises. In a very close study in terms of CSR and sustainability, Pucheta-Martinez and Gallego-Alvarez (2020) found that women on the boards of directors of enterprises induce CSR performance of enterprises. Although none of these studies are carried out in the tourism sector, they can be referenced.

Further Studies

Qualitative studies are used as an effective tool in research types such as exploratory and grounded theory. Qualitative studies are very functional to reveal the level of awareness of any phenomenon and to predict the relationships between variables. As seen in this study, women's empowerment and corporate social responsibility awareness in tourism enterprises are very low. In the following studies, grounded theory studies can be carried out by using interviews and interviews with the owners of the enterprises operating in the tourism sector. In this way, the perceptions of business owners and managers can be analyzed. Remedial measures can be planned for these perceptions.

The tourism sector is a service sector where female employees have a higher density than male employees. In this sense, although the model of the study is very original and remarkable, satisfactory results have not emerged. It is thought that the same study model may produce more remarkable results in western countries where CSR and sustainability culture are widespread.

Limitations of Work

Enterprises were selected using the tourism business classifications in Borsa İstanbul. All tourism sector enterprises operating in Borsa Istanbul were evaluated in this study. Although it is considered a tourism enterprise, the existence of enterprises focusing on different related-unrelated business lines raises doubts about the effectiveness of this classification. Studies to be carried out only in enterprises operating in tourism business lines can give healthier results to policymakers or practitioners. The data on the companies' pages do not give healthy information about the focus business lines.

Final Remarks

Organizations in general and turist sector in particular should support research to establish supportive policies and practices that promote work-life balance, such as flexible working arrangements, parental leave, and childcare facilities. These measures could help women board members manage their professional and personal responsibilities more effectively, fostering their empowerment and retention. To establish transparent reporting mechanisms to track progress in gender diversity and the empowerment of women board members can be done by reporting on metrics, such as the representation of women in leadership positions, pay equity, and diversity initiatives, demonstrates accountability and encourages ongoing improvement.

Recognizing and rewarding the contributions of women board members should ensure that performance evaluations and compensation systems are fair, transparent, and free from gender bias, thereby acknowledging the achievements and value that women bring to the boardroom, also involves advocating for gender equality and diversity in corporate leadership beyond the organization. Companies can participate in industry initiatives, public policy discussions, and partnerships with organizations that promote women's empowerment, contributing to systemic change and broader gender equality efforts.

Investing in leadership development programs tailored to women board members in tourism can strengthen their skills, knowledge, and confidence. Providing opportunities for training, workshops, and executive education can help women enhance their leadership capabilities and prepare for higher-level roles.

Highlighting successful women board members as role models can inspire other women to aspire to leadership positions. Encouraging the visibility and recognition of women's achievements in the boardroom can help shift societal perceptions and stereotypes, encouraging more women to pursue leadership roles. By actively promoting the empowerment of women board members, organizations can drive positive change, enhance decision-making processes, and foster a more inclusive and equitable corporate culture.

REFERENCES

Abdalla, F., Samah, A., Hashim, A., & Rosnon, M. R. (2022). Relationship between Participation and Women Empowerment in the Tourism Industry; a Structural Equation Modeling Approach. *South Asian Research Journal of Business and Management.*, *4*(1), 11–27. doi:10.36346arjbm.2022.v04i01.002

Ahmadi, A., Nakaa, N., & Bouri, A. (2018). Chief Executive Officer Attributes, oard Structures, Gender Diversity and Performance among French. *Research in International Business and Finance*, *44*(21), 218–226. doi:10.1016/j.ribaf.2017.07.083

Aluchna & Krejner-Nowecka. (2016). Why do we need women on boards? A synthesis of theories. *Przegl ąd Organizacji, 10*, 72–80.

Brammer, S., & Millington, A. (2008). Does it pay to be different? An analysis of the relationship between corporate social and financial performance. *Strategic Management Journal*, *29*(12), 1325–1343. doi:10.1002mj.714

Camilleri, M. A. (2016). Responsible tourism creates shared value among stakeholders. *Tourism Planning & Development*, *13*(2), 219–235. doi:10.1080/21568316.2015.1074100

Camilleri, M. A. (2020). Strategic corporate social responsibility in tourism and hospitality. *Sustainable Development (Bradford)*, *1–3*(3), 504–506. Advance online publication. doi:10.1002d.2059

Chang, H.-H. (2017). Consumer socially sustainable consumption: The perspective toward corporate social responsibility, perceived value, and brand loyalty. *J. Econ. Manag.*, *13*, 167–191.

Chilufya, A., Hughes, E., & Scheyvens, R. (2019). Tourists and community development: Corporate social responsibility or tourist social responsibility? *Journal of Sustainable Tourism*, *27*(10), 1513–1529. doi:10.1080/09669582.2019.1643871

Dutta, N., & Bhakta, K. (2017). *Women Empowerment: Dimensions, Needs and Hurdles in India IMS Business School, Doctoral Colloquium*. Retrieved From https://www.managejournal.com/up/conference/20170210152711.pd

European Commission. (2019). *Gender Balance on Company Boards-Information from the Commission.* Available online: https://data.consilium.europa.eu/doc/document/ST-9823-2019-INIT/en/pdf

Eyben, R. (2011). Supporting pathways of women's empowerment: a brief guide for international development organizations. Institute of Development Studies (IDS).

Garay, L., & Font, X. (2012). Doing good to do well? Corporate social responsibility reasons, practices, and impacts in small and medium accommodation enterprises. *International Journal of Hospitality Management*, *31*(2), 329–337. doi:10.1016/j.ijhm.2011.04.013

Goetz, A. M., & Musembi, C. N. (2008). Voice and Women's Empowerment: Mapping a Research Agenda. Brighton, UK: Institute of Development Studies (IDS).

Gruszczynski, M. (2020). Women on Boards and Firm Performance: A Microeconomic Search for Connection. *Journal of Risk and Financial Management.*, *13*(9), 2018. doi:10.3390/jrfm13090218

Hadj, T. B. (2020). Effects of corporate social responsibility towards stakeholders and Environmental management on responsible Innovation and competitiveness. *Journal of Cleaner Production*, *250*, 119490. doi:10.1016/j.jclepro.2019.119490

Halliday, C. S., Paustian-Underdahl, S. C., & Fainshmidt, S. (2020). Women on boards of directors: A meta-analytic examination of the roles of organizational leadership and national context for gender quality. *Journal of Business and Psychology*.

Heesup, H. (2021). Consumer behavior and environmental sustainability in tourism and hospitality: A review of theories, concepts, and latest research. *Journal of Sustainable Tourism*, *29*(7), 1021–1042. doi:10.1080/09669582.2021.1903019

Huis, M. A., Hansen, N., Otten, S., & Lensink, R. A. (2017). Three-Dimensional Model of Women's Empowerment: Implications in the Field of Microfinance and Future Directions. *Front Psychol.*, *8*(1678). doi:10.3389/fpsyg.2017.01678

Ibrahim, S., & Alkire, S. (2007). *Agency and Empowerment: A Proposal for Internationally Comparable Indicators*. Routledge.

International Labor Organization. (2023). *Achieving gender equality at work*. International labour conference, 111th Session, Geneve.

Jenkins, H. A. (2004). Critique of conventional CSR theory: An SME perspective. *Journal of General Management*, *2004*(29), 37–57. doi:10.1177/030630700402900403

Kotler, P., & Lee, N. (2002). *Corporate social responsibility: Doing the most good for your company and your cause*. John Wiley & Sons.

Leavy, P. (2014). *The Oxford Handbook of Qualitative Research*. Oxford University Press. doi:10.1093/oxfordhb/9780199811755.001.0001

Lee, Y., & Jin-Li, H. (2018). Integrated Approaches for Business Sustainability: The Perspective of Corporate Social Responsibility. *Sustainability (Basel)*, *10*(7), 2318. doi:10.3390u10072318

Lund-Durlacher, D. (2015). Corporate social responsibility and tourism. In P. Moscardo & G. Benckendorff (Eds.), *Education for Sustainability in Tourism—A Handbook of Processes, Resources, and Strategies* (pp. 59–72). Springer. doi:10.1007/978-3-662-47470-9_4

Madanaguli, A., Srivastava, S., Ferraris, A., & Dhir. A. (2021). *Corporate social responsibility and sustainability in the tourism sector: A systematic literature review and future outlook*. doi:10.1002/sd.2258

Malik, S. (2018). Corporate Social Responsibility: A Tool for Woman Empowerment in Hospitality Sector. *IJRAR*, *5*, 4.

Martínez M.C.V., & Rambaud, S. C. (2019). Women on corporate boards and ˛rm™s ˛nancial performance. *Women's Stud Int Forum*. doi:.wsif.2019.102251 doi:10.1016/j

Mayer, M., & Vogt, L. (2016). Economic effects of tourism and its influencing factors: An overview focusing on the spending determinants of visitors. *Zeitschrift für Tourismuswissenschaft*, *8*(2), 169–198. doi:10.1515/tw-2016-0017

Mihalic, T. (2016). Sustainable-responsible tourism discourse–Towards 'responsustable' tourism. *Journal of Cleaner Production*, *111*, 461–470. doi:10.1016/j.jclepro.2014.12.062

Nugraheni, A. I. P., Priyambodo, T. K., Kusworo, H. A., & Sutikno, B. (2019). The Social Dimension Of Sustainable Development: Defining Tourism Social Sustainability. *ICESC. EAI. Proceedings of the 1st International Conference on Engineering, Science, and Commerce, ICESC 2019*. 10.4108/eai.18-10-2019.2289855

Pansardi, P., & Bindi, M. (2021). The new concepts of power? Power-over, power-to and power-with. *Journal of Political Power*, *14*(1), 51–71. doi:10.1080/2158379X.2021.1877001

Qiu, H., Fan, D. X. F., Lyu, J., Lin, P. M. C., & Jenkins, C. L. (2019). Analyzing the Economic Sustainability of Tourism Development: Evidence from Hong Kong. *Journal of Hospitality & Tourism Research (Washington, D.C.)*, *43*(2), 226–248. doi:10.1177/1096348018777046

Ristić, D., Vukoičić, D., & Milinčić, M. (2019). Tourism and sustainable development of rural settlements in protected areas - Example NP Kopaonik (Serbia). *Land Use Policy*, *89*, 104231. doi:10.1016/j.landusepol.2019.104231

Roberts, C., Reynolds, J., & Dolasinski, M. J. (2022). Meta-Analysis of Tourism Sustainability Research: 2019–2021. *Sustainability (Basel)*, *14*(6), 3303. doi:10.3390u14063303

Sığrı, U. (2018). *Qualitative Research Methods*. Beta Publishing.

Simionescu, L. N., Ghergina, S. C., Sheikha, H. T., & Sheikha, Z. (2021). Does board gender diversity affect firm performance? Empirical evidence from Standard & Poor's 500 Information Technology Sector. *Financial Innovation*, *7*(52), 52. doi:10.118640854-021-00265-x

Tengland, P. A. (2007). Empowerment: A Conceptual Discussion. In *Health and Society*. Springer Science+Business Media, LLC. doi:10.1007/s10728-007-0067-3

Van Eerdewijk, A., Wong, F., Vaast, C., Newton, J., Tyszler, M., & Pennington, A. (2017). *A conceptual model of women's and girls' empowerment*. KIT Royal Tropical Institut. https://ec.europa.eu/commission/presscorner/detail/en/IP_12_1205

Yachin, J. M. (2018). The 'customer journey': Learning from customers in tourism experience encounters. *Tourism Management Perspectives, 28*, 201-210. https://www.sciencedirect.com/science/article/pii/S2211973618300795 doi:10.1016/j.tmp.2018.09.002

ENDNOTE

1 https://www.unwto.org/tourism-and-covid-19-unprecedented-economic-impacts

Chapter 23
Measuring Career Anchor Perceptions of Female Tourist Guide Candidates:
A Pilot Study

Seraynur Bektaş
Afyon Kocatepe University, Turkey

Harun Özbey
Afyon Kocatepe University, Turkey

Özcan Zorlu
Afyon Kocatepe University, Turkey

ABSTRACT

Career anchors are essential for the success of business life. Gender may also be a critical factor that determines the way of career orientation as well as career anchors. At this point, we assume that focusing on the career anchors of female tourist guide candidates will contribute to understanding the nature of career decisions. From this point of view, this study aims to reveal the career anchors of female tourist guide candidates studying at university. The data was gathered from female tourist guidance program students via a questionnaire. Results of the study revealed that female tourist guides perceive technical career anchors most, while the security anchor is perceived as least. Additionally, their career anchor perceptions significantly differ due to their career goals. Participants who want to be tourist guides in the future perceive higher autonomy as career anchors and service anchors, which are also inherent in the tourist guidance profession.

DOI: 10.4018/978-1-6684-8417-3.ch023

INTRODUCTION

The term career, derived from carrus in the Latin language (Etmyonline, 2023), infers the period of time that an individual spends working or doing a particular thing. Vocationally, it refers to a series of works that a person has in a particular area of work (Oxford Advanced Learner's Dictionary, 2023). Yarnall (2008) also defines a career as work and non-work experiences that shape an individual's path in life. Since the career encompasses a long period consisting of work and non-work elements, it could affect an individual's life in a significant proportion. Thus, it should be managed or guided through specific goals (Türkay, 2015). However, the individual and the organization should always endeavor for the career process to succeed. While the individual and the organization undertake joint responsibilities in career management, the employee's job performance will be much better, and the work satisfaction of the individual, whose talents and skills are discovered at a young age, will increase (Kong et al., 2015).

Although career is a broad term, management of the career is shaped based on the relevant job context. Thus, each profession involves different career goals and career paths. Therefore, the career expectations of individuals vary due to the essential characteristics of the work. Career expectations express an individual's career choices and goals (Rojewski, 2005). During the formation of career expectations, both the individuals and the social environment (family, friends, teachers, sector managers, etc.) have various preliminary inferences about the individual and the individual's profession. Therefore, career expectations begin to take shape before the career starts and include various predictions for the individual to have a promising career in the future. Career expectations are an essential element that determines what the working life will be like in adulthood (Olenik-Shemesh et al., 2018). Considering that if the career expectations of the individual are not met, he/she will evaluate other career opportunities, career expectations should be well understood, negotiated well, and measured regularly (Knight et al., 2006).

Secondary and graduate education highly affects individuals' career expectations. Hence, if an individual wants to do high-skilled jobs/professions in the future, he/she should first develop the skills and abilities needed and take a university education beyond the high school level (Kucharvy, 2014). In this context, higher education institutions are undeniable institutions in shaping careers. Universities are also the institutions where people's CA begins to be determined. In his pioneering study, Edgar H. Schein (1978: 125-126) has grounded CA as self-perceived talents & abilities, motives & needs, and attitudes & motives. He also stated that CA is broader than the typical concepts of work motivation & job value. According to Schein, it is almost impossible to predict CA via tests, and it cannot be observed unless a person has work experience. CA concept emphasizes significant interactions among abilities, motives, and values, and it expresses the person's stability while considering an individual's change or growth (Schein, 1978). However, Schein indicates the necessity of work experience to observe CA; graduate students could have some CA perceptions due to the existence of applied courses and training opportunities. At this point, stress on CA perceptions of female tourist guide candidates studying for the graduate degree.

THEORETICAL BACKGROUND

A career is defined as the individually perceived sequence of attitudes and behaviors associated with work-related experiences and activities throughout the person's life (Hall, 1976). The individual and the work organization are the two main components of this definition. As a result, career development can be thought of as the activities performed over time by both the individual and the work organization

Table 1. Schein's career anchors

Career Anchor	Explanation
Technical/Function Competence	Primarily motivated by the work's content; only wants to advance in his or her technical or functional area of expertise; dislikes and fears general management as being too political.
Managerial Competence	Primarily motivated by chance to analyze and solve problems in the face of uncertainty and incomplete information; enjoys bringing people together to accomplish shared objectives; stimulated rather than drained by the crisis.
Security/Stability	Primarily motivated by job security and loyalty to a single organization over time; able to adjust and thoroughly mingle into an association's qualities and standards; dislikes moving and traveling frequently.
Entrepreneurial creativity	Primarily motivated by the need to construct or create something entirely on their own, they are quickly bored and enjoy switching projects, more interested in starting new businesses than overseeing existing ones.
Autonomy/independence	Primarily motivated to seek employment opportunities as free of organizational constraints as possible; wants to work at her own pace and schedule; is willing to give up promotions in exchange for more freedom.
Service/Dedication to a Cause	Primarily motivated to make a difference in the world in some way, it aims to combine personal values of helping the community with work activities; it is more concerned with their skills than their values regarding finding work.
Pure Challenge	Primarily motivated by overcoming significant challenges, resolving almost insoluble issues, or prevailing over extremely tough opponents, they define their careers in terms of competition or daily combat, where winning is everything; they are highly determined and intolerant of people whose goals are different from theirs.
Lifestyle	Primarily motivated to strike a balance between work and life; extremely concerned about things like paternity and maternity leave, day-care options, and so on; looks for organizations that have programs and strong values that support families

Source: Feldman and Bolino (1996, p. 92)

(Beck, 1996). In this sense, activities performed by individuals encompass career goals, career choices, career orientations, career improvement efforts, career path decisions, and CA. Indeed, CA guides the career decisions of individuals (Cerdin &Pargneux, 2010). CA reflects the individual's career aspirations, motivations, and newly discovered abilities based on experience. Based on a person's perceived strengths and weaknesses, a CA develops over time, and it has an impact on career choices, decisions to switch jobs, personal goals, views of the future, the choice of particular occupations and work settings, and affective reactions to experiences at work (Coetzee & Schreuder, 2014). The more individuals get work experience and build careers, the more they have an understanding and learn about themselves. However, these understandings and learning, namely CA, somehow affect their career decisions. Once a person notices the CA that shapes his/her career, he/she also recognizes that relevant CA will stay with him/her throughout the career (Evans, 1996). Hence, a person without knowledge of his/her CA can always be misguided or oriented wrong, suffer from unsatisfactory work experiences, and feel "this is not really me" (Schein, 1990).

Although CA plays a crucial role in a successful career, current literature substantially takes as a reference to Schein's typology and explanations. In 1990, Edgar H. Schein identified eight distinctive CA follow as technical-functional competence, general managerial competence, security/stability, entrepreneurial creativity, autonomy/independence, service/dedication to a cause, pure challenge, and lifestyle (Schein, 1990: 20). Feldman & Bolino (1996), in their study, summarize those CA types in Table 1, as referring Schein's original study conducted in 1990.

Individuals with different CA have different career paths (Gubler et al., 2015) and perform better in their specialized competencies. For instance, someone with technical, functional competence would be more successful in a narrower professional position, while someone with managerial competence performs better in comprehensive senior positions (Jacobs, 1987). People with technical CA make career decisions by considering opportunities for developing technical competencies. On the other hand, people with managerial CA insist on realizing their primary career goal, which means becoming a general manager. People with security CA are likely to continue at work positions due to their willingness to stay in their current company or geographic location. Entrepreneurial CA expresses the career decisions mainly characterized by business setup. Defining the content of a job and carrying it out operationally identifies autonomy CA (Beck & La Lopa, 2001). Those five CA are the prominent anchors that somehow affect individuals' career success and career paths. Additionally, service CA concentrates on values and willingness to help others. People with service CA substantially want to help others, make an effort to solve environmental problems or add value to something. People characterized by pure challenge CA like to deal with seemingly unsolvable problems and have job satisfaction when they overcome difficult obstacles. Finally, lifestyle CA reflects the integration of personal needs, family needs, and requirements (Yarnall, 1998). When an individual ultimately develops CA, he/she will not abandon a job even if forced to make a difficult choice (Hardin et al., 2001).

Schein asserts that individuals are more likely to achieve positive career outcomes such as job effectiveness, job satisfaction, and stability when they achieve congruence between their work and career anchor. Indeed, people with congruence are generally more likely to succeed in their careers than those without (Danziger & Valency, 2006). A distinct CA is generally established at the mid-career or early late-career stages, and CA patterns are grown over time (Van Vuuren & Fourie, 2000). In his series of studies about CA, Schein states that each individual can only have one steadfast career anchor, and this CA is the result of life and work experiences. However, most careers require different CAs simultaneously. For example, a professor would have autonomy, security, technical/functional competence, and service CA the succeed in their career (Weber &Ladkin, 2009). One of those professions is tourist guiding, which requires more than one CA to perform adequately.

World Federation of Tourist Guides Association (WFTGA) defines the tourist guide as "*a person who guides visitors in the language of their choice and interprets the cultural and natural heritage of an area which person normally possesses an area-specific qualification usually issued and/or recognized by the appropriate authority*" (WFTGA, 2003). Tourist guides are the experts in interpretation and come to the forefront with their specific knowledge (Manning, 2016). Hence, they play a variety of essential roles in their work. Zhang & Chow (2004) summarize these roles as actors, ambassadors, buffers, caretakers, catalysts, culture brokers, information-givers, intermediaries, interpreters, translators, leaders, mediators, organizers, salespeople, shamans, and teachers. On the other hand, Tsaur & Teng (2017) assert that tourist guides play six distinctive leadership roles: instrumental, social, interactional, communicative, dealing with the emergency, and caring. Pond (1993) also defines a series of tourist guide traits such as enthusiasm, self-confidence, having an outgoing nature, proactive behavior, sensitivity, flexibility, authenticity, exhibiting a pleasant and professional appearance, a sense of humor, being knowledgeable, good communication skills, organization talent, and decisiveness. Roles and traits of tourist guides manifest that the tourist guiding profession is multidirectional, and people performing as tourist guides are miscellaneous. Therefore, a tourist guide could have more than one CA due to achieve career goals and career path. At this point, the tourist guiding education gains importance since many CAs of individuals begin to shape with training and courses at the university.

The countries that attract the most tourists worldwide and have gained momentum in the tourism sector attach importance to tourist guide training. It is seen that the most critical common condition in guidance education and in all countries required to be a guide is to speak at least one foreign language fluently other than the mother tongue. In training or courses held in different countries, it is seen that guide candidates should have communication skills and be knowledgeable about first aid. In many countries, candidates receiving tourist guide training must have at least a high school or associate degree. The specialties of the tourist guide profession have expanded and diversified. For this reason, countries that provide tourist guidance training should be constantly renewed by considering this development and differentiation. From this point of view, training tourist guides who catch up with the changing tourism trends will contribute to the image and advertising of the country. Furthermore, a comprehensive tourist guide education provides new insights into candidates, and they can tackle difficulties in the profession thanks to the booming CA at this stage.

METHODOLOGY

This pilot study aims to determine female tourist guide candidates' career anchor perceptions (CAP) through an adapted career orientation inventory (CHI). The CAP used in the study is a short form of a 41-item inventory determined by İgbaria & Baroudi (1993). İgbaria & Baroudi (1993), in their research, measured CA with a test-retest approach and developed a 25-item CHI for further studies. That 25-item CAP has been used in this study to measure the perceptions of female tourist guide candidates. The survey was conducted from 15th to 30th December with the Tourism Guidance Department of Tourism Faculty and the Tourism Guidance students of the Institute of Social Sciences of Afyon Kocatepe University. An online questionnaire form was delivered to the research population. The data was gathered from 70 female students as a pilot study.

Data analyses of the study were started with validity and reliability tests. In the ongoing process, descriptive analyses were performed to measure participants' demographic variables such as class, group of ages, and primary career goals. Afterward, the arithmetic means and standard deviation measured each item's participation level. Discriminant analyses followed descriptive analyses to reveal significant differences among groups.

Data Analyses

Within the scope of data analyses, factor analysis is performed to measure the data set's validity, and the data's reliability is tested with Cronbach's Alpha coefficient for each factor. The results in Table 2 indicate that the KMO value is above 0,60, and Bartlett's test for sphericity is 0,000. Kaiser (1974) stated that values between 0,50 and 0,70 are considered mediocre for factor analyses (Field, 2005). Therefore, the data set is valid, and factor analysis is appropriate due to Bartlett's test result.

Table 2 indicates that this study has the same factors as İgbaria & Baroudi (1993) extracted in their research. The primary factor for CAP is the service with 6,754 eigenvalues, which explains approximately %23 of the total variance. Other factors in the CAP also have reliable eigenvalues, whereas all items of CAP have loadings above 0,40. The lowest loading is 0,478 at PC3 item, "*I feel successful only if I am constantly challenged by a tough problem or a competitive situation,*" and the highest value is 0,894 at E2 item indicating "*I am always on the lookout for ideas that would permit me to start and build my*

Table 2. Validity and reliability of data set

Factors	Items	Loadings	Eigenvalue	% Variance	% Cumulative Variance	Reliability	Scale Reliability
Service	S1	0,89	6,754	12,909	12,909	,813	,859
	S2	0,89					
	S3	0,65					
Autonomy	A1	0,86	3,106	10,746	23,654	,878	
	A2	0,81					
	A3	0,87					
Technical	T1	0,79	2,144	9,817	33,471	,812	
	T2	0,75					
	T3	0,79					
Managerial	M1	0,74	1,716	8,304	41,775	,754	
	M2	0,84					
	M3	0,52					
Security(Job tenure)	JT1	0,88	1,543	8,187	49,962	,874	
	JT2	0,89					
Entrepreneurship	E1	0,84	1,361	7,984	57,946	,637	
	E2	0,89					
	E3	0,83					
Lifestyle	L1	0,83	1,217	7,598	65,544	,606	
	L2	0,69					
	L3	0,49					
Pure challenge	PC1	0,85	1,083	7,588	73,132	,725	
	PC2	0,86					
	PC3	0,48					
Security (geographic)	SG1	0,75	1,009	6,596	79,728	,607	
	SG2	0,81					
Kaiser-Meyer-Olkin (KMO) Measure of Sampling Adequacy: 0,657							
Bartlett's Test of Sphericity: x^2: 1044,386 Sig: 0,000							

own enterprise." On the other hand, CAP factors explain %79,728 of the total, and each factor has a reliable Cronbach's alpha coefficient. Moreover, the overall reliability of CAP is 0,859, which confirms high reliability.

The demographics of respondents, such as the class, age group, and career goal, are presented in Table 3. 19 of 70 respondents are beginners in undergraduate education with a rate of %24,17, while %30 of the total is in the 4th class. In other words, 21 respondents are potential business life employees who are in milestones for career decisions. Eight respondents are students in graduate study as likely academicians. More than half of the respondents (n: 39, %55,71) are 18-21, while the rest are 22-25 ages (n: 31, %44,49).

Table 3. Demographic variables

Group	*Variables*	*n*	*%*	*Group*	*Variables*	*n*	*%*
Class	1st	19	27,14	***Career goal***	To be a tourist guide.To work in tourism enterprises. To be a civil servant.To build my own business.To be an academician. Others.	35271268	50,002,8610,0017,148,5711,43
	2nd	14	20,00				
	3th	12	17,14				
	4th	21	30,00				
	Master	8	5,71				
Age Group	18-21 ages	39	55,71				
	22-25 ages	31	44,29				

Table 4. Descriptive of CAP items and factors

Factor	**Items**	$\bar{x}$	**S.d**	$\bar{x}$	**Factor**	**Items**	$\bar{x}$	**S.d**	$\bar{x}$
Managerial	M1	4,171	1,021	4,105	**Technical**	T1	4,500	0,757	4,410
	M2	3,643	1,130			T2	4,357	0,869	
	M3	4,500	0,913			T3	4,371	0,981	
Autonomy	A1	4,186	1,081	4,343	**Pure Challenge**	PC1	2,900	1,331	3,481
	A2	4,514	0,830			PC2	3,157	1,315	
	A3	4,329	1,059			PC3	4,386	0,889	
Service	S1	3,743	1,348	3,467	**Lifestyle**	L1	4,729	0,658	4,162
	S2	3,571	1,325			L2	4,600	0,710	
	S3	3,086	1,380			L3	3,157	1,304	
Security (Job tenure)	JT1	4,386	0,906	4,307	**Entrepreneurship**	E1	4,057	1,034	3,890
	JT2	4,229	1,038			E2	3,843	1,235	
***Security (Geographic)**	SG1	3,471	1,282	3,343		E3	3,771	1,253	
	SG2	3,214	1,382		**CA (TOTAL)**				**3,900**

* Reverse coded, $\bar{x}$: *Mean*, **S.d.**: *Standard deviation*

The career goals presented in Table 3 reveal that half of the respondents want to be tourist guides in the future. Entrepreneurs of the sample constitute another significant group in career goals with %17,14. On the other hand, only two respondents said they wanted to work in any position in tourism enterprises. The career goals of others were listed as living abroad, being a tattooist, being a make-up artist, being a teacher, working at a private firm, and matriculating again.

The ongoing data analysis phase determines how respondents perceive CAP items and factors. Namely, their participation level in each item and the arithmetic means of factors have been evaluated in this phase. Results in Table 4 prove that the respondents highly agree with CAP items ($\bar{x}$: 3,90), and the most participated factor of CAP is Technical, with a 4,41 mean. In other words, respondents are motivated by the work's content and only want to advance in the technical or functional expertise. They also wish for autonomy ($\bar{x}$: 4,34) and job tenure ($\bar{x}$: ≈ 4,31) in their future career. However, they do not want to stay in any city or place in case of better career opportunities ($\bar{x}$: ≈2,66).

Table 5. Discriminant analyses within the context of respondents' classes

		Sum of Squares	df	Mean Square	F	Sig.
CA	**Between Groups**	,719	4	,180	,632	,641
	Within Groups	18,493	65	,285		
Managerial	Between Groups	4,848	4	1,212	1,802	,139
	Within Groups	43,717	65	,673		
Autonomy	Between Groups	4,340	4	1,085	1,389	,247
	Within Groups	50,765	65	,781		
Service	Between Groups	5,898	4	1,475	1,118	,356
	Within Groups	85,746	65	1,319		
Security(Job tenure)	Between Groups	3,902	4	,976	1,169	,333
	Within Groups	54,244	65	,835		
Security (Geographic)	Between Groups	3,197	4	,799	,622	,649
	Within Groups	83,574	65	1,286		
Technical	Between Groups	3,300	4	,825	1,534	,203
	Within Groups	34,961	65	,538		
Pure challenge	Between Groups	5,264	4	1,316	1,464	,223
	Within Groups	58,433	65	,899		
Lifestyle	Between Groups	1,551	4	,388	,963	,434
	Within Groups	26,170	65	,403		
Entrepreneurship	Between Groups	4,844	4	1,211	1,554	,197
	Within Groups	50,650	65	,779		

The most participated item of CAP is L1 ($\bar{x}$: ≈4,73), indicating that "*I want to develop a career that permits me to continue to pursue my own lifestyle.*" The second most participated item in the CAP scale is also listed under the Lifestyle factor as L2 ($\bar{x}$: 4,60), and it refers to "*I think a career is worthwhile only if it enables me to lead my life in my way.*" Contrary to these, the least participated item is PC1 ($\bar{x}$: 2,90), referring that "*I want a career working on problems that are almost insoluble.*"

Although the descriptive analyses concluded that the respondents highly perceive CA, it is necessary to perform discriminant analyses to reveal any potential differences among groups. Therefore, Table 5, Table 6, and Table 7 present the results of discriminant analyses for each group in order starting from the classes of respondents. In discriminant analyses, 0,05 is accepted as a critical value for significance; values under 0,05 indicate a significant difference between groups. In this case, the Ryan-Einot-Gabriel-Welsch Range test determines which group significantly differs from others.

Table 5 presents the discriminant analyses performed to assign any potential group differences regarding respondents' classes. As seen in the Table, respondents similarly perceive CAP items and the sub-dimensions of CAP (Sig.>0,05). Thus, there are no significant differences between groups regarding respondents' classes.

Table 6 presents the results of the independent samples T-test for the age group of respondents. It is clear from the table that there are no significant differences in perceiving CAP items when we consider

Table 6. Discriminant analyses within the context of respondents' age groups

Categories		Levene's Test for Equality of Variances		t-Test for Equality of Means						
		F	Sig.	t	df	Sig. (2-tailed)	Mean Difference	Std. Error Difference	95% Confidence Interval of the Difference	
									Lower	Upper
CA	**1**	,000	,989	-,865	68	,390	-,11001	,12720	-,36383	,14382
	2	--	--	-,871	66,028	,387	-,11001	,12631	-,36219	,14218
Managerial	**1**	,568	,454	-1,175	68	,244	-,23656	,20132	-,63828	,16516
	2	--	--	-1,193	67,264	,237	-,23656	,19835	-,63245	,15933
Autonomy	**1**	1,904	,172	-,726	68	,470	-,15660	,21577	-,58717	,27397
	2	--	--	-,752	67,494	,454	-,15660	,20814	-,57200	,25879
Service	**1**	,444	,507	,791	68	,432	,22002	,27806	-,33485	,77488
	2	--	--	,781	60,766	,438	,22002	,28175	-,34343	,78346
Security(Job tenure)	**1**	,034	,855	,266	68	,791	,05914	,22239	-,38463	,50291
	2	--	--	,266	64,188	,791	,05914	,22265	-,38564	,50392
Security (Geographic)	**1**	,001	,971	-1,433	68	,156	-,38379	,26780	-,91817	,15060
	2	--	--	-1,426	63,161	,159	-,38379	,26916	-,92163	,15406
Technical	**1**	,219	,641	-1,069	68	,289	-,19134	,17899	-,54852	,16583
	2	--	--	-1,072	65,095	,288	-,19134	,17853	-,54788	,16520
Pure challenge	**1**	,042	,839	-1,195	68	,236	-,27543	,23048	-,73534	,18447
	2	--	--	-1,198	64,982	,235	-,27543	,22999	-,73476	,18390
Lifestyle	**1**	,608	,438	-,244	68	,808	-,03750	,15357	-,34393	,26894
	2	--	--	-,250	67,993	,803	-,03750	,14971	-,33623	,26124
Entrepreneurship	**1**	,015	,904	-,105	68	,916	-,02288	,21735	-,45661	,41084
	2	--	--	-,104	62,471	,917	-,02288	,21899	-,46058	,41481

1: Equal variances assumed, **2:** Equal variances not assumed

the age groups of respondents (Sig.> 0,05). Additionally, the significance values for sub-dimensions of CAP are all above 0.05, which is the critical value for discriminants. Therefore, it is concluded that respondents perceive CAP items and sub-dimensions similarly regarding their group ages. This result is reasonable since the respondents' ages are very close and they are university students of the same generation.

Respondents' career anchor perceptions were tested with discriminant analyses based on their career goals in Table 7. Results in the Table indicate that the career anchor perceptions significantly differ according to career goals (Sig.: 0,001). Additionally, it is concluded that the autonomy (Sig.: 0,007), service (Sig.: 0,000), and entrepreneurship (Sig.: 0,000) perceptions of respondents significantly differ based on career goals.

After determining significant differences at CAP within the context of career goals, post-hoc tests were performed to reveal different groups. Results in Table 8 indicate that participants who want to be

Table 7. Discriminant analyses within the context of respondents' career goals

Categories		Sum of Squares	df	Mean Square	F	Sig.
CA	**Between Groups**	5,051	5	1,010	4,566	**,001***
	Within Groups	14,161	64	,221		
Managerial	Between Groups	4,603	5	,921	1,340	,259
	Within Groups	43,962	64	,687		
Autonomy	Between Groups	11,851	5	2,370	3,507	**,007***
	Within Groups	43,253	64	,676		
Service	Between Groups	40,451	5	8,090	10,114	**,000***
	Within Groups	51,194	64	,800		
Security(Job tenure)	Between Groups	3,108	5	,622	,723	,609
	Within Groups	55,038	64	,860		
Security (Geographic)	Between Groups	6,288	5	1,258	1,000	,425
	Within Groups	80,483	64	1,258		
Technical	Between Groups	5,168	5	1,034	1,999	,091
	Within Groups	33,092	64	,517		
Pure challenge	Between Groups	5,501	5	1,100	1,210	,315
	Within Groups	58,196	64	,909		
Lifestyle	Between Groups	2,991	5	,598	1,548	,188
	Within Groups	24,730	64	,386		
Entrepreneurship	Between Groups	16,791	5	3,358	5,553	**,000***
	Within Groups	38,702	64	,605		

tourist guides ($\bar{x}$: ≈4,54) and want to build their businesses ($\bar{x}$: ≈4,63) significantly differ from others ($\bar{x}$: ≈3,25). However, participants who wish to be academicians, want to be civil servants, and work in tourism enterprises similarly perceive the CA. Table 8 also presents participants who wish to be academicians perceive significantly far less autonomy ($\bar{x}$: ≈3,28) than those whose career goals are to be civil servants ($\bar{x}$: ≈4,67), to build their businesses ($\bar{x}$: ≈4,67) and to be tourist guides ($\bar{x}$: ≈4,48).

When Table 8 is considered in terms of the Service sub-dimension, it is seen that the participants who want to be tourist guides perceive the Service much more ($\bar{x}$: 4,00) than the two groups [(Others $\bar{x}$: ≈1,92) and Civil Servant candidates $\bar{x}$: ≈2,62)]. Those groups significantly differ in perceiving the Service dimension. Finally, participants who want to build their businesses perceive much more entrepreneurship ($\bar{x}$: ≈4,53) than those whose career goals vary ($\bar{x}$: ≈2,99). There are no significant differences in other groups of entrepreneurship sub-dimension. Career goals could be an important antecedent for shaping CA and sub-dimensions.

CONCLUSION

Tourist guidance is a dynamic process requiring high performance and good service in multifarious atmospheres. Regardless of tour types, such as cultural routes, ecotours, or urban tours, tourist guides

Table 8. Post-hoc tests (Ryan-Einot-Gabriel-Welsch F)

Categories	Career Goals	Mean
CA	*Other*	**3,2450a**
	To be an academician.	3,6267ab
	To work in tourism enterprises.	3,8400ab
	To be a civil servant.	3,8514ab
	To be a tourist guide.	**4,0537b**
	To build my own business.	**4,0633b**
Autonomy	*To be an academician.*	**3,2778a**
	Other	3,7917ab
	To work in tourism enterprises.	4,3333ab
	To be a civil servant.	**4,6667b**
	To build my own business.	**4,6667b**
	To be a tourist guide.	**4,4762b**
Service	*Other*	**1,9167a**
	To be a civil servant.	**2,6190a**
	To be an academician.	2,6667ab
	To work in tourism enterprises.	3,1667abc
	To build my own business.	3,8611bc
	To be a tourist guide.	**4,0095c**
Entrepreneurship	*Other*	**2,9583a**
	To be an academician.	3,1667ab
	To work in tourism enterprises.	3,6667abc
	To be a civil servant.	3,6190abc
	To be a tourist guide.	4,0762bc
	To build my own business.	**4,5270c**

simultaneously deal with the guests, control them, and interpret touristic attractions. Hence, tourist guidance is generally accepted as challenging work with many specific roles. Being a personal assistant, having management capability, being an expert, knowing local issues, communicating with the group members peer to peer, and taking responsibility are some of these specific roles observed at tourist guidance (Manning, 2016). Besides, tourist guiding is heavily based on physical strength as an interdisciplinary profession. Thus, people who want to be tourist guides need to evaluate the conditions and content of the work in advance. This evaluation process gains more importance and meaning with university education, where tourist guide candidates become familiar with the job through courses and training. Herein, they also start to build up their career anchors based on experiences and narratives related to the job. Some candidates focus on the autonomy anchor, while others take cognizance of the service anchor, and some are concerned with the job as a challenge anchor.

Although the absence of prior studies about tourist guides CA perceptions, many factors, such as personal characteristics and demographics, conceivably affect CA growth. Gender can also be an essential factor shaping CA. Thus, this study aims to evaluate female tourist guide candidates with descriptive analyses. Seventy female participants were the sample of this study, and it is concluded that half of this sample has chosen the primary career goal of being a tourist guide. However, participants have high scores in perceiving career anchors, technical/ function competence is the highest perceived anchor among others. From this, we can assume that female candidates are motivated by the content of tourist guidance and want to be experts in this profession. On the other hand, as expected, they perceive geographic security less than other CA. Namely, they desire a job that enables freedom to travel. This fact is also a part of the tourist guidance since the members of this profession perform their job by traveling to various touristic attractions and interpreting them. Another study finding is that CA perceptions of female tourist guide candidates do not differ significantly based on their age group and classes. However, CA perceptions significantly differ based on their career goals. It is determined that autonomy, service, and entrepreneurship are the differentiating anchors.

The results of this study present some important evidence regarding career management and career path through CA perceptions of female tourist guide candidates. Above all, the perceptions about CA of participants are congruent with the nature of the job. Additionally, participants are highly aware of CA, and it seems they harmonized CA with their career goals. However, these results are obtained from a small sample within the context of a pilot study. Thus, more comprehensive studies still need to be conducted with larger samples. More control variables must also be added to receive more detailed results, and studies using qualitative and quantitative research techniques should be conducted in the future. By the way, tourist guide candidates should be more informed about career anchors. They also need to be guided by academicians with theoretical explanations and scenarios of role model plays.

REFERENCES

Beck, J., & La Lopa, J. M. (2001). An exploratory application of Schein's career anchors inventory to hotel executive operating committee members. *International Journal of Hospitality Management*, *20*(1), 15–28. doi:10.1016/S0278-4319(00)00035-9

Beck, J. A. (1996). *Career anchors, organizational commitment, and job plateaus: An analysis of hotel executive operating committee members' career development* [Doctor of Philosophy Thesis]. Purdue University Publications.

Cerdin, J. L., & Pargneux, M. L. (2010). Career anchors: A comparison between organization-assigned and self-initiated expatriates. *Thunderbird International Business Review*, *52*(4), 287–299. doi:10.1002/tie.20350

Coetzee, M., & Schreuder, D. (2014). Career anchors as a meta-capacity in organizational career development. In M. Coetzee (Ed.), *Psycho-social Career Meta-capacities* (pp. 139–154). Springer. doi:10.1007/978-3-319-00645-1_8

Danziger, N., & Valency, R. (2006). Career anchors: Distribution and impact on job satisfaction, the Israeli case. *Career Development International*, *11*(4), 293–303. doi:10.1108/13620430610672513

Etmyonline. (2023). *Career*. https://www.etymonline.com/word/career

Evans, C. (1996). A review of career anchors in use. *European Journal of Work and Organizational Psychology*, *5*(4), 609–615. doi:10.1080/13594329608414883

Gubler, M., Biemann, T., Tschopp, C., & Grote, G. (2015). How career anchors differentiate managerial career trajectories: A sequence analysis perspective. *Journal of Career Development*, *42*(5), 412–430. doi:10.1177/0894845315572891

Hardin, J. R., Stocks, M. H., & Graves, O. F. (2001). The effect of match or mismatch between the career anchors and the job settings of CPAs: An empirical analysis. *Advances in Accounting*, *18*, 119–148. doi:10.1016/S0882-6110(01)18008-9

Igbaria, M., & Baroudi, J. J. (1993). A short-form measure of career orientations: A psychometric evaluation. *Journal of Management Information Systems*, *10*(2), 131–154. doi:10.1080/07421222.1993.11518003

Jacobs, R. C. (1987). Identifying Career anchors and team perceptions: A new approach to diagnosing management team-building needs. *Asia Pacific Journal of Human Resources*, *25*(3), 35–50. doi:10.1177/103841118702500303

Knight, D. K., Crutsinger, C., & Kim, H. (2006). The impact of retail work experience, career expectation, and job satisfaction on retail career intention. *Clothing & Textiles Research Journal*, *24*(1), 1–14. doi:10.1177/0887302X0602400101

Kong, H., Wang, S., & Fu, X. (2015). Meeting Career Expectation: Can It Enhance Job Satisfaction of Generation Y? *International Journal of Contemporary Hospitality Management*, *27*(1), 1–27. doi:10.1108/IJCHM-08-2013-0353

Kucharvy, T. (2014). *Creating a pathway to your dream: Career designing and controlling a career around your life goals*. Business Expert Press.

Olenik-Shemesh, D., Heiman, T., & Keshet, N. S. (2018). The role of career aspiration, self-esteem, body esteem, and gender in predicting sense of well-being among emerging adults. *The Journal of Genetic Psychology*, *179*(6), 343–356. doi:10.1080/00221325.2018.1526163 PMID:30332342

Oxford Advanced Learner's Dictionary. (2023). *Definition of Career Noun from the Oxford Advanced Learner's Dictionary*. https://www.oxfordlearnersdictionaries.com/definition/english/career_1?q=career

Pond, K. L. (1993). *The professional guide: Dynamics of tour guiding*. Van Nostrand Reinhold Company.

Rojewski, J. W. (2005). Occupational aspirations: constructs, meaning, and application. In S. D. Brown & R. W. Lent (Eds.), *Career development and counseling: Putting theory and research to work*. John Wiley & Sons.

Schein, E. H. (1990). *Career anchors: Discovering your real values* (revised edition). Pfeiffer & Company.

Schein, E. H. (1996). Career anchors revisited: Implications for career development in the 21st century. *The Academy of Management Perspectives*, *10*(4), 80–88. doi:10.5465/ame.1996.3145321

Tsaur, S.-H., & Teng, H.-Y. (2017). Exploring tour guiding styles: The perspective of tour leader roles. *Tourism Management*, *59*, 438–448. doi:10.1016/j.tourman.2016.09.005

Türkay, O. (2015). Turizm İşletmelerinde Kariyer Yönetimi. In E. Pelit (Ed.), *Turizm İşletmelerinde İnsan Kaynakları Yönetimi*. Grafiker Yayınları.

Van Vuuren, L. J., & Fourie, C. (2000). Career anchors and career resilience: Supplementary constructs? *SA Journal of Industrial Psychology*, *26*(3), 15–20. doi:10.4102ajip.v26i3.714

Weber, K., & Ladkin, A. (2009). Career anchors of convention and exhibition industry professionals in Asia. *Journal of Convention & Event Tourism*, *10*(4), 243–255. doi:10.1080/15470140903356411

WFTGA. (2003). *What is a tourist guide?* https://wftga.org/about-us/what-is-a-tourist-guide/

Yarnall, J. (1998). Career anchors: Results of an organisational study in the UK. *Career Development International*, *3*(2), 56–61. doi:10.1108/13620439810207536

Zhang, H. Q., & Chow, I. (2004). Application of importance-performance model in tour guides' performance: Evidence from mainland Chinese outbound visitors in Hong Kong. *Tourism Management*, *25*(1), 81–91. doi:10.1016/S0261-5177(03)00064-5

Chapter 24
Women in Tourism:
Gender Bias and Constraints

Donsy Joy
CHRIST University (Deemed), India

Havilah Prarthana Johnson
CHRIST University (Deemed), India

R. Kavitha
CHRIST University (Deemed), India

ABSTRACT

Tourism could be spending time away from home in search of leisure, relaxation, and pleasure while utilizing the commercial provision of services. The capacity of tourism to improve livelihoods by reducing poverty, guarantee or enhance environmental sustainability, and promote gender equality and women's empowerment are some of its strongest points. "Women's empowerment" can mean a variety of things, such as respecting women's opinion, making an effort to find them, and improving women's status through training, awareness, and education. They may have the chance to redefine gender norms and other similar roles, giving them more flexibility to pursue their objectives. The main aim of the chapter is to identify the major challenges faced by women in the tourism sector, the ways to overcome constraints, gender stereotypes, and the social stigmas while they juggle between multiple responsibilities. With this chapter, the authors aim to gain more knowledge more specifically on how women can be empowered in the tourism and hospitality sector.

INTRODUCTION

Tourism is a social, cultural, and economic phenomenon that refers to traveling outside of one's normal area for personal, business, or professional reasons. These people are called visitors, and tourism has to do with their activities, some of which involve spending money on tourism. The tourism sector is made up of a variety of industries, including lodging, food and beverage service, passenger transportation

DOI: 10.4018/978-1-6684-8417-3.ch024

(by rail, road, water, and air), rental of transportation equipment, travel agencies, and other reservation services, as well as cultural, sporting, and recreational activities, retail trade of goods specifically for the tourism industry of a nation, and other activities.

The tourist sector is one of the most crucial for boosting employment, reducing poverty, and empowering women. Giving women access to earning opportunities that only raise their level of economic empowerment is a primary focus of the bulk of tourism-related projects aimed at eliminating poverty. However, this usually doesn't lead to general advancement. Power disparities in society, intricate cultural norms, and conventional gender roles are to blame for this.

Thanks to tourism, women now have more employment options, which boosts their freedom and self-reliance. Women in rural developing nations are moving away from agriculture and towards tourism since it needs less physical labour, which is proof of this(Spenceley & Meyer, 2012). In addition, because they might improve service delivery, women find respect, thanks, and recognition in the tourism sector. Women have typically been expected to take care of their family duties at home rather than leave their comfort zones to look for independent jobs (Sahay, 1998b). Additionally, historically, women have been viewed as being primarily in charge of reproduction.

Women have the capacity to influence economic and social development when given the chance to actively participate in decisions that affect their life. Giving women the tools they need to lead within their communities, regions, and countries will help achieve this goal. Women can achieve their life goals through empowerment, such as raising their income on their own to meet the expanding financial, health care, and educational needs of their children.

It has been demonstrated that tourism can open doors to empowerment, and it is important to take full advantage of this possibility. Women's ability to fully participate to the tourism industry is now unrealized due to their concentration in lower status and lower paying jobs.

Through the growth of tourism, the promotion of sustainability, and inclusive development, UNWTO aims to improve the quality of women's lives. "Tourism opens doors for women" was the subject chosen for World Tourism Day 2007, bringing up gender concerns and supporting women's empowerment and gender equality. Secretary-General Taleb Rifai stated in the UNWTO report from 2010 that "more must be done to achieve gender equality" and that "tourism plays a critical role in empowering women politically, socially, and economically."

The UN Convention on Human Rights states that all women should have the same rights as males. However, programmes and initiatives for tourism development that are designed and put into action are typically gender-blind and adhere to the patterns of gender relations that already exist, not addressing the concerns of gender disparity (Longwe, 2002a). Women confront more difficulties since they have less access to education, exposure, and social support. Government initiatives should support women's overall development so that instead of being viewed as welfare recipients, they take an active role in constructing the country. Development should take into account social and human variables in addition to economic ones (Varghese, 2011)

The tourist sector offers both men and women a variety of job opportunities . Consequently, one of the largest sectors in the world is tourism . The tourist sector needs more women to participate . The National Tourism Development Master Plan's Article 28 (b) controls the participation of women in the industry and identifies maximising gender mainstreaming through tourism development as one of the community empowerment strategies in the industry(Tristanti et al., 2022).

The social and cultural norms that frequently give rise to gender discrimination against women make it crucial to raise public awareness of these issues and draw attention to the working circumstances that women face in the tourism sector. However, women also need to be motivated, supported, and empowered in order for them to establish their legitimate demands and actively participate in tourism. In this regard, providing adequate training for women and girls is one of the most important factors(Tristanti et al., 2022).

The emphasis should be on the tourism industry's sustainable expansion, in which all participants engage equally, obtain comparable advantages, have an equal opportunity to find employment, and are paid fairly(Tristanti et al., 2022).

OBJECTIVE AND PURPOSE

The chapter aims to seek in-depth analysis of the bottlenecks of women empowerment, comparison in trends relating to gender stereotypes, changes in economy that's evidently seen the upper hand in women empowerment during the past years in the tourism sector. The theme emphasized in this paper allows us to observe the advancement of women empowerment, the problems and obstacles encountered, and the gendered treatment of women in the tourism sector. It furnishes the target audience with adequate knowledge about the current scenario of women empowerment and their projected growth trajectory.

GENDER BIAS

Gender Bias is a situation wherein a person receives different treatment because of their perceived gender identity. Unconscious or implicit bias frequently takes the form of gender bias.

One of our "core" identities is gender. It begins early in life and carries on all the way through. While "sex" refers to the biological factors of being male or female, "gender" refers to the social, cultural, and psychological components of men's and women's lives. The features of gender change over time and differ between cultures (based on values, beliefs, behavioural expectations, etc.) (Sabina & Nicolae, 2013). It is a learned identity that evolves over time and differs greatly within and between cultures. According to the International Research and Training Institute for the Advancement of Women (INSTRAW) of the United Nations, Gender refers to "a range of socially constructed roles and relationships, personality traits, attitudes, behaviours, values, relative power and influence that society ascribes to the two sexes on a differential basis." (Sabina & Nicolae, 2013)

(Kabeer, 2005a) defined empowerment as "the capacity to make decisions." Since women are frequently denied the right to vote and are marginalised in practically every country, this concept significant. It is crucial to establishing women's empowerment as it gives women control over their lives and promotes their development.

The tourist sector is one of the most significant areas where gender-based discrimination is present in the workplace. Even though there are nearly twice as many women working in the industry as there are males, which is made up entirely of businesses that provide various services, women still face discrimination and differing working conditions because of their gender. (Ramchurjee & Paktin, 2011) in his study mentions that 'Tourism' has demonstrated potential for creation of job opportunities and growth that helps in supporting the local communities in destination areas. However, less attention has been paid to the unequal ways the benefits of tourism are spread to men and women especially in the developing world. Opportunities and obstacles for gender equality and women's empowerment are both present in the tourism industry. Despite the fact that women are underrepresented in management and leadership positions, they have recently made a greater contribution to the economic world. However, even though there are a lot of women working in the tourism sector, the majority of their positions are low-skilled and poorly paid.

(Ramchurjee & Paktin, 2011) also mentions the tourism industry's cyclical nature which is another prevalent feature that causes some to claim that it can create a good condition for women to manage their many duties, but this necessarily cannot be seen as a general rule. In many circumstances, mothers and women cannot count on a consistent salary to serve as the foundation of the household income and must instead create income on their own throughout the year. (Ramchurjee & Paktin, 2011) opines that, women are more likely to work in jobs that are common and expected of them, such as cooking, cleaning, and hospitality which happens because of discrimination and gender stereotypes. However, if a strong gender perspective is incorporated into the planning and execution processes, tourism can be used as a platform to promote gender equality and women's empowerment at the household, community, national, and international levels.

The phrase "glass ceiling" refers to an invisible barrier that keeps a particular population from moving up the social hierarchy beyond a certain level. It is typically used to refer to women. No matter how subtly it is portrayed, the glass ceiling is actually a challenging barrier to overcome. In simple words, it refers to the situation in which a qualified individual who wants to move up an organizational hierarchy is hindered from doing so at a greater level due to prejudice, most frequently based on sexism or racism.

Additionally, according to traditional gender norms, women are primarily responsible for taking care of children, the elderly, and household chores. As a result, women frequently have to pick or in other words, 'forced to choose' between seasonal and informal jobs. The concept of "glass floor" that tends to exempt women from many of the unpleasant characteristics of a hierarchy's lower levels is generally not considered a problem however it seeks to highlight or raise awareness of the discriminatory inequity. (Ramchurjee & Paktin, 2011)

Although the concept of glass ceiling was true for the majority of other industries, the hospitality sector's glass ceilings were substantially higher (Janta, 2011). The glass ceilings for women who are immigrants or from other nations are considerably greater for them (Janta, 2011). The migrant women encounter additional difficulties because they are unaware of the local laws and labour practises that have been implemented.(Rasanjali et al., 2021)

In the study conducted by (Ramchurjee & Paktin, 2011), it is observed that, the vertical segregation of the labour market in the administrative/managerial and service sectors is a reflection of the state of the labour markets generally. More than 40% of the world's workforce is now made up of women, who have attained higher levels of education than ever before. However, only a small percentage of them succeed in breaking past thc "glass cciling," and their representation in executive positions continues to be unacceptably low.

GENDER DISCRIMINATION/STEREOTYPING

(Kogovsek & Kogovsek, 2015) states that organizations and larger communities miss out on opportunities by ignoring female potential and fail to take action to overcome existing gender prejudices. Along with income inequality, occupational gender segregation practises that are typically based on customs and prejudices are where gender discrimination is most openly present.

Gender issues have an impact on sexual objectification, gender roles, women's rights, and women's employment and participation in tourist planning and management. Tourism-related exploitation can take many different forms: Poor pay, hazardous working conditions, sexual harassment and prostitution, as well as slavery and human trafficking. (Sabina & Nicolae, 2013)

Women make up between 60 and 70 percent of the tourism workforce as per International-Labour-Office, 2001 (Singhania et al., 2021). However, there is a significant gender gap in the labour division in the tourism sector (Kinnaird & Hall, 1996). Opportunities offered to men and women are not dispersed equitably (Swain, 1993). When compared to men who are hired for more senior roles like managers and owners, women are typically employed in temporary, less trained, and seasonal professions like cleaning or cooking and are paid less. Even though they require skills, the professions that women have are not classed as skilled, which further devalues them if they are performing menial labour. The prevailing cultural stereotypes against women in society are to blame. (Singhania et al., 2021)

Additionally, women frequently miss out on opportunities to participate in official employment and instead work mostly in informal, gender-biased jobs(Scheyvens, 2000). The jobs associated with women and performed by them in the tourism sector are viewed as unskilled, whereas the identical tasks performed by males are regarded seriously and viewed as skilled for them(Elmas, 2007). For instance, many males in the hotel industry are choosing to work in the traditionally female-dominated field of cooking. Even non-traditional employment for women, including tour guides and drivers, are forbidden since they are seen as a man's job and not appropriate for women. (Wilkinson & Pratiwi, 1995)

The fact that women are primarily concentrated in informal jobs that are primarily extensions of their domestic household activities, such as housework, cooking, etc., has led to the acceptance of female participation in the tourism industry in male-dominated countries (Elmas, 2007). (Swain, 1993), in their study states that, Kuna women in Panama participated in ethnic mola art but needed men to market it because they were not permitted to work outside of their community. Similarly, Tucker asserted in her study that it is only appropriate for men to interact with visitors, and that women should not do so (Tucker & Boonabaana, 2012).

The United Nations (2015) reports that working women lack access to their own income, particularly in developing nations. Even in their own houses, they are not permitted to make financial decisions. It stands for the hierarchies of power that prevail in society (Singhania et al., 2021). Women are still unable to rise to positions of leadership and social prestige even after becoming financially independent. In order to understand how changes occur over time as a result of tourism development, it is essential to look beyond women's economic empowerment and comprehend the complexities of the gendered nature of tourism employment and the power relationships involved (Tucker & Boonabaana, 2012).

The term "occupational gender segregation," which refers to the unequal distribution of men and women across various jobs, is widespread throughout the world. As a result, promotion chances for women are constrained. Men's perceptions of women and even women's perceptions of themselves are affected by occupational gender segregation, which results in an overrepresentation of women in lower professional categories, lower prestige, and lower pay (Jordan, 1997). Occupation gender segregation arises because a sizable fraction of businesses base their hiring decisions on stereotypes about certain jobs that are exclusively suitable for men or women (Kogovsek & Kogovsek, 2015)

The tourism industry's labour market exhibits significant horizontal and vertical gender segregation, where, women are employed as waitresses/servers, cleaners, travel agency sales representatives, flight attendants, etc., whereas men are employed as gardeners, construction workers, drivers, pilots, etc on a horizontal level. The traditional "gender pyramid" is evident vertically in the tourism industry, with males predominating in important managerial positions whereas lower levels of employment with limited chances for professional advancement are dominated by women.

Despite the fact that women make up the majority of the hospitality workforce, they are still underrepresented in management positions as a result of formal and informal exclusionary practises and stereotypes that successful people must possess certain traits that are associated with men. Women are still under higher pressure after they reach management positions (Kogovsek & Kogovsek, 2015).

STEREOTYPICAL IMAGE OF WOMEN

Women have traditionally been expected to stay indoors and take care of their family responsibilities without leaving the confines of their homes to pursue independent employment (Sahay, 1998a). Women have also traditionally been seen as being largely responsible for reproduction (Singhania et al., 2021). Women are routinely excluded from positions of authority and responsibility within the sector, and are concentrated in low-skilled and low-paying jobs, objectified as part of the tourism "package," and permitted to continue in their traditional roles within a sector that thrives on such stereotypical images (Ramchurjee & Paktin, 2011). According to Cukier (1996), in his research on gendered effects on the tourism development and employment conducted in Bali, Indonesia, it was found that opportunities for female employees were constrained because they had no access to male jobs (e.g., security guards, drivers, groundskeeping), and the need to care for children prevents them from working the graveyard shift. Only 7% of women hold the well-paid, respected job of tour guide.

While tourism generates jobs, women are less likely to be able to access them due to family responsibilities, religious convictions, cultural norms on gender roles, the ability to travel farther for work, and childrearing. As a result, women supplement their income with side jobs (such baking or cleaning) to support their families. They also rely on social media for help with childcare and domestic duties, as well as money from relatives who live abroad. (Cukier, 1996)

Gender discrepancies are most widespread in developing nations where it is the duty of a women to look after domestic chores while the men handle the finances/cash. Women are viewed as physically frail and are restricted by cultural norms hence, they are not permitted to perform male vocations (Elmas, 2007). If women had to get involves in the outside work like tourism activities, they are subjected to being judged based on their moral character by society(Hosni & Lundberg, 2005).

Women are unable to negotiate better terms for themselves with their family, society, or the government since they have always been sidelined. This limits their talents and leaves them vulnerable and poor (Singhania et al., 2021). Women who are dependent on others for survival frequently lose their cultural identity, pride, and self-worth. Even working women who are the sole providers for their families may not have adequate assistance or access to resources to enable them to frequently continue working in challenging circumstances (Wallace & March, 1991).

Women have little opportunity to consider if they could engage in travel because their roles as wives and mothers were already predetermined. It was socially unacceptable since it deviated from the accepted standard. Many women have encountered issues related to how other people perceived them because their friends, family, and co-workers showed disapproval of their decision to travel alone. Many of the female passengers traveling alone have often received blatant displays of social impoliteness from others, which effectively limited their freedom to leave or travel solo. The perception of others, according to many women, is that traveling alone is only acceptable while you're young (Wilson & Little, 2005)

It takes time for women to negotiate gender roles in order to promote equality, but even then, these advances are not immune from social backlash(Díaz-Carrión & Vizcaino, 2022).

JOB ROLES

In the tourism industry, women are more likely to work in cleaning, customer service, and other lower-level positions than males are in maintenance, bartending, gardening, and higher-level positions, indicating that societal norms continue to play a significant role in patterns of segregation. In reality, there is a common vertical gender pyramid where men hold the higher and management positions and women often occupy the lower levels, with a few notable exceptions (Kogovsek & Kogovsek, 2015).

Although it is established that women are well-represented in the workforce, it is seen that they typically work in service or secretarial roles rather than professional or decision-making ones. It was shown that women are almost twice as likely to be business owners in the tourist sector than in other industries, and they frequently hire more women than men (Ramchurjee & Paktin, 2011). Women's leadership in tourism has lot of potential to grow, however, women still make up barely one-fifth of all tourism ministers and tourist board chairs. Although tourism frequently aids in community development and offers chances for women to work for themselves, the study discovers that women contribute to a large amount of unpaid work in tourism family business, when compared to other industries (Ramchurjee & Paktin, 2011).

Most women who start enterprises are limited to producing agarbatti, papad and pickles, embroidery, handicrafts, catering services, running snack bars, small retail stores, beauty salons, etc. However, this has not entirely improved the position of women. Women continue to have issues with money, culture, technology, marketing, and access to resources. Many Indian women business owners reported that, although owning the company, males predominate in decision-making (Bharathi, 2019).

Their average salary is typically lower than men. Unskilled women typically work in subpar conditions, and the time spent on paid employment frequently conflicts with women's obligations to their families and homes. Women commonly work in part-time occupations, which are underpaid and limit their access to management and skilled labour roles. Often the cultural stereotypes/preconceptions that originate in communities act as a constraining factor for women leadership because they tend to limit or restrict a woman's aspiration or goals (Sabina & Nicolae, 2013).

PATRIARCHAL SOCIETY

Women are not permitted to do anything against men's preferences because men have all of the power and authority in the home and even outside of it (Longwe, 2002b). Major choices are still made by men in the home even when women are participating in decision-making. They need men's approval before working outside the home or opening their own business, and males are the only ones who determine what jobs women should perform. Additionally, it is not permitted for women to interact with men outside of the group because it is not acceptable. As a result, women working in the tourism industry are frequently restricted to positions that do not need them to contact visitors(Panta & Thapa, 2018a) .

It is evident from this literature and some official reports that the patriarchal gender roles that are in place influence both hosts and guests. As a result, women who work in the tourism industry typically hold seasonal, part-time, unofficial, unskilled, and low-paying jobs in the hospitality, cleaning, or sex industries. Some scholars have also observed that employment in the tourism sector, whether paid or unpaid, does not necessarily translate into social, political, or cultural empowerment (Jiménez-Esquinas, 2017).

Working conditions for women in the tourist sector are gendered practises that often continue and maybe worsen gender disparities by placing them in a global labour force that is exploited and dominated by capital and patriarchal social and economic institutions (Kinnaird & Hall, 1996).

LOWER WAGE PAYMENT

Despite the fact that laws promoting gender equality have been in place for many years, gender discrimination still occurs because women are statistically paid less than men for the same job because the value of women's work is discounted (Foubert et al., 2010).

Even though 54% of the world's tourist workers are women, there exist pay gaps (World Tourism Organization, 2019). This difference is worse in developing nations. In the workforce, there are inequities that are both at horizontal and vertical for women (Rinaldi & Salerno, 2020). The issue of gender wage disparities is significant worldwide and is particularly serious in the hotel industry. Women are paid less than their male counterparts not only when they first enter the workforce but also after they achieve their career peak positions. In other words, even though they perform the same type of work with the same degree of responsibility, they are paid less. The researchers draw the conclusion that lower wages received by women in the sample of the Spanish hospitality workforce are the result of gender segregation as a result of the disproportionate concentration of women in the lowest-paying jobs, functional areas, jobs with lower levels of responsibility, and least desirable positions (Purcell, 1996).

EMOTIONAL CONTEXT

Feminist theory has significantly influenced social studies by recognising experience and subjectivity in the examination of gender issues and paying close attention to the role of emotions in cultural values and beliefs as well as in the creation of social relations and hierarchies (Dilley & Scraton, 2010).

The nature of such job and the existing gender expectations and stereotypes generate emotions with varying effects at the individual and social level, despite the fact that rural women tend to appreciate the positive features of tourism work (Ratten & Dana, 2017) . Participants who engaged in masculinized activities in particular spoke of the need to constantly compare themselves to their male peers or the pressure to prove that they could perform just as well as men, if not better: "you feel tired of always having to prove yourself... or show that you are better than men, sometimes it is exhausting... "(Laura, 34, single, adventure tourist guide, Jalcomulco, 2016). In addition, they had to contend with the stereotypes that "women don't know the monte" (the hills or the outdoors) and that "their place is in the kitchen." Therefore, women had to carry an extra weight of demonstrating their competence in order to get higher paying positions in rural tourism, such as tourist guides (Ratten & Dana, 2017)

Shame is related to an insecurity that has 2 meanings: a lack of self-confidence and a certain level of dread of others. Participants from Mexico had also started to "undo" the emotion of shame, gradually overcoming the sociocultural norms that attach shame to women's interactions with strangers (primarily male tourists or other male stakeholders, such as consultants, researchers, suppliers, or trainers), in order to possibly work more comfortably in tourism environments. This is similar to the experiences of Göreme women in Turkey (Tucker, 2007) . Furthermore, some of the women in our study engaged in affective performances to overcome their shame and turn it into pride to negotiate gender norms. As women encountered the limitations of gendered employment or gendered places in tourism industry, responses associated with rage, despair, or sadness also emerged as a result of the presence of gender stereotypes. Young women who developed an interest in activities that were often associated with men were more likely to experience this predicament (Díaz-Carrión & Vizcaino, 2022).

STRATEGIES TO IMPROVE WOMEN'S PARTICIPATION

- From a study conducted in Ella, Srinlanka, it was observed that the majority of those who were active in the tourism sector (43%) think workshops and training programmes would be beneficial. This makes sense given that those already employed in the field are aware of its demands and may require this training to hone specific abilities. Promotional techniques on social media (54%) are thought to be effective in luring more women into the sector, according to those who were not involved in it (Rasanjali et al., 2021).
- Additionally, previous studies have suggested that in order to empower women, an appropriate institutional framework is required (Herawati et al., 2014) . This might be accomplished through the involvement of cooperatives, businesses, or even the government. Once a system is in place, it is very simple to remove the negative stigma associated with this sector. The coordination role could also be assumed by municipalities (Rasanjali et al., 2021) .When organisations encourage women to work in the tourism sector, their ability to generate revenue increases, which boosts their self-worth and bargaining strength (Rinaldi & Salerno, 2020). As women are exposed to more stakeholders through institutionalisation, many other abilities, including transferrable personal and professional skills, also develop (Gajic & Cvijanović, 2020).
- Additionally, quick training in hospitality management would give these women the chance to meet the urgent demands of the sector while also enhancing their linguistic and social capital. This would help the people' outdated way of thinking to disappear. Women should be encouraged to work in rural tourism because these businesses can serve both domestic and international tourists, which will benefit the vulnerable populations that live in poverty (Rasanjali et al., 2021)
- Researchers have occasionally emphasized that the hospitality and tourist industries are oppressive to work-life balance because seasonality changes cause the demand cycle to fluctuate and necessitate long hours. Although there are significant disparities across nations and between various segments, working conditions in the hospitality and tourism sectors are usually difficult, with employees dealing with a tremendous workload and constant pressure (Kogovsek & Kogovsek, 2015). Policies that support appropriate childcare options and make sure women stay in touch with the company while on maternity leave are necessary to create family-friendly workplaces (Baum, 2013).
- In order to use the female talent that is being wasted due to gender discrimination, policy makers and management must take steps towards effective solutions and establish equal opportunity grounds, offering them status positions and salaries that are commensurate with their knowledge and capabilities (Kogovsek & Kogovsek, 2015).

CONSTRAINTS

In today's world there have been many initiatives in the tourism industry to foster the development of women entrepreneurs. However, a majority of factors still exist that hamper and adversely affect the journey. These pain points are deep rooted and prove to be a challenge in the pathway of success.

According to (Apostolopoulos & Sönmez, 2001), although women's travel behaviour reflects the norms and values of the societies in which they are immersed, socioeconomic constraints and cultural or religious restrictions pertaining to their social roles are more likely to affect the types of travel that they are capable of making(Khan, 2011).When women engage in leisure and recreational activities outside of the home or in the great outdoors, restrictions like feelings of vulnerability and concern for personal safety seem to be amplified (Deem, 1996; Gilmartin, 1997; Little, 2002; Mason, 1988; Virden & Walker, 1999) According to market research on women's domestic travel in South Australia (South Australian Tourism Commission, 1996), women perceive obstacles like cost, time, a lack of knowledge about where to go, a lack of companions, and obligations to their homes, jobs, and families as well as obstacles like these. In addition, a lot of these South Australian women expressed worry about being afraid of being attacked or raped, especially while visiting isolated locations.

Due to the prevalence of crimes against women, safety is regularly mentioned as one of the top concerns among female travellers(Ratthinan & Selamat, 2019). Obtaining permission from their guardians is one of the challenges that female travellers encounter while planning journeys as solo travellers or in female-only groups. A single female under the age of 18 who is not married must receive permission from her guardian, but a married woman must obtain permission from her husband (Tavakoli & Mura, 2021). Depending on the country a lady comes from, this rule may be different.

The constraint categories of personal, sociocultural, Cultural barriers, Leisure constraints, Constrained by fear, Practical constraints, Spatial constraints, Sexual harassment, Loneliness, Concern for safety, Endanger women's physical health, Family and work conflict, Abusing women, Cultural transmutation, Dual responsibility, Lack of education and skills, Lack of Access to Information are combined in studies of women's barriers to adventure travel, tourism, and adventure leisure, and subcategories within each category are identified depending on major barriers. The discussion that follows will demonstrate how many restrictions from all three categories can be felt at once and how many are thought to be related since they can inform and impact one another. Additionally, women may experience limitations at various points along their adventure tourism journey. Pre-travel restrictions may affect a woman's decision to travel but may not stop her from doing so. Once the pre-travel restrictions have been worked out, women frequently encounter new restrictions when engaging in adventure tourism, which can significantly affect this experience. Each restriction will therefore be explored in relation to other restrictions as well as those that women face both before and during their adventure tourism experience(Doran, 2016).

PERSONAL CONSTRAINTS

Women's self-perceptions, beliefs, and attitudes serve as the foundation for their personal limitations. Although it is asserted that personal restrictions are the strongest of all restrictions, fewer women have mentioned personal restrictions than socio-cultural and practical restrictions. This is because personal restrictions affect a person's motivation to act and influence their decision to participate. The research has identified four major subcategories of personal constraints: guilt, self-doubt, fear, and the idea that one is not adventurous. The idea of leaving their husbands and children behind while they plan their personal trips might cause intense guilt. Women who experience this shame may reassess their commitment to altruism, revaluate their plans for adventurous travel, and prioritize the needs of others. The conflict between selfishness (wanting to take part in adventure tourism) and duty (ethic of caring) "complicates for women the dilemmas of choice, leaving them poised between an ideal of selflessness and the reality of their own agency and desires," according to the study. Other significant barriers for women include uncertainty about their athletic prowess, their capacity to learn the necessary skills for the activity, their familiarity with the surroundings, and their capacity to work in isolated areas(Doran, 2016).

(Wilson & Little, 2008)contend that female tourists' fears of male violence and harassment are out of proportion to what they actually encounter. They contend that fear is socially and culturally ingrained in women, who are taught not to go alone, to be afraid of strangers, to be afraid of the dark, and to remain in or close to the safety of their homes.

SOCIOCULTURAL

Women may endure social and cultural limitations before and during their travels.For women, other major deterrents include a lack of companions and not knowing anyone who engages in the activity of choice. Women's concerns about safety, loneliness, and self-confidence can all be exacerbated by being alone. Australian women who have travelled for adventure tourism have reported that unwelcome male attention and harassment has limited their trip plans, especially for women travelling alone and without male partners (Wilson & Little, 2005, 2008). This might increase women's already-present concerns about personal safety. According to the women in these studies, they were followed, grabbed, fondled, verbally abused, fondled, and forced to sexual behaviours because they were considered to be sexually "available" (Doran, 2016).

CULTURAL BARRIERS

The traditional values in various cultural locales may have a significant impact on women's participation in the tourism industry (Papadopoulou, 2021). The definitions of culture encompass more than just artistic creations, musical traditions, or traditional ways of living; they also cover the spiritual and intellectual traits of a certain social group of individuals. Due to the fact that social values are ingrained in societal conventions, practises, and institutions, it may have an impact on an individual's conduct (Deen et al., 2016; Jolly, 2002)

Another significant restriction was discovered to be sexualized male attention. As a result of the harassment they experienced from men while travelling, the women developed a fear of being attacked or mistreated, which reduced their utilisation of the leisure facilities available at the location (Seow & Brown, 2018).

LEISURE CONSTRAINTS

Women have historically encountered barriers due to their gender when trying to pursue hobbies. Women are nonetheless limited by cultural conventions regarding gendered behaviour, and stigma still exists to varied degrees in various regions of the world. However, there were means for women to travel, especially during the 19th century, which is regarded as the "golden age of travel" (Robinson, 1994). Although marketers failed to see women's potential in the 1970s, their changing roles had a huge impact on the travel industry (Bartos, 1982)and public leisure remained predominantly a male domain (Rybczynski & Glancy, 1992)

The experiences of the traveller are influenced by cultural elements, including gender, making tourism a socio-cultural phenomenon (Seow & Brown, 2018)The term "leisure limitations" refers to the conditions that make it difficult for someone to engage in, commit to, and enjoy their leisure activities (Jackson & Henderson, 1995).

The majority of research on the challenges experienced by female travellers has been centred on western women's experiences.

Young white and South Asian women in England were surveyed about their leisure activities in Green and Singleton's (2006) study, and the results revealed that Asian respondents felt more threatened by accessing public spaces for leisure than did white respondents. According to studies, Asian women are frequently portrayed as afraid and dependant, making them unsuitable (Teo & Leong, 2006) and unwilling (Cai & Combrink, 2000) for independent travel.

CONSTRAINED BY FEAR

Participants mentioned fear as a factor that prevented them from fully appreciating their vacation. They were afraid since they were Asian ladies travelling alone and believed that they were vulnerable. Participants described harassment in detail, including groping and attempted robbery (Seow & Brown, 2018).

Travel constraints are regarded as major deterrents to starting or finishing a journey (Kerstetter et al., 2005).

Travel constraints refer to "factors that inhibit continued travel, cause inability to begin travel, result in the inability to maintain or increase frequency of travel, and lead to negative effects on the quality of travel" (Hung & Petrick, 2010).

PRACTICAL CONSTRAINTS

This kind of constraint takes into account the difficulties and obstacles that women who travel alone must deal with on a day-to-day basis, such as a lack of funds and time, unfamiliarity with the destination's culture, and the stress and exhaustion that come with it. A lack of time and money could, historically speaking, impede women's travel plans and access to travel. The women's accounts of stress and exhaustion associated with their position as single females were closely related to anxiety and social expectations. During their travels, some women described feeling restricted and constricted due to their lack of familiarity with the local language, customs, and geography of a region or destination. For many women, being unable to communicate in their native tongue was a cause of dissatisfaction that limited their engagement with locals (Wilson & Little, 2005).

SPATIAL CONSTRAINTS

Women's freedoms and movements inside tourist destinations, spaces, and places are limited and restricted by spatial constraints. This category includes problems including restrictions on destination options and mobility in tourist environments. The constraint that came out of the inductive analysis of the women's experiences was their perception of their ability to move around freely and safely in tourist environments. Spatial restrictions are those elements that limit solitary women's freedoms and movements inside tourist contexts or that affect their decision-making about where they can go. These elements are strongly related to the effects of the first three categories. Pre-travel space limitations could limit the available destinations. Women reported having fewer options for travel and access to particular locations because they are single females (Wilson & Little, 2005).

SEXUAL HARASSMENT

From the study, factors identified as limitations on the women's solo travel experiences were unwanted attention and harassment. Due to the fact that they were traveling alone and without a male companion, many women believed that they were seen as being "available" sexually. The women's experiences frequently included details of male harassment, including being touched, catcalled, verbally insulted, followed, and subjected to other sexual actions. Numerous testimonials from the women attest to the persistence of unwanted male harassment during their solitary excursions .Travel freedoms were restricted due to the fear of harassment and the perceptions that lone female travellers were more vulnerable (Wilson & Little, 2005).

LONELINESS

The women's experiences travelling alone were also said to have been hampered by their feelings of loneliness. Even though many choose to travel alone, they nevertheless indicated a great yearning for social interaction while on the road. At times, being alone could cause them to feel lonely, and loneliness could amplify any existing feelings of vulnerability and fear (Wilson & Little, 2005).

CONCERN FOR SAFETY

Several women who had travelled reported feeling somewhat constrained in their movements in some tourist locations, particularly in remote and dark areas. Women frequently thought that they should remain within this zone of modesty for their own safety, as if there were an illusory "safety sphere" that surrounded big tourist destinations. Women's movements were also restricted by the time of day, with the night being seen as particularly dangerous for them. As a result, many reduced their activities and decided to stay close to their lodging rather than take the chance of being harassed (Wilson & Little, 2005).

ENDANGER WOMEN'S PHYSICAL HEALTH

Most tourism occupations require travel to various locations, and they are typically stressful. Women's physical status being more fragile than men's (Hoseini, 2000), working in such tasks will exhaust them more than males do, harming their physical health as a result.

FAMILY AND WORK CONFLICT

The scope of the family would be impacted in various ways by the employment of women in the tourism sector. Any stressful occupations in the tourism sector will impact women's families and interfere with their domestic responsibilities. Married women's time spent on family and home duties may be reduced by fatigue from job and frequent travel, and they are unable to set aside enough time to raise children and check on husband chores (Bagheri & Tayebi Khorami, 2014).

ABUSING WOMEN

During their working hours, women may come into contact with a lot of foreign males, which could lead to sexual problems such sexual assault and harassment as well as covert relationships that are more harmful to women than men (Bagheri & Tayebi Khorami, 2014).

CULTURAL TRANSMUTATION

Since women working in the tourism sector interact with people from many cultures and nations, they could draw comparisons between them and the residents of their own neighbourhood. They might change their approach to their religious, social, and cultural beliefs in the long run if they come to the conclusion that tourists are generally superior to their native population (Bagheri & Tayebi Khorami, 2014).

DUAL RESPONSIBILITY

A woman has a dual role where she plays two parts: one as a working woman for which she is paid, and the other as a stay-at-home mother for which she is not paid (Sahay, 1998b).Women who work in the tourism sector struggle to strike a balance between social reproduction and professional responsibilities since, while working outside the home, they are still expected to care for their families. Women who work in family tourist enterprises and support their husbands are typically regarded to be performing a task that is not compensated or acknowledged. Women work more hours than men do due to having to manage two responsibilities, which can occasionally cause stress and impair their health (Panta & Thapa, 2018b)

They frequently miss out on sleep in order to complete their obligations and do not have time for themselves. Additionally, it decreased their social interactions with their families and neighbours. They frequently struggle to find aid because of their poor pay (Gentry, 2007). They do not, however, get an equal position in society, and their achievements are frequently ignored.

LACK OF EDUCATION AND SKILLS

Although girls and women have the opportunity to work in the tourism business, the majority of them are engaged there in low-paying, low-skilled occupations due to their lower educational attainment and lack of experience (Kabeer, 2005b). Since the tourism profession requires interaction with international visitors, which necessitates language competency and communication skills, uneducated or less educated women confront several difficulties (Panta & Thapa, 2018b). Women's understanding of their status and rights is enabled by meaningful access to education and skills. They become more self-assured and are better equipped to compete in the tourism business (Radović-Marković & Živanović, 2019) and fully participate in the development process (UNFPA, 1994).

LACK OF ACCESS TO INFORMATION

Information access is a must for the growth and empowerment of the tourism industry. Given that it is a fundamental right, information about the perks and disadvantages of tourism should be equally given to all people. However, there are gender inequalities in the way information is dispersed. Statistics demonstrate that women have less access to information than men (Narayan-Parker, 2002). Women were found to be less knowledgeable and informed than men to a greater extent, according to a study on gender disparities in rural tourism in Vojvodina. Women are better able to take advantage of opportunities that present themselves when they have access to adequate, timely, and pertinent information.

Although the Botswana government has made a few revisions to traditional laws to advance women, male dominance still predominates in rural areas. Women experience prejudice, are denied the chance to reach their full potential, and continue to be underpowered. Through the creation of jobs and business opportunities, the government promoted community-based cultural tourism while also establishing a tourist policy that supported it. Women were given the opportunity to participate in a variety of projects, including pottery making, running an art and craft business, and cooking and also engaged in other unconventional jobs such as tour guide and site officers. According to the study, women have a sense of freedom whether they are employed or self-employed. They were liberated from dependence on men, depravity, and a sense of despair as they discovered some meaning in their lives through work and community service. They gained social, psychological, and economic dominance. They gained the freedom to act independently, make purchases, contribute to the family income, improve their children's education, travel, seek extra business interests, and engage in confident interactions with international tourists. Due to tourism-related activities, women were unable to pursue formal education; however, they did pursue vocational training to improve the skills needed to be independent. In addition, they are teaching their children, particularly their girl child, which will empower future generations. The government's initiative to promote cultural tourism in Botswana has given women more options. But there are still challenges that women must overcome, including limited access to cash, low wages, low levels of education, and a lack of participation in designing and implementing tourist initiatives. Women were able to overcome these obstacles and a patriarchal society, moving from passive to active participation (Singhania et al., 2021).

CONCLUSION

Women have been actively involved in tourism activities in some communities, supported by their families and societies, whereas in other places, even when they are given permission to work outside, they receive less support. Despite these problems, women continued to labour and take any possibilities that presented themselves. They were guided into empowerment through tourism, which served as a facilitator. They were able to break the pattern of believing that women are only homemakers because of tourism, which supported them in questioning long-held stereotypes about gender roles. They now have a higher social status and a higher level of living attributable to tourism. It also gave them a sense of pride in themselves, respect for themselves, happiness, and independence, all of which contributed to the empowerment of women. Various studies also adhered to previous literature and limited itself to life stage, nationality, economic status, and education, which help to understand why young women travel, how they perceive risks and restrictions, and how they think and feel about their destination. These factors also help to predict how often they will travel in the future.

Women play an informal role in the tourism industry, according to an examination of the literature on the subject. Low-paying jobs are available for women. Due to a number of challenges, women are also constrained in how they can perform their duties. Access, involvement, power, and benefits are all part of what women do in the tourism industry. In the family, women play a more significant role in decision-making, but this role does not exist in group decision-making.

REFERENCES

Apostolopoulos, Y., & Sönmez, S. (2001). Working producers, leisured consumers: Women's experiences in developing regions. *Women as Producers and Consumers of Tourism in Developing Regions*, 3–18.

Bagheri, M., & Tayebi Khorami, M. (2014). Reasons, Opportunities, Challenges and Solutions for Women's Employment in the Tourism Industry of Islamic Republic of Iran. *Journal of Tourism Hospitality Research*, *3*(2), 91–121.

Bartos, R. (1982). Women and travel. *Journal of Travel Research*, *20*(4), 3–9. doi:10.1177/004728758202000402

Baum, T. (2013). *International perspectives on women and work in hotels, catering and tourism*. Academic Press.

Bharathi, P. S. (2019). Woman entrepreneurship and its scope in tourism sector in Talangana. *Ictact Journal on Management Studies*, *5*(1), 912–915.

Cai, L. A., & Combrink, T. E. (2000). Japanese female travelers-a unique outbound market. *Asia Pacific Journal of Tourism Research*, *5*(1), 16–24. doi:10.1080/10941660008722055

Cukier, J. (1996). Tourism and employment in Bali: trends and implications. *Tourism and Indigenous Peoples*, 49–75.

Deem, R. (1996). No time for a rest? An exploration of women's work, engendered leisure and holidays. *Time & Society*, *5*(1), 5–25. doi:10.1177/0961463X96005001001

Deen, A., Harilal, V., & Achu, F. N. (2016). Women in tourism: Experiences and challenges faced by owners of small accommodation establishments. *African Journal of Hospitality, Tourism and Leisure*, *5*(3), 223–814.

Díaz-Carrión, I. A., & Vizcaino, P. (2022). Mexican women's emotions to resist gender stereotypes in rural tourism work. *Tourism Geographies*, *24*(2–3), 244–262. doi:10.1080/14616688.2020.1867886

Dilley, R. E., & Scraton, S. J. (2010). Women, climbing and serious leisure. *Leisure Studies*, *29*(2), 125–141. doi:10.1080/02614360903401927

Doran, A. (2016). Empowerment and women in adventure tourism: A negotiated journey. *Journal of Sport & Tourism*, *20*(1), 57–80. doi:10.1080/14775085.2016.1176594

Elmas, S. (2007). Gender and tourism development: a case study of the Cappadoccia Region of Turkey. In *Tourism and gender: Embodiment, sensuality and experience* (pp. 302–314). CABI Wallingford UK. doi:10.1079/9781845932718.0302

Foubert, P., Burri, S., & Numhauser-Henning, A. (2010). *The gender pay gap in Europe from a legal perspective*. Publications Office of the European Union Luxembourg.

Gajic, T., & Cvijanović, D. (2020). The Level of Engagement of the Female Workforce in the Rural Tourism Development of Serbia. *Journal of Tourism Leisure and Hospitality*, *2*(1), 36–42.

Gentry, K. M. (2007). Belizean women and tourism work: Opportunity or impediment? *Annals of Tourism Research*, *34*(2), 477–496. doi:10.1016/j.annals.2006.11.003

Gilmartin, P. (1997). The dangers of independent travel: A century of advice for 'lady travellers.'. *Journal of Interdisciplinary Gender Studies*, *2*(1), 1–13.

Herawati, A., Purwaningsih, A., Pudianti, A., & Surya, R. V. (2014). Rural tourism community empowerment based on local resources for improving community welfare: Case on Pentingsari village, Yogyakarta, Indonesia. *Review of Integrative Business and Economics Research*, *3*(2), 88.

Hoseini, S. H. (2000). *Biological difference: man and woman: body, spirit and brain*. Critics Books.

Hosni, D., & Lundberg, C. (2005). The status of women revisited: The case of Nepal. *Equal Opportunities International*, *24*(3/4), 73–85. doi:10.1108/02610150510788088

Hung, K., & Petrick, J. F. (2010). Developing a measurement scale for constraints to cruising. *Annals of Tourism Research*, *37*(1), 206–228. doi:10.1016/j.annals.2009.09.002

Jackson, E. L., & Henderson, K. A. (1995). Gender-based analysis of leisure constraints. *Leisure Sciences, 17*(1), 31–51. doi:10.1080/01490409509513241

Janta, H. (2011). Polish migrant workers in the UK hospitality industry: Profiles, work experience and methods for accessing employment. *International Journal of Contemporary Hospitality Management, 23*(6), 803–819. doi:10.1108/09596111111153484

Jiménez-Esquinas, G. (2017). "This is not only about culture": On tourism, gender stereotypes and other affective fluxes. *Journal of Sustainable Tourism, 25*(3), 311–326. doi:10.1080/09669582.2016.1206109

Jolly, S. (2002). Gender and cultural change. *The Bridge.*

Jordan, F. (1997). An occupational hazard? Sex segregation in tourism employment. *Tourism Management, 18*(8), 525–534. doi:10.1016/S0261-5177(97)00074-5

Kabeer, N. (2005b). Gender equality and women's empowerment: A critical analysis of the third millennium development goal 1. *Gender and Development, 13*(1), 13–24. doi:10.1080/13552070512331332273

Kerstetter, D. L., Yen, I.-Y., & Yarnal, C. M. (2005). Plowing uncharted waters: A study of perceived constraints to cruise travel. *Tourism Analysis, 10*(2), 137–150. doi:10.3727/108354205776259295

Khan, S. (2011). Gendered leisure: Are women more constrained in travel for leisure? *Tourismos, 6*(1), 105–121.

Kinnaird, V., & Hall, D. (1996). Understanding tourism processes: A gender-aware framework. *Tourism Management, 17*(2), 95–102. doi:10.1016/0261-5177(95)00112-3

Kogovsek, M., & Kogovsek, M. (2015). Hospitality and tourism gender issues remain unsolved: A call for research. *Quaestus, 6*, 194.

Little, D. E. (2002). Women and adventure recreation: Reconstructing leisure constraints and adventure experiences to negotiate continuing participation. *Journal of Leisure Research, 34*(2), 157–177. doi:10.1080/00222216.2002.11949967

Longwe, S. H. (2002b). Addressing rural gender issues: a framework for leadership and mobilisation. *III World Congress for Rural Women.*

Mason, J. (1988). No peace for the wicked: older married women and leisure. *No Peace for the Wicked: Older Married Women and Leisure*, 75–85.

Narayan-Parker, D. (2002). *Empowerment and poverty reduction: A sourcebook.* World Bank Publications.

Panta, S. K., & Thapa, B. (2018b). Entrepreneurship and women's empowerment in gateway communities of Bardia National Park, Nepal. *Journal of Ecotourism, 17*(1), 20–42. doi:10.1080/14724049.2017.1299743

Papadopoulou, G. (2021). Promoting gender equality and women empowerment in the tourism sector. In *Handbook of Research on the Role of Tourism in Achieving Sustainable Development Goals* (pp. 152–174). IGI Global. doi:10.4018/978-1-7998-5691-7.ch009

Purcell, K. (1996). The relationship between career and job opportunities: Women's employment in the hospitality industry as a microcosm of women's employment. *Women in Management Review*, *11*(5), 17–24. doi:10.1108/09649429610122618

Radović-Marković, M., & Živanović, B. (2019). Fostering green entrepreneurship and women's empowerment through education and banks' investments in tourism: Evidence from Serbia. *Sustainability (Basel)*, *11*(23), 6826. doi:10.3390u11236826

Ramchurjee, N., & Paktin, W. (2011). *Tourism a vehicle for women's empowerment: prospect and challenges*. University of Mysore.

Rasanjali, C., Sivashankar, P., & Mahaliyanaarachchi, R. P. (2021). Women Participation in Rural Tourism: A Case of Ella, Sri Lanka. *AGRARIS: Journal of Agribusiness and Rural Development Research*, *7*(2), 256–269. doi:10.18196/agraris.v7i2.11294

Ratten, V., & Dana, L.-P. (2017). Gendered perspective of indigenous entrepreneurship. *Small Enterprise Research*, *24*(1), 62–72. doi:10.1080/13215906.2017.1289858

Ratthinan, S. P., & Selamat, N. H. (2019). Negotiating travel constraints via technology: A study of Malay Muslim women through a hierarchical constraint model perspective. *Asian Journal of Business Research*, *9*(2), 55. doi:10.14707/ajbr.190060

Rinaldi, A., & Salerno, I. (2020). The tourism gender gap and its potential impact on the development of the emerging countries. *Quality & Quantity*, *54*(5–6), 1465–1477. doi:10.100711135-019-00881-x

Robinson, J. (1994). *Unsuitable for ladies: An anthology of women travellers*. OUP Oxford.

Rybczynski, W., & Glancy, M. (1992). Waiting for the Weekend. *Journal of Leisure Research*, *24*(3), 295–296. doi:10.1080/00222216.1992.11969895

Sabina, J. M., & Nicolae, J. C. (2013). Gender trends in tourism destination. *Procedia: Social and Behavioral Sciences*, *92*, 437–444. doi:10.1016/j.sbspro.2013.08.698

Sahay, S. (1998a). *Women and empowerment: Approaches and strategies*. Discovery Publishing House.

Scheyvens, R. (2000). Promoting women's empowerment through involvement in ecotourism: Experiences from the Third World. *Journal of Sustainable Tourism*, *8*(3), 232–249. doi:10.1080/09669580008667360

Seow, D., & Brown, L. (2018). The solo female Asian tourist. *Current Issues in Tourism*, *21*(10), 1187–1206. doi:10.1080/13683500.2017.1423283

Singhania, O., Swain, S., & George, B. (2021). Women empowerment in a traditional masculine industry: Tourism in context. *Economics. Management and Sustainability*, *6*(2), 60–71. doi:10.14254/jems.2021.6-2.5

Spenceley, A., & Meyer, D. (2012). Tourism and poverty reduction: Theory and practice in less economically developed countries. *Journal of Sustainable Tourism*, *20*(3), 297–317. doi:10.1080/09669582.2012.668909

Swain, M. B. (1993). Women producers of ethnic arts. *Annals of Tourism Research*, *20*(1), 32–51. doi:10.1016/0160-7383(93)90110-O

Tavakoli, R., & Mura, P. (2021). Muslim women travellers' constraints: A critical review. *Women in Tourism in Asian Muslim Countries*, 25–40.

Teo, P., & Leong, S. (2006). A postcolonial analysis of backpacking. *Annals of Tourism Research*, *33*(1), 109–131. doi:10.1016/j.annals.2005.05.001

Tristanti, T., Nurhaeni, I. D. A., Mulyanto, M., & Sakuntalawati, R. D. (2022). The Role of Women in Tourism: A Systematic Literature Review. *KnE Social Sciences*, 545–554.

Tucker, H. (2007). Undoing shame: Tourism and women's work in Turkey. *Journal of Tourism and Cultural Change*, *5*(2), 87–105. doi:10.2167/jtcc089.0

Tucker, H., & Boonabaana, B. (2012). A critical analysis of tourism, gender and poverty reduction. *Journal of Sustainable Tourism*, *20*(3), 437–455. doi:10.1080/09669582.2011.622769

Varghese, T. (2011). Women empowerment in Oman: A study based on Women Empowerment Index. *Far East Journal of Psychology and Business*, *2*(2), 37–53.

Virden, R. J., & Walker, G. J. (1999). Ethnic/racial and gender variations among meanings given to, and preferences for, the natural environment. *Leisure Sciences*, *21*(3), 219–239. doi:10.1080/014904099273110

Wallace, T., & March, C. (1991). *Changing perceptions: Writings on gender and development*. Oxfam GB.

Wilkinson, P. F., & Pratiwi, W. (1995). Gender and tourism in an Indonesian village. *Annals of Tourism Research*, *22*(2), 283–299. doi:10.1016/0160-7383(94)00077-8

Wilson, E., & Little, D. E. (2005). A "relative escape"? The impact of constraints on women who travel solo. *Tourism Review International*, *9*(2), 155–175. doi:10.3727/154427205774791672

Wilson, E., & Little, D. E. (2008). The solo female travel experience: Exploring the 'geography of women's fear.'. *Current Issues in Tourism*, *11*(2), 167–186. doi:10.2167/cit342.0

Chapter 25
Female Entrepreneurship in Tourism:
Hurdles, Challenges, and Issues With Respect to Pakistan

Muhammad Faisal Sultan
Khadim Ali Shah Bukhari Institute of Technology, Pakistan

Raza Ali Khan
NED University, Karachi, Pakistan

Nazia Abdul Rehman
Federal Urdu University, Karachi, Pakistan

Erum Shaikh
Shaheed Benazir Bhutto University, Pakistan

ABSTRACT

Female entrepreneurship is perceived as the new beam of light for the growth of the world economy. The phenomenon is especially important for the developing world. However, female entrepreneurs are always associated with challenges and hurdles and similar is valid even for the developed world. However, due to the segregation of gender roles, female entrepreneurship is especially difficult in the Islamic world and also in Pakistan. Several studies highlighted the importance of the phenomenon and also place light on hurdles that may impair female entrepreneurs to cultivate their own businesses in the tourism industry. However, rear work has been conducted on female entrepreneurship in the tourism industry with reference to Pakistan. Thus, the importance of this study is massive for a diverse set of stakeholders. Therefore, this chapter has been written by linking the hurdles, issues, and challenges pertaining to female entrepreneurship with female entrepreneurship in the tourism industry of Pakistan.

DOI: 10.4018/978-1-6684-8417-3.ch025

BACKGROUND

The origin of entrepreneurship is based on the French word "Entreprende", which means "Undertake". In fact, entrepreneurship has been used as a term by different disciplines e.g., economics, finance, sociology, anthropology, and psychology, etc. (Cem *et al.,* 2019). However, the use of the term i.e., "entrepreneurship" might be replaced by any synonym word to have the appropriate relationship with the concerned discipline. Although generically we may define entrepreneurship as an opportunity for discovery, analysis, and marketing of any tangible or intangible item that is not existing (Cem *et al.,* 2019). In fact, nurturing the term will provide us with the broadest scope of men's freedom to understand their abilities through independent & unrestricted comprehension of ideas. Hence the population of entrepreneurs is increasing day after another whether in the form of startups or in the form of entrepreneurs working in larger companies. However, the ratio of women entrepreneurs is still on the lower side significantly either for company ownership or as the corporate entrepreneur (Serafimova & Petrevska, 2018). These indications are aligned with patriarchal philosophies that believe females are inferior as compared to men. Hence, the numbers of men-owned startups are much higher than the number of women-owned startups (Yaqoob, 2020).

On the other side women entrepreneurs are perceived as an un-tapped source that may foster economic growth. The research uses the word of the new engines for female entrepreneurs due to their ability to foster the growth of economies. The tendency to contribute in the development of the economy is found to be much higher for developing sides of the world. However, their importance can also be legitimized in developed parts of the world. World Economic Forum in its annual general meeting for the year 2012 used the word "***Way Forward***", for explaining the importance of female entrepreneurs (Serafimova & Petrevska, 2018). However, most of the time women are forced towards entrepreneurship through severe challenges related to occupation. Thus, most of the studies in this domain are focused on differences between male and female entrepreneurs or personal and economic aspects that lead to female entrepreneurship. On the other side tourism sector is massively dominated by male entrepreneurs and female has to face several problems (Yildrim et al., 2020).

Hence females are mostly associated with minor or lower-level entrepreneurial activities e.g., accommodation, food & beverage, souvenir, production, and presentation of local products food & beverages, accommodations & production or presentation of local products, etc. Hence, women are also found to have difficulties in initiating entrepreneurial ventures in the tourism sector. In fact, support from the government may not be found effective to foster entrepreneurial efforts and venture from females (Yildrim et al., 2020).

INTRODUCTION

The debate is still in progress regarding how to push more females toward entrepreneurship. The debate is especially fruitful for developing sides of the world where the surge in women's participation in business and social activities has increased phenomenally. However, due to gender differences in business decisions, there is a need for motivation to make females become more inclined toward entrepreneurship. One way to do this is to provide entrepreneurial education to females. However, female entrepreneurship in the informal tourism sector is flourishing and most of the activities related to the informal tourism sector are owned by female entrepreneurs. Thus females are becoming more independent and have more control over their earnings without relying on any other source. Hence legitimate to declare an increase in women in management and entrepreneurial activities will also enhance economic growth & development (Tajeddini *et al.,* 2022). Although women without formal education have been mostly associated with food or those forms of businesses that do not require advanced skills. Contrary to this females with formal education are found to be associated with businesses like event organization and selling of handicrafts etc., that needs advanced skills e.g., planning and organizing, etc (Ojong et al., 2021).

However, a surge of female entrepreneurship has been observed in the informal tourism sector & therefore several females from reputed families are now trying to build their brands in the informal tourism sector. This is also due to the fact that the hospitality industry is making females earn and build in a socially acceptable way through the extension of traditional gender-based roles (Tajeddini *et al.,* 2022).

Women Entrepreneurship and Islam

Islamic economic system is a holistic economic system that promotes entrepreneurial activities, social-wellbeing as well as economic justice (Muhammad *et al.,* 2020). Similarly, there is no restriction in Islam i.e., either from the Holy Book Quran and Hadith (Sayings & Acts of Prophet Mohammad), that may prevent females from working either from home or outside the home (Roomi & Parrott, 2008).

In fact, there is no religious evidence that may hinder in the way of female entrepreneurship as it is allowed for Muslim women and is also reflected in the Quran and Hadith. In fact, it is also evidenced by the example of Hazrat Khadija R.A, who as successful trader, and hence nothing is wrong if Muslim Women is pursuing entrepreneurship. However, Islam has provided some guidelines and Sharia principles for daily engagements of females & through following these rules and principles females may increase the level of their spirituality. Hence females are allowed to be engaged in any business that is not-prohibited or Haram and can also take advantage of modalities like ***Mudarabah*** & ***Musharikah*** etc. Hence Muslim women may use their basic skills to get involved in entrepreneurial activities through working products e.g., food, etc from their home and may utilize their education and advanced skills to get involved in advanced-level entrepreneurship activities through online businesses, etc. Although due to the absence of regulatory authority for entrepreneurial activities, women may avoid Islamic teaching. This may happen through charging more prices for providing goods at the doorstep, especially for products that do not have a fixed price & also by avoiding tax-related payment due to non-registration with tax authorities. Similarly devoting more time to business may create work-family issues as due to time-related issues it is not possible for a female to perform their familial roles of daughter, mother and wife, etc. The issue is especially critical for married women (Muhammad *et al.,* 2020).

Women and Pakistan

Women's population in Pakistan constitutes almost half of the total population of Pakistan which formulates a healthy proportion of the total workforce that is committed and enthusiastic to contribute in the economic development of the country (Riffat & Ul-Haque, 2016).

Therefore, women's entrepreneurship is common in Pakistan although it has not attained the desired appreciation due to the cultural myths and negative perceptions regarding the level of female productivity. Thus, the country ranked lowest in terms of female entrepreneurship which is around 1% which is severely low as compared to the rate of their counterparts i.e., 21%. However, female entrepreneurship is perceived as a keystone for the development of small and medium-sized enterprises (SMEs). In fact, in Pakistan Small and Medium Sized Enterprises Development Authority (SMEDA) is trying hard for the development and growth of the sector. Thus, the role of women entrepreneurship has been realized and SMEDA also signifies its importance for the development and economic growth of the industry (Shah *et al.,* 2021).

Although it is also a fact that there are several obstacles in the way women entrepreneurs in Pakistan and therefore exploration of the female entrepreneurial segment is still in progress. Difficulties raised due to conservative thinking that are part of Pakistan's culture create a hindrance to women's participation in private as well as public sectors. In fact, as worldwide employment rate for Pakistan's women has decreased significantly. On the other side women empowerment and female-owned small-scale enterprises are now getting common that are the evidence for future economic growth (Riffat & Ul-Haque, 2016).

RESEARCH GAP

Low-income economies have more tendency for entrepreneurship as jobs are not desirably attractive in these locations (Yildrim et al., 2020).

Although in Pakistan the opportunities for female entrepreneurship are much lower than the opportunities for male entrepreneurship. Hence, the numbers of female-owned entrepreneurial ventures are negligible even in the era that demands individuals and groups to become job providers rather than job-seekers (Yaqoob, 2020).

Some of the studies claimed that hurdles encountered by female entrepreneurs from Pakistan are deep-rooted in the structurally-enforced lower status of women in Islamic society. These are the clear indications of gender biases that are a part of the regional, tribal & feudal culture of Islamic societies (Roomi & Parrott, 2008). These sorts of challenges are common for female entrepreneurs, especially from Pakistan where absence of research that may explore the impact of female entrepreneurship on society or the economy (Riffat & Ul-Haque, 2016).

THEORIES AND PHILOSOPHIES ASSOCIATED WITH THE RESEARCH GAP

Shah et al (2021) indicated major feminist theories that may be effectively related to issues and hurdles faced by female entrepreneurs from Pakistan and the Muslim sides of the world.

Liberal Feminist Theory

According to this theory, females are not inferior to their counterparts and if we can supplement women with equal resources, education, and work experience then they will be found equally effective to their counterparts. Although less access to business, resources, and education is making them inferior to males. The theory also claimed that male dominant society, discrimination against women, and structural barriers are also hampering the rate of female entrepreneurship.

Social Feminist Theory

The indications of this theory are based upon the way to observe the world by different genders. In fact, the theory does not discuss the level of female productivity & emphasizes the point that the socialization of females creates different goals, priorities, perspectives, and options.

Thus, the major purpose of the theory is to discuss the selection of different approaches that may cause lesser productivity of females than their counterparts.

Psychoanalytical Theory

The theory highlighted that male domination in society has been reflected in the negative feelings of females. Moreover, past experiences also have a significant impact on the personality and thinking process of females. Thus, risk-aversion is one of the prominent characteristics of female nature due to which they take fewer risks as compared to their counterparts.

SIGNIFICANCE OF THE STUDY

There is a severe need for attention towards female entrepreneurship in developing sides of the world (Tajeddini *et al.,* 2022). Especially in Pakistan where the research on the economic impact of female entrepreneurship has a lackluster trend (Riffat & Ul-Haque, 2016). On the other side female entrepreneurship is perceived as the major tool for economic progression for developed & developing ideas of the world (Serafimova & Petrevska, 2018). In fact, female entrepreneurship may also result in a decrease in poverty, and female entrepreneurship in the informal tourism sector in Islamic countries like Indonesia is exemplary. However, in Pakistan, the number of female-owned startups is severely low (Yaqoob, 2020) and the reasons are cultural norms and lower-social status (Roomi & Parrott, 2008). Hence the recent study is impactful in making researchers, academicians, entrepreneurs, and policymakers understand the hurdles that female entrepreneurs are facing. Thus, special attention may be given to female entrepreneurship in the tourism sector which may enhance further research and better policy-making required in this regard.

OBSTACLES FACED BY FEMALE ENTREPRENEURS

International Labor Organization (ILO), highlighted there are several obstacles in the way of female entrepreneurship. However, lacking of required training and development is the prime obstacle & in association with the absence of adequate financial resources creates a massive hindrance for female entrepreneurs. However, hindrances are especially high for inexperienced entrepreneurs who are trying to enter in the segment and facing issues like credit history, collateral and tracking of history from banks (Riffat & Ul-Haque, 2016).

Although according to Lockyer and George (2012), barriers faced by experienced and experienced women entrepreneurs are almost the same. A list of barriers highlighted by the study is given in Table 1 and important barriers & obstacles encountered by female entrepreneurs are under:

Table 1. Barriers faced by female entrepreneurs (Lockyer & George, 2012)

Barriers Perceived or Actual	Business	Non-Business
Work Commitment	31	46
Children Cost	17	16
Flexibility of Childcare	25	10
Caring of elderly or ill	06	10
Study Commitment	12	12
Lack of Time	21	28
Too Old	00	07
Too Young	02	06
Lack of Business Idea	02	26
Business Management Skills	33	41
Financial Stability Issues	02	12
Others	23	10

a) **Customs & Local Culture:** Several studies indicated the lower presentation of females in male-dominated and non-traditional occupations. Male-dominated professions require technical backgrounds, tough physical expectations along with high-level of self-efficacy. Therefore, females may not be found consistent with the requirements due to prevailing cultural norms and expectations (Riffat & Ul-Haque, 2016).
Similarly has been indicated by Shah et al (2021), women's entrepreneurship is suffering due to a lack of support for religious and cultural norms and beliefs. The situation is especially valid for developing parts of the world and therefore entrepreneurship has been a forte of males rather than females.

b) **Gender Discrimination with Females:** Gender-related discriminations are not only creating hindrances to women's growth as employees but also creating resistance to women's entrepreneurship. In fact, women may not be able to attain stability in their careers as they have to follow the career path permitted or directed by their husbands. Thus vivid to write that females are also forced to focus on short-term growth as most of the time they are seems to be job-oriented rather than career-oriented. There are several reasons for the short-term orientation e.g., family responsibilities, career paths correlated with the career path and permission of spouse, and discretion of male supervisors (Riffat & Ul-Haque, 2016).

c) **Lack of Confidence:** Lack of confidence and communication skills also perceive as a major hurdle for female entrepreneurs (Riffat & Ul-Haque, 2016). Similar has also been reflected in Naguib (2022) as females are mainly evaluated on the bases of their family and household responsibilities. Therefore, it is recommended to supplement females with the required social and psychological skills in order to excel the economic and social development.

d) **Lack of financial resources:** Lack of financial resources is one of the prime hurdles in the way of female entrepreneurship. Although females who haven't started their own business view a lack of financial resources as one of the prime hurdles in entering to the field of entrepreneurship (Lockyer & George, 2012)

Table 2. Skills gap of female entrepreneurs (Lockyer & George, 2012)

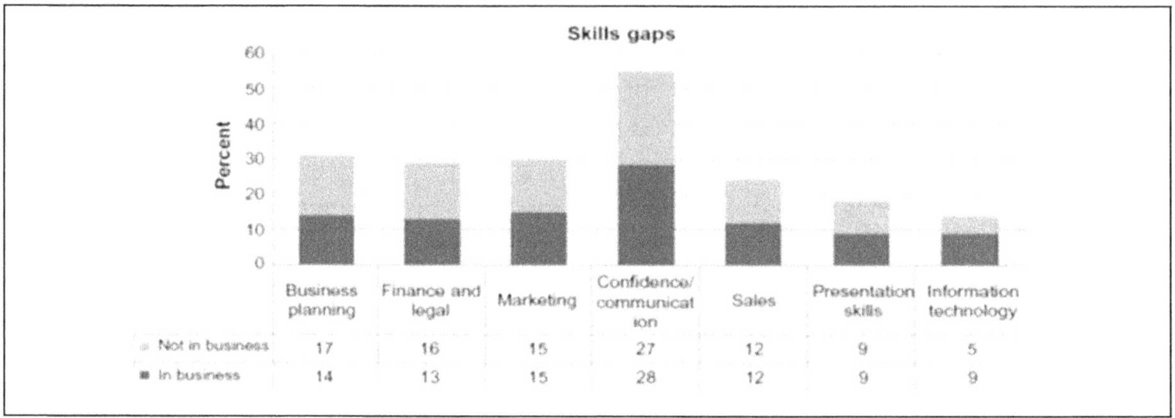

e) **Skills Gap:** Skill Gap in comparison to the market demand is also one of the prime factors that hindered the way of female entrepreneurs. The major skills required for entrepreneurship are to understand the potential of business and use of effective marketing and sales strategies are found to be on the lower side of females. A deeper understanding might be gained through table 2 which is used to reflect skills and their perceive gap in female entrepreneurs (Riffat & Ul-Haque, 2016). The study conducted by Shah et al (2021) also highlighted lacking skills, knowledge, and expertise in female entrepreneurs is making it difficult for female entrepreneurs to become established or to gain momentum.

f) **Personal Barriers:** Personal Barriers like family care and responsibilities are always formulated hurdles for female entrepreneurship. Due to these responsibilities' females cannot invest time in the required skills for entrepreneurship like skills required for making the business flourish in log-run etc (Riffat & Ul-Haque, 2016).

g) **Harassments:** Gender-related harassment creates a negative impact on the intentions of female entrepreneurs. This phenomenon is quite common in developing sides of the world and also creates restrictions on female traveling and the creation of contacts (Shah *et al.,* 2021)

h) **Lack of Access to loans:** Male and female entrepreneurs have a significant difference in access to loans. The challenge still prevails even in the 21st century. The same is also valid for developed sides of the world like the USA as female entrepreneurs are only able to achieve 4% of loans allocated to the SME sector. Moreover, female entrepreneurs also faced issues like shorter time for loan repayment and the need for strong collaterals, etc. (Shah *et al.,* 2021).

i) **Lack of Access to Policy Making:** Female entrepreneurs face difficulties in accessing policy makers The access gets even more difficult in developing sides of the world and therefore there are very few policies that may favor businesses owned by females. On the contrary, males have easy access to the policymakers as policy-making is mainly associated with males which increases the probability of access to the policymakers & also aids in policy formulation that favors male-owned businesses and larger firms (Shah *et al.,* 2021).

METHODOLOGY

The methodology used for writing this study is based on parameters given by Saunders et al (2007); Saunders et al (2015) and Sekaran and Bougie (2016). These three authors are used in order to leverage the study with the concepts of research onion and research tree. The philosophy used to write this chapter is epistemology as it is the philosophy of knowledge and the purpose of this chapter is to highlight major obstacles and challenges associated with Muslim female entrepreneurship which has received rear focus from Muslim and Non-Muslim scholars. The research approach is archival as it is based upon past data that was published in scholarly research articles while the mono-qualitative method as the data has only been collected through observation (Saunders et al., 2007 & Saunders et al., 2015). Thus, the purpose of the study is descriptive and the time horizon for conducting the study is cross-sectional (Sekaran & Bougie, 2016) & follows qualitative content analysis according to Creswell and Creswell (2017).

REFERENCES

Brush, C. G. (1992). Research on women business owners: Past trends, a new perspective and future directions. *Entrepreneurship Theory and Practice*, *16*(4), 5–30. doi:10.1177/104225879201600401

Cem, I. Ş. I. K., Küçükaltan, E. G., Çelebi, S. K., Çalkin, Ö., Enser, İ., & Çelik, A. (2019). Tourism and entrepreneurship: A literature review. *Journal of Ekonomi*, *1*(1), 1–27.

Creswell, J. W., & Creswell, J. D. (2017). *Research design: Qualitative, quantitative, and mixed methods approaches*. Sage publications.

Faizan, R. & Ul-Haque, A. (2016). *The relationship between societal attributes, feminine leadership & management style: Responses from Pakistan's urban region female-owned businesses*. Academic Press.

Lockyer, J., & George, S. (2012). What women want: Barriers to female entrepreneurship in the West Midlands. *International Journal of Gender and Entrepreneurship*, *4*(2), 179–195. doi:10.1108/17566261211234661

Muhammad, S., Ximei, K., Sharif, I., & Haq, Z. (2020). An overview of women entrepreneurship from Islamic perspective. *Review of Economics and Development Studies*, *6*(4), 857–866. doi:10.47067/reads.v6i4.285

Naguib, R. (2022). Motivations and barriers to female entrepreneurship: Insights from Morocco. *Journal of African Business*, 1–28. doi:10.1080/15228916.2022.2053400

Ojong, N., Simba, A., & Dana, L. P. (2021). Female entrepreneurship in Africa: A review, trends, and future research directions. *Journal of Business Research*, *132*, 233–248. doi:10.1016/j.jbusres.2021.04.032

Roomi, M. A., & Parrott, G. (2008). Barriers to development and progression of women entrepreneurs in Pakistan. *The Journal of Entrepreneurship*, *17*(1), 59–7. doi:10.1177/097135570701700105

Saunders, M., Lewis, P., & Thornhill, A. (2007). Research methods. In Business Students (4th ed.). Pearson Education Limited.

Saunders, M. N. K., Lewis, P., Thornhill, A., & Bristow, A. (2015). *Understanding research philosophies and approaches: Research methods for business students*. Pearson.

Sekaran, U., & Bougie, R. (2016). *Research methods for business: A skill building approach*. John Wiley & sons.

Serafimova, M., & Petrevska, B. (2018). Female entrepreneurship in tourism: A strategic management perspective in Macedonia. *Journal of Applied Economics and Business*, *6*(1), 21–32.

Shah, F. U. H., Fatima, U., Ahmad, M., Ahad, A., Saeed, R., Riaz, M. U., & Ghafoor, F. (2021). Factors affecting women entrepreneurship and its impact on SMEs development. *International Journal of Innovation. Creativity and Change*, *15*(7), 228–252.

Yaqoob, S. (2020). The Emerging trend women entrepreneurship in Pakistan. *Journal of Arts & Social Sciences*, *7*(2), 217–230. doi:10.46662/jass-vol7-iss2-2020(217-230)

Yildirim, G., Tarinç, A., & Kilinç, C. C. (2020). Women entrepreneurship in tourism: The case of Turkey. *Journal of Tourism and Gastronomy Studies*, *8*(4), 2462–2477. doi:10.21325/jotags.2020.721

Chapter 26
Tourism Social Entrepreneurship and Roles in Activating SDGs

Ali Raza
Near East University, Cyprus

Laiba Asif
KBSAS, Sindh University, Naushahro Feroze, Pakistan

Erum Shaikh
Shaheed Benazir Bhutto University, Sanghar, Pakistan

Waleed M. Alhanhanah
Istanbul Okan University, Sisli Mecidiyekoy, Turkey

ABSTRACT

Sustainable development along with social entrepreneurship is providing a new approach to the tourism industry. Tourism social entrepreneurship is all about how to take beneficial steps that reduce the negative effects in the tourism industry. It is the responsibility of the entrepreneurs to actively play their role in maintaining sustainability and taking advanced, innovative, and bold steps. It depends upon them how they fulfill the current needs without sacrificing future needs, as this is a key concept of sustainable development goals. Corporate socially responsible firms are sustainable firms. Those industries which are not applying these strategies are now suffering because there is a limited understanding of the fusion of SDGs and tourism social entrepreneurship. How do they apply the SDGs strategies as a part of their strategic planning industry? This chapter contributes to the emerging role of tourism social entrepreneurs and also tells how they achieve sustainability with a combination of 17 strategies and fusions of SDGs.

DOI: 10.4018/978-1-6684-8417-3.ch026

SOCIAL ENTREPRENEURSHIP

Through the primary support of organizations, social sustainability pursues productivity expansion to accomplish progressive change without damaging the surroundings (Correia & Dolnicar, 2021). The united nations have already worked on a variety of global projects to achieve this (Movono & Hughes, 2022). To advance socio, economical, ecological, and political components, 193 nations endorsed vision 2030 in September 2015 as just a hypothesis that includes 17 Sustainable Development Objectives that must be achieved by 2030 (Diaz & Ariza, 2022). According to this concept, Social entrepreneurship's (SE) multifaceted strategy aids in incorporating business operations into a blueprint for sustainable development (Bansal et al., 2019). The process of jointly generating benefits for society & community that go beyond simple profitability is known as Social Entrepreneurship (Olmedo et al., 2019). Social entrepreneurship is becoming increasingly important in addressing today's societal & environmental issues (Suriyankietkaew et al., 2022). A study of over a hundred environmental firms in 9 OECD nations in 2015 revealed sales above Euro 600 million as well as the development of over 6 million jobs, including about 500,000 from vulnerable communities (Boluk et al., 2019). In Australia as well as the European Union, social companies represent 3% to 8% of something like the respective economies' economic growth. However, Canada would contribute 4 percent of the total GDP during the upcoming ten years (Europe, 2018). The assessment of social effect continues to be an open question that is unresolved & prevents determining the exact importance of social businesses in the accomplishment of social development goals, although the establishment of many indexes and methodologies (Pothipala et al., 2021). About 40% of public companies analyze the SDGs within their evaluation of overall effect objectives, according to the European Sustainable Business Monitoring Document 2020-2021 (Ciccarino et al., 2022). Just under 60% of community organizations evaluate their effect aims. The overall impact of a firm's operations on social progress as well as the accomplishment of such sustainable development goals, nevertheless, really aren't assessed (Kowszyk & Vanclay, 2020). Our chapter intends to develop a new evaluation system based on vision 2030 to be used with various forms of social innovation to close the gap (Diaz & Ariza, 2022). Contrary to a widespread belief that charitable businesses are a relic of industry, this inquiry offers a chance to further the discussion on the importance of social innovation as a vital part of corporate operations toward SDGs (Foncubierta, 2021; Abbasi et al., 2022). Three main studies (Schumpeter & Redvers, 1934; Kirzner, 1973; Buzinde et al., 2017) are largely responsible for the present understanding of entrepreneurialism in the economy. When describing both industrial and non-economic (such as social) entrepreneurialism, Swedberg (2009), for instance, adopts Schumpeterian terminology and describes it as "the pressing through or effective adoption of a novel mixture based on previously existent stuff and processes". Therefore, Schumpeterian entrepreneurialism would support the economic, societal, and environmental objectives of sustainable development. Parallel to this, (Buzinde et al., 2017) contrast the market consequences of corporate innovation in the Schumpeterian & Buzinde schools towards the societal & communal impacts. Many scholars and many others base their comprehension of social innovation here on the sophisticated and well-established contemporary conception of entrepreneurship in economists instead of developing novel models of social entrepreneurialism. The involvement of conventional enterprises has thus far been found to be inadequate, despite the attempts put forth by the United Nations to involve both government and private entities in achieving the 2030 Agenda. As a result, social entrepreneurship has become works better at utilizing business opportunities to address social issues. Nevertheless, there are currently no particular methods available to assess how social companies affect the achievement of Sustainable Development Goals (SDGs). Alternative strategies for tourist growth are made possible by the industry's ongoing search for ethical and environmentally friendly methods. Localities frequently play a prominent role in these creative tourism businesses and growth plans. The aim of involving and effectively improving underprivileged and underprivileged people through the sector is refocused on the expansion of social entrepreneurs working inside the tourism sector. Like a business technique to solve social issues while maximizing the positive effects and limiting the bad impacts that tourism could have on host societies, Tourism Social Entrepreneurship (TSE) has been proposed. The potential role of TSE as a driver of long-term societal development is yet poorly understood. This research seeks to fill this information vacuum by conceiving TSE as an all-encompassing approach to long-term development initiatives. The dynamism in the environment has brought a bundle of changes each day. Almost in every sector, industry, or field, so how it is possible the tourism industry will not be a part of this? The dynamics for the development of the tourism industry, this study has been carried out. The function of organizations is crucial to every type of innovation, especially social entrepreneurship. The significance of organizations seems indication in the functioning of a Kirznerian equilibrium & Schumpeterian innovation ideas of entrepreneurship for each of which isn't instantly obvious. In several words, the tourism industry plays a crucial part in entrepreneurial behavior by organizing the possibilities where income may be realized or where innovative pairings can be taken out, according to the Kirznerian and Schumpeterian concepts of entrepreneurship. If tourism industries are the source of entrepreneurship chances, therefore for, philanthropic, & government sector organizations alike will all experience Kirznerian profitable opportunities also Schumpeterian opportunities for entrepreneurship which result in structural transformation.

SOCIAL ENTREPRENEURSHIP IN TOURISM INDUSTRY

Tourism is one of the most essential economic activities and has great importance for the economies. The dynamism in the environment has completely changed everything and now it has affected the world tourism industry as well. Although tourism has now become the most developing industry and highly contributing to economic growth (Buckley, 2011). Tourism has great importance nowadays. Tourism creates opportunities for countries to boost their performance of the countries at double rates (Goffi et al., 2019). Currently, the tourism industry has been affected by the pandemic and there has been no development made in this period. The world has suffered a lot almost and every economic activity is almost paralyzed in that worst conditions. After the pandemic, it has been a hot topic of interest for the researcher. Tourism is the platform that promotes the beauty of the countries and attracts foreigners to visit historical places and enjoy the beauty of other nations (Roy & Roy, 2015). It is a real economic asset for the countries that not only promote the places in the world but also support and enhance growth (Abbasi et al., 2022).

Sustainable Tourism Development (STD) is tourism that fully considers the economical, societal, and ecological effects it has today and tomorrow, fulfilling the requirements of travelers, businesses, the ecosystem, and community partners. The world has been suffering from many environmental challenges and most of the economic resources have been destroyed that were the place of attraction for tourists, and that is the reason environmental sustainability and tourism are interrelated with each other (Jamal & Higham, 2021). Governments are taking steps for the protection of resources. Tourism is concerned with the development of natural resources and their maintenance is necessary. Tourism and the environment are sharing the closest bond (Hole & Snehal, 2019). And tourism is playing a vital role in the working performance of the economy. If we are talking about "tourism" it is not just a word it now becomes a need of individuals as well. The 21st century is full of stress and most people are suffering from anxiety and depression (Agyeiwaah, 2020). Traveling is the key source for them to overcome and forget their worries visiting historical places and enjoying the beauty of the world enables them to feel refreshed. Tourism is the combination of places, people protection, and power (Kim et al., 2019). This enables individuals to meet with different people and visit the places and government regulations offer them security and power to enjoy the whole natural beauty. Most of the economies their all economic activities are just dependent upon tourism they have no other sources so it represents how important this sector is. Tourism is an enjoyable activity almost every minute each tourist visits their favorite place some of them are lucky they happily and conveniently enjoy their trips and some of them have to face certain challenges due to a lack of proper management by the local government of them (Belli et al., 2020).

Tourism is not just an economic activity it offers plenty of opportunities for countries and enables them to utilize their natural resources for revenue generation and also very help full to maintain their old, ancient historical places clean and attractive so that tourists pay more attention. It is also provided several benefits to the inhabitants of the countries and local people. It is now considered the platform for generating opportunities (Abbasi et al., 2022). Every region has its specifications and regional characteristics that need proper management for the protection of the resources. Tourism and environmental sustainability are two linked things that gave birth to the concept of social entrepreneurship and nowadays social entrepreneurship practices are adopted by organizations rapidly. This world has been suffering from many challenges and these challenges threaten the world and organizations (Belli et al., 2020). Those organizations are socially active and their priority is to enhance the quality of the environment and to facilitate their stakeholders. Different authors, researchers, and scholars have their concepts regarding social entrepreneurship. Most prominently it is defined as the process and the practices that are taken for the welfare of humanity and the protection of the environment and their productions are used for the social cause. This need of the current situation and its practices are highly effective in the development of sustainable development goals (SDGs). As the concept of entrepreneurship is all about creativity and uniqueness and its combination with sociality presents a different perspective. And contributing to developing society and the environment currently, the main focus of most businesses and organizations is only to maximize their wealth and generate large amounts of revenues and create value. Sustainability is the primary concern and now it is necessary for the future of the world. Social entrepreneurship is playing a pivotal role in the growth of the tourism industry and also takes an active part in maintaining innovativeness with sustainability practices (Huang et al., 2020).

If we generally discuss the term social entrepreneurship it is quite simple to understand its characteristics it is the systematic way of analyzing the environment finding the gap and filling this gap with creativity and innovation and also taking care of the environmental protections and social causes. The main objectives of social entrepreneurs are to enhance the quality of the environment. They enthusiastically work for the environment (Tseng et al., 2019). When they start working in the environment bundle of challenges is faced by them. These challenges are not minor, these are globally unsolved problems and their capability to accept the challenges and turn them into opportunities is the real potential of social entrepreneurs. Social entrepreneurship is the ultimate solution to the most critical social problems that have been faced the all over world. Their strategies are designed according to the current problems and how they can play an active role in the solution of the problems. The other most important myth about social entrepreneurship is that it is providing the road map to the unsolved market conditions as well that have been brought by the mismanagement and carelessness of persons who are just profit oriented. Bill Drayton, the founder of Asoka, is quoted as saying:

Social Entrepreneurs are not content just to give a fish, or teach how to fish. They will not rest until they have revolutionized the fishing industry.

Social entrepreneurs are highly active and supremely talented and they know how they can enhance social and environmental conditions with their innovative thinking and capabilities. Social entrepreneurship implementation is popular in business organizations and now it has been implemented in practical life as well (Gabriel & Drayton, 2016).

Social entrepreneurship is a broader approach and it has different models. It enhanced the performance of the tourism industry (von der Weppen & Cochrane, 2012). The old tourism methods are just profit-oriented and their motive is to just maximize the wealth of the country. But with the change in the environmental conditions and the dynamism in the environment, everything is now totally changed and now social entrepreneurship has revolutionized the overall procedures and it brought innovation in the tourism industry as well. Tourism social entrepreneurship is all about the current trends and has developed such procedures and techniques in the industry so that new and innovatively providing benefits to the social environment maintain the sustainability of the tourism industry (Zeyen et al., 2013). Social entrepreneurship is the most advanced and useful approach for the development of the economy and enables the country to be recognized the among the most developing economies. And those social threats that they have to deal with are the most critical challenges that are poverty, lack of education, human rights violations, poor sanitation and public health, unemployment, environmental problems, and other societal needs that are unmet by the public and private sectors (Defourny & Nyssens, 2010). As the main concern of the public sector organization is to just use the assets of the government for their personal use there is no maintenance of anything. And resources are available but not utilized properly. That is the reason the local people are suffering from poverty. They don't have enough food for survival and do not have the proper education facilities as well. And most of the youth are suffering from anxiety and depression because of the lack of job opportunities. There are such economies in the world that are updated and working with strategies and the concept of social sustainability and social entrepreneurship (Aquino et al., 2022). And are enjoying a better standard of living. They are successful in providing all the basic facilities to their people. The basic motive of social entrepreneurship is to maximize the rate of social welfare and minimize such factors that have a bad influence on society and eagerly work for the development of society. According to the current situation, there has been a need for social enterprises that work more effectively and efficiently for the social development factor rather than generating profit. Most of the firms are highly working day and night for maximizing the value of the firms. Not concerned with the quality and the other environmental aspects. Compromising on the health of the individuals (Boluk et al., 2019).

The establishment of social enterprises proved very effective in vanishing such factors from society. And that is very necessary because they develop business strategies innovatively along with providing several other benefits to boost economic growth as well. They are highly skilled a know how to use scarce resources for maximum benefits. Their strategies are designed in a way that enables the features of less input and more output. They believe in equality and are highly concerned about that they know that how they can distribute social and economic benefits simultaneously. When the united nation has presented the concept of the SDGs which are commonly known as sustainable development goals and are considered as the agenda 2030 for sustainable development. And governments are actively adopting such strategies and start putting efforts into that as well. he bold agenda sets out a global framework to end extreme poverty, fight inequality and injustice, and fix climate change until 2030. The world has been drastically changed and we are living in an era where the world is experiencing so many changes that have influenced human life and survival in the era very difficult because of the rapid arising of environmental issues (Ketprapakorn & Kantabutra, 2019). Some uncertainties have proven very terrible for human life and after influenza that infected the world in 1918, the COVID-19 pandemic has affected humanity badly and paralyzed all human activities it has been considered big trouble for the world since 1918. The world has been shifting towards sustainability rapidly. In the last decades the world has been experiencing a bundle of challenges organizations and businesses are now transforming themselves towards sustainability and innovativeness the big challenge for them to manage the dynamism and the environmental change (De Lange & Dodds, 2017). It has been observed that most of the researchers mainly focused on evaluating the firm's performance in context with the internal management and external factors and how they affect the firm's performance but some other variables have also impacted the firm's financial performance. The United Nations accepted the Agenda for Sustainable Development 2030 on September 25, 2015. The 17 Sustainable Development Goals (SDGs) including 169 worldwide objectives inside the 2030 Sustainable development agenda seek to enhance everyone's lives, safeguard the environment, to assure a promising future for everyone (United Nations). The United Nations have established the 17 goals agenda to achieve for all nations so that the living standards of the people can be changed and everyone has a better way of living. These now become the fundamental key factors for the development of the nations and the success of the businesses as well (Munsters & Niesten, 2013). Its implementations in business activities have promoted economic growth and provided a bundle of social benefits to the world. Globalization has also increased the rate of the implementation process as to remain in the global market it is also necessary to successfully implement them but despite them, some other factors force the firms to adopt them more rapidly as the current changes in the climate conditions the consumers and the non-governmental organizations are also putting pressure on the organizations. Tourism is a fundamental industry that can prove very helpful for the promotion of sustainable development goals and the 17 goals of sustainable development highly contribute to the development of natural resources and their protection (Osinubi et al., 2023). Social entrepreneurship along with tourism plays an incredible role in the achievement of sustainable development goals. Tourism social entrepreneurship is playing an active role in activating the SDGs. And this unique combination of the duo can facilitate the world and reduce the level of arising challenges. And we have categorized each role differently and mentioned them below. The implementation of these goals is not easy it takes proper planning and strategy formulation (Naderi et al., 2019).

ROLE OF TOURISM SOCIAL ENTREPRENEURSHIP IN ACTIVATING THE SUSTAINABLE DEVELOPMENT GOALS

The role that the tourism industry plays in promoting sustainably has been expressly acknowledged by the UNO. Declaring 2017 the Worldwide Year of Sustainable Tourism for Development and then emphasizing the significance of the tourist industry in supporting 3 of the seventeen 2030 Agenda for (SDGs): SDG 12 Fully accountable Production and Consumption United nations sustainable development goals eight reasonably good Job and Economic Progress, and SDG 14 Existence Below Fluid.

Nevertheless, little focus has been given to exploring how the tourism industry may successfully utilize the SDGs to improve tourism's capacity for sustainability because the SDGs were announced in 2015. Although tourism is only briefly mentioned in the UN vision 2030 for Environmental Sustainability resolutions (three times), the SDGs have emerged as the main focus of research on tourism's capability to support sustainable development. By urging academics to use "analytical reasoning to realize sustainable in tourist networks," the first special edition on the SDGs to be released in the sector of tourism in the journal JoST (Boluk et al., 2019) began to close a gap in the body of research. For instance, two publications that addressed SDG 1 No Poverty, (Scheyvens & Hughes, 2019) criticized perceptions of tourism's contribution to eradicating poverty. In one study responding to Sdgs 10—Reduced Disparities the researchers discussed the importance of transnational corporations in the adoption of equitable and equitable practices—aligned with this study (Zapata et al., 2018). The purpose of this present chapter, which builds on earlier research on tourism and the SDGs, would have been to release a compilation of research and industry standards that would expand our knowledge of how travel and hospitality contribute to the advancement of the SDGs. We requested analytical studies to provide light on the difficulties, fallacies, and limitations of using tourism to tackle sustainability issues. We believed that new and innovative paths for investigation may be opened up through a more specific investigation of the role which tourism plays in attaining SDGs at the call for papers was written. Their examination of Cairo, Dubai, & Muscat, (Alhowaish, 2016) focused on gender disparity (SDG 5), noting the significance of views as determinants of the degree of female empowerment in tourism activities. Dubai was placed #1 among the study locations for women's rights according to the writers' input. Given that women are all still underrepresented in the tourism sector and that SDG 5 is backed by the Saudi Vision 2030, (Sarwar, 2022) research was a model and inspiration. As per the findings, the inclusion of women inside the tourism industry mediates both STD & emotional, legal, and societal development. By using Netherlands Pride as a lens, (Eijgelaar et al., 2022; Hummel & van der Duim, 2012; Pojani & Stead, 2015) cutting-edge study illuminates capitalist amusement while highlighting SDGs 5 and 12 for sustainable development and consumption.

Goal 1: No Poverty

One of the most popular and highly growing across the world and also very problematic for the survival of humans. Tourism has the potential to play a significant role in the reduction of poverty in the world. It can play an active role in reducing the poverty. It can several opportunities for the local people and their involvement enables them to earn their living. From the perspective of social entrepreneurship, this can further enhance the capabilities of the entrepreneur as well. Entrepreneurs have the potential to avail the of minor opportunities and they creatively involve minorities in the process of developing small jobs. And utilize the resources properly. It attracts foreign investors as well which will be beneficial for the economy as well (Wang et al., 2020).

Goal 2: Zero Hunger

When social entrepreneurs actively create unique and innovative products from the local production of the country and deliver them to tourist destinations can promote the economy and agriculture sector. They provide benefits to the farmers and other local communities as well. Tourists also enjoy the experience of new things (Barbosa, 2020).

Goal 3: Good Health and Well-Being

When new and advanced procedures are applied to the tourism industry can lead to better standards of living for individuals. When the tourism industry generates more revenue it will directly enhance economic growth and that will enhance the quality of the healthcare services in the economy. Increasing the level of better medical facilities enables the government to purchase new and advanced medical technology for the hospitals and advanced procedures and medication provide better facilities as well (Goyal et al., 2021).

Goal 4: Quality Education

When tourism social entrepreneurship developed the sector the impact also create career developing opportunities for women and other individuals to become part of such programs so that the better services lead to attracting more customers and creating more opportunities for individuals to acquire a quality education and become more productive in their respective field (Cagarman et al., 2020; Gunzel et al., 2020).

Goal 5: Gender Equality

Entrepreneurs' mindset promotes gender equality and creates the same level of opportunities for males and females. It has the potential to empower women in a variety of ways, particularly by giving them access to employment and other earnings possibilities in both small and large-scale tourism and accommodation businesses. Being one of the industries with the largest percentage of female employees and business owners, tourism social entrepreneurship could be a resource enabling women to achieve their full potential to take the lead in all aspects of society (Punzi & Werner, 2020; McKague & Harrison, 2019).

Goal 6: Clean Water and Sanitation

Tourism social entrepreneurship is also efficient for maintaining the quality of the water their change and advanced models enable them to acquire pure water measures for the tourist destination and for over the country. The country's resources are sacred and utilized properly for the economic development role. All the funds that are being generated from the industry are also reused for the development of the sector (Sultana & Rahman, 2019; Goyal et al., 2021).

Goal 7: Affordable and Clean Energy

Currently, many environmental issues are arising and one of the most hazardous is carbon dioxide emissions that are harmful to the lives of individual's tourism social entrepreneurship practices has increased the utilization of renewable energy and also these practices are promoted all over the world. Energy consumption is high for most economies and their economic growth is dependent upon them. And as alternative renewable energy utilization is safe secure and affordable and efficient for CO2 reduction (Ranabahu, 2020).

Goal 8: Decent Work and Economic Growth

An industry that is highly contributing to economic growth is the tourism industry opportunities that are created from the industry are decent and provide better services to tourists as well. This is one other major role of tourism social entrepreneurship for achieving sustainable development goals. And there have been several other benefits that social tourism can provide to the overall world. These efforts boost the level of the economy fastest and are considered the profitable sector as well (Vázquez-Maguirre, 2020).

Goal 9: Industry, Innovation, and Infrastructure

Tourist destinations are sometimes privately owned and sometimes public the infrastructure and other development work are under their supervision when other key factors enter the industry this enhances the maintenance of the resources along with the creativity and innovation that attract customers and provide proper maintenance to economic assets as well. As order to draw visitors as well as other types of foreign direct investment, the industry could also encourage nation states to modernize their facilities & transform its industries so that they are more green, source of energy, & hygienic (Eichler, & Schwarz, 2019; Cagarman et al., 2020).

Goal 10: Reduced Inequalities

Tourism social entrepreneurship engages all the communities in the work and provides an equal chance for all the individuals to play their role in the development and also provides a platform to showcase their skills as well. It will reduce the differentiation between the genders and also encourage work enthusiasm when they get equal opportunities to perform. Tourism can contribute to urban renewal and rural development and reduce regional imbalances by allowing communities to prosper in their place of origin. That will increase the peace and harmony in the economy and also each sector's development leading to achieving the other sustainable development goals (Macassa, 2021; Pothipala et al., 2021).

Goal 11: Sustainable Cities and Communities

The place which is not safe and secure for the persons who lived there how will be the ideal tourist destination. Tourism social entrepreneurship supports the preservation of sustainability of the locations so that they can become the perfect popular attractions, promotes the rehabilitation of regions in degradation, & preserves the natural and cultural heritage, resources over which tourism depends. Increased investment in environmentally friendly infrastructures (e.g., more effective transportation systems, lower pollution levels, preservation of historic buildings and open spaces) should lead to wiser, healthier communities that benefit both locals and visitors (Goyal et al., 2021).

Goal 12: Responsible Consumption and Production

Social entrepreneurship has promoted the concept of the sustainable tourism sector that adopts sustainable consumption and production practices can play a significant role in accelerating the global shift towards sustainability (Ranabahu, 2020).

Goal 13: Climate Action

Tourism social entrepreneurship contributes to climate change activities It is, therefore, in the sector's interest to play a leading role in the global response to climate change. By lowering energy consumption and shifting to renewable energy sources, especially in the transport and accommodation sector, tourism can help tackle one of the most pressing challenges of our time. And also creating some other ways for the development in the sector along with taking care of climate change (Barry & Hoyne, 2021).

Goal 14: Life Below Water

The financial effects of appropriate the use of ocean biodiversity, particularly via appropriate governance of fisheries, farming, and tourism, are increased for SIDS & LDCs in 2030. This is because social entrepreneurship and tourism are promoting the preservation of life underneath the surface. It is highly concerned with it (Costales & Zeyen, 2022).

Goal 15: Life on Land

Its primary draws for visitors are frequently grand vistas, unspoiled jungles, abundant biodiversity, & historical landmarks. Due to these initiatives to reduce pollution & consumerism, preserve native flora and fauna, & make people aware, eco-tourism may play a key role not just in maintaining diversity but in honoring natural environments. (Wang et al., 2020)

Goal 16: Peace and Justice

As billions of encounters between people from various experiences occur every year as a result of tourism, the industry can promote intercultural and interreligious compassion and understanding, establishing the groundwork for an even more stable world. Sustainable tourism, which enriches and involves communities, may also serve as an income source, build individual cultures, and encourage entrepreneurship development, helping to deter the escalation of conflicts & violence and to solidify calm in post-war countries (Sottini & Ciambotti, 2019).

Goal 17: Partnerships for The Goals

Tourism has emerged the businesses across the world with different sectors of people working in the partnership whether it is private or public sector partnership it has been creating a bundle of job opportunities and internationally and nationally different cultures different departments working together for the achievement of the common goal. Indeed, public/public cooperation and public/private partnerships are a necessary and core foundation for tourism development, as is an increased awareness of the role of tourism in the implementation of sustainable development goals (Wang et al., 2020; Goyal et al., 2021). Even despite being located in cosmopolitan cities, the sub-epithelial landscapes lack essential services like waste disposal, cleanliness, and access to drinking water. We investigate social entrepreneurial approaches to deal with these problems and, in doing so, provide examples of approaches that could be utilized to achieve SDGs.

CONCLUSION

The study of SE & STD is topical and important, especially in light of the fact of societies in several countries are becoming unwilling or incapable to address those problems, whether those issues are caused by ecological or economic factors. As a result, this then will become more and more important to evidenced-oriented tourist studies that concentrate on and challenge existing SE as a unit of investigation to be situated at the core of the knowledge of various types of sustainable-focused tourism firms. Contemporary tourism scholars now have a wider range of ways to participate in the global conversation about social reform and advance the theory of tourism growth attributable to the idea of social businesses. Therefore, in the chapter, we suggest two research areas for tourism researchers to consider: (i) the function of SE as crucial players in the growth of STD; and (ii) the relationships involving social entrepreneurialism & 17 sustainable development goals for the tourism industry. This chapter per the authors are looking for advanced and innovative methods & techniques for tourism social entrepreneurship and their roles in activating SDGs. Currently, sustainability is the key factor, that is considered the main source for developing the sector and enhancing the capabilities, of the humans contributing in the field for the development of the industry. It has been necessary to formulate such strategies, that are socially beneficial for all sustainability. Their practices in businesses are now increasing day by day. There have been many social organizations that are actively contributing to tourism and promoting the concept of social development goals (SDGs).

REFERENCES

Abbasi, A. R., Raza, A., & Shaikh, H. (2022). A Nexus Between Political Instability & International Tourism Demand. *International Journal of Social Science & Entrepreneurship*, *2*(2), 297–312. doi:10.58661/ijsse.v2i2.32

Agyeiwaah, E. (2020). Over-tourism and sustainable consumption of resources through sharing: The role of government. *International Journal of Tourism Cities*, *6*(1), 99–116. doi:10.1108/IJTC-06-2019-0078

Alhowaish, A. K. (2016). Is tourism development a sustainable economic growth strategy in the long run? Evidence from GCC countries. *Sustainability (Basel)*, *8*(7), 605. doi:10.3390u8070605

Aquino, R. S., Lück, M., & Schänzel, H. A. (2022). Mapping the outcomes of social entrepreneurship and tourism on host communities: A three-dimensional approach. *Journal of Sustainable Tourism*, *30*(8), 1799–1820. doi:10.1080/09669582.2021.1986515

Bansal, S., Garg, I., & Sharma, G. D. (2019). Social entrepreneurship as a path for social change and driver of sustainable development: A systematic review and research agenda. *Sustainability (Basel)*, *11*(4), 1091. doi:10.3390u11041091

Barbosa, L. (2020). Social entrepreneurship and digital technology-case study of the blockchain for zero hunger initiative. *Economic and Social Development: Book of Proceedings*, 193-202.

Barry, D., & Hoyne, S. (2021). Sustainable measurement indicators to assess impacts of climate change: Implications for the New Green Deal Era. *Current Opinion in Environmental Science & Health*, *22*, 100259. doi:10.1016/j.coesh.2021.100259

Belli, L., Cilfone, A., Davoli, L., Ferrari, G., Adorni, P., Di Nocera, F., Dall'Olio, A., Pellegrini, C., Mordacci, M., & Bertolotti, E. (2020). IoT-enabled smart sustainable cities: Challenges and approaches. *Smart Cities*, *3*(3), 1039–1071. doi:10.3390martcities3030052

Boluk, K. A., Cavaliere, C. T., & Higgins-Desbiolles, F. (2019). A critical framework for interrogating the United Nations Sustainable Development Goals 2030 Agenda in tourism. *Journal of Sustainable Tourism*.

Buckley, R. (2011). Tourism and environment. *Annual Review of Environment and Resources*, *36*(1), 397–416. doi:10.1146/annurev-environ-041210-132637

Buzinde, C., Shockley, G., Andereck, K., Dee, E., & Frank, P. (2017). Theorizing social entrepreneurship within tourism studies. *Social entrepreneurship and tourism: Philosophy and practice*, 21-34.

Cagarman, K., Kratzer, J., von Arnim, L. H., Fajga, K., & Gieseke, M. J. (2020). Social entrepreneurship on its way to significance: The case of Germany. *Sustainability (Basel)*, *12*(21), 8954. doi:10.3390u12218954

Ciccarino, I., da Silva, J. F., & Rodrigues, S. (2022). Changes in the Portuguese social economy identified by a social entrepreneurial taxonomy. *Journal of Cleaner Production*, *376*, 134268. doi:10.1016/j.jclepro.2022.134268

Correia, A., & Dolnicar, S. (2021). Tourism social entrepreneurs and their role in activating the SDGs–Contributions by Karla Boluk. *Women's Voices in Tourism Research.*

Costales, E., & Zeyen, A. (2022). Social Entrepreneurship, Grand Challenges and Crisis: What We Know So Far. In *Social Entrepreneurship and Grand Challenges: Navigating Layers of Disruption from COVID-19 and Beyond* (pp. 17–32). Springer International Publishing. doi:10.1007/978-3-031-07450-9_2

De Lange, D., & Dodds, R. (2017). Increasing sustainable tourism through social entrepreneurship. *International Journal of Contemporary Hospitality Management*, *29*(7), 1977–2002. doi:10.1108/IJCHM-02-2016-0096

Defourny, J., & Nyssens, M. (2010). Conceptions of social enterprise and social entrepreneurship in Europe and the United States: Convergences and divergences. *Journal of Social Entrepreneurship*, *1*(1), 32–53. doi:10.1080/19420670903442053

Diaz-Sarachaga, J. M., & Ariza-Montes, A. (2022). The role of social entrepreneurship in the attainment of the sustainable development goals. *Journal of Business Research*, *152*, 242–250. doi:10.1016/j.jbusres.2022.07.061

Eichler, G. M., & Schwarz, E. J. (2019). What sustainable development goals do social innovations address? A systematic review and content analysis of social innovation literature. *Sustainability (Basel)*, *11*(2), 522. doi:10.3390u11020522

Eijgelaar, E., Peeters, P., & Reinecke, T. (2022). Netherlands: Review of sustainable tourism policies, strategies and instruments. *European SME Going Green 2030 Report. Review and analysis of policies, strategies and instruments for boosting sustainable tourism in Europe*, 316-326.

Europe, I. (2018). *A policy brief from the policy learning platform on low-carbon economy*. Academic Press.

Foncubierta-Rodríguez, M. J. (2021). Influence of the entrepreneur's personal values in business governance style and their relationship with happiness at work. *Corporate Governance (Bradford)*, *22*(3), 592–617. doi:10.1108/CG-05-2021-0197

Gabriel, P., & Drayton, B. (2016). *Social entrepreneurship and innovation: International case studies and practice*. Kogan Page Publishers.

Goffi, G., Cucculelli, M., & Masiero, L. (2019). Fostering tourism destination competitiveness in developing countries: The role of sustainability. *Journal of Cleaner Production*, *209*, 101–115. doi:10.1016/j.jclepro.2018.10.208

Goyal, S., Agrawal, A., & Sergi, B. S. (2021). Social entrepreneurship for scalable solutions addressing sustainable development goals (SDGs) at BoP in India. *Qualitative Research in Organizations and Management*, *16*(3/4), 509–529. doi:10.1108/QROM-07-2020-1992

Günzel-Jensen, F., Siebold, N., Kroeger, A., & Korsgaard, S. (2020). Do the United Nations' Sustainable Development Goals matter for social entrepreneurial ventures? A bottom-up perspective. *Journal of Business Venturing Insights*, *13*, e00162. doi:10.1016/j.jbvi.2020.e00162

Hole, Y., & Snehal, P. (2019). The significance of pilgrimage tourism to sustainable development with special reference to the Indian context. *African Journal of Hospitality, Tourism and Leisure*, *8*(3), 1–11.

Huang, X., Dai, S., & Xu, H. (2020). Predicting tourists' health risk preventative behaviour and travelling satisfaction in Tibet: Combining the theory of planned behaviour and health belief model. *Tourism Management Perspectives*, *33*, 100589. doi:10.1016/j.tmp.2019.100589

Hummel, J., & van der Duim, R. (2012). Tourism and development at work: 15 years of tourism and poverty reduction within the SNV Netherlands Development Organisation. *Journal of Sustainable Tourism*, *20*(3), 319–338. doi:10.1080/09669582.2012.663381

Jamal, T., & Higham, J. (2021). Justice and ethics: Towards a new platform for tourism and sustainability. *Journal of Sustainable Tourism*, *29*(2-3), 143–157. doi:10.1080/09669582.2020.1835933

Ketprapakorn, N., & Kantabutra, S. (2019). Sustainable social enterprise model: Relationships and consequences. *Sustainability (Basel)*, *11*(14), 3772. doi:10.3390u11143772

Kim, S., Whitford, M., & Arcodia, C. (2019). Development of intangible cultural heritage as a sustainable tourism resource: The intangible cultural heritage practitioners' perspectives. *Journal of Heritage Tourism*, *14*(5-6), 422–435. doi:10.1080/1743873X.2018.1561703

Kirzner, I. M. (1973). *Competition and entrepreneurship*. The University of Chicago Press.

Kowszyk, Y., & Vanclay, F. (2020). The possibilities and limitations regarding the use of impact evaluation in corporate social responsibility programs in Latin America. *Corporate Governance (Bradford)*, *21*(2), 279–293. doi:10.1108/CG-01-2020-0038

Macassa, G. (2021). Social enterprise, population health and sustainable development goal 3: A public health viewpoint. *Annals of Global Health*, *87*(1), 52. doi:10.5334/aogh.3231 PMID:34221905

McKague, K., & Harrison, S. (2019). Gender and health social enterprises in Africa: A research agenda. *International Journal for Equity in Health, 18*(1), 1–10. doi:10.118612939-019-0994-2 PMID:31221156

Movono, A., & Hughes, E. (2022). Tourism partnerships: Localizing the SDG agenda in Fiji. *Journal of Sustainable Tourism, 30*(10), 2318–2332. doi:10.1080/09669582.2020.1811291

Munsters, W., & Niesten, M. (2013). The cultural tourism sustainability mix applied to the development of contemplative tourism in Limburg, the Netherlands. In *Cultural tourism* (pp. 140–150). CABI. doi:10.1079/9781845939236.0140

Naderi, A., Nasrolahi Vosta, L., Ebrahimi, A., & Jalilvand, M. R. (2019). The contributions of social entrepreneurship and transformational leadership to performance: Insights from rural tourism in Iran. *The International Journal of Sociology and Social Policy, 39*(9/10), 719–737. doi:10.1108/IJSSP-06-2019-0124

Olmedo, L., van Twuijver, M., & O'Shaughnessy, M. (2019). *Community-based social enterprises fostering inclusive development in peripheral European rural areas. The United Nations Inter-Agency Task Force on Social and Solidarity Economy*. UNTFSSE.

Osinubi, T., Adedoyin, A., Olufemi, O., & Ajide, F. (2023). Does tourism affect sustainable development in MINT countries? *Global Journal of Emerging Market Economies, 15*(1), 72–92. doi:10.1177/09749101211064388

Pojani, D., & Stead, D. (2015). Going Dutch? The export of sustainable land-use and transport planning concepts from the Netherlands. *Urban Studies (Edinburgh, Scotland), 52*(9), 1558–1576. doi:10.1177/0042098014562326

Pothipala, V., Keerasuntonpong, P., & Cordery, C. (2021). Alleviating social and economic inequality? The role of social enterprises in Thailand. *Journal of Accounting & Organizational Change, 17*(1), 50–70. doi:10.1108/JAOC-09-2020-0127

Punzi, M. C., & Werner, M. (2020). Challenging the menstruation taboo one sale at a time: The role of social entrepreneurs in the period revolution. The Palgrave handbook of critical menstruation studies, 833-851.

Ranabahu, N. (2020). 'Wicked'solutions for 'wicked'problems: Responsible innovations in social enterprises for sustainable development. *Journal of Management & Organization, 26*(6), 995–1013. doi:10.1017/jmo.2020.20

Roy, S. C., & Roy, M. (2015). Tourism in Bangladesh: Present status and future prospects. *International Journal of Management Science and Business Administration, 1*(8), 53–61. doi:10.18775/ijmsba.1849-5664-5419.2014.18.1006

Sarwar, S. (2022). Impact of energy intensity, green economy and blue economy to achieve sustainable economic growth in GCC countries: Does Saudi Vision 2030 matters to GCC countries. *Renewable Energy, 191*, 30–46. doi:10.1016/j.renene.2022.03.122

Scheyvens, R., & Hughes, E. (2019). Can tourism help to "end poverty in all its forms everywhere"? The challenge of tourism addressing SDG1. *Journal of Sustainable Tourism*.

Schumpeter, J. A., & Redvers, O. P. I. E. (1934). *Theorie Der Wirtschaftlichen Entwicklung. The theory of economic development. An inquiry into profits, capital, credit, interest, and the business cycle... translated... by redvers opie*. Academic Press.

Sottini, A., & Ciambotti, G. (2020). Social Entrepreneurship Toward a Sustainable Peacebuilding. *Peace, Justice and Strong Institutions*, 1-11.

Sultana, N., & Rahman, A. (2019). Social business for sustainable development: a developing country perspective. In The Components of Sustainable Development: Engagement and Partnership (pp. 179-192). Springer Singapore. doi:10.1007/978-981-13-9209-2_12

Suriyankietkaew, S., Krittayaruangroj, K., & Iamsawan, N. (2022). Sustainable Leadership practices and competencies of SMEs for sustainability and resilience: A community-based social enterprise study. *Sustainability (Basel)*, *14*(10), 5762. doi:10.3390u14105762

Swedberg, R. (2009). *Principles of economic sociology*. Princeton University Press. doi:10.2307/j.ctvcm4g75

Tseng, M. L., Lin, C., Lin, C. W. R., Wu, K. J., & Sriphon, T. (2019). Ecotourism development in Thailand: Community participation leads to the value of attractions using linguistic preferences. *Journal of Cleaner Production*, *231*, 1319–1329. doi:10.1016/j.jclepro.2019.05.305

Vázquez-Maguirre, M. (2020). Building sustainable rural communities through indigenous social enterprises: A humanistic approach. *Sustainability (Basel)*, *12*(22), 9643. doi:10.3390u12229643

von der Weppen, J., & Cochrane, J. (2012). Social enterprises in tourism: An exploratory study of operational models and success factors. *Journal of Sustainable Tourism*, *20*(3), 497–511. doi:10.1080/09669582.2012.663377

Wang, X., Yuen, K. F., Wong, Y. D., & Li, K. X. (2020). How can the maritime industry meet Sustainable Development Goals? An analysis of sustainability reports from the social entrepreneurship perspective. *Transportation Research Part D, Transport and Environment*, *78*, 102173. doi:10.1016/j.trd.2019.11.002

Zapata Campos, M. J., Hall, C. M., & Backlund, S. (2018). Can MNCs promote more inclusive tourism? Apollo tour operator's sustainability work. *Tourism Geographies*, *20*(4), 630–652. doi:10.1080/14616688.2018.1457074

Zeyen, A., Beckmann, M., Mueller, S., Dees, J. G., Khanin, D., Krueger, N., Murphy, P. J., Santos, F., Scarlata, M., Walske, J., & Zacharakis, A. (2013). Social entrepreneurship and broader theories: Shedding new light on the 'Bigger Picture'. *Journal of Social Entrepreneurship*, *4*(1), 88–107. doi:10.1080/19420676.2012.725422

Chapter 27
Women's Entrepreneurship in Halal Tourism:
A Game Changer for Developing Economies

Muhammad Omer Rafique
Muhammad Ali Jinnah University, Pakistan

ABSTRACT

'Halal' travelers are a rising market niche for tourism, and they look for locations that accommodate their dietary, sartorial, and ritual needs. "Halal" implies lawful in accordance with Islamic teachings. Gambling, drinking, eating pork, and nudity are prohibited. By 2022, it is anticipated that Muslim tourists would spend $220 billion, up from 121 million in 2016, and they will number 156 million. The chapter is intended to present the concept of the role of women in halal tourism and highlight the opportunity to increase entrepreneurship in developing countries through a robust model. The majority of the developing or underdeveloped countries are Islamic, with extraordinary tourism destinations. However, the tourism business in these countries is still monopolized by males. The discussed topic promotes SDG 5 (Gender Equality) and SDG 8 (Economic Growth). Encouraging women's entrepreneurship in Halal tourism shall attract an inflow of finance into the local economy and reduce the unemployment level while maintaining the cultural limitations and securing the belief systems.

INTRODUCTION

Islam is a religion that primarily draws its inspiration from the Quran and the Prophet Muhammad's Sunnah (the Hadith) (SAW). The terms halal (allowed) and haram are found in the Shariah (forbidden). The phrase halal tourism, sometimes known as Islamic tourism, developed from this idea of halal. The newest item on the market and a sector that is expanding rapidly is halal tourism. Its net value, excluding Umrah and Hajj pilgrimages, is currently above $126 billion and is expected to reach $192 billion by 2020. (Dinar Standard and CrescentRating LLC, 2012).

DOI: 10.4018/978-1-6684-8417-3.ch027

Even though the acceptability of halal tourism has significantly increased over the past ten years, the notion still lacks clarity. The concept's lack of standardisation occasionally even causes Muslim tourists to get perplexed. The majority of the study done so far has been concentrated on one location, or in certain cases, two locations have been compared from the perspective of halal tourism. The results of this study revealed that, in many cases, halal tourism locations did not adhere to the core principles of Shariah (Islamic teaching). It is important to note that halal tourism is one of the world's most prestigious subsectors of travel. Consequently, any problems with halal tourism might potentially have a bad effect on the worldwide tourist industry(Rasul, 2019).

Explaining Halal Tourism

Any activity or thing that is acceptable to utilize in the tourist sector according to Islamic teachings may be summed up as halal tourism. The acceptance of Islamic teachings and practises in all facets of tourist operations must thus serve as a guide for the development and promotion of Halal tourism destinations (El-Gohary, 2016).

Mohsin et al. (2016) commented that the Arabic halal is a general term that applies to all aspects of life, not just food. It said it started integrating the concept of halal into the tourism sector after noticing the popularity and acceptance of the concept. Consolidation of Sharia law in the banking and investment sectors. "Halal tourism is one of the new phenomena emerging from the growth of the halal industry". In general, "halal tourism can be regarded as any object or activity for which the use or involvement of the tourism industry is permissible according to Islamic teachings" (Battour & Ismail, 2016, p. 1). After 11 September 2001, the term "Islamic tourism" was introduced by the quarterly Anglo-Arab Islamic Tourism magazine on the impact of tourism across the Islamic world". The focus of halal tourism seems to be largely on Middle Eastern demographics, especially Gulf families who are willing to follow and follow conservative customs and traditions (Syathir et al., 2021).

Halal tourism is still a fairly young subject of research, as the literature review revealed. Despite this and the seeming uncertainty, halal tourism may be defined as the provision of travel services that are tailored to Muslim visitors' demands in line with their religious duties (Vargas-Sánchez & Moral-Moral, 2020).

Ingredients of Halal Tourism

Ingredients of Halal tourism were found in the literature (El-Gohary, 2016).

As a result, certain non-Muslim nations including Japan, the Philippines, Singapore, and Brazil now provide Muslim-friendly tourism amenities to ensure that Muslims visiting these nations have no trouble following their religious tenets.

For instance, the Chamber of Commerce in Japan and the Philippine Travel Agency Associations held training seminars to ensure that Muslim visitors to those countries would experience a Muslim-friendly atmosphere. The seminar's objectives included both promoting Muslim-friendly halal tourism and ensuring Muslim visitors' degree of satisfaction while they were there (Rahman et al. 2019).

Sourcing These Ingredients

In order to ensure that tourism activities and services comply with halal requirements, the ingredients of Halal tourism are typically charted and sourced through these measures.

Figure 1. Ingredients of Halal tourism

The first step is to obtain halal certification for the tourism establishment or service provider. This certification is typically issued by recognized halal certification bodies or organizations that verify and confirm that the products or services meet the halal standards. The certification process involves rigorous inspections, audits, and compliance checks. Moreover, Tourism businesses need to establish a halal-compliant supply chain. This involves sourcing ingredients, products, and services from suppliers and vendors that adhere to halal guidelines. The supply chain should ensure that all ingredients used in food, beverages, cosmetics, and other products meet the halal requirements (Manzoor et al., 2019).

For food and beverage services, the ingredients used should be halal-certified or sourced from suppliers who provide halal-certified products. This includes ensuring that meat, poultry, seafood, and other animal-derived products are slaughtered according to Islamic dietary laws and that non-halal ingredients or additives are avoided. Halal tourism establishments, especially hotels and resorts, should provide designated prayer facilities for Muslim guests. These facilities should be equipped with prayer mats, Qur'ans, and appropriate ablution areas to facilitate Muslim guests' religious practices (Yanes et al., 2019).

Halal tourism establishments should offer a variety of non-alcoholic beverage options, including halal-certified soft drinks, juices, and mocktails. This ensures that Muslim travelers can enjoy refreshments without compromising their religious beliefs. Halal tourism aims to provide an environment that respects the principles of modesty. This includes offering separate swimming pools or designated swimming times for men and women, providing private changing rooms, and offering modest clothing options for guests who may require them. Halal tourism activities should be designed to align with Islamic values. This may involve offering family-friendly entertainment, cultural activities, and recreational options that are free from activities that contradict Islamic principles (Wood and Al-Azri 2019).

When organizing halal travel packages or tours, destination selection is crucial. Women entrepreneurs in the Halal travel industry should consider destinations that provide halal-friendly environments, such as countries with a significant Muslim population or destinations that offer halal-certified services. It is important for women entrepreneurs in the Halal travel industry to conduct thorough research, establish partnerships with halal-certified suppliers, and maintain regular communication with halal certification bodies to ensure the ingredients of Halal tourism are charted and sourced appropriately. By doing so, they can provide authentic halal experiences to Muslim travelers and meet their specific needs (Goffi et al., 2019).

Potential in Tourism for Developing Countries (SDG-8)

Because inbound tourism is contributing more and more to long-term economic growth, its significance has increased dramatically. It promotes industrial development, creates jobs and thereby increases income, inbound tourism also generates positive externalities, and finally, as the economy expands, one can argue that growth in GDP could lead to further increase in international tourism. All of these factors contribute to economic growth (Scheyvens & Hughes, 2019).

The number of jobs created by tourism in many different sectors is substantial. These occupations are not only part of the tourism sector, but also the agricultural sector, the telecommunications sector, the health sector, and the education sector. Many tourists travel to learn about the host country's culture, different traditions, and gastronomy. This is very beneficial for local restaurants, malls, and shops (Wood and Al-Azri 2019). The population of Melbourne, Australia is heavily influenced by tourism. It has a population of about 4 million and employs about 22,000 residents in the tourism sector alone (Goffi et al., 2019).

One of the primary causes seems to be that demand for international travel has been relatively income elastic, especially in the big nations that produce the majority of the world's tourists (Davis, 1968). Travel-related costs rise even more quickly as national incomes rise (Yanes et al., 2019).

In South Asia, the tourist industry contributed 8.9% of the region's GDP ($281.6 billion) in 2017, and experts predict an additional 7.2% rise in 2018. According to Figure 1, it will be 9.0% of GDP ($301.8 billion), with auxiliary income increasing by 6.2% by 2028 to reach 9.4% of GDP ($579.6 billion). With a projected rise of 3.0% or a total share of 7.6% of the job market in 2018, tourism is expected to provide 7.6% of jobs to the employment sector, up from its 7.5% share (49,931,500 jobs) in 2017. (51,436,500 jobs). The employment impact of tourism is anticipated to reach 7.8% of the total workforce by 2028. (63,006,000 jobs) (Manzoor et al., 2019).

Role of Women Entrepreneurship in Halal Tourism (SDG-3)

The role of women in entrepreneurship has been the subject of several prior studies. According to Minniti & Naudé's (2010) research, fewer women own firms than males. Women run micro and small companies at a rate of just 10.28%. Women company owners are less likely to launch new enterprises. If women possess a high level of knowledge, entrepreneurship will be their passion. For women to become entrepreneurs, there are basic issues with knowledge, skills, and talents (Bachri et al., 2022).

Figure 2. Common jobs for female engagements in tourism in developing countries
Source: Hutchings et al. (2020)

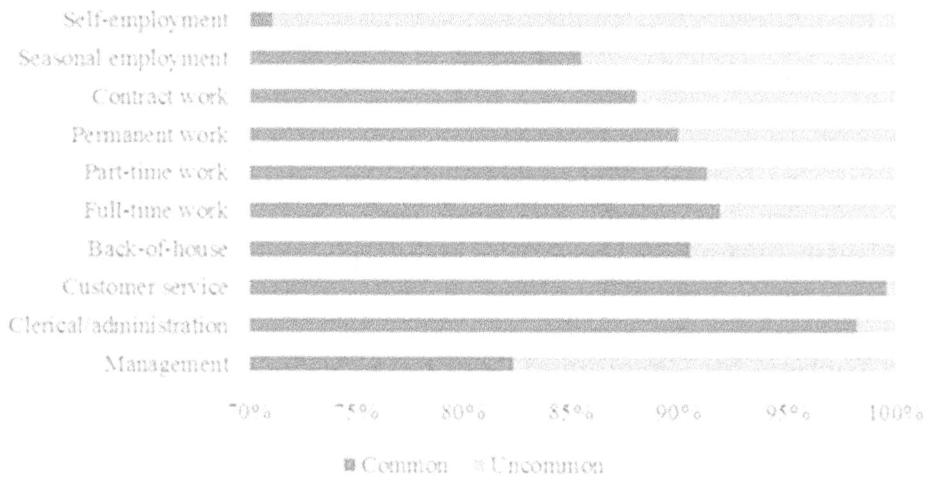

The United Nations World Tourism Organization (UNWTO) reported that "women in tourism are still underpaid, under-utilized, under-educated, and under-represented" in its Global Report on Women in Tourism, which was published in 2010 (UNWTO, 2017, p. ii). However, the UNWTO asserted in the same study that tourism remained one of the greatest ways for women to achieve economic empowerment, particularly in developing nations when other sectors are underperforming (Pastore et al. 2021).

In comparison to other industries, the percentage of women working in the tourist industry is greater, but it is still below average, and according to World Bank data, women working in the industry often earn 35–40% less than males do. Studies have already shown that there is a gender pay difference in the tourist business in numerous countries, which is not only explained by differing production features but also because of discrimination, namely because men and women are paid differently for the same attributes (Hutchings et al. 2020).

Women encounter several difficulties when working in the tourist industry due to their exposure to commercial procedures and the social world more generally outside of their families, which frequently prevents them from being empowered. Compared to the farther-flung cities of Chilas and Abbottabad, Karachi has a more developed tourist industry and, as a port city, has more external effects on its character. As a result, more women independently contribute to the tourism industry in Karachi. Although this may contribute to their empowerment, many of the women still encounter obstacles as tourist producers from people of both sexes who adhere to rigid gendered roles (Sayira and Andrews 2023).

Since labour tends to be a reproduction or extension of women's domestic labor, such as cooking and cleaning, tourism offers strong opportunities for women to enter the informal sector of employment through self-employment and entrepreneurship (Hutchings et al. 2020). Such household labour may include producing and selling goods to visitors, cooking, and traditional crafts like the billum in Papua New Guinea. Although not all self-employment happens in the informal sector, the activity may be seen as appropriate in cultures where women often do not work outside the house or have few employment opportunities (Mohammed et al. 2022).

Women in developing countries are able to work while circumventing cultural hurdles and may choose entrepreneurial activities that are seen as an extension of their wives' role (del Alonso-Almeida, 2012). Gender and race or gender and class or gender and race and class intersectionality can affect some women. In particular, women belonging to certain ethnic groups and women living in economically disadvantaged rural and remote areas who lack the infrastructure and educational facilities of urban areas may suffer a double or even triple disadvantage (Valek and Almuhrzi 2021).

Challenges for Females

Challenges that women entrepreneurs may face in the Halal travel industry:

Sociocultural Constraints: In some societies, women may face cultural and societal expectations that limit their participation in the business world or restrict their mobility, which can pose challenges for women entrepreneurs in the Halal travel industry.

- Access to Funding and Resources: Women entrepreneurs often face difficulties in accessing funding and resources necessary to start or grow their businesses. They may encounter biases or stereotypes that make it harder for them to secure financial support from investors or traditional banking institutions.
- Limited Networking Opportunities: Women entrepreneurs may have limited access to professional networks and mentoring opportunities, which are crucial for business growth and development. This can make it challenging to establish connections with industry experts and potential partners.
- Balancing Work and Family Responsibilities: Women often have to juggle multiple roles and responsibilities, including family and household duties. Balancing these demands with the demands of running a business can be challenging and may require effective time management and support systems.
- Overcoming Gender Bias: Women entrepreneurs in any industry may face gender bias, including stereotypes, prejudices, or discriminatory practices that can hinder their progress. Overcoming these biases and proving their capabilities can be a significant challenge.
- Despite these challenges, the Halal travel industry presents promising opportunities for women entrepreneurs to tap into a growing market and cater to the specific needs of Muslim travelers. With determination, strategic planning, and a supportive network, women can overcome these obstacles and thrive in the industry.

Islamic Finance and Women's Halal Tourism

Islamic finance (IF) and the halal business have grown together, particularly since the 1960s and 1970s, respectively. A vast range of economic activities, including food, fashion, medicines, cosmetics, travel, media, and entertainment, total around 2.2 trillion dollars in the global Halal market. According to Mastercard-prediction, CrescentRating's which states that Muslims would spend roughly $300 billion by 2026, halal tourism (HT) is one of the industries with the quickest growth rates. However, the Islamic Financial Services Board has assessed IF assets to be less than 1% of all financial assets globally, or around 2.19 trillion dollars (Martín et al. 2020).

Halal consumption and business is more than just halal ingredients and manufacturing methods. All aspects of Halal and Haram in Islam must be considered. Islam should be practiced as a whole and not just as a part of it. In this case, halal products should not only be viewed from the production part, but the entire process from procurement of raw materials to receipt by the consumer should be monitored. Ramli (2014) provide a synthesis-based discussion of this industry. Based on this argument, Islam is an inclusive way of life where guidance is given to guide people. Halal compliance should cover broader aspects rather than a selective scope.

Some of the key drivers for achieving financial inclusion, particularly for women, low-income earners, and rural poor people, include reliable technology, microcredit and microfinance services, legal and regulatory commitment from regulators and policymakers of Islamic financial institutions, widespread public awareness of Islamic financial services and products, financial proficiency and literacy, and financial infrastructure (Aliyu et al. 2019; Mahadi et al. 2019).

Halal Tourism Economic Model for Businesswomen

Figure 3. Halal tourism economic model for businesswomen

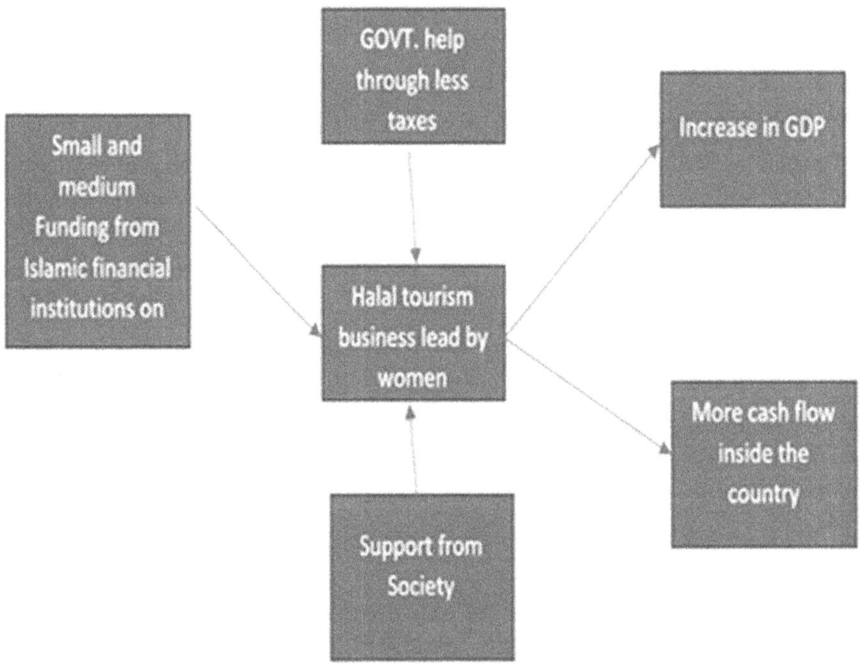

The above model describe the story how Halal tourism business can influence the economy of a developing country firstly it is proposed that small and medium funding from Islamic Financial Institutions shall be increased and guided towards the halal to is in businesses lead by local female they need the assistant from government in less taxes and easy facilities moreover the support from a Muslim society is also mandatory because in many cases the society norms and cultural situations do not allow female entrepreneurs to initiate their business however in the modern society the female entrepreneurs are encouraged with in the families if all these supports are given to this business it is very highly expected that it will increase the inflow of cash and other currencies like dollars which is a great dilemma of the developing countries that they don't have sufficient dollars to import their needs moreover it will decrease the unemployment rate as well as it shall empower the local female within the society while keeping all the religious limits imposed through Quran and sunnah.

Halal Tourism Promotes Sustainability

Women entrepreneurs in halal tourism can play a significant role in promoting sustainability within the industry. Here are some ways they can contribute:

Women entrepreneurs can establish eco-friendly halal accommodations by incorporating sustainable practices. This may include using renewable energy sources, implementing water and energy conservation measures, utilizing eco-friendly materials in construction and operations, and promoting waste management and recycling programs. Women entrepreneurs can educate and encourage their customers to engage in responsible tourism practices. This includes promoting cultural and environmental awareness, respecting local customs and traditions, minimizing waste generation, supporting local communities, and participating in activities that contribute positively to the destination's socio-economic development.

Women entrepreneurs can prioritize sustainable and locally sourced food and beverage options in their halal tourism businesses. This can involve partnering with local farmers and suppliers to provide organic and seasonal produce, reducing food waste through proper planning and portion control, and offering plant-based or vegetarian menu choices to minimize the environmental impact of meat consumption. Women entrepreneurs can support conservation efforts and protect biodiversity in their halal tourism ventures. This can include collaborating with local conservation organizations, promoting nature-based activities that respect and preserve the natural environment, and organizing educational programs to raise awareness about the importance of biodiversity conservation.

Women entrepreneurs can contribute to the sustainable development of local communities by actively involving and empowering them in their halal tourism businesses. This can include providing employment opportunities, supporting local artisans and handicrafts, promoting cultural heritage preservation, and initiating community-based tourism initiatives that directly benefit the local population. Women entrepreneurs can organize workshops, seminars, and training programs to educate both staff and guests about sustainable practices in halal tourism. By raising awareness and sharing knowledge, they can inspire individuals to adopt sustainable behaviors and make informed choices while traveling. Moreover, Women entrepreneurs can collaborate with like-minded organizations and stakeholders to promote sustainability in halal tourism. This can involve joining industry associations or networks focused on sustainable tourism, partnering with local environmental and social organizations, and sharing best practices and resources to collectively advance sustainable initiatives.

By integrating sustainability into their halal tourism businesses, women entrepreneurs can demonstrate their commitment to responsible and ethical practices. This not only benefits the environment but also enhances the overall experience for Muslim travelers who seek halal options aligned with their values.

Recommendations for Promoting Women Halal Tourism

There are several ideas that can be initiated by the local females as entrepreneurs to promote Halal tourism and thus participate in the economic development of the country.

Halal Travel Agencies

Launching a travel agency that specializes in organizing halal-friendly travel packages, including flights, accommodations, transportation, and sightseeing activities that comply with Islamic principles can be a very good solution for females.

Modest Fashion and Apparel

Creating a clothing brand that focuses on modest fashion and offers a wide range of stylish and fashionable options for Muslim women travelers is also an easy option.

Halal Travel Apps and Technology

Literate females can Develop mobile applications or online platforms that offer halal travel guides, prayer time notifications, halal food recommendations, and other travel-related services specifically designed for Muslim travelers.

Muslim Friendly Hotels

Female entrepreneurs can easily establish a Muslim-friendly hotel with limited funds and government subsidies.

Halal Snack Stalls

Moreover, local female entrepreneurs can also initiate Halal snacks stalls specially in the seasons of tourism.

Halal Beverages Stalls

This can also be a good option for businesswomen even to introduce local halal beverage stalls for tourists with a very limited budget and resources.

Female Drivers for Long-Route Transport

Females can also learn how to drive HTVs (heavy vehicles) and get a license for this. This freelancing or getting a job in a transportation company is a good option for female workers keeping within the limits of Halal.

Female Tourist Guide

Various female tourists need a guide but they are not able to find a female for this service and they feel uncomfortable with the male guide. This business can be initiated with zero investment.

Handicraft Shop

Local handicrafts are less costly and generate a big profit. Female entrepreneurs shall consider this idea for start-up businesses.

Training

Additionally, service programs from colleges may teach female entrepreneurs about successful, intelligent, and strategic entrepreneurship. Training aid from NGOs, financial assistance, and business equipment from the government may help female entrepreneurs work autonomously.

After receiving training in business management and receiving excellent service, it is necessary to support the women business owners for six months, up until they are independent and prepared to handle a variety of business issues, in order to ensure prime success and optimal business sustainability. This procedure is crucial for company control.

Fundings

Without assistance from the municipal government, Islamic banks, and private parties/large firms in the City, women entrepreneurs will not be able to effectively carry out the process of women's entrepreneurship. To retain their creativity and company viability, governments might provide working cash and commercial investment. Additionally, women's business owners require marketing support, such as promoting travel through various media including radio, billboards, television, private television, TVRI, and other forms of advertising.

CONCLUSION

This book chapter entails the concept of introducing the role of a female entrepreneur in promoting Halal to a reason in developing countries like Pakistan the chapter provides a robust and comprehensive model which can be further researched empirically this chapter refers to the relevant literature on Halal to the origin and how it can be mapped with female participation in business this idea can be a game changer if promoted and facilitated by the government as well as a funding agency to new startups through this economy of a developing country can prosper and cash flow may come inside the country for the betterment of local people.

The Halal sectors are divided, and integration for all types of goods, commodities, and services, including IF, should be encouraged(Budi et al. 2020). As a result, it is recommended that the IF play the function of a catalyst in the growth of the HT as a means of using Islamic finance as a source of cash to launch an integrated Halal enterprise. In order to engage in HT development and more effectively utilise the cultural Arab riches of specific Spanish regions, several stakeholders must become visible in this regard. FinTech firms that assist emerging businesses in offering a broad range of cutting-edge goods and services in the expanding Halal niche market can also promote this growth.

REFERENCES

Bachri, N., Wardhiah, W., Roni, M., & Nur, M. M. (2022). Women's entrepreneurship in supporting halal tourism in lhokseumawe city. *International Journal of Economic, Business, Accounting, Agriculture Management and Sharia Administration*, *2*(1), 49–56.

Battour, M., & Ismail, M. N. (2016). Halal tourism: Concepts, practises, challenges and future. *Tourism Management Perspectives*, *19*, 150–154. doi:10.1016/j.tmp.2015.12.008

Davis, H. D. (1968). Potentials for tourism in developing countries. *Finance and Development*. https://search.proquest.com/openview/58c2cd1bf137b001c599a40c2fec3ae7/1?pq-origsite=gscholar&cbl=1819673

El-Gohary, H. (2016). Halal tourism, is it really Halal? *Tourism Management Perspectives*, *19*, 124–130. doi:10.1016/j.tmp.2015.12.013

Goffi, G., Cucculelli, M., & Masiero, L. (2019). Fostering tourism destination competitiveness in developing countries: The role of sustainability. *Journal of Cleaner Production*, *209*, 101–115. doi:10.1016/j.jclepro.2018.10.208

Manzoor, F., Wei, L., Asif, M., Haq, M. Z., & Rehman, H. (2019). The Contribution of Sustainable Tourism to Economic Growth and Employment in Pakistan. *International Journal of Environmental Research and Public Health*, *16*(19), 3785. doi:10.3390/ijerph16193785 PMID:31597376

Mohsin, A., Ramli, N., & Alkhulayfi, B. A. (2016). Halal tourism: Emerging opportunities. *Tourism Management Perspectives*, *19*, 137–143. doi:10.1016/j.tmp.2015.12.010

Rasul, T. (2019). The trends, opportunities, and challenges of halal tourism: A systematic literature review. *Tourism Recreation Research*, *44*(4), 434–450. doi:10.1080/02508281.2019.1599532

Scheyvens, R., & Hughes, E. (2019). Can tourism help to "end poverty in all its forms everywhere"? The challenge of tourism addressing SDG1. *Journal of Sustainable Tourism*, *27*(7), 1061–1079. doi:10.1080/09669582.2018.1551404

Syathir, S. A., Abror, A., Wardy, P. T., Muslihati, M., Sofyan, S., Sirajuddin, S., Nasri, K. M., & Zulfikar, D. A. (2021). Crisis and disaster management for halal tourism: A systematic review. *Tourism Review*, *77*(1), 129–145.

Vargas-Sánchez, A., & Moral-Moral, M. (2020). Halal tourism: Literature review and experts' view. *Journal of Islamic Marketing*, *11*(3), 549–569. doi:10.1108/JIMA-04-2017-0039

Yanes, A., Zielinski, S., Diaz Cano, M., & Kim, S.-I. (2019). Community-Based Tourism in Developing Countries: A Framework for Policy Evaluation. *Sustainability (Basel)*, *11*(9), 2506. doi:10.3390u11092506

Chapter 28
Women Voluntourism for the Empowerment of Refugee Women

Selen Subaşı
https://orcid.org/0000-0003-0893-9330
University of Szeged, Hungary

ABSTRACT

The term "voluntourism" refers to vacations in which participants give their time to charitable organizations. Humanitarian crises, either natural causes like natural disasters or human causes such as armed conflicts and civil wars, caused millions of people to be displaced. According to the United Nations Refugee Agency (UNHCR), global forced displacement has reached 103 million, with 53.2 million of those internally displaced and 32.5 million refugees. Sustainable Development Goals (SDGs) call on the global community to "leave nobody behind," which has resulted in the formation of dozens of new organizations and motivated thousands of individuals to participate in volunteer travel. In addition, non-governmental organizations have been actively looking for volunteers to aid displaced people. Women hailing from a wide range of nations have been taking on the mantle of a refugee and making the journey to a number of conflict zones in order to establish organizations and/or lend a hand in the process of resettling refugees.

INTRODUCTION

The emergence of the refugee crisis as a consequence of the Syrian civil war has underscored the importance of providing assistance and fostering empowerment among refugees, with particular emphasis on women and children. Women-led organizations, including those run by voluntourists, have stepped up to provide essential support for women who often face significant challenges such as language barriers, discrimination, and lack of access to education and employment opportunities. Through their programs and services, these organizations are providing a lifeline for vulnerable populations, helping them to rebuild their lives and contribute to their new communities.

DOI: 10.4018/978-1-6684-8417-3.ch028

In Turkey, a number of outstanding community-based initiatives are run by women voluntourists, including Small Projects Istanbul (SPI), ONAR Istanbul, MADRE, Knitstanbul, and Team International Assistance for Integration (TIAFI), C@rma Foundation as well as Jana Impact. These organizations play a critical role in supporting Syrian refugee women in Turkey, providing them with access to education, healthcare, and vocational training, as well as psychological support and childcare. These initiatives not only provide skills that can lead to employment opportunities but also empower refugee women to take control of their lives and become more independent.

Moreover, these voluntourist women-led organizations provide an opportunity for women from around the world to get involved in making a difference in the lives of Syrian refugee women. By volunteering with these organizations, women can help support the education, healthcare, and psychological well-being of Syrian women refugees. These voluntourist women-led organizations are not only making a difference in the lives of Syrian women refugees but also in the larger struggle for gender equality and social justice.

Overall, the efforts of these organizations demonstrate the power of community-based, voluntourist women-led initiatives to create positive change in the lives of those most in need. As Turkey continues to host millions of Syrian refugees, it is vital that the government and larger aid organizations prioritize support for these grassroots efforts and work to ensure that the voices and experiences of refugee women are centered in all aspects of the response.

The work of women-led organizations in Turkey is not only commendable for the assistance they provide but also for the broader impact they have on promoting gender equality and social justice.

Voluntourism: Volunteer Tourism

Voluntourism, also known as volunteer tourism or volunteer travel, is a growing trend in the tourism industry where travellers participate in voluntary work while on vacation in a foreign country (Wearing, 2001). This type of tourism is typically short-term, with volunteers engaging in a range of activities such as conservation work, community development, and teaching. The voluntourism industry has grown rapidly in recent years, with an estimated 10 million volunteers participating in various projects worldwide (CBI-Ministry of Foreign Affairs, 2020).

The motivations behind voluntourism can vary, with some volunteers seeking to make a difference and contribute to the local community, while others may be driven by personal development or cultural immersion (Daldeniz, & Hampton, 2010). However, there is ongoing debate about the potential positive and negative impacts of voluntourism on host communities, as well as on the volunteers themselves.

Studies have highlighted the benefits of voluntourism for both volunteers and host communities, including personal development, increased intercultural awareness, and positive impacts on the environment and local economies (Tomazos & Butler, 2009). However, it is also important to consider the potential negative impacts of voluntourism on host communities, such as the displacement of local workers or cultural disruption (Guttentag, 2009). This is especially important when working with vulnerable populations, such as refugees, who require sensitive and thoughtful approaches to their needs.

The Benefits and Positive Impacts on Volunteers and Host Communities

Voluntourism, a combination of volunteering and tourism, has gained significant attention in recent years due to the potential benefits it offers to both volunteers and the communities they serve. Numerous studies have highlighted the diverse benefits that can be gained from voluntourism, including personal development, increased intercultural awareness, and positive impacts on host communities.

The positive outcomes of voluntourism for host communities have been extensively examined in empirical studies. For instance, the study of McGehee and Andereck (2004) study found that residents of a rural community in Arizona were generally supportive of volunteer tourism and believed that it could have positive economic and social impacts on their community. Similarly, Wearing and McGehee (2013) reviewed existing literature and found that voluntourism can provide much-needed labour and resources to host communities, promote cross-cultural understanding, and contribute to sustainable development. Orams (1995) conducted a study that examined the impact of tourism on a host community and found that ecotourism, of which volunteer tourism is a subset, can have positive impacts on both the environment and the local community. Wearing (2001) stated that volunteer tourism can provide host communities with important economic benefits, such as increased employment and investment opportunities. The study also found that host communities view volunteer tourism as a way to build relationships with people from other countries and learn about their cultures.

Voluntourism can also have positive impacts on the volunteers themselves. The psychological benefits of voluntourism have been reported in several studies. For example, voluntourism can increase intercultural awareness and understanding, leading to personal growth, increased self-esteem, and improved psychological well-being (Wearing, 2001). Waterman (2011) found that voluntourism can enhance volunteers' psychological well-being and efforts toward self-actualization. Lepp (2008) discovered that volunteer tourism can promote personal growth and the discovery of self, leading to enhanced psychological well-being.

Voluntourism can also enhance volunteers' personal and professional development by increasing interpersonal, problem-solving, and communication skills (Jones, 2005), identity development (Matthews, 2008), and the discovery and development of self (Wearing et al., 2008). Overall, voluntourism not only brings positive impacts to the communities and causes being supported, but it also offers numerous benefits to the volunteers themselves.

While these studies indicate positive outcomes, it is crucial that volunteer programs are carefully designed and implemented to ensure their sustainability and cultural appropriateness. In conclusion, voluntourism can have positive impacts on both volunteers and host communities when it is thoughtfully designed and implemented in a culturally sensitive and sustainable manner.

The Potential Negative Impacts of Volunteer Tourism on Host Communities

Although volunteer tourism, or voluntourism, is often viewed as a way to do good and help people in need, there is some evidence to suggest that it can have negative impacts on host communities. While empirical data on this topic is limited and often mixed, some studies have identified concerns such as a lack of local ownership and control over volunteer projects, as well as the potential for volunteers to displace local workers or create dependency on foreign aid. Several studies have found that voluntourism can lead to dependence on external support and undermine local initiatives for development. For instance, McGehee and Andereck (2008) pointed out that voluntourism can lead to the commodification of poverty, where the host communities are objectified, and their poverty is presented as a product to be consumed by tourists. Other studies have noted that voluntourism can create unrealistic expectations among the host communities and contribute to a "poverty tourism" mentality, where the local people are objectified, and their lives are displayed as a spectacle for the tourists (Wearing & McGehee, 2013;

Mostafanezhad, 2013). This can create a power dynamic where the volunteers are seen as superior and the host communities as inferior, leading to a lack of respect for local culture and customs. In addition to social impacts, some studies have highlighted the potential negative effects of voluntourism on the local environment. For example, voluntourism activities such as building construction or wildlife conservation can have unintended consequences on the local ecosystem and biodiversity (Mostafanezhad, 2013). Similarly, Liu and Leung (2019) noted that voluntourism activities aimed at wildlife conservation can have unintended consequences on the local ecosystem and biodiversity, such as the displacement of native species and the introduction of non-native species. These studies highlight the importance of carefully planning and managing voluntourism activities to minimize their negative impact on the environment. Furthermore, volunteer tourism can create conflicts between local residents and volunteers, due to differences in cultural norms and expectations. For instance, Tomazos and Butler (2009) found that volunteer tourism programs in some destinations have led to conflicts between local residents and volunteers. The researchers also found that volunteer projects may not always align with the needs and priorities of local communities and that the volunteer programs can sometimes be exploitative of the host communities. In conclusion, while the empirical data on the negative impacts of volunteer tourism on host communities is mixed, it is important to consider the specific context of each program and destination. It is also crucial to implement volunteer programs in a way that is culturally sensitive, sustainable, and respectful of the needs and priorities of the host communities. As the research in this area is still limited, more studies are needed to fully understand the impact of voluntourism on host communities. Additionally, it is crucial that volunteer programs are carefully designed and implemented with the involvement and consultation of the host communities to ensure that they are sustainable, culturally appropriate, and beneficial to both volunteers and the host communities.

Voluntourism for Refugees: A Tool for Empowerment

Humanitarian crises caused by natural disasters, armed conflicts, and civil wars have led to millions of people being displaced. According to the United Nations Refugee Agency ([UNHCR], 2021a), the number of forcibly displaced people has reached 103 million globally, with 53.2 million internally displaced persons and 32.5 million refugees. The Sustainable Development Goals (SDGs) call for the global community to "leave nobody behind," resulting in the formation of new organizations and motivating thousands of individuals to participate in volunteer travel (United Nations Development Programme, n.d.)

Non-governmental organizations (NGOs) have been actively seeking volunteers to aid displaced people, and women from a range of nations have been traveling to conflict zones to establish organizations and lend a hand in resettling refugees (United Nations Volunteers, 2018; UNHCR, n.d.). Voluntourism for refugees refers to the practice of traveling to refugee camps or communities and volunteering to support refugees. This can involve providing direct assistance such as teaching, building infrastructure, distributing aid, or engaging in more indirect activities like fundraising, awareness-raising, or advocacy (ibid.).

Empowerment occurs when people take greater control over the decisions, assets and policies, Institutions, and Processes that affect their livelihoods (UNHCR, 2005). Refugee empowerment is to enable them to shape their futures, improve their environment and help overcome limitations in service provision by giving refugees the tools, resources, and opportunities needed to take control of their own lives and achieve self-sufficiency (ibid.). The goal of refugee empowerment is to help refugees transition from being passive recipients of aid to active participants in their own recovery and resettlement so that they can build new lives for themselves and their families.

Voluntourism can be a means to empower refugees in several ways. First, it can provide education and job skills training to refugees, helping them acquire new skills and knowledge that can improve their economic prospects. Second, through voluntourism, refugees can develop leadership skills, increase their confidence, and become more active in their communities. Third, voluntourism can increase awareness about the challenges faced by refugees and promote advocacy for their rights and needs. Fourth, voluntourism projects can provide refugees with access to healthcare services that they might not otherwise have access to. Finally, voluntourism can support community development by providing resources and support for refugee-led initiatives and projects (UNHCR, n.d.).

Overall, voluntourism can empower refugees by providing them with the education, skills, resources, and support they need to improve their lives and communities. It can also increase awareness about the challenges faced by refugees and promote advocacy for their rights and needs. However, it is crucial to approach voluntourism with cultural sensitivity, appropriate training, and a long-term perspective that prioritizes building the capacity of the local community rather than creating unsustainable dependency on short-term volunteers. By doing so, voluntourism can be a powerful tool for refugee empowerment and a meaningful way for individuals to contribute to the global effort to “leave nobody behind”.

Syrian Civil War and Refugees

The Syrian Civil War, which began in 2011, has caused immeasurable harm to the people of Syria. The conflict has resulted in the displacement of more than half of Syria’s prewar population, leaving millions of people without homes, basic necessities, and protection. The war has created one of the most pressing humanitarian crises of the 21st century, with the plight of Syrian refugees becoming a global issue.

The conflict has resulted in a massive exodus of people from Syria, with over 6.8 million refugees fleeing to neighboring countries (Internal Displacement Monitoring Centre [IDMC], 2019). These refugees have sought asylum in countries such as Turkey, Lebanon, Jordan, Iraq, and Egypt. While some refugees live in camps, the majority of them reside in urban areas, facing numerous challenges in accessing essential services and protection (IDMC, 2019). Jordan is currently hosting 661,670 registered refugees, with only 20 percent of them residing in camps (UNHCR, 2021b). Similarly, Lebanon is accommodating 1.5 million Syrian refugees, with 814,715 officially registered (ibid.). In Iraq, over 260,000 refugees are registered, and the UNHCR provides shelter and aid to more than 140,000 Syrian refugees in Egypt (ibid.). Turkey has been accommodating the largest population of refugees, and the government has granted them “Temporary Protection” status, which provides them with lawful residence, defends them from prosecution for illegal entry, and promises their nonrefoulement (UNHCR Turkey, 2022). As of 2021, 3.7 million Syrians are receiving temporary protection, and over 320,000 refugees and asylum seekers are receiving international protection in Turkey (ibid.).

The Syrian conflict has caused a massive displacement crisis, with millions of people forced to flee their homes and seek refuge in neighboring countries. The host countries are facing numerous challenges in providing essential services, protection, and assistance to the displaced population. Among these, Turkey is accommodating the largest population of Syrian refugees, providing them with temporary protection and basic necessities. However, the crisis continues to pose significant challenges for the host countries and the international community, highlighting the need for sustained efforts to address the root causes of the conflict and support the displaced population.

Syrian Women Refugees in Turkey

Syrian women refugees in Turkey face a multitude of challenges and barriers in accessing basic services, healthcare, education, and employment opportunities. According to a report by the United Nations Entity for Gender Equality and the Empowerment of Women (UN Women, 2018), the majority of Syrian refugees with official temporary protection status (93%) live outside of camp settings, where only 15% of women have access to income-generating jobs. This is largely due to language barriers and limited access to quality housing and employment opportunities.

To better understand the challenges facing Syrian women refugees in Turkey, a comprehensive needs assessment was conducted by UN Women and the Association for Solidarity with Asylum Seekers and Immigrants (ASAM), which was funded by the Government of Iceland. The study involved structured and in-depth interviews with 1,291 Syrian women and girls across seven cities in Turkey *and* revealed that language barriers, limited access to quality housing, and few employment opportunities were among the most prominent challenges faced by Syrian women in Turkey *(UN Women,2018)*.

In addition, the study found that only 15% of Syrian women work in income-generating jobs, and the vast majority (73%) do not know where to seek assistance for violence or harassment. Many are also not aware of where to seek assistance for their children, and a significant proportion *is* not aware of free legal counseling, psychosocial support, home care, or childcare services*(ibid)*.

Access to healthcare is also a significant barrier for Syrian women refugees in Turkey, as they face language barriers, lack of health insurance, limited financial resources, and discrimination. This can lead to misdiagnosis or mistreatment, according to a study by Alawa et al. (2019).

The lack of employment opportunities and restrictive gender norms make it challenging for Syrian women in Turkey to access employment or start their own businesses. In fact, only 15% of Syrian women work in income-generating jobs, with approximately 24% living below the national extreme poverty line (UN Women, 2017).

Syrian refugee women in Turkey also experience high levels of trauma, depression, and anxiety, with most women reporting at least two mental health symptoms (A*l*pak et al., 2015). They tend to have post-traumatic stress disorder and report low levels of well-being, increased risk of depression, and lower life satisfaction compared to men (Sagır, 2018), underscoring the need for psychosocial services.

Access to education and skills development programs is also a significant challenge for Syrian women refugees in Turkey due to language barriers, financial constraints, and cultural norms. The Norwegian Refugee Council (2019) found that many Syrian children, particularly girls, are not attending school in Turkey, with factors such as language barriers, discrimination, and financial constraints contributing to the problem.

Child marriage is also a significant problem among Syrian refugee communities in Turkey, with many girls being married off at a young age and being unable to continue their education (*Spencer*, 2015). Syrian women and girls in Turkey are at high risk of experiencing gender-based violence, with many facing violence from family members or in intimate partner relationships (Dogutas, 2019).

In conclusion, the conditions of Syrian women refugees in Turkey are challenging, with numerous barriers and challenges affecting their physical safety and security, access to basic services, economic insecurity, and mental health concerns. The studies highlight the need for increased support and resources to address these issues, including better access to healthcare, economic empowerment, education and skills development, child marriage prevention, and gender-based violence prevention. It is essential to address the needs of Syrian women refugees to ensure their well-being and safety and to help them rebuild their lives in a new country.

Voluntarist Women-Led Organizations in Turkey to Support Syrian Women Refugees

The Syrian civil war and the refugee crisis have brought attention to the importance of supporting and empowering refugees, including women and children. Voluntourist women-led organizations in Turkey have emerged as a significant force in the voluntourism industry, working to provide education, vocational training, psychological support, healthcare, and childcare for Syrian refugee women.

These women-led initiatives provide essential support for Syrian refugee women, who often face significant challenges such as language barriers, discrimination, and lack of access to education and employment opportunities. By providing the resources and support they need to succeed, voluntourism can be a powerful tool for empowering refugee women to build a better future for themselves and their families.

In this chapter, we explore outstanding community-based initiatives run by women in places of violence and conflict, with a particular focus on Turkey. These initiatives include SPI, ONAR Istanbul, MADRE, Knitstanbul, TIAFI, C@rma Foundation, and JANA Impact. These organizations are led by women and work to empower and support women and children who have been displaced by war, conflict, and persecution. Through their programs and services, they are providing a lifeline for vulnerable populations, helping them to rebuild their lives, and contribute to their new communities.

Karyn Thomas is the expatriate woman founder of SPI, a non-profit organization based in Istanbul, Turkey, that provides support to Syrian refugees through various initiatives, including education and skills training programs, women's empowerment initiatives, and community events (Small Projects Istanbul-, n.d). The organization runs a community center and education program that empowers refugee women by providing them with language classes, vocational training, and opportunities to develop new skills. These programs aim to increase the social and economic self-sufficiency of refugee women, helping them to integrate into their new communities and become more independent. SPI also provides a range of other services to support the well-being of refugee women and their families, including emergency assistance, health care services, and childcare. The organization works closely with local partners and community groups to ensure that its programs are responsive to the needs of the community and that the voices of refugee women are heard and represented. Through its programs and services, SPI has empowered many refugee women to take control of their lives and build a brighter future for themselves and their families.

ONAR Istanbul is another women-led and women-aimed NGO based in Istanbul, Turkey that aids Syrian refugees in the country. ONAR Istanbul is an organization founded by women to provide support to refugees, especially women, and children. The organization offers various programs and services to Syrian women refugees in Turkey, including language classes, vocational training, psychological support, and health services. The role of women in ONAR Istanbul is to lead and implement these programs and services, as well as to provide support and mentorship to the women refugees who participate in them. Women in ONAR Istanbul also work to raise awareness and advocate for the needs and rights of refugees, particularly women, and children. Through its programs and services, ONAR Istanbul aims to empower Syrian women refugees by providing them with the skills and resources they need to build a better future for themselves and their families in Turkey. In addition to emergency assistance, education, healthcare, and psycho-social support, ONAR offers various programs to empower Syrian women. These programs include vocational training and job placement services, which aim to improve the economic and social status of refugee women. Through these initiatives, ONAR strives to enhance the well-being and agency of refugee women, as well as promote their active participation in society. ONAR works closely with other organizations and local communities to provide coordinated support to Syrian refugees in Istanbul. Their mission is to provide ongoing assistance and relief to those in need, with a focus on empowering refugees and helping them to rebuild their lives. (ONAR Istanbul, n.d.)

MADRE is a women-led organization that works to empower women, promote gender equality, and advance human rights. The organization believes that women have a critical role to play in creating a more just and peaceful world, and therefore works to ensure that women's voices are heard, and their needs are addressed in all aspects of its programs (MADRE, n.d.). MADRE works with women-led groups and organizations around the world, including in Turkey, to provide support and resources for women's empowerment and to advance women's rights. In the context of Syrian refugee women in Turkey, MADRE has partnered with local women-led organizations to provide psychosocial support, legal aid, and other services to women and girls. The organization recognizes that women are particularly vulnerable in crisis situations and works to ensure that their needs and rights are not overlooked.

"Knitstanbul" is run by a team of Syrian refugees and British volunteers in Istanbul and London (Knitistanbul, n.d.). "Knitstanbul" is an inspiring social enterprise that empowers refugee women by providing them with opportunities for skill-building, community connection, and economic independence. The organization is run by a team of Syrian refugees and British volunteers who work together to create hand-knitted products, such as scarves, hats, and gloves, which are sold through various channels, including online platforms and pop-up markets. Through "Knitstanbul", refugee women are given access to paid employment and training in knitting, as well as language and communication skills. This not only helps them to build their confidence and self-esteem but also provides them with the financial resources they need to support themselves and their families. "Knitstanbul" also facilitates social connections between refugees and locals, which helps to break down barriers and foster understanding and empathy. In addition to the economic and social benefits provided to refugee women, "Knitstanbul" also contributes to sustainability by promoting the use of natural and locally sourced materials. The organization also emphasizes the importance of ethical production practices and fair-trade principles. Overall, Knitstanbul is a wonderful example of how social enterprises can make a positive impact on the lives of vulnerable populations, while also promoting sustainability and community-building.

TIAFI is an NGO founded in 2017 by Anne O'Rorke, an Irish woman who moved to Turkey to improve the living conditions of women refugees, and children (Team International Assistance for Integration-TIAFI, n.d.). TIAFI's mission is to support refugees who have been displaced by war, conflict, and persecution, with a particular focus on empowering women and children. TIAFI's work is centred around three main areas: education, healthcare, and community support. The organization operates a school in Istanbul that provides free education to children from refugee backgrounds. The school is staffed by qualified teachers, many of whom are also refugees, and provides a safe and nurturing environment for children to learn and grow. TIAFI also provides healthcare services to refugees, including medical consultations, vaccinations, and mental health support. In addition, the organization offers community support through a range of programs and activities designed to promote social integration and cultural exchange. One of TIAFI's key initiatives is the Women's Empowerment Program, which aims to provide refugee women with the skills and resources they need to support themselves and their families. The program includes training in areas such as language and computer skills, as well as vocational training in areas such as tailoring and hairdressing. TIAFI also provides micro-loans to women who wish to start their own businesses and offers mentorship and support to help them succeed. TIAFI's work is crucial in a context where many refugees face significant barriers to integration and social mobility. By providing access to education, healthcare, and community support, as well as opportunities for economic empowerment, TIAFI is helping to improve the lives of refugees and their families and promote greater social inclusion and cohesion in Turkey.

Istanbul & I is an organization in which ex-pat women work to empower marginalized women in Istanbul, including Syrian refugees (Istanbul & I, n.d.). They offer a variety of programs, including language classes, vocational training, and mentorship opportunities, with the aim of promoting gender equality and social inclusion. Their language classes and vocational training programs aim to help women overcome some of the challenges they face in accessing education and employment opportunities. The mentorship opportunities also provide a supportive network for women and help them build their confidence and skills. By empowering marginalized women, Istanbul & I is contributing to the social and economic integration of Syrian refugees and helping to build more inclusive and resilient communities in Istanbul.

C@rma Foundation is a women-led NGO based in Turkey that strives to bridge the gap between Turkish professionals who are eager to give back to the community and civic organizations in need of volunteers (C@rma Foundation, n.d.). Founded by Sandrine Ramboux, Margarita Mecheva, Isil Arisoy Kaya, Amelie Gaucherand, and Julie Ramboux, the organization launched a platform in 2014 to connect individuals willing to volunteer with organizations looking for volunteers.

One of the non-profit organizations that the C@rma Foundation partners with is JANA Impact, which is also a women-initiated organization that focuses on empowering refugee women in Turkey. JANA Impact's "Women Empowerment for Inclusive Growth" project provides mentorship, training, and seed funding to support women entrepreneurs (JANA Impact., n.d.). The five-month project in Gaziantep selects 15 refugee women to receive support to establish and sustain their businesses, helping them to become more self-sufficient and economically stable.

C@rma Foundation and JANA Impact are both volunteer women-initiated organizations that share the belief in the power of women to drive positive change in their communities. They are committed to creating a more equitable and just world for all, with a particular focus on supporting women and marginalized groups. Through their efforts, they hope to create a sustainable and inclusive future for all those they serve.

C@rma Foundation's platform is a testament to the power of connecting individuals who want to give back to organizations that need help. By providing a bridge between volunteers and non-profit organizations, they enable individuals to make a difference in their communities while also supporting the efforts of organizations working towards positive change. C@rma Foundation and JANA Impact's work in empowering women and promoting social justice highlights the importance of supporting women-led organizations and the power of collective action toward creating a better future for all.

Karam Foundation is an NGO that was founded in 2007 by a group of Syrian American women (Karam Foundation, n.d.). The organization provides aid and resources to Syrian refugees, with a particular focus on supporting women and children. Karam Foundation offers a range of programs, including education and health care initiatives, to support Syrian women refugees and help them to rebuild their lives. Karam Foundation is a non-profit organization that provides aid and resources to people affected by the Syrian conflict, with a focus on empowering women and girls. The organization's work focuses on three main areas: education, humanitarian aid, and sustainable development. Additionally, the Women's Empowerment Programs aim to support Syrian women and girls, including through initiatives such as job training, literacy programs, and access to health care.

Syrian refugee women in Turkey face a multitude of challenges, including limited access to education, healthcare, and employment opportunities. They often struggle to adapt to a new culture and language, and many face discrimination and social isolation. In this context, organizations play a crucial role in providing much-needed support and resources to these women. By offering language classes, vocational training, and mentorship opportunities, these organizations not only help refugee women gain practical skills and knowledge, but also provide a sense of community and empowerment. Moreover, the involvement of ex-pat women in these organizations highlights the potential for cross-cultural collaboration and solidarity, as well as the importance of leveraging diverse skills and resources for the greater good. Overall, these organizations demonstrate the power of grassroots initiatives to effect positive change, particularly in the context of complex and challenging humanitarian crises such as the Syrian refugee crisis. These organizations offer important support and resources to Syrian refugee women in Turkey and are examples of how ex-pat women can use their skills and resources to make a difference in their communities.

CONCLUSION

The Syrian civil war and the refugee crisis have highlighted the need to support and empower refugees, including women and children. Women-led organizations provide essential support for women who often face significant challenges such as language barriers, discrimination, and lack of access to education and employment opportunities. Through their programs and services, women-led organizations are providing a lifeline for vulnerable populations, helping them to rebuild their lives and contribute to their new communities.

The outstanding community-based initiatives run by women in Turkey, include SPI, ONAR İstanbul, MADRE, Knitstanbul, TIAFI, C@rma foundation as well as Jana impact.

The paper emphasizes the important role these organizations led by ex-pat women have played in supporting Syrian refugee women in Turkey.

One of the main challenges faced by Syrian refugee women in Turkey is accessing education and employment opportunities. Many of these women have had to leave school or work due to the conflict and have limited resources to pursue further education or vocational training. Women-led organizations such as SPI and Knitstanbul have set up programs to provide vocational training in fields such as textile production, sewing, and handicrafts. These initiatives not only provide skills that can lead to employment opportunities but also empower refugee women to take control of their lives and become more independent.

In addition to education and vocational training, women-led organizations in Turkey also provide psychological support and healthcare to Syrian refugee women. Many of these women have experienced trauma and loss as a result of the conflict and require counselling and mental health services to cope with their experiences. Organizations such as ONAR Istanbul and MADRE provide counselling and psychosocial support, while TIAFI offers healthcare services and access to medical professionals.

Furthermore, childcare is another major concern for Syrian refugee women in Turkey, as many women are unable to work or attend school due to the lack of childcare options. Women-led organizations such as TIAFI provide daycare services, allowing women to pursue education or work opportunities while ensuring that their children are in a safe and supportive environment.

Jana Impact and Knitistanbul are dedicated to promoting women's economic empowerment and creating sustainable solutions for women entrepreneurs.

Through their programs and services, these women-led organizations empower refugee women to take control of their lives and build a brighter future for themselves and their families. They provide essential support for vulnerable populations, helping them to rebuild their lives and become more independent. Overall, this article highlights the critical role that women-led organizations play in providing support and resources to Syrian refugee women in Turkey, and providing Syrian refugee women with the skills and resources they need to build a better future for themselves and their families in Turkey, thus promoting their well-being, agency, and active participation in society.

The work of women-led organizations in Turkey to support Syrian women refugees is commendable not only for the assistance they provide but also because they empower women from all around the world to take an active role in helping others. The involvement of women who are themselves refugees or members of marginalized communities is critical because they can better understand the needs and challenges faced by Syrian women refugees. These women also have unique insights into the cultural, social, and economic barriers that limit access to resources and services.

By empowering women to take an active role in supporting other women, these organizations are contributing to the larger struggle for gender equality and women's empowerment. Women's empowerment is essential for sustainable development and the achievement of gender equality. By supporting and empowering women, organizations can help women develop the skills and resources needed to participate fully in society and contribute to their communities.

Moreover, the work of these organizations challenges the traditional gender roles and expectations that have limited women's participation in public life. By providing opportunities for women to take leadership roles, these organizations are challenging the gendered power dynamics that have long dominated society. By working towards gender equality and women's empowerment, these organizations are not only making a difference in the lives of Syrian women refugees but also in the larger struggle for gender equality and social justice.

Overall, the efforts of these organizations demonstrate the power of community-based, women-led initiatives to create positive change in the lives of those most in need. As Turkey continues to host millions of Syrian refugees, it is vital that the government and larger aid organizations prioritize support for these grassroots efforts and work to ensure that the voices and experiences of refugee women are centred in all aspects of the response.

REFERENCES

Alawa, J., Zarei, P., & Khoshnood, K. (2019). Evaluating the provision of health services and barriers to treatment for chronic diseases among Syrian refugees in Turkey: A review of literature and stakeholder interviews. *International Journal of Environmental Research and Public Health, 16*(15), 2660. doi:10.3390/ijerph16152660 PMID:31349639

Alpak, G., Unal, A., Bulbul, F., Sagaltici, E., Bez, Y., Altindag, A., Dalkilic, A., & Savas, H. A. (2015). Post-traumatic stress disorder among Syrian refugees in Turkey: A cross-sectional study. *International Journal of Psychiatry in Clinical Practice, 19*(1), 45–50. doi:10.3109/13651501.2014.961930 PMID:25195765

CBI Ministry of Foreign Affairs. (2020). *The European market potential for SAVE tourism.* https://www.cbi.eu/market-information/tourism/save-tourism/market-potential

C@rma Foundation. (n.d.). Retrieved February 26, 2023, from https://www.4carma.com/en

Daldeniz, B., & Hampton, M. P. (2010). 3 VOLUNtourists versus volunTOURISTS. *Volunteer tourism: Theoretical frameworks and practical applications*, 30.

Dogutas, A. (2019). Gender Based Violence against Syrian Refugee Women in Turkey. *Border Crossing*, *9*(2), 113–124. doi:10.33182/bc.v9i2.811

Edmonds, K. (2013). Beyond good intentions: The structural limitations of NGOs in Haiti. *Critical Sociology*, *39*(3), 439–452. doi:10.1177/0896920512437053

Guttentag, D. A. (2009). The possible negative impacts of volunteer tourism. *International Journal of Hospitality & Tourism Administration*, *10*(4), 329–345.

ImpactJ. A. N. A. (n.d.). *Home*. Retrieved from https://jana-impact.org/

Internal Displacement Monitoring Centre. (2019). *Global Report on Internal Displacement 2019*. Norwegian Refugee Council.

Istanbul, O. N. A. R. (n.d.). *ONAR İstanbul Yabanci Uyruklu Kadinlari Destekleme Derneği (ONAR İstanbul).* Retrieved February 9, 2023, from https://www.4carma.com/en/ngo/ONARİstanbul

Istanbul & I. (n.d.). *Empowering Women in Istanbul*. Retrieved February 21, 2023, from https://istanbulandi.org.tr/

Jones, A. M. (2005). Assessing international youth service programmes in two low-income countries: Voluntary Action. *The Journal of the Institute for Volunteering Research*, *7*(2), 87–100.

Karam Foundation. (n.d.). *Karam Foundation*. Retrieved February 9, 2023, from https://karamfoundation.org/

Knitistanbul. (n.d.). *Our Team*. Retrieved February 9, 2023, from https://knitstanbul.com/about/team/

Lepp, A. (2008). Discovering self and discovering others through the Taita Discovery Centre volunteer tourism programme, Kenya. In S. Wearing & N. G. McGehee (Eds.), *International volunteer tourism: Integrating travelers and communities* (pp. 86–100). CABI. doi:10.1079/9781845933807.0086

Liu, T. M., & Leung, K. K. (2019). Volunteer tourism, endangered species conservation, and aboriginal culture shock. *Biodiversity and Conservation*, *28*(1), 115–129. doi:10.100710531-018-1639-2

Lyons, K. D., & Wearing, S. (2008). Volunteer tourism as alternative tourism: Journeys beyond otherness. In S. Wearing & N. G. McGehee (Eds.), *Journeys of discovery in volunteer tourism: International case study perspectives* (pp. 3–11). CABI. doi:10.1079/9781845933807.0003

MADRE. (n.d.). *Care In Crisis for Women and Girls Syria*. Retrieved February 9, 2023, from https://www.madre.org/projects/care-crisis-women-and-girls

Matthews, A. (2008). Negotiated selves: Exploring the impact of local-global interactions on young volunteer travellers. In K. Lyon & S. Wearing (Eds.), Journeys of discovery in volunteer tourism (pp. 101–117). CABI.

McGehee, N. G., & Andereck, K. (2008). 'Pettin'the critters': Exploring the complex relationship between volunteers and the voluntoured in McDowell County, West Virginia, USA, and Tijuana, Mexico. In Journeys of discovery in volunteer tourism: International case study perspectives (pp. 12–24). CABI.

McGehee, N. G., & Andereck, K. L. (2004). Factors predicting rural residents' support of tourism. *Journal of Travel Research*, *43*(2), 131–140. doi:10.1177/0047287504268234

McGehee, N. G., & Santos, C. A. (2005). Social change, discourse, and volunteer tourism. *Annals of Tourism Research*, *32*(3), 760–779. doi:10.1016/j.annals.2004.12.002

Mostafanezhad, M. (2013). The politics of aesthetics in volunteer tourism. *Annals of Tourism Research*, *43*, 150–169. doi:10.1016/j.annals.2013.05.002

Orams, M. B. (1995). Towards a more desirable form of ecotourism. *Tourism Management*, *16*(1), 3–8. doi:10.1016/0261-5177(94)00001-Q

Sagır, Z. (2018). *Cultural adaptation, mental health, and religion among Syrian women refugees* [Suriyeli kadın mültecilerde kültürel uyum, ruh sağlığı ve din] [Unpublished PhD dissertation]. Istanbul University, Istanbul, Turkey.

Şimşek, Z., Yentur Doni, N., Gül Hilali, N., & Yildirimkaya, G. (2018). A community-based survey on Syrian refugee women's health and its predictors in Şanliurfa, Turkey. *Women & Health*, *58*(6), 617–631. doi:10.1080/03630242.2017.1321609 PMID:28430082

Small Projects Istanbul. (n.d.). *Community development*. Retrieved February 9, 2023, from https://www.smallprojectsistanbul.org/community-development/

Spencer, D. (2015). 'To Protect Her Honour': Child Marriage in Emergencies – the Fatal Confusion between Protecting Girls and Sexual Violence. *CARE Gender and Protection in Humanitarian Contexts*. https://insights.careinternational.org.uk/media/k2/attachments/CARE_Child-marriage-in-emergencies_2015.pdf

Team International Assistance for Integration-TIAFI. (n.d.). *About TIAFI*. Retrieved February 9, 2023, from https://tiafi.org/

The Norwegian Refugee Council. (2019). *Left Behind: The Education Crisis for Syrian Refugee Children in Turkey*. Retrieved from https://www.nrc.no/resources/reports/left-behind-the-education-crisis-for-syrian-refugee-children-in-turkey/

Tomazos, K., & Butler, R. (2009). Volunteer tourism: Working on holiday or playing at work? *Tourismos*, *4*(4), 331–349.

Turkey, U. N. H. C. R. (2022). *UNHCR Turkey Fact Sheet - February 2022*. ReliefWeb. Retrieved February 24, 2023, from https://reliefweb.int/report/turkey/unhcr-turkey-fact-sheet-february-2022-entr

UNHCR. (2005). *Handbook of self-reliance*. Retrieved from https://www.unhcr.org/44bf7b012.pdf

UNHCR. (2021a). *Figures at a Glance*. UNHCR. https://www.unhcr.org/figures-at-a-glance.html

UNHCR. (2021b). *Syria Regional Refugee Response*. Retrieved February 24, 2023, from https://data2.unhcr.org/en/situations/syria/location/113

UNHCR. (n.d.). *How volunteers help refugees*. Retrieved from https://www.unhcr.org/en-my/how-volunteers-help-refugees.html

United Nations Development Programme. (n.d.). *Sustainable Development Goals*. https://www.undp.org/content/undp/en/home/sustainable-development-goals.html

United Nations Volunteers. (2018). *State of the World's Volunteerism Report 2018: The Thread that Binds*. Retrieved from https://unv-swvr2018.org/files/51692_UNV_SWVR_2018_WEB_OVERVIEW.pdf

Waterman, A. S. (2011). Eudaimonic identity theory: Identity as self-discovery. In S. J. Schwartz, K. Luyckx, & V. L. Vignoles (Eds.), Handbook of identity theory and research (pp. 357–379). Springer Science + Business Media.

Wearing, S. (2001). *Volunteer tourism: Experiences that make a difference*. CABI. doi:10.1079/9780851995335.0000

Wearing, S., Deville, A., & Lyons, K. (2008). The volunteer's journey through leisure into the self. In S. Wearing & N. G. McGehee (Eds.), *Journeys of discovery in volunteer tourism: International case study perspectives* (pp. 65–71). CABI. doi:10.1079/9781845933807.0063

Wearing, S., & McGehee, N. G. (2013). Volunteer tourism: A review. *Tourism Management*, *38*, 120–130. doi:10.1016/j.tourman.2013.03.002

Women, U. N. (2017). *Syrian Women in Turkey: Investing in the Untapped Potential. Economic empowerment*. Available at: https://eca.unwomen.org/en/digital-library/publications/2017/06/syrian-women-in-turkey-investing-in-the-untapped-potential

Women, U. N. (2018). *Needs assessment of Syrian women and girls under temporary protection status in Turkey*. Retrieved from: https://www2.unwomen.org/-/media/field%20office%20eca/attachments/publications/country/turkey/the%20needs%20assessmentengwebcompressed.pdf?la=en&vs=3139

Women Inspiring Change. (2020). *Karyn Thomas*. Available at: https://orgs.law.harvard.edu/womeninspiringchange/2016-honorees/karyn-thomas/

Yılmaz, E., Usta, J., Akyüz, A., & Özbay, F. (2018). Livelihoods and Coping Strategies of Syrian Women Refugees in Turkey: A Gender-Sensitive Analysis. *Journal of International Women's Studies*, *19*(4), 27–44.

Chapter 29
Interest of Women in the Tourism Sector and How to Empower Them

Areeba Ashfaq
Mohammad Ali Jinnah University, Pakistan

Atiqa Asif
Mohammad Ali Jinnah University, Pakistan

Hawa Imran
Mohammad Ali Jinnah University, Pakistan

Maryum Ikhlaque
Mohammad Ali Jinnah University, Pakistan

Neha Nehal
Mohammad Ali Jinnah University, Pakistan

Naveeda Katper
University of Sindh, Jamshoro, Pakistan

Mahrukh Shamim
Mohammad Ali Jinnah University, Pakistan

ABSTRACT

This study aims to determine the interest of women working in the tourism industry and their empowerment in the industry through inclusion and supporting entrepreneurial activities and to study the different perceptions and beliefs that women have about working in the tourism industry through assessment of their feelings toward the working environment in this industry. The main focus of the study was two major areas: (1) interest of women in employment tourism industry and (2) interest of women in entrepreneurial activities in the tourism industry. This study also considers whether the uncertainties of the tourism industry and its inclusion practices deform the interest of women to work in the tourism industry. Data were collected from 20 female respondents representing the interest of women to work in the tourism industry. The results indicate a positive response of 70.6% female respondents, suggesting their interest in the tourism industry, while 29.4% of the female respondents showed lack of interest in the industry.

DOI: 10.4018/978-1-6684-8417-3.ch029

1. INTRODUCTION

Tourism industry is one of the highest growing industries in the world. From entrepreneurial perspective tourism industry has so much potential for nascent entrepreneurs yielding greater employment opportunities for both men and women and greater contribution in country's economic output (Memon, & Tunio, 2023; Sultan, & Tunio, 2022; Abdullah, et al., 2020). The generation of employment for women in the tourism industry is one of the most important factors for innovation and rapid growth of the industry (Afshan, et al., 2021; Alshareef, et al., 2022; Aurangzeb, et al., 2021). According to United nations world Tourism organization, women contribute to the majority of the tourism workforce, in most parts of the world (UNWTO, 2022). Women make up the 54% of the total tourism workforce all around the world (UNWTO, 2022). But only 20% of that group hold the management positions due to the gender discrimination (Tunio, et al., 2022; Chaudhry, et al., 2021; Gilal, et al., 2021; Tunio, et al., 2021). The inclusion of women in the tourism can reduce the gender inequality by substantial amount. As a part of society, it's the very right of every woman to actively participate in the tourism activities (Memon, et al., 2021; Tunio, et al., 2021; Mushtaq, et al., 2021; Tunio, et al., 2021). Both men and women are the resources to develop the tourism sector and its is necessary that the two should have equal job opportunities (Shaikh, and Tunio, 2020; Tunio and Shaikh, 2020; Shaikh, et al., 2021; Tunio, et al., 2021). The inequality and lack of inclusion of women leads to underappreciation of women and hence cause them to have low motivation and productivity levels. This inequality results in wage gap which causes Women to earn 14.7 per cent less than men in tourism sector (UNWTO, 2022).

Tourism is a tool for economic, social and political empowerment of women. It brings important employment opportunities to women and in most of the case income generation either as the main source of income or as the additional side income (Shaikh, et al., 2021; Tunio, et al., 2021; Shaikh, et al., 2022). These jobs are often flexible and can be done from the workplace, household or communities (Katpar., et al., 2020; Shaikh, et al., 2022; Tunio; 2020). Tourism can offer opportunities for women to run their own businesses (Jucan, 2013). The inclusion and equal opportunities for women make them feel empowered to work in any industry of workplace (Sultan, et al., 2022; Tunio, 2020; Katper, et al., 2017). The gender diversity leads to innovation and greater flexibility of workforce of workforce. Inclusion of women lead to their empowerment which causes them to start their own venture in the tourism industry.

This paper focuses mainly on the analysis of women in tourism from two major perspectives (i) Interest of women in Employment tourism industry; (ii)interest of women in entrepreneurial activities in the tourism industry. We have also studied the various factors which distorts the interest of women in tourism industry. These factors include the inequality at workplace, inclusion of women in the workplace, gender diversity, discrimination faced by women entrepreneurs, work timings and family obligations.

2. RESEARCH QUESTIONS

In order to analyse the women's interest in working in the tourism industry this paper examines the following research questions:

1) How much women are interested to work in the tourism industry?
2) Are women willing to participate in entrepreneurial activities/ start their own venture in the tourism industry?

3) What are the factors that deforms the interest of the women to work in the tourism industry?

3. RESEARCH METHODOLOGY

To explore the women's interest in working in tourism industry, we used a Qualitative interview design. The interviewers reported their feelings and beliefs related to working in the tourism industry based on their working experiences or prior knowledge of the tourism industry. The questionnaire consisted of eight questions which was returned by 25 female respondents with work experience and knowledge regarding the tourism industry.

4. LITERATURE REVIEW

4.1 Role of Women in the Tourism Industry

The role of women in tourism industry and regulation is also emphasized in the Regulation of the Minister of Culture and Tourism of 2011 which suggests that the women community is mostly involved in all the activities from the planning, implementation, supervision, maintenance and utilization processes, which results in broad opportunities for active participation from women. (Ismi, 2022) On the basis of some evidence, tourism sector is segmented both horizontally and vertically along the gender line (Sinclair and Stabler, 1997; Sultan, et al., 2022).

4.2 Workplace Inclusion of Women in the Tourism Sector

Although, there are some evidences that suggests some jobs are disproportionately taken by women which is referred to as "occupational crowding" which is due to differences across the occupations of employer's perception of discrimination has been explained as the result of differences across occupations in employers' or consumers' taste for discrimination (Neumark, 1996; Tunio, et al., 2017; Gilal. et al., 2020).Women experience barriers in almost every aspect of work – including whether tey have paid work at all, the type of work they obtain or are excluded from, the availability of support services such as childcare, their pay, benefits and conditions of work, the insecurity of their jobs or enterprises, (ILO, 2009), their access to vocational training (Kabeer, 2008; Tunio, et al., 2020). Human resource departments must concentrate on how to add value to the department, coordinate, enhance their effectiveness, and create new skills in order to successfully face new challenges. The higher the degree of administrative expert role cognition of human resource practitioners, the more positive the impact on job performance (Tunio, et al., 2021). The findings indicated that human resource administrative expert role is positively related to the job performance and, further, achievement motivation exists mediating effect and affection is completely mediating effect between human resource administrative expert and the job performance. (Sultan, et al., 2023). Individual job commitment is driven by achievement motivation, which is a vital internal driving force. We may assess candidates' achievement motivation through personality tests during the recruiting and selection phase in business practice. Employees with a higher level of achievement motivation will predict better job performance in the future than those with a lower level of achievement motivation.

4.3 Entrepreneur Women in the Tourism Industry

Entrepreneur's context acts as an "asset" by providing the opportunities to entrepreneurs and also "liability" by setting actions under boundaries (Welter, 2011). Women entrepreneurs also face difficulties in building and managing social networks than their male counterparts (Benschop, 2009). The context constitutes culture, history, social networks, skills and abilities, technology and education as few important elements that set conditions for entrepreneurs. Additionally, entrepreneur's perception including gender differences are largely defined by the context. When it comes to the job market, women have to face socio-cultural barriers besides economic inequality, discrimination, work–life imbalance and harassment. (Katper, et al., 2017). In this regard, several independent bodies have been working, such as the small and medium enterprise development authority (SMEDA), women chamber of commerce and industry (WCCI) and first women bank (FWB). These regulatory bodies help women entrepreneurs' access microfinance, exploit market opportunities and administrative support to run the business successfully (Gul, et al., 2021).

The graduates of universities from business administration, information technology, and other disciplines are found in the entry registers for entrepreneurship. This tract of technological progress can be a turning point for the youth in Pakistan to transform unemployment into self-employment and for more opportunities (Sagath, et al., 2019;)

4.4 Empowerment of Women in the Tourism Industry

Women's empowerment refers to the ability of women to transform economic and social development when empowered to fully participate in the decisions that affect their lives through leadership training, coaching, consulting, and the provision of enabling tools for women to lead within their communities, regions, and countries (Wiser Earth, 2005). Aspects of limiting the roles of women and men in tourism which include access, participation, control, and benefits, are manifestations of gender justice (Najih MA.,2017).

As noted above, policy interventions aimed at fostering female participation in employment, management and ownership of tourist activities are expected to a better chance of success (Ferguson, 2011; Tunio, 2020).

4.5 Survey

Name
Age
1. Do you work well with people from different backgrounds and diverse personalities
Yes
No
2. Can you tackle challenges on daily basis
Yes
No
3. Are you interested to work in tourism industry
Yes
No

4. Reason to choose yes
Travelling
Good pay/benefits
Like working with people
All of the above
None of the above(choose this option if you've selected no as your answer in question 3)
5. Reason to choose no
Family restrictions
Lack of interest in the field
Can't travel for work
All of the above
None of the above(choose this option if you've selected yes as your answer in question 3)
6. Will you be willing to learn about different topics to be a better tour guide to customers
Yes
No
7. Do you think women will face higher level of gender discrimination if they pursue tourism as a career as compared to other fields
Yes
No
8. which issue affecting women in tourism world you say is most pressing
Flexible work hours
Child Care
Remuneration
Gender Discrimination

FINDINGS AND RESULTS

A quantitative survey was conducted where 20 respondents participated. The survey consisted of 8 multiple choice questions. The participants' age ranged from 14-40 with the median age being 22. The first questions were to find if surveyed individuals can work with people with diverse backgrounds or can tackle challenges on daily basis. Most of the answers were yes. Followed by the question which was about if women are interested in working in the tourism sector, 70.6% of responses were positive with followed by 29.4% responses being negative. The biggest reason behind those positive answers was travelling, which weighed around 45%, followed by pay/benefits at 12%. For the respondents who chose no as their answer to the third question, the main reasons are family restrictions 20% and lack of interest being 25%. 85% of surveyed individuals, when presented with the question, if they think the gender discrimination would be higher as compared to other fields, said yes. The last question which asked about issues affecting women in the tourism world, 40% of people selected gender discrimination as the main issue, followed by flexible work hours at 50% and remuneration and childcare being at 5%.

Graphical Representation of Data:

Figure 1.

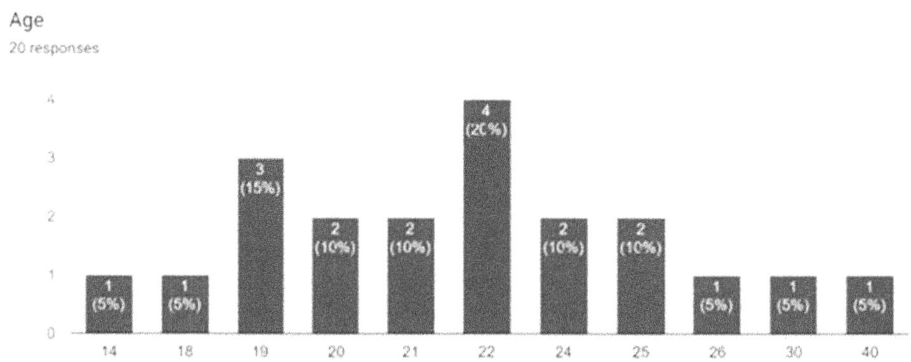

Figure 2.

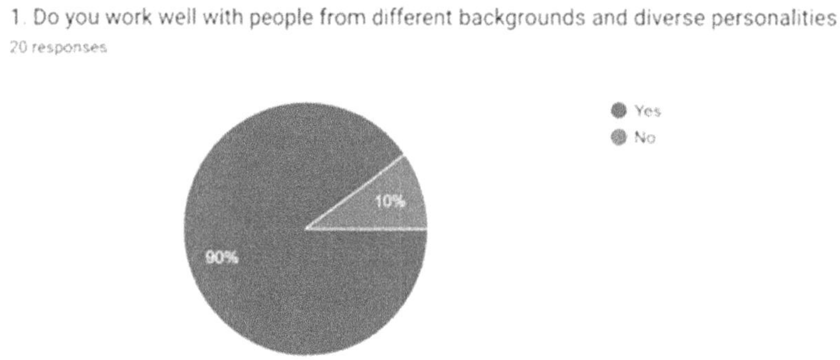

Figure 3.

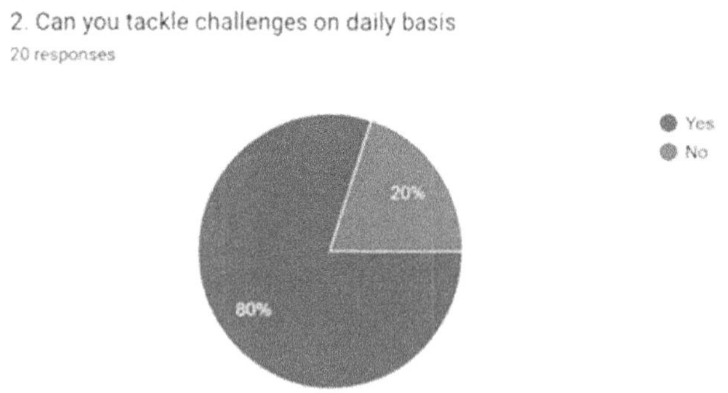

Figure 4.

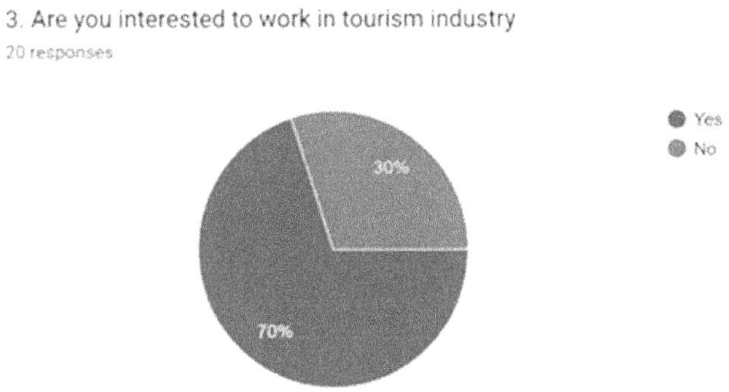

Figure 5.

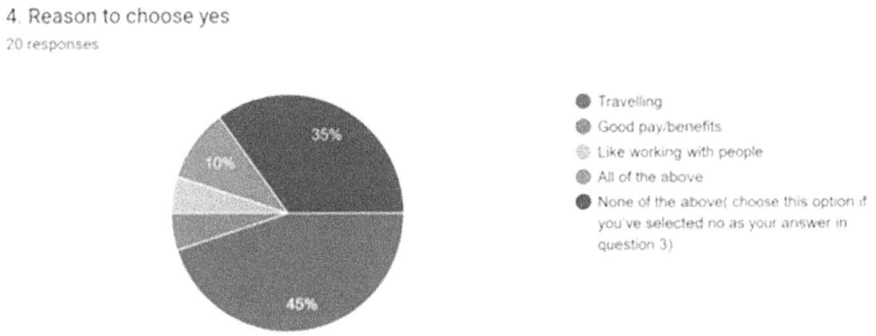

Figure 6.

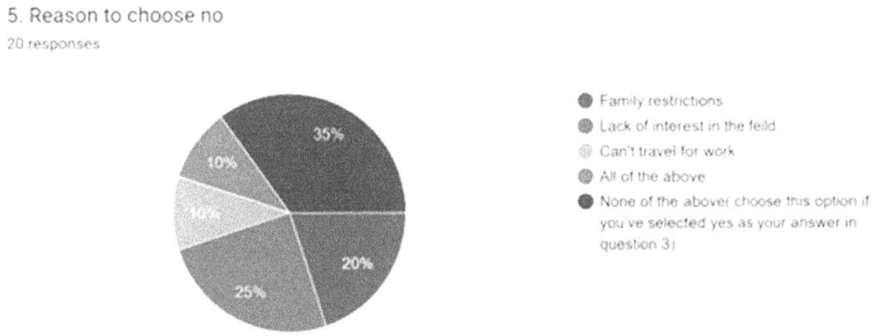

Figure 7.

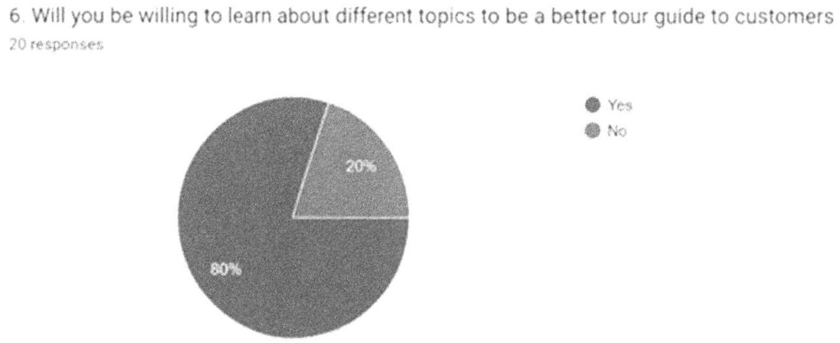

Figure 8.

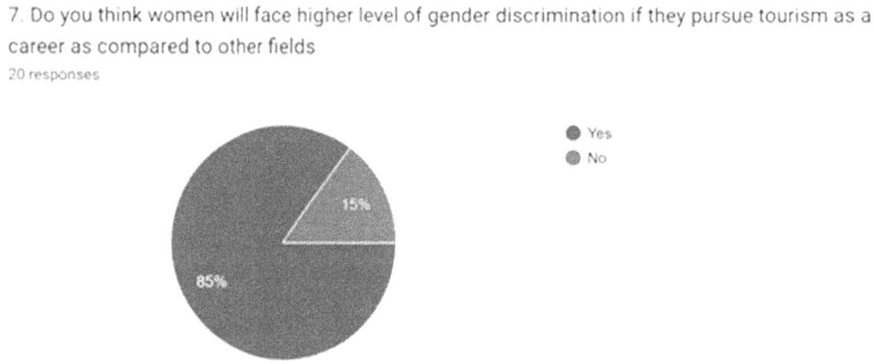

Figure 9.

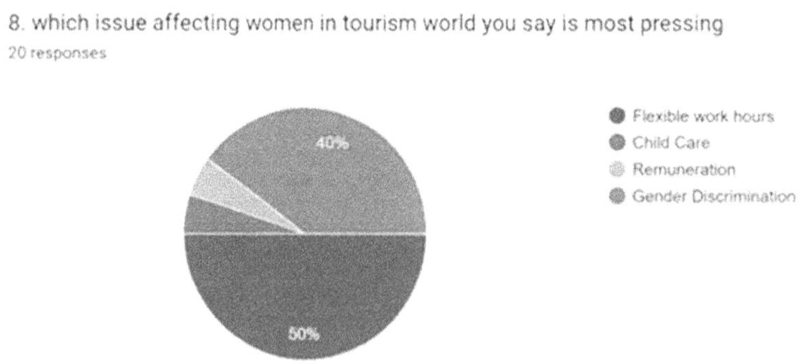

4.6 Woman Empowerment and Tourism

Tourism play an important role in achieving the objectives and goals to empower woman of sustainable development. Tourism is considered as a significant sector to empower woman in the developing world. It create employment opportunities for the woman, reduce poetry in country, foreign investments, inflow of foreign exchange earnings and financial improvements. Tourism is documented in some financial sectors. The specialized UN agency is UNWTO (United Nation World Tour Organization) to enhance the positive impact of tourism in development in woman lives. They address some issue. Tourism has its potential to contribute in growth of economy and the empowerment of woman. If woman were to be given the opportunity to work in the tourism industry they will obtain higher managerial posts to access to economic resources and have opportunity to participate in decision making as a result to gain employment. Through tourism woman are given the opportunity to use their natural and cultural resources as well as their knowledge and skills. Tourism is the primary source of income in developing countries. Women mainly get the opportunities like to visit the destination, run by local communities, the involvement in tourism projects and tour guides.

4.7 Women's Empowerment Through Tourism

After analyzing the academic literature on women empowerment in tourism, it is also useful to look at the more practical implications of it. During a conference in Berlin in March 2008 the United Nations World Travel Organization (UNWTO) implemented an Action Plan to Empower Women through Tourism. The objectives of the plan are derived from the United Nations Millennium Development Goals, which aims are to benefit poor (also through tourism), protect the environment and empower women. Tourism is an important part of world's GDP (more than 10%) and creates around 8% of world's employment. Of all people employed in the tourism sector 60-70% are women . There are predictions of huge annual growth in total demand coming year. It offers many opportunities to both developed and developing countries in rural, urban, regional or cultural areas. Tourism development can be a great incentive for development and thus for the empowerment of women, which makes the plan of the WTO a reasonable motivation to empower women through this sector. The UN defines empowerment as a critical aspect of gender equality, and defines both concepts as following (UNFPA, 2008): Gender equality implies a society in which women and men enjoy the same opportunities, outcomes, rights and obligations in all spheres of life. Equality between men and women exists when both sexes are able to share equally in the distribution of power and influence; have equal opportunities for financial independence through work or through setting 3 12 up businesses; enjoy equal access to education and the opportunity to develop personal ambitions. A critical aspect of promoting gender equality is the empowerment of women, with a focus on identifying and redressing power imbalances and giving women more autonomy to manage their own lives. Women's empowerment is vital to sustainable development and the realization of human rights for all.'

Tourism is seen as one of many tools that can contribute to the achievement of the MDG's. The strength of tourism is that it can help to improve livelihoods by alleviating poverty, ensure or improve environmental sustainability, and promote gender equality and empower women. There is no region in the developing world where women seem to be equal to men in terms of legal, social and economics' rights. This gender gap influences the access to and control of resources, political voice, in economic opportunities, and power relations. Ultimately, the costs of these inequalities cut broadly across the society and ultimately harm everyone. Increasing power and control over natural resources. Women in developing nations are usually in charge of securing water, food and fuel and of overseeing family health and diet. For this reason, they tend to put in practice whatever they have learned about preserving the environment and natural resources. Tourism development could be a tool to introduce environmental conscious practices within livelihoods by stressing the importance of conservation and sustainable practices within the environment, so that it can function as any kind of tourism supply and thus as a new income generator. Further, by learning and improving techniques of crop planting and other types of production, women can become suppliers within the value chain production for the tourism market. - Economic empowerment: More women than men live in poverty (UNFPA, 2008). Much work within families and communities is unpaid and often falls on the shoulders of women. These women have to deal with discrimination in the economic sphere every day, or are struggling for some economic independence. Narrowing the gap between men and women in the economic sphere, for instance 'We know that poverty is not just about lack of money; it is also about lack of choice. This is particularly true for women. Today, many women cannot make their own choices about pregnancy and childbearing, they cannot make their own choices about seeking medical care. These choices are made for them and, in the worst cases, there simply are no choices. Tourism can function as an instrument for economic empowerment by opening up new markets for small and medium enterprises. Educational empowerment: About two thirds of the illiterate adults in the world are female (UN, 2008). The level of education of women is strongly associated with levels of infant mortality and fertility, as well as the level of education and the economic opportunities for their offspring. Tourism can be used as a sector where women from all hierarchies in society, with all types of educational background, can be trained and educated for certain skills within the tourism industry. Women can engage in different forms of creative entrepreneurship which gives women a whole range of skills, like speaking different languages, hospitality skills, guiding tourists, managing project, and so on. Political empowerment, the equality of women to men is still not guaranteed in many social and legal institutions. This influences the quality and implementation of their basic human rights, the access and/or control over land, in employment and earnings, and in participation in decision making processes in general. During the implementation or management of tourism projects this problem could be acknowledged by including women in the planning and organizational process, with the aim of giving them a voice in the process. Tourism could empower women by giving them more power and control over what happens to their community, their livelihood.

4.8 How to Empower Women in the Tourism Industry

Woman in leadership position is the growing evidence on tourism industry.as the gender gap ay, the tourism sector got fewer woman in management and executive position. We have to inspired woman as:

- They have to provide them the fairly work, respecting and supporting nondiscrimination human rights.
- The tourism industry provide every woman their safety, health and well-being.
- To empower they have to provide tourism education, training and professional development for woman.
- Implement employment opportunities supply chain, management, marketing, sales and enterprise development that empower woman.
- Support and empower woman that will benefit to your tourism sector.

- The tourism sector has to empower woman by giving house, medical and transport facilities.

4.9 Economic Empowerment in the Tourism Industry

The issue of female economic empowerment in the Middle East and the North Africa region (MENA) has attracted significant attention; however, its potential in the tourism sector has not been fully explored. This study aims to identify the steps involved in conceptualizing, designing, and conducting a customized tourism training program to economically empower women in the United Arab Emirates (UAE).

4.10 Status of Women in the Tourism Sector

Currently, although women represent 55% of the workforce in the tourism sector, only 20% of women hold management positions. Even more striking is the fact that only 8% hold positions of high responsibility. However, it is women who are the decision-makers in 70% of cases when booking a holiday.

In June 2020, pilot project was launched supporting tourism professionals in Gilgit-Baltistan as they navigated impacts of the COVID-19 pandemic. Over 190 professionals from 12 tourism hotspots were trained.

For Root Network's next project, the lack of women's representation in Pakistan's tourism industry while responding to needs identified by the community during the training sessions, including:

- Industry-specific training
- Support in accessing larger market base
- Financial support
- Networking support from experienced professionals

The Women in Tourism Fellowship will build off skills and knowledge of new female graduates and young industry professionals to further support their goals within the sector. We will collaborate with local partners to conduct multi-week tourism trainings to ensure women are well-prepared to succeed in the tourism industry, and mentorship opportunities to establish support networks as women grow in their careers.

4.11 Women's Self-Employment in Tourism

Women's entrepreneurship is found as an opportunity for economic empowerment.

Involvement of women in entrepreneurship is strongly connected with improving the status of women, family and community well-being.

Conventional entrepreneurship refers to innovations, new business venturing and models as strong commercial objectives. On the other side, social entrepreneurship is more ethical in entrepreneurial activity, with a precise social change outline, influenced by social and environmental values.

The factors that influence development of woman in entrepreneurship in tourism sector.

1. It improve the financial condition, personal interest, earning money, deliberately selfemployment, knowledge, self-employment and confidence.
2. Comprise of higher demand, infrastructure, leadership, low investment and easy for marketing.

3. It contains social projection, degree of social acceptance and respect from society.

4.12 Empowering Through Culture Tourism

Cultural heritage is widely recognized as a significant resource for heritage tourism. Integration of cultural heritage into tourism development could strengthen the authenticity of tourism offerings and facilitate a deeper understanding of a destination's culture, thereby enhancing attractiveness and competitiveness. However, it has also been argued that heritage tourism brings changes to the meaning of cultural heritage as it comes to be treated as a commodity for touristic consumption. To ensure authenticity in the process of tourism development, it is strongly advised that the cultural and socio-economic values of heritage should be safeguarded when it is used in tourism. Considering the evolving nature of culture itself, called for a more open approach to authenticity with the understanding that heritage tourism supports the continuous revival and representation of the past in the present with selection and imagination.

Much intangible cultural heritage has been practiced and transmitted by communities for generations and continuously contributes to their vitality and wellbeing .It is incomplete if only the material format is presented for the community has played an important role in sustaining the values of such heritage, and this should also be the case in its use for tourism . As such, use of cultural heritage for tourism inevitably also generates impacts for local communities, for good and ill, particularly for the practitioners of such heritage . A community's intention for the tourism use of cultural heritage should be understood, respected, and reflected in the tourism development process and community control and benefit acquisition from tourism should be safeguarded as residents' perspectives and involvement help to foster greater sustainability and cultural integrity for heritage tourism.

Traditional crafts and artifacts, such as weaving, carving, embroidery and mask making, are common types of cultural heritage that are widely used in tourism When modified to showcase local culture and serve as souvenirs for tourists, local arts and crafts become the material embodiment of tourist experience and support cross-cultural understanding and meaning making . Changes in the presentation, the design, the meaning, the techniques and the quality of traditional crafts and artifacts have been widely documented when adapting them to tourist needs . For example, in the case of Maori carvings in New Zealand, Ryan and Crotts (1997) documented changes in the use of paint, carving techniques and production mode to reduce the cost to fit tourists' willingness to pay. Changes in the design and product development to suit tourists' tastes and needs for modern daily use have also been widely recognized, such as new designs of Miao and Naxi ethnic embroidery and silver ornaments in south-western China and weaving products in Chinchero, Peru.

CONCLUSION

Viewing the status quo of women in the tourism industry, it is important that women should be aware of both the positives and negatives of the profession by educating themselves about other people with experience. To truly bring about a change, challenges that the tourism industry might face should be predicted and solutions should be created in foresight. This might not only help private sectors around the world but also public sectors, international organizations, and civil societies.

REFERENCES

Abdullah, M., Katper, N. K., Chaudhry, N. I., & Tunio, M. N. (2020). An Impact of Workaholics on Creativity: The mediating role of Negative Mood and moderating role of Supervisor Support. *Sukkur IBA Journal of Management and Business, 7*(2).

Afshan, G., Ilyas, S., Tunio, M. N., & Kalhoro, M. (2021). CSR actions and post-COVID-19 consequences in the hotel industry: A conceptual framework. *International Journal of Strategic Change Management, 7*(4), 275–289. doi:10.1504/IJSCM.2021.122845

Alshareef, N., & Tunio, M. N. (2022). Role of Leadership in Adoption of Blockchain Technology in Small and Medium Enterprises in Saudi Arabia. *Frontiers in Psychology, 13*, 2284. doi:10.3389/fpsyg.2022.911432

Aurangzeb, M. T., Tunio, M. N., Rehman, Z., & Asif, M. (2021). Influence of administrative expertise on human resources practitioners on the job performance: Mediating role of achievement motivation. *International Journal of Management, 12*(4), 408–421.

Chaudhry, I. S., Paquibut, R. Y. & Tunio, M. N. (2021). Do workforce diversity, inclusion practices, & organizational characteristics contribute to organizational innovation? Evidence from the U.A.E. *Cogent Business & Management, 8*(1), 1947549. . doi:10.1080/23311975.2021.1947549

Gilal, F. G., Gilal, N. G., Channa, N. A., Gilal, R. A., Gilal, R. G., & Tunio, M. N. (2020). Towards an integrated model for the transference of environmental responsibility. *Business Strategy and the Environment, 29*(6), 1–10. doi:10.1002/bse.2524

Gilal, F. G., Gilal, N. G., Gilal, R. G., Gon, Z., Gilal, W. G., & Tunio, M. N. (2021). The Ties That Bind: Do Brand Attachment and Brand Passion Translate Into Consumer Purchase Intention? *Central European Management Journal, 29*(1), 14–38. doi:10.7206/cemj.2658-0845.39

Gul, A., Subhan, S., & Tunio, M. N. (2021). Learning experiences of women entrepreneurs amidst COVID-19. *International Journal of Gender and Entrepreneurship, 13*(2), 1756–6266. doi:10.1108/IJGE-09-2020-0153

Katpar., N. K., Chaudhry, N. I., Tunio, M. N., & Ali, M. A. (2020). Impact of Leadership Style and Organizational Culture on Organizational Commitment. *Sukkur IBA Journal of Management and Business, 7*(1), 92-106.

Katper, N. K. Medan, A., Syed, K. B. S., & Tunio, M. N. (2017). Determinants of Debt Maturity Structure in Shariah and Non-Shariah Firms in Pakistan: A comparative Study. *Journal of Applied Economic Sciences, 12*(4), 1210-1225.

Memon, A.B., Meyer, K., & Tunio, M. N. (2021). Toward collaborative networking among innovation laboratories: a conceptual framework. *International Journal of Innovation Science.* doi:10.1108/IJIS-04-2021-0069

Memon, A. B., & Tunio, M. N. (2023). Assessing the Potential of ICT Education at Secondary Schools in Pakistan: A Comparative Study. *Pakistan Journal of Educational Research and Evaluation, 10*(2), 16–33.

Mushtaq, T., Tunio, M. N., Akbar, Z., & Jariko, M. (2021). Green Organizational identity: Antecents and consequences: An Emprical Study. *Contemporary Issues in business and Government, 27*(3), 2056-2069.

Najih, M.A. (2017). Gender dan kemajuan teknologi pemberdayaan perempuan pendidikan dan Keluarga. *HARKAT: Media Komunikasi Islam Tentang Gebder dan Anak, 12*(2), 1826.

Shaikh, E., & Tunio, M. N. (2020). Customer satisfaction and Customer loyalty: An empirical case study on the impact of benefits generated through Smartphone applications. *International Journal of Public Sector Performance Management.*

Shaikh, E., Tunio, M. N., Khoso, W. M., Brahmi, M., & Rasool, S. (2022). The COVID-19 Pandemic Overlaps Entrepreneurial Activities and Triggered New Challenges: A Review Study. *Managing Human Resources in SMEs and Start-ups: International Challenges and Solutions*, 155-182.

Shaikh, E., Tunio, M. N., & Qureshi, F. (2021). Finance and women's entrepreneurship in DETEs: A literature review. Entrepreneurial Finance. *Innovation and Development*, 191–209.

Shaikh, E., Watto, W. A., & Tunio, M. N. (2022). Impact of Authentic Leadership on Organizational Citizenship Behavior by Using The Mediating Effect of Psychological Ownership. *ETIKONOMI, 21*(1), 89–102. doi:10.15408/etk.v21i1.18968

Shaikh, S., Sultan, M. F., Mushtaque, T., & Tunio, M. N. (2021). Impact of COVID-19 on GDP: A serial mediation effect on international tourism and hospitality. *International Journal of Management, 12*(84), 422–430.

Sultan, M. F., Shaikh, S. K., Tunio, M. N., & Asim, M. (2022). Sustainable Leadership: A Divine Form to Lead and Succeed in a Holistic Way. In Sustainable Development of Human Resources in a Globalization Period (pp. 111-118). IGI Global. doi:10.4018/978-1-6684-4981-3.ch007

Sultan, M. F., & Tunio, M. N. (2022). Gauging the COVID-19 on Academic Stress on Tertiary Level Female Students on the eve of COVID-19. *KASBIT Business Journal*, *15*(3), 1–10.

Sultan, M. F., Tunio, M. N., Aziz, A., & Shaikh, S. K. (2023). Social Innovation and Social Entrepreneurship in the Wake of COVID-19: A Perspective From the Developing Side of the World. In Frugal Innovation and Social Transitions in the Digital Era (pp. 113-118). IGI Global.

Tunio, M. N. (2020) Role of ICT in Promoting Entrepreneurial Ecosystems in Pakistan. *Journal of Business Ecosystems, 1*(2).

Tunio, M. N. (2020). Role of ICT in promoting entrepreneurial ecosystems in Pakistan. *Journal of Business Ecosystems*, *1*(2), 1–21. doi:10.4018/JBE.2020070101

Tunio, M. N., Chaudhry, I. S., Mughal, F., & Shaikh, E. (2022). Marketing Mode and Survival of the Entrepreneurial Activities of Nascent Entrepreneurs. In Big Data Analytics (pp. 1-18). Auerbach Publications. doi:10.1201/9781003307761-1

Tunio, M. N., Chaudhry, I. S., Shaikh, S., Jariko, M. A., & Brahmi, M. (2021). Determinants of the Sustainable Entrepreneurial Engagement of Youth in Developing Country—An Empirical Evidence from Pakistan. *Sustainability (Basel)*, *13*(14), 7764. doi:10.3390u13147764

Tunio, M. N., Jariko, M. A., Børsen, T., Shaikh, S., Mushtaque, T., & Brahmi, M. (2021). How Entrepreneurship Sustains Barriers in the Entrepreneurial Process—A Lesson from a Developing Nation. *Sustainability (Basel), 13*(20), 1–18. doi:10.3390u132011419

Tunio, M. N., & Shaikh, E. (2020). Nascent entrepreneurs and challenges in digital market in developing countries. *International Journal of Public Sector Performance Management.*

Tunio, M. N., Shaikh, E., Niaz, S., & Katper, N. S. (2021). Multifaceted perils of the Covid-19 and implications: A Review. *Estudios de Economía Aplicada*. Advance online publication. doi:10.25115/eea.v39i2.3957

Tunio, M. N., Soomro, A. A., & Bogenhold, D. (2017). The Study of Self-employment at SMEs Level with Reference to Poverty in Developing Countries. Business and Management Research, 6(2). doi:10.5430/bmr.v6n2p33

Tunio, M. N., Yusrini, L., Shah, Z. A., Katper, N., & Jariko, M. A. (2021). How Hotel Industry Cope up with the COVID-19: An SME Perspective. *Etikonomi*, *20*(2), 213–224. doi:10.15408/etk.v20i2.19172

Tunio, M. N., Yusrini, L., Shah, Z. A., Katper, N., & Jariko, M. L. (2021). How Hotel Industry Copes Up with the COVID-19: An SME Perspective. *Etikonomi, 20*(2), 213 – 224.

Tunio, M. N., Yusrini, L., & Shoukat, G. (2021). Corporate social responsibility (CSR) in Hotels in Austria, Pakistan, and Indonesia: small and medium Enterprise spillover of COVID-19. In Handbook of research on entrepreneurship, innovation, sustainability, and ICTs in the post-COVID-19 era (pp. 263-280). IGI Global.

Chapter 30
Women Empowerment in Human Resource Management

Balaach Kamran
Mohammad Ali Jinnah University, Pakistan

Osama Anwar
Mohammad Ali Jinnah University, Pakistan

Syed Asad Zaidi
Mohammad Ali Jinnah University, Pakistan

Hafiz Qasim Naseer
Mohammad Ali Jinnah University, Pakistan

Saad Yousaf
Mohammad Ali Jinnah University, Pakistan

Muhammad Asif Qureshi
Mohammad Ali Jinnah University, Pakistan

ABSTRACT

Researchers all over the globe have showed increased research attention in management literature towards discrimination at the workplace. The research has revealed that human resource management practices play a critical role in shaping the extent of discrimination that occurs within an organization as they influence the access that women employees have to opportunities within the workplace regarding training, fair performance appraisals, career advancement, equal rewards, and developmental assignments.

DOI: 10.4018/978-1-6684-8417-3.ch030

1. INTRODUCTION

Human resource management (HRM) is the strategic and operational approach to managing an organization's most valuable asset – its people. It involves recruiting, managing, training and developing people within the organization to ensure that the organization is able to meet its goals and objectives. HRM is a critical function of any organization and is responsible for the success or failure of the organization (Boselie, Van Harten, & Veld, 2021). HRM plays a vital role in developing, maintaining, and improving organizational structure, culture, and employee engagement. It also has a major impact on the organization's bottom line, as it is responsible for managing employee compensation, benefits, and other incentives. Additionally, HRM is responsible for ensuring legal compliance with labor laws, as well as monitoring and evaluating employee performance (Greer, 2021; Memon, A. B., & Tunio, M. N. 2023; Sultan, & Tunio, 2022; Abdullah, et al., 2020).

The focus of HRM is to ensure the organization's success by finding, developing, and retaining the right people with the right skills. HRM implements strategies to identify, recruit, and retain qualified employees, and develop and maintain a competitive workforce. It is responsible for creating and implementing policies, procedures, and strategies to ensure compliance with labor laws and regulations, and to ensure a safe and healthy work environment(Boselie, Van Harten, & Veld, 2021). HRM also works to improve employee morale, motivation, and engagement by implementing programs such as employee recognition, rewards, and incentives. In addition, HRM is responsible for managing employee relationships, resolving conflicts, and providing guidance and support on employee-related matters. Finally, HRM is responsible for creating and implementing strategies to ensure that the organization is competitive in the labor market(Greer, 2021).

1.1 Concept of Empowerment

Empowerment is the process of giving people the authority, resources, and confidence to take initiative and make decisions that improve their lives and the lives of those around them. It can involve both individuals and groups, and can take place at the organizational, community, and individual levels(Elsenhans, 2021). At the organizational level, empowerment can involve creating a culture of trust and shared responsibility, developing an environment in which employees feel they can make decisions and take risks without fear of punishment or failure, and allowing them to take ownership of their work and act on it in a way that meets their goals. It also involves giving employees access to resources, training, and support that help them become more effective and successful (Afshan, et al., (2021; Alshareef, et al., 2022; Aurangzeb, et al., 2021).

At the community level, empowerment involves creating a sense of ownership among citizens and encouraging them to take an active role in their own development (Tunio, et al., 2022; Chaudhry, et al., 2021). This can include providing access to resources and support, advocating for policy changes that give citizens a voice in decision-making, and creating spaces where community members can come together to learn, share, and work together (Gilal, et al., 2021; Tunio, et al., 2021). At the individual level, empowerment entails giving people the skills, knowledge, and resources they need to make decisions that are beneficial to their own lives and the lives of those around them (Memon, et al., 2021; Tunio, et al., 2021). This can include access to education and training, developing a sense of self-efficacy, and creating a supportive environment that promotes individual growth and development (Elsenhans, 2021).

Empowerment is an ongoing process that requires commitment and ongoing effort. It requires a shift in mindset, both on the part of individuals and organizations, to ensure that everyone is given the opportunity to make decisions and take action that improves their lives and the lives of those around them.

1.2 Women Empowerment in Human Resource Management

Women empowerment in human resource management is the process of creating an environment where women can reach their full potential in the workplace (Mushtaq, et al., 2021; Tunio, et al., (2021). It is about enabling them to use their skills, knowledge, and experience to contribute to the success of the organization. This involves creating a supportive work environment for women, providing career guidance, and engaging in activities that promote gender equality (Shaikh, and Tunio, 2020; Tunio and Shaikh, 2020). Women empowerment in HRM also involves developing policies and practices that support the advancement of women in the workplace(Ihalainen, et al., 2021).

This includes offering equal pay and promotion opportunities, providing flexible work arrangements, and empowering women by giving them a voice in decision-making (Shaikh, et al., 2021; Tunio, et al., 2021). Additionally, it is important to create a culture of inclusivity and respect, one which values the contributions of both men and women(Ihalainen, et al., 2021). Finally, it is essential to ensure that there are no forms of discrimination in the workplace, such as sexual harassment and gender-based violence. By implementing measures to empower women in the workplace, organizations can enjoy the benefits of increased productivity, employee loyalty, and a diverse workforce.

2. LITERATURE REVIEW

Women empowerment in HRM is an issue that has been gaining increasing attention from employers and employees alike. Despite advances in gender equality, there are still many issues that can hinder women from achieving their full potential in the workplace (Shaikh, et al., 2021; Tunio, et al., 2021).

2.1 Lack of Representation in Management Positions

The lack of representation of women in management positions is an ongoing challenge for women's empowerment in HRM (Shaikh, et al., 2022; Katpar., et al., 2020). Women have the potential to bring a unique perspective to the workplace and to better understand the issues that women in the workplace face (Shaikh, et al., 2022; Tunio, 2020). However, they are not given the same opportunities as men to take on leadership roles, resulting in a lack of representation in management positions (Casad, et al., 2021).

One of the primary impacts of this lack of representation is that women are not given a platform to share their views and concerns (Sultan, et al., 2022; Tunio, 2020). This means that their voices are not heard, which can lead to a lack of recognition for the valuable contributions that women can make (Katper, et al., 2017; Tunio, et al., 2017). It can also mean that important decisions are made without considering the perspectives of female workers (Gilal. et al., 2020; Tunio, et al., 2020). In addition, research has shown that having more women in leadership positions can result in a more inclusive culture and more gender-balanced teams (Casad, et al., 2021).

Furthermore, the lack of representation of women in management positions can also lead to gender discrimination in the workplace. Studies have found that women who are in positions of authority are more likely to be subjected to gender discrimination than their male counterparts (Gul, et al., 2021; Tunio, et al., 2021). This can include micro aggressions such as being interrupted more often or being talked over, as well as more overt forms of discrimination such as being passed over for promotions or being paid less than men (Casad, et al., 2021).

Finally, the lack of representation of women in management positions can also contribute to the gender pay gap. This is because women often face a glass ceiling when it comes to reaching high-level positions. This can result in women having less earning potential than their male counterparts, as they are not given the same opportunities to climb the corporate ladder. It is a significant challenge for women's empowerment in HRM. It can lead to a lack of recognition for their contributions, gender discrimination, and an unequal wage gap. As such, it is essential that organizations make an effort to increase the representation of women in senior roles in order to create a more equal and inclusive workplace (Casad, et al., 2021).

2.2 Unconscious Bias

Unconscious bias is a form of implicit bias that occurs without conscious awareness or intention. Unconscious bias is a challenge for the women empowerment in HRM because it can lead to decisions and actions that are unfair and discriminatory. Unconscious bias can manifest itself in a number of ways, such as making assumptions about someone's capabilities, disregarding the opinions of certain genders, and providing unequal access to resources (Ventura, Morillas, Martins-Loução, & Cruz, 2021).

In HRM, unconscious bias can lead to hiring and promotion decisions that favor male applicants over female applicants, regardless of qualifications. It can lead to the exclusion of women from training and development opportunities, and can even lead to the creation of a hostile work environment. Unconscious bias can be especially damaging to the advancement of women in the workplace, as it can limit the visibility of female employees, limit their potential for promotion, and lead to gender pay gaps (Ventura, Morillas, Martins-Loução, & Cruz, 2021). Unconscious bias can also influence the way people perceive performance reviews and appraisals, leading to biased assessments and evaluations. This can lead to women being judged more harshly than men and receiving lower scores on performance reviews.

2.3 Gender-Based Pay Gaps

Gender-based pay gaps can be a major challenge for the women empowerment in HRM. This is because the wage gap continues to be a major source of discrimination against women in the workplace. Women typically earn less than their male counterparts in the same job and with similar qualifications. The wage gap is exacerbated by the undervaluation of traditionally female-dominated professions. In addition, women are more likely to work part-time and/or take on more family responsibilities, leading to lower wages than their male counterparts (Kurchenko, Kolomiyets-Ludwig, & Ilnytskyy, 2021).

Furthermore, gender-based pay gaps can lead to a lack of job security for women in HRM. Women may be more likely to experience job insecurity due to the lower wages they receive. This can lead to them feeling less empowered in the workplace and feeling less secure in their positions. Finally, gender-based pay gaps can contribute to the overall gender inequality in the workplace. Women are often seen as being less capable and valuable than their male counterparts. This can lead to women feeling powerless in their jobs and not having the same level of career advancement as their male counterparts (Kurchenko, Kolomiyets-Ludwig, & Ilnytskyy, 2021).

This contributes to the gender gap in the workplace and can lead to women feeling less empowered in the workplace overall. Overall, gender-based pay gaps can be a major challenge for the women empowerment in HRM. The wage gap leads to greater job insecurity for women and contributes to the overall gender inequality in the workplace. This can lead to women feeling less empowered in the workplace and less secure in their positions (Kurchenko, Kolomiyets-Ludwig, & Ilnytskyy, 2021).

2.4 Hostile Work Environment

Hostile work environment can be a major challenge for women empowerment in Human Resource Management (HRM). It is defined as an environment where employees are subjected to unwelcome and offensive behavior that is based on their gender, race, age or other protected characteristics. This type of environment can lead to an atmosphere of fear, intimidation and humiliation and can have a significant impact on women's ability to perform their jobs effectively (Errami, Ochir, & Peppoloni, 2021).

Women are often subjected to discrimination and harassment, which can contribute to hostile work environment. In addition, women may also experience gender-based micro aggressions, which are subtle forms of discrimination that can have a negative impact on their self-esteem and career development (Errami, Ochir, & Peppoloni, 2021). This type of environment can lead to lower morale and increased stress levels, which can further impede women's ability to advance in their careers.

The lack of female representation in executive and managerial positions can also be a challenge for women empowerment in HRM. Women are often underrepresented in these positions, which can limit their opportunities for career advancement and create an atmosphere of gender bias. Furthermore, women may not have access to the same resources and support as their male counterparts, which can lead to a feeling of isolation and exclusion (Errami, Ochir, & Peppoloni, 2021).

Finally, the prevalence of sexual harassment in the workplace can be a major challenge for women empowerment in HRM. Sexual harassment is a form of gender-based discrimination that can have a negative impact on women's mental and physical health, as well as their career prospects. It is important for employers to take action to prevent and address sexual harassment in the workplace, as it can create an environment of fear and intimidation that limits women's advancement opportunities (Errami, Ochir, & Peppoloni, 2021).

3. METHODOLOGY

3.1 Solution for the Lack of Representation in Management Positions

1. One of the best ways to increase representation in management positions is to expand the talent pool. Companies should ensure that they are actively recruiting and interviewing qualified women, and that they are equally represented in leadership and recruitment processes. Additionally, companies should look for qualified female candidates from a variety of educational institutions, and also consider non-traditional sources such as boot camps and professional associations.
2. Mentorship programs are a great way to increase representation in leadership positions. Companies should create mentor relationships between senior executives and up-and-coming female employees. This will not only provide female employees with mentors who can provide them with guidance and support, but also helps increase their visibility and presence in the organization.

3. Flexible working arrangements, such as job sharing and telecommuting, are a great way to increase representation in management positions. Flexible working arrangements make it easier for women to juggle work and family responsibilities, and also help reduce the gender gap in management positions.
4. Companies should also create initiatives to support and promote women in leadership positions. These initiatives should focus on providing resources and training to help women succeed and thrive in their roles. Companies can also create initiatives that focus on networking and career advancement opportunities for women.

3.2 Resolving Unconscious Bias

In order to improve gender equality and women empowerment in HRM, organizations need to address unconscious bias and create a more inclusive workplace. Here are some steps that can be taken to reduce unconscious bias in HRM:

- Organizations should ensure that all HR staff, managers, and executives are aware of the issues related to unconscious bias and its potential negative effects on workplace equality. This can be done through training programs and resources that emphasize the importance of diversity, inclusion, and fairness.
- Organizations should implement initiatives that encourage diversity and inclusion in the workplace. These initiatives should focus on recruiting and hiring from a diverse pool of candidates, and should also promote equal opportunities for advancement and compensation.
- Organizations should also ensure that performance reviews are conducted objectively, with a focus on skills and job qualifications rather than on gender or race. Additionally, organizations should strive to create a culture of feedback and open dialogue, so that any potential biases can be identified and addressed.
- Organizations should also create policies and procedures that clearly outline expectations for fair and equitable treatment of all employees. These policies should also be regularly monitored and updated as needed.

By taking these steps, organizations can reduce unconscious bias and create a more equitable and inclusive workplace. This can ultimately lead to greater gender equality and improved women empowerment in HRM.

3.3 Reducing Gender-Based Pay Gaps

To address this issue, organizations must take a proactive approach to resolving gender-based pay gaps and improving women's empowerment in HRM.

- Organizations must ensure that women are paid the same as their male counterparts for performing the same job. This can be done by implementing a fair and equitable salary structure without taking into account any gender factors and making sure that everyone is paid the same amount for the same job.

- Organizations should establish transparent pay policies that clearly define the process of setting salaries and wages. This will help ensure that all employees are aware of the salary and wage structure in the organization and that the right amount of pay is given to all employees regardless of gender.
- Unconscious bias in the workplace can lead to gender-based pay gaps. Organizations must work to eliminate any unconscious bias that may exist in their workplace. This can be done by conducting regular training programs on gender equality and awareness and making sure that all employees are aware of the importance of eliminating gender-based discrimination in the workplace.
- Organizations should provide flexible working opportunities to their employees, especially women. This will allow them to balance their work and personal lives and avoid having to take a pay cut if they need to take time off for family or personal reasons.
- Organizations should provide equal opportunities for career advancement for both men and women in the workplace. This will not only help to bridge the gender pay gap, but will also create a more diverse and inclusive workplace.

By taking these steps, organizations can work towards bridging the gender pay gap and improving the overall empowerment of women in the workplace. This will not only help to create a more equal and fairer workplace, but will also benefit the organization in the long run by improving its overall efficiency and productivity.

3.4 Creating Safe Work Environment

- The first step in resolving a hostile work environment issue to improve women empowerment in HRM is to establish a zero-tolerance policy. This policy should clearly outline unacceptable behaviors and the consequences of engaging in them. This policy should also include language that makes it clear that any form of discrimination or harassment based on gender, race, age, religion, sexual orientation, and other protected characteristics will not be tolerated.
- Training is a key component to resolving hostile work environment issues and improving women empowerment in HRM. Employees should be trained on the zero-tolerance policy and how to recognize and respond to incidents of hostile behavior. This training should be ongoing and should be tailored to the unique needs of the organization.
- When a complaint is made that an employee is experiencing a hostile work environment, it is important to investigate the complaint thoroughly and promptly. During the investigation, it is important to maintain confidentiality and to be impartial. It is also important to ensure that the employee making the complaint is protected from any form of retaliation.
- Once the investigation is complete, it is important to take appropriate action. Depending on the severity of the incident, this could range from a verbal warning to termination. It is important to ensure that all employees are held accountable for their actions and that appropriate steps are taken to prevent further incidents.
- To ensure a workplace free of hostile behavior, it is important to foster an inclusive environment. This can be done by creating an open dialogue between employees, encouraging diversity in the workplace, and creating a culture of respect. Additionally, it is important to ensure that all employees feel safe and respected.

By implementing these steps, organizations can ensure that hostile work environment issues are addressed and that women empowerment in HRM is improved. By taking the necessary steps to create a safe and respectful workplace, organizations can create a work environment that is conducive to success and the advancement of women.

REFERENCES

Abdullah, M., Katper, N. K., Chaudhry, N. I., & Tunio, M. N. (2020). An Impact of Workaholics on Creativity: the mediating role of Negative Mood and moderating role of Supervisor Support. *Sukkur IBA Journal of Management and Business, 7*(2).

Afshan, G., Ilyas, S., Tunio, M. N., & Kalhoro, M. (2021). CSR actions and post-COVID-19 consequences in the hotel industry: A conceptual framework. *International Journal of Strategic Change Management, 7*(4), 275–289. doi:10.1504/IJSCM.2021.122845

Alshareef, N., & Tunio, M. N. (2022). Role of Leadership in Adoption of Blockchain Technology in Small and Medium Enterprises in Saudi Arabia. *Frontiers in Psychology, 13*, 2284. doi:10.3389/fpsyg.2022.911432

Aurangzeb, M. T., Tunio, M. N., Rehman, Z., & Asif, M. (2021). Influence of administrative expertise on human resources practitioners on the job performance: Mediating role of achievement motivation. *International Journal of Management, 12*(4), 408–421.

Boselie, P., Van Harten, J., & Veld, M. (2021). A human resource management review on public management and public administration research: Stop right there...before we go any further.... *Public Management Review, 23*(4), 483–500. doi:10.1080/14719037.2019.1695880

Casad, B., Franks, J., Garasky, C., Kittleman, M., Roesler, A., Hall, D., & Petzel, Z. (2021). Gender inequality in academia: Problems and solutions for women faculty in STEM. *Journal of Neuroscience Research, 99*(1), 13–23. doi:10.1002/jnr.24631 PMID:33103281

Chaudhry, I. S., Paquibut, R. Y. & Tunio, M. N. (2021). Do workforce diversity, inclusion practices, & organizational characteristics contribute to organizational innovation? Evidence from the U.A.E. *Cogent Business & Management, 8*(1), 1947549. . doi:10.1080/23311975.2021.1947549

Elsenhans, H. (2021). *Capitalism, Development and Empowerment of Labour: A Heterodox Political Economy*. Routledge. doi:10.4324/9781003182511

Errami, E., Ochir, G., & Peppoloni, S. (2021). Achieve gender equality and empower all women and girls. *Geosciences and the Sustainable Development Goals*, 105-126.

Gilal, F. G., Gilal, N. G., Channa, N. A., Gilal, R. A., Gilal, R. G., & Tunio, M. N. (2020). Towards an integrated model for the transference of environmental responsibility. *Business Strategy and the Environment, 29*(6), 1–10. doi:10.1002/bse.2524

Gilal, F. G., Gilal, N. G., Gilal, R. G., Gon, Z., Gilal, W. G., & Tunio, M. N. (2021). The Ties That Bind: Do Brand Attachment and Brand Passion Translate Into Consumer Purchase Intention? *Central European Management Journal, 29*(1), 14–38. doi:10.7206/cemj.2658-0845.39

Greer, C. (2021). *Strategic human resource management.* Pearson Custom Publishing.

Gul, A., Subhan, S., & Tunio, M. N. (2021). Learning experiences of women entrepreneurs amidst COVID-19. *International Journal of Gender and Entrepreneurship, 13*(2), 1756–6266. doi:10.1108/IJGE-09-2020-0153

Ihalainen, M., Shaikh, S., Mujawamariya, G., Mayanja, S., Adetonah, S., Tavenner, K., & Elias, M. (2021). Promise and contradiction: Value chain participation and women's empowerment. *Advancing gender equality through agricultural and environmental research: Past, present and future*, 147-188.

Katpar., N. K., Chaudhry, N. I., Tunio, M. N., & Ali, M. A. (2020). Impact of Leadership Style and Organizational Culture on Organizational Commitment. *Sukkur IBA Journal of Management and Business, 7*(1), 92-106.

Katper, N. K. Medan, A., Syed, K. B. S.., Tunio, M. N. (2017). Determinants of Debt Maturity Structure in Shariah and Non-Shariah Firms in Pakistan: A comparative Study. *Journal of Applied Economic Sciences, 12*(4), 1210-1225.

Kurchenko, L., Kolomiyets-Ludwig, E., & Ilnytskyy, D. (2021). Women's empowerment as a tool for sustainable development of higher education and research in the digital age. *Stagnancy Issues and Change Initiatives for Global Education in the Digital Age*, 141-172.

Memon, A.B., Meyer, K., & Tunio, M. N. (2021). Toward collaborative networking among innovation laboratories: a conceptual framework. *International Journal of Innovation Science.* doi:10.1108/IJIS-04-2021-0069

Memon, A. B., & Tunio, M. N. (2023). Assessing the Potential of ICT Education at Secondary Schools in Pakistan: A Comparative Study. *Pakistan Journal of Educational Research and Evaluation, 10*(2), 16–33.

Mushtaq, T., Tunio, M. N., Akbar, Z., & Jariko, M. (2021). Green Organizational identity: Antecents and consequences: An Emprical Study. *Contemporary Issues in Business and Government, 27*(3), 2056-2069.

Shaikh, E., & Tunio, M. N. (2020). Customer satisfaction and Customer loyalty: An empirical case study on the impact of benefits generated through Smartphone applications. *International Journal of Public Sector Performance Management.*

Shaikh, E., Tunio, M. N., Khoso, W. M., Brahmi, M., & Rasool, S. (2022). The COVID-19 Pandemic Overlaps Entrepreneurial Activities and Triggered New Challenges: A Review Study. *Managing Human Resources in SMEs and Start-ups: International Challenges and Solutions*, 155-182.

Shaikh, E., Tunio, M. N., & Qureshi, F. (2021). Finance and women's entrepreneurship in DETEs: A literature review. Entrepreneurial Finance. *Innovation and Development*, 191–209.

Shaikh, E., Watto, W. A., & Tunio, M. N. (2022). Impact of Authentic Leadership on Organizational Citizenship Behavior by Using The Mediating Effect of Psychological Ownership. *ETIKONOMI, 21*(1), 89–102. doi:10.15408/etk.v21i1.18968

Shaikh, S., Sultan, M. F., Mushtaque, T., & Tunio, M. N. (2021). Impact of COVID-19 on GDP: A serial mediation effect on international tourism and hospitality. *International Journal of Management, 12*(84), 422–430.

Sultan, M. F., Shaikh, S. K., Tunio, M. N., & Asim, M. (2022). Sustainable Leadership: A Divine Form to Lead and Succeed in a Holistic Way. In Sustainable Development of Human Resources in a Globalization Period (pp. 111-118). IGI Global. doi:10.4018/978-1-6684-4981-3.ch007

Sultan, M. F., & Tunio, M. N. (2022). Gauging the COVID-19 on Academic Stress on Tertiary Level Female Students on the eve of COVID-19. *KASBIT Business Journal*, *15*(3), 1–10.

Sultan, M. F., Tunio, M. N., Aziz, A., & Shaikh, S. K. (2023). Social Innovation and Social Entrepreneurship in the Wake of COVID-19: A Perspective From the Developing Side of the World. In Frugal Innovation and Social Transitions in the Digital Era (pp. 113-118). IGI Global.

Tunio, M. N. (2020) Role of ICT in Promoting Entrepreneurial Ecosystems in Pakistan. *Journal of Business Ecosystems, 1*(2).

Tunio, M. N. (2020). Role of ICT in promoting entrepreneurial ecosystems in Pakistan. *Journal of Business Ecosystems*, *1*(2), 1–21. doi:10.4018/JBE.2020070101

Tunio, M. N., Chaudhry, I. S., Mughal, F., & Shaikh, E. (2022). Marketing Mode and Survival of the Entrepreneurial Activities of Nascent Entrepreneurs. In Big Data Analytics (pp. 1-18). Auerbach Publications. doi:10.1201/9781003307761-1

Tunio, M. N., Chaudhry, I. S., Shaikh, S., Jariko, M. A., & Brahmi, M. (2021). Determinants of the Sustainable Entrepreneurial Engagement of Youth in Developing Country—An Empirical Evidence from Pakistan. *Sustainability (Basel)*, *13*(14), 7764. doi:10.3390u13147764

Tunio, M. N., Jariko, M. A., Børsen, T., Shaikh, S., Mushtaque, T., & Brahmi, M. (2021). How Entrepreneurship Sustains Barriers in the Entrepreneurial Process—A Lesson from a Developing Nation. *Sustainability (Basel), 13*(20), 1–18. doi:10.3390u132011419

Tunio, M. N., & Shaikh, E. (2020). Nascent entrepreneurs and challenges in digital market in developing countries. *International Journal of Public Sector Performance Management.*

Tunio, M. N., Shaikh, E., Niaz, S., & Katper, N. S. (2021). Multifaceted perils of the Covid-19 and implications: A Review. *Estudios de Economía Aplicada*. Advance online publication. doi:10.25115/eea.v39i2.3957

Tunio, M. N., Soomro, A. A., & Bogenhold, D. (2017). The Study of Self-employment at SMEs Level with Reference to Poverty in Developing Countries. Business and Management Research, 6(2). doi:10.5430/bmr.v6n2p33

Tunio, M. N., Yusrini, L., Shah, Z. A., Katper, N., & Jariko, M. A. (2021). How Hotel Industry Cope up with the COVID-19: An SME Perspective. *Etikonomi*, *20*(2), 213–224. doi:10.15408/etk.v20i2.19172

Tunio, M. N., Yusrini, L., & Shoukat, G. (2021). Corporate social responsibility (CSR) in Hotels in Austria, Pakistan, and Indonesia: small and medium Enterprise spillover of COVID-19. In Handbook of research on entrepreneurship, innovation, sustainability, and ICTs in the post-COVID-19 era (pp. 263-280). IGI Global.

Tunio, M. N., Yusrni, L., Shah, Z. A., Katper, N. K., & Jariko, M. A. (2021). How hotel industry cope up with the COVID-19: An SME Perspective. *Etikonomi, 20*(2).

Ventura, A., Morillas, L., Martins-Loução, M., & Cruz, C. (2021). Women's Empowerment, Research, and Management: Their Contribution to Social Sustainability. *Sustainability (Basel)*, *13*(22), 12754. doi:10.3390u132212754

Chapter 31
Women in Tourism Opportunities for Women in Remote Areas

Muhammad Affan Javed
Mohammad Ali Jinnah University, Pakistan

Rimsha Brohi
Mohammad Ali Jinnah University, Pakistan

Aisha Aqeel Ahmed
Mohammad Ali Jinnah University, Pakistan

Asad Ahmed Khan
Mohammad Ali Jinnah University, Pakistan

Mubashir Zafar Iqbal
Mohammad Ali Jinnah University, Pakistan

Muhammad Bilal Aslam
Mohammad Ali Jinnah University, Pakistan

Areej Razzaq
Mohammad Ali Jinnah University, Pakistan

ABSTRACT

In this chapter, the authors are going to look into tourism regarding women as it is known gender equality is a priority for national governments and international organizations, as evidenced by their adherence to a number of agreements, including the Beijing Declaration and Platform for Action, the Sustainable Development Goals, and the Convention on the Elimination of All Forms of Discrimination Against Women (CEDAW). Just and equitable societies must priorities gender equality and women's empowerment.

DOI: 10.4018/978-1-6684-8417-3.ch031

INTRODUCTION

Tourism is a region that has generated many job possibilities for both men and women. This makes tourism one in every of the largest industries within the global. The involvement of women is an important aspect in tourism. The lifestyles of girls in tourism is regulated in Ministerial Regulation No. 50 of 2011 regarding the National Tourism Development Master Plan Article 28 which states that one of the community empowerments in tourism includes optimizing gender mainstreaming via tourism development. Optimizing the function of girls in tourism is also regulated inside the Regulation of the Minister of Culture and Tourism of 2011 which explains that the community is actively worried in sports starting from the making plans, implementation, supervision, renovation and utilization techniques, by means of supplying large possibilities for lively participation from girls. Women as part of society have the right and capability to participate in tourism activities. The involvement of ladies in tourism can reduce gender inequality in improvement. The promoting of equality and girls' empowerment became one of the agendas within the Third Millennium Development Goals (MDGs) which turned into proclaimed by using the United Nations in 2000 primarily based on the consequences of a joint settlement with 189 heads of UN member states consisting of Indonesia.

In development, efforts want to be made to boom the ability and participation of present resources to maximize the improvement. Both women and men, as resources in society, basically have the same opportunity to have the ability to participate in development. From the aforementioned policies, women have an identical role as men in the tourism area.

However, the practice within the field illustrates various things concerning the role of ladies in tourism mentioning the extraordinary roles of males and females in tourism work. The outcomes of display that women in tourism have less function in agencies however most effective play a position in nearby dances, agriculture, homestays, food and beverage services, and making souvenirs. It seems that there are variations in the roles of men and women in tourism jobs.

LITERATURE VIEW

The tourism sector definitely provides various entry points for women's employment and opportunities for creating self-employment in small and medium sized income generating activities. The importance of women in the tourism sector, analyze and evaluate the reasons for and constraints to women's vertical mobility, and addresses the issue of how tourism affects their activities. More specifically it investigates issues concerning the access to natural resources by women by Nichola Ramchurjee (2011).

According to the findings, there are differences in the roles of women and men in tourism. Women have a role in supporting the tourism sector, namely in maintaining culture and tourism development. However, women's roles are more concentrated in low-skill and low-paid areas. Women's skills are more dominant in the accommodation sector which includes homestay services, food and beverage services, making souvenirs, opening stalls, hawking trade, cooking, cleaning public places and working in the hospitality sector. This is due to women's limited access to education, skills in tourism, family background, physical condition, language, and ability to set their criteria. The results of the literature review were used to analyze the role of women in the tourism sector in the tourist village of Nglanggeran in Patuk, Gunugkidul, Yogyakarta in (Tristanti et. al, 2022).

RESEARCH METHODOLOGY

The technique in this study we use is to writing survey by utilizing a complete system. The main strategy is to look for diaries both broadly globally and locally through the EBSCO data set, Science Direct, and Scopus. Writing survey is an efficient, express and reproducible strategy for distinguishing, assessing and combining research works and thoughts that have been delivered by scientists and experts. The reason for the writing survey is to examine and bring existing information connected with the point to be read up to view holes for the exploration as done. The indexed lists for articles that are relevent to the examination title are 10 articles, both public and worldwide articles. The articles were analyzed in light of targets, reasonableness of subjects, strategies, and exploration results.

OPPORTUNITIES IN TOURISM

Industries are continually disrupted via new trends and new innovations, and the is not any exception to this. Keeping pace with those emerging tourism developments can assist companies to stay aggressive and cater to the desires of customers. Nevertheless, with COVID as a backdrop, a number of those wishes have advanced, priorities have shifted, and entirely new requirements have emerged.

Figure 1.

KEEPING UP WITH THE LATEST TOURISM TRENDS

Anything part of the travel industry zone your venture is worried about, you want to keep up with your finger at the beat. Previously preferred standards and items exit of style, to be supplanted by utilizing more noteworthy cutting-edge components that wind up catching extra of the market. New objections, new innovations and ways of conveyance have been welcomed on essential movements inside the undertaking. Early reception of ongoing advancements is significant.

1. The Factors Responsible for Upcoming Tourism Trends

A fashion can be described as a generalized change in situation or behavior or a fashionable path in which things are transpiring. Within tourism, a number of exclusive trends have brought about adjustments in patron behavior and enterprise techniques, which means they're responsible for new tourism traits rising. One of these developments has been a essential exchange in the manner technology is used, specifically in regards to artificial intelligence and system gaining knowledge of. This has led to the multiplied adoption of robotics and AI-powered generation in inns. It has also brought about modifications in purchaser behaviors through the upward push of voice manage and voice search.

The COVID-19 pandemic located a focal point on hygiene, protection, and neighborhood in preference to international services, at the same time as additionally accelerating the usage of contactless payments and remote working. Meanwhile, environmental concerns have delivered to the nearby-first approach of many customers, and have given rise to traits associated with natural food and eco-travel.

It is crucial to observe the cutting-edge tourism traits in response to growing patron movements due to the coronary pandemic. However, maximum styles have arisen from more wellknown shifts in consumer behaviors.

2. Tourism Trends Related With COVID-19

The trends which affect tourism in the period of covid are as follows:

Safety and Hygiene Tourism Trends

Whether it travels through air, inns, cafes, or bars since the episode of Coronavirus, security and cleanliness norms have been totally principal. Considering this, there are various travel industry drifts that are connected with this, for example, expanded cleaning, socially separated seating, giving hand gel, and implementing veils in certain settings.

COVID-19 SOP's

The coronavirus has forced many nations to take on limitations, meanwhile many organizations are empowering representatives are working from home through many platforms. Also, business occasions have been especially getting impacted and one of the reasons the travel industry patterns has been a switch in the center towards recreation clients.

Shifting From International to Local Clients

The different travel limitations and many individuals traveling abroad have meant many in the travel industry are not focused on international clients, as opposed to global ones. This doesn't mean abandoning global clients completely, yet it is probably going to require an adjustment of your center-promoting techniques.

Growth of Payments

Contact less payments has been a staple with regards to for quite a while, yet the development of choices using local e-transfers has assisted with taking this to a higher level, meaning clients don't have to worry about credit cards or Mastercard to pay for dinners, lodging stays, transport, and different administrations.

GLOBAL UPCOMING TOURISM TRENDS

There are several changes going to be affect the tourism globally and will make an new way to have more opportunities and jobs for the employment.

Virtual Reality Effect on Tourism:

VR affects is one more significant level that travel industry patterns disturbing the business and gaining by the innovation can give you an edge over rivals who have not yet taken on it. Through internet-based VR visits, clients can encounter inn insides, eatery insides, outside vacation spots and that's just the beginning, all from their home.

Solo Travelling:

Relaxation heading out used to be a family issue or something that people faced together. While that is for some individuals are deciding to strike out all alone. Going alone on a trip is as of now not really strange and traveler drifts progressively have shown this. The needs of solo travelers are different. Some just need to go without interruption of anyone. Others are youthful singles searching for social exercise. A few old or seniors even utilize their lives in and stay in a luxury hotel. These travel industry patterns are set to develop growth.

Eco Travel:

Tourism trends are heavily influenced by the concerns and mores of the customer base. As a new generation becomes increasingly relevant in the marketplace, the ideals driving their purchasing decisions create new tourism trends. Eco travel is just one example of these tourism trends, reflecting a growing concern among today's travellers for ethical and sustainable tourism options. Eco travel includes simple changes, such as the availability of carbon credits when booking a flight or the option to rent an electric instead of a conventional vehicle. More sophisticated examples might include tourism with a volunteer element, perhaps working on a nature reserve or engaging in conservation work.

Local Experience:

Today's tourists do not want to be protected from the places they visit, a culture. They want to engage with people in the local culture. From enjoying local rituals like regional festivals and holidays, local experiences are the top tourist trends to watch.

Personalization:

You are probably familiar with those ads that pop up on social media and certain other websites, that you've looked at or purchased online. Today consumer expect experiences that closely match their personal preferences, from destinations to accommodation and the kinds of activities they'll engage in. The more closely an experience can be tailored to a client's desires and expectations, the more likely they are to return and to use the same service again.

TOURISM'S COMPARATIVE ADVANTAGE FOR WOMEN

I. Women Entrepreneur in Tourism

The travel industry additionally has another superpower: most of its labor force is female. With such a wide geological reach, and a tremendous scope of abilities required, the travel industry has the chance to haul ladies out of poverty, furnish them with abilities and give noble, manageable work. In small countries, where the ladies struggle most, as any similarity to uniformity falls away. Women and young ladies are impeded from the very beginning by an disadvantage in training, education, social rights regarding the lady's part in the home, and innate orientation gender baised in the working environment.

So - the travel industry holds a gigantic measure of potential. However, as the UN World The travel industry Association's (UNWTO, 2017) report uncovered, the subtleties behind the insights are not exactly so equivalent. Most of these ladies are gathered in lower-level positions - cooking, cleaning, serving and administrative work, and for these jobs they are paid on normal 10 to 15 percent not exactly male staff. Moreover, a ton of ladies embrace neglected jobs inside the travel industry - supporting privately-run companies as undetectable client care suppliers, servants, PAs, etc.

II. Leadership in Women

Most of individuals working in the travel industry are ladies - yet the higher up the professional bureaucracy you go, the less and further between they are. Our manual for ladies' business visionaries in the travel industry investigates why there are scarcely any female forerunners in the travel industry and the social and social difficulties they experience. We likewise hear from a few moving female business visionaries on how they fired up their travel industry organizations - and how they are utilizing their abilities and impact to work on the possibilities for different ladies.

Tourism is considered a key source of poverty reduction and inclusive economic growth in poor countries. In general, as a labor-intensive sector, a large share of unskilled or semi-skilled workers, including youth and women are employed in the tourism sector. In KP tourism employs thousands of small and medium enterprises and increases economic activity in rural locations with widespread poverty. Estimates suggest that the total number of domestic tourists visiting KP per year is about 8.8 million, which is 19 percent of total national domestic tourist traffic. Geography has provided KP with a competitive edge over other provinces in attracting tourists so this is another sector with tremendous potential for growth. 126 "The current institutional setup for tourism management and promotion in KP includes the Tourism Corporation Khyber Pakhtunkhwa (TCKP) and the Department of Tourists Services (DTS). The main responsibilities of TCKP are to promote local and foreign tourism, to monitor and manage all tourism activities and to play the role of catalyst by providing a level playing field for the private sector. In addition, Pak-Austrian Institute of Tourism and Hotel Management (PAITHOM) is responsible for producing skilled human resources in the hospitality sector. All these organizations are collectively working under the Department of Sports, Tourism, Youth Affairs, Archaeology & Museum (DoT).

III. Employment for Women

The World The travel industry Association (UNWTO), in a joint effort with UN Ladies, The German Culture for Global Participation (GIZ), World Bank Gathering and Amadeus report submitted on 6 November 2019.

The Worldwide Report on Ladies in the travel industry 2010 gave the main look and a gauge concentrate on the circumstance of ladies in the travel industry area. The second release of the report thinks about how the circumstance has advanced and gives a careful evaluation of the travel industry's commitment towards Joined Countries Supportable Improvement Objective 5 to accomplish orientation correspondence and ladies' strengthening.

The report has a lengthy geological degree covering created and non-industrial nations, extra top to bottom industry examination, and a few contextual investigations that show how ladies all over the planet are involving the travel industry as a vehicle for strengthening and improvement.

Figure 2.

RESULT AND DISCUSSION

Research on gender, covid and the travel industry is distinguished for three reasons. To begin with, the most common way of including ladies and men in the travel industry is formed by orientation philosophy in the public arena. Second, gender relations add to the elements of the ease of male and female character in the travel industry. Third, the covid affect viewpoint is a key for figuring out the connection among the travel industry and social cycles.

This is in accordance with the 2017 UNWTO statement which positions the travel industry as a device to propel the 2030 general plan for practical improvement including the 17 manageable advancement objectives (SDGs). For this situation, the job of ladies in the travel industry means quite a bit to study.

In different examinations that have been done, the job of ladies isn't just as a spouse and mother, in particular regular and organic capabilities showing up as ladies, as "family chiefs" specifically completing daily schedule and exhausting family tasks. In view of these capabilities, the significant investment that ladies give here are much of the time neglected, frequently seen as only a commitment, considered non-work and have no monetary worth. On the off chance that that job is supplanted by another lady, obviously there is a money related esteem in the work.

The association of ladies in the travel industry is beginning to increase. Report proof that ladies' cooperation in the proprietorship and the board of the travel industry exercises has expanded emphatically in numerous nations, particularly created nations like the European Association.

Some roles of women are as follows:

The contribution of ladies in the travel industry is beginning to increment. report proof that ladies' support in the possession and the board of the travel industry exercises has expanded decisively in numerous nations, particularly created nations like the European Association.

Women contribute 66% of services b. work, but only gets 10% of his wages. Women are responsible for 50% of world food production, but only control 1% of existing material goods.

Ladies appreciate not as much as men as c. its commitment to public creation; the typical time-based compensation is lower than that of men; legitimate security for them is deficient with regards to, the job of ladies is restricted to low-paid or neglected work and admittance to creation sources is tiny.

The affect of covid also changes the landscape of the industry in term of technology or in term of any other proprietorship. The covid affect the world and the industry of tourism to moved toward the locally participates and enjoy the rituals and festivals along with the individuals living in the country. The more focused now a days is for those people who is becoming a self-oriented and prefer to be in home and exploring around the world the tourism takes another change when this affect the world as the technology designs for those targeted audience to gain more attention and be in a hot topic for the world.

Getting the nerves in control and to get out of exhaustion the people or individuals around the world decides to go for tour or vacation in order to get rid of the situations they are facing either couple or solo. They use to go for the eco travel and exploring either on the locally or internationally.

From the quest for articles connected with the job of ladies in the travel industry, it turned into the reason for specialists to direct research connected with the job of ladies in the travel industry in the vacationer town of Nglanggeran, Patuk, Gunungkidul, Yogyakarta. The examination depends on the job of the local area in Nglanggeran the travel industry. The fundamental job of the nearby local area (people) as program implementers is the way to the travel industry achievement (Manaf, et al, 2018). In any case, the job of ladies is as yet restricted, specifically that ladies don't partake in the Nglanggeran the travel industry town local area. Ladies are engaged with culinary gatherings, homestay supplier gatherings, and cocoa maker gatherings. While men for the most part involve key positions (Putri and Adinia, 2018). There are various jobs among ladies and men in the travel industry in Nglanggeran.

CONCLUSION

We come to the conclusion that just though gender segregation in the labour market is a general issue, that doesn't imply it shouldn't be addressed when talking about tourism. The essential adjustments must be brought about via efforts from all sectors. However, it appears that the tourist sector is a particularly strong "candidate" for participation in initiatives for the progress of women. The tourist business has a tremendous amount of flexibility because of its size, its expansion, and its incredibly diverse and dynamic character. So that other industries can use efforts and tactics in the tourism sector as examples for their own development, this can help the industry build important initiatives for the progress of women.

The writing audit of articles applicable to the job of ladies in the travel industry shows that ladies possess a casual job in the travel industry and also the article admitted the change in the travel industry depending on the technology and what is the landscape in the tourism industry has in the world. Ladies have work positions with low wages. Ladies additionally have constraints in doing their jobs because of a few hindrances. The job of ladies in the travel industry incorporates access, support, control, and advantages The job of ladies is more conspicuous in the field of dynamic in the family, however this job doesn't happen in collective choice making. The consequences of this study will be utilized as a reason for exploration to be done in regards to the job of ladies in the travel industry area.

The change in industry the broadly accept in the change in the industry of tourism affect worldwide and the families, individuals or spouses learn more about the tourism by ads or promotion which for the people to have more knowledge about the industry of tourism. The change in technology also affect the industry and the industry of tourism is board to be educated to everyone.

REFERENCES

Manaf, A., Purbasari, N., Damayanti, M., Aprilia, N., & Astuti, W.Community-Based Rural Tourism in Inter-Organizational Collaboration. (2018). How Does It Work Sustainably? Lessons Learned from Nglanggeran Tourism Village, Gunungkidul Regency, Yogyakarta, Indonesia. *Sustainability (Basel), 10*(7), 2142. doi:10.3390u10072142

Putri, F. A., & Adinia, N. C. (2018). The Role of Communication in Sustainable Development Tourism: A Case Study on Community-based Tourism (Pokdarwis) in Nglanggeran Village. *Jurnal Komunikasi Indonesia*, *7*(2), 5. Advance online publication. doi:10.7454/jki.v7i2.9795

Ramchurjee, N. (2011). *"Tourism" a vehicle for women's empowerment: Prospect and challenges.* https://www.researchgate.net/publication/256505214_Tourism_a_Vehicle_for_Women's_Empowerment_Prospect_and_Challenges

Tristanti, Nurhaeni, Mulyanto, & Sakuntalawati. (2022). The Role of Women in Tourism: A Systematic Literature Review. In *2021 Annual Conference of Indonesian Association for Public Administration*. KnE Social Sciences. doi:10.18502/kss.v7i5.10576

UNWTO. (2017). *Gender and tourism*. https://www.unwto.org/gender-and-tourism

Chapter 32
Bibliometric Analysis of Conducted Studies Related to Women's Entrepreneurship in Turkey

Nuray Yıldız
Department of Tourist Guiding, Cappadocia University, Turkey

Zerrin Yalçın
Department of Tourism Management, Aydin Adnan Menderes University, Turkey

ABSTRACT

The main purpose of this research is to determine the way in which the concept of entrepreneurship, the importance of which is increasing day by day, is discussed in the national tourism literature, to reveal the current characteristics of the studies and to examine the subject. In addition, it is aimed to contribute to the literature by compiling studies on a national scale in which the topic of entrepreneurship research is discussed within the framework of the tourism discipline. Starting from this point, between 2019 and 2022, Google Academic, ULAKBIM, and National Dissertation/Thesis Center (YOK) are reviewed for words such as women's entrepreneurship, entrepreneurship, tourism sector. Within the scope of the qualitative method, the academic papers obtained will be reviewed firstly through document analysis and then via bibliometric analysis.

INTRODUCTION

Müftüoğlu & Durukan (2004) state that entrepreneur; identify opportunities in existing and interpret, organize, and with the ability to manage business-related risks he took responsibility for, and desire to create innovation in a time of uncertainty showcases the skill is defined as the person (Hisrich & Peters, 2002; Ferrante, 2005; Bozdemir et al., 2021). It is the entrepreneurial ability that that person possesses that differentiates a labor resource from others. Some important factors such as risk-taking tendency, in-

DOI: 10.4018/978-1-6684-8417-3.ch032

novation, need for success, social control center, uncertainty tolerance and self-confidence, which guide the entrepreneur's abilities, indicate the character traits of the entrepreneur (Koh, 1996). Assuming the risk indicates that a person makes a decision for a situation under certain conditions, without knowing where and how it will happen, by making predictions through probabilities (Wakkee et al., 2010). If there is no element of risk, similar conditions will exist for the entire workforce, and the characteristics characteristic of the entrepreneur will remain ineffective in terms of realizing the enterprise. Therefore, the tendency to tolerate ambiguity and take risks is among the factors that make a difference for the workforce (Alpkan et al., 2010; Bourne, 2011; Comeche & Loras, 2010; Ga´mez-Gonza´lez et al., 2010; Hormiga et al., 2011; Kautonen & Palmroos, 2010; Lin et al., 2010; Mainardes et al., 2011; Sebora & Theerapatvong, 2010; Wakkee et al., 2010; Akehurst et al., 2012; Goyal & Parkash, 2011; Okafor & Amalu, 2010; Bozdemir et al., 2021). Entrepreneurship is one of the most effective factors in ensuring economic, social, cultural and technological development (Nişancı & Ödemiş, 2022). Entrepreneurs who carry out these activities produce goods and services by bringing other production factors together, contribute to employment and realize economic development (Sarfaraz et al., 2014; Gündüz & Karadeniz, 2022; Saçak Gündüz, 2021). Women entrepreneurship is the fulfillment of entrepreneurial activities by women. Entrepreneurial activities benefit women to go beyond their traditional roles and to be present and active in the global business environment (Sharma, 2013). Work-life balance means that an individual balances their daily work activities with their non-work activities. Work-life balance is critical for employees to be able to carry on their work life and private lives without any problem (Nişancı & Ödemiş, 2022). The increase in the number of female and male entrepreneurs in a country accelerates the economic development of that country and raises the country to the level of developed countries. Since women entrepreneurship is one of the biggest reasons underlying the development of developed countries, it is important to support women entrepreneurs. As a result of the social changes experienced with the developing technologies today, it is extremely important for women to be active in working life every day, both economically and socially (Kısa & Şahin, 2022). Today, encouraging women to become entrepreneurs by including policies that support women entrepreneurs is an important investment for both women and the country's economy (Sarfaraz et al., 2014). Therefore, women entrepreneurs in Turkey should be supported, encouraged, informed about the stages of starting and running a business related to entrepreneurial activities. However, there are many social and economic problems in our country that prevent women from taking part in working life. With the elimination of these problems, it will be easier for women to participate in the working life, which will positively affect production and employment (Gündüz & Karadeniz, 2022). Investments benefited by women entrepreneurs in Turkey increased in the first months of 2022. In this context, one out of every 4 investments was granted to women entrepreneurs in the first quarter of 2022. During this period, a total of 23 venture investments were made, and 6 investments were allocated to women entrepreneurs (www.haberturk.com). COVID-19 it has affected many consumer behaviours, such as staying at home, avoiding using public transport, and paying attention to savings (www.tursab.org.tr). The main purpose of this research is to determine the way in which the concept of entrepreneurship, the importance of which is increasing day by day, is discussed in the national tourism literature, to reveal the current characteristics of the studies and to examine the subject. In addition, it is aimed to contribute to the literature by compiling studies on a national scale in which the topic of entrepreneurship research is discussed within the framework of the tourism discipline. Starting from this point; between 2019 and 2022, Google Academic, ULAKBIM and National dissertation/thesis center (YOK) aim to review for words such as women's entrepreneurship, entrepreneurship, tourism sector on these platforms. Within the scope of the qualitative method, the academic papers obtained will be reviewed firstly through document analysis and then via bibliometric analysis.

ENTREPRENEURSHIP

The first article about women's entrepreneurship was published in 1970. In the pioneering article titled "Entrepreneurship, A New Female Frontier" published in 1976, the motivations, characteristics and attitudes of women entrepreneurs were examined (Casson et al., 2009). Individual values are fixed, even if they are relative, and individuals related to the socialization process have different it emerges with preferences (Bulut, 2012).

Schwartz has grouped the concept of individual value under ten dimensions based on the individual. These are Power, (Dominance over people and resources or social status, reputation and control), Success, (Personal success that comes through proving their abilities in accordance with social standards), Hedonism, (Emotional satisfaction and pleasure), Excitement, (Innovation, excitement and challenge in life), Self-Direction, (Creation, discovery and independent thinking), Universalism, (Protection, tolerance, appreciation and understanding for nature and the well-being of all humanity), Benevolence, (Increasing and protecting the well-being of all people, as well as people with frequent contact), Traditionalism (accepting, being connected and respecting the thoughts and traditions provided by faith and traditional culture), Compliance (Protecting the environment, being integrated with nature and having a world full of beauty), Security, (Continuation, harmony and security of one's own existence, relationships and society) can be expressed (Pazarcık, 2010).

The word enterprise is a word of French origin that appeared in the fifteenth century. As a meaning, it is defined as 'to start, initiate. The first meaning was used to describe upper-class individuals. In addition, it is explained as someone who is aware of and evaluates the opportunities for these people, is brave, is ready to complete a job (Halıcı, 2005; Açıkgöz, 2021). According to the Turkish Language Institute, the concept of initiative has been used in the sense of 'undertaking or attempting a job' (Türk Dil Kurumu, [TDK], 2022). Şimşek (1998) states that the initiative aims to achieve its goals through a realistic approach with conscious planning for products and services by bringing production factors such as capital, labor, labor force and raw materials together. In addition, it should not be forgotten that it is an economic, social and technical unit while benefiting from these service resources (Açıkgöz, 2021).

Tokat & Şerbetçi (1998) emphasize that the concept of enterprise is seen as an economic unit established to produce economic goods and services that have value in the market to meet needs and demands, or to create income for an entrepreneur, to perform commercial and economic activities or to achieve specified goals (Açıkgöz, 2021; Goyal & Parkash, 2011). According to Akdoğan & Cingöz (2006), the definition of an entrepreneur is a person seeing the opportunities he has and can turn these opportunities into a business idea, take risks, and turn the idea into action with the limited resources at his/her disposal. İrmiş et al. (2010) state that an entrepreneur has a number of qualifications such as an innovative, creative, confidence, leadership (Okafor & Amalu, 2010; Ca´ceres, Guzma´n, & Rekowski, 2011; Cavalcante et al., 2011; Goktan & Miles, 2011; Hotho & Champion, 2011; Huang et al., 2010; Huarng & Yu, 2011; Naranjo-Valencia et al., 2011; Zhang & Duan, 2010; Akehurst et al., 2012) role model person due to his character. Entrepreneurship is generally seen as the process of creating new ideas or products with added value that provide economic gains and personal satisfaction by combining time and effort and taking on many psychological, economic and social risks (Hirsch & Peter, 2002). Types of entrepreneurship; among genuine entrepreneurship, corporate entrepreneurship, internal entrepreneurship, professional entrepreneurship, opportunity entrepreneurship, entrepreneurship, initiative entrepreneurship, environmental entrepreneurship, innovative entrepreneurship, social entrepreneurship, female entrepreneurship are mentioned (Açıkgöz, 2021).

The analysis and study of women entrepreneurs can be viewed from the perspective of a number of scientific fields (Abu-Rabia-Queder, 2010; Collins & Low, 2010; Elam & Brush, 2010; Essers et al., 2010; Leoni & Falk, 2010; Pardo-del-Val, 2010; Rashid, 2010; Sadi & Al-Ghazali, 2010). Ecevit (1993) as noted, women entrepreneurs, in an operation that has established itself outside of the house, alone or with someone else who partner with business workers working in the establishment phase as a business owner, contacted with private or public institutions, planning the future of established businesses, earnings and investments that is authoritative (Lee, 1997) for, and is the person who undertakes all the risks of the enterprise. (Açıkgöz, 2021). Liman (1993) expression in the context of in the 1970s and 1980s, women's desire to be active in business life, their liberation and desire to use their talents, their desire for flexible working hours led to obstacles and made them turn to the idea of starting their own business with their savings instead of working for a wage. Towards the middle of the 1990s, women's entrepreneurship emerged and the number of women entrepreneurs increased. Over time, thanks to scientific research, there has been an increase in female entrepreneurship (Kelly et al., 2011; Sarfaraz, 2014; Hoxha & Krasniqi, 2008).

The female entrepreneur not only builds the business, but also plans her own future and career with the business. Innovative, visionary, risk-taking, courageous female characteristics are emerging. Women have the ability to solve all the problems they face at this stage, who have goals and the intelligence and desire for freedom necessary to achieve these goals (Moore, 2000). The push factors that lead women to entrepreneurship are poverty in the country, the place of women in business life, inequality and discrimination (Minniti & Arenius, 2003; Du Rietz & Henrekson, 2000) in business life due to being a woman, women have tried to exist in many areas due to reasons such as. Besides, the freedom of women pull factors desires, want to be your own boss, want to work in a flexible environment, family and work life balanced between wanting to make a transition occurs (Akehurst et al., 2012; Soysal, 2010; Okafor & Amalu, 2010; Goyal & Parkash, 2011; Sarfaraz et al., 2014). A person's cultural value is shaped by the society to which she belongs (Brush et al., 2009). The level of economic development reflects the entrepreneurial spirit of the society (Tambunan, 2009; Kumbhar, 2013; Georgeta, 2012; Durak, 2010). Among the types of female entrepreneurs; There are varieties such as traditional female entrepreneur, familial female entrepreneur, innovative female entrepreneur, original female entrepreneur (Kutanis & Hancı, 2004). Coulter (2001) states that the importance of entrepreneurship can be emphasized in three areas. These are (Yirik & Ilgaz Yıldırım, 2014b):

- Innovation and inventiveness,
- The emergence and growth of new businesses,he apparence and growth of new businesses,
- It is the creation of new business areas.

Factors such as increasing income distribution inequalities, regional economic differences, increasing poverty and unemployment, inability to create an effective competitive environment have increased the importance of entrepreneurship (Soysal, 2010). Entrepreneurs are defined as innovative, risk-taking, fast-making decisions and, most importantly, determined and selfless people in the work they will do (Yirik & Ilgaz Yıldırım, 2014a). Similarly, researchers have emphasized in studies that in order for a female entrepreneur to be successful, they must also have the following characteristics. These are characteristics; the ability to make the right decision, the ability to plan the future, the ability to evaluate time well are also qualifications (Şahin, 2006; Karakaş & Gökmen, 2016).

Yeniçeri & İnce (2005) state that entrepreneurship, which is fed by traditional and natural resources in the information society, industrial society, has fallen into the second place by keeping labor and capital constant. Production is provided thanks to knowledge. Knowledge, on the other hand, comes about only with the contributions of people. Therefore, the other important input in the information society is the human factor (Yirik & Ilgaz Yıldırım, 2014b). Today, many interrelated factors, together with laws and practices that ensure the protection and support of working women around the world, have been instrumental in increasing the number of women entrepreneurs. Tokol (1999) states that the most important of these factors are (Yirik & Ilgaz Yıldırım, 2014b):

- Demographic developments,
- The increase of primary education opportunities,
- The formation and dissemination of non-standard forms of work,
- Increasing the small family structure,
- Decrease in marriage rates and increase in divorce rates,
- Improvement of attitudes towards the image of working women in different societies,
- Improving services in the field of children's care and growth.

Although an increase in the number of women entrepreneurs has been observed due to the developments in the economic and social fields, it is understood that the situation does not offer significant results compared to developed countries in Turkey. There was a sudden increase in the female labor force in Turkey with 72% in 1955. This rate decreased to 30.9% in the 1990s and was stated as 25.4% as of 2004 (İlter, 2008; Yirik & Ilgaz Yıldırm, 2014b).

Bego & Gehrels (2014) suggest that there are two main factors that lead women to entrepreneurship in their study. One of them consists of "mandatory" factors and the other consists of "attractive" factors. In addition to their income, women's creation of individual work opportunities is evaluated within the scope of attractive factors. Changing living conditions and increasing needs have pushed women to explore ways to earn more income; creating an environment where they can work individually is considered very attractive for women. Women are exploring business lines such as self-employment where they can increase their income. It is understood that women as entrepreneurs have the desire to achieve more economic freedom, to work as owners and managers, to make decisions, to obtain more economic power than others (Bego & Gehrels, 2014; Erkol Bayram, 2018).

At this point, the development of rural tourism is mostly realized with the activity and activity of women. For this reason, various projects are being carried out in some European countries to support women's entrepreneurship in order to develop rural tourism. On the other hand, it is aimed to develop rural tourism in Turkey with the five-year development plan (Fidan & Nam, 2012). Tarlak (2007) states that rural tourism is a tourism based on women's labor as a type, it supports social development by creating employment for women. Women are in the position of both producing and selling products in rural tourism. Within the framework of family management, a woman is the most competent person in rural tourism. The fact that women engaged in such handicrafts in villages have a very large contribution will contribute to the family budget by providing an important opportunity for rural women's employment. In many European countries, initiatives, agricultural policies and other policies focus on activities that support women's initiatives in rural areas, these practices are especially focused on mountainous and it is a priority for disadvantaged regions (Koutsou et al., 2009; Fidan & Nam, 2012).

Women entrepreneurs offer employment opportunities to women again in the enterprises they have established themselves. In this context, women's role conflict factor is again defined as the recognition of tolerance by women managers, the more effective perception and attempts to solve the psychological and sexual violence situation that women will experience by a woman, the unity and togetherness of women within the social structure makes them feel stronger (Soysal, 2010). Research shows that it is known that one of the types of tourism in which women are most successful in tourism is rural tourism. With the rural tourism activities of women, the appreciation of local people and tourists comes to the fore. The problems experienced by women entrepreneurs who want to carry out tourism activities in rural areas are; insufficient attention to activities by the people of the region, the adoption of an urban entrepreneurship approach instead of an entrepreneurship approach unique to the region, the lack of acceptance of women as an active tourism entrepreneur, the micro-level of their activities, the length of incentive and credit processes, gender-based perceptions can be expressed as a lack of education (Soysal, 2013; Kantar, 1999; Erkol Bayram, 2018).

THE EFFECTS OF COVID-19 PANDEMIC IN TOURISM

In some periods, people may go through some difficult processes. All these difficult periods can be closely linked to economic, political, cultural and public health situations. These negative processes in question bring about some changes in human behavior (Temizkan et al., 2021). The Covid-19 pandemic, which originated in Wuhan province of China in 2019 and spread rapidly all over the world, was put on the pandemic diseases list by the World Health Organization on March 11 (Öztürk & Tankuş, 2020). Today, pandemic conditions have disrupted not only the tourism industry, but also many industrial activities (Baker et al., 2020). Tourism is a very sensitive sector and is very quickly affected by the negative factors developing around it. Even a slight risk related to human health or safety; such as natural disasters, terrorist incidents, epidemics, it is enough not to choose a tourist area (Lee & Chen, 2011). There is a negative relationship between the tourism activities of the countries and the epidemic diseases (Karim vd., 2020). Therefore, the COVID-19 pandemic has negatively affected the tourism sector all over the world and in Turkey (Bahar & İlal, 2020; Yazıcı Ayyıldız, 2021).

In this context, among the studies in the literature (Zhang et al. 2019) emphasized that during the SARS epidemic, tourists prefer to eat alone more and individuality comes to the fore. Therefore, it has been suggested that businesses offering open buffet services should take appropriate measures. In addition, Zheng et al. (2012) suggested that the most appropriate service to prevent the spread of the virus in food and beverage establishments with a capacity of more than fifty in America is a takeaway service.

Acar (2020) investigated the possible effects of the COVID-19 pandemic on the tourism sector in the current and future times, emphasizing that the virus has negative effects that will be felt for many years, and the tourism sector will also be affected by these negativities. Bahar & İlal (2020) examined the economic effects of the COVID-19 pandemic on the tourism sector and stated that the demand shocks caused by the pandemic will lead to income losses in the tourism sector. Bakar & Rosbi (2020), COVID-19 in their study on the impact to the tourism industry, according to the balance of the market theory of supply and demand, the price of the tourism sector will continue to decline in parallel with the decrease in demand, declining demand in the tourism industry should develop a preventive mechanism for the prevention of government stated. Furthermore, they stated that in the absence of preventive action by governments, the tourism industry will be exposed to October economic collapse.

İbiş (2020) conducted interviews with 11 travel agency officials operating in Istanbul to reveal how the COVID-19 pandemic has affected travel agencies in Turkey. In the study, it was stated that travel agencies were greatly affected by the epidemic and their income sources were cut off. Despite this, it has been revealed that many fixed expenses continue, and agencies need more support. Kıvılcım (2020) stated that a new era will start in the tourism sector after COVID-19, contactless digital transactions, social distancing tours and isolated holidays will be at the forefront in his study on the effects of the COVID-19 pandemic on the tourism sector.

When looking at other literature studies on women's entrepreneurship, it is emphasized that the level of education of women entrepreneurs in general is a very important factor in their individual values in the women's entrepreneurship study called the example of Antalya (Yirik & Ilgaz Yıldırım, 2014a; 2014b). In another study, it was concluded that the perspective of the social and cultural environment affects women's entrepreneurial life (Sayın, 2011). Bedük (2015) claims that the basis of the problems that women face in their working life is due to some failures in the economic, social and cultural structure of society.

Soysal (2010) emphasizes that it is known that women entrepreneurs should increase their education levels and have a professional equipment within the scope of growing their businesses and contributing more to the economy (Soysal, 2010).

METHODOLOGY

In the study, one of the qualitative research methods, the document review technique was used. Document review is the study of written materials containing information about events or phenomena being examined (Yıldırım & Şimşek, 2021; Çatlak et al., 2015). It is defined as a scientific research method based on the collection, study and interpretation of various documents as primary source of research data. In addition, while it is usually used as a support for various research methods in literature, it is also preferred as a single method (Yıldırım & Şimşek, 2021). Bibliometric studies are playing role about determining the current status of scientific publications in their field also deficiencies and effectiveness (Şen, 2020).

RESULTS

In the table below, the articles regarding conducted studies related to women entrepreneurship since the Covid-19 virus in Turkey, are examined according to the author's name, year, article name, purpose of the study.

In the table above, academic works are examined according to the years. Within the scope of bibliometric analysis, it was determined that there were the most single-authored studies with 11 authors. When viewed by years, it is seen that the most studies are in the years 2019 and 2022. In this case, it is undoubtedly understood that the effect of the pre-pandemic period and the recovery processes after the pandemic are effective. It is the qualitative method with a significant difference that the authors prefer the most as a method. Although there is an increase, it can be said that the qualitative research method is preferred as a data collection method, especially since the number of women entrepreneurs in Turkey is relatively few. Studies have generally focused on women entrepreneurs in rural areas in terms of their

Table 1.

Author(s)	Year	Article Name	Method	Purpose
Açıkgöz, C.	2019	The Role and Importance of the Women Entrepreneur in Rural Tourism: Case of Muğla	Qualitative	This study that aims to investigate the role and the importance of women entrepreneurs in Muğla province employed qualitative research methods.
Akca, Ç.	2019	Determining The Women Entrepreneurs Entrepreneurial Intent And Its Relations With Self-Confidence: A Research Of Food And Beverage Sector	Quantitave	The aim of the study is determinig the women entrepreneurs entrepreneurial intent and its relations with self-confidence in food and beverage sector.
Aylan, F. N., Sarı Gök, H. & Şalvarcı, S.	2019	The Role of Woman Entrepreneurship in Reduction of Rural Poverty: A Case of Levander Smelling Village	Qualitative	The aim of this study is to reveal the role of tourism sector and women entrepreneurship activities in rural poverty reduction.
Bakay, M. E., Müftüoğlu, M., Nalbantoğlu, A. & Çoçan, E.	2019	A Qualitative Research on Women Entrepreneurship in Rural Areas and Problems Encountered by Entrepreneurial Women: The Case of Manisa-Demirci	Qualitative	In this study, it was aimed to identify the challenges and problems faced by women entrepreneurs in the Demirci district of Manisa, to provide a variety of suggestions in line with the findings and to analyze the identity of rural women's entrepreneurship.
Başaran, H. & Irmak, E.	2022	The importance of cooperatives in local product production and local women's entrepreneurship in Devrek	Qualitative	In this study, analyzes and inferences were made by examining the activities of two separete cooperatives that transform local products into local delicacies with the production and marketing of walking stick, women's entrepreneurship and local products in Devrek District of Zonguldak Province.
Bayram, A. T.	2019	Bibliometric Analysis Of Women Studies On Turkish Tourism Literature	Qualitative	The aim of the research from this perspective; It is the examination of the bibliometric characteristics of the works published with the keyword in woman on Turkish Tourism literature.
Bozdemir, M., Gülbuçuk, B. & Bayramoğlu, Z.	2021	Determination Of Entrepreneur Qualifications And Entrepreneurship Skills Of Women In The Rural Area: A Research Province Konya	Qualitative	This study was carried out to determine the entrepreneurial qualifications and entrepreneurship skills of women who have a significant labor force potential in the countryside.
Demiral, N. Ö. & Hassan, A.	2020	Women Entrepreneurs and Social Media: A Research on Instagram Pages of Bozcaada Accommodation Businesses	Qualitative	This study, which will be evaluated in the triangle of women, social media and tourism sector, aims to determine the intensity of the women entrepreneurs of Bozcaada accommodation enterprises to use Instagram, which is a social media tool.
Gündüz, A. Y. & Karadeniz, B.	2022	Historical Development Of Entrepreneurship And Women's Entrepreneurship: A Review Of The Studies In Malatya	Qualitative	In the study, the participation processes of women entrepreneurs in Turkey and the difficulties they experince are mentioned and the similarity levels of the research carried out in Malatya with the literature studies are examined by giving place to the studies in the literature results have been achieved.
İncili, İ. İ.	2022	Financial Policies For Women Entrepreneurs: Case Of Denizli	Quantitave	In the study, the factors that lead women to entrepreneurship were examined, the problems they faced during the stages of starting a business and running the business were analyzed; The effect of the support given by public institutions and organizations to women entrepreneurs has been examined.
Kaleözü, S.	2021	Motivation Factors Of Female Entrepreneurs In Tourism Sector: A Typology Proposal For Turkey	Qualitative	The aim of the study is to evaluate the concept of motivation and women's entrepreneurship together, to reveal the motivation factors that lead women to become entrepreneurs and to understand the importance of the motivation factors on women's entrepreneurship.
Kemer, E. & Aslan, Z.	2021	Women in Hospitality Businesses: Vignette Technique Application	Qualitative	This study aims to analyze and evaluate the experiences of women in choosing to work in the tourism sector and coming to managerial positions.
Kısa, M. C. & Şahin, D.	2022	Being a Female Manager in Hotel Businesses: The Case of Ordu Province	Qualitative	In this study, it was aimed to determine a number of problems encountered by women at the executive level in the tourism sector, which allows women to be employed at a high rate compared to other sector, and it was carried out in 4- and 5- star accommodation enterprises with operation certificates in Ordu.
Kurtay, E.	2021	Evaluation Of Women's Entrepreneurship In The Ecotourism Sector Through The Practical Examples In Duzce Province	Qualitative	With this study, the role of women entrepreneurs in the ecotourism sector in Düzce and the participation rate of women in ecotourism in the region will be determined and evaluated in terms of female entrepreneur candidates performing their commercial activities using the fields and spaces they own.
Nişancı, Z. N. & Ödemiş, G.	2022	Women Entrepreneurs And Work-Life Balance: An Analysis In Terms Of Demographic Factors	Quantitative	The aim of this research is to reveal the perceptions of women entrepreneurs on work-life balance and whether work-life balance changes according to demographic variables. Questionnaire technique was used as data collection method in the research.
Özden, C.	2019	Financial Problems of Women Entrepreneurs in Rural Areas: Milas (Muğla)- Bydgoszcz (Poland) Comparison	Quantitative	This study was carried out to compara the female entrepreneurship activites carried out in the rural district of Milas in the province of Mugla and in Bydgoszcz, a rural area of Poland.
Özen, İ. A., Karadeniz, G. & Zaro, E.	2022	Women Entrepreneur Restaurants and Determination of Tourist Preferences: The Case of Cappadocia	Qualitative	In the research, it is aimed to determine the reasons for the preference of tourists in restaurants run by women entrepreneurs and serving local dishes in the Cappadocia Region.

Author(s)	Year	Article Name	Method	Purpose
Saçak Gündüz, S.	2021	Women's Entrepreneurship Literature in Turkey During The COVID-19 Pandemic	Qualitative	Within the scope of the study, a literature review was conducted for academic publications which issues women entrepreneurship and published in 2020 and in the first half of 2021, when the COVID-19 pandemic prevailed.
Serinikli, N.	2019	The Attitudes of The Micro Female Entrepreneurs in The Rural Area Towards Becoming A Cooperative: Agro-Toruism Woman Cooperative	Quantitative	The research has been carried out to ascertain the current entrepreneurial activities of the female entrepreneurs who live in the rural area and to reveal the attitudes of the women towards becoming a cooperative.
Tamer, M.	2021	Women Entrepreneurs In The Cıttaslow Movement: A Case Study Of Seferihisar	Qualitative	The objective of this study is to examine the Cittaslow movement in Seferihisar in the context of women.
Tamer, M. & Alaeddinoğlu, F.	2022	Women Entrepreneurs in the Cittaslow Movement: The Case of Seferihisar	Qualitative	This study, which scrutinizes the Cittaslow movement in Seferihisar by focusing on women entrepreneurs, aims to reveal the impacts of the Cittaslow movement on women entrepreneurs and how women are affected by this process.
Toptaş, A.	2020	The Impact of Cultural Tourism on the Innovations of Women Entrepreneurs: Investigation of Boğatepe Village by Case Study Analysis	Qualitative	The purpose of this study is to show these innovation studies as an example to entrepreneur women and to encourage women living in different regions to invest in cultural tourism.

Figure 1.

economic contribution. Women's entrepreneurship in accommodation or catering establishments has been investigated, but it has been women's entrepreneurship in rural areas that has been particularly emphasized in tourism.

In the above figure, the most repeated words in academic works were analyzed and it is seen that there are mostly expressions of women, enterprise, entrepreneur, entrepreneurship, tourism, pandemic and Covid-19.

In the graph above, the distribution of academic studies by years is given. it is understood that there are 7 studies in 2019 and 2022, followed by the year 2021 and the year 2020. It is understood that the impact of the pandemic has made itself felt especially in 2020 with restrictions and bans, and in this context, it has changed its business models.

In the graph above, the maximum number of authors seen in academic studies is given. In this context, 11 studies were prepared with one author, 7 studies with two authors, 3 studies with three authors and 1 study with four authors.

In the graph above, the most commonly used research method in academic studies is a relatively limited topic and the qualitative method has been preferred in terms of data collection. Accordingly, the

Figure 2.

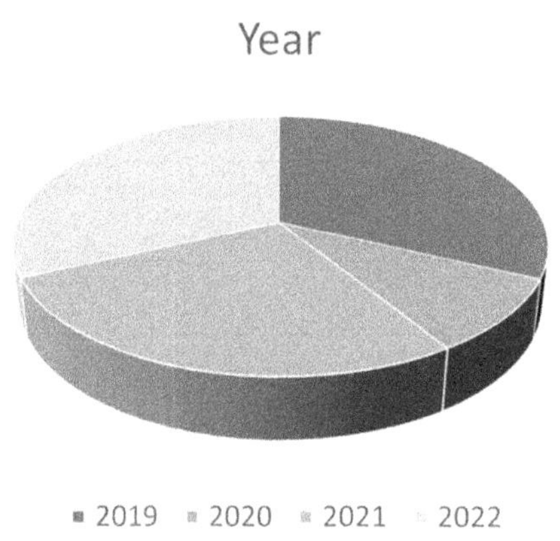

Figure 3.

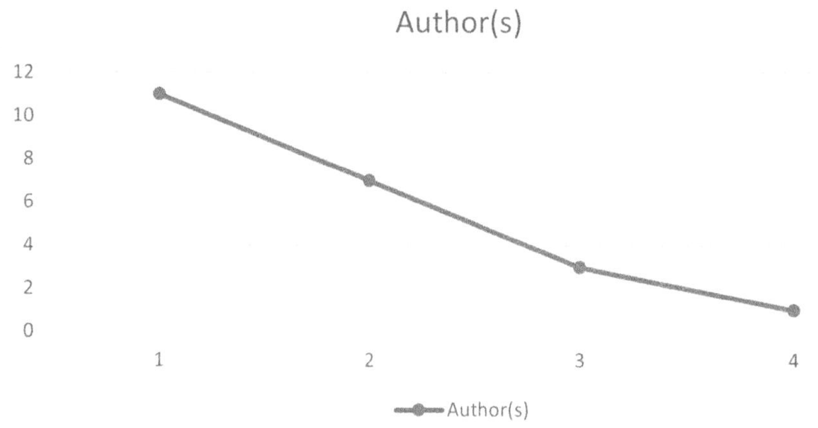

Figure 4.

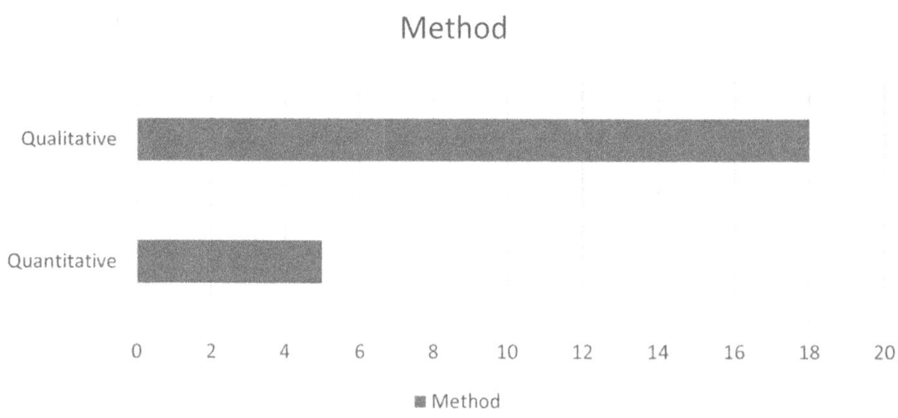

qualitative method was used in 18 studies and the quantitative method was used in 5 studies, and it is understood that there is no study that uses the mixed method.

LIMITATIONS

Among the limitations of the study, the first is time. In addition, conducted studies related to women entrepreneurship since the covid-19 virus in Turkey in the literature, qualitative method was used. However, the subject can be enriched by using quantitative and different qualitative method techniques in future studies.

CONCLUSION

Within the scope of the study results in the literature, it is understood that women entrepreneurs are around 30, 40 and 50 years old and have about 2, 4, 6 or 9 years of experience. In this context, the length of the working period shows the expertise and continuity in the profession. It is understood that women entrepreneurs mainly aim to generate economic income and be free, to create an independent working environment (Karakaş & Gökmen, 2016; Erkol Bayram, 2018; Yirik & Ilgaz Yıldırım, 2014a; 2014b). The results of academic studies on women's entrepreneurship in Turkey reveal that women are taking an active role in rural tourism more. In addition, he reveals that while some of them receive support from their family regarding women's entrepreneurship, some of them do not receive support. In this context, due to women's responsibilities, the opportunity to achieve status and rise in business life becomes more difficult.

However, it is very promising that women are supported in entrepreneurship in society and that local governments offer incentives in this regard. While there are projects supporting women's entrepreneurship in European countries, there are five-year development plans in Turkey concerning women's entrepreneurship (Fidan & Nam, 2012). In the modern era, the vast majority of women want to start their own business, and the desire for entrepreneurship is directly proportional to the level of education (Esen, 2009; Kaan & Gökmen, 2016). Moreover, it has been revealed as a result of research that women entrepreneurs mostly want intelligent and talented employees in small and medium-sized enterprises (Yirik & Ilgaz Yıldırım, 2014b). The vast majority of women entrepreneurs in Turkey want to start their own business in order to benefit society and individuals. There are many women who have started their own business with the incentives within the scope of the European Union. It is also understood that the tourism sector is advantageous for women. Furthermore, it is understood that women entrepreneurs in Turkey generally attach importance to concepts such as universality and compatibility (Yirik & Ilgaz Yıldırım, 2014a).

The fact that the tourism industry offers income and employment opportunities to the national economies is very important for the world. However, negative outcomes are inevitable as a result of unforeseen crises in the sector from a regional or global point of view. The epidemic that spread in China and affected the whole world in a short time has not only affected tourism, but has pushed it into a recession process.

For this reason, there has been a decrease in tourism revenues. It should not be forgotten that tourism is a labor-intensive sector, but it consists of a human-oriented service process. In this context, among the results of the study, it was understood that there was an increase, especially in the pre- and post-pandemic process. Women's entrepreneurship in Turkey has been among the topics discussed for years. State-sponsored organizations support women's employment and various financial supports are offered for women entrepreneurs in this context. However, compared to developing countries, women's employment has just started to increase in our country. This has been caused by the fact that women have become active in business life with technological developments and want to start their own business with the opportunities they have in this context.

Especially in rural regions, the phenomenon of women's employment and entrepreneurship is very important. It provides economic benefits to the region. Women have many roles in home and business life. In this context, women entrepreneurs who have various roles today need support in business life. Today, women's entrepreneurship is among the most popular topics, while its difficulties are ignored. In particular, the factor of female empowerment comes to the fore.

Strong women and entrepreneurs are needed for strong societies to exist. In this context, the study focused on women's entrepreneurship related to women's empowerment. The reason why the most used method in the findings obtained is the qualitative research method according to the years is that the sector needs to develop. Similarly, due to the restrictions in the pandemic period, the way of doing business has changed and there has been a loss of income. In this case, financial support for women entrepreneurs has brought with it a number of difficulties, such as the recovery process.

REFERENCES

Abu-Rabia-Queder, S. (2010). Independence and entrepreneurship among Arab Muslim rural and Bedouin women in Israel. *Gender, Place and Culture*, *17*, 123–126.

Acar, Y. (2020). Yeni Koronavirüs (COVİD-19) Salgını ve Turizm Faaliyetlerine Etkisi [The New Coronavirus (COVID-19) Pandemic and Its Impact on Tourism Activities]. *Güncel Turizm Araştırmaları Dergisi*, *4*(1), 7–21. doi:10.32572/guntad.703410

Açıkgöz, C. (2019). *Kırsal Turizmde Kadın Girişimcinin Yeri ve Önemi Muğla Örneği* [The Role and Importance of the Women Enrepreneur in Rural Tourism: Case of Muğla] [Master Thesis]. Muğla Sıtkı Koçman University.

Akca, Ç. (2019). *Kadın Girişimcilerin Girişimcilik Niyetleri ile Kendine Güven İlişkisinin Belirlenmesi: Yiyecek İçecek Sektöründe Bir Araştırma* [Determining the Women Entrepreneurs Entrepreneurial Intent and Its Relations With Self-Confidence: A Research of Food and Beverage Sector] [Master Thesis]. Ankara Hacı Bayram Veli University.

Akehurt, G., Simarro, E., & Mastur, A. (2012). Women entrepreneurship in small service firms: Motivations, barriers and performance. *Service Industries Journal*, *32*(15), 2489–2505. doi:10.1080/02642069.2012.677834

Alpkan, L., Bulut, C., Gunday, G., Ulusoy, G., & Kilic, K. (2010). Organizational support for intrapreneurship and its interaction with human capital to enhance innovative performance. *Management Decision*, *48*(5), 732–755. doi:10.1108/00251741011043902

Aylan, F. K., Sarı Gök, H., & Şalvarcı, S. (2019). Kırsal Yoksulluğun Giderilmesinde Kadın Girişimciliğin Rolü: Lavanta Kokulu Köy Örneği [The Role of Woman Entrepreneurship in Reduction of Rural Poverty: A Case of Levander Smelling Village]. *Journal of Tourism Gastronomy Studies*, *7*(2), 1271–1289. doi:10.21325/jotags.2019.420

Bahar, O., & Çelik İlal, N. (2020). Coronavirüsün (COVİD-19) Turizm Sektörü Üzerindeki Ekonomik Etkileri [The Economic Effects of Coronavirus (COVID-19) on the Tourism Sector]. *International Journal of Social Sciences and Education Research*, *6*(1), 125–139. doi:10.24289/ijsser.728121

Bakar, N. A., & Rosbi, S. (2020). Effect of Coronavirus Disease (COVİD-19) to Tourism Industry. *International Journal of Advanced Engineering Research and Science*, *7*(4), 189–193. doi:10.22161/ijaers.74.23

Bakay, M. E., Müftüoğlu, M., Nalbantoğlu, A., & Çoçan, E. (2019). Kırsal Kesimlerdeki Kadın Girişimciliği ve Girişimci Kadınların Karşılaştıkları Sorunlar Üzerine Nitel Bir Araştırma: Manisa-Demirci Örneği [A Qualitative Research on Women Entrepreneurship in Rural Areas and Problems Encountered by Entrepreneurial Women: The Case of Manisa-Demirci]. *Journal of Yasar University*, *15*(Special Issue), 295–306.

Başaran, H., & Irmak, E. (2022). Devrek'te Yerel Ürün Üretimi ve Yerelde Kadın Girişimciliğinde Kooperatiflerin Yeri [The importance of cooperatives in local product production and local women's entrepreneurship in Devrek]. *Ahi Evran Akademi*, *3*(1), 19–31.

Bayram, A. T. (2019). Türk Turizm Alan Yazınında Kadınlar Üzerine Yapılan Çalışmaların Bibliyometrik Analizi [Bibliometrik Analysis of Women Studies On Turkish Tourism Literature]. *Turizm ve Araştırma Dergisi*, *8*(2), 56–77.

Bedük, A. (2005). Women Working in Turkey and Women's Entrepreneurship. *Electronic Journal of Social Sciences*, *3*(12), 106–117.

Beqo, I., & Gehrels, A. S. (2014). Women entrepreneurship in developing countries: A European example. *Research in Hospitality Management*, *4*(1-2), 97–103. doi:10.1080/22243534.2014.11828320

Bourne, L. (2011). Advising upwards: Managing the perceptions and expectations of senior management stakeholders. *Management Decision*, *49*(6), 1001–1023. doi:10.1108/00251741111143658

Bozdemir, M., Gülçubuk, B. & Bayramoğlu, Z. (2021). Kırsal Alandaki Kadınların Girişimci Niteliklerinin ve Girişimcilik Becerilerinin Belirlenmesi: Konya İli Araştırması [Determination of Entrepreneur Qualifications and Entrepreneurship Skills of Women in the Rural Area: A Research Province Konya]. *Trakya Üniversitesi Sosyal Bilimler Dergisi, 23*, 87-106. doi:10.26468/trakyasobed.871643

Brush, C. G., Bruin, A., & Welter, F. (2009). A gender-aware framework for women's entrepreneurship. *International Journal of Gender and Entrepreneurship*, *1*(1), 8–24. doi:10.1108/17566260910942318

Ca'ceres, R., Guzma'n, J., & Rekowski, M. (2011). Firms as source of variety in innovation: Influence of size and sector. *The International Entrepreneurship and Management Journal*, *7*(3), 357–372. doi:10.100711365-011-0198-8

Çatlak, Ş., Tekdal, M., & Baz, F. Ç. (2015). Scratch yazılımı ile programlama öğretiminin durumu: bir doküman inceleme çalışması [The status of programming teaching with Scratch software: a document review study]. *Journal of Instructional Technologies ve Teacher Education*, *4*(3), 13–25.

Cavalcante, S., Kesting, P., & Ulhøi, J. (2011). Business model dynamics and innovation: (Re)establishing the missing linkages. *Management Decision*, *49*(8), 1327–1342. doi:10.1108/00251741111163142

Collins, J., & Low, A. (2010). Asian female immigrant entrepreneurs in small and medium-sized businesses in Australia. *Entrepreneurship and Regional Development*, *22*(1), 97–111. doi:10.1080/08985620903220553

Comeche, J. M., & Loras, J. (2010). The influence of variables of attitude on collective. Entrepreneurship. *The International Entrepreneurship and Management Journal*, *6*(1), 23–38. doi:10.100711365-009-0131-6

Demiral, N. Ö., & Hassan, A. (2020). Kadın Girişimciler ve Sosyal Medya: Bozcaada Konaklama İşletmelerinin Instagram Sayfalarına Yönelik Bir Araştırma [Women Entrepreneurs and Social Media: A Research on Instagram Pages of Bozcaada Accommodation Businesses]. *AHBVÜ Turizm Fakültesi Dergisi*, *23*(2), 359–373. doi:10.34189/tfd.23.02.007

Du Rietz, A., & Henrekson, M. (2000). Testing the female underperformance hypothesis. *Small Business Economics*, *14*(1), 1–10. doi:10.1023/A:1008106215480

Elam, A., & Brush, C. G. (2010). Women, entrepreneurship and social capital: A dialogue and construction. *Organization Studies*, *31*(2), 245–249. doi:10.1177/0170840610363530

Erkol Bayram, G. (2018). Women Entrepreneurs and Tourism: An Investıgation on the Current Situation and Problems in Sinop Province. *Journal of Business Research*, *10*(2), 56–88.

Essers, C., Benschop, Y., & Doorewaard, H. (2010). Female ethnicity: Understanding Muslim immigrant businesswomen in The Netherlands. *Gender, Work and Organization*, *17*(3), 320–339. doi:10.1111/j.1468-0432.2008.00425.x

Ferrante, F. (2005). Revealing entrepreneurial talent. *Small Business Economics*, *25*(2), 159–174. doi:10.100711187-003-6448-6

Fidan, F., & Nam, D. (2012). New Dynamics in Rural Tourism: Women Entrepreneus-Sample of Taraklı. *KMU Journal of Social and Economic Research*, *14*(23), 51–57.

Ga'mez-Gonza'lez, J., Rondan-Catalun˜a, F. J., Diez-de Castro, E. C., & Navarro-Garcia, A. (2010). Toward an international code of franchising. *Management Decision*, *48*(10), 1568–1595. doi:10.1108/00251741011090333

Georgeta, I. (2012). Women entrepreneurship in the current international business environment. *Cogito-Multidisciplinary Research Journal*, *1*, 122–131.

Goktan, A. B., & Miles, G. (2011). Innovation speed and radicalness: Are they inversely related? *Management Decision*, *49*(4), 533–547. doi:10.1108/00251741111126477

Goyal, M., & Parkash, J. (2011). Women Entrepreneurship in India- Problems and Prospects. *ZENITH International Journal of Multidisciplinary Research*, *1*(5), 195–207.

Gündüz, A. Y., & Karadeniz, B. (2022). Girişimcilik Olgusunun Tarihsel Gelişimi ve Kadın Girişimciliği: Malatya İlinde Yapılan Çalışmanın İncelenmesi [Historical Development Of Entrepreneurship And Women's Entrepreneurship: A Review Of The Studies In Malatya]. *Bingöl University Journal of Social Sciences Institute*, *24*(24), 96–112. doi:10.29029/busbed.1119269

Haberturk. (2022). *Kadın Girişimcilerin Ağırlığı Artıyor* [The Influence of Women Entrepreneurs is Growing]. Available at: https://www.haberturk.com/kadin-girisimcilerin-agirligi-artiyor-3370908-ekonomi

Halıcı, E. (2005). *Girişimcilik ve teknoparklar: Girişimciliğin eski ve yeni yüzü.* Available at: https://www.emo.org.tr/etkinlikler/egitim/etkinlik_bildirileri_detay.php?etkinlikkod=19&bilkod=350

Hisrich, D. R., & Peters, M. P. (2002). *Entrepreneurship* (5th ed.). McGraw Hill Higher Education.

Hormiga, E., Batista-Canino, R. M., & Sa'nchez-Medina, A. (2011). The role of intellectual capital in the success of new ventures. *The International Entrepreneurship and Management Journal*, *7*(1), 71–92. doi:10.100711365-010-0139-y

Hotho, S., & Champion, K. (2011). Small businesses in the new creative industries: Innovation as a people management challenge. *Management Decision*, *49*(1), 29–54. doi:10.1108/00251741111094428

Hoxha, D., & Krasnıq, B. (2008). Female Entrepreneurship in Kosova: Does Gender Difference Hypothesis Hold? *International Journal of Business and Management Science*, *1*(1), 49–65.

Huang, J. Y., Chou, T. C., & Lee, G. G. (2010). Imitative innovation strategies: Understanding resource management of competent followers. *Management Decision*, *48*(6), 952–975. doi:10.1108/00251741011053488

Huarng, K. H., & Yu, T. H. K. (2011). Entrepreneurship, process innovation and value creation by a non-profit SME. *Management Decision*, *49*(2), 284–296. doi:10.1108/00251741111109160

İbiş, S. (2020). COVİD-19 Salgınının Seyahat Acentaları Üzerine Etkisi. (The Impact of the COVID-19 Pandemic on Travel Agencies). *Safran Kültür ve Turizm Araştırmaları Dergisi*, *3*(1), 85–98.

İlter, B. (2008). *Girişimcilik Sürecinde Kadın Girişimcilerin Karşılaştıkları Sorunların Analizi: Kagider Örneği* [Analysis of The Problems Faced by Women Entrepreneurs in the Entrepreneurship Process: The Case of Kagider] [Doctoral Thesis]. Afyon Kocatepe University.

İncili, İ. İ. (2022). *Girişimci Kadınlara Yönelik Mali Politikalar: Denizli Örneği* [Financial Policies for Women Entrepreneurs: Case of Denizli] [Master Thesis]. Pamukkale University.

Kaleözü, S. (2021). *Turizm Sektöründe Kadın Girişimcilerin Motivasyon Faktörleri: Türkiye için Bir Tipoloji Önerisi* [Motivation Factors of Femalae Entrepreneurs in Tourism Sector: A Typology Proposal for Turkey] [Master Thesis]. Ankara Hacı Bayram Veli University.

Kantar, M. (1999). Entrepreneurship and Rural Women. *Journal of Agricultural Economics*, *4*(1/2), 29–42.

Karakaş, A., & Gökmen, G. (2016). Studying Woman Entrepreneurs Profile In Tourism Sector. *Journal of Recreation and Tourism Research*, *3*(3), 18–25.

Karim, W., Haque, A., Anis, Z., & Ulfy, M. A. (2020). The Movement Control Order (MCO) for COVİD-19 Crisis and its Impact on Tourism and Hospitality Sector in Malaysia. *International Tourism and Hospitality Journal*, *3*(2), 1–7.

Kautonen, T., & Palmroos, J. (2010). The impact of a necessity-based start-up on subsequent entrepreneurial satisfaction. *The International Entrepreneurship and Management Journal*, *6*(3), 285–300. doi:10.100711365-008-0104-1

Kelley, D. J., Brush, C. G., Greene, P. G., & Litovsky, Y. (2013). *Global Entrepreneurship Monitor: 2012 Women's Report*. The Center for Women's Leadership at Babson College and London Business School.

Kemer, E., & Aslan, Z. (2021). Konaklama İşletmelerinde Kadın Yönteici Olmak: Vignette Tekniği Uygulaması [Women in Hospitality Businesses: Vignette Technique Application]. *Turizm Akademik Dergisi*, *9*(1), 39–53.

Kısa, M. C., & Şahin, D. (2022). Otel İşletmelerinde Kadın Yönetici Olmak: Ordu İli Örneği [Being a Female Manager in Hotel Enterprises: The Example of Ordu Province]. *ODÜSOBİAD*, *12*(2), 1061–1088. doi:10.48146/odusobiad.1102336

Kıvılcım, B. (2020). COVİD-19 (Yeni Koronavirüs) Salgınının Turizm Sektörüne Muhtemel Etkileri [Possible Effects of the COVID-19 (Novel Coronavirus) Pandemic on the Tourism Sector]. *USOBED Uluslararası Batı Karadeniz Sosyal ve Beşeri Bilimler Dergisi*, *4*(1), 17–27. doi:10.46452/baksoder.727208

Koh, C. H. (1996). Testing hypotheses of entrepreneurial characteristics a study of Hong Kong. *Journal of Managerial Psychology*, *11*(3), 12–25. doi:10.1108/02683949610113566

Koutsou, S., Notta, O., Samathrakis, V., & Partalidou, M. (2009). Women's Entrepreneurship and Rural Tourism in Greece: Private Enterprises and Cooperatives. *South European Society & Politics*, *14*(2), 191–209. doi:10.1080/13608740903037968

Kumbhar, V. (2013). Some Critical Issues of Women Entrepreneurship in Rural India. *European Academic Research*, *1*(2), 192–200.

Kurtay, E. (2021). Ekoturizm Sektöründe Kadın Girişimciliğinin Düzce İli Uygulama Örnekleri Üzerinde Değerlendirilmesi [Evaluation of Women's Entrepreneurship in the Ecotourism Sector Through the Practical Examples in Düzce Province]. *PEYZAJ- Eğitim, Bilim. Kültür ve Sanat Dergisi*, *3*(2), 93–105. doi:10.53784/peyzaj.1025166

Kutanis, R. Ö. & Hancı, A. (2004). Kadın Girişimcileri Kişisel Özgürlük Algılamaları [Women Entrepreneurs' Perceptions of Personal Freedom]. *3. Ulusal Bilgi, Ekonomi ve Yönetim Kongresi, Osmangazi Üniversitesi İİBF,* 457- 464.

Lee, C. C., & Chen, C. J. (2011). The reaction of elderly Asian tourists to avian influenza and SARS. *Tourism Management*, *32*(6), 1421–1422. doi:10.1016/j.tourman.2010.12.009 PMID:32287735

Lee, J. (1996). The motivation of women entrepreneurs in Singapore. *Women in Management Review*, *11*(2), 18–29. doi:10.1108/09649429610112574

Leoni, T., & Falk, M. (2010). Gender and field of study as determinants of self-employment. *Small Business Economics*, *34*(2), 167–185. doi:10.100711187-008-9114-1

Lin, E., Lin, T. M. Y., & Lin, B. W. (2010). New high-tech venturing as process of resource accumulation. *Management Decision*, *48*(8), 1230–1246. doi:10.1108/00251741011076762

Mainardes, E. W., Alves, H., & Raposo, M. (2011). Stakeholder theory: Issues to resolve. *Management Decision*, *49*(2), 226–252. doi:10.1108/00251741111109133

Minniti, M., & Arenius, P. (2003). *Women in Entrepreneurship*. Presented at the Symposium of the Entrepreneurial Advantage of Nations, First Annual Global Entrepreneurship Symposium, New York, NY.

Moore, P. (2000). Transitions-Milestones On The Route To Success, USASBE Proceeding. *İstanbul Kadın Araştırmaları Dergisi, 18*, 1-5.

Naranjo-Valencia, J. C., Jime'nez-Jime'nez, D., & Sanz-Valle, R. (2011). Innovation or imitation? The role of organizational culture. *Management Decision*, *49*(1), 55–72. doi:10.1108/00251741111094437

Nişancı, Z. N., & Ödemiş, G. (2022). Kadın Girişimciler ve İş-Yaşam Dengesi: Demografik Faktörler Açısından Bir Analiz [Women Entreprenurs and Work-life Balance: An Analysis in terms of Demographic Factors]. *Journal of Business, Innovation and Governance*, *5*(1), 44–72.

Okafor, C., & Amalu, R. (2010). Entrepreneurial Motivations as Determinants of Women Entrepreneurship Challenges. *Petroleum-Gas University of Ploiesti BULLETIN Economic Sciences Series*, *62*(2), 67–77.

Özden, C. (2019). *Kırsal Alanlarda Kadın Girişimcilerin Karşılaştıkları Finansal Sorunlar: Milas (Muğla) – Bydgoszcz (Polonya) Karşılaştırması* [Financial Problems of Women Entrepreneurs in Rural Areas: Milas (Muğla)- Bydgoszcz (Poland) Comparison] [Master Thesis]. Muğla Sıtkı Koçman University.

Özen, İ. A., Karadeniz, G., & Zaro, E. (2022). Kadın Girişimci Restoranlar ve Turist Tercihlerinin Belirlenmesi: Kapadokya Örneği [Women Entrepreneur Restaurants and Determination of Tourist Preferences: The Case of Cappadocia]. *Journal of Tourism and Gastronomy Studies*, *10*(3), 2314–2326. doi:10.21325/jotags.2022.1093

Öztürk, İ., & Tankuş, E. (2020). Covid-19'un Potansiyel Turist Profili Oluşturan Tüketici Davranışı Üzerindeki Etkilerinin Çözüm Önerileri Perspektifinde Değerlendirilmesi [Evaluation of the Effects of Covid-19 on Consumer Behavior That Creates a Potential Tourist Profile in the Perspective of Solution Proposals]. *Sivas Cumhuriyet Üniversitesi İktisadi ve İdari Bilimler Dergisi*, *21*(2), 103–121.

Pardo-del-Val, M. (2010). The Services supporting female entrepreneurs. *Service Industries Journal*, *30*(9), 1479–1498. doi:10.1080/02642060802626840

Pazarcık, Y. (2010). *Eğitimin Bireysel Değerler ve Yöneticilerin İş Görme Anlayışına Etkisi: Türkiye'de Yabancı Dilde Eğitim Veren Orta Eğitim Kurumları Mezunları Üzerine Bir Araştırma* [The Impact of Education on Individual Values And Managers' Understanding of Working: A Research on Graduates of Secondary Educational Institutions Providing Education in a Foreign Language in Turkey] [Doctoral Thesis]. Marmara University.

Rashid, A. T. (2010). Development through social entrepreneurship: Perspectives and evidence from Bangladesh. *Canadian Journal of Development Studies*, *30*(3-4), 441–455. doi:10.1080/02255189.2010.9669311

Saçak Gündüz, S. (2021). COVID-19 Salgını Döneminde Türkiye'de Kadın Girişimcilik Yazını [Women's Entrepreneurship Literature in Turkey During The COVID-19 Pandemic]. *Journal of Research in Entrepreneurship Innovation and Marketing*, *5*(10), 148–165.

Sadi, M. A., & Al-Ghazali, B. M. (2010). Doing business with impudence: A focus on women entrepreneurship in Saudi Arabia. *African Journal of Business Management*, *4*, 1–11.

Şahin, E. (2006). *Kadın Girişimcilik ve Konya İlinde Kadın Girişimcilik Profili Üzerine Bir Uygulama* [An Application on Women Entrepreneurship and the Profile of Women Entrepreneurship in Konya Province] [Master Thesis]. Selcuk University.

Sarfaraz, L., Faghih, N., & Majd, A. A. (2014). The relationship between women entrepreneurship and gender equality. *Journal of Global Entrepreneurship Research*, *2*(6), 1–11. doi:10.1186/2251-7316-2-6

Sayın, E. (2011). Descriptive Analysis of The Problems of Women Entrepreneurs. *Journal of Organization and Management Sciences*, *3*(1), 1309–8039.

Sebora, T. C., & Theerapatvong, T. (2010). Corporate entrepreneurship: A test of external and internal influences on managers' idea generation, risk taking, and proactiveness. *The International Entrepreneurship and Management Journal*, *6*(3), 331–350. doi:10.100711365-009-0108-5

Şen, Ö. (2020). Bibliyometrik Analiz Yöntemi ile Ağızdan Ağıza İletişim (Wom) Konusunun İncelenmesi. *Journal of International Social Sciences Studies*, *6*(54), 1–10. doi:10.26449ssj.1919

Serinikli, N. (2019). Kırsal Alandaki Mikro Kadın Girişimcilerin Kooperatifleşmeye Karşı Tutumları: Agro-Turizm Kadın Kooperatifleri [The Attitudes of The Micro Female Entrepreneurs in The Rural Area Towards Becoming A Cooperative: Agro-Toruism Woman Cooperative]. *Girişimcilik ve Kalkınma Dergisi*, *14*(1), 45–57.

Sharma, P. (2013). Women Entrepreneurship Development in India. *Global Journal of Management and Business Studies*, *3*(4), 371–376.

Soysal, A. (2010). Characteristics of Women Entrepreneurs, the Problems They Face and Suggestions for Women who will Start a Business: A Research in Kahramanmaraş Province. *Journal of Eskisehir Osmangazi University IIBF*, *5*(1), 71–95.

Tambunan, T. (2009). Women entrepreneurship in Asian developing countries: Their development and main constraints. *Journal of Development and Agricultural Economics*, *1*(2), 27–40.

Tamer, M. (2021). *Cittaslow Hareketinde Kadın Girişimciler: Seferihisar Örneği* [Women Entrepreneurs in the Cittaslow Movement: A Case Study of Seferihisar] [Master Thesis]. Van Yüzüncü Yıl University.

Tamer, M., & Alaeddinoğlu, F. (2022). Cittaslow Hareketinde Kadın Girişimciler: Seferihisar Örneği [Women Entrepreneurs in the Cittaslow Movement: The Case of Seferihisar]. *Coğrafi Bilimler Dergisi/ Turkish Journal of Geographical Sciences*, *20*(2), 500–527. doi:10.33688/aucbd.1127083

TDK. (2022). *Girişim* [Enterprise]. Available at: https://sozluk.gov.tr/

Temizkan, V., Güven, E. Ö., Yılmazer, A., & Andsoy, C. (2021). Covid-19 ile Gerçekleşen Tüketici Davranışları ve Eğilimlerine Yönelik Bir Araştırma [A Research on Consumer Behaviors and Trends That Occur with Covid-19]. *Elektronik Sosyal Bilimler Dergisi*, *20*(78), 1311–1327. doi:10.17755/esosder.839996

Toptaş, A. (2020). Kültür Turizminin Kadın Girişimcilerin Yaptıkları Yeniliklere Etkisi: Boğatepe Köyü'nün Örnek Vaka Analizi ile İncelenmesi [The Impact of Cultural Tourism on the Innovations of Women Entrepreneurs: Investigation of Boğatepe Village by Case Study Analysis]. *Journal of Turkish Tourism Research*, *4*(4), 3378–3391. doi:10.26677/TR1010.2020.568

TURSAB. (2022). *Sektörel Raporlar* [Sectoral reports]. Available at: https://www.tursab.org.tr/turkiye-turizm-degerlendirmesi/sektorel-raporlar

Wakkee, I., Elfring, T., & Monaghan, S. (2010). Creating entrepreneurial employees in traditional service sectors. *The International Entrepreneurship and Management Journal*, *6*(1), 1–21. doi:10.100711365-008-0078-z

Yazıcı Ayyıldız, A. (2020). COVİD-19 Salgın Döneminde Otellerin Pazarlama Stratejileri: Kuşadası Örneği [Marketing Strategies of Hotels During the COVID-19 Pandemic Period: The Case of Kusadasi]. *BMIJ*, *8*(3), 3328–3358. doi:10.15295/bmij.v8i3.1587

Yıldırım, A., & Şimşek, H. (2021). Sosyal Bilimlerde Nitel Araştırma Yöntemleri [Qualitative Research Methods in Social Sciences]. Ankara: Seçkin Yayıncılık.

Yirik, Ş., & Ilgaz Yıldırım, B. (2014a). Field Study of Relations Between Personal Values of Women Entrepreneurs In Tourism Sector and Their Perception of Risk and Uncertainty Perceptions: The Case of Antalya. *Journal of Niğde University IIBF*, *7*(2), 97–111.

Yirik, Ş., & Ilgaz Yıldırım, B. (2014b). Analysis of Personal Values of Women Entrepreneurs of Tourism Sector In Terms of Their Demographic Characteristics: The Case of Antalya. *Mustafa Kemal University Journal of Graduate School of Social Sciences*, *11*(26), 361–378.

Zhang, J., & Duan, Y. (2010). The impact of different types of market orientation on product innovation performance: Evidence from Chinese manufacturers. *Management Decision*, *48*(6), 849–867. doi:10.1108/00251741011053433

Zhang, K., Chen, Y., & Li, C. (2019). Discovering the tourists' behaviors and perceptions in a tourism destination by analyzing photos' visual content with a computer deep learning model: The Case of Beijing. *Tourism Management*, *75*, 595–608. doi:10.1016/j.tourman.2019.07.002

Zheng, Y., Goh, E. & Wen, J. (2020). The effects of misleading media reports about COVID-19 on Chinese tourists' mental health: a perspective article. *Anatolia: An International Journal of Tourism and Hospitality Research*. . doi:10.1080/13032917.2020.1747208

Chapter 33
Glass Ceilings and Glass Elevators as Gender Segregation in Female Employment in the Tourism Sector:
Placing a Lens on the Issue With Reference to the Muslim World

Muhammad Faisal Sultan
Khadim Ali Shah Bukhari Institute of Technology, Pakistan

Muhammad Nawaz Tunio
University of Modern Science and Sufism, Pakistan

Afifa Sardar
Bahria University, Karachi, Pakistan

Aamir Firoz Shamsi
Bahria University, Karachi, Pakistan

ABSTRACT

There is no need to highlight the importance of women as religious and cultural history is full of roles that women have played in religious, cultural, as well as business activities. Starting from Hazrat Khadija R.A., Muslim women are always admired and well-known for their acts and responsibilities. Although due to some adopted cultural, familial, political, and social barriers, professional growth of women is always hindered. Similar is the case in the Islamic world where women have to encounter several barriers in their professional growth. However, growth in the tourism industry is a special case due to the perception attached to the industry and working practices. The situation is not getting improved not only due to cultural and religious barriers but also due to inability of the educational sector to train and groom women as employees and entrepreneurs. Therefore, this chapter has been written specifically with reference to the hurdles that Muslim women are experiencing in their professional growth with special emphasis on barriers in the tourism industry.

DOI: 10.4018/978-1-6684-8417-3.ch033

INTRODUCTION

There is no one who can ignore the women's contribution to economic growth. In fact, the contribution is much more significant as in the industrial sector female representation is 25% & in the agriculture sector, it's around 40%However, females earn much lesser as compared to their counterparts & legal protection has also been found on the lower side for females. Similar is found true even for those sectors or areas where females contributed equally to their counterparts like world food production where 50% of work hangs upon the shoulders of females. However, their earning is much lesser than the males. Similar is the case of the tourism industry where women are holding around 46% of jobs although the representation is not free from sectorial and occupational separation. The postulate is true as most of the women are serving in lower-level jobs like cooking, cleaning, and hospitality positions and their representation in high-level jobs is severely low (Tristanti *et al.,* 2022).

However, tourism may also provide growth to women's careers through self-employment and entrepreneurial ventures. In fact, the leverage of women's leadership & entrepreneurship in the tourism sector is almost twice as compared to the other sectors (Hutchings et al., 2020).

ISSUES AND CHALLENGES FACED BY WOMEN WORKFORCE IN THE TOURISM INDUSTRY

Issues Faced by Women Workforce:

A study by Mohanty et al (2018), indicated that women hold a fair share of the work activities associated with the tourism industry. Although several issues still need to be resolved

a) Lack of attention on female efforts and entrepreneurial work resulted in lesser appreciation and acceptance of women's intrapreneurship and efforts.
b) Family-owned businesses in the tourism industry are heavily dependent upon the role of women. However, the role has rarely been noticed and appreciated.
c) Lack of family support also reflected negatively upon women's intrapreneurship and efforts.
d) Male domination also causes women to work in lower-level jobs even in the tourism industry. Thus, their work and efforts go unnoticed.
e) Stereotyping and perceptual barriers regarding women's capabilities are restricting women to lower-level jobs and salary scales.
f) Sexual and Professional harassment of women is raising in the tourism industry.
g) Tourism is the most attractive sector for females and has a larger share of women. Although serious efforts are required to remove gender disparity from the tourism industry.

Challenges Associated With Women's Workforce

Mohanty et al (2018), also mentioned that the participation of the women workforce in tourism is essential for the sustainable development of the industry. Although there are several challenges that may intervene in the way of women's participation and sustainable growth of the tourism industry that are as under:

a) Level of literacy of women is one of the major challenges that interrupt women's role and intrapreneurship in the tourism industry.
b) Orthodox mentality regarding women's role is hindering in the way of women's intrapreneurship and participation in the tourism industry.
c) Women are earning lower salaries than their counterparts even in similar positions. This is creating a severe negative impact on women's career development and paths.
d) Religious Problems is also creating hindrance in the way of women's intrapreneurship.

RESEARCH GAP AND GRAY AREA

Women's job is not limited to household work as it goes much beyond the boundaries of our houses. Although the work is invisible in nature and society also does not acknowledge that in the intended way (Tristanti *et al.,* 2022). Similar is the case of the tourism industry where women are mainly employed for lower-level or part-time jobs, where the jobs are not only insecure but also accompanied by violence, poor working conditions, stress, exploitation, sexual harassment, etc (Tristanti *et al.,* 2022). However, in recent times there is an increase in focus on the role of different genders in association with the tourism industry. Although the focus is not on gender segregation and therefore the research on gender with respect to occupation in the tourism industry is still lacking and not part of mainstream research (Hutchings et al., 2020). On the other side, there is also a lack of research on the area of Muslim women and tourism as initial work in this vein was started in 2010. Hence there are fewer research papers that are based on Muslim women and the tourism industry and the least of these articles are based on South-East Asia (Feldbauer & Jaffrey, 2021).

Thus, legitimate to believe Sayira et al (2021) indicated that the absence of women from the literature on tourism activities might be because of their exclusion or limited participation in tourism activities. These exclusions and restrictions are more dominant in the Muslim world (Sheikhi & Imirizaldu, 2021) based on cultural, religious, and political reasons. Thus, research work related to issues and challenges faced by women (Mohanty et al., 2018), must be conceptualized for tourism i.e., labor-intensive sector (Hutchings *et al*., 2020 & Sayira *et al.,* 2021), with reference to the Muslim women participation as host (Feldbauer & Jaffrey, 2021).

LITERATURE REVIEW

Women in Tourism Industry

Tourism is included in the list of the world's largest industries and therefore it is also labor-intensive in nature as other than full-time work roles individuals may join as a part-time worker or casual etc. In fact, due to the labor-intensive nature of the industry individuals with limited formal education or with on-the-job training may be indulged in the workforce (Tristanti *et al.,* 2022). However, a review of the literature highlighted that tourism also appears as an industry with a high level of segregation with most of the higher-level jobs being performed by men (Hutchings *et al.,* 2020). Hence the involvement of women in tourism activities is an important component for the tourism industry not only to decrease gender inequality but also for assistance in marching towards (MDGs) Third Millennium Development

Goals. Although recent findings indicated that most of the women in the tourism industry are involved in lower-level jobs like regional dances, making souvenirs, food, and beverage services& agriculture, etc. Hence, it is legitimate to reflect that the roles of men and women are also different in the tourism industry (Tristanti *et al.,* 2022).

Other than role segregation it has also been evidence that male perception regarding women is stereotyped and therefore women are not only operating on the lower levels but also on lower-paid jobs. In fact, discrimination against women in the workplace has no boundaries as it extended to sexual harassment and workplace violence. This is found true even for societies and economies with sex discrimination and anti-discrimination laws. Hence in light of the literature, two forms of segregation i.e., horizontal segregation and vertical segregation have been associated with women's prospects in the workplace. Horizontal segregation is the term that indicated women's selection for a particular form of job that is female-oriented and with fewer prospects. On the other hand, vertical segregation is related to the advancement of women at managerial level positions (Hutchings et al., 2020). Hence there is a need to focus on the development and growth of females in order to provide equal opportunities as well as to make them play role in the growth & development of industry (Tristanti *et al.,* 2022).

In fact, women are more associated with nature and therefore the better companion of the tourism industry. Moreover, they may also perform part-time jobs, i.e., after completing their household work. In fact, tourism yields several forms of value chains that may provide formal as well as informal employment to women. In fact, there are some forms of tourism that are mostly associated with the environment like eco-tourism; agro-tourism & rural tourism may be more effectively handled by women (Mohanty *et al.,* 2018). Hence women have the chance to provide their opinion and also to occupy better positions as compared to the other industries. However, still, there are several issues that are associated with women's occupation in the tourism industry that is still unsolved and needs to be addressed effectively (Mohanty *et al.,* 2018). A similar has been highlighted by Alrwajfah et al (2020) indicating the presence of several cultural, political, and family restrictions & difficulties in managing time for family, etc.

Barriers Faced by Muslim Women in Tourism Industry

A tourism system cannot be developed without close interaction between host and guest. In fact, the performance of the tourism system is not only dependent upon the satisfaction of the host (Sheikhi & Imirizaldu, 2021). Thus, in order to analyze barriers faced by Muslim women in the tourism industry there is a need to look into the research work that has been based on Muslim areas as hosts. Trends indicated that there are very few studies that are based on tourism with reference to the Muslim sides of the world. However, limited available literature associated with Muslim women as the host indicated barriers associated with religion, family life, and societal and cultural norms are creating hindrances in the way of Muslim women. In fact, Muslim women have encountered a combination of social, cultural, and religious barriers; especially for employment in the tourism industry, & these barriers are creating significant hurdles (Feldbauer & Jaffrey, 2021).

On the other side women's participation in developing countries is always preferred and it is perceived as the important factor behind the development of the country. Contrary to this Muslim world evidences significantly low women participation in economic, political, social, and cultural activities (Sheikhi & Imirizaldu, 2021). In fact, studies also believe Muslim traditions and beliefs are following old traditions and values (Masa'deh *et al.,* 2019) Studies like Feldbauer and Jaffrey (2021) and Sheikhi and Imirizaldu, (2021) provided a significant review of hurdles and challenges Muslim women faced in the tourism industry that are as under:

Barriers Due to Religious Values

Religious values are among the top-ranked predictor of gender segregation in various societies and cultures. In fact, these values even affect employment as well as employability (Feldbauer & Jaffrey, 2021). Similar sort of findings is found from the research on women working in the tourism sector of Maldives as religion was found to be a major obstacle for women working in the tourism industry, especially in enclave tourism operations (Sheikhi & Imirizaldu, 2021).

In fact, studies also believe that Islam provides men liberty to access resources and to take a decision but restricted women to domestic work and is surrounded by the pressure of household responsibility and the pressure of marriages (Feldbauer & Jaffrey, 2021). Adding to these studies also indicated that historical and political barriers create ambiguity in the understanding of religion in the desired way and therefore there is a significant gender gap in work-related roles for women in Middle-East & Arab countries (Koburtay *et al.,* 2020).

Barriers Due to Cultural Values

In Muslim countries, women are allowed to work only in selected sectors that may fall under cultural, social, and religious values (Shaban et al., 1995). However, Muslim women employed in the tourism industry have to face discrimination on the bases of cultural values from expatriates working in the tourism industry of Muslim sides (Sadi & Henderson, 2005). Although on the bases of moral guidelines from the Quran and Hadith, we must treat both genders differently but not unequally. However, in some of the Muslim parts of the world women are still perceived as house managers and therefore they are required to fulfill the duties of housekeepers or mothers. Hence there is a severe glass ceiling in the way of Muslim women to get employed (Koburtay *et al.,* 2018). In fact, due to social stigma females are perceived as inferior to their counterparts and therefore barriers always lie in front of females either in the form of barriers of mobility or barriers to meeting and commuting with strangers, etc (Alrwajfah *et al.,* 2020)

Barriers due to Marriage

In Islamic societies, marriage contracts may also have their roots in patriarchy as people with these values may treat women as perishable products. Hence, working women are believed as demanding personalities and are subjected to social scrutiny. Hence women are barred to do jobs (Feldbauer & Jaffrey, 2021).

Barriers due to Familial Duties and Responsibilities

It is also a critical factor that may restrict Muslim women's employment in the tourism industry. In fact, due to male domination, women are not allowed to perform jobs freely. Moreover, they are also subjected to the threat of family loss due to their role as working women. In fact, for women working in the tourism industry the conflict between job and family life is a usual thing that always resulted in an increase in stress and causes women to quit their jobs, especially after getting married (Feldbauer & Jaffrey,

2021). A similar sort of hurdle is indicated by Alrwajfah et al (2020) as the study indicated due to time management issues women cannot excel and grow in the tourism sector especially after getting married.

Lack of Women-Skilled Workers

There is not much difference in the education facilities available to males and females working in the tourism sector. However, the lack of skilled female employees in the tourism sector is also one of the reasons for the absence of females from top positions (Alwajifah et al., 2020)

Inefficient Educational Strategies for Women's Empowerment in Tourism

Lacking educational strategies that may empower women in the labor market is a severe issue and hinders women's empowerment as well as growth in the tourism industry (Alwajifah et al., 2020)

Empirical Evidence From the Muslim World

There are very fewer women participating in the tourism industry of Saudi Arabia regardless of the massive push that has been given to tourism by the Saudi government in recent times. Although the increase in the involvement of women's participation in the tourism industry of Saudi Arabia will not only increase women's empowerment but also produces a favorable impact on the Saudi Arab economy. In fact, the Saudi Arab Government also wants to increase women's participation in the tourism industry through its 2030 program. Although studies are still in progress to understand the way that how the government will overcome these challenges and will achieve a desirable work environment in Saudi Arabia (Almathami et al., 2022).

One of the early studies from Saudi Arabia Indicated that females have to face various personal and job-related factors, especially working with women in the tourism sector is not easy and people perceive tourism as an unfit industry for Muslim women. On the other side, it has also been noticed that working in tourism is supplemented with a lower salary structure as compared to the oil and gas sector. Therefore, in addition to familial and cultural barriers women in Saudi Arabia are reluctant to prefer the tourism sector due to religious as well as job-related issues (Al-Ismail et al., 2019).

Similar sort of findings was found from the perception of female students from UAE which indicated that cultural hurdles are the major theme associated with future job prospects in the industry. Although students also believe that extensive work experience in the field of tourism may reduce the cultural hurdles from the way of Muslim women in the tourism industry. However, lower reward expectations from the tourism industry are also making cultural hurdles more strong and deviating students towards other occupations (Griffin *et al.*, 2021). Malaysia with a Muslim majority is also a country where religion has a significant role in shaping occupation choices for females. Gender differences are also on the higher side in Malaysia and therefore working in the tourism sector is not preferred for females. In fact, tourism is not been preferred by female students for advanced-level education. Faculty associated with advanced courses in tourism education is also and females are restricted to travel abroad without a mahram (Jafari & Scot, 2019).

Alrwajfah et al (2020), also indicated that women's participation in Jordan's job market is severely low as it is around 17.3% in 2017 and due to a decrease in the rate of employment the rate of women's unemployment reaches 31.2%. The major reasons for the major women's unemployment in Jordan are driven by low-level of education, gender-based biases for leadership roles & women's responsibility towards family & home. However, women from rural areas are mostly interested in personal development, grooming, and growth. On the other side females from rural areas believes that tourism does not create a positive economic impact as well as desired job opportunities. Thus, to increase women's participation in the tourism industry there is a need to socially accept women's participation in the tourism sector and also to increase training activities for women to leverage their skills for the required roles. One of the other studies from Jordon also indicated several challenges for Muslim women's participation in the tourism industry as the part-time nature of work and social obligations are found to be creating major hurdles. However, cultural barriers, stereotyping and working for long hours, and night timing also create a considerable hindrance to women's participation in the tourism sector of Jordon (Koburtay *et al.,* 2020).

METHODOLOGY

The study is qualitative in nature which is based upon qualitative content analysis (Creswell & Creswell, 2017) and its methodology is developed through the incorporation of research and research design parameters highlighted by well-known authors like Creswell and Creswell (2017); Saunders et al (2007); Saunders et al (2015) and Sekaran and Bougie (2016) etc. The philosophy associated with the study is epistemology which is treated as the philosophy that is used to build or create knowledge and approach for conducting study is deductive. The methodological choice is mono-qualitative and research strategy is archival (Saunders et al., 2015) while purpose of research is descriptive & time horizon is cross-sectional in nature. All the above-mentioned points are valid as the study is written specifically with reference to the Muslim world with data collection is made from previous scholarly publications.

Implications and Discussion

The feminist group in the Islamic world is trying to revamp the perception and attitudes of both genders. In fact, some of the feminist groups in the Islamic world are behaving like western feminist groups that are demanding equality and more rights for women (Tavakoli & Mura, 2021). Although most of the feminist groups in the Islamic world are working for more freedom for women within the Islamic laws that are consistent with women's rights and status in Muslim Societies (Tavakoli & Muras, 2021). However, getting success in a demanding industry like tourism is much more complex for women from any side or religion of the world, especially with the responsibilities of wife, home manager, mother, etc. In fact, some of the studies argued that most of the time women got professional success prior to getting married. This is evident in Islamic as well as non-Islamic cultures but the issue is more observed in Muslim countries where several challenges and hurdles exist for married women, especially for those who are operating in the tourism industry (Nagar, 2021).

REFERENCES

Al-Ismail, S., Carmichael, F., & Duberley, J. (2019). Female employment in hotels in Saudi Arabia and UAE. *Gender in Management*, *34*(7), 554–576. doi:10.1108/GM-11-2017-0151

Almathami, R., Khoo-Lattimore, C., & Yang, E. C. L. (2022). Exploring the challenges for women working in the event and festival sector in the Kingdom of Saudi Arabia. *Tourism Recreation Research, 47*(1), 47–61. doi:10.1080/02508281.2020.1821329

Alrwajfah, M. M., Almeida-García, F., & Cortés-Macías, R. (2020). Females' perspectives on tourism's impact and their employment in the sector: The case of Petra, Jordan. *Tourism Management, 78*, 104069. doi:10.1016/j.tourman.2019.104069

Creswell, J. W., & Creswell, J. D. (2017). *Research design: Qualitative, quantitative, and mixed methods approaches*. Sage publications.

Feldbauer, I. G., & Jeffrey, H. L. (2021). Disempowered Hosts? A Literature Review of Muslim Women and Tourism. *Women in Tourism in Asian Muslim Countries*, 13-23.

Griffin, M., Hodgson, R., & Sivam, S. (2021). Career incentives, career deterrents, and cultural blocks: An investigation of factors impacting female Emirati students' perceptions of tourism. *Journal of Human Resources in Hospitality & Tourism, 20*(3), 472–496. doi:10.1080/15332845.2021.1923945

Hutchings, K., Moyle, C. L., Chai, A., Garofano, N., & Moore, S. (2020). Segregation of women in tourism employment in the APEC region. *Tourism Management Perspectives, 34*, 100655. doi:10.1016/j.tmp.2020.100655

Jafari, J., & Scott, N. (2014). Muslim world and its tourisms. *Annals of Tourism Research, 44*, 1–19. doi:10.1016/j.annals.2013.08.011

Koburtay, T., Syed, J., & Haloub, R. (2020). Implications of religion, culture, and legislation for gender equality at work: Qualitative insights from Jordan. *Journal of Business Ethics, 164*(3), 421–436. doi:10.100710551-018-4036-6

Masa'deh, R. E., Alananzeh, O., Jawabreh, O., Alhalabi, R., Syam, H., & Keswani, F. (2019). The association among employees' communication skills, image formation and tourist behaviour: Perceptions of hospitality management students in Jordan. *International Journal of Culture, Tourism and Hospitality Research, 13*(3), 257–272. doi:10.1108/IJCTHR-02-2018-0028

Mohanty, P., Swain, S. K., & Besra, S. (2018). Women at work: Exploring the issues and challenges of women employees in travel and tourism. *Indian Journal of Economics and Development, 6*(1), 1–5.

Nagar, K. (2021). Representation of women managers in hospitality and tourism: A content analysis of related magazine articles. *Gender in Management, 36*(3), 329–348. doi:10.1108/GM-10-2019-0180

Saunders, M., Lewis, P., & Thornhill, A. (2007). Research methods. Business Students, 6(3), 1-268

Saunders, M. N. K., Lewis, P., Thornhill, A., & Bristow, A. (2015). *Understanding research philosophies and approaches: Research methods for business students*. Pearson.

Sayira, T., Andrews, H., & Ayesha, Q. U. A. (2021). Are Women Empowered Producers of Tourism in Pakistan? In *Women in Tourism in Asian Muslim Countries* (pp. 81–96). Springer. doi:10.1007/978-981-33-4757-1_6

Sekaran, U., & Bougie, R. (2016). *Research methods for business: A skill building approach*. John Wiley & sons.

Sheikhi, A. R., & Imirizaldu, I. A. (2021). Women's Footprint in Traditional Muslim Ethnic Communities: The Case of World Pottery Village in Iran. In *Women in Tourism in Asian Muslim Countries* (pp. 63–80). Springer. doi:10.1007/978-981-33-4757-1_5

Tavakoli, R., & Mura, P. (2021). Muslim women travellers' constraints: A critical review. *Women in tourism in Asian Muslim countries*, 25-40.

Tristanti, T., Nurhaeni, I. D. A., Mulyanto, M., & Sakuntalawati, R. D. (2022). The Role of Women in Tourism: A Systematic Literature Review. *KnE Social Sciences*, 545-554.

Chapter 34
The Contribution of Women in Latin American Tourism Research:
CIET as a Succesfull Study Case

Maximiliano Emanuel Korstanje
University of Palermo, Argentina

ABSTRACT

In recent years, some voices have alerted on the gender inequalities not only in the tourism industry, a private sphere where the authority seems to be monopolized by men, but also in academic research. These emerging studies have emphasized the role of women in scientific knowledge production and distribution. This chapter goes in that direction, interrogating the first Latin American center fully specialized in tourism and hospitality research, CIET. Founded by Regina Schluter and her colleagues, this center inspired thousands of professional researchers in the region standing for a leading platform for tourism research in Spanish. CIET has been a source of consult and excellence in research in Argentina and Latin America over more than three decades.

INTRODUCTION

In general, the tourism industry [so to speak from its inception] has been a great employer for women in different positions (Bownell 1994; Gentry 2007; Acharya & Halpenny, 2013). Over recent years, some voices have called attention to these positions that were subordinated to the authority of men. Under other circumstances, women even earned low wages while being subject to excessive working conditions (Duffy et al 2015; Cave & Killic 2010; Norrild & Korstanje 2021). Having said this, women have starred significant contributions and innovation not only to the tourism industry but also to tourism research [as well as epistemology] (Caton 2012; Chambers & Rakic 2015). Interesting studies in (Caton 2012) & Belhassen & Caton (2009) have applauded the role of women as a democratic agent of deconstruction that lays the foundations towards a new epistemology in tourism. As these works alert, the expansion

DOI: 10.4018/978-1-6684-8417-3.ch034

of the colonial European powers in former centuries has been legitimated by a positivist version of the science. Some voices have been placed into a peripheral position or treated as irrational (child-like) actors. The dismantling of European Empires -in the mid of the twentieth century- has created a good opportunity for the rise of a deconstruction process where this relegated voices and knowledge gained further recognition. Today, tourism research has the opportunity to put women in a good position to deal with the limitations of an exhausted -if not old-fashioned- epistemology (Manwa, 2008). What seems to be more important, there is a belated recognition of women in applied tourism research, which needs to be discussed (Tucker 2009). For some reason, which is very hard to precise here, women have been historically relegated not only in the labor market but also in academic careers (Hjalager, 2003; Costa et al 2013; Bolles, 1997). As Kinnaird & Hall (2000) put it, the gendered perspective has been systematically discriminated from the epistemological discussion over decades. The opposite is equally true, the horizons of gender were monopolized by some radicalized discourses placing asunder some mix-balanced discussions. Aside from this, tourism research should incorporate the question of gender as a valid instrument to generate new theories and fresh alternatives to the problems the industry often faces (Pritchard, 2014). In the same line of inquiry, women earn lower levels of wages in tourism-related universities in comparison to men. At the same time, although women are included in the circles of professional tourism research, they are excluded from occupying chair positions. Some studies have widely shown how women are not only discriminated against but also, they are subject to material inequalities. Having said this, the obtained outcome reveals two important assumptions. Women are often limited to low-paid jobs or positions, while men monopolize the academic market. On another, men place a privileged position in research whereas women are displaced to teaching (Basurto-Barcia & Ricaurte-Quijano, 2017).

As the previous backdrop, the present book chapter explores the role played by women in Latin America during the 80s decade in the foundation of the first Center of Studies fully specialized in tourism research. Needless to say that this is the case of CIET [founded by Regina Schluter and colleagues] which historically gravitated furtherly on a shared agenda for scientific research in the constellations of hospitality and tourism for the region. CIET [Centro de Investigacion y Estudios en Turismo] is a leading institution that took the lead for tourism production knowledge in Latin America. Henceforth, the first section focuses on the question of gender in tourism, at best as it has been discussed by the literature. Under-developing economies, which develop a dependence on tourism have higher risks to encourage patriarchal practices that ultimately relegate women from the labor marketplace. As a result of this, gender inequalities are mainly marked by local cultures and traditions, as some scholars conclude. We open the discussion in another direction while introducing new critical studies to expand the current understanding of the connection between gender asymmetries and tourism. The second section puts the discussion through the lens of gender and tourism research. Over the years, women have shed light on the methodological problems of the discipline laying the foundations for the maturation of tourism research. Last but not least, we narrate the case of CIET, founded by Regina Schluter, a research center which illuminated the pathways of many professional fieldworkers in Latin America (Norrild & Korstanje, 2017).

THE QUESTION OF GENDER IN TOURISM

As stated in the introductory section, the question of gender and tourism is inevitably entwined (Abou-Shouk. 2023). Not only the industry is the main employer of women but also most of them play a leading role in different levels of service sectors (Chambers 1997; Goodse 2003). It is safe to say that women occupied a central position in the expansion of tourism and hospitality industries worldwide (Cole 2018; Gentry 2007; Duffy et al. 2015; Nassani et al 2019). As Fergusson (2011) notes, the feminist claims are related to the fact women not only are discriminated they play a game in the tourist marketplace in inferior conditions than men. Methodologically speaking, gender inequalities are measured by the MDG3 (Millennium Development Goal) with some accuracy. The results suggest that tourism development may reduce women´s inequalities but further investigation is needed to calibrate their potential in the industry for the years to come. Similar concerns and discussion can be furtherly found in (Sony 2012; Moswete & Lacey, 2015; Boley et al. 2017; Siamak-Seyfi 2023)

In consonance with this, Regina Schluter (2008) holds that the changes accelerated by the First World War enlarged women's autonomy and liberty to enter the tourism industry. Women planned their academic careers though trapped in strict protocols and gender inequalities. At a closer look, Gentry (2007) alerts that the tourism industry offers [at least in some peripheral destinations] a houseification of labor mainly based on gender segregation and low-paid wages. Although women gravitate to different positions in the industry, their participation has been minimized in the specialized literature. At the same time, women`s experience tends to be packaged and unilaterally labelled in mechanical stereotypes by the housification of labor. Anyway, this gives a further opportunity for women to gain autonomy from the local norms. What is more important, the gender inequalities created by tourism has been increased just after four decades of business expansion. Mass tourism is often seen as a perpetuating force that cements gender inequalities through the subordination and exploitation of women who are classified by their domestic skills. As the author explains, gender inequalities can be deepened by tourism or not, but it depends upon cultural mainstream values. In those societies where gender inequality culturally becomes acute and clear tourism -far from reducing the salary gap- aggravates the problem. This reminds us that segregation is culturally enrooted in a much deep-seated ideology. Small destination paradoxically provides women with new opportunities to negotiate their jobs and working conditions while cementing their subordination to the status quo. To put things simply, much of the machismo can be explained by the gender relations molded in the post-colonial period (Duffy et al. 2015). To wit, K Butler (1995) argues convincingly that host guests' encounter accelerates some long-dormant revolution [even in the question of gender]. This means that Western tourists often have the power of emancipating the local [female] workers through letters and publications while interrogating the local patriarchal order. Western female tourists, in this way, exhibit a fertile ground which needs to be investigated. In a seminal entry, Donna Chambers (2021) questions the relationship between colonialism, ethnicity and sexism. Mass migration offers a positive landscape that adjusts the exploitative colonial relations between the imperial center and its periphery. Mass migration and human diasporas allow former colonies to deliver their citizens to improve their lives. The political imbalances which are culturally engulfed in the colonial period, put all citizens in equal conditions in a free labor market. At first glimpse, the ethnic boundaries are revitalized by racism and sexism while affirming the exploitative mechanism and ideologies to preserve the ruling elite. Per Chambers, through international tourism, new forms of imperialism surface. The tourism industry is the main employer [economic driver] that recruits relegated economic factors such as migrants, and asylum-seekers [if not women] who are subject to hard working conditions and low salaries. Paradoxically, the main European metropolises that centuries ago have been the epicenters of colonial powers, nowadays receive thousands of students coming from former colonies. Higher tourism education opens the doors to valuable possibilities for overseas pre-graduate students.

As the previous argument is given, the question of segregation in tourism seems to be far from being new. Kate Hutchings et al. (2020) lament that gender asymmetries are a neglected fact, that is very hard to be empirically measured. At first glimpse, authorities and corporations deny women are segregated or subordinated to be paid lower salaries than men. Secondly and most importantly, women are over-represented in low-paid positions and employment. Even if there is no consensus in publications about gender discrimination, no less true seems to be that women are daily labelled under specific stereotypes which curtail their employment opportunities in tourism and hospitality. The introduction of anti-discrimination laws and policies makes it subtle but does not fix the problem. Hence, some studies have emphasized the role of higher tourism education to reverse the gender inequalities in the industry (Carvalho et al. 2014; Costa, Carvalho & Breda 2011; Bakas, et al 2018). In consonance with the literature, Rinaldi & Salerno (2020) suggest that the growth of tourism globally offers good opportunities for women to break with local paternal hegemony while negotiating new forms of authority and independence, but rarely does the industry revolves the vertical gender segregation. Mitra, Chattopadhyay & Chatterjee (2022) bring some reflection to what extent tourism growth aggravates or revolves the gender inequalities. Despite the classic literature suggesting that tourism consumption is a valuable poverty relief for relegated actors such as migrants, women or ethnic minorities, the sector needs skilled abilities these groups often lack. Therefore, the access of women and migrants to high-skilled employment is previously determined by strong socio-economic barriers. Regardless of the numerous opportunities and legal efforts to establish gender parity in the higher education system, women are rejected to be recruited in high-level education. It is important to discuss that women represent 70% of the labor force in tourism and hospitality. The obtained empirical evidence is far from being decisive. While tourism shows to be a key driver for reducing gender inequalities, leading women to psychological achievements, their income varies on country and GPD. In low-income countries which are more dependent on tourism, women are paid lower wages in comparison to men.

Lastly, this coincides with an ethnography conducted by Juana Norrild & Maximiliano Korstanje who discuss the excessive working hours often main and housekeepers face in Argentina. The consulted key informants reveal that luxury hotels are structured in a strong hierarchal pyramid where women are placed in a peripheral position. A whole portion of women employed in luxury hotels is low-skilled staff or persons who lack higher education. At the time the organization is marked by high-status consumption, [like the case of luxury hotels which are offered to high-purchasing power segments] less flexible is the access of women to better job opportunities. This suggests that gender asymmetries and capital are inextricably intertwined.

GENDER AND TOURISM RESEARCH

Although there is a considerable number of publications denouncing the inequalities of women in the tourism or hospitality industries, less attention was given to their roles in applied tourism research or the fields of tourism epistemology (Li 2003). One of the pioneering of gender and tourism research harks back to Margaret B. Swine`s paper Gender in Tourism. In this seminal work, she holds the thesis that although a majority of researchers discussing gender inequalities are simply women, they run the risk of legitimate the stereotypes of a patriarchal society. As Swine notably observes, scholars enthusiastically embrace feminist theories to explain gender`s asymmetries fabricated and potentiated by tourism consumption, but rarely, do these theories inspect the fields of academic circles. This creates a paradoxical situation, where the number of women accepted to take part in academic staff does not crystalize into better opportunities for women in jobs or labor market. As the author goes on to write: "One danger here has been the construction of an essentialist universal female experience as a converse to the universal male, without taking into account race, ethnicity, age, education, and other characteristics of individual women. Such an approach has limited use beyond generalized description, while gender studies promise an understanding of how relationships work and change" (Swine 1995: 253).

Coincident with the above-cited excerpt, Kinnaird & Hall (2000) calls attention to the urgency to theorize on gender issues. Per their viewpoint, the gendered-based approach helps in providing a valuable conceptual instrument to decipher the subordination of women in the labor marketplace. In this vein, Fiona Bakas (2017) gives a snapshot of how feminism may contribute very well to the adoption of new methods and epistemologies in tourism research. The feminist ideals are fertile to cement tourism entrepreneurship research to overcome the current setbacks. The gender-based theory allows more reciprocity and encourages the engagement of professional researchers as well as academicians with new topics that help the academia to reverse the present tourism research crisis. Heather Gibson (2001) discusses critically to what extent a gender-aware framework is necessary for tourism fields. The liberal-gendered perspective alludes to feminist theory to gain further emancipation of women respecting men. But the limitation of these theories lies in the fact that women already are part of society while taking active participation in public life. The problem appears to be those women`s contributions are often silenced by the ruling classes. To fill this gap, recently two influential voices Antonia Correia & Sara Dolnicar (2021) edited a more than interesting volume entitles *Women`s Voices in Tourism Research.* This ambitious volume contains different chapters authored by women that narrate their contribution to the current tourism research. Of course, not all of them are epistemologists or theorists of tourism. This editorial project articulates a wide invitation to well-skilled researchers who enthusiastically describe the direct impact of women in academia. Having said this, the book is not in opposition to the knowledge production or legacy left by men, even recognizing the influence of men's tourism research. Additionally, chapters toy with the belief that there is restrictive access to what has been investigated by women in the past. To make these contributions mainly overt to the academic community, the book gives a spotlight on knowledge production/distribution in women's hands. As Ines Carvalho (2021) eloquently points out, women`s career advancement describes with much accuracy how gender interacts in the bridge between higher education and the real obstacles to women in the industry. Women's mobilities comprise not only upward progression in the different jobs but also learning experiences, intrinsic rewards as well as a net of contacts. Women often come across hinders that prevent their advancement. These hindrances include sexism, problems dealing with men in workplaces and structural economic asymmetries based on gender. Carvalho recognizes that gender flexibility gives further possibilities towards an "ideal tourism worker", which means "the availability-related flexibility" that describes how women are ultimately perceived by managers. Per her outcomes, women are mistakenly considered less flexible than men or less available for some jobs because they should prefer their families overworking. These types of stereotypes play a negative role in gender configuration. In consonance with this, Vizcaino-Suarez & Diaz-Carrion (2019) examine the contribution and gravitation of gender in tourism research in Latin America. Based on bibliometric analysis, they hold that the leading countries in tourism research include Brazil, Mexico and Argentina. At the same time, the majority of produced knowledge is associated with empirical research but focuses on a local scope. Not surprisingly, this production is far from being determined by marketing and a managerial perspective, rather they are structured within the fields of humanities and social sciences -sociology and anthropology to be more exact-. Having said this, serious challenges are posited for these scholars to integrate a global platform jointly with women from the US and Europe towards a cosmopolitan spirit. At the same time, as authors lament, Latin American universities are promoting patriarchal lines of authority that daily exclude women from applied research. For public universities, tourism syllabuses or programs in Latin America are managed by social science faculties instead of the business schools which monopolize tourism education in the Anglophone world. Beyond the importance of women`s engagement in tourism research, our chapter reflects on the first steps, problems, methodological limitations and opportunities of CIET, which is situated as the first research center headed by Regina Schluter and other women to boost tourism research in Latin America. Doubtless, this center was a cornerstone not only for future tourism researchers but also for higher tourism education in the entire region.

CIET AS A SUCCESSFUL STUDY CASE (CENTER OF RESEARCH AND TOURISM STUDIES)

As debated in earlier sections, CIET [Centro de Investigacion y Estudios Turistiticos] is a research center geographically located in Buenos Aires, Argentina. Its history leads us back to 1982, a special moment when a group of professors headed by Regina Schluter, decided to create an academic platform [center] with the primordial end to promote scientific research. All these pioneering voices had tourism as their main object of study. The different problems the industry daily faced, as well as the urgency to adopt the scientific method to understand tourism dynamism inspired these young professionals and researchers. Abroad, some scholars [like Jafar Jafari, Christ Cooper, and Neil Leiper] emphasized the need of developing accurate methods to understand tourism as an all-encompassing institution. Professor Schluter pivoted in the creation of a local center because of the lack of a theoretical framework that ultimately confronts the economic-based paradigm. Educated in tourism and Social Sciences [Social psychology to be exact], Professor Schluter devoted part of her life to training future researchers concerned with the future of tourism in the region. The economy-based theory defined unilaterally tourism as a commercial activity, ignoring the emergent contributions of sociology, social psychology or even anthropology in the field. From its outset, CIET has held in Buenos Aires, through the time it inaugurated venues in different parts of the country. In this respect, Regina has been followed by brilliant colleagues such as Judith Carrillo, Fernando Monedero Galvez, Arnoldo Neiff, Adela Puig and Alicia Tagliorette [only to name a few]. CIET has taken active steps in the training of professional researchers in different Latin American countries as well as active participation in international congresses and seminars. A foundational event in the history of CIET is mainly marked by the organization of Congreso Internacional CIET-AMFORT [CIET-AMFORT International Conference] held in 1989 in Buenos Aires, Argentina. In consequence, the center financially supported the creation of the first Latin American journal [inspired by the example of Annals of Tourism Research], Estudios y Perspectivas en Turismo [Studies and Perspectives in Tourism] which monopolized a whole portion of leading studies, research and publications in tourism for more than three decades.

It is noteworthy that in the 70s decade, though some universities incorporated tourism in their curricula, the career was certainly subordinated to business and management. Pre-graduate students, who envisaged tourism as a promising career, dreamt to work as tour operators, agencies, hotels or owners of their undertakings. Neither the interest in tourism associated with business and management nor the professional publications was marginal [if not very poor] in this period. The main scientist research center overtly neglected the idea that tourism has been conceived as a serious object of study. Meanwhile, the creation of Annals of Tourism Research [in the US] as well as the works of AIEST- Association of Scientific Experts in Tourism [in Switzerland and Germany] have been germane for the first tourism-related publications worldwide. Tourism studies in Spanish have been not only inexistent but also of poor quality. Without any doubt, this was a major challenge posed to Professor Schluter and her colleagues who decided to adopt Spanish as a main shared language. This was the main reason behind the creation of the center which was designed to edit the first leading books in Latin America. The project was rapidly complemented by the creation of the journal Estudios y Perspectivas en Turismo. As the previous backdrop, the journal has taken the lead in discussing publications with a focus on gender inequalities and tourism. Since the turn of the century, the journal was a pioneering voice (in Latin America) exploring the inequalities suffered by women in the tourism industry as well as in the constellations of professional research and education. Mendoza-Ontiveros & Chapullin Carillo (2015) conduct an interesting ethnography on 21 women working in the tourism industry. The general line of inquiry says overtly that women are trapped in a flexible net in the working condition determined by excessive working hours and low salaries. The jobs accessed by female workers are low-skilled or with little probability of social upward. Anyway, interviewees manifested their opportunity to bolster fluid dialogue with tourists or persons in quest of better opportunities. Similar reflections have been made by Huete, Brotons, & Singuenza, 2016). Their analysis is based on the Spanish market. Although women represent a major portion of the recruited workforce, they are low-skilled and consequently low-paid. There is a much deep wage gap between gender which is associated with fragile working conditions, and vertical segregation, which go beyond the current pro-gender policies.

The journal accepted papers written in other languages than Spanish, but all pieces were translated and edited into Spanish. Of course, with the digital technological breakthrough, the center was subject to different setbacks and barriers which alludes to the adoption of English as the main lingua franca. The lack of interest of English native readers in Spanish adjoined to the radical shifts experienced by the industry at the turn of the century has re-founded the aims and goals of the center, returning to the original spirit. CIET still works in the high-quality edited book while inviting well-recognized experts and specialists to publish their advances in Spanish. It is also argued that CIET represents a successful study case that reflects the academic itinerary of Regina Schluter and many other brilliant women who inspired tourism research in Latin America. This center not only filled the curricula of many universities but also was an academic reference for Latin American countries.

CONCLUSION

It is tempting to say that feminist voices have struggled to incorporate women into equal labor conditions to men over the years. Of course, beyond the current material inequalities between women and men, specialists suggest that the question of gender plausibly sparks detractors, supporters, and if not, controversies. Needless to say that the tourism fields seem not to be an exception. Over the decades, countless studies have emphasized the problem often came across women in the tourism and hospitality industries. These problems include excessive working hours, low-paid salaries, as well as harassment or abuse of power in luxury hotels. Having said this, an emerging dearth of publications have approached the role of women not only in the different levels of service sectors but also in tourism research [without mentioning the higher education system]. This behoves us to put the example of CIET [Centro de Investigacion y Estudios turisticos] a leading and well-renowned Research center that for more than four decades shed the paths of tourism researchers, epistemologists, policy-makers and practitioners interested in the deep understanding of tourism in Latin America. The present book chapter synthetizes and describes the barriers, opportunities and concerns that ultimately faced Regina Schluter and other leading women while managing this center. At the same time, this does not represent a biographical chapter, but rather a simple description of how women have eloquently permeated, impacted and interrogated furtherly the status quo in Latin America Tourism research. It is worth mentioning that women have directly engaged and contributed to the advance of the discipline in the region. In a nutshell, CIET not only occupied a central position in the academic debates revolving around the future of tourism but also was a fertile ground for thousands of colleagues, researchers and academicians interested in the constellations of tourism and hospitality continent-wide. The first section reviews the literature with a focus on gender inequalities in the tourism and labor marketplace. The literature overly acknowledges that the tourism industry democratizes the labor market, but often it does not suffice to reverse the local gender inequalities. Some outcomes reveal those under-developed economies, which are dependent on international tourism, develop strong gender relations which impede women to get better Jobs [at least in comparison with developed economies]. Nevertheless of this fact, the tourism industry offers a fertile ground for women to negotiate their identities locally while interrogating the status quo. In some cases, western female tourists plausibly have a positive influence on local working staff through a type of acculturation process. If the question of gender inequality is very hard to measure, and explain, further obstacles show the discrimination of women in professional tourism research. This is the topic that draws readers` attention in the second section. Lastly, the third section brings some reflections on Regina Schluter`s first steps as well as the creation of CIET.

REFERENCES

Abou-Shouk, M., Zoair, N., Elbaz, A. M., & Abdel-Jalil, M. (2023). Local tourists' perceptions of tourist destinations' competitiveness: A comparative study of the United Arab Emirates, Egypt, and Oman. *Worldwide Hospitality and Tourism Themes*, *15*(2), 143–154. doi:10.1108/WHATT-10-2022-0117

Acharya, B. P., & Halpenny, E. A. (2013). Homestays as an alternative tourism product for sustainable community development: A case study of women-managed tourism product in rural Nepal. *Tourism Planning & Development*, *10*(4), 367–387. doi:10.1080/21568316.2013.779313

Bakas, F. E. (2017). 'A beautiful mess': Reciprocity and positionality in gender and tourism research. *Journal of Hospitality and Tourism Management*, *33*, 126–133. doi:10.1016/j.jhtm.2017.09.009

Bakas, F. E., Costa, C., Breda, Z., & Durão, M. (2018). A critical approach to the gender wage gap in tourism labor. *Tourism, Culture & Communication*, *18*(1), 35–49. doi:10.3727/109830418X15180180585167

Basurto-Barcia, J., & Ricaurte-Quijano, C. (2017). Women in tourism: Gender (in) equalities in university teaching and research. *Anatolia*, *28*(4), 567–581. doi:10.1080/13032917.2017.1370781

Belhassen, Y., & Caton, K. (2009). Advancing understandings: A linguistic approach to tourism epistemology. *Annals of Tourism Research*, *36*(2), 335–352. doi:10.1016/j.annals.2009.01.006

Boley, B. B., Ayscue, E., Maruyama, N., & Woosnam, K. M. (2017). Gender and empowerment: Assessing discrepancies using the resident empowerment through tourism scale. *Journal of Sustainable Tourism*, *25*(1), 113–129. doi:10.1080/09669582.2016.1177065

Bolles, A. L. (1997). Women as a category of analysis in scholarship on tourism: Jamaican women and tourism employment. *Tourism and culture: An applied perspective*, 77-92.

Brownell, J. (1994). Women in hospitality management: General managers' perceptions of factors related to career development. *International Journal of Hospitality Management*, *13*(2), 101–117. doi:10.1016/0278-4319(94)90032-9

Butler, K. L. (1995). Independence for Western Women Through Tourism. *Annals of Tourism Research*, *22*(2), 487–489. doi:10.1016/0160-7383(94)00101-4

Carvalho, I. (2021). Gender and Tourism. In A. Correia & S. Dolnicar (Eds.), *Women's Voices in Tourism Research* (pp. 1–15). University of Queens Press.

Carvalho, I., Costa, C., Lykke, N., & Torres, A. (2014). An analysis of gendered employment in the Portuguese tourism sector. *Journal of Human Resources in Hospitality & Tourism*, *13*(4), 405–429. doi:10.1080/15332845.2014.888509

Caton, K. (2012). Taking the moral turn in tourism studies. *Annals of Tourism Research*, *39*(4), 1906–1928. doi:10.1016/j.annals.2012.05.021

Cave, P., & Kilic, S. (2010). The role of women in tourism employment with special reference to Antalya, Turkey. *Journal of Hospitality Marketing & Management*, *19*(3), 280–292. doi:10.1080/19368621003591400

Chambers, D. (2021). *Tourism and (re)presentation: beyond colonialism.* In A. Correia & S. Dolnicar (Eds.), *Women's Voices in Tourism Research* (pp. 25–32). University of Queens Press.

Chambers, D., & Rakic, T. (2015). *Tourism Research Frontiers: beyond the boundaries of knowledge*. Emerald Group. doi:10.1108/S1571-5043201520

Chambers, E. (Ed.). (1997). *Tourism and culture: an applied perspective*. SUNY Press.

Chant, S. (2005). Gender and tourism employment in Mexico and the Philippines. In *Gender, work and tourism* (pp. 119–177). Routledge.

Cole, S. (Ed.). (2018). *Gender equality and tourism: Beyond empowerment*. CABI. doi:10.1079/9781786394422.0000

Correia, A., & Dolnicar, S. (2021). *Women's voices in tourism research: Contributions to knowledge and letters to future generations*. University of Queensland Press. doi:10.14264/817f87d

Costa, C., Carvalho, I., & Breda, Z. (2011). Gender inequalities in tourism employment: The Portuguese case. *Revista Turismo & Desenvolvimento (Aveiro)*, (15), 39–54.

Costa, H. A., Costa, C., Breda, Z., Caçador, S., Carvalho, I., & Vieira, R. (2013). Women in Brazil with higher education in tourism: Preliminary results of the Gentour Project. *Revista Turismo & Desenvolvimento (Aveiro)*, (19), 53–66.

Duffy, L. N., Kline, C. S., Mowatt, R. A., & Chancellor, H. C. (2015). Women in tourism: Shifting gender ideology in the DR. *Annals of Tourism Research*, *52*, 72–86. doi:10.1016/j.annals.2015.02.017

Frew, E. A., & Shaw, R. N. (1999). The relationship between personality, gender, and tourism behavior. *Tourism Management*, *20*(2), 193–202. doi:10.1016/S0261-5177(98)00081-8

Gentry, K. M. (2007). Belizean women and tourism work: Opportunity or impediment? *Annals of Tourism Research*, *34*(2), 477–496. doi:10.1016/j.annals.2006.11.003

Ghodsee, K. (2003). State support in the market: Women and tourism employment in post-socialist Bulgaria. *International Journal of Politics Culture and Society*, *16*(3), 465–482. doi:10.1023/A:1022364630327

Gibson, H. (2001). *Gender in tourism: Theoretical perspectives. In Women as producers and consumers of tourism in developing regions (pp. 19-43).* Praeger.

Hjalager, A. M. (2003). Global tourism careers? Opportunities and dilemmas facing higher education in tourism. *Journal of Hospitality, Leisure, Sport and Tourism Education*, *2*(2), 26–37. doi:10.3794/johlste.22.35

Hutchings, K., Moyle, C. L., Chai, A., Garofano, N., & Moore, S. (2020). Segregation of women in tourism employment in the APEC region. *Tourism Management Perspectives*, *34*, 100655. doi:10.1016/j.tmp.2020.100655

Kinnaird, V., & Hall, D. (2000). Theorizing gender in tourism research. *Tourism Recreation Research*, *25*(1), 71–84. doi:10.1080/02508281.2000.11014901

Li, J. (2003). Playing upon fantasy: Women, ethnic tourism and the politics of identity construction in contemporary Xishuang Banna, China. *Tourism Recreation Research*, *28*(2), 51–65. doi:10.1080/02508281.2003.11081404

Long, V. H., & Kindon, S. L. (2005). Gender and tourism development in Balinese villages. In *Gender, work and tourism* (pp. 89–117). Routledge.

Manwa, H. (2008). Enhancing participation of women in tourism. In *Building community capacity for tourism development* (pp. 116–122). CABI. doi:10.1079/9781845934477.0116

Mendoza Ontiveros, M. M., & Chapulín Carrillo, J. D. V. (2015). Turismo, trabajo femenino y empoderamiento de las mujeres en bahías de Huatulco, Oaxaca-México. *Estudios y Perspectivas en Turismo*, *24*(2), 316–335.

Mitra, S. K., Chattopadhyay, M., & Chatterjee, T. K. (2022). Can Tourism Development Reduce Gender Inequality? *Journal of Travel Research*.

Moswete, N., & Lacey, G. (2015). "Women cannot lead": Empowering women through cultural tourism in Botswana. *Journal of Sustainable Tourism*, *23*(4), 600–617. doi:10.1080/09669582.2014.986488

Nakata, M., & Momsen, J. D. (2010). Gender and tourism: Gender, age and mountain tourism in Japan. *Geografia*, *6*(2), 1–15.

Norrild, J., & Korstanje, M. E. (2017). Regina Schlüter–an entire life devoted to tourism research. *Anatolia*, *28*(3), 493–500. doi:10.1080/13032917.2017.1324226

Norrild, J. A., & Korstanje, M. E. (2021). Maids and housekeepers at luxury hotels: Life stories in hotels of Buenos Aires, Argentina. *International Journal of Tourism Anthropology*, *8*(2), 129–147. doi:10.1504/IJTA.2021.116091

Norris, J., & Wall, G. (1994). Gender and tourism. *Progress in Tourism, Recreation and Hospitality Management, 6*, 57-78.

Pritchard, A. (2014). Gender and feminist perspectives in tourism research. *The Wiley Blackwell companion to tourism*, 314-324.

Richter, L. K. (2012). Exploring the political role of gender in tourism research. In *Global tourism* (pp. 426–439). Routledge.

Rinaldi, A., & Salerno, I. (2020). The tourism gender gap and its potential impact on the development of the emerging countries. *Quality & Quantity*, *54*(5), 1465–1477. doi:10.100711135-019-00881-x

Schlüter, R. G. (2008). Mujer y turismo. Vestimenta e interacción social en los centros turísticos de Argentina durante la década de 1930 [Women and tourism: dressing and social interaction at the tourist destinations in Argentina during the 30s decade]. *Cuadernos de Turismo*, (21), 181–199.

Siamak Seyfi, I. D. (2023). *The gendered effects of statecraft on women in tourism: economic sanctions, women's disempowerment and sustainability? In Gender and Tourism Sustainability*. Routledge.

Sony, K. C. (2012). Roles and challenges of women in tourism sector of western Nepal: A micro-ethnographic study. *Nepal Tourism and Development Review*, *2*(1), 32–59.

Swain, M. B. (1995). Gender in tourism. *Annals of Tourism Research*, *22*(2), 247–266. doi:10.1016/0160-7383(94)00095-6

Swain, M. B. (1995). Gender in tourism. *Annals of Tourism Research*, *22*(2), 247–266. doi:10.1016/0160-7383(94)00095-6

Tucker, H. (2009). Recognizing emotion and its postcolonial potentialities: Discomfort and shame in a tourism encounter in Turkey. *Tourism Geographies*, *11*(4), 444–461. doi:10.1080/14616680903262612

Vizcaino-Suárez, L. P., & Díaz-Carrión, I. A. (2019). Gender in tourism research: Perspectives from Latin America. *Tourism Review*, *74*(5), 1091–1103. doi:10.1108/TR-02-2017-0021

Wall, G., & Norris, J. (2003). Gender and tourism. *Classic reviews in tourism*, 235-264.

Wilkinson, P. F., & Pratiwi, W. (1995). Gender and tourism in an Indonesian village. *Annals of Tourism Research*, *22*(2), 283–299. doi:10.1016/0160-7383(94)00077-8

Woyo, E., & Venganai, H. (2022). Gender and tourism. In *Encyclopedia of tourism management and marketing* (pp. 387–390). Edward Elgar Publishing. doi:10.4337/9781800377486.gender.and.tourism

Xu, H., & Gu, H. (2018). Gender and tourism development in China. *Journal of China Tourism Research*, *14*(4), 393–404. doi:10.1080/19388160.2018.1539426

KEY TERMS AND DEFINITIONS

Empowerment: It is a term used to obtain further opportunities for marginalized or low skilled people.

Gender Inequalities: They are material or financial asymmetries based on question of gender.

Tourism: Is a set of services oriented to travel and leisure practices which include different subsectors.

Tourism Research: Is the combination of technique, methods and instrument aimed at producing tourism knowledge.

Unskilled Workforce: It is a work manpower mainly marked by low levels of knowledge and skills.

Chapter 35
The Leadership of the Higher Education Institutions in Empowering Female Entrepreneurship

Shakeel Ahmed
Greenwich University, Pakistan

Erum Shaikh
Shaheed Benazir Bhutto University, Sanghar, Pakistan

ABSTRACT

The aim of this study is to find out the leadership of the higher education institutions in empowering female entrepreneurship. Qualitative methodology is used in this study in which semi-structured interviews are conducted from the female candidates who are pursuing entrepreneurship and relying on the higher education institutions and their leadership in initiating and maintaining support system for the females. Findings of the study show that female entrepreneurs are more determined for the entrepreneurship as they are mostly necessity driven in the developing countries. Their leadership skills enable them to tackle the various challenges and drive the situation in the order to carry on with their entrepreneurial motives.

1. INTRODUCTION

Female entrepreneurship has come under more and more scrutiny in recent years, given the verifiable data demonstrating the significance of new firm formation for economic growth and development. Beyond only creating jobs and boosting the economy, female-owned small enterprises are essential to society (Abdullah, et al., 2020; Afshan, et al., 2021). Despite the fact that governments and academia have taken note of the rise in the proportion of female entrepreneurs in developing countries, women still own or manage roughly 30% of all small enterprises in industrialized countries (Datta & Gailey, 2012). Additionally, programs or policies targeted at encouraging and developing female entrepreneur-

DOI: 10.4018/978-1-6684-8417-3.ch035

ship have been started by donors, international public organizations, national and local governments, NGOs, private corporations, charities, knowledge institutes, and business groups (Vossenberg, 2013; Alshareef, et al., 2022; Aurangzeb, et al., 2021).

This study tries to identify the components of the support environment for the female entrepreneurs in the developing country like Pakistan.

2. LITERATURE REVIEW

In the upcoming years, female labor force participation is anticipated to rise at an accelerated rate. This development is a result of the high rates of female education, with more women currently holding university degrees than males (Chaudhry, et al., 2021; Gilal, et al., 2021). The Kingdom of Bahrain has started several initiatives, projects, and training and rehabilitation programs targeted at empowering women by giving them the knowledge and skills necessary to start and run small businesses or to pursue entrepreneurship (Alexandre & Kharabsheh, 2019). One of the most important aspects of entrepreneurship is the underlying traits of the business owners themselves (Tunio, et al. 2021; Memon, et al., 2021). Male and female entrepreneurs often have different natural traits (Tunio, et al., 2021; Shaikh, et al., 2022; Shaikh, et al., 2022).

Valeri & Katsoni. (2021) discussed that women entrepreneurs are equally capable of co-founding businesses, occupying positions of leadership in the industry, and taking part in and making decisions on the operational aspects of such businesses, either by themselves or in partnership with others (Mushtaq, et al., 2021; Tunio, et al., 2021). Pioneering spirit, tenacity, attentive service, attention to staff (general knowledge/education and authorization), authors emphasis on social goals such as: quality, credit, charity activities, pursuit of expansion and profit, and a sense of accomplishment of self-realization are among the traits of women entrepreneurs (Shaikh, and Tunio, 2020; Tunio and Shaikh, 2020).

More women than ever before are in public office today. However, promoting more female leadership will contribute to achieving more gender equality (Shaikh, et al., (2021; Tunio, et al., 2021). In order to place businesses in the markets where they are selling their goods and services, company management must thus search for skilled collaborators whose competences are appropriate for managerial roles, adding to image, reputation, and profitability (Mabula & Mwakapala, 2020; Shaikh, et al., 2021; Tunio, et al., 2021). By eschewing the outdated exclusivity of organizational management, which is often held by males, managerial leadership adds to the perfection of its missionary processes (Shaikh, et al., 2022; Katpar., et al., 2020). Therefore, dominance, strength, robustness, intellect, perseverance, and responsibility are growing feminine qualities. These traits foster a sense of equality in managerial decision-making (Demirovic et al., 2020; Shaikh, et al., 2022; Tunio (2020)).

According to Biswas and Stainback. (2021) assert that the leader role is necessary in any sector since the collaborators are guided by the leader's activities to reach shared goals. Small and medium-sized businesses are similarly affected by these cutting-edge practises for preserving market sustainability and stability (Acevedo-Duque et al., 2021; Tunio, 2020; Katper, et al., 2017), particularly in Colombia.

Any institution's educational leadership is an essential element. Leaders at universities set the standard for how academic staff and students conduct themselves. Successful administrators say that using a variety of leadership behaviors, not just one, can improve academic results (Guterresa et al., 2020). Higher education was created to adhere to the highest standards of excellence and has raised the bar for using different leadership philosophies. Leadership is concerned in an academic setting not only with the demands of the organization but also with the national mission (Tunio, et al., 2017; Gilal. et al., 2020). University administrators have a considerable impact on academic programs and are accountable for the accomplishments or failures of their institutions or organizations. The success of leaders in advancing and ensuring the welfare of the university's numerous stakeholders may have a big influence on the academic programmed. Leadership is characterized by several qualities, characteristics, and behaviors, such as a clear vision, action, setting an example, ethical links, congruence, trustworthiness, and teamwork (Shaari et al., 2022).

The positive effects that an increase in the percentage of female entrepreneurs generates on the economic conditions and social well-being of a country are the reason for the growing interest among policy makers and academics in the role of entrepreneurial women (Martínez-Rodríguez et al., 2022), both in developed and underdeveloped countries It is especially pertinent in the case of developing nations since it may serve as an effective way to fight poverty. Women are becoming more educated, politically powerful, and granted economic liberties, which has dramatically increased the number of female entrepreneurs globally in recent years (Tunio, et al., 2020; Gul, et al., 2021).

Universities have placed an increasing emphasis on providing technical assistance to industry, disseminating technical information, and doing more applied research since the 1970s. Collaboration between universities and their local stakeholders is increasingly important for innovation, entrepreneurship, and regional development, especially in high-tech areas (Huggins & Johnston, 2009; Tunio, et al., 2021). Since the 1980s, there have been an exceptionally large number of programs teaching entrepreneurship. These university-based entrepreneurship centers serve as a crucial platform for a variety of services and initiatives that promote entrepreneurship and economic growth. These organizations organize, coordinate, encourage and direct academic participation with small and growing enterprises, bridging universities and communities (Blesia et al., 2021).

A university-based entrepreneurship center's contributions to entrepreneurship and business growth include incubators, chances for pitching, training, consulting, easing access to finance, and helping to prepare business plans, to name a few examples (Wang, 2021). Entrepreneurship centers at universities may be a great source of funding, promote student-faculty interaction, and involve stakeholders in building networks with many partners for outreach and economic growth. However, the activities of these entrepreneurship centers seldom ever explicitly support marginalized populations (Wang, 2021). Riebe (2012) made the case that these centers offer distinctive and significant benefits for women learners and local communities, as well as innovative and potent opportunities for Higher Education Institutes to advance their educational missions, using the experiences of two university-sponsored centers for women entrepreneurs in the U.S. as an exceptional case. However, initiatives to address possibilities for various groups connected to economic and social fairness in impoverished communities have often gone unnoticed (Nabi et al. 2017).

3. METHODOLOGY

Qualitative research method is used in this study in which semi-structured interviews are conducted from the female candidates who are under the influence of the higher education institutions (Shaikh, E., et al., 2021; Shaikh, et al., 2022). Respondents are pointed out and approached through snow ball approach, snow ball approach is a process to find out and reach the participants through well known people and

Table 1. Demographics

No.	Respondents	Gender	Age	Marital Status	Qualification
1.	Razia	F	28	Married	MS
2.	Koomal	F	27	Single	MBA
3.	Shaista	F	29	Married	BBA
4.	Naheed	F	25	Married	MCS
5.	Moomal	F	30	Single	MA
6.	Rabia	F	31	Single	MBA
7.	Abida	F	32	Married	MS
8.	Zakia	F	33	Separated	BBA
9.	Shazia	F	30	Widow	BBA
10.	Nadia	F	29	Single	MBA

social circle (Shaikh, et al., 2021; Memon, et al., 2019). After reaching participants, consent was secured to conducted interview, record the interview, and use the data for the research purpose only (Tunio, et al., 2022; Shaikh, et al., 2019).

In order to meet the saturation, 10 interviews were conducted from the different female candidates. After completing interviews, transcription was developed. After completing transcription, data was reviewed again to rectify any mistake, spelling error, or grammatical mistake (Govindarajo, et al., 2021; Tunio, et al., 2022).

In the next step, codes were generated from the transcription, selected codes were categorized in the separate categories, from the selected categories, themes are generated, those themes were labeled with respect to the nature of the data/ information. Thus, these themes are the final findings of the study (Shaikh, et al., 2022; Raza, et al., 2022).

4. FINDINGS

4.1. Sustainable Leadership

Razia: "It appears that leaders in Higher education institutions are trying to utilize their own concepts of sustainability to adjust and continually improve overall performance."

Koomal: "SL embodies the promise of higher education evolution towards a more equitable and wealthier world in which the academic environment and higher education achievements are preserved for generations to come."

Shaista: "Higher education institutions (HEIs) need a sustainable leadership that are responsible for creating a sustainable future for the entrepreneurship and innovation"

Naheed: "Sustainable leadership of the higher education institutions can play vital role in the development of the entrepreneurship, which requires changes in many organizational fields, including leadership and innovation"

Figure 1. Themes

4.2. Family Businesses

Naheed: Family hierarchy is somehow challenging when it comes to the female entrepreneurship because there are so many things which are not socially allowed for the female like easy and freely navigation to the market and communication with the customers.

4.3. Resilient Leadership

Razia: Education enables us to handle the challenges and cope up in the right way at the right time.

Koomal: Everyone has many issues to handle at the same time but to handle them in the proper way is a suitable way through the entrepreneurial approach.

Shaista: Social and economic fluctuations causes many crises but the leadership of the educated female can deal with all sort of the issues and challenges.

1. Indigenous Culture:

Naheed: our family structure and native culture is very much concerned about the females and movements of the females at home and outside the home. However, family support can make it possible for the females to go easily with their business interest and work in the limited way.

4.4. Higher Education

Razia: It is obvious that female entrepreneurs can differentiated from the rest of the population by their actions and activities. It is a leadership role of the females to initiate a business with the creative idea and maintain it in the disciplined way.

Rabia: There is great importance of the education which enables the normal female as leader in any field of the life, thus, in the specific way, when you talk about the business, they can become best leaders in the business with support of the family. In this case, family support is very important.

Abida: I want to do more better in the business, but due to the social constraints, I can not work independently because I do not have any male or family support. However, I feel that education institutions can play important role to fill this gape in supporting female entrepreneurs who are lacking family support.

4.5. Leadership Skills

Razia: As you know that females are naturally leaders and have more leading skills in the administration of the domestic and family affairs. They can be proved more important

Koomal: females are the business leaders because they have power to deal with the challenging tasks.

Shaista: I can do business because of my high qualification and support of the university.

Naheed: High education institution is a major source of support for me in my business initiative and maintained of the business.

5. DISCUSSION

Higher education institutions play important role in nurturing leadership skills in the females by providing them education and skills. Education attended in the higher education provides dynamic ideas and plans to the females for the feasible business. Sustainable business is hot topic now a days in research as well as education institutions. In the high education institutions, concept of the sustainability is giving more importance in order to create economic and social value.

High education institutions experience different challenges in the process of the development and progress; however, challenges may be like leadership, finance, staffing, curriculum, students' affairs, community support, trust issues, different kinds of the conflict, social, regional and tribal issues and so on. Thus, every section offers new kind of the challenges for the educational institutions but even though, they play their important and tangible role in the performing their duties. Education female entrepreneurs who have attended higher education from the reputable institutions indicate higher interest in entrepreneurship. It is seemed that most of the female entrepreneurs are necessity driven due to the different contextual factors.

6. CONCLUSION

Sustainable leadership of the higher education institutions plays vital role in developing entrepreneurial ecosystem in the region for the female entrepreneurs of all ages. This study entails that family plays important role in supporting females for the entrepreneurial activities, however, other social constrains resist them and create more challenges. Self-reliant nature of the females enable them to be resilience in the unfavorable situations in order to focus on the entrepreneurial motives and keep moving forward in their mission.

REFERENCES

Abdullah, M., Katper, N. K., Chaudhry, N. I., & Tunio, M. N. (2020). An Impact of Workaholics on Creativity: the mediating role of Negative Mood and moderating role of Supervisor Support. *Sukkur IBA Journal of Management and Business, 7*(2).

Acevedo-Duque, Á., Gonzalez-Diaz, R., Vargas, E. C., Paz-Marcano, A., Muller-Pérez, S., Salazar-Sepúlveda, G., Caruso, G., & D'Adamo, I. (2021). Resilience, leadership and female entrepreneurship within the context of smes: Evidence from latin america. *Sustainability (Basel), 13*(15), 8129. doi:10.3390u13158129

Afshan, G., Ilyas, S., Tunio, M. N., & Kalhoro, M. (2021). CSR actions and post-COVID-19 consequences in the hotel industry: A conceptual framework. *International Journal of Strategic Change Management, 7*(4), 275–289. doi:10.1504/IJSCM.2021.122845

Alexandre, L., & Kharabsheh, R. (2019). The evolution of female entrepreneurship in the Gulf Cooperation Council, the case of Bahrain. *International Journal of Gender and Entrepreneurship, 11*(4), 390–407. doi:10.1108/IJGE 02 2019 0041

Alshareef, N., & Tunio, M. N. (2022). Role of Leadership in Adoption of Blockchain Technology in Small and Medium Enterprises in Saudi Arabia. *Frontiers in Psychology, 13*, 2284. doi:10.3389/fpsyg.2022.911432

Aurangzeb, M. T., Tunio, M. N., Rehman, Z., & Asif, M. (2021). Influence of administrative expertise on human resources practitioners on the job performance: Mediating role of achievement motivation. *International Journal of Management, 12*(4), 408–421.

Biswas, P. K., Roberts, H., & Stainback, K. (2021). Does women's board representation affect nonmanagerial gender inequality? *Human Resource Management, 60*(4), 659–680. doi:10.1002/hrm.22066

Blesia, J. U., Iek, M., Ratang, W., & Hutajulu, H. (2021). Developing an entrepreneurship model to increase students' entrepreneurial skills: An action research project in a higher education institution in Indonesia. *Systemic Practice and Action Research, 34*(1), 53–70. doi:10.100711213-019-09506-8

Chaudhry, I. S., Paquibut, R. Y. & Tunio, M. N. (2021). Do workforce diversity, inclusion practices, & organizational characteristics contribute to organizational innovation? Evidence from the U.A.E. *Cogent Business & Management, 8*(1), 1947549. . doi:10.1080/23311975.2021.1947549

Datta, P. B., & Gailey, R. (2012). Empowering women through social entrepreneurship: Case study of a women's cooperative in India. *Entrepreneurship Theory and Practice*, *36*(3), 569–587. doi:10.1111/j.1540-6520.2012.00505.x

Demirović Bajrami, D., Radosavac, A., Cimbaljević, M., Tretiakova, T. N., & Syromiatnikova, Y. A. (2020). Determinants of residents' support for sustainable tourism development: Implications for rural communities. *Sustainability (Basel)*, *12*(22), 9438. doi:10.3390u12229438

Gilal, F. G., Gilal, N. G., Channa, N. A., Gilal, R. A., Gilal, R. G., & Tunio, M. N. (2020). Towards an integrated model for the transference of environmental responsibility. *Business Strategy and the Environment*, *29*(6), 1–10. doi:10.1002/bse.2524

Gilal, F. G., Gilal, N. G., Gilal, R. G., Gon, Z., Gilal, W. G., & Tunio, M. N. (2021). The Ties That Bind: Do Brand Attachment and Brand Passion Translate Into Consumer Purchase Intention? *Central European Management Journal*, *29*(1), 14–38. doi:10.7206/cemj.2658-0845.39

Govindarajo, N. S., Kumar, D., Shaikh, E., Kumar, M., & Kumar, P. (2021). Industry 4.0 and business policy development: Strategic imperatives for SME performance. *Etikonomi*, *20*(2), 239–258. doi:10.15408/etk.v20i2.20143

Gul, A., Subhan, S., & Tunio, M. N. (2021). Learning experiences of women entrepreneurs amidst COVID-19. *International Journal of Gender and Entrepreneurship*, *13*(2), 1756–6266. doi:10.1108/IJGE-09-2020-0153

Guterresa, L. F. D. C., Armanu, A., & Rofiaty, R. (2020). The role of work motivation as a mediator on the influence of education-training and leadership style on employee performance. *Management Science Letters*, *10*(7), 1497–1504. doi:10.5267/j.msl.2019.12.017

Huggins, R., & Johnston, A. (2009). The economic and innovation contribution of universities: A regional perspective. *Environment and Planning. C, Government & Policy*, *27*(6), 1088–1106. doi:10.1068/c08125b

Katpar, N. K., Chaudhry, N. I., Tunio, M. N., & Ali, M. A. (2020). Impact of Leadership Style and Organizational Culture on Organizational Commitment. *Sukkur IBA Journal of Management and Business, 7*(1), 92-106.

Katper, N. K. Medan, A., Syed, K. B. S., & Tunio, M. N. (2017). Determinants of Debt Maturity Structure in Shariah and Non-Shariah Firms in Pakistan: A comparative Study. *Journal of Applied Economic Sciences, 12*(4), 1210-1225.

Mabula, J. B., Dongping, H., & Mwakapala, L. Y. (2020). SME's use of ICT and financial services on innovation performance: The mediating role of managers' experience. *Human Systems Management*, *39*(3), 427–439. doi:10.3233/HSM-190790

Martínez-Rodríguez, I., Quintana-Rojo, C., Gento, P., & Callejas-Albinana, F. E. (2022). Public policy recommendations for promoting female entrepreneurship in Europe. *The International Entrepreneurship and Management Journal*, *18*(3), 1235–1262. doi:10.100711365-021-00751-9

Memon, A.B., Meyer, K., & Tunio, M. N. (2021). Toward collaborative networking among innovation laboratories: a conceptual framework. *International Journal of Innovation Science*. doi:10.1108/IJIS-04-2021-0069

Memon, A. R., Shaikh, E., & Khan, M. S. (2019). Determination of Customer Satisfaction of Hyderabad Restaurants. *Irish Interdisciplinary Journal of Science and Research*, *3*(3), 11–18.

Mushtaq, T., Tunio, M. N., Akbar, Z., & Jariko, M. (2021). Green Organizational identity: Antecents and consequences: An Emprical Study. *Contemporary Issues in Business and Government, 27*(3), 2056-2069.

Nabi, G., Liñán, F., Fayolle, A., Krueger, N., & Walmsley, A. (2017). The impact of entrepreneurship education in higher education: A systematic review and research agenda. *Academy of Management Learning & Education*, *16*(2), 277–299. doi:10.5465/amle.2015.0026

Raza, A., Shaikh, E., Tursoy, T., & Almashaqbeh, H. A. (2022). Economics and Business Perspectives of Sustainable HRM. In *Sustainable Development of Human Resources in a Globalization Period* (pp. 36–48). IGI Global. doi:10.4018/978-1-6684-4981-3.ch003

Riebe, M. (2012). A place of her own: The case for university-based centers for women entrepreneurs. *Journal of Education for Business*, *87*(4), 241–246. doi:10.1080/08832323.2011.591846

Shaari, R., Kamarudin, D., Ju, S. O. O. N., & Zakaria, H. (2022). Effects of Leadership Types on Job Satisfaction Among Malaysian Higher Education Institutions. *Asya Öğretim Dergisi*, *10*(1), 54–70. doi:10.47215/aji.1020324

Shaikh, E., Azhar, H., Brahmi, M., & Zehra, N. (2022). The impact of monetary and non-monetary motivation on employees' performance: A case study of Hyderabad Electric Supply Company. *International Journal of Technology Transfer and Commercialisation*, *19*(1), 127–141. doi:10.1504/IJTTC.2022.123088

Shaikh, E., Brahmi, M., Thang, P. C., Watto, W. A., Trang, T. T. N., & Loan, N. T. (2022). Should I Stay or Should I Go? Explaining the Turnover Intentions with Corporate Social Responsibility (CSR), Organizational Identification and Organizational Commitment. *Sustainability (Basel)*, *14*(10), 6030. doi:10.3390u14106030

Shaikh, E., Khoso, I., & Chandio, F. (2019). Effects of Corporate Social Responsibility on Organizational Performance: A Conceptual and Literature Review. *Journal of Grassroot*, *53*(1).

Shaikh, E., Mishra, V., Ahmed, F., Krishnan, D., & Dagar, V. (2021). Exchange rate, stock price and trade volume in US-China trade war during COVID-19: An empirical study. *Estudios de Economía Aplicada*, *39*(8). Advance online publication. doi:10.25115/eea.v39i8.5327

Shaikh, E., & Tunio, M. N. (2020). Customer satisfaction and Customer loyalty: An empirical case study on the impact of benefits generated through Smartphone applications. *International Journal of Public Sector Performance Management*.

Shaikh, E., Tunio, M. N., Khoso, W. M., Brahmi, M., & Rasool, S. (2022). The COVID-19 Pandemic Overlaps Entrepreneurial Activities and Triggered New Challenges: A Review Study. *Managing Human Resources in SMEs and Start-ups: International Challenges and Solutions*, 155-182.

Shaikh, E., Tunio, M. N., & Qureshi, F. (2021). Finance and women's entrepreneurship in DETEs: A literature review. Entrepreneurial Finance. *Innovation and Development*, 191–209.

Shaikh, E., Watto, W. A., & Tunio, M. N. (2022). Impact of Authentic Leadership on Organizational Citizenship Behavior by Using The Mediating Effect of Psychological Ownership. *ETIKONOMI*, *21*(1), 89–102. doi:10.15408/etk.v21i1.18968

Shaikh, S., Sultan, M. F., Mushtaque, T., & Tunio, M. N. (2021). Impact of COVID-19 on GDP: A serial mediation effect on international tourism and hospitality. *International Journal of Management*, *12*(84), 422–430.

Tunio, M. N. (2020). Role of ICT in Promoting Entrepreneurial Ecosystems in Pakistan. *Journal of Business Ecosystems, 1*(2).

Tunio, M. N. (2020). Role of ICT in promoting entrepreneurial ecosystems in Pakistan. *Journal of Business Ecosystems*, *1*(2), 1–21. doi:10.4018/JBE.2020070101

Tunio, M. N., Chaudhry, I. S., Mughal, F., & Shaikh, E. (2022). Marketing Mode and Survival of the Entrepreneurial Activities of Nascent Entrepreneurs. In *Big Data Analytics* (pp. 1–18). Auerbach Publications. doi:10.1201/9781003307761-1

Tunio, M. N., Chaudhry, I. S., Shaikh, S., Jariko, M. A., & Brahmi, M. (2021). Determinants of the Sustainable Entrepreneurial Engagement of Youth in Developing Country—An Empirical Evidence from Pakistan. *Sustainability (Basel)*, *13*(14), 7764. doi:10.3390u13147764

Tunio, M. N., Jariko, M. A., Børsen, T., Shaikh, S., Mushtaque, T., & Brahmi, M. (2021). How Entrepreneurship Sustains Barriers in the Entrepreneurial Process—A Lesson from a Developing Nation. *Sustainability (Basel)*, *13*(20), 1–18. doi:10.3390u132011419

Tunio, M. N., Shah, S. M. M., Qureshi, M. A., Tunio, A. N., & Shaikh, E. (2022). Career Predilections and Options to Opt Occupation for the Youth in Pakistan. In *Developing Entrepreneurial Ecosystems in Academia* (pp. 156–170). IGI Global. doi:10.4018/978-1-7998-8505-4.ch009

Tunio, M. N., & Shaikh, E. (2020). Nascent entrepreneurs and challenges in digital market in developing countries. *International Journal of Public Sector Performance Management.*

Tunio, M. N., Shaikh, E., Niaz, S., & Katper, N. S. (2021). Multifaceted perils of the Covid-19 and implications: A Review. *Estudios de Economía Aplicada*. Advance online publication. doi:10.25115/eea.v39i2.3957

Tunio, M. N., Soomro, A. A., & Bogenhold, D. (2017). The Study of Self-employment at SMEs Level with Reference to Poverty in Developing Countries. Business and Management Research, 6(2). doi:10.5430/bmr.v6n2p33

Tunio, M. N., Yusrini, L., Shah, Z. A., Katper, N., & Jariko, M. A. (2021). How Hotel Industry Cope up with the COVID-19: An SME Perspective. *Etikonomi*, *20*(2), 213–224. doi:10.15408/etk.v20i2.19172

Tunio, M. N., Yusrini, L., & Shoukat, G. (2021). Corporate social responsibility (CSR) in Hotels in Austria, Pakistan, and Indonesia: small and medium Enterprise spillover of COVID-19. In Handbook of research on entrepreneurship, innovation, sustainability, and ICTs in the post-COVID-19 era (pp. 263-280). IGI Global.

Tunio, M. N., Yusrni, L., Shah, Z. A., Katper, N. K., & Jariko, M. A. (2021). How hotel industry cope up with the COVID-19: An SME Perspective. *Etikonomi, 20*(2).

Vossenberg, S. (2013). Women Entrepreneurship Promotion in Developing Countries: What explains the gender gap in entrepreneurship and how to close it. *Maastricht School of Management Working Paper Series, 8*(1), 1-27.

Wang, Q. (2021). Higher education institutions and entrepreneurship in underserved communities. *Higher Education*, *81*(6), 1273–1291. doi:10.100710734-020-00611-5

Compilation of References

A global report on women in tourism. (2019). UNWTO. Retrieved from https://www.unwto.org/publication/global-report-women-tourism-2-edition

Abbasi, A. R., Raza, A., & Shaikh, H. (2022). A Nexus Between Political Instability & International Tourism Demand. *International Journal of Social Science & Entrepreneurship*, *2*(2), 297–312. doi:10.58661/ijsse.v2i2.32

Abbas, S., Isaac, N., Zia, M., Zakar, R., & Fischer, F. (2021). Determinants of women's empowerment in Pakistan: Evidence from Demographic and Health Surveys, 2012–13 and 2017–18. *BMC Public Health*, *21*(1), 1328. doi:10.118612889-021-11376-6 PMID:34229626

Abdalla, F., Samah, A., Hashim, A., & Rosnon, M. R. (2022). Relationship between Participation and Women Empowerment in the Tourism Industry; a Structural Equation Modeling Approach. *South Asian Research Journal of Business and Management.*, *4*(1), 11–27. doi:10.36346arjbm.2022.v04i01.002

Abdillah, Y., Supriono, S., & Supriyono, B. (2022). Change and innovation in the development of Balinese dance in the garb of special interest tourism. *Cogent Social Sciences*, *8*(1), 2076962. doi:10.1080/23311886.2022.2076962

Abdullah, M., Katper, N. K., Chaudhry, N. I., & Tunio, M. N. (2020). An Impact of Workaholics on Creativity: the mediating role of Negative Mood and moderating role of Supervisor Support. *Sukkur IBA Journal of Management and Business, 7*(2).

Abdullah, M., Katper, N. K., Chaudhry, N. I., & Tunio, M. N. (2020). An Impact of Workaholics on Creativity: The mediating role of Negative Mood and moderating role of Supervisor Support. *Sukkur IBA Journal of Management and Business, 7*(2).

Abou-Shouk, M. A., Mannaa, M. T., & Elbaz, A. M. (2021). Women's empowerment and tourism development: A cross-country study. *Tourism Management Perspectives*, *37*, 100782. doi:10.1016/j.tmp.2020.100782

Abou-Shouk, M., Zoair, N., Elbaz, A. M., & Abdel-Jalil, M. (2023). Local tourists' perceptions of tourist destinations' competitiveness: A comparative study of the United Arab Emirates, Egypt, and Oman. *Worldwide Hospitality and Tourism Themes*, *15*(2), 143–154. doi:10.1108/WHATT-10-2022-0117

About European Family Businesses - European Family. (2023). *European Family*. https://europeanfamilybusinesses.eu/about-european-family-businesses/

Abu-Rabia-Queder, S. (2010). Independence and entrepreneurship among Arab Muslim rural and Bedouin women in Israel. *Gender, Place and Culture*, *17*, 123–126.

Acar, Y. (2020). Yeni Koronavirüs (COVİD-19) Salgını ve Turizm Faaliyetlerine Etkisi [The New Coronavirus (COVID-19) Pandemic and Its Impact on Tourism Activities]. *Güncel Turizm Araştırmaları Dergisi*, *4*(1), 7–21. doi:10.32572/guntad.703410

Acevedo-Duque, Á., Gonzalez-Diaz, R., Vargas, E. C., Paz-Marcano, A., Muller-Pérez, S., Salazar-Sepúlveda, G., Caruso, G., & D'Adamo, I. (2021). Resilience, leadership and female entrepreneurship within the context of smes: Evidence from latin america. *Sustainability (Basel)*, *13*(15), 8129. doi:10.3390u13158129

Acharya, B. P., & Halpenny, E. A. (2013). Homestays as an alternative tourism product for sustainable community development: A case study of women-managed tourism product in rural Nepal. *Tourism Planning & Development*, *10*(4), 367–387. doi:10.1080/21568316.2013.779313

Açıkgöz, C. (2019). *Kırsal Turizmde Kadın Girişimcinin Yeri ve Önemi Muğla Örneği* [The Role and Importance of the Women Enrepreneur in Rural Tourism: Case of Muğla] [Master Thesis]. Muğla Sıtkı Koçman University.

Advisor, N. B. (2015). Focus social media outreach on cultivating 'influencers' among your followers. *Nonprofit Business Advisor Report*, *6–8*. Advance online publication. doi:10.1002/nba

Afshan, G., Ilyas, S., Tunio, M. N., & Kalhoro, M. (2021). CSR actions and post-COVID-19 consequences in the hotel industry: A conceptual framework. *International Journal of Strategic Change Management*, *7*(4), 275–289. doi:10.1504/IJSCM.2021.122845

Afshan, G., Shahid, S., & Tunio, M. N. (2021). Learning experiences of women entrepreneurs amidst COVID-19. *International Journal of Gender and Entrepreneurship*, *13*(2), 162–186. doi:10.1108/IJGE-09-2020-0153

Agenic.io. (2022). *5 Influencer Marketing Campaign Examples In Travel Industry*. https://agenic.io/5-influencer-marketing-campaign-examples-in-travel-industry/

Aghazamani, Y., & Hunt, C. A. (2017). Empowerment in tourism: A review of peer-reviewed literature. *Tourism Review International*, *21*(4), 333–346. doi:10.3727/154427217X15094520591321

Aghazamani, Y., Kerstetter, D., & Allison, P. (2020). Women's perceptions of empowerment in Ramsar, a tourism destination in northern Iran. *Women's Studies International Forum*, *79*, 102340. doi:10.1016/j.wsif.2020.102340

Agrawal, R. (2018). Constraints and challenges faced by women entrepreneurs in emerging market economy and the way forward. *JWEE*, (3-4), 1–19. doi:10.28934/jwee18.34.pp1-19

Agüera, F. O., Cuadra, S. M., López-Guzmán, T., & Morales, P. C. (2017). Estudio de la demanda existente en torno al oleoturismo. El caso de Andalucía. *Cuadernos de Turismo*, (39), 437–453. doi:10.6018/turismo.39.290641

Aguilar, M. J. C., & García, E. M. (2019). Olive oil tourism and rural development: Advances and challenges in the case of the province of jaén (Andalusia, Spain). *Journal of Tourism and Development*, (32), 255–264.

Agyeiwaah, E. (2020). Over-tourism and sustainable consumption of resources through sharing: The role of government. *International Journal of Tourism Cities*, *6*(1), 99–116. doi:10.1108/IJTC-06-2019-0078

Ahl, H., & Marlow, S. (2021). Exploring the false promise of entrepreneurship through a postfeminist critique of the enterprise policy discourse in Sweden and the UK. *Human Relations*, *74*(1), 41–68. doi:10.1177/0018726719848480

Ahmadi, A., Nakaa, N., & Bouri, A. (2018). Chief Executive Officer Attributes, oard Structures, Gender Diversity and Performance among French. *Research in International Business and Finance*, *44*(21), 218–226. doi:10.1016/j.ribaf.2017.07.083

Akbolat, M., Yılmazer, A., & Tutar, H. (2014). Konaklama işletmeleri çalışanlarının algıladıkları mobbingin iş tatmini ve işten ayrılma niyetine etkisi. *Manas Sosyal Araştırmalar Dergisi*, *3*(3), 1–17.

Akca, Ç. (2019). *Kadın Girişimcilerin Girişimcilik Niyetleri ile Kendine Güven İlişkisinin Belirlenmesi: Yiyecek İçecek Sektöründe Bir Araştırma* [Determining the Women Entrepreneurs Entrepreneurial Intent and Its Relations With Self-Confidence: A Research of Food and Beverage Sector] [Master Thesis]. Ankara Hacı Bayram Veli University.

Akdağ, G. & Üçyıldız, U. H. (2021). Seyahat acentası çalışanlarının cam tavan sendromu algılamaları: Ege bölgesi seyahat acentalarında bir uygulama. *İşletme Araştırmaları Dergisi, 12*(3), 2540–2654.

Akehurt, G., Simarro, E., & Mastur, A. (2012). Women entrepreneurship in small service firms: Motivations, barriers and performance. *Service Industries Journal*, *32*(15), 2489–2505. doi:10.1080/02642069.2012.677834

Akgül, S. Ö. (2020). Turizmde Toplumsal Cinsiyet Eşitliği ve Kadınların Güçlendirilmesi ile Mutluluk Endeksi Arasındaki İlişki [The Relationship Between Gender Equality and Women's Empowerment in Tourism and Happiness Index.] *Turar Turizm ve Araştırma Dergisi*, *9*(2), 160–178.

Akgun, A. E., Koseoglu, M. A., & Dogru, T. (2021). COVID-19 pandemic and the Turkish hospitality industry: Evaluating the impact on family-owned hotels. *Journal of Hospitality and Tourism Management*, *48*, 32–40. doi:10.1016/j.jhtm.2021.01.004

Akinci, Z., & Kasalak, M. A. (2016). Management of Special Interest Tourism in Terms of Sustainable Tourism. In Global Issues and Trends in Tourism. St. Kliment Ohridski University Press.

Akoğlan, M. (1996). Konaklama endüstrisinde kadının konumu. *Anatolia Turizm Araştırmaları Dergisi*, *7*(3-4), 16–23.

Alawa, J., Zarei, P., & Khoshnood, K. (2019). Evaluating the provision of health services and barriers to treatment for chronic diseases among Syrian refugees in Turkey: A review of literature and stakeholder interviews. *International Journal of Environmental Research and Public Health*, *16*(15), 2660. doi:10.3390/ijerph16152660 PMID:31349639

Albayrak, A. (2019). *Yiyecek içecek endüstrisinde kadının yeri ve sorunu. In Turizmde Kadın Olmak*. Değişim Yayınları.

Albuz, N. (2018). *Turist Rehberliğine Giriş, Profesyonel Turist Rehberlerinin Turizm Sektörü içindeki Yeri ve Önemi*. Detay Yayıncılık.

Alcañiz, M., Bigné, E., & Guixeres, J. (2019). Virtual Reality in Marketing: A Framework, Review, and Research Agenda. *Frontiers in Psychology*, *10*, 1530. doi:10.3389/fpsyg.2019.01530 PMID:31333548

Alcántara-Alcover, E., Artacho-Ramírez, M. A., Martínez-Guillamóna, N., & Campos-Sorianoa, N. (2013). Purpose of stay and willingness to stay as dimensions to identify and evaluate hotel experiences. *International Journal of Hospitality Management*, *33*, 357–365. doi:10.1016/j.ijhm.2012.10.006

Al-Dajani, H., & Marlow, S. (2013). Empowerment and entrepreneurship: A theoretical framework. *International Journal of Entrepreneurial Behaviour & Research*, *19*(5), 503–524. doi:10.1108/IJEBR-10-2011-0138

Aldırmaz Akkaya, F. (2018). *Türkiye'de Kadın Kooperatiflerinin Sürdürülebilir Kırsal Turizmdeki Önemi ve Finansal Performanslarının Analizi*. Doktora Tezi, İnönü Üniversitesi, Sosyal Bilimler Enstitüsü.

Alexandre, L., & Kharabsheh, R. (2019). The evolution of female entrepreneurship in the Gulf Cooperation Council, the case of Bahrain. *International Journal of Gender and Entrepreneurship*, *11*(4), 390–407. doi:10.1108/IJGE-02-2019-0041

Alhas, F. (2020). Kadınların üst düzey yönetici olmasına engel olan nedenler cam uçurum, cam yürüyen merdiven ve cam asansör arasındaki ilişkinin incelenmesi. *Ağrı İbrahim Çeçen Üniversitesi Sosyal Bilimler Enstitüsü Dergisi*, *6*(2), 101–113.

Alhowaish, A. K. (2016). Is tourism development a sustainable economic growth strategy in the long run? Evidence from GCC countries. *Sustainability (Basel)*, *8*(7), 605. doi:10.3390u8070605

Ali, M. A., & Salisu, Y. (2019). Women Entrepreneurship and Empowerment Strategy for National Development. *Journal of Economics. Management and Trade*, *22*(3), 1–13. doi:10.9734/JEMT/2019/44828

Al-Ismail, S., Carmichel, F., & Duberley, J. (2019). Female employment in hotels in Saudi Arabia and UAE. *Gender in Management*, *34*(7), 554–576. doi:10.1108/GM-11-2017-0151

Alkan, Ö., Yılmaz, F. & Abar, H. (2020). Türkiye'de kadına yönelik aile içi şiddeti etkileyen faktörlerin sıralı probit regresyon modeliyle belirlenmesi. *İnsan ve Toplum Bilimleri Araştırmaları Dergisi*, *9*(5), 3338-3360.

Alkan, Z., & Ulas, S. (2022). Practices of social media and influencer: A research based on influencer's followers. *MEDIAJ*, *5*(2), 187–205.

Alkan, Z., & Ulas, S. (2023). Trust in social media influencers and purchase intention: An empirical analysis. *Online Journal of Communication and Media Technologies*, *13*(1), e202301. doi:10.30935/ojcmt/12783

Alkhaled, S., & Berglund, K. (2018). 'And now I'm free': Women's empowerment and emancipation through entrepreneurship in Saudi Arabia and Sweden. *Entrepreneurship and Regional Development*, *30*(7-8), 877–900. doi:10.1080/08985626.2018.1500645

Almathami, R., Khoo-Lattimore, C., & Yang, E. C. L. (2022). Exploring the challenges for women working in the event and festival sector in the Kingdom of Saudi Arabia. *Tourism Recreation Research*, *47*(1), 47–61. doi:10.1080/02508281.2020.1821329

Alonso-Almeida, M. del M. (2013). Influence of gender and financing on tourist company growth. *Journal of Business Research*, *66*(5), 621–631. doi:10.1016/j.jbusres.2012.09.025

Alonso, P., Moscoso, S., & Salgado, J. F. (2017). Structured Behavioral Interview as a Legal Guarantee for Ensuring Equal Employment Opportunities for Women: A Meta-Analysis. *The European Journal of Psychology Applied to Legal Context*, *9*(1), 15–23. doi:10.1016/j.ejpal.2016.03.002

Aloulou, W. (2019). Entrepreneurship and innovation in the digitalization era. *Advances in E-Business Research*, 179–203. doi:10.4018/978-1-5225-7262-6.ch011

Alpak, G., Unal, A., Bulbul, F., Sagaltici, E., Bez, Y., Altindag, A., Dalkilic, A., & Savas, H. A. (2015). Post-traumatic stress disorder among Syrian refugees in Turkey: A cross-sectional study. *International Journal of Psychiatry in Clinical Practice*, *19*(1), 45–50. doi:10.3109/13651501.2014.961930 PMID:25195765

Alparslan, A. M., Çetinkaya, B. Ö., & Özgöz, A. (2015). İşletmelerde cinsiyet ayrımcılığı ve kadın çalışanların sorunları. *MAKÜ İktisadi ve İdari Bilimler Fakültesi Dergisi*, *2*(3), 66–81.

Alpkan, L., Bulut, C., Gunday, G., Ulusoy, G., & Kilic, K. (2010). Organizational support for intra-preneurship and its interaction with human capital to enhance innovative performance. *Management Decision*, *48*(5), 732–755. doi:10.1108/00251741011043902

Alrwajfah, M. M., Almeida-García, F., & Cortés-Macías, R. (2020). Females' perspectives on tourism's impact and their employment in the sector: The case of Petra, Jordan. *Tourism Management*, *78*, 104069. doi:10.1016/j.tourman.2019.104069

Alshareef, N., & Tunio, M. N. (2022). Role of Leadership in Adoption of Blockchain Technology in Small and Medium Enterprises in Saudi Arabia. *Frontiers in Psychology*, *13*, 2284. doi:10.3389/fpsyg.2022.911432

Altunışık, R., Boz, H., Gegez, E., Koç, E., Sigri, U., Yildiz, E., & Yuksel, A. (2022). *Sosyal Bilimlerde Araştırma Yöntemleri: Yeni Perspektifler*. Seçkin Yayıncılık.

Ambardar, M. R., & Singh, S. (2017). Empowerment and socio-economic upliftment of rural women through tourism: A case study of Bhaderwah, Jammu & Kashmir. *International Journal of Scientific Research*, *6*(5), 346–352.

Anafarta, N., Sarvan, F., & Yapıcı, N. (2008). Konaklama işletmelerinde kadın yöneticilerin cam tavan algısı: Antalya ilinde bir araştırma. *Akdeniz İ.İ.B.F. Dergisi*, *15*, 111–137.

Anggadwita, G., Luturlean, B. S., Ramadani, V., &Ratten, V. (2017). Socio-cultural environments and emerging economy entrepreneurship: Women entrepreneurs in Indonesia. *Journal of Entrepreneurship in Emerging Economies*.

Ap, J., & Wong, K. K. (2001). Case Study on Tour Guiding: Professionalism, Issues and Problems. *Tourism Management*, *22*(5), 551–563. doi:10.1016/S0261-5177(01)00013-9

Apostolopoulos, Y., & Sönmez, S. (2001). Working producers, leisured consumers: Women's experiences in developing regions. *Women as Producers and Consumers of Tourism in Developing Regions*, 3–18.

Aquino, R. S., Lück, M., & Schänzel, H. A. (2022). Mapping the outcomes of social entrepreneurship and tourism on host communities: A three-dimensional approach. *Journal of Sustainable Tourism*, *30*(8), 1799–1820. doi:10.1080/09669582.2021.1986515

Ardıç Yetiş, Ş., & Çalışkan, N. (2020). Turizm sektöründe kadın istihdamı: Mevcut duruma ilişkin bir değerlendirme. *Manisa Celal Bayar Üniversitesi Sosyal Bilimler Dergisi*, *18*(3), 105–119.

Ardrey, W. J. IV, Pecotich, A., & Shultz, C. J. II. (2006). Entrepreneurial women as catalysts for socioeconomic development in transitioning Cambodia, Laos, and Vietnam. *Consumption Markets & Culture*, *9*(4), 277–300. doi:10.1080/10253860600921811

Ar, H., & Çelik Uğuz, S. (2017). Küresel Sürdürülebilir Kalkınma Hedeflerinde Turizmin Rolü: Türkiye Örneği. *Uluslararası Sosyal Araştırmalar Dergisi*, *10*(49), 521–530.

Arora, M., & Sharma, R. L. (2021). Neutralizing Maleficent Effects of COVID-19 Through Entrepreneurship: Peeping Through the Lens of Communication. In Effective Strategies for Communicating Insights in Business (pp. 67-86). IGI Global.

Arora, M., & Sharma, R. L. (2021). Repurposing the Role of Entrepreneurs in the Havoc of COVID-19. In Entrepreneurship and Big Data (pp. 229-250). CRC Press. doi:10.1201/9781003097945-16

Arora, M., & Singh, S. (2018). Impact Assessment of Self-Help Group Bank Linkage Programme on Women Empowerment in the State of Himachal Pradesh. Academic Press.

Arora, A., Bansal, S., Kandpal, C., Aswani, R., & Dwivedi, Y. (2019). Measuring social media influencer index- insights from facebook, Twitter and Instagram. *Journal of Retailing and Consumer Services*, *49*, 86–101. doi:10.1016/j.jretconser.2019.03.012

Arora, M. (2016). Creative dimensions of entrepreneurship: A key to business innovation. *Pacific Business Review International*, *1*(1), 255–259.

Arora, M., & Sharma, R. L. (2022). Coalescing skills of gig players and fervor of entrepreneurial leaders to provide resilience strategies during global economic crises. In *COVID-19's Impact on the Cryptocurrency Market and the Digital Economy* (pp. 118–140). IGI Global. doi:10.4018/978-1-7998-9117-8.ch008

Arora, M., & Singh, S. (2018). Microfinance, Women Empowerment, and Transformational Leadership: A Study of Himachal Pradesh. *International Journal on Leadership*, *6*(2), 23.

Arora, M., & Singh, S. (2023). Microfinance for achieving sustainable development goals: pondering over Indian experiences for the preservation of magnificent african natural resources. In *Research Anthology on Microfinance Services and Roles in Social Progress* (pp. 378–394). IGI Global.

Arroyo, C., Barbieri, C., Sotomayor, S., & Knollenberg, W. (2019). Cultivating women's empowerment through agritourism: Evidence from Andean communities. *Sustainability*, *11*(11), 1–14.

Arshad, M., Qasim, N., Farooq, O., & Rice, J. (2022). Empowering leadership and employees' work engagement: A social identity theory perspective. *Management Decision*, *60*(5), 1218–1236. doi:10.1108/MD-11-2020-1485

Arslan, A., & Şimşek, G. (2018). Kadın Turist Rehberlerinin Yaşadıkları Mesleki Sorunlar: Aydın Turist Rehberleri Örneği. *Journal of Travel and Tourism Research*, (13), 23–49.

Arslan, A., & Şimşek, G. (2018). Kadın turist rehberlerinin yaşadıkları mesleki sorunlar: Aydın turist rehberleri örneği. *Journal of Travel and Tourism Research*, *13*, 23–49.

Asan, K. (2022). Measuring the impacts of travel influencers on bicycle travelers. *Current Issues in Tourism*, *25*(6), 978–994. doi:10.1080/13683500.2021.1914004

Asbollah, A. Z. B., Lade, C., & Michael, E. (2013). The tourist's gaze: From the perspective of a Muslim woman. *Tourism Analysis*, *18*(6), 677–690. doi:10.3727/108354213X13824558188703

Ashforth, B. E. (1989). The experience of powerlessness in organizations. *Organizational Behavior and Human Decision Processes*, *43*(2), 207–242. doi:10.1016/0749-5978(89)90051-4

Ataünal, L., & Aybars, A. (2018). Yönetim Kurullarındaki Cinsiyet Çeşitliliği ile Çeşitliliğin Finansal Performans Etkisi: Gelişmekte Olan Piyasa Örneği. *Maliye Finans Yazıları*, (109).

Ateljevic, I., Pritchard, A., & Morgan, N. (2018). Gender and tourism: A feminist perspective. In The Routledge Handbook of Transport, Tourism and Mobilities (pp. 183-195). Routledge.

Ateş, M., Kazan, İ., & Soybalı, H. H. (2022). Turist Rehberliğinde Ayrımcılık: Kadın Turist Rehberlerine Yönelik Bir Uygulama. *Turist Rehberliği Dergisi*, *5*(1), 1–16. doi:10.34090/tured.1064528

Audrezet, A., de Kerviler, G., & Moulard, J. G. (2018). Authenticity under threat: When social media influencers need to go beyond self-presentation. *Journal of Business Research*, *18*(July). Advance online publication. doi:10.1016/j.jbusres.2018.07.008

Aurangzeb, M. T., Tunio, M. N., Rehman, Z., & Asif, M. (2021). Influence of administrative expertise on human resources practitioners on the job performance: Mediating role of achievement motivation. *International Journal of Management*, *12*(4), 408–421.

Avais, M. A., Wassan, A., & Shah, S. (2014). A case study on problems of working women in city Sukkur. *Academic Research International*, *5*(2), 325–333.

Avcı, E. & Bilgili, B. (2020). Sosyal Medya Fenomen Özelliklerinin Takipçilerin Destinasyonu Ziyaret Etme Niyeti Üzerindeki Etkisi. *Tourism and Recreation*, *2*(1), 83-92.

Avcı, N. (2018). *Turizm sektöründe çalışan kadınların karşılaştıkları engeller ve bu engellere yönelik algıları* [Unpublished Master Thesis]. Muğla Sıtkı Koçman Üniversitesi, Muğla.

Avcı, C. (2021). Turizm İşletmelerinde Cinsiyet Ayrımcılığı: Kavramsal Bir Değerlendirme. *Turizm ve İşletme Bilimleri Dergisi*, *1*(2), 1–14.

Aydın Tükeltürk, Ş., & Şahin Perçin, N. (2008). Turizm sektöründe kadın çalışanların karşılaştıkları kariyer engelleri ve cam tavan sendromu: Cam tavanı kırmaya yönelik stratejiler. *Yönetim Bilimleri Dergisi*, *6*(2), 113–128.

Aydoğdu, G. (2021, March 11). *Toplumsal Cinsiyet ve Turizm*. https://kalkinmaguncesi.izka.org.tr/index.php/2021/03/11/toplumsal-cinsiyet-ve-turizm/

Ay, E., İpek, K., Özdağ, N. B., Özekici, E., & Alvarez, M. D. (2019). Travel bloggers as influencers: What compels them to blog. In N. Kozak & M. Kozak (Eds.), *Tourist destination management* (pp. 159–175). Springer International Publishing. doi:10.1007/978-3-030-16981-7_9

Aykaç, A. (2009). *Yeni işler, yeni işçiler*. İletişim Yayınları.

Aylan, F. K., Sarı Gök, H., & Şalvarcı, S. (2019). Kırsal Yoksulluğun Giderilmesinde Kadın Girişimciliğin Rolü: Lavanta Kokulu Köy Örneği [The Role of Woman Entrepreneurship in Reduction of Rural Poverty: A Case of Levander Smelling Village]. *Journal of Tourism Gastronomy Studies*, *7*(2), 1271–1289. doi:10.21325/jotags.2019.420

Aytaç, S., & Dursun, S. (2013). Çalışma hayatında kadına yönelik şiddet: Sağlık sektöründe bir uygulama. *Journal of Istanbul University Law Faculty*, *71*(1), 67–78.

Ayyıldız, T., Ayyıldız, A. Y., & Turan, O. K. U. L. (2019). Turizm Rehberliği Öğrencilerinde Türkiye İmajı Algılamaları: Aydın Adnan Menderes Üniversitesi Turizm Fakültesi Örneği. *Journal of Travel and Tourism Research*, (14), 93–100.

Bachri, N., Wardhiah, W., Roni, M., & Nur, M. M. (2022). Women's entrepreneurship in supporting halal tourism in lhokseumawe city. *International Journal of Economic, Business, Accounting, Agriculture Management and Sharia Administration*, *2*(1), 49–56.

Backaler, J. (2018). *Digital influence: Unleash the power of influencer marketing to accelerate your global business*. Palgrave Macmillan. doi:10.1007/978-3-319-78396-3

Badoc-Gonzales, B. P., Mandigma, M. B. S., & Tan, J. J. (2022). SME resilience as a catalyst for tourism destinations: A literature review. *Journal of Global Entrepreneurship Research*, *12*(1), 1–22. doi:10.100740497-022-00309-1

Bagheri, F., Ghaderi, Z., Abdi, N., & Hall, C. M. (2022). Female entrepreneurship, creating shared value, and empowerment in tourism; the neutralizing effect of gender-based discrimination. *Current Issues in Tourism*, 1–18. doi:10.1080/13683500.2022.2126749

Bagheri, M., & Tayebi Khorami, M. (2014). Reasons, Opportunities, Challenges and Solutions for Women's Employment in the Tourism Industry of Islamic Republic of Iran. *Journal of Tourism Hospitality Research*, *3*(2), 91–121.

Bagilhole, B. (2002). *Women in non-traditional occupations: Challenging men*. Springer. doi:10.1057/9780230501102

Bahar, O., & Çelik İlal, N. (2020). Coronavirüsün (COVİD-19) Turizm Sektörü Üzerindeki Ekonomik Etkileri [The Economic Effects of Coronavirus (COVID-19) on the Tourism Sector]. *International Journal of Social Sciences and Education Research*, *6*(1), 125–139. doi:10.24289/ijsser.728121

Bakar, N. A., & Rosbi, S. (2020). Effect of Coronavirus Disease (COVİD-19) to Tourism Industry. *International Journal of Advanced Engineering Research and Science*, *7*(4), 189–193. doi:10.22161/ijaers.74.23

Bakas, F. E. (2017). 'A beautiful mess': Reciprocity and positionality in gender and tourism research. *Journal of Hospitality and Tourism Management*, *33*, 126–133. doi:10.1016/j.jhtm.2017.09.009

Bakas, F. E., Costa, C., Breda, Z., & Durão, M. (2018). A critical approach to the gender wage gap in tourism labor. *Tourism, Culture & Communication*, *18*(1), 35–49. doi:10.3727/109830418X15180180585167

Bakay, M. E., Müftüoğlu, M., Nalbantoğlu, A., & Çoçan, E. (2019). Kırsal Kesimlerdeki Kadın Girişimciliği ve Girişimci Kadınların Karşılaştıkları Sorunlar Üzerine Nitel Bir Araştırma: Manisa-Demirci Örneği [A Qualitative Research on Women Entrepreneurship in Rural Areas and Problems Encountered by Entrepreneurial Women: The Case of Manisa-Demirci]. *Journal of Yasar University*, *15*(Special Issue), 295–306.

Baker, M. A., & Nelson, R. E. (2005). Creating something from nothing: Resource construction through entrepreneurial bricolage. *Administrative Science Quarterly*, *50*(3), 329–366. doi:10.2189/asqu.2005.50.3.329

Baker, T. E., Aldrich, H., & Nina, L. (1997). Invisible entrepreneurs:the neglect of women business owners by mass media and scholarly journals in the USA. *Entrepreneurship and Regional Development*, *9*(3), 221–238. doi:10.1080/08985629700000013

Balcı, E., & Aslan, E. B. (2022). Turizmde çalışan kadın sorunları: Antalya'da 5 yıldızlı otellerin yiyecek-içecek departmanında çalışan kadınlar. *Journal of Tourism and Gastronomy Studies*, *10*(3), 1674–1698.

Balkız, Ö. I., & Öztürk, E. (2013). Neo-liberal gelişme anlayışı ve kadın: Mikro finans uygulamaları kadınları güçlendiriyor mu? *Mediterranean Journal of Humanities*, *3*(2), 1–21. doi:10.13114/MJH/201322451

Bandiera, O., Buehren, N., Burgess, R., Goldstein, M., Gulesci, S., Rasul, I., & Sulaiman, M. (2020). Women's empowerment in action: Evidence from a randomized control trial in Africa. *American Economic Journal. Applied Economics*, *12*(1), 210–259. doi:10.1257/app.20170416

Bansal, S., Garg, I., & Sharma, G. D. (2019). Social entrepreneurship as a path for social change and driver of sustainable development: A systematic review and research agenda. *Sustainability (Basel)*, *11*(4), 1091. doi:10.3390u11041091

Baran, M. (2023). *Under the influence*. https://www.travelweekly.com/Travel-News/Travel-Technology/social-media-influencers

Baratta, R., Vigolo, V., & Simeoni, F. (2021). Perceived risks and future travel intentions: The experience of solo female travelers….*: Restarting tourism*… https://aisberg.unibg.it/retrieve/c04b9d53-fa1b-4e1c-804e-9860dfef2296/Baratta%20et%20al.%202021_EA_Tourman.pdf

Baratta, R., Vigolo, V., & Simeoni, F. (2021). Perceived risks and future travel intentions: The experience of solo female travelers. *Tourman 2021 Book of*… https://aisberg.unibg.it/retrieve/c04b9d53-fa1b-4e1c-804e-9860dfef2296/Baratta%20et%20al.%202021_EA_Tourman.pdf

Barbosa, L. (2020). Social entrepreneurship and digital technology-case study of the blockchain for zero hunger initiative. *Economic and Social Development: Book of Proceedings*, 193-202.

Barnes, S. J., Rutter, R., Mattsson, J., & Sørensen, F. (2020). Patron Sentiment of Employee–Customer Interaction: Exploring Hotel Customer Reviews through Machine Learning: An Abstract. In S. Wu, F. Pantoja, & N. Krey (Eds.), *Marketing Opportunities and Challenges in a Changing Global Marketplace. AMSAC 2019. Developments in Marketing Science: Proceedings of the Academy of Marketing Science*. Springer. doi:10.1007/978-3-030-39165-2_161

Barrett, R., Jones, T., & McEvoy, D. (2016). "How do we get from there to here?": An interpretive phenomenological analysis of the lived experiences of women family business founders. *International Small Business Journal*, *34*(8), 1060–1080.

Barry, D., & Hoyne, S. (2021). Sustainable measurement indicators to assess impacts of climate change: Implications for the New Green Deal Era. *Current Opinion in Environmental Science & Health*, *22*, 100259. doi:10.1016/j.coesh.2021.100259

Bartos, R. (1982). Women and travel. *Journal of Travel Research*, *20*(4), 3–9. doi:10.1177/004728758202000402

Başaran, H., & Irmak, E. (2022). Devrek'te Yerel Ürün Üretimi ve Yerelde Kadın Girişimciliğinde Kooperatiflerin Yeri [The importance of cooperatives in local product production and local women's entrepreneurship in Devrek]. *Ahi Evran Akademi*, *3*(1), 19–31.

Bastian, B. L., Sidani, Y. M., & El Amine, Y. (2018). Women entrepreneurship in the Middle East and North Africa: A review of knowledge areas and research gaps. *Gender in Management*, *33*(1), 14–29. doi:10.1108/GM-07-2016-0141

Bastida, M., Pinto, L. H., Olveira Blanco, A., & Cancelo, M. (2020). Female entrepreneurship: Can cooperatives contribute to overcoming the gender gap? A Spanish first step to equality. *Sustainability (Basel)*, *12*(6), 1–17. doi:10.3390u12062478

Basurto-Barcia, J., & Ricaurte-Quijano, C. (2017). Women in Tourism: Gender (in) Equalities in University Teaching and Research. *Anatolia*, *28*(4), 567–581. doi:10.1080/13032917.2017.1370781

Bates, T. (1995). Self-employment entry across industry groups. *Journal of Business Venturing*, *10*(2), 143–156. doi:10.1016/0883-9026(94)00018-P

Batliwala, S. (1994). The meaning of women's empowerment: New concept from action. Population Policies Reconsidered: Health, Empowerment and Rights, 127-138.

Battour, M. (2017). *Muslim Travel Behavior in Halal Tourism, Mobilities, Tourism and Travel Behavior-Contexts and Boundaries*. Intech Open. Available online: https://www. intechopen. com/books/mobilities-tourism-and-travel-behavior-contexts-and-boundaries/muslimtravel-behavior-in-halal-tourism

Battour, M., & Ismail, M. N. (2016). Halal tourism: Concepts, practises, challenges and future. *Tourism Management Perspectives*, *19*, 150–154. doi:10.1016/j.tmp.2015.12.008

Battour, M., Ismail, M. N., & Battor, M. (2011). The impact of destination attributes on Muslim tourist's choice. *International Journal of Tourism Research*, *13*(6), 527–540. doi:10.1002/jtr.824

Baum, T. (2013). *International perspectives on women and work in hotels, catering and tourism*. Academic Press.

Baum, T. (2013). *International perspectives on women and work in hotels, catering and tourism*. International Labour Organisation (No. 289). Gender Working Paper 1/2013, Sector Working Paper.

Bayeh, E. (2016). The role of empowering women and achieving gender equality to the sustainable development of Ethiopia. *Pacific Science Review B. Humanities and Social Sciences*, *2*(1), 37–42.

Bayram, A. T. (2019). Bıblıometrıc Analysıs of Women Studıes on Turkısh Tourısm Lıterature. *Journal of Tourism And Research*, *8*(2), 55–77.

Bayram, A. T. (2019). Türk Turizm Alan Yazınında Kadınlar Üzerine Yapılan Çalışmaların Bibliyometrik Analizi [Bibliometrik Analysis of Women Studies On Turkish Tourism Literature]. *Turizm ve Araştırma Dergisi*, *8*(2), 56–77.

Bazyar, A. (2014). Social Sustainable Development and Economic Transformation, OIDA. *International Journal of Sustainable Development*, *7*(5), 40–50.

Beard, J., & Ragheb, M. (1983). Measuring leisure motivation. *Journal of Leisure Research*, *15*(3), 219–228. doi:10.1080/00222216.1983.11969557

Beck, J. A. (1996). *Career anchors, organizational commitment, and job plateaus: An analysis of hotel executive operating committee members' career development* [Doctor of Philosophy Thesis]. Purdue University Publications.

Beck, J., & La Lopa, J. M. (2001). An exploratory application of Schein's career anchors inventory to hotel executive operating committee members. *International Journal of Hospitality Management*, *20*(1), 15–28. doi:10.1016/S0278-4319(00)00035-9

Bedük, A. (2005). Women Working in Turkey and Women's Entrepreneurship. *Electronic Journal of Social Sciences*, *3*(12), 106–117.

Belhassen, Y., & Caton, K. (2009). Advancing understandings: A linguistic approach to tourism epistemology. *Annals of Tourism Research*, *36*(2), 335–352. doi:10.1016/j.annals.2009.01.006

Belias, D., Velissariou, E., Kyriakou, D., Varsanis, K., Vasiliadis, L., Mantas, C., Sdrolias, L., & Koustelios, A. (2018). Tourism consumer behavior and alternative tourism: the case of agrotourism in Greece. In *Innovative Approaches to Tourism and Leisure* (pp. 465–478). Springer. doi:10.1007/978-3-319-67603-6_35

Bell, D. E., & Staw, B. M. (1989). People as sculptors versus sculpture: The roles of personality and personal control in organizations. In L. L. Cummings & B. M. Staw (Eds.), *Research in Organizational Behavior* (Vol. 11, pp. 1–39). JAI Press. doi:10.1017/CBO9780511625459.014

Belli, L., Cilfone, A., Davoli, L., Ferrari, G., Adorni, P., Di Nocera, F., Dall'Olio, A., Pellegrini, C., Mordacci, M., & Bertolotti, E. (2020). IoT-enabled smart sustainable cities: Challenges and approaches. *Smart Cities*, *3*(3), 1039–1071. doi:10.3390martcities3030052

Beqo, I., & Gehrels, A. S. (2014). Women entrepreneurship in developing countries: A European example. *Research in Hospitality Management*, *4*(1-2), 97–103. doi:10.1080/22243534.2014.11828320

Berdychevsky, L., & Gibson, H. J. (2015). Sex and risk in young women's tourist experiences: Context, likelihood, and consequences. *Tourism Management*. https://www.sciencedirect.com/science/article/pii/S0261517715000850

Bernard, S., Rahman, I., & McGehee, N. G. (2022). Breaking barriers for Bangladeshi female solo travelers. *Tourism Management Perspectives*. https://www.sciencedirect.com/science/article/pii/S2211973621001458

Bernard, S., Rahman, I., & McGehee, N. G. (2022). Breaking barriers for Bangladeshi female solo travelers. *Tourism Management Perspectives*, *41*, 100932. doi:10.1016/j.tmp.2021.100932

Bernhard, I., & Olsson, A. K. (2020). Network collaboration for local and regional development - the case of Swedish women entrepreneurs. *International Journal of Entrepreneurship and Small Business*, *41*(4), 539. doi:10.1504/IJESB.2020.111578

Berno, T. (2011). Sustainability on a Plate: Linking Agriculture and Food in the Fiji Island Tourism Industry. In Tourism and Agriculture: New Geographies of Consumption, Production and Rural Restructuring (pp. 87-103). London: Routledge.

Bharathi, P. S. (2019). Woman entrepreneurship and its scope in tourism sector in Talangana. *Ictact Journal on Management Studies*, *5*(1), 912–915.

Bilsland, C., Nagy, H., & Smith, P. (2020). Virtual internships and work-integrated learning in hospitality and tourism in a post-COVID-19 world. *International Journal of Work-Integrated Learning*, *21*(4), 425–437.

Birdir, K., Derinalp, S., & Çanakçı, T. (2013). Otel mutfaklarında çalışan iş görenlerin sorunları, 14. In Ulusal Turizm Kongresi (pp. 980-997). Kayseri.

Bisnath, S. (2001). *Poverty in a globalizing world at different stages of women's life cycle. Içinde Globalization, Poverty and Women's Empowerment*. Sayı November.

Biswas, P. K., Roberts, H., & Stainback, K. (2021). Does women's board representation affect non-managerial gender inequality? *Human Resource Management*, *60*(4), 659–680. doi:10.1002/hrm.22066

Biswas, R., & Cassel, C. (1996). Strategic HRM and gendered division of labour in the hotel industry: A case study. *Personnel Review*, *25*(2), 19–34. doi:10.1108/00483489610147870

Blau, G., & Tatum, D. (2000). Correlates of perceived gender discrimination for female versus male medical technologists. *Sex Roles*, *43*(1/2), 105–118. doi:10.1023/A:1007043813257

Blesia, J. U., Iek, M., Ratang, W., & Hutajulu, H. (2021). Developing an entrepreneurship model to increase students' entrepreneurial skills: An action research project in a higher education institution in Indonesia. *Systemic Practice and Action Research*, *34*(1), 53–70. doi:10.100711213-019-09506-8

Blood, R. O., & Wolfe, D. M. (1960). *Husbands and Wives: The Dynamics of Married Living*. The Free Press.

Blumstein, P. W., & Schwartz, P. (1983). *American Couples: Money, Work, Sex*. William Morrow.

Boden, R. J. Jr, & Nucci, A. R. (2000). On the survival prospects of men's and women's new business ventures. *Journal of Business Venturing*, *15*(4), 347–362. doi:10.1016/S0883-9026(98)00004-4

Bodosca, S., & Diaconescu, M. D. (2015). Tourism Development After the Implementation of Sustainable Strategies in Neamt County. *Procedia: Social and Behavioral Sciences*, *188*, 230–236. doi:10.1016/j.sbspro.2015.03.378

Boğan, E. (2020). Halal tourism: The practices of halal hotels in Alanya, Turkey. *Journal of Tourism and Gastronomy Studies*, *8*(1), 29–42. doi:10.21325/jotags.2020.534

Boley, B. B., Ayscue, E., Maruyama, N., & Woosnam, K. M. (2016). Gender and empowerment: Assessing discrepancies using the resident empowerment through tourism scale. *Journal of Sustainable Tourism*, *25*(1), 113–129. doi:10.1080/09669582.2016.1177065

Boley, B. B., & Gard McGehee, N. G. (2014). Measuring empowerment: Developing and validating the resident empowerment through tourism scale (RETS). *Tourism Management*, *45*, 85–94. doi:10.1016/j.tourman.2014.04.003

Bolles, A. L. (1997). Women as a category of analysis in scholarship on tourism: Jamaican women and tourism employment. *Tourism and culture: An applied perspective*, 77-92.

Boluk, K. A., Cavaliere, C. T., & Higgins-Desbiolles, F. (2019). A critical framework for interrogating the United Nations Sustainable Development Goals 2030 Agenda in tourism. *Journal of Sustainable Tourism*.

Bond, M. (1997). *Women Travellers: A New Growth Market*. Pacific Asia Travel Association.

Booth, N., & Matic, J. A. (2011). Mapping and leveraging influencers in social media to shape corporate brand perceptions. *Corporate Communications*, *16*(3), 184–191. doi:10.1108/13563281111156853

Bor, H. M., & Erten, A. (2019). Dijital Çağın Mesleği-Nasıl Influencer Olunur? İstanbul: Hürriyet Kitap.

Boselie, P., Van Harten, J., & Veld, M. (2021). A human resource management review on public management and public administration research: Stop right there…before we go any further…. *Public Management Review*, *23*(4), 483–500. doi:10.1080/14719037.2019.1695880

Bosma, N., & Kelley, D. (2019). *Global entrepreneurship monitor 2018/2019 global report. Global Entrepreneurship Research Association*. GERA.

Boudrias, J. S., Gaudreau, P., & Laschinger, H. K. S. (2004). Testing the structure of psychological empowerment: Does gender make a difference? *Educational and Psychological Measurement*, *64*(5), 861–877. doi:10.1177/0013164404264840

Bouhazzama, M., & Guenaoui, A. (2020). The Moroccan women's cooperative in response to lasting impacts: Social cohesion, solidarity and inclusion. *International Journal of Management*, *11*(5), 195–202. doi:10.34218/IJM.11.5.2020.01

Bourne, L. (2011). Advising upwards: Managing the perceptions and expectations of senior management stakeholders. *Management Decision*, *49*(6), 1001–1023. doi:10.1108/00251741111143658

Bozdemir, M., Gülçubuk, B. & Bayramoğlu, Z. (2021). Kırsal Alandaki Kadınların Girişimci Niteliklerinin ve Girişimcilik Becerilerinin Belirlenmesi: Konya İli Araştırması [Determination of Entrepreneur Qualifications and Entrepreneurship Skills of Women in the Rural Area: A Research Province Konya]. *Trakya Üniversitesi Sosyal Bilimler Dergisi, 23*, 87-106. doi:10.26468/trakyasobed.871643

Bozkurt, Ö. (2019). Kariyer engelleri ve kariyer sorunları. In Kariyer, Temel Kavramlar, Yönetimi, Güncel Konular (pp. 63-75). Konya: Eğitim Yayınevi.

Brammer, S., & Millington, A. (2008). Does it pay to be different? An analysis of the relationship between corporate social and financial performance. *Strategic Management Journal*, *29*(12), 1325–1343. doi:10.1002mj.714

Branzei, O. (2012). *Social change agency under adversity: how relational processes (re)produce hope in hopeless settings*. Routledge.

Breda, Z., Santos, A., Kliček, T., & Dinis, G. (2020). Profile, motivations, and experiences from Portuguese solo female travelers. *... management approaches to ...* https://www.igi-global.com/chapter/profile-motivations-and-experiences-from-portuguese-solo-female-travelers/251672

Brobbey, V. A., Asiedu, R., Mensah, P. O., Antwi, F. A., Sie, E. F., & Kumah, A. (2022). Employees' Perception of Sexual Harassment and Its Impacts on Employee Performance in Ghana. *Social Values and Society*, *4*(2), 41–45. doi:10.26480vs.02.2022.41.45

Brotherton, B., & Himmetoglu, B. (1997). Beyond destinations— special interest tourism. *Anatolia: An International Journal of Tourism and Hospitality Research, 8*(3), 11–30.

Brotherton, B., & Himmetoğlu, B. (1997). Beyond Destinations—Special Interest Tourism. *Anatolia*, *8*(3), 11–30. doi:10.1080/13032917.1997.9687118

Brown, D., & Hayes, N. (2008). *Influencer marketing: Who really influences your customers?* Routledge. doi:10.4324/9780080557700

Brownell, J. (1994). Women in hospitality management: General managers' perceptions of factors related to career development. *International Journal of Hospitality Management*, *13*(2), 101–117. doi:10.1016/0278-4319(94)90032-9

Brush, C. G. (1992). Research on women business owners: Past trends, a new perspective and future directions. *Entrepreneurship Theory and Practice*, *16*(4), 5–30. doi:10.1177/104225879201600401

Brush, C. G., De Bruin, A., & Welter, F. (2009). A gender-aware framework for women's entrepreneurship. *International Journal of Gender and Entrepreneurship*, *1*(1), 8–24. doi:10.1108/17566260910942318

Bucher, R. D. (2000). Diversity Consciousness. Academic Press.

Buckley, R. (2011). Tourism and environment. *Annual Review of Environment and Resources*, *36*(1), 397–416. doi:10.1146/annurev-environ-041210-132637

Buhalis, D., & Law, R. (2008). Progress in information technology and tourism management: 20 years on and 10 years after the Internet—The state of eTourism research. *Tourism Management*, *29*(4), 609–623. doi:10.1016/j.tourman.2008.01.005

Bune, K. (2008, June 4). *Queen Bee Syndrome Produces Stinging Victimisation*. https://www.lawofficer.com/queen-bee-syndrome-produces-stinging-victimization/

Burns, P. M., & Novelli, M. (2008). *Tourism development: Growth, myths, and inequalities*. CABI Publishing. doi:10.1079/9781845934255.0000

Butler, K. L. (1995). Independence for Western Women Through Tourism. *Annals of Tourism Research*, *22*(2), 487–489. doi:10.1016/0160-7383(94)00101-4

Butler, R. W. (1990). Alternative tourism: Peons hope or Trojan Horse? *Journal of Travel Research*, *28*(3), 40–45. doi:10.1177/004728759002800310

Buzinde, C., Shockley, G., Andereck, K., Dee, E., & Frank, P. (2017). Theorizing social entrepreneurship within tourism studies. *Social entrepreneurship and tourism: Philosophy and practice*, 21-34.

C@rma Foundation. (n.d.). Retrieved February 26, 2023, from https://www.4carma.com/en

Ca'ceres, R., Guzma'n, J., & Rekowski, M. (2011). Firms as source of variety in innovation: Influence of size and sector. *The International Entrepreneurship and Management Journal*, *7*(3), 357–372. doi:10.100711365-011-0198-8

Cagarman, K., Kratzer, J., von Arnim, L. H., Fajga, K., & Gieseke, M. J. (2020). Social entrepreneurship on its way to significance: The case of Germany. *Sustainability (Basel)*, *12*(21), 8954. doi:10.3390u12218954

Cai, L. A., & Combrink, T. E. (2000). Japanese female travelers-a unique outbound market. *Asia Pacific Journal of Tourism Research*, *5*(1), 16–24. doi:10.1080/10941660008722055

Çakır, Ö. (2008). Türkiye'de kadının çalışma yaşamından dışlanması. *Erciyes Üniversitesi İktisadi İdari Bilimler Fakültesi*, *31*, 25–47.

Çakır, P. G., Barakazı, M., & Barakazı, E. (2017). Turizm Sektöründe Çalışan Kadınların Karşılaştığı Sorunları Değerlendirmeye Yönelik Bir Araştırma. *Journal of Academic Social Science Studies*, *61*, 461–474.

Calas, M., Smircich, L., & Bourne, K. (2009). Extending the boundaries: Reframing 'entrepreneurship as social change' through feminist perspectives. *Academy of Management Review*, *34*(3), 552–569. doi:10.5465/amr.2009.40633597

Cambronero-Saiz, B. (2013). Gender policies and advertising and marketing practices that affect women's health. *Global Health Action*, *6*(1), 1–7. doi:10.3402/gha.v6i0.20372 PMID:23806207

Camilleri, M. A. (2016). Responsible tourism creates shared value among stakeholders. *Tourism Planning & Development*, *13*(2), 219–235. doi:10.1080/21568316.2015.1074100

Camilleri, M. A. (2020). Strategic corporate social responsibility in tourism and hospitality. *Sustainable Development (Bradford)*, *1–3*(3), 504–506. Advance online publication. doi:10.1002d.2059

Campón-Cerro, A. M., Folgado-Fernández, J. A., & Hernández-Mogollón, J. M. (2017). Rural destination development based on olive oil tourism: The impact of residents' community attachment and quality of life on their support for tourism development. *Sustainability (Basel)*, *9*(9), 1624. doi:10.3390u9091624

Campos-Soria, J. A., García-Pozo, A., & Sánchez-Ollero, J. L. (2015). Gender wage inequality and labour mobility in the hospitality sector. *International Journal of Hospitality Management*, *49*, 73–82. doi:10.1016/j.ijhm.2015.05.009

Campos-Soria, J. A., Marchante-Mera, A., & Ropero-García, M. A. (2011). Patterns of occupational segregation by gender in the hospitality industry. *International Journal of Hospitality Management*, *30*(1), 91–102. doi:10.1016/j.ijhm.2010.07.001

Can, M. (2022). *Meryem Can Instagram Hesabı*. https://www.instagram.com/meeryemcan/?hl=tr

Çanga, A. Ç., Kutlu, T., & Çalışkan, H. (2018). Applıcatıons of Agrıcultural Tourısm n the World and Turkey. *International Journal of Tourism. Economic and Business Sciences*, *2*(2), 450–457.

Cano, M. (2019). *Masculinity in the Kitchen: Gender Performance in the Culinary Arts Industry* [Master's Thesis]. University of Texas, El Paso.

Carlson, D., Kacmar, K. M., Zivnuska, S., Ferguson, M., & Whitten, D. (2011). Work-family enrichment and job performance: A constructive replication of affective events theory. *Journal of Occupational Health Psychology, 16*(3), 297–312. doi:10.1037/a0022880 PMID:21728437

Carvalho, I. (2021). Gender and Tourism. In A. Correia & S. Dolnicar (Eds.), *Women's Voices in Tourism Research* (pp. 1–15). University of Queens Press.

Carvalho, I., Costa, C., Lykke, N., & Torres, A. (2014). An analysis of gendered employment in the Portuguese tourism sector. *Journal of Human Resources in Hospitality & Tourism, 13*(4), 405–429. doi:10.1080/15332845.2014.888509

Carvalho, I., Costa, C., Lykke, N., & Torres, A. (2019). Beyond the glass ceiling: Gendering tourism management. *Annals of Tourism Research, 75*, 79–91. doi:10.1016/j.annals.2018.12.022

Carvalho, I., Costa, C., Lykke, N., Torres, A., & Wahl, A. (2018). Women at the top of tourism organizations: Views from the glass roof. *Journal of Human Resources in Hospitality & Tourism, 17*(4), 397–422. doi:10.1080/15332845.2018.1449551

Carvalho, S., & Carvalho, J. V. (2020). The Implications of Digital Marketing and e-Commerce in the Tourism Sector Growth. In Á. Rocha, A. Abreu, J. de Carvalho, D. Liberato, E. González, & P. Liberato (Eds.), *Advances in Tourism, Technology and Smart Systems. Smart Innovation, Systems and Technologies* (Vol. 171). Springer. doi:10.1007/978-981-15-2024-2_15

Casad, B., Franks, J., Garasky, C., Kittleman, M., Roesler, A., Hall, D., & Petzel, Z. (2021). Gender inequality in academia: Problems and solutions for women faculty in STEM. *Journal of Neuroscience Research, 99*(1), 13–23. doi:10.1002/jnr.24631 PMID:33103281

Casado-Díaz, J. M., Driha, O., & Simón, H. (2020). The gender wage gap in hospitality: New evidence from Spain. *Cornell Hospitality Quarterly, 63*(3), 399–417. doi:10.1177/1938965520971273

Castro, M. P., & Zermeño, M. G. G. (2020). Being an entrepreneur post-COVID-19–resilience in times of crisis: A systematic literature review. *Journal of Entrepreneurship in Emerging Economies, 13*(4), 721–746. doi:10.1108/JEEE-07-2020-0246

Castro, O., & Zermeño, M. G. (2020). Gender perspective in crisis and disaster management: Entrepreneurship and tourism. *Journal of Business Research, 117*, 454–462.

Cater, C. I. (2010). Steps to space; opportunities for astro-tourism. *Tourism Management, 31*(6), 838–845. doi:10.1016/j.tourman.2009.09.001

Çatlak, Ş., Tekdal, M., & Baz, F. Ç. (2015). Scratch yazılımı ile programlama öğretiminin durumu: bir doküman inceleme çalışması [The status of programming teaching with Scratch software: a document review study]. *Journal of Instructional Technologies ve Teacher Education, 4*(3), 13–25.

Caton, K. (2012). Taking the moral turn in tourism studies. *Annals of Tourism Research, 39*(4), 1906–1928. doi:10.1016/j.annals.2012.05.021

Cavalcante, S., Kesting, P., & Ulhøi, J. (2011). Business model dynamics and innovation: (Re)establishing the missing linkages. *Management Decision, 49*(8), 1327–1342. doi:10.1108/00251741111163142

Cave, P., & Kılıç, S. (2010). The role of women in tourism employment with special reference to Antalya, Turkey. *Journal of Hospitality Marketing & Management, 19*(3), 280–292. doi:10.1080/19368621003591400

Çayır, C., & Çetin, Ö. (2011). Din ve şiddet üzerine psikolojik bir yaklaşım. *Dicle Üniversitesi İlahiyat Fakültesi Dergisi*, *13*(1), 1–34.

CBI Ministry of Foreign Affairs. (2020). *The European market potential for SAVE tourism.* https://www.cbi.eu/market-information/tourism/save-tourism/market-potential

Çelen, O., & Tuna, M. (2021). Konaklama işletmelerindeki kadın çalışanların kraliçe arı sendromu algılarının ölçülmesine yönelik bir ölçek geliştirme çalışması. *Türk Turizm Araştırmaları Dergisi*, *5*(3), 2127–2148.

Çelik Uğuz, S., & Topbaş, F. (2016). Turizmde kadın istihdamı ve ücret ayrımcılığı: Karşılaştırmalı bir analiz. *Anatolia: Turizm Araştırmaları Dergisi*, *27*(1), 62–78.

Çelik Uğuz, S., & Topbaş, T. (2014). Toplumsal cinsiyet ve kadın emeği: Turizm sektörüne betimsel bir yaklaşım. *Akademik Sosyal Araştırmalar Dergisi*, *2*(7), 487–509.

Çelik, M.B. & Akar Şahingöz, S. (2018). İş yaşamında cinsiyet ayrımcılığı: Kadın aşçılar örneği. *Journal of Tourism and Gastronomy Studies, 6*(3), 370-383.

Cem, I. Ş. I. K., Küçükaltan, E. G., Çelebi, S. K., Çalkin, Ö., Enser, İ., & Çelik, A. (2019). Tourism and entrepreneurship: A literature review. *Journal of Ekonomi*, *1*(1), 1–27.

Cerdin, J. L., & Pargneux, M. L. (2010). Career anchors: A comparison between organization-assigned and self-initiated expatriates. *Thunderbird International Business Review*, *52*(4), 287–299. doi:10.1002/tie.20350

Çetinkaya, M.Y., & Öter, Z. (2021). Turizm Toplum Kültür, Turist Rehberlerinin Turizm, Toplum ve Kültür Etkileşimi Üzerine Yansımaları. *Nobel Yayıncılık.*

Cevher, F. M. (2020). *Influencer marketing ve sosyal medya fenomenleri*. Nobel Bilimsel Eserler.

Chakraborty, U., & Biswal, S. K. (2022). Psychological empowerment of women entrepreneurs: A netnographic study on twitter. *Management Research Review*, *45*(6), 717–734.

Chalermchaikit, V., & Kozak, M. (2021). The practice of triple bottom line in the sustainability of tourism: A gender perspective. *Gender and Tourism*, 285–300. doi:10.1108/978-1-80117-322-320211016

Chambers, D. (2021). *Tourism and (re)presentation: beyond colonialism.* In A. Correia & S. Dolnicar (Eds.), *Women's Voices in Tourism Research* (pp. 25–32). University of Queens Press.

Chambers, D., & Rakic, T. (2015). *Tourism Research Frontiers: beyond the boundaries of knowledge*. Emerald Group. doi:10.1108/S1571-5043201520

Chambers, E. (Ed.). (1997). *Tourism and culture: an applied perspective*. SUNY Press.

Chang, H.-H. (2017). Consumer socially sustainable consumption: The perspective toward corporate social responsibility, perceived value, and brand loyalty. *J. Econ. Manag.*, *13*, 167–191.

Chant, S. (1997). Gender and tourism employment. In Gender, Work and Tourism (pp. 120-180). London: Routledge.

Chant, S. (2005). Gender and tourism employment in Mexico and the Philippines. In *Gender, work and tourism* (pp. 119–177). Routledge.

Chatterjee, S., Dutta Gupta, S., & Upadhyay, P. (2018). Empowering women and stimulating development at bottom of pyramid through micro-entrepreneurship. *Management Decision*, *56*(1), 160–174. doi:10.1108/MD-04-2017-0296

Chatzigeorgiou, C. & Christou, E. (2020). Adoption of social media as distribution channels in tourism marketing: A qualitative analysis of consumers' experiences. *Journal of Tourism, Heritage & Services Marketing*, *6*(1), 25-32. doi:10.5281/zenodo.3603355

Chaudhry, I. S., Paquibut, R. Y. & Tunio, M. N. (2021). Do workforce diversity, inclusion practices, & organizational characteristics contribute to organizational innovation? Evidence from the U.A.E. *Cogent Business & Management, 8*(1), 1947549. . doi:10.1080/23311975.2021.1947549

Chawla, M., Rajan, S., & Sahni, P. (2020). *Powering the economy with her: Women entrepreneurship in India*. Bain& Company, Google.

Cheema, S. N., & Jamal, W. N. (2022). An Empirical Study on Gender Based Discrimination at Pakistani Workplaces: Determination of the Causes of Gender Based Discrimination in Pakistan's Private Service Sector Workplaces. *Sustainable Business and Society in Emerging Economies*, *4*(2), 327–338. doi:10.26710bsee.v4i2.2272

Chen, B. Y., Wang, Y., Wang, D., & Lam, W. H. (2019). Understanding travel time uncertainty impacts on the equity of individual accessibility. *Transportation Research Part D, Transport and Environment*, *75*, 156–169. doi:10.1016/j.trd.2019.08.027

Chen, C. C., & Petrick, J. F. (2016). The Roles of Perceived Travel Benefits, Importance, and Constraints in Predicting Travel Behavior. *Journal of Travel Research*, *55*(4), 509–522. doi:10.1177/0047287514563986

Chen, H.-J., Chen, P.-J., & Okumus, F. (2013). The relationship between travel constraints and destination image: A case study of Brunei. *Tourism Management*, *35*, 198–208. doi:10.1016/j.tourman.2012.07.004

Chen, X. (2021). Research on the Influencing Factors of Hotel "Green and Low Carbon" Behavior on Consumer Experience. *IOP Conference Series. Earth and Environmental Science*, *692*(3), 032083. doi:10.1088/1755-1315/692/3/032083

Chia, K. W., & And Xiong, J. (2023). Once upon a time in quarantine: Exploring the memorable quarantine hotel experiences of Chinese student returnees during the COVID-19 pandemic. *Tourism and Hospitality Research*, *23*(1), 72–87. doi:10.1177/14673584221085209

Chiang, C. Y., & Jogaratnam, G. (2006). Why do women travel solo for purposes of leisure? *Journal of Vacation Marketing*, *12*(1), 59–70. doi:10.1177/1356766706059041

Chiao, H.-M., Yu-Li, C., & Wei-Hsin, H. (2018). Examining the usability of an online virtual tour-guiding platform for cultural tourism education. Journal of Hospitality, Leisure, Sports and Tourism Education, 23, 29–38.

Chilufya, A., Hughes, E., & Scheyvens, R. (2019). Tourists and community development: Corporate social responsibility or tourist social responsibility? *Journal of Sustainable Tourism*, *27*(10), 1513–1529. doi:10.1080/09669582.2019.1643871

Chinomona, E., & Maziriri, E. T. (2015). Women in action: Challenges facing women entrepreneurs in the Gauteng Province of South Africa. *The International Business & Economics Research Journal*, *14*(6), 835. doi:10.19030/iber.v14i6.9487

Chou, C. M., Chen, Y. L., & Chen, C. J. (2021). Using Social Media to Overcome the COVID-19 Pandemic: The Influence of Social Media Use on Hotel Performance. *Journal of Travel Research*, *60*(3), 490–504.

Chuah, S. H.-W., & Yu, J. (2021). The future of service: The power of emotion in human-robot interaction. *Journal of Retailing and Consumer Services*, *61*(102551), 1–8. doi:10.1016/j.jretconser.2021.102551

Chung, N., & Koo, C. (2015). The use of social media in travel information search. *Telematics and Informatics*, *32*(2), 215–229. doi:10.1016/j.tele.2014.08.005

Ciccarino, I., da Silva, J. F., & Rodrigues, S. (2022). Changes in the Portuguese social economy identified by a social entrepreneurial taxonomy. *Journal of Cleaner Production*, *376*, 134268. doi:10.1016/j.jclepro.2022.134268

Çiçen, C., Boyacıoğlu, E. Z. & Oğuzhan, A. (2020). Turizmde kadın işgücünün cinsiyet ayrımcılığı algısı: Edirne ilinde bir araştırma. *Tourism and Recreation, 2*(1), 54-63.

Ciğer, A., & Uyar, G. F. (2016). Konaklama işletmelerinde kadın muhasebecilerin cam tavan algısına yönelik bir araştırma: Antalya ili örneği. *MANAS Sosyal Araştırmalar Dergisi*, *5*(2), 43–58.

Cinar, K., Akyuz, S., Ugur-Cinar, M., & Onculer-Yayalar, E. (2021). Faces and phases of women's empowerment: The case of women's cooperatives in Turkey. *Social Politics*, *28*(3), 778–805. doi:10.1093p/jxz032

Clark, P. J., Newcomer, J. M., & Jones, A. M. (2015). Overcoming gender barriers in aircraft maintenance: women's perceptions in the United States. *The Collegiate Aviation Review International, 33*(2).

Clevenger, L., & Singh, N. (2013). Exploring barriers that lead to the glass ceiling effect for women in the US hospitality industry. *Journal of Human Resources in Hospitality & Tourism*, *12*(4), 376–399. doi:10.1080/15332845.2013.790258

Cobanoglu, C., Kasim, A., & Nadiri, H. (2019). The impact of women's empowerment on destination loyalty and word of mouth: The mediating role of destination satisfaction. *Journal of Travel Research*, *58*(1), 111–126.

Cocobolinho. (2022). *Cocobolinho Insgram Hesabı*. https://www.instagram.com/cocobolinho/?hl=tr

Coetzee, M., & Schreuder, D. (2014). Career anchors as a meta-capacity in organizational career development. In M. Coetzee (Ed.), *Psycho-social Career Meta-capacities* (pp. 139–154). Springer. doi:10.1007/978-3-319-00645-1_8

Cohen, E. (2018). *How Nisha Katona's Mowgli Street Food became a social media sensation*. Evening Standard. Retrieved from https://www.standard.co.uk/insider/fooddrink/how-nisha-katona-s-mowgli-street-food-became-a-social-media-sensation-a3824181.html

Cohen, E. (1985). The Tourist Guide: The Origins, Structure and Dynamics of a Role. *Annals of Tourism Research*, *12*(1), 5–29. doi:10.1016/0160-7383(85)90037-4

Cole, S. (2018). *Gender Equality and Tourism: Beyond Empowerment*. CABI. doi:10.1079/9781786394422.0000

Collins, J., & Low, A. (2010). Asian female immigrant entrepreneurs in small and medium-sized businesses in Australia. *Entrepreneurship and Regional Development*, *22*(1), 97–111. doi:10.1080/08985620903220553

Comeche, J. M., & Loras, J. (2010). The influence of variables of attitude on collective. Entrepreneurship. *The International Entrepreneurship and Management Journal*, *6*(1), 23–38. doi:10.100711365-009-0131-6

Company.shopltk.com. (2023). *The ultimate guide to travel influencer marketing*. https://company.shopltk.com/ultimate-guide-influencer-marketing-travel-brands

Connell, J., & Rugendyke, B. (2016). Gender and tourism: Women 's experiences in tropical North Queensland. In Tourism and Gender (pp. 17-31). Channel View Publications.

Co-operativesU. K. (2022). https://www.uk.coop/understanding-co-ops/what-co-op/quick-facts-about-co-ops

Correia, A., & Dolnicar, S. (2021). Tourism social entrepreneurs and their role in activating the SDGs–Contributions by Karla Boluk. *Women's Voices in Tourism Research.*

Correia, A., & Dolnicar, S. (2021). *Women's voices in tourism research: Contributions to knowledge and letters to future generations*. University of Queensland Press. doi:10.14264/817f87d

Costa, C., Breda, Z., Bakas, F. E., Durão, M., & Pinho, I. (2016). Through the gender looking-glass: Brazilian tourism entrepreneurs. *International Journal of Gender and Entrepreneurship*, *8*(3), 282–306. doi:10.1108/IJGE-07-2015-0023

Costa, C., Carvalho, I., & Breda, Z. (2011). Gender inequalities in tourism employment: The Portuguese case. *Revista Turismo & Desenvolvimento (Aveiro)*, (15), 39–54.

Costa, H. A., Costa, C., Breda, Z., Caçador, S., Carvalho, I., & Vieira, R. (2013). Women in Brazil with higher education in tourism: Preliminary results of the Gentour Project. *Revista Turismo & Desenvolvimento (Aveiro)*, (19), 53–66.

Costales, E., & Zeyen, A. (2022). Social Entrepreneurship, Grand Challenges and Crisis: What We Know So Far. In *Social Entrepreneurship and Grand Challenges: Navigating Layers of Disruption from COVID-19 and Beyond* (pp. 17–32). Springer International Publishing. doi:10.1007/978-3-031-07450-9_2

Çötok, N. A. (2020). *Şiddet, çalışma yaşamı ve medya bağlamında kadın*. Pegem Akademi.

Craney, T. A., & Surles, J. G. (2002). Model-dependent variance inflation factor cutoff values. *Quality Engineering*, *14*(3), 391–403. doi:10.1081/QEN-120001878

Crawford, D. W., & Godbey, G. (1987). Reconceptualizing barriers to family leisure. *Leisure Sciences*, *9*(2), 119–127. doi:10.1080/01490408709512151

Crawford, D. W., Jackson, E. L., & Godbey, G. (1991). A hierarchical model of leisure constraints. *Leisure Sciences*, *13*(4), 309–320. doi:10.1080/01490409109513147

CreatorDen. (2018). *Tüm Pazarlamacıların Bilmesi Gereken 3 Influencer Türü*. https://creatorden.com/tr/tum-pazarlamacilarin-bilmesi-gereken-3-influencer-turu/

Creswell, J. W., & Creswell, J. D. (2017). *Research design: Qualitative, quantitative, and mixed methods approaches*. Sage Publication.

Creswell, J. W., & Creswell, J. D. (2017). *Research design: Qualitative, quantitative, and mixed methods approaches*. Sage publications.

Crompton, J. (1979). Motivation for pleasure vacation. *Annals of Tourism Research*, *6*(4), 408–424. doi:10.1016/0160-7383(79)90004-5

Crowe, S., Cresswell, K., Robertson, A., Huby, G., Avery, A., & Sheikh, A. (2011). The case study approach. *BMC Medical Research Methodology*, *11*(1), 1–9. doi:10.1186/1471-2288-11-100 PMID:21707982

Cukier, J. (1996). Tourism and employment in Bali: trends and implications. *Tourism and Indigenous Peoples*, 49–75.

D'Auria, T. (2012). *2013 winter star party*. Southern Cross Astronomical Society (SCAS) Inc. Retrieved from http://www.scas. org/wsp.html

D'Espallier, B., Guérin, I., & Mersland, R. (2011). Women and repayment in microfinance: A global analysis. *World Development*, *39*(5), 758–772. doi:10.1016/j.worlddev.2010.10.008

Daengbuppha, J., Hemmington, N., & Wilkes, K. (2006). Using grounded theory to model visitor experiences at heritage sites Methodological and practical issues. *Qualitative Market Research*, *9*(4), 367–388. doi:10.1108/13522750610689096

Dahles, H., & Susilowati, T. P. (2015). Business resilience in times of growth and crisis. *Annals of Tourism Research*, *51*, 34–50. doi:10.1016/j.annals.2015.01.002

Dahles, H., & Susilowati, T. P. (2015). The role of tourism in coping with poverty in rural Indonesia. *Journal of Sustainable Tourism*, *23*(4), 554–573.

Dahl, S. (2018). *Social Media Marketing Theories & Applications* (2nd ed.). SAGE Publications Ltd.

Dal Mas, F., & Paoloni, P. (2020). A relational capital perspective on social sustainability; the case of female entrepreneurship in Italy. *Measuring Business Excellence*, *24*(1), 114–130. doi:10.1108/MBE-08-2019-0086

Daldeniz, B., & Hampton, M. P. (2010). 3 VOLUNtourists versus volunTOURISTS. *Volunteer tourism: Theoretical frameworks and practical applications*, 30.

Dalkıranoğlu, T. (2006). *Çalışma yaşamında kadın işgücü ve cinsiyet ayrımcılığı: konaklama işletmelerinde bir uygulama* [Unpublished Master Thesis]. Anadolu University.

Danish, A. Y., & Smith, H. L. (2012). Female entrepreneurship in Saudi Arabia: Opportunities and challenges. *International Journal of Gender and Entrepreneurship*.

Danziger, N., & Valency, R. (2006). Career anchors: Distribution and impact on job satisfaction, the Israeli case. *Career Development International*, *11*(4), 293–303. doi:10.1108/13620430610672513

Das, A. (2020). *Impact of travel journalism on female solo travelers' experience*. https://www.academia.edu/download/63370019/AnantaDas_NtccDissertation20200520-123320-1q2l0ro.pdf

Datta, P. B., & Gailey, R. (2012). Empowering women through social entrepreneurship: Case study of a women's cooperative in India. *Entrepreneurship Theory and Practice*, *36*(3), 569–587. doi:10.1111/j.1540-6520.2012.00505.x

Davidsson, P., & Wiklund, J. (1997). Values, beliefs and regional variations in new firm formation rates. *Journal of Economic Psychology*, *18*(2-3), 179–199. doi:10.1016/S0167-4870(97)00004-4

Davis, H. D. (1968). Potentials for tourism in developing countries. *Finance and Development*. https://search.proquest.com/openview/58c2cd1bf137b001c599a40c2fec3ae7/1?pq-origsite=gscholar&cbl=1819673

Dayour, F., & Adam, A. A. (2022). Impact of Covid-19 on women entrepreneurs in tourism: The case of women tour operators in Ghana. *African Journal of Hospitality, Tourism and Leisure*, *11*(2), 1–22.

Dayour, F., & Adam, I. (2022). Entrepreneurial motivations among COVID-19 induced redundant employees in the hospitality and tourism industry. *Journal of Human Resources in Hospitality & Tourism*, *21*(1), 130–155. doi:10.1080/15332845.2022.2015239

De Bruin, A., Brush, C., & Welter, F. (2006). Introduction to the Special Issue: Towards building cumulative knowledge on women's entrepreneurship. *Entrepreneurship Theory and Practice*, *30*(5), 585–593. doi:10.1111/j.1540-6520.2006.00137.x

De Lange, D., & Dodds, R. (2017). Increasing sustainable tourism through social entrepreneurship. *International Journal of Contemporary Hospitality Management*, *29*(7), 1977–2002. doi:10.1108/IJCHM-02-2016-0096

De Veirman, M., Cauberghe, V., & Hudders, L. (2016). Marketing through Instagram influencers: Impact of number of followers and product divergence on brand attitude. *International Journal of Advertising*, *36*(5), 798–828. doi:10.1080/02650487.2017.1348035

Deem, R. (1996). No time for a rest? An exploration of women's work, engendered leisure and holidays. *Time & Society*, *5*(1), 5–25. doi:10.1177/0961463X96005001001

Deen, A., Harilal, V., & Achu, F. N. (2016). Women in tourism: Experiences and challenges faced by owners of small accommodation establishments. *African Journal of Hospitality, Tourism and Leisure*, *5*(3), 223–814.

Defourny, J., & Nyssens, M. (2010). Conceptions of social enterprise and social entrepreneurship in Europe and the United States: Convergences and divergences. *Journal of Social Entrepreneurship*, *1*(1), 32–53. doi:10.1080/19420670903442053

Degreef, A. (2019). *Rakuten Marketing 2019 Influencer Marketing Global Survey*. https://blog.rakutenadvertising.com/insights/rakuten-marketing-2019-influencer-marketing-global-survey-report/

Deming, A. (1996). The edge of the civilized world: Tourism and the hunger for wild Places. *Orion*, *15*, 28–35.

Demiral, N. Ö., & Hassan, A. (2020). Kadın Girişimciler ve Sosyal Medya: Bozcaada Konaklama İşletmelerinin Instagram Sayfalarına Yönelik Bir Araştırma [Women Entrepreneurs and Social Media: A Research on Instagram Pages of Bozcaada Accommodation Businesses]. *AHBVÜ Turizm Fakültesi Dergisi*, *23*(2), 359–373. doi:10.34189/tfd.23.02.007

Demirdağ, Ş. A. (2022). Otel işletmelerinde karşılaşılan fiziksel sorunların örgütsel bağlılık ve iş tatmini üzerindeki etkisi: Kadın çalışanlar üzerine bir araştırma. *KADEM Kadın Araştırmaları Dergisi*, *8*(1), 95–121.

Demirel, E. T. (2019). Tokenizm teorisinin Türkiye bağlamında işlerliğine yönelik bir ölçek geliştirme çalışması. *Fırat Üniversitesi Uluslararası İktisadi ve İdari Bilimler Dergisi*, *3*(2), 1–34.

Demirkol, Ş., Fidan, F., & Pelit, E. (2004). Turizm sektöründeki bayan işgörenlerin karşılaştıkları sorunlar ve otel işletmelerinde bir uygulama. *Abant İzzet Baysal Üniversitesi Sosyal Bilimler Enstitüsü Dergisi*, *1*(8), 71–88. doi:10.11616/AbantSbe.128

Demir, M. (2011). İş yaşamında ayrımcılık: Turizm sektörü örneği. *Uluslararası İnsan Bilimleri Dergisi*, *8*(1), 760–784.

Demirović Bajrami, D., Radosavac, A., Cimbaljević, M., Tretiakova, T. N., & Syromiatnikova, Y. A. (2020). Determinants of residents' support for sustainable tourism development: Implications for rural communities. *Sustainability (Basel)*, *12*(22), 9438. doi:10.3390u12229438

Derks, B., Van Laar, C., & Ellemers, N. (2016). The Queen Bee Phenomenon: Why Women Leaders Distance Themselves From Junior Women. *The Leadership Quarterly*, *27*(3), 456–469. doi:10.1016/j.leaqua.2015.12.007

Díaz-Carrión, I. A., & Vizcaino, P. (2022). Mexican women's emotions to resist gender stereotypes in rural tourism work. *Tourism Geographies*, *24*(2–3), 244–262. doi:10.1080/14616688.2020.1867886

Diaz-Sarachaga, J. M., & Ariza-Montes, A. (2022). The role of social entrepreneurship in the attainment of the sustainable development goals. *Journal of Business Research*, *152*, 242–250. doi:10.1016/j.jbusres.2022.07.061

Dilley, R. E., & Scraton, S. J. (2010). Women, climbing and serious leisure. *Leisure Studies*, *29*(2), 125–141. doi:10.1080/02614360903401927

Dinçer, F. İ., Akova, O., Ertuğral, S. M., & Çiftçi, M. A. (2016). Woman Labour Force in Tourism Industry in Turkey: Opportunities and Barriers. Eurasian Academy of Sciences Social Sciences Journal, 379-395.

Djafarova, E., & Rushworth, C. (2017). Exploring the credibility of online celebrities' Instagram profiles in influencing the purchase decisions of young female users. *Computers in Human Behavior*, *68*, 1–7. doi:10.1016/j.chb.2016.11.009

Doğan, M. (2017). The role of cookery vocation in the tourism sector on women employment. *The Online Journal of Science and Technology*, *7*(1), 127–132.

Dogutas, A. (2019). Gender Based Violence against Syrian Refugee Women in Turkey. *Border Crossing*, *9*(2), 113–124. doi:10.33182/bc.v9i2.811

Dolezal, C., & Novelli, M. (2022). Power in community-based tourism: Empowerment and partnership in Bali. *Journal of Sustainable Tourism*, *30*(10), 2352–2370. doi:10.1080/09669582.2020.1838527

Dönmez, B., & Dağaşan, G. (2017). Kadın aşçıların dikkatine! Profesyonel mutfakta yemek yapmak evde iki kap yemek yapmaya benzemezmiş. In Uluslararası Turizmin Geleceği Kongresi (pp. 1053-1064). Mersin.

Doran, A. (2016). Empowerment and women in adventure tourism: A negotiated journey. *Journal of Sport & Tourism*, *20*(1), 1–24. doi:10.1080/14775085.2016.1176594

Douglas, N., Douglas, N., & Derret, R. (Eds.). (2001). *Special Interest Tourism*. Wiley.

Dreher, G. (2003). Breaking the glass ceiling: The effects of sex ratios and work–life programs on female leadership at the top. *Human Relations*, *56*(5), 541–562. doi:10.1177/0018726703056005002

Du Rand, G. E., & Heath, E. (2006). Towards a framework for food tourism as an element of destination marketing. *Current Issues in Tourism*, *9*(3), 206–234. doi:10.2167/cit/226.0

Du Rietz, A., & Henrekson, M. (2000). Testing the female underperformance hypothesis. *Small Business Economics*, *14*(1), 1–10. doi:10.1023/A:1008106215480

Duan, W., Yu, Y., Cao, Q., & Levy, S. (2016). Exploring the Impact of Social Media on Hotel Service Performance: A Sentimental Analysis Approach. *Cornell Hospitality Quarterly*, *57*(3), 282–296. doi:10.1177/1938965515620483

Duarte Alonso, A., & Nyanjom, J. (2017). Local stakeholders, role and tourism development. *Current Issues in Tourism*, *20*(5), 480–496. doi:10.1080/13683500.2015.1078782

Duffy, B. E., & Hund, E. (2015). "Having it all" on social media: Entrepreneurial femininity and self-branding among fashion bloggers. *Social Media + Society*, *1*(2). doi:10.1177/2056305115604337

Duffy, L. N., Kline, C. S., Mowatt, R. A., & Chancellor, H. C. (2015). Women in Tourism: Shifting Gender Ideology in the DR. *Annals of Tourism Research*, *52*, 72–86. doi:10.1016/j.annals.2015.02.017

Duguid, F., Durutaş, G., & Wodzicki, M. (2015). *The current state of women's co-operatives in Turkey*. World Bank. doi:10.1596/25977

Duman, T. (2012). The value of Islamic tourism: Perspectives from the Turkish experience. *ICR Journal*, *3*(4), 718–739. doi:10.52282/icr.v3i4.513

Duruoğlu, T. (2007). Emek piyasasında cinsiyetçi ücret ayrımı: Bursa Organize Sanayi bölgesinde bir araştırma. *İletişim Kuram ve Araştırma Dergisi*, *24*, 61-76.

Dutta, N., & Bhakta, K. (2017). *Women Empowerment: Dimensions, Needs and Hurdles in India IMS Business School, Doctoral Colloquium*. Retrieved From https://www.managejournal.com/up/conference/20170210152711.pd

Duve.com. (2022). *The ultimate guide to working with travel influencers*. https://duve.com/guide-to-travel-influencers/

Du, Z., Liu, J., & Wang, T. (2022). Augmented Reality Marketing: A Systematic Literature Review and an Agenda for Future Inquiry. *Frontiers in Psychology*, *13*, 925963. doi:10.3389/fpsyg.2022.925963 PMID:35783783

Edgell, D. L. (2016). *Managing Sustainable Tourism: A Legacy for the Future*. Routledge. doi:10.4324/9781315688343

Edmonds, K. (2013). Beyond good intentions: The structural limitations of NGOs in Haiti. *Critical Sociology*, *39*(3), 439–452. doi:10.1177/0896920512437053

Eichler, G. M., & Schwarz, E. J. (2019). What sustainable development goals do social innovations address? A systematic review and content analysis of social innovation literature. *Sustainability (Basel)*, *11*(2), 522. doi:10.3390u11020522

Eid, R., & El-Gohary, H. (2015). The role of Islamic religiosity on the relationship between perceived value and tourist satisfaction. *Tourism Management*, *46*, 477–488. doi:10.1016/j.tourman.2014.08.003

EIGE. (2016). *European Institute for Gender Equality(EIGE)'s Gender Equality Glossary & Thesaurus*. https://eige.europa.eu/taxonomy/term/1102

Eijgelaar, E., Peeters, P., & Reinecke, T. (2022). Netherlands: Review of sustainable tourism policies, strategies and instruments. *European SME Going Green 2030 Report. Review and analysis of policies, strategies and instruments for boosting sustainable tourism in Europe*, 316-326.

Elam, A., & Brush, C. G. (2010). Women, entrepreneurship and social capital: A dialogue and construction. *Organization Studies*, *31*(2), 245–249. doi:10.1177/0170840610363530

Elder, S., & Smith, A. (2010). Women in labour markets: Measuring progress and identifying challenges. Ginebra, Suiza: International Labour Office.

El-Gohary, H. (2016). Halal tourism, is it really Halal? *Tourism Management Perspectives*, *19*, 124–130. doi:10.1016/j.tmp.2015.12.013

Elli, D. M. (2017). *The phenomenon and rise of Influencer Marketing and how it affects customer opinion and helps or damages brands* [Master Thesis]. International Hellenic University.

Elmas, S. (2007). Gender and tourism development: a case study of the Cappadoccia Region of Turkey. In *Tourism and gender: Embodiment, sensuality and experience* (pp. 302–314). CABI Wallingford UK. doi:10.1079/9781845932718.0302

Elsenhans, H. (2021). *Capitalism, Development and Empowerment of Labour: A Heterodox Political Economy*. Routledge. doi:10.4324/9781003182511

Elshaer, I. A., Moustafa, M. M., Sobaih, A. E. E., Aliedan, A. M., & Azazz, A. M. (2021). The mediating role of psychological empowerment in the relationship between transformational leadership and women's job performance. *Journal of Organizational Change Management*, *34*(2), 541–560.

Elshaer, I., Moustafa, M., Sobaih, A. E., Aliedan, M., & Azazz, A. M. (2021). The impact of women's empowerment on sustainable tourism development: Mediating role of tourism involvement. *Tourism Management Perspectives*, *38*, 100815. doi:10.1016/j.tmp.2021.100815

Enoch, Y., & Grossman, R. (2010). Blogs of Israel and Danish backpackers to India. *Annals of Tourism Research*, *37*(2), 520–536. doi:10.1016/j.annals.2009.11.004

Erdoğmuş, N. (2003). *Kariyer geliştirme: kuram ve uygulama*. Nobel Yayınları.

Erikli, S. (2020). Çalışma yaşamında toplumsal cinsiyet ayrımcılığının görünümü. *Hacettepe Üniversitesi Sosyal Bilimler Dergisi*, *2*(1), 39–60.

Erkol Bayram, G. (2018). Women Entrepreneurs and Tourism: An Investıgation on the Current Situation and Problems in Sinop Province. *Journal of Business Research*, *10*(2), 56–88.

Erkol Bayram, G., Bayram, A. T., & Sürücü, Ö. A. (2018). *Turizm ve Kadın*. Detay Yayıncılık.

Erkol, G. B. (2022). Kadınların turizm sektöründe karşılaşmış olduğu engellerin demografik değişkenler kapsamında incelenmesi: Kadın rehberler üzerine bir araştırma, *Yönetim. Ekonomi ve Pazarlama Araştırmaları Dergisi*, *6*(3), 144–161.

Errami, E., Ochir, G., & Peppoloni, S. (2021). Achieve gender equality and empower all women and girls. *Geosciences and the Sustainable Development Goals*, 105-126.

Ertac, M., & Tanova, C. (2020). Flourishing Women through Sustainable Tourism Entrepreneurship. *Sustainability (Basel)*, *12*(5643), 1–17. doi:10.3390u12145643

Eryılmaz, G. (2022). Restoran işletmelerinde istihdam edilen kadın çalışanların karşılaştıkları sorunlar ve çözüm önerileri: Örnek olay çalışması, *Journal of Social. Humanities and Administrative Sciences*, *8*(56), 1056–1071. doi:10.29228/JOSHAS.64314

Essers, C., Benschop, Y., & Doorewaard, H. (2010). Female ethnicity: Understanding Muslim immigrant businesswomen in The Netherlands. *Gender, Work and Organization, 17*(3), 320–339. doi:10.1111/j.1468-0432.2008.00425.x

Etmyonline. (2023). *Career.* https://www.etymonline.com/word/career

Ettl, K., & Welter, F. (2010). Gender, context and entrepreneurial learning. *International Journal of Gender and Entrepreneurship, 2*(2), 108–129. doi:10.1108/17566261011050991

Europe, I. (2018). *A policy brief from the policy learning platform on low-carbon economy.* Academic Press.

European Commission. (2019). *Gender Balance on Company Boards-Information from the Commission.* Available online: https://data.consilium.europa.eu/doc/document/ST-9823-2019-INIT/en/pdf

European Commission. (2020). *Supporting entrepreneurship.* Internal Market, Industry, Entrepreneurship and SMEs. from https://single-marketeconomy.ec.europa.eu/smes/supporting-entrepreneurship_en

European Institute for Gender Equality (EIGE). (2016). *Gender Equality Index.* Author.

European Institute for Gender Equality. (2016). *Gender in tourism.* Luxembourg: EIGE. Publications Office of the European Union. Retrieved from: https://eige.europa.eu/

Eurostat. (1998). *Community Methodology on Tourism Statistics.* European Commission.

Evans, C. (1996). A review of career anchors in use. *European Journal of Work and Organizational Psychology, 5*(4), 609–615. doi:10.1080/13594329608414883

Expedia. (2019). *2019 Expedia Group travel trends report.* https://www.expediagroup.com/travel-research/travel-trends-report-2019

Eyben, R. (2011). Supporting pathways of women's empowerment: a brief guide for international development organizations. Institute of Development Studies (IDS).

Eyel, C. Ş. & Şen, B. (2020). Influencer pazarlama ve tüketicilerin influencer'lara yönelik tutumlarinin satin alma niyeti üzerindeki etkisi. *Atlas Sosyal Bilimler Dergisi, Atlas Sosyal Bilimler Dergisi,* 7-29.

Facebook. (2019). *Facebook IQ: Driving revenue for travel brands.* https://www.facebook.com/iq/insights-to-go/driving-revenue-for-travel-brands

Fagence, M. (2000). Ethnic Tourism in Developed Countries: Special Interest or Specialized Mass Tourism? *Tourism Recreation Research, 25*(2), 77–87. doi:10.1080/02508281.2000.11014913

Faizan, R. & Ul-Haque, A. (2016). *The relationship between societal attributes, feminine leadership & management style: Responses from Pakistan's urban region female-owned businesses.* Academic Press.

Family Business Institute. (2016). *Global Family Business Survey 2016.* Retrieved from http://fambizpwc.com/2018/03/global-family-business-survey-2016-results-announced/

Faraj, S., Pachidi, S., & Sayegh, K. (2018). Working and organizing in the age of the learning algorithm. *Information and Organization, 28*(1), 62–70. doi:10.1016/j.infoandorg.2018.02.005

Faten Ali, M. H. (2014). Challenges facing female employees in quick service restaurants: Egypt as a case study. *Journal of Association of Arab Universities for Tourism and Hospitality, 11*(3), 59–74. doi:10.21608/jaauth.2014.57689

Fayos-Sola, E., Marin, C., & Jafari, J. (2014). Astro-tourism: No requiem for meaningful travel. *Pasos (El Sauzal), 12*(4), 663–671. doi:10.25145/j.pasos.2014.12.048

Feldbauer, I. G., & Jeffrey, H. L. (2021). Disempowered Hosts? A Literature Review of Muslim Women and Tourism. *Women in Tourism in Asian Muslim Countries*, 13-23.

Fennell, D. (2015). Ecotourism in Costa Rica: Environmental Impacts and Management. *Journal of Sustainable Tourism*, *23*(2), 169–189.

Ferguson, L. (2013). Promoting Gender Equality and Empowering Women? Tourism and the Third Millennium Development Goal. In Tourism and the Millennium Development Goals (pp. 73-87). Routledge.

Fernández, Z., Nieto, M. J., & Ramos, A. (2013). Family involvement and internationalization in family firms: A review of the literature. *Journal of Family Business Strategy, 4*(3), 157-166.

Fernández, J. I. P., Montilla, J. C., & Hidalgo, I. C. (2019). Introducing olive-oil tourism as a special interest tourism. *Heliyon*, *5*, 12. PMID:31872132

Fernández, J. I. P., Montilla, J. C., Hidalgo, I. C., & Fernández, M. C. P. (2022). Evaluating olive oil tourism experiences based on the segmentation of demand. *International Journal of Gastronomy and Food Science*, 27.

Fernández, Z., Nieto, M. J., & Ramos, C. (2013). Internationalization and innovation: An analysis of complementarity and substitutability. *International Business Review*, *22*(2), 371–386.

Ferrante, F. (2005). Revealing entrepreneurial talent. *Small Business Economics*, *25*(2), 159–174. doi:10.100711187-003-6448-6

Fidan, F., & Nam, D. (2012). New Dynamics in Rural Tourism: Women Entrepreneus-Sample of Taraklı. *KMU Journal of Social and Economic Research*, *14*(23), 51–57.

Figueroa-Domecq, C., Anna de Jong, A., & Williams, A. M. (2020). Gender, tourism & entrepreneurship: A critical review. *Annals of Tourism Research*, *84*(102980), 1–13. doi:10.1016/j.annals.2020.102980

Filimonau, V., Matyakubov, U., Matniyozov, M., Shaken, A., & Mika, M. (2022). Women entrepreneurs in tourism in a time of a life event crisis. *Journal of Sustainable Tourism*, 1–23. Advance online publication. doi:10.1080/09669582.2022.2091142

Flashpack. (2023). *Group Solo Travel in Turkey*. Retrieved from https://www.flashpack.com/destinations/turkey/

Foncubierta-Rodríguez, M. J. (2021). Influence of the entrepreneur's personal values in business governance style and their relationship with happiness at work. *Corporate Governance (Bradford)*, *22*(3), 592–617. doi:10.1108/CG-05-2021-0197

Foubert, P., Burri, S., & Numhauser-Henning, A. (2010). *The gender pay gap in Europe from a legal perspective*. Publications Office of the European Union Luxembourg.

Francis, B., & Alagas, M. (2020). Structural empowerment and psychological empowerment: Effects on innovative behavior among police officers. *Journal of Police and Criminal Psychology, 35*(4), 441–450.

Franzke, S., Wu, J., Froese, F. J., & Chan, Z. X. (2022). Female entrepreneurship in Asia: A critical review and future directions. *Asian Business & Management*, *21*(3), 343–372. doi:10.105741291-022-00186-2

Freberg, K., Graham, K., McGaughey, K., & Freberg, L. A. (2010). Who are the social media influencers? A study of public perceptions of personality. *Public Relations Review*, *37*(1), 90–92. doi:10.1016/j.pubrev.2010.11.001

Frederick, B. (2023). *The Top 100 Social Media Influencers Worldwide*. https://www.searchenginejournal.com/top-social-media-influencers/475776/#close

Fredman, P., & Tyrväinen, L. (2010). Frontiers in nature-based tourism. *Scandinavian Journal of Hospitality and Tourism, 10*(3), 177–189. doi:10.1080/15022250.2010.502365

Frew, E. A., & Shaw, R. N. (1999). The relationship between personality, gender, and tourism behavior. *Tourism Management, 20*(2), 193–202. doi:10.1016/S0261-5177(98)00081-8

Ga'mez-Gonza'lez, J., Rondan-Catalun˜a, F. J., Diez-de Castro, E. C., & Navarro-Garcia, A. (2010). Toward an international code of franchising. *Management Decision, 48*(10), 1568–1595. doi:10.1108/00251741011090333

Gabriel, P., & Drayton, B. (2016). *Social entrepreneurship and innovation: International case studies and practice.* Kogan Page Publishers.

Gajic, T., & Cvijanović, D. (2020). The Level of Engagement of the Female Workforce in the Rural Tourism Development of Serbia. *Journal of Tourism Leisure and Hospitality, 2*(1), 36–42.

Galbreath, J. (2017). Drivers of Green Innovations: The impact of export intensity, women leaders, and Absorptive Capacity. *Journal of Business Ethics, 158*(1), 47–61. doi:10.100710551-017-3715-z

Gallo, M. A. (2018). Family businesses and entrepreneurship: A literature review. *Journal of Family Business Management, 8*(2), 141–157.

Gannon, M., Rasoolimanesh, S. M., & Taheri, B. (2021). Assessing the mediating role of residents' perceptions toward tourism development. *Journal of Travel Research, 60*(1), 149–171. doi:10.1177/0047287519890926

Garanti, Z. (2022). Alternative and special interest tourism to mitigate the effects of tourism seasonality: The debate from Cyprus. *Worldwide Hospitality and Tourism Themes, 14*(5), 451–460. doi:10.1108/WHATT-07-2022-0083

Garay, L., & Font, X. (2012). Doing good to do well? Corporate social responsibility reasons, practices, and impacts in small and medium accommodation enterprises. *International Journal of Hospitality Management, 31*(2), 329–337. doi:10.1016/j.ijhm.2011.04.013

Garcia-Ramon, M. D., Canoves, G., & Valdovinos, N. (1995). Farm tourism, gender and the environment in Spain. *Annals of Tourism Research, 22*(2), 267–282. doi:10.1016/0160-7383(94)00096-4

García-Sánchez, I. M., Cuadrado-Roura, J. R., & Martínez-Fernández, M. T. (2017). Determinants of women entrepreneurship in Spain: An institutional approach. *Journal of Business Research, 76*, 1–10.

García-Sánchez, I. M., Cuadrado-Roura, J. R., & Martínez-Fernández, M. T. (2017). Gender and family business in tourism: An exploratory study of the Spanish case. *Journal of Family Business Management, 7*(1), 50–69.

Garrigos, J. A., Haddaji, M., Segovia, P. G., & Signes, A. P. (2019). Gender Differences in the Evolution of Haute Cuisine Chef's Carerr. *Journal of Culinary Science & Technology*. Advance online publication. doi:10.1080/15428052.2019.1640156

Gatewood, E. J., Brush, C. G., Carter, N. M., Greene, P. G., & Hart, M. M. (2009). Diana: a symbol of women entrepreneurs' hunt for knowledge, money, and the rewards of entrepreneurship. *Small Business Economics, 32*, 129-144.

Gebelek, G. (2008). *The new international migration from a gender perspective: A case study of post-Soviet servants of tourism in Antalya* [Unpublished Master Thesis]. Koç University.

Ge, J., & Gretzel, U. (2018). A taxonomy of value co-creation on Weibo – A communication perspective. *International Journal of Contemporary Hospitality Management, 30*(4), 2075–2092. doi:10.1108/IJCHM-09-2016-0557

Genç, R. (2018). The Impact of Tourism in The Reduction of Gender Inequality. *Trakya Üniversitesi İktisadi ve İdari Bilimler Fakültesi Dergisi, 7*(1), 20–35.

Gentry, K. M. (2007). Belizean women and tourism work: Opportunity or impediment? *Annals of Tourism Research*, *34*(2), 477–496. doi:10.1016/j.annals.2006.11.003

Georgeta, I. (2012). Women entrepreneurship in the current international business environment. *Cogito-Multidisciplinary Research Journal*, *1*, 122–131.

Geyser, W. (2023). *What is an influencer? – Social Media Influencer Defined.* https://influencermarketinghub.com/what-is-an-influencer/#toc-0

Ghodsee, K. (2003). State support in the market: Women and tourism employment in post-socialist Bulgaria. *International Journal of Politics Culture and Society*, *16*(3), 465–482. doi:10.1023/A:1022364630327

Ghouse, S. M., Durrah, O., & McElwee, G. (2021). Rural women entrepreneurs in Oman: Problems and opportunities. *International Journal of Entrepreneurial Behaviour & Research*, *27*(7), 1674–1695. doi:10.1108/IJEBR-03-2021-0209

Giantari, I. G., Surya, I. B., Yasa, N. N., & Yasa, I. B. (2018). Development and revitalization strategies for traditional markets in Bali. *International Journal of Social Economics*, *45*(7), 1058–1070. doi:10.1108/IJSE-09-2017-0414

Gibson, H. (2001). *Gender in tourism: Theoretical perspectives. In Women as producers and consumers of tourism in developing regions (pp. 19-43).* Praeger.

Giddens, A. (2012). *Sosyoloji.* Kırmızı Yayınları.

Gilal, F. G., Gilal, N. G., Channa, N. A., Gilal, R. A., Gilal, R. G., & Tunio, M. N. (2020). Towards an integrated model for the transference of environmental responsibility. *Business Strategy and the Environment*, *29*(6), 1–10. doi:10.1002/bse.2524

Gilal, F. G., Gilal, N. G., Gilal, R. G., Gon, Z., Gilal, W. G., & Tunio, M. N. (2021). The Ties That Bind: Do Brand Attachment and Brand Passion Translate Into Consumer Purchase Intention? *Central European Management Journal*, *29*(1), 14–38. doi:10.7206/cemj.2658-0845.39

Gilmartin, P. (1997). The dangers of independent travel: A century of advice for 'lady travellers.'. *Journal of Interdisciplinary Gender Studies*, *2*(1), 1–13.

Gist, M. E. (1987). Self-efficacy: Implications for organizational behavior and human resource management. *Academy of Management Review*, *12*(3), 472–485. doi:10.2307/258514

Global CompactU. N. (2022). https://unglobalcompact.org/take-action/action/womens-principles

Global Entrepreneurship Monitor, G. E. M. (2022). *Gem Global Entrepreneurship Monitor. 2021/22 Women's Entrepreneurship Report: From Crisis to Opportunity*. GEM. https://www.gemconsortium.org/

Goeldner, C. R., & Ritchie, J. R. B. (2009). *Tourism: Principles, Practices, Philosophies*. John Wiley & Sons.

Goetz, A. M., & Musembi, C. N. (2008). Voice and Women's Empowerment: Mapping a Research Agenda. Brighton, UK: Institute of Development Studies (IDS).

Goffi, G., Cucculelli, M., & Masiero, L. (2019). Fostering tourism destination competitiveness in developing countries: The role of sustainability. *Journal of Cleaner Production*, *209*, 101–115. doi:10.1016/j.jclepro.2018.10.208

Goktan, A. B., & Miles, G. (2011). Innovation speed and radicalness: Are they inversely related? *Management Decision*, *49*(4), 533–547. doi:10.1108/00251741111126477

Gömleksiz, M. N., & Kan, A. Ü. (2009). Sosyal Bilgiler Dersi Öğretim Programının Eleştirel Düşünme, Yaratıcı Düşünme ve Girişimcilik Becerilerini Kazandırmadaki Etkililiğinin Belirlenmesi (Diyarbakır İli Örneği). *Doğu Anadolu Bölgesi Araştırmaları*, *8*(1), 39–49.

Gönüllü Atakan, A., & Çoban, A. (2022). *Türkiye'de Kadınların Kooperatifler Yoluyla Güçlenmesi: Yasal ve Yapısal Stratejiler, Uluslararası Çalışma Örgütü*. ILO Türkiye Ofisi.

Govindarajo, N. S., Kumar, D., Shaikh, E., Kumar, M., & Kumar, P. (2021). Industry 4.0 and business policy development: Strategic imperatives for SME performance. *Etikonomi*, *20*(2), 239–258. doi:10.15408/etk.v20i2.20143

Goyal, M., & Parkash, J. (2011). Women Entrepreneurship in India- Problems and Prospects. *ZENITH International Journal of Multidisciplinary Research*, *1*(5), 195–207.

Goyal, S., Agrawal, A., & Sergi, B. S. (2021). Social entrepreneurship for scalable solutions addressing sustainable development goals (SDGs) at BoP in India. *Qualitative Research in Organizations and Management*, *16*(3/4), 509–529. doi:10.1108/QROM-07-2020-1992

Grandy, G., Cukier, W., & Gagnon, S. (2020). (In) visibility in the margins: COVID-19, women entrepreneurs and the need for inclusive recovery. *Gender in Management*, *35*(7/8), 667–675. doi:10.1108/GM-07-2020-0207

Grandy, G., Montagno, R. V., Campbell, J. E., & Nedic, O. (2020). Women entrepreneurs and the Covid-19 pandemic: A systematic literature review and research agenda. *Journal of Business Venturing Insights*, *14*, e00184.

Greenberg, J., Ashton-James, C. E., & Ashkanasy, N. M. (2007). Social comparison processes in organizations. *Organizational Behavior and Human Decision Processes*, *102*(1), 22–41. doi:10.1016/j.obhdp.2006.09.006

Green, R. T., & Cunningham, I. C. M. (1976). Empowerment Status, Feminine Role Perception and Family Purchasing Decisions. *Journal of Business Research*, *4*(4), 313–324. doi:10.1016/0148-2963(76)90041-2

Greer, C. (2021). *Strategic human resource management*. Pearson Custom Publishing.

Greer, G. (1999). *The Whole Woman*. Anchor.

Gretzel, U. (2018). Influencer marketing in travel and tourism. In M. Sigala & U. Gretzel (Eds.), *Advances in Social Media For Travel, Tourism And Hospitality: New Perspectives* (pp. 147–156). Routledge.

Gretzel, U. (2018). Tourism And Social Media Sage Handbook Of. *Tourism Management*, 1–23.

Gretzel, U., & Yoo, K. H. (2013). Premises and promises of social media marketing in tourism. In S. McCabe (Ed.), *The Routledge handbook of tourism marketing* (pp. 491–504). Routledge.

Griffin, T., Giberson, J., Lee, S. H., Guttentag, D., Kandaurova, M., Sergueeva, K., & Dimanche, F. (2017). Virtual Reality and Implications for Destination Marketing. *Travel and Tourism Research Association: Advancing Tourism Research Globally*, 29. https://scholarworks.umass.edu/ttra/2017/Academic_Papers_Oral/29

Griffin, M., Hodgson, R., & Sivam, S. (2021). Career incentives, career deterrents, and cultural blocks: An investigation of factors impacting female Emirati students' perceptions of tourism. *Journal of Human Resources in Hospitality & Tourism*, *20*(3), 472–496. doi:10.1080/15332845.2021.1923945

Grönroos, C. (2007). *Service management and marketing: customer management in service competition* (3rd ed.). John Wiley.

Gruszczynski, M. (2020). Women on Boards and Firm Performance: A Microeconomic Search for Connection. *Journal of Risk and Financial Management.*, *13*(9), 2018. doi:10.3390/jrfm13090218

Gubler, M., Biemann, T., Tschopp, C., & Grote, G. (2015). How career anchors differentiate managerial career trajectories: A sequence analysis perspective. *Journal of Career Development*, *42*(5), 412–430. doi:10.1177/0894845315572891

Güçer, E., Keleş, Y., Demirdağ, S. A., & Çelikkanat, N. (2018). Kadın isgörenlerin karşılaştığı sorunların örgütsel sessizlikleri üzerindeki etkisi: Bodrum'daki otel işletmelerinde bir araştırma. *Afyon Kocatepe Üniversitesi Sosyal Bilimler Dergisi*, *20*(2), 1–18. doi:10.32709/akusosbil.414693

Gudani, A. M. I. (2020). *Travel influencer.* https://www.academia.edu/43198447/Travel_Influencer?email_work_card=reading-history

Gündüz, A. Y., & Karadeniz, B. (2022). Girişimcilik Olgusunun Tarihsel Gelişimi ve Kadın Girişimciliği: Malatya İlinde Yapılan Çalışmanın İncelenmesi [Historical Development Of Entrepreneurship And Women's Entrepreneurship: A Review Of The Studies In Malatya]. *Bingöl University Journal of Social Sciences Institute*, *24*(24), 96–112. doi:10.29029/busbed.1119269

Günzel-Jensen, F., Siebold, N., Kroeger, A., & Korsgaard, S. (2020). Do the United Nations' Sustainable Development Goals matter for social entrepreneurial ventures? A bottom-up perspective. *Journal of Business Venturing Insights*, *13*, e00162. doi:10.1016/j.jbvi.2020.e00162

Gupta, D. G., Shin, H., & Jain, V. (2022). Luxury experience and consumer behavior: A literature review. Marketing Intelligence & Planning. doi:10.1108/MIP-12-2021-0438

Guterresa, L. F. D. C., Armanu, A., & Rofiaty, R. (2020). The role of work motivation as a mediator on the influence of education-training and leadership style on employee performance. *Management Science Letters*, *10*(7), 1497–1504. doi:10.5267/j.msl.2019.12.017

Guttentag, D. A. (2009). The possible negative impacts of volunteer tourism. *International Journal of Hospitality & Tourism Administration*, *10*(4), 329–345.

Haberturk. (2022). *Kadın Girişimcilerin Ağırlığı Artıyor* [The Influence of Women Entrepreneurs is Growing]. Available at: https://www.haberturk.com/kadin-girisimcilerin-agirligi-artiyor-3370908-ekonomi

Hadj, T. B. (2020). Effects of corporate social responsibility towards stakeholders and Environmental management on responsible Innovation and competitiveness. *Journal of Cleaner Production*, *250*, 119490. doi:10.1016/j.jclepro.2019.119490

Haibo, C., Ayamba, E. C., Udimal, T. B., Agyemang, A. O., & Ruth, A. (2020). Tourism and sustainable development in China: A review. *Environmental Science and Pollution Research International*, *27*(31), 39077–39093. doi:10.100711356-020-10016-7 PMID:32638313

Hair, J. F. Jr, Hult, G. T. M., Ringle, C., & Sarstedt, M. (2016). *A primer on partial least squares structural equation modeling (PLS-SEM).* Sage Publications.

Halıcı, E. (2005). *Girişimcilik ve teknoparklar: Girişimciliğin eski ve yeni yüzü.* Available at: https://www.emo.org.tr/etkinlikler/egitim/etkinlik_bildirileri_detay.php?etkinlikkod=19&bilkod=350

Halkias, D., Nwajiuba, C., Harkiolakis, N., & Caracatsanis, S. M. (2011). Challenges facing women entrepreneurs in Nigeria. *Management Research Review*, *34*(2), 221–235. doi:10.1108/01409171111102821

Hall, C. M., & Lew, A. A. (2009). *Understanding and Managing Tourism Impacts: An Integrated Approach.* Routledge. doi:10.4324/9780203875872

Halliday, C. S., Paustian-Underdahl, S. C., & Fainshmidt, S. (2020). Women on boards of directors: A meta-analytic examination of the roles of organizational leadership and national context for gender quality. *Journal of Business and Psychology.*

Hamid, S., Ali, R., Azhar, M., & Khan, S. (2021). Solo travel and well-being amongst women: An exploratory study. *Indonesian Journal of Tourism and Leisure*, *2*(1), 1–13. doi:10.36256/ijtl.v2i1.125

Han, H., Lee, S., Ariza-Montes, A., Al-Ansi, A., Tariq, B., Vega-Muñoz, A., & Park, S. H. (2021). Muslim travelers' inconvenient tourism experience and self-rated mental health at a non-Islamic country: Exploring gender and age differences. *International Journal of Environmental Research and Public Health*, *18*(2), 758. doi:10.3390/ijerph18020758 PMID:33477400

Hania, U., Rachmaniaa, I., Setyaningsih, S., & Putri, R. (2012). Patterns of Indonesian women entrepreneurship. *Procedia Economics and Finance*, 4.

Hanmer, J., & Saunders, S. (1984). *Well-founded Fear: A Community Study of Violence to Women*. Hutchinson.

Hardin, J. R., Stocks, M. H., & Graves, O. F. (2001). The effect of match or mismatch between the career anchors and the job settings of CPAs: An empirical analysis. *Advances in Accounting*, *18*, 119–148. doi:10.1016/S0882-6110(01)18008-9

Harris, H. (2004). Global careers: Work-Life Issues and the Adjustment of Women International Managers. *Journal of Management Development*, *23*(9), 818–832. doi:10.1108/02621710410558431

Harron, S., & Weıler, B. (1992). Ethnic tourism. In B. Weiler & C. M. Hall (Eds.), *Special Interest Tourism*. Belhaven Press.

Hashemi, S. M., & Schuler, S. R. (1993). Defining and studying empowerment of women: A research note from Bangladesh. *JSI Working Paper, 3*.

Hassan, A. (2012). Developing Country Studies 'Package Eco-tour' as Special Interest Tourism Product-Bangladesh Perspective. *Developing Country Studies*, *2*(1), 1–9.

Hassan, O. H. M., Sobaih, A. E. E., & Elshaer, I. A. (2022). The Impact of Women's Empowerment on Their Entrepreneurship Intention in the Saudi Food Industry. *Journal of Risk and Financial Management*, *15*(12), 571. Advance online publication. doi:10.3390/jrfm15120571

Hassan, S. B., & Damir, A. K. (2022). *Sustainable Development and Gender Equality: Empowerment Through Solo Female Travel Experiences. In Planning and Managing Sustainability in Tourism*. Springer. doi:10.1007/978-3-030-92208-5_12

Haugh, H. M., & Talwar, A. (2014). Linking social entrepreneurship and social change: The mediating role of Empowerment. *Journal of Business Ethics*, *133*(4), 643–658. doi:10.100710551-014-2449-4

Hechavarria, D., Bullough, A., Brush, C., & Edelman, L. (2019). High growth women's entrepreneurship: Fueling social and economic development. SSRN *Electronic Journal*. doi:10.2139/ssrn.3460214

Heesup, H. (2021). Consumer behavior and environmental sustainability in tourism and hospitality: A review of theories, concepts, and latest research. *Journal of Sustainable Tourism*, *29*(7), 1021–1042. doi:10.1080/09669582.2021.1903019

Heimtun, B., & Abelsen, B. (2013). Singles and solo travel: Gender and type of holiday. *Tourism Culture & Communication*. https://www.ingentaconnect.com/content/cog/tcc/2013/00000013/00000003/art00002

Henderson, K. A., & Allen, K. R. (1991). The ethic of care: Leisure possibilities and constraints for women. *Society and Leisure*, *14*(1), 97–113. doi:10.1080/07053436.1991.10715374

Henderson, K. A., & Dialeschki, M. D. (1991). A sense of entitlement to leisure as constraint and empowerment for women. *Leisure Sciences*, *13*(1), 51–65. doi:10.1080/01490409109513124

Hennekam, S., & Shymko, Y. (2020). Coping strategies of female entrepreneurs in times of crisis: The impact of Covid-19. *International Journal of Entrepreneurship and Small Business*, *40*(2), 187–202.

Hennekam, S., & Shymko, Y. (2020). Coping with the COVID-19 crisis: Force majeure and gender performativity. *Gender, Work and Organization*, *27*(5), 788–803. doi:10.1111/gwao.12479 PMID:32837010

Hennessy, B. (2022). *Influencer Sosyal Medya Çağında Marka Yaratmak*. Nova Kitap: Kadiköy-İstanbul.

Henry, C., Hill, F., & Leitch, C. (2005). Entrepreneurship education and training: Can entrepreneurship be taught? Part I. *Education + Training*, *47*(2), 98–111. doi:10.1108/00400910510586524

Henseler, J., Ringle, C. M., & Sarstedt, M. (2015). A new criterion for assessing discriminant validity in variance-based structural equation modeling. *Journal of the Academy of Marketing Science*, *43*(1), 115–135. doi:10.100711747-014-0403-8

Hepola, S. (2019). *Ice cream queen Jeni Britton Bauer on building her brand through social media*. Inc. Retrieved from https://www.inc.com/sonya-hepola/jeni-britton-bauer-social-media-marketing-brand-building.html

Herawati, A., Purwaningsih, A., Pudianti, A., & Surya, R. V. (2014). Rural tourism community empowerment based on local resources for improving community welfare: Case on Pentingsari village, Yogyakarta, Indonesia. *Review of Integrative Business and Economics Research*, *3*(2), 88.

Hillman, W., & Radel, K. (2023). The social, cultural, economic and political strategies extending women's territory by encroaching on patriarchal embeddedness in tourism in Nepal. *Gender and Tourism Sustainability*, 303–324. doi:10.4324/9781003329541-19

Hisrich, D. R., & Peters, M. P. (2002). *Entrepreneurship* (5th ed.). McGraw Hill Higher Education.

Hjalager, A. M. (2003). Global tourism careers? Opportunities and dilemmas facing higher education in tourism. *Journal of Hospitality, Leisure, Sport and Tourism Education*, *2*(2), 26–37. doi:10.3794/johlste.22.35

Hogg, M. A., van Knippenberg, D., & Rast, D. E. III. (2012). The social identity theory of leadership: Theoretical origins, research findings, and conceptual developments. *European Review of Social Psychology*, *23*(1), 258–304. doi:10.1080/10463283.2012.741134

Hole, Y., & Snehal, P. (2019). The significance of pilgrimage tourism to sustainable development with special reference to the Indian context. *African Journal of Hospitality, Tourism and Leisure*, *8*(3), 1–11.

Hookway, N. (2008). Entering the blogosphere': Some strategies for using blogs in social research. *Qualitative Research*, *8*(1), 91–113. doi:10.1177/1468794107085298

Hootsuite. (2019). *The social media trends shaping the travel industry*. https://www.hootsuite.com/pages/the-social-media-trends-shaping-the-travel-industry

Hormiga, E., Batista-Canino, R. M., & Sa'nchez-Medina, A. (2011). The role of intellectual capital in the success of new ventures. *The International Entrepreneurship and Management Journal*, *7*(1), 71–92. doi:10.100711365-010-0139-y

Horodnic, I. A., & Williams, C. (2019). Tackling undeclared work in the European Union. SSRN *Electronic Journal*. doi:10.2139/ssrn.3437406

Hoseini, S. H. (2000). *Biological difference: man and woman: body, spirit and brain*. Critics Books.

Hosni, D., & Lundberg, C. (2005). The status of women revisited: The case of Nepal. *Equal Opportunities International*, *24*(3/4), 73–85. doi:10.1108/02610150510788088

Hotels - Europe | Statista Market Forecast. (2023). *Statista*. https://www.statista.com/outlook/mmo/travel-tourism/hotels/europe

Hotho, S., & Champion, K. (2011). Small businesses in the new creative industries: Innovation as a people management challenge. *Management Decision*, *49*(1), 29–54. doi:10.1108/00251741111094428

Howard, L. (2022). *Best International Destinations for Solo Travelers in 2022*. Retrieved from https://www.forbes.com/uk/advisor/travel-insurance/best-international-destinations-for-solo-travellers-in-2022/

Hoxha, D., & Krasnıq, B. (2008). Female Entrepreneurship in Kosova: Does Gender Difference Hypothesis Hold? *International Journal of Business and Management Science*, *1*(1), 49–65.

Huang, J. Y., Chou, T. C., & Lee, G. G. (2010). Imitative innovation strategies: Understanding resource management of competent followers. *Management Decision*, *48*(6), 952–975. doi:10.1108/00251741011053488

Huang, L., Guo, Z., Deng, B., & Wang, B. (2023). Unlocking the relationship between environmentally specific transformational leadership and employees' green behaviour: A cultural self-representation perspective. *Journal of Cleaner Production*, *382*, 134857. doi:10.1016/j.jclepro.2022.134857

Huang, S., Hsu, C. H., & Chan, A. (2010). Tour Guide Performance and Tourist Satisfaction: A Study of the Package Tours in Shanghai. *Journal of Hospitality & Tourism Research (Washington, D.C.)*, *34*(1), 3–33. doi:10.1177/1096348009349815

Huang, X., Dai, S., & Xu, H. (2020). Predicting tourists' health risk preventative behaviour and travelling satisfaction in Tibet: Combining the theory of planned behaviour and health belief model. *Tourism Management Perspectives*, *33*, 100589. doi:10.1016/j.tmp.2019.100589

Huang, Y. C., Backman, K. F., Backman, S. J., & Chang, L. L. (2016). Exploring the Implications of Virtual Reality Technology in Tourism Marketing: An Integrated Research Framework. *International Journal of Tourism Research*, *18*(2), 116–128. doi:10.1002/jtr.2038

Huarng, K. H., & Yu, T. H. K. (2011). Entrepreneurship, process innovation and value creation by a non-profit SME. *Management Decision*, *49*(2), 284–296. doi:10.1108/00251741111109160

Hubbard, E. E. (2004). *The Manager's Pocket Guide to Diversity Management*. Human Resource Development.

Huggins, R., & Johnston, A. (2009). The economic and innovation contribution of universities: A regional perspective. *Environment and Planning. C, Government & Policy*, *27*(6), 1088–1106. doi:10.1068/c08125b

Hughes, K. D., Saunders, C., & Denier, N. (2022). Lockdowns, Pivots & triple shifts: Early challenges and opportunities of the COVID-19 pandemic for women entrepreneurs. *Journal of Small Business and Entrepreneurship*, *34*(5), 483–501. doi:10.1080/08276331.2022.2042657

Huis, M. A., Hansen, N., Otten, S., & Lensink, R. A. (2017). Three-Dimensional Model of Women's Empowerment: Implications in the Field of Microfinance and Future Directions. *Front Psychol.*, *8*(1678). doi:10.3389/fpsyg.2017.01678

Hummel, J., & van der Duim, R. (2012). Tourism and development at work: 15 years of tourism and poverty reduction within the SNV Netherlands Development Organisation. *Journal of Sustainable Tourism*, *20*(3), 319–338. doi:10.1080/09669582.2012.663381

Hung, K., & Petrick, J. F. (2010). Developing a measurement scale for constraints to cruising. *Annals of Tourism Research*, *37*(1), 206–228. doi:10.1016/j.annals.2009.09.002

Hutchings, K., Moyle, C. L., Chai, A., Garofano, N., & Moore, S. (2020). Segregation of women in tourism employment in the APEC region. *Tourism Management Perspectives*, *34*, 100655. doi:10.1016/j.tmp.2020.100655

Iakovidou, O., (2010). Agro-Tourism In Greece: The Case of Women Agro-Tourism Co-operatives of Ambelakia. *Medit W*, *1*(97).

İbiş, S. (2020). COVİD-19 Salgınının Seyahat Acentaları Üzerine Etkisi. (The Impact of the COVID-19 Pandemic on Travel Agencies). *Safran Kültür ve Turizm Araştırmaları Dergisi*, *3*(1), 85–98.

Ibrahim, S., & Alkire, S. (2007). *Agency and Empowerment: A Proposal for Internationally Comparable Indicators.* Routledge.

Igbaria, M., & Baroudi, J. J. (1993). A short-form measure of career orientations: A psychometric evaluation. *Journal of Management Information Systems*, *10*(2), 131–154. doi:10.1080/07421222.1993.11518003

Ihalainen, M., Shaikh, S., Mujawamariya, G., Mayanja, S., Adetonah, S., Tavenner, K., & Elias, M. (2021). Promise and contradiction: Value chain participation and women's empowerment. *Advancing gender equality through agricultural and environmental research: Past, present and future*, 147-188.

İkiz, A. N. (2020). Turizm İşletmelerinde Cinsiyet Ayrımcılığı ve Kadın Çalışanlar: Yöneticilerinin Konuya İlişkin Tutum ve Uygulamaları. *Yüzüncü Yıl Üniversitesi Sosyal Bilimler Enstitüsü Dergisi*, (50), 373–396.

ILO Şiddet ve Taciz Sözleşmesi (C190). (2021). *Çalışma yaşamında şiddet ve taciz*. Friedrich Ebert Stiftung.

ILO. (2015). *Women in business and management: gaining momentum. International Labour Organisation. International Labour Office*. ILO.

İlter, B. (2008). *Girişimcilik Sürecinde Kadın Girişimcilerin Karşılaştıkları Sorunların Analizi: Kagider Örneği* [Analysis of The Problems Faced by Women Entrepreneurs in the Entrepreneurship Process: The Case of Kagider] [Doctoral Thesis]. Afyon Kocatepe University.

ImpactJ. A. N. A. (n.d.). *Home*. Retrieved from https://jana-impact.org/

İncili, İ. İ. (2022). *Girişimci Kadınlara Yönelik Mali Politikalar: Denizli Örneği* [Financial Policies for Women Entrepreneurs: Case of Denizli] [Master Thesis]. Pamukkale University.

Influence4You. (2022). *Top 10 Travel Influencers in 2022*. https://blogen.influence4you.com/the-top-10-travel-influencers/

Influencermarketinghub.com. (2022). *Must-Have Features To Look for In An Influencer Marketing Platform*. https://influencermarketinghub.com/must-have-features-to-look-for-in-an-influencer-marketing-platform/#:~:text=Many%20influencer%20marketing%20platforms%20offer,your%20sales%20and%20engagement%20rates

Internal Displacement Monitoring Centre. (2019). *Global Report on Internal Displacement 2019*. Norwegian Refugee Council.

International Labor Organization. (2023). *Achieving gender equality at work*. International labour conference, 111th Session, Geneve.

Irawan, N., & Nara, V. (2020). Managing women empowerment through participation in sustainable tourism development in Kampong Phluk, Siem Reap, Cambodia. *International Journal of Economics, Business and Accounting Research*, *4*(2).

Isaac, R. K., & Eid, T. A. (2019). Tourists' destination image: An exploratory study of alternative tourism in Palestine. *Current Issues in Tourism*, *22*(12), 1499–1522. doi:10.1080/13683500.2018.1534806

Islam, M. S., & Kirillova, K. (2021). Nonverbal communication in hotels as a medium of experience co-creation. *Tourism Management*, *87*, 104363. doi:10.1016/j.tourman.2021.104363

Istanbul & I. (n.d.). *Empowering Women in Istanbul*. Retrieved February 21, 2023, from https://istanbulandi.org.tr/

Istanbul, O. N. A. R. (n.d.). *ONAR İstanbul Yabanci Uyruklu Kadinlari Destekleme Derneği (ONAR İstanbul)*. Retrieved February 9, 2023, from https://www.4carma.com/en/ngo/ONARİstanbul

İstanbullu Dinçer, F. İ., Akova, A., Muğan Ertuğral, S., & Aydoğan Çifçi, M. (2016). Türkiye'de turizm sektöründe kadın istihdamı: Imkanlar ve engeller. *Eurasian Academy of Sciences Social Sciences Journal*, *1*, 379–395.

Itchfeet. (2023). Retrieved from http://charukesi.com/itchyfeet/category/europe/turkey/

Ito, E., & Higham, J. (2020). Supplemental tourism activities: A conceptual framework to maximise sport tourism benefits and opportunities. *Journal of Sport & Tourism*, *24*(4), 269–284. doi:10.1080/14775085.2020.1850322

Iwu, C. G., Nsengimana, S., & Robertson, T. K. (2019). The factor contributing to the low numbers of women entrepreneurs in kigali. *Acta Universitatis Danubius Economica*, *15*(6).

Jackson, E. L. (1991). Leisure Constraints/Constrained Leisure: Special Issue Introduction. *Journal of Leisure Research*, *23*(4), 279–285. doi:10.1080/00222216.1991.11969860

Jackson, E. L., & Henderson, K. A. (1995). Gender-based analysis of leisure constraints. *Leisure Sciences*, *17*(1), 31–51. doi:10.1080/01490409509513241

Jackson, G., & Morpeth, N. (1999). Local Agenda 21 and community participation in tourism policy and planning: Future or fallacy. *Current Issues in Tourism*, *2*(1), 1–38. doi:10.1080/13683509908667841

Jacobs, R. C. (1987). Identifying Career anchors and team perceptions: A new approach to diagnosing management team-building needs. *Asia Pacific Journal of Human Resources*, *25*(3), 35–50. doi:10.1177/103841118702500303

Jafari, J., & Scott, N. (2014). Muslim world and its tourisms. *Annals of Tourism Research*, *44*, 1–19. doi:10.1016/j.annals.2013.08.011

Jalilvand, M. R., Ebrahimi, A., & Samiei, N. (2013). Electronic word of mouth effects on tourists' attitudes toward Islamic destinations and travel intention: An empirical study in Iran. *Procedia: Social and Behavioral Sciences*, *81*, 484–489. doi:10.1016/j.sbspro.2013.06.465

Jamali, D. (2009). Constraints and opportunities facing women entrepreneurs in developing countries. *Gender in Management*, *24*(4), 232–251. doi:10.1108/17542410910961532

Jamal, T., & Higham, J. (2021). Justice and ethics: Towards a new platform for tourism and sustainability. *Journal of Sustainable Tourism*, *29*(2-3), 143–157. doi:10.1080/09669582.2020.1835933

Jamshaid, S., Bahadar, N., Jamshed, K., Rashid, M., Imran Afzal, M., Tian, L., Umar, M., Feng, X., Khan, I., & Zong, M. (2023). Pre-and Post-Pandemic (COVID-19) Mental Health of International Students: Data from a Longitudinal Study. *Psychology Research and Behavior Management*, *16*, 431–446. doi:10.2147/PRBM.S395035 PMID:36814636

Janta, H. (2011). Polish migrant workers in the UK hospitality industry: Profiles, work experience and methods for accessing employment. *International Journal of Contemporary Hospitality Management*, *23*(6), 803–819. doi:10.1108/09596111111153484

Javadian, G., & Singh, R. P. (2012). Examining successful Iranian women entrepreneurs: An exploratory study. *Gender in Management*, *27*(3), 148–164. doi:10.1108/17542411211221259

Jenkins, H. A. (2004). Critique of conventional CSR theory: An SME perspective. *Journal of General Management*, *2004*(29), 37–57. doi:10.1177/030630700402900403

Jennings, J. E., & Brush, C. G. (2013). Research on women entrepreneurs: Challenges to (and from) the broader entrepreneurship literature? *The Academy of Management Annals*, *7*(1), 663–715. doi:10.5465/19416520.2013.782190

Jeong, E., Ryu, I., & Brown, A. (2018). Moderating effect of sense of community on the relationship between psychological empowerment and tourism policy participation of local residents. *Global Business & Finance Review*, *23*(1), 36–46. doi:10.17549/gbfr.2018.23.1.36

Jia, X. (2018, May). Consideration of the role of tourism environmental protection on the sustainable development of Tourism. In *2018 7th International Conference on Energy, Environment and Sustainable Development (ICEESD 2018)* (pp. 1642-1646). Atlantis Press. 10.2991/iceesd-18.2018.296

Jiménez-Esquinas, G. (2017). "This is not only about culture": On tourism, gender stereotypes and other affective fluxes. *Journal of Sustainable Tourism*, *25*(3), 311–326. doi:10.1080/09669582.2016.1206109

Jin, S. V., Muqaddam, A., & Ryu, E. (2019). Instafamous and social media influencer marketing. *Marketing Intelligence & Planning*, *37*(5), 567–579. Advance online publication. doi:10.1108/MIP-09-2018-0375

Johanson, M. (2017). *How this woman turned her Instagram account into a $4,000-a-week job*. Business Insider. Retrieved from https://www.businessinsider.com/how-this-woman-turned-her-instagram-account-into-a-4000-a-week-job-2017-1

Jolly, S. (2002). Gender and cultural change. *The Bridge*.

Jones, A. M. (2005). Assessing international youth service programmes in two low-income countries: Voluntary Action. *The Journal of the Institute for Volunteering Research*, *7*(2), 87–100.

Jongmeewasin, S. (2016). *Religious Tourism, Pilgrimage, and Cultural Tourism*. Silpakorn University. Retrieved from https://www.academia.edu/12252237/Religious_Tourism_Pilgrimage_and_Cultural_Tourism

Jordan, F., & Gibson, H. (2005). "We're not stupid... But we'll not stay home either": Experiences of solo women travelers. *Tourism Review International*. https://www.ingentaconnect.com/content/cog/tri/2005/00000009/00000002/art00006

Jordan, F. (1997). An occupational hazard? Sex segregation in tourism employment. *Tourism Management*, *18*(8), 525–534. doi:10.1016/S0261-5177(97)00074-5

Jordan, F., & Gibson, H. (2005). 'We're not stupid . . . but we'll not stay home either': Experiences of solo women travellers. *Tourism Review International*, *9*(2), 195–212. doi:10.3727/154427205774791663

JourneyWomen. (2022). *Safety Resources for Women Travelers*. Retrieved from https://journeywoman.com/solo-travel-safety/

Kaba, B., & Emekli, G. (2018). A rising Trend in Tourism: Solo Women Travelers (The Case of Turkey). *Aegean Geographical Journal*, *27*(2), 111–126.

Kabeer, N. (1999). Resources, agency, achievements: Reflections on the measurement of women's empowerment. *Development and Change*, *30*(3), 435–464. doi:10.1111/1467-7660.00125

Kabeer, N. (2005). Gender equality and women's empowerment: A critical analysis of the third millennium development goal 1. *Gender and Development*, *13*(1), 13–24. doi:10.1080/13552070512331332273

Kabeer, N. (2005). Is microfinance a 'magic bullet' for women's empowerment? Analysis of findings from South Asia. *Economic and Political Weekly*, 4709–4718.

Kabil, E. N. (2022). Toplumsal cinsiyet eşitsizliklerinin kadınların motivasyon ve verimliliği üzerindeki rolü: Mutfak departmanı çalışanları üzerine bir araştırma. *Journal of New Tourism Trends*, *3*(1), 70–85.

Kaiser, M. (2016). *Best Female Travel Blogs: Top Influential Views*. Retrieved from https://www.creditdonkey.com/female-travel-blogs.html

Kaleözü, S. (2021). *Turizm Sektöründe Kadın Girişimcilerin Motivasyon Faktörleri: Türkiye için Bir Tipoloji Önerisi* [Motivation Factors of Femalae Entrepreneurs in Tourism Sector: A Typology Proposal for Turkey] [Master Thesis]. Ankara Hacı Bayram Veli University.

Kalinowski, K. M., & Weiler, B. (1992). Review. Educational Travel. In C. M. Hall & B. Weiler (Eds.), *Special Interest Tourism*. John Wiley & Sons, Inc.

Kamışlı, M., & Akın, F. (2009). The Effects Of Tourism Sector To Economic Development In Emerging And Developing Countries: An Assessment On Turkish Tourism. *Anadolu Bil Vocational School Journal*, (16), 16–29.

Kantar, M. (1999). Entrepreneurship and Rural Women. *Journal of Agricultural Economics*, *4*(1/2), 29–42.

Kanter, R. M. (2003). *Men and women of the corporation*. Basic Books.

Karabulut, E. (2022, April 19). *Turizmde çalışan kadın Sayısı.* https://www.turizmdatabank.com/turizm-istatistikleri/turizmde-kadin-istihdami/

Kara, D., Kim, H., & Uysal, M. (2018). The effect of manager mobbing behaviour on female employees' quality of life. *Current Issues in Tourism*, *21*(13), 1453–1467. doi:10.1080/13683500.2015.1078298

Karakaş, A., & Gökmen, G. (2016). Studying Woman Entrepreneurs Profile In Tourism Sector. *Journal of Recreation and Tourism Research*, *3*(3), 18–25.

Karam Foundation. (n.d.). *Karam Foundation*. Retrieved February 9, 2023, from https://karamfoundation.org/

Karaset, Ş.B., & Yiğit, T. (2017). Turizm sektöründe kadın istihdamı ve sorunları. III. In *Uluslararası Girişimcilik, İstihdam ve Kariyer Kongresi* (pp. 1114-1130). Muğla.

Karasoula, K. (2010). *The competitiveness of Greece in the touristic sector. The case of prefecture Achaea and the development of the specific and alternative forms of tourism* [MBA thesis]. University of Patras.

Karim, W., Haque, A., Anis, Z., & Ulfy, M. A. (2020). The Movement Control Order (MCO) for COVİD-19 Crisis and its Impact on Tourism and Hospitality Sector in Malaysia. *International Tourism and Hospitality Journal*, *3*(2), 1–7.

Katpar, N. K., Chaudhry, N. I., Tunio, M. N., & Ali, M. A. (2020). Impact of Leadership Style and Organizational Culture on Organizational Commitment. *Sukkur IBA Journal of Management and Business, 7*(1), 92-106.

Katpar., N. K., Chaudhry, N. I., Tunio, M. N., & Ali, M. A. (2020). Impact of Leadership Style and Organizational Culture on Organizational Commitment. *Sukkur IBA Journal of Management and Business, 7*(1), 92-106.

Katper, N. K. Medan, A., Syed, K. B. S., & Tunio, M. N. (2017). Determinants of Debt Maturity Structure in Shariah and Non-Shariah Firms in Pakistan: A comparative Study. *Journal of Applied Economic Sciences, 12*(4), 1210-1225.

Katper, N. K. Medan, A., Syed, K. B. S.., Tunio, M. N. (2017). Determinants of Debt Maturity Structure in Shariah and Non-Shariah Firms in Pakistan: A comparative Study. *Journal of Applied Economic Sciences, 12*(4), 1210-1225.

Kattara, H. (2005). Career challenges for female managers in egyptian hotels. *International Journal of Contemporary Hospitality Management*, *17*(3), 238–251. doi:10.1108/09596110510591927

Kautonen, T., & Palmroos, J. (2010). The impact of a necessity-based start-up on subsequent entrepreneurial satisfaction. *The International Entrepreneurship and Management Journal*, *6*(3), 285–300. doi:10.100711365-008-0104-1

Kaya, İ., & Atçı, D. (2015). Türk turizm sektöründe çalışma saatlerinin analizi ve değerlendirilmesi. *Cag University Journal of Social Sciences*, *12*(2), 23–47.

Kaya, İ., & Topbaş, F. (2019). Turizm sektöründe cinsiyete dayalı ücret ayrımcılığı. *Izmir Democracy University Social Sciences Journal*, *2*(1), 74–87.

Kaya, Ş. (2017). Turizmde Kadın Emeği. *Karatahta İş Yazıları Dergisi*, *9*, 1–22.

KEDV & CCA. (2015). Türkiye'de Kadın Kooperatifleri: Ayakta Kalma Mücadelesinden Büyümeye. KEDV Yayınları.

KEİG. (2015). *Türkiye'de Kadın Kooperatifleşmesi: Eğilimler ve İdeal Tipler*. Kayhan Matbaa.

KEİG. (2018). *Kadın Kooperatifleri Kılavuzu*. Hazal Matbaa.

Kelley, D. J., Brush, C. G., Greene, P. G., & Litovsky, Y. (2013). *Global Entrepreneurship Monitor: 2012 Women's Report*. The Center for Women's Leadership at Babson College and London Business School.

Kemer, E., & Aslan, Z. (2021). Konaklama İşletmelerinde Kadın Yönteici Olmak: Vignette Tekniği Uygulaması [Women in Hospitality Businesses: Vignette Technique Application]. *Turizm Akademik Dergisi*, *9*(1), 39–53.

Kensbock, S., Jennings, G., Bailey, J., & Patiar, A. (2013). The lowest rung': Women room attendants' perceptions of five star hotels' operational hierarchies. *International Journal of Hospitality Management*, *35*, 360–368. doi:10.1016/j.ijhm.2013.07.010

Kerstetter, D. L., Yen, I.-Y., & Yarnal, C. M. (2005). Plowing uncharted waters: A study of perceived constraints to cruise travel. *Tourism Analysis*, *10*(2), 137–150. doi:10.3727/108354205776259295

Ketprapakorn, N., & Kantabutra, S. (2019). Sustainable social enterprise model: Relationships and consequences. *Sustainability (Basel)*, *11*(14), 3772. doi:10.3390u11143772

Khan, S. (2011). Gendered leisure: Are women more constrained in travel for leisure? *Tourismos*, *6*(1), 105–121.

Khan, S. (2011). Gendered Leısure: Are Women More Constraıned in Travel for Leısure? *TOURISMOS: An Internatıonal Multıdıscıplınary Journal of Tourısm*, *6*(1), 105–121.

Khatiwada, L. K., & Silva, J. A. (2015). Mitigating Gender Inequality in Rural Regions: The Effects of Tourism Employment in Namibia. *International Journal of Tourısm Research*, *17*(5), 442–450. doi:10.1002/jtr.2010

Kimbu, A. N., & Ngoasong, M. Z. (2016). Women as vectors of Social Entrepreneurship. *Annals of Tourism Research*, *60*, 63–79. doi:10.1016/j.annals.2016.06.002

Kim, H., Lee, J. S., & Yoo, J. J. (2015). The impact of social media on consumers' acculturation and purchase intentions in tourism-related industries. *Journal of Travel & Tourism Marketing*, *32*(5), 564–586.

Kim, S., Whitford, M., & Arcodia, C. (2019). Development of intangible cultural heritage as a sustainable tourism resource: The intangible cultural heritage practitioners' perspectives. *Journal of Heritage Tourism*, *14*(5-6), 422–435. doi:10.1080/1743873X.2018.1561703

Kinnaird, V., & Hall, D. (1996). Understanding Tourism Processes: A Gender-Aware Frame Work. *Tourism Management*, *17*(2), 95–102. doi:10.1016/0261-5177(95)00112-3

Kinnaird, V., & Hall, D. (2000). Theorizing gender in tourism research. *Tourism Recreation Research*, *25*(1), 71–84. doi:10.1080/02508281.2000.11014901

Kiráľová, A., & Pavlíčeka, A. (2015). Development of social media strategies in tourism destination. *Procedia: Social and Behavioral Sciences*, *175*, 358–366. doi:10.1016/j.sbspro.2015.01.1211

Kirzner, I. M. (1973). *Competition and entrepreneurship*. The University of Chicago Press.

Kısa, M. C., & Şahin, D. (2022). Otel İşletmelerinde Kadın Yönetici Olmak: Ordu İli Örneği [Being a Female Manager in Hotel Enterprises: The Example of Ordu Province]. *ODÜSOBİAD*, *12*(2), 1061–1088. doi:10.48146/odusobiad.1102336

Kıvılcım, B. (2020). COVİD-19 (Yeni Koronavirüs) Salgınının Turizm Sektörüne Muhtemel Etkileri [Possible Effects of the COVID-19 (Novel Coronavirus) Pandemic on the Tourism Sector]. *USOBED Uluslararası Batı Karadeniz Sosyal ve Beşeri Bilimler Dergisi*, *4*(1), 17–27. doi:10.46452/baksoder.727208

Knight, D. K., Crutsinger, C., & Kim, H. (2006). The impact of retail work experience, career expectation, and job satisfaction on retail career intention. *Clothing & Textiles Research Journal*, *24*(1), 1–14. doi:10.1177/0887302X0602400101

Knitistanbul. (n.d.). *Our Team*. Retrieved February 9, 2023, from https://knitstanbul.com/about/team/

Koburtay, T., Syed, J., & Haloub, R. (2020). Implications of religion, culture, and legislation for gender equality at work: Qualitative insights from Jordan. *Journal of Business Ethics*, *164*(3), 421–436. doi:10.100710551-018-4036-6

Kocacık, F. & Gökkaya, B.V. (2005). Türkiye'de çalışan kadınlar ve sorunları. *C.Ü. İktisadi ve İdari Bilimler Dergisi, 6*(1), 195-219.

Koç, B., & Uşaklı, A. (2022). Kadın çalışanların turizm sektöründen ayrılma nedenleri: Keşifsel bir araştırma. *Journal of Tourism and Gastronomy Studies*, *10*(1), 589–609.

Koc, E. (2020). Do Women Make Better in Tourism and Hospitality? A Conceptual Review from a Customer Satisfaction and Service Quality Perspective. *Journal of Quality Assurance in Hospitality & Tourism*, *21*(4), 402–429. doi:10.1080/1528008X.2019.1672234

Koc, E., & Ayyildiz, A. Y. (2021). Culture's Influence on the Design and Delivery of the Marketing Mix Elements in Tourism and Hospitality. *Sustainability (Basel)*, *13*(21), 11630. doi:10.3390u132111630

Kodagoda, T., & Jayawardhana, S. (2022). Forced or voluntary reluctance or voluntary preference to work? Women in hotel industry: Evidence from Sri Lanka. *Gender in Management*, *37*(4), 524–534. doi:10.1108/GM-01-2021-0011

Kogovsek, M., & Kogovsek, M. (2015). Hospitality and tourism gender issues remain unsolved: A call for research. *Quaestus*, *6*, 194.

Koh, C. H. (1996). Testing hypotheses of entrepreneurial characteristics a study of Hong Kong. *Journal of Managerial Psychology*, *11*(3), 12–25. doi:10.1108/02683949610113566

Komunte, M. (2015). Usage of Mobile Technology in Women Entrepreneurs: A case Study of Uganda. *The African Journal of Information Systems*, *7*(3), 3. https://digitalcommons.kennesaw.edu/ajis/vol7/iss3/3

Kong, H., Wang, S., & Fu, X. (2015). Meeting Career Expectation: Can It Enhance Job Satisfaction of Generation Y? *International Journal of Contemporary Hospitality Management*, *27*(1), 1–27. doi:10.1108/IJCHM-08-2013-0353

Köroğlu, Ö., Yıldırım, H. U., & Kılıç, A. (2020). Kadın Turist Rehberlerinin Mesleki Sorunlarının Belirlenmesi. *Mehmet Akif Ersoy Üniversitesi Sosyal Bilimler Enstitüsü Dergisi*, (31), 26–40. doi:10.20875/makusobed.586054

Köse, Z. (2014). *Turizmde Kadın İstihdamı ve Kadın Girişimciliği Beypazarı Örneği*. Hacettepe Üniversitesi Sosyal Bilimler Enstitüsü Antropoloji Anabilim Dalı.

Koskela, H. (1997). 'Bold walk and breakings': Women's spatial confidence versus fear of violence. *Gender, Place and Culture*, *4*(3), 301–319. doi:10.1080/09663699725369

Kothari, T. (2017). Women entrepreneurs' path to building venture success: Lessons from India. *South Asian Journal of Business Studies*, *6*(2), 118–141. doi:10.1108/SAJBS-03-2016-0021

Kotler, N. (2020). *How Christina Tosi is keeping Milk Bar sweet during the pandemic*. Forbes. Retrieved from https://www.forbes.com/sites/nataliekotler/2020/05/12/how-christina-tosi-is-keeping-milk-bar-sweet-during-the-pandemic/?sh=66c5ccbc58dc

Kotler, P., & Lee, N. (2002). *Corporate social responsibility: Doing the most good for your company and your cause*. John Wiley & Sons.

Koutsou, S., Notta, O., Samathrakis, V., & Partalidou, M. (2009). Women's Entrepreneurship and Rural Tourism in Greece: Private Enterprises and Cooperatives. *South European Society & Politics*, *14*(2), 191–209. doi:10.1080/13608740903037968

Kowszyk, Y., & Vanclay, F. (2020). The possibilities and limitations regarding the use of impact evaluation in corporate social responsibility programs in Latin America. *Corporate Governance (Bradford)*, *21*(2), 279–293. doi:10.1108/CG-01-2020-0038

Koyuncu, M. B., & Sevin, H. D. (2017). Konaklama işletmelerinde çalışanların psikolojik şiddet (mobbing) algısının iş tatminine etkisi. *Yönetim. Ekonomi ve Pazarlama Araştırmaları Dergisi*, *1*(3), 33–52.

Kozinets, R. V. (2002). The field behind the screen: Using netnography for marketing research in online communities. *JMR, Journal of Marketing Research*, *39*(1), 61–72. doi:10.1509/jmkr.39.1.61.18935

Kramer, B. (2012). *Eclipse chasers log. Eclipse chasers log summary*. Retrieved from http://www.eclipsechasers.com/tseChaserLogSums.php

Kraus, S., Palmer, C., Kailer, N., Kallinger, F. L., & Spitzer, J. (2018). Digital Entrepreneurship. *International Journal of Entrepreneurial Behavior & Research*. doi:10.1108/IJEBR-06-2018-0425

Krco, M. (2005). *Curious about astronomy? Ask an Astronomer*. Retrieved from http://curious.astro.cornell.edu/question. php?number=657

Kreitz, P. A. (2008). Best Practices for Managing Organizational Diversity. *Journal of Academic Librarianship*, *34*(2), 101–120. doi:10.1016/j.acalib.2007.12.001

Krishnamurthy, L., Rajasekar, A., & Jebasingh, D. (2019). An Empirical Study on Role of Travel Start-Ups and Entrepreneurship in E-Tourism. *Journal of Service Science and Management*, *12*(5), 620–627. doi:10.4236/jssm.2019.125042

Kruja, D., & Gjyrezi, A. (2011). The Special Interest Tourism Development and the Small Regions. *Turizam*, *15*(2), 77–89. doi:10.5937/Turizam1102077K

Ktoturkiye.com. (2022). *Meryem Can's Korea Trip*. https://ktoturkiye.com/meryem-can-ile-kore-seyahati

Kucharvy, T. (2014). *Creating a pathway to your dream: Career designing and controlling a career around your life goals*. Business Expert Press.

Küçükaltan, D. (2009). Genel bir yaklaşımla girişimcilik. *Girişimcilik Gündemi*, *4*(1), 21–28.

Kulkarni, R., & Harshitha.Rao, R. (2022). A Study on Challenges and Opportunities of Women Entrepreneurs in Tourism Sector. *Journal of Positive School Psychology*, *6*(5), 6732–6743.

Kumar, A. (2017). Cultural and heritage tourism: A tool for sustainable development. *Global Journal of Commerce & Management Perspective*, *6*(6), 56–59. doi:10.24105/gjcmp.6.6.1709

Kumar, A., & Singh, G. (2019). Seasonal effect on tourism in India. *Journal of Financial Economics*, *7*(2), 48–51.

Kumar, N., & Singh, D. L. K. (2021). Status of Women-entrepreneur in Indian Startups. *International Journal of Engineering Technology and Management Sciences*, *5*(2), 1–12. doi:10.46647/ijetms.2021.v05i02.001

Kumbhar, V. (2013). Some Critical Issues of Women Entrepreneurship in Rural India. *European Academic Research*, *1*(2), 192–200.

Kunene, B., Mashau, P., & And Nyawo, J. (2022). Ascertaining The Challenges Faced By Women Entrepreneurs in Ethekwini Municipality, Kwazulu-Natal Province. *Gender & Behaviour*, *20*(1), 19060–19083.

Kunjurman, V., & Hussin, R. (2016). Women participation in ecotourism development: Are they empowered. *World Applied Sciences Journal*, *34*(12), 1652–1658.

Kurchenko, L., Kolomiyets-Ludwig, E., & Ilnytskyy, D. (2021). Women's empowerment as a tool for sustainable development of higher education and research in the digital age. *Stagnancy Issues and Change Initiatives for Global Education in the Digital Age*, 141-172.

Kurtay, E. (2021). Ekoturizm Sektöründe Kadın Girişimciliğinin Düzce İli Uygulama Örnekleri Üzerinde Değerlendirilmesi [Evaluation of Women's Entrepreneurship in the Ecotourism Sector Through the Practical Examples in Düzce Province]. *PEYZAJ- Eğitim, Bilim. Kültür ve Sanat Dergisi*, *3*(2), 93–105. doi:10.53784/peyzaj.1025166

Kusumawardani, K. A., Widyanto, H. A., & Deva, I. P. (2020). Understanding the entrepreneurial intention of female entrepreneurs in the Balinese tourism industry. *International Journal of Research in Business and Social Science, 9*(1), 63–79. doi:10.20525/ijrbs.v9i1.611

Kutanis, R. Ö. & Hancı, A. (2004). Kadın Girişimcileri Kişisel Özgürlük Algılamaları [Women Entrepreneurs' Perceptions of Personal Freedom]. *3. Ulusal Bilgi, Ekonomi ve Yönetim Kongresi, Osmangazi Üniversitesi İİBF,* 457- 464.

Kutay, T. (2022). Kırsal Kalkınmada Kadın Kooperatiflerinin Önemi: Türkiye Özelinde Bir Değerlendirme. *Politik Ekonomik Kuram*, *6*(1), 119–150. doi:10.30586/pek.1020843

Kuzgun, Y., & Sevim, S. A. (2004). Kadınların çalışmasına karşı tutum ve dini yönelim arasındaki ilişki. *Ankara Üniversitesi Eğitim Bilimleri Fakültesi Dergisi*, *37*(1), 14–27.

La Lopa, J. M., & Gong, Z. (2020). Sexual harassment of hospitality interns. *Journal of Hospitality & Tourism Education*, *32*(2), 88–101. doi:10.1080/10963758.2020.1726767

Leavy, P. (2014). *The Oxford Handbook of Qualitative Research*. Oxford University Press. doi:10.1093/oxfordhb/9780199811755.001.0001

Lebrun, A.-M., Su, C.-J., & Bouchet, P. (2021). A More Sustainable Management of Domestic Tourists in Protected Natural Parks: A New Trend in Sport Tourism after the Covid-19 Pandemic? *Sustainability (Basel)*, *13*(14), 7750. doi:10.3390u13147750

Lee, C. C., & Chen, C. J. (2011). The reaction of elderly Asian tourists to avian influenza and SARS. *Tourism Management*, *32*(6), 1421–1422. doi:10.1016/j.tourman.2010.12.009 PMID:32287735

Lee, C. K., & Kang, S. (1998). Measuring earnings inequality and median earnings in the tourism industry. *Tourism Management*, *19*(4), 341–348. doi:10.1016/S0261-5177(98)00033-8

Lee, H., & Hyun, S. S. (2015). The impact of social media on tourist behavior. *Journal of Travel & Tourism Marketing*, *32*(5), 527–540.

Lee, J. (1996). The motivation of women entrepreneurs in Singapore. *Women in Management Review*, *11*(2), 18–29. doi:10.1108/09649429610112574

Lee, Y., & Jin-Li, H. (2018). Integrated Approaches for Business Sustainability: The Perspective of Corporate Social Responsibility. *Sustainability (Basel)*, *10*(7), 2318. doi:10.3390u10072318

Leoni, T., & Falk, M. (2010). Gender and field of study as determinants of self-employment. *Small Business Economics*, *34*(2), 167–185. doi:10.100711187-008-9114-1

Lepp, A., & Gibson, H. (2003). Tourist roles, perceived risk and international tourism. *Annals of tourism research*. https://www.sciencedirect.com/science/article/pii/S0160738303000240

Lepp, A. (2008). Discovering self and discovering others through the Taita Discovery Centre volunteer tourism programme, Kenya. In S. Wearing & N. G. McGehee (Eds.), *International volunteer tourism: Integrating travelers and communities* (pp. 86–100). CABI. doi:10.1079/9781845933807.0086

Leymann, H. (1996). The content and development of mobbing at work. *European Journal of Work and Organizational Psychology*, *5*(2), 165–184. doi:10.1080/13594329608414853

Liberato, P., Alen, E., & Liberato, D. (2018). Smart tourism destination triggers consumer experience: The case of Porto. *European Journal of Management and Business Economics*, *27*(1), 6–25. doi:10.1108/EJMBE-11-2017-0051

Li, C., Guo, S., Wang, C., & Zhang, J. (2019). Veni, vidi, vici: The impact of social media on virtual acculturation in tourism context. *Technological Forecasting and Social Change*, *145*, 513–522. doi:10.1016/j.techfore.2019.01.013

Li, H., Xie, K. L., & Zhang, Z. (2020). The effects of consumer experience and disconfirmation on the timing of online review: Field evidence from the restaurant business. *International Journal of Hospitality Management*, *84*(102344), 1–11. doi:10.1016/j.ijhm.2019.102344

Li, J. (2003). Playing upon fantasy: Women, ethnic tourism and the politics of identity construction in contemporary Xishuang Banna, China. *Tourism Recreation Research*, *28*(2), 51–65. doi:10.1080/02508281.2003.11081404

Li, L., & Leung, R. W. (2001). Female managers in Asian hotels: Profile and career challenges. *International Journal of Contemporary Hospitality Management*, *13*(4), 189–196. doi:10.1108/09596110110389511

Li, M., Lehto, X., & Li, H. (2020). 40 Years of Family Tourism Research: Bibliometric Analysis and Remaining Issues. *Journal of China Tourism Research*, *16*(1), 1–22. doi:10.1080/19388160.2020.1733337

Lin, E., Lin, T. M. Y., & Lin, B. W. (2010). New high-tech venturing as process of resource accumulation. *Management Decision*, *48*(8), 1230–1246. doi:10.1108/00251741011076762

Linke, S., Turak, E., & Nel, J. (2010). Freshwater Conservation Planning: The case for systematic approaches. *Freshwater Biology*, *56*(1), 6–20. doi:10.1111/j.1365-2427.2010.02456.x

Little, D. E. (2002). Women and adventure recreation: Reconstructing leisure constraints and adventure experiences to negotiate continuing participation. *Journal of Leisure Research*, *34*(2), 157–177. doi:10.1080/00222216.2002.11949967

Liu, T. M., & Leung, K. K. (2019). Volunteer tourism, endangered species conservation, and aboriginal culture shock. *Biodiversity and Conservation*, *28*(1), 115–129. doi:10.100710531-018-1639-2

Li, X., Liang, X., Li, Y., & Liu, J. (2020). Family tourism research: A bibliometric review. *Current Issues in Tourism*, *23*(6), 673–693.

Lockyer, J., & George, S. (2012). What women want: Barriers to female entrepreneurship in the West Midlands. *International Journal of Gender and Entrepreneurship*, *4*(2), 179–195. doi:10.1108/17566261211234661

Lok Sabha Secretariat. (2013). *Tourism Sector in India.* Retrieved from http://164.100.47.134./internet/Tourismsectorin India.pdf

Long, V. H., & Kindon, S. L. (2005). Gender and tourism development in Balinese villages. In *Gender, work and tourism* (pp. 89–117). Routledge.

Longwe, S. H. (2002b). Addressing rural gender issues: a framework for leadership and mobilisation. *III World Congress for Rural Women.*

Lou, C., & Yuan, S. (2018). Influencer marketing: How message value and credibility affect consumer trust of branded content on social media. *Journal of Interactive Advertising*, 1–45.

Loureiro, P. R. A., Carneiro, F. C., & Sachsida, A. (2004). Race and gender discrimination in the labor market: An urban and rural sector analysis for Brazil. *Journal of Economic Studies (Glasgow, Scotland)*, *31*(2), 129–143. doi:10.1108/01443580410527114

Loureiro, S. (2014). The role of rural tourism experience economy in place attachment and behavioral intentions. *International Journal of Hospitality Management*, *40*, 1–9. doi:10.1016/j.ijhm.2014.02.010

Lund-Durlacher, D. (2015). Corporate social responsibility and tourism. In P. Moscardo & G. Benckendorff (Eds.), *Education for Sustainability in Tourism—A Handbook of Processes, Resources, and Strategies* (pp. 59–72). Springer. doi:10.1007/978-3-662-47470-9_4

Luo, F., Moyle, B. D., Moyle, C. L. J., Zhong, Y., & Shi, S. (2020). Drivers of carbon emissions in China's tourism industry. *Journal of Sustainable Tourism*, *28*(5), 747–770. doi:10.1080/09669582.2019.1705315

Lyons, K. D., & Wearing, S. (2008). Volunteer tourism as alternative tourism: Journeys beyond otherness. In S. Wearing & N. G. McGehee (Eds.), *Journeys of discovery in volunteer tourism: International case study perspectives* (pp. 3–11). CABI. doi:10.1079/9781845933807.0003

Mabula, J. B., Dongping, H., & Mwakapala, L. Y. (2020). SME's use of ICT and financial services on innovation performance: The mediating role of managers' experience. *Human Systems Management*, *39*(3), 427–439. doi:10.3233/HSM-190790

Macassa, G. (2021). Social enterprise, population health and sustainable development goal 3: A public health viewpoint. *Annals of Global Health*, *87*(1), 52. doi:10.5334/aogh.3231 PMID:34221905

MacRobert, A. (2012). How to start right in astronomy. *Sky and Telescope Magazine*. Retrieved from http://www. skyandtelescope.com/howto/basics/How_to_Start_Right_in_Astronomy.html

Madanaguli, A., Srivastava, S., Ferraris, A., & Dhir. A. (2021). *Corporate social responsibility and sustainability in the tourism sector: A systematic literature review and future outlook.* doi:10.1002/sd.2258

MADRE. (n.d.). *Care In Crisis for Women and Girls Syria*. Retrieved February 9, 2023, from https://www.madre.org/projects/care-crisis-women-and-girls

Mahdavi Zafarghandi, M. (2016). *Iranian Female Tour Guides' Perceptions of Working in the Tourism Industry*. Academic Press.

Mahmoud, M., Abou-Shouk, M., & Fawzy, N. (2019). The role of governmental authorities in supporting entrepreneurship in tourism industry. International Journal of Heritage. *Tourism and Hospitality*, *13*(2), 107–123.

Mahmud, S., Shah, N. M., & Becker, S. (2012). Measurement of Women's Empowerment in Rural Bangladesh. *World Development*, *40*(3), 610–619. doi:10.1016/j.worlddev.2011.08.003 PMID:23637468

Mainardes, E. W., Alves, H., & Raposo, M. (2011). Stakeholder theory: Issues to resolve. *Management Decision*, *49*(2), 226–252. doi:10.1108/00251741111109133

Malik, S. (2018). Corporate Social Responsibility: A Tool for Woman Empowerment in Hospitality Sector. *IJRAR*, *5*, 4.

Manaf, A., Purbasari, N., Damayanti, M., Aprilia, N., & Astuti, W.Community-Based Rural Tourism in Inter-Organizational Collaboration. (2018). How Does It Work Sustainably? Lessons Learned from Nglanggeran Tourism Village, Gunungkidul Regency, Yogyakarta, Indonesia. *Sustainability (Basel)*, *10*(7), 2142. doi:10.3390u10072142

Mandipaka, F. (2014). An investigation of the challenges faced by women entrepreneurs in developing countries: A case of king williams' town, South Africa. *Mediterranean Journal of Social Sciences*. Advance online publication. doi:10.5901/mjss.2014.v5n27p1187

Mannaa, M. T. (2020). Halal food in the tourist destination and its importance for Muslim travellers. *Current Issues in Tourism*, *23*(17), 2195–2206. doi:10.1080/13683500.2019.1616678

Manthiou, A., Kang, J., Sumarjan, N., & Tang, L. R. (2016). The Incorporation of Consumer Experience into the Branding Process: An Investigation of Name-Brand Hotels. *International Journal of Tourism Research*, *18*(2), 105–115. doi:10.1002/jtr.2037

Manwa, H. (2008). Enhancing participation of women in tourism. In *Building community capacity for tourism development* (pp. 116–122). CABI. doi:10.1079/9781845934477.0116

Manzoor, F., Wei, L., Asif, M., Haq, M. Z., & Rehman, H. (2019). The Contribution of Sustainable Tourism to Economic Growth and Employment in Pakistan. *International Journal of Environmental Research and Public Health*, *16*(19), 3785. doi:10.3390/ijerph16193785 PMID:31597376

Markowska, M., Ahl, H., & Naldi, L. (2018, January 1). *Paths to entrepreneurship:life stories of Swedish Mumpreneurs*. Lancaster EPrints. Retrieved April 26, 2023, from https://eprints.lancs.ac.uk/id/eprint/127670/

Martínez M.C.V., & Rambaud, S. C. (2019). Women on corporate boards and ˛rm™s ˛nancial performance. *Women's Stud Int Forum*. doi:.wsif.2019.102251 doi:10.1016/j

Martínez-Rodríguez, I., Quintana-Rojo, C., Gento, P., & Callejas-Albiñana, F.-E. (2021). Public policy recommendations for promoting female entrepreneurship in Europe. *The International Entrepreneurship and Management Journal*, *18*(3), 1235–1262. doi:10.100711365-021-00751-9

Masa'deh, R. E., Alananzeh, O., Jawabreh, O., Alhalabi, R., Syam, H., & Keswani, F. (2019). The association among employees' communication skills, image formation and tourist behaviour: Perceptions of hospitality management students in Jordan. *International Journal of Culture, Tourism and Hospitality Research*, *13*(3), 257–272. doi:10.1108/IJCTHR-02-2018-0028

Masadeh, M. (2013). Women in the hotel industry: What's missing from this picture? *International Journal of Academic Research in Business & Social Sciences*, *3*(1), 573–580.

Maser, B., & Weiermair, K. (1998). Travel Decision-Making: From the Vintage Point of Perceived Risk and Information Preferences. *Journal of Travel & Tourism Marketing*, *7*(4), 107–121. doi:10.1300/J073v07n04_06

Mashapure, R., Nyagadza, B., Chikazhe, L., Msipa, N., Ngorora, G. K., & Gwiza, A. (2022). Challenges hindering women entrepreneurship sustainability in rural livelihoods: Case of Manicaland Province. *Cogent Social Sciences*, *8*(1), 2132675. Advance online publication. doi:10.1080/23311886.2022.2132675

Mason, J. (1988). No peace for the wicked: older married women and leisure. *No Peace for the Wicked: Older Married Women and Leisure*, 75–85.

Matthew Scott, J., & Harrison, T., R., Hussain, J., & Millman, C. (2014). The role of *guanxi* networks in the performance of women-led firms in China. *International Journal of Gender and Entrepreneurship*, *6*(1), 68–82. doi:10.1108/IJGE-03-2013-0014

Matthews, A. (2008). Negotiated selves: Exploring the impact of local-global interactions on young volunteer travellers. In K. Lyon & S. Wearing (Eds.), Journeys of discovery in volunteer tourism (pp. 101–117). CABI.

Mayer, M., & Vogt, L. (2016). Economic effects of tourism and its influencing factors: An overview focusing on the spending determinants of visitors. *Zeitschrift für Tourismuswissenschaft*, *8*(2), 169–198. doi:10.1515/tw-2016-0017

Maziriri, E. T., Nyagadza, B., & Chuchu, T. (2022). Innovation conviction, innovation mindset and innovation creed as precursors for the need for achievement and women's entrepreneurial success in South Africa: entrepreneurial education as a moderator. *European Journal of Innovation Management*.

McCulley, K. (2022). *Solo Female Travel in Turkey — Is Turkey Safe?* Retrieved from https://www.adventurouskate.com/solo-female-travel-turkey-safe/

McGehee, N. G., & Andereck, K. (2008). 'Pettin'the critters': Exploring the complex relationship between volunteers and the voluntoured in McDowell County, West Virginia, USA, and Tijuana, Mexico. In Journeys of discovery in volunteer tourism: International case study perspectives (pp. 12–24). CABI.

McGehee, N. G., & Andereck, K. L. (2004). Factors predicting rural residents' support of tourism. *Journal of Travel Research*, *43*(2), 131–140. doi:10.1177/0047287504268234

McGehee, N. G., & Santos, C. A. (2005). Social change, discourse, and volunteer tourism. *Annals of Tourism Research*, *32*(3), 760–779. doi:10.1016/j.annals.2004.12.002

McKague, K., & Harrison, S. (2019). Gender and health social enterprises in Africa: A research agenda. *International Journal for Equity in Health*, *18*(1), 1–10. doi:10.118612939-019-0994-2 PMID:31221156

McNamara, K. E., & Prideaux, B. (2010). *A typology of solo independent women travellers. International Journal of Tourism*. doi:10.1002/jtr.751

Meera, S. (2014, March). Challenges of Women Entrepreneurs: With Specific Reference to Tourism Industry, TMIMT. *International Journal (Toronto, Ont.)*, *1*(1), •••. www.tmimtjournal.org

Mehrabian, A., & Russell, J. A. (1974). *An approach to environmental psychology*. The MIT Press.

Memon, A.B., Meyer, K., & Tunio, M. N. (2021). Toward collaborative networking among innovation laboratories: a conceptual framework. *International Journal of Innovation Science*. doi:10.1108/IJIS-04-2021-0069

Memon, A. B., & Tunio, M. N. (2023). Assessing the Potential of ICT Education at Secondary Schools in Pakistan: A Comparative Study. *Pakistan Journal of Educational Research and Evaluation*, *10*(2), 16–33.

Memon, A. R., Shaikh, E., & Khan, M. S. (2019). Determination of Customer Satisfaction of Hyderabad Restaurants. *Irish Interdisciplinary Journal of Science and Research*, *3*(3), 11–18.

Mendoza Ontiveros, M. M., & Chapulín Carrillo, J. D. V. (2015). Turismo, trabajo femenino y empoderamiento de las mujeres en bahías de Huatulco, Oaxaca-México. *Estudios y Perspectivas en Turismo*, *24*(2), 316–335.

Mentzer, J. T., & Kahn, K. B. (1995). A framework of logistics research. *Journal of Business Logistics*, *16*(4), 231–250.

Meredith, J. (1993). Theory building through conceptual methods. *International Journal of Operations & Production Management*, *13*(5), 3–11. doi:10.1108/01443579310028120

Messiaen, J. (2017). *Influencer Marketing How The Popularity Threshold Of Istara Influencers Impacts Consumer Behavıour: The Moderating Role Of Purchase Involvement* [Master's Thesis]. Universiteit Gent.

Mihalic, T. (2016). Sustainable-responsible tourism discourse–Towards 'responsustable' tourism. *Journal of Cleaner Production, 111*, 461–470. doi:10.1016/j.jclepro.2014.12.062

Milgrath, L. (1989). An inquiry in to valves for a sustainable society: a personal Statement. In L. Milgrath (Ed.), *Envesconing a sustainable society*. SUNY Press.

Millan, M. G., Amador, L., & Arjona, J. M. (2015). El oleoturismo: una alternativa para preservar los paisajes del olivar y promover el desarrollo rural y regional de Andalucía (Espana). *Rev. Geogr. Norte Gd., 60*, 195–214.

Minniti, M., & Arenius, P. (2003). *Women in Entrepreneurship*. Presented at the Symposium of the Entrepreneurial Advantage of Nations, First Annual Global Entrepreneurship Symposium, New York, NY.

Mitchell, M. A., & Mitchell, S. J. (2001). Consumer Experience Tourism in the Nonprofit and Public Sectors. *Journal of Nonprofit & Public Sector Marketing, 9*(3), 21–34. doi:10.1300/J054v09n03_02

Mitra, S. K., Chattopadhyay, M., & Chatterjee, T. K. (2022). Can Tourism Development Reduce Gender Inequality? *Journal of Travel Research*.

Mitten, D. (1992). Empowering Girls and Women in the Outdoors. *Journal of Physical Education, Recreation & Dance, 63*(2), 56–60. doi:10.1080/07303084.1992.10604117

Mizrahi, R., & Aracı, H. (2010). A Resaerch On Women Managers And Glass Ceiling Syndrome. *Organizasyon ve Yönetim Bilimleri Dergisi, 2*(1), 149–156.

Modernspeak. (2022). *How travel influencer marketing can impact branding*. https://www.modernspeak.co/stories/how-travel-influencer-marketing-can-positively-impact-your-branding

Mohamad, N. H., & Hamzah, A. (2013). Tourism cooperative for scaling up community-based tourism. *Worldwide Hospitality and Tourism Themes, 5*(4), 315–328. doi:10.1108/WHATT-03-2013-0017

Mohanty, P., Kumar Swain, S., & Besra, S. (2018). Women at work: Exploring the issues and challenges of women employees in travel and tourism. *Indian Journal of Economics and Development, 6*(1), 1–5.

Mohsin, A., Ramli, N., & Alkhulayfi, B. A. (2016). Halal tourism: Emerging opportunities. *Tourism Management Perspectives, 19*, 137–143. doi:10.1016/j.tmp.2015.12.010

Mooney, S., & Ryan, I. (2009). A woman's place in hotel management: Upstairs or downstairs? *Gender in Management, 24*(3), 195–210. doi:10.1108/17542410910950877

Moore, P. (2000). Transitions-Milestones On The Route To Success, USASBE Proceeding. *İstanbul Kadın Araştırmaları Dergisi, 18*, 1-5.

More, J. S., & Lingam, C. (2017). *A SI model for social media influencer maximization*. Applied Computing and Informatics. doi:10.1016/j.aci.2017.11.001

Morrison, A., Breen, J., & Ali, S. (2003). Small business growth: Intention, ability, and opportunity. *Journal of Small Business Management, 41*(4), 417–425. doi:10.1111/1540-627X.00092

Mostafanezhad, M. (2013). The politics of aesthetics in volunteer tourism. *Annals of Tourism Research, 43*, 150–169. doi:10.1016/j.annals.2013.05.002

Moswete, N., & Lacey, G. (2015). "Women cannot lead": Empowering women through cultural tourism in Botswana. *Journal of Sustainable Tourism, 23*(4), 600–617. doi:10.1080/09669582.2014.986488

Mota, A., Braga, V., & Ratten, V. (2019). Entrepreneurship motivation: Opportunity and necessity. *Contributions to Management Science*, 139–165. doi:10.1007/978-3-030-12342-0_8

Moufakkir, O. (2015). The stigmatized tourist. *Annals of Tourism Research*, *53*, 17–30. doi:10.1016/j.annals.2015.04.004

Moufakkir, O. (2020). Experience of Arab/Muslim women visiting relatives in the West and the management of stigma by association. *Tourism Management*, *78*, 104073. doi:10.1016/j.tourman.2019.104073

Movono, A., & Dahles, H. (2017). Female empowerment and tourism: A focus on businesses in a Fijian village. *Asia Pacific Journal of Tourism Research*, *22*(6), 681–692. doi:10.1080/10941665.2017.1308397

Movono, A., & Hughes, E. (2022). Tourism partnerships: Localizing the SDG agenda in Fiji. *Journal of Sustainable Tourism*, *30*(10), 2318–2332. doi:10.1080/09669582.2020.1811291

Mowforth, M., & Munt, I. (2015). *Tourism and Sustainability: Development, Globalisation and New Tourism in the Third World*. Routledge. doi:10.4324/9781315795348

Mrema, A. (2015). Tourism and women empowerment in monduli district, arusha-Tanzania. *African Journal of Hospitality, Tourism and Leisure*, *4*(2), 1–14.

Mrema, A. A. (2015). Tourism and women empowerment in moduli district, arusha-Tanzania. *African Journal of Hospitality, Tourism and Leisure*, *4*(2), 1–14.

Muhammad, S., Ximei, K., Sharif, I., & Haq, Z. (2020). An overview of women entrepreneurship from Islamic perspective. *Review of Economics and Development Studies*, *6*(4), 857–866. doi:10.47067/reads.v6i4.285

Munoz-Bullon, F. (2009). The gap between male and female pay in the Spanish tourism industry. *Tourism Management*, *30*(5), 638–649. doi:10.1016/j.tourman.2008.11.007

Munsters, W., & Niesten, M. (2013). The cultural tourism sustainability mix applied to the development of contemplative tourism in Limburg, the Netherlands. In *Cultural tourism* (pp. 140–150). CABI. doi:10.1079/9781845939236.0140

Muravyev, A., Talavera, O., & Schäfer, D. (2009). Entrepreneurs' gender and financial constraints: Evidence from International Data. *Journal of Comparative Economics*, *37*(2), 270–286. doi:10.1016/j.jce.2008.12.001

Muresan, I. C., Oroian, C. F., Harun, R., Arion, F. H., Porutiu, A., Chiciudean, G. O., Todea, A., & Lile, R. (2016). Local residents' attitude toward sustainable rural tourism development. *Sustainability (Basel)*, *8*(1), 100. doi:10.3390u8010100

Mushtaq, T., Tunio, M. N., Akbar, Z., & Jariko, M. (2021). Green Organizational identity: Antecents and consequences: An Emprical Study. *Contemporary Issues in business and Government, 27*(3), 2056-2069.

Mushtaq, T., Tunio, M. N., Akbar, Z., & Jariko, M. (2021). Green Organizational identity: Antecents and consequences: An Emprical Study. *Contemporary Issues in Business and Government, 27*(3), 2056-2069.

Nabi, G., Liñán, F., Fayolle, A., Krueger, N., & Walmsley, A. (2017). The impact of entrepreneurship education in higher education: A systematic review and research agenda. *Academy of Management Learning & Education*, *16*(2), 277–299. doi:10.5465/amle.2015.0026

Naderi, A., Nasrolahi Vosta, L., Ebrahimi, A., & Jalilvand, M. R. (2019). The contributions of social entrepreneurship and transformational leadership to performance: Insights from rural tourism in Iran. *The International Journal of Sociology and Social Policy*, *39*(9/10), 719–737. doi:10.1108/IJSSP-06-2019-0124

NadwornyK. (2013). *Turkey Posts*. Retrieved from https://katrinkaabroad.com/turkey/

Nagar, K. (2021). Representation of women managers in hospitality and tourism: A content analysis of related magazine articles. *Gender in Management*, *36*(3), 329–348. doi:10.1108/GM-10-2019-0180

Naguib, R. (2022). Motivations and barriers to female entrepreneurship: Insights from Morocco. *Journal of African Business*, 1–28. doi:10.1080/15228916.2022.2053400

Najih, M.A. (2017). Gender dan kemajuan teknologi pemberdayaan perempuan pendidikan dan Keluarga. *HARKAT: Media Komunikasi Islam Tentang Gebder dan Anak, 12*(2), 1826.

Nakata, M., & Momsen, J. D. (2010). Gender and tourism: Gender, age and mountain tourism in Japan. *Geografia*, *6*(2), 1–15.

Nalçacı İkiz, A. (2020). Turizm işletmelerinde cinsiyet ayrımcılığı ve kadın çalışanlar: Yöneticilerinin konuya ilişkin tutum ve uygulamaları. *Van Yüzüncü Yıl Üniversitesi Sosyal Bilimler Enstitüsü Dergisi*, *50*, 373–396.

Naldi, L., Rasmussen, E., & Colombo, M. G. (2018). Social media in family business: What do we know and where do we go? *Journal of Family Business Management*, *8*(2), 126–147.

Nambisan, S., Wright, M., & Feldman, M. (2019). The Digital Transformation of Innovation and Entrepreneurship: Progress, challenges and key themes. *Research Policy*, *48*(8), 103773. doi:10.1016/j.respol.2019.03.018

Naranjo-Valencia, J. C., Jime'nez-Jime'nez, D., & Sanz-Valle, R. (2011). Innovation or imitation? The role of organizational culture. *Management Decision*, *49*(1), 55–72. doi:10.1108/00251741111094437

Narayan, D., Stern, N., Nankani, G., Page, J., & Jorgensen, S. (2002). Empowerment and poverty reduction; a source book, the world bank. *PREM.*

Narayan-Parker, D. (2002). *Empowerment and poverty reduction: A sourcebook.* World Bank Publications.

Nassani, A., Aldakhil, A., Abro, M., Islam, T., & Zaman, K. (2019). The impact of tourism and finance on women empowerment. *Journal of Policy Modeling*, *41*(2), 234–254. doi:10.1016/j.jpolmod.2018.12.001

National Bureau of Statistics of China. (2020). *China Statistical Yearbook 2020.* China Statistics Press.

Nayyar, A., Mahapatra, B., Le, D., & Suseendran, G. (2018). Virtual Reality (VR) & Augmented Reality (AR) technologies for tourism and hospitality industry. *International Journal of Engineering & Technology*, *7*(2.21), 156-160.

Nazariadli, S., Morais, D. B., Bunds, K., Baran, P., & Supak, S. (2019). Rural tourism microentrepreneurs' self-representation through photography: A counter-hegemonic approach. *Rural Society*, *28*(1), 29–51. doi:10.1080/10371656.2019.1576294

Ngoasong, M. Z., & Kimbu, A. N. (2016). Informal microfinance institutions and development-led tourism entrepreneurship. *Tourism Management*, *52*, 430–439. doi:10.1016/j.tourman.2015.07.012

Nguyen, D. T. A., & Hsu, L. (2022). Exploring visit intention to India among Southeast Asian solo female travelers. *Journal of Destination Marketing & Management*. https://www.sciencedirect.com/science/article/pii/S2212571X22000373

Nguyen, D.T.A., & Hsu, L. (n.d.). Travel experience of Southeast Asian solo female travelers. *Journal of Indonesian Tourism, Hospitality and …* https://ejournal.upi.edu/index.php/Jithor/article/view/51021

Nguyen, H. N. (2018). *An investigation into female solo travelers: Solo travel exposure and perceived constraints, personal development, authentic personality and self-esteem* (Order No. 10981186). Available from ProQuest Dissertations & Theses Global. (2269043425). Retrieved from https://www.proquest.com/dissertations-theses/investigation-into-female-solo-travelers-travel/docview/2269043425/se-2

Nguyen, H. N. (2018). *An Investigation into Female Solo Travelers: Solo Travel Exposure and Perceived Constraints, Personal Development, Authentic Personality and Self-Esteem.* https://search.proquest.com/openview/5f27aff116766f5334371c4bc4baba90/1?pq-origsite=gscholar&cbl=18750&diss=y

Ngwira, C., Tse, S., & Vongvisitsin, T. (2020). Negotiation strategies and constraints for solo female travelers in Africa. *Tourism Culture & …* https://www.ingentaconnect.com/content/cog/tcc/2020/00000020/00000001/art00004

Ngwira, C., Tse, S., & Vongvisitsin, T. B. (2020). Negotiation Strategies and Constraints for Solo Female Travelers in Africa. *Tourism, Culture & Communication, 20*(1), 35–47. doi:10.3727/109830420X15859302993081

Nikjoo, A., Markwell, K., & Nikbin, M. (2021). The flag-bearers of change in a patriarchal Muslim society: Narratives of Iranian solo female travelers on Instagram. *Tourism Management.* https://www.sciencedirect.com/science/article/pii/S2211973621000301

Nişancı, Z. N., & Ödemiş, G. (2022). Kadın Girişimciler ve İş-Yaşam Dengesi: Demografik Faktörler Açısından Bir Analiz [Women Entreprenurs and Work-life Balance: An Analysis in terms of Demographic Factors]. *Journal of Business, Innovation and Governance, 5*(1), 44–72.

Nisha, F., & Cheung, C. (2022). Locating Muslimah in the travel and tourism research. *Tourism Management Perspectives.* https://www.sciencedirect.com/science/article/pii/S2211973622000058

Nizami, M. (2023). *13 Top Travel Influencers.* https://smallbiztrends.com/2022/05/top-travel-influencers.html?utm_content=cmp-true

Nooripoor, M., Khosrowjerdi, M., Rastegari, H., Sharifi, Z., & Bijani, M. (2021). The role of tourism in rural development: Evidence from Iran. *GeoJournal, 86*(4), 1705–1719. doi:10.100710708-020-10153-z

Norrild, J. A., & Korstanje, M. E. (2021). Maids and housekeepers at luxury hotels: Life stories in hotels of Buenos Aires, Argentina. *International Journal of Tourism Anthropology, 8*(2), 129–147. doi:10.1504/IJTA.2021.116091

Norrild, J., & Korstanje, M. E. (2017). Regina Schlüter–an entire life devoted to tourism research. *Anatolia, 28*(3), 493–500. doi:10.1080/13032917.2017.1324226

Norris, J., & Wall, G. (1994). Gender and tourism. *Progress in Tourism, Recreation and Hospitality Management, 6*, 57-78.

Nugraheni, A. I. P., Priyambodo, T. K., Kusworo, H. A., & Sutikno, B. (2019). The Social Dimension Of Sustainable Development: Defining Tourism Social Sustainability. *ICESC. EAI. Proceedings of the 1st International Conference on Engineering, Science, and Commerce, ICESC 2019.* 10.4108/eai.18-10-2019.2289855

Nunkoo, R., Thelwall, M., Ladsawut, J., & Goolaup, S. (2020). Three decades of tourism scholarship: Gender, collaboration and research methods. *Tourism management.* https://www.sciencedirect.com/science/article/pii/S0261517719302547

Nurhaeni, I. D. A., Popping, R., Sugiarti, R., Nugroho, R. A., &Pratiwi, R. D. (2019). The dynamics of gender differences on tourism governance in Indonesia. *BISNIS & BIROKRASI: JurnalIlmuAdministrasi dan Organisasi, 26*(3).

Nurhandayani, A., Syarief, R.. & Najib, M. (2019). The impact of social media influencer a brand images to purchase intention. *Journal of Applied Management, 4*(17), 650-661.

Nyaupane, G. P., Morais, D. B., & Graefe, A. R. (2004). Nature Tourism Constraints: A Cross-Activity Comparison. *Annals of Tourism Research, 31*(3), 540–555. doi:10.1016/j.annals.2004.01.006

Obermayer, N., Kővári, E., Leinonen, J., Bak, G., & Valeri, M. (2022). How social media practices shape family business performance: The wine industry case study. *European Management Journal, 40*(3), 360–371. doi:10.1016/j.emj.2021.08.003

OECD. (2017). *Policy brief on women's entrepreneurship.* Publications Office of the EU.

OECD. (2017). *Report on the implementation of the OECD gender recommendations—Some progress on gender equality but much left to do*. OECD.

Oh, H., Fiore, A. M., & Jeoung, M. (2007). Measuring experience economy concepts: Tourism applications. *Journal of Travel Research, 46*(2), 119–132. doi:10.1177/0047287507304039

Ojong, N., Simba, A., & Dana, L. P. (2021). Female entrepreneurship in Africa: A review, trends, and future research directions. *Journal of Business Research, 132*, 233–248. doi:10.1016/j.jbusres.2021.04.032

Okafor, C., & Amalu, R. (2010). Entrepreneurial Motivations as Determinants of Women Entrepreneurship Challenges. *Petroleum-Gas University of Ploiesti BULLETIN Economic Sciences Series, 62*(2), 67–77.

Öktem, Ş., Kubat, G., & Kızıltan, B. (2018). Otel İşletmelerinde Kadın İşgören Davranışlarını Etkileyen Algılara İlişkin Bir Araştırma. *Anatolia: Turizm Araştırmaları Dergisi, 29*(2), 197–208.

Okumuş, F., Sarıışık, M., & Naipaul, S. (2010). Understanding why women work in five-star hotels in a developing country and their work-related problems. *International Journal of Hospitality & Tourism Administration, 11*(1), 76–105. doi:10.1080/15256480903539693

Olenik-Shemesh, D., Heiman, T., & Keshet, N. S. (2018). The role of career aspiration, self-esteem, body esteem, and gender in predicting sense of well-being among emerging adults. *The Journal of Genetic Psychology, 179*(6), 343–356. doi:10.1080/00221325.2018.1526163 PMID:30332342

Olmedo, L., van Twuijver, M., & O'Shaughnessy, M. (2019). *Community-based social enterprises fostering inclusive development in peripheral European rural areas. The United Nations Inter-Agency Task Force on Social and Solidarity Economy*. UNTFSSE.

Olsson, A. K., & Bernhard, I. (2020). Keeping up the pace of digitalization in small businesses–women entrepreneurs' knowledge and use of social media. *International Journal of Entrepreneurial Behaviour & Research, 27*(2), 378–396. doi:10.1108/IJEBR-10-2019-0615

Orams, M. B. (1995). Towards a more desirable form of ecotourism. *Tourism Management, 16*(1), 3–8. doi:10.1016/0261-5177(94)00001-Q

Orhan, G. (2009). *Ücret ayrımcılığının önlenmesinde iş değerlemenin rolü* [Unpublished Master Thesis]. Marmara University.

Orido, C. O. (2017). *Challenges faced by female chefs in the kenyan hospitality industry: a study through an african oral tradition of storytelling* [Doctoral dissertation]. Auckland University of Technology.

Osinubi, T., Adedoyin, A., Olufemi, O., & Ajide, F. (2023). Does tourism affect sustainable development in MINT countries? *Global Journal of Emerging Market Economies, 15*(1), 72–92. doi:10.1177/09749101211064388

Osman, H., Brown, L., & Phung, T. M. T. (2020). The travel motivations and experiences of female Vietnamese solo travellers. *Tourist Studies*. doi:10.1177/1468797619878307

Otegui-Carles, A., Araújo-Vila, N., & Fraiz-Brea, J. A. (2022). Solo Travel Research and Its Gender Perspective: A Critical Bibliometric Review. *Tourism and Hospitality*. https://www.mdpi.com/2673-5768/3/3/45

Otto, J. E., & Ritchie, J. R. B. (1996). The service experience in tourism. *Tourism Management, 17*(3), 165–174. doi:10.1016/0261-5177(96)00003-9

Oxford Advanced Learner's Dictionary. (2023). *Definition of Career Noun from the Oxford Advanced Learner's Dictionary*. https://www.oxfordlearnersdictionaries.com/definition/english/career_1?q= career

Özden, C. (2019). *Kırsal Alanlarda Kadın Girişimcilerin Karşılaştıkları Finansal Sorunlar: Milas (Muğla) – Bydgoszcz (Polonya) Karşılaştırması* [Financial Problems of Women Entrepreneurs in Rural Areas: Milas (Muğla)- Bydgoszcz (Poland) Comparison] [Master Thesis]. Muğla Sıtkı Koçman University.

Özen, İ. A., Karadeniz, G., & Zaro, E. (2022). Kadın Girişimci Restoranlar ve Turist Tercihlerinin Belirlenmesi: Kapadokya Örneği [Women Entrepreneur Restaurants and Determination of Tourist Preferences: The Case of Cappadocia]. *Journal of Tourism and Gastronomy Studies*, *10*(3), 2314–2326. doi:10.21325/jotags.2022.1093

Öz, H., Dönmez, B., & Akdağ, G. (2020). İş yerinde cinsel taciz ile iş doyumu ve işten ayrılma niyeti ilişkileri: Kadın mutfak çalışanları örneği. *Türk Turizm Araştırmaları Dergisi*, *4*(4), 3842–3854. doi:10.26677/TR1010.2020.594

Özkan, G. S. & Özkan, B. (2010). Kadın çalışanlara yönelik ücret ayrımcılığı ve kadın ücretlerinin belirleyicilerine yönelik bir araştırma. *Çalışma ve Toplum, 1*(24), 91-104.

Ozkazanc-Pan, B., & Susan Clark Muntean, S. C. (2018). Networking towards (in)equality: Women entrepreneurs in technology. *Gender, Work and Organization*, *25*(4), 379–400. doi:10.1111/gwao.12225

Öztürk, İ., & Tankuş, E. (2020). Covid-19'un Potansiyel Turist Profili Oluşturan Tüketici Davranışı Üzerindeki Etkilerinin Çözüm Önerileri Perspektifinde Değerlendirilmesi [Evaluation of the Effects of Covid-19 on Consumer Behavior That Creates a Potential Tourist Profile in the Perspective of Solution Proposals]. *Sivas Cumhuriyet Üniversitesi İktisadi ve İdari Bilimler Dergisi*, *21*(2), 103–121.

Padilla-Meléndez, A., & del Águila-Obra, A. R. (2013). Web and social media usage by museums: Online value creation. *International Journal of Information Management*, *33*(5), 892–898. doi:10.1016/j.ijinfomgt.2013.07.004

Palaz, S. (2003). Türkiye'de cinsiyet ayırımcılığı analizinde neo klasik yaklaşıma karşı kurumcu yaklaşım: Eşitliği sağlayıcı politika önerileri. *Balıkesir Üniversitesi Sosyal Bilimler Enstitüsü Dergisi*, *6*(9), 87–109.

Panda, S. (2018). Constraints faced by women entrepreneurs in developing countries: Review and ranking. *Gender in Management*, *33*(4), 315–331. doi:10.1108/GM-01-2017-0003

Pansardi, P., & Bindi, M. (2021). The new concepts of power? Power-over, power-to and power-with. *Journal of Political Power*, *14*(1), 51–71. doi:10.1080/2158379X.2021.1877001

Pantano, E., & Priporas, C.-V. (2016). The effect of mobile retailing on consumers' purchasing experiences: A dynamic perspective. *Computers in Human Behavior*, *61*, 548–555. doi:10.1016/j.chb.2016.03.071

Panta, S. K., & Thapa, B. (2017). Entrepreneurship and women's empowerment in gateway communities of Bardia National Park, Nepal. *Journal of Ecotourism*, 1–23. doi:10.1080/14724049.2017.1299743

Papadopoulou, G. (2021). Promoting gender equality and women empowerment in the tourism sector. In *Handbook of Research on the Role of Tourism in Achieving Sustainable Development Goals* (pp. 152–174). IGI Global. doi:10.4018/978-1-7998-5691-7.ch009

Pardo-del-Val, M. (2010). The Services supporting female entrepreneurs. *Service Industries Journal*, *30*(9), 1479–1498. doi:10.1080/02642060802626840

Passportandplates. (2021a). Retrieved from https://passportandplates.com/destination-guides/intrepid-travel-turkey-review/

Passportandplates. (2021b). Retrieved from https://passportandplates.com/turkey-travel-guide/

Passportandplates. (2023). Retrieved from https://passportandplates.com/aboutme/

Patiar, A., Ma, E., Kensbock, S., & Cox, R. (2017). Students' perceptions of quality and satisfaction with virtual field trips of hotels. *Journal of Hospitality and Tourism Management*, *31*, 134–141. doi:10.1016/j.jhtm.2016.11.003

Pazarcık, Y. (2010). *Eğitimin Bireysel Değerler ve Yöneticilerin İş Görme Anlayışına Etkisi: Türkiye'de Yabancı Dilde Eğitim Veren Orta Eğitim Kurumları Mezunları Üzerine Bir Araştırma* [The Impact of Education on Individual Values And Managers' Understanding of Working: A Research on Graduates of Secondary Educational Institutions Providing Education in a Foreign Language in Turkey] [Doctoral Thesis]. Marmara University.

Pechlaner, H., & Innerhofer, E. (2021). Sustainability and digitalization in family-owned hotels in the post-COVID era. *Journal of Tourism Futures*, *7*(1), 92–101. doi:10.1108/JTF-11-2020-0092

Peeters, L., & Ateljevic, I. (2009). Women empowerment entrepreneurship nexus in tourism: processes of social innovation. In J. Ateljevic & S. J. Page (Eds.), *Tourism and Entrepreneurship*. Routledge.

Pelit, E., Güçer, E., & Demírdağ, Ş. A. (2016). Kadın işgörenlerin karşılaştığı sorunların iş bırakma eğilimlerine etkisi: Otel işletmelerinde bir araştırma. *Mehmet Akif Ersoy Üniversitesi Sosyal Bilimler Enstitüsü Dergisi*, *8*(15), 43–65. doi:10.20875b.16565

Pencarelli, T. (2020). The digital revolution in the travel and tourism industry. *Information Technology & Tourism*, *22*(3), 455–476. doi:10.100740558-019-00160-3

Pereira, A., & Silva, C. (2018). Women solo travellers: motivations and experiences. *Millenium*. https://repositorio.ipv.pt/handle/10400.19/5257

Pereira, A., & Silva, C. (2018). Women solo travellers: Motivations and experiences. *Millenium*, *6*(6), 99–106. doi:10.29352/mill0206.09.00165

Perkov, D., Primorac, D., & Perkov, M. (2015). Position of female managers in Croatian tourism. *Economic and Social*. https://www.esd-conference.com/upload/book_of_proceedings/Book%20of%20Proceedings_Belgrade%202015.pdf#page=107

Phadnis, A. (2023, February 10). *Foreign tourists quadrupled to 6.19 mn in 2022, but 43% lower than 2019*. Business Standard News. Retrieved March 24, 2023, from https://www.business-standard.com/article/current-affairs/foreign-tourists-quadrupled-to-6-19mn-in-2022-but-43-lowerthan

Pine, B. J., & Gilmore, J. H. (1999). *The experience economy, work is theatre and every business a stage*. Harvard Business School Press.

Pojani, D., & Stead, D. (2015). Going Dutch? The export of sustainable land-use and transport planning concepts from the Netherlands. *Urban Studies (Edinburgh, Scotland)*, *52*(9), 1558–1576. doi:10.1177/0042098014562326

Pond, K. L. (1993). *The professional guide: Dynamics of tour guiding*. Van Nostrand Reinhold Company.

Ponthieux, S., & Meurs, D. (2015). Gender Inequality. In Handbook of Income Distribution (Vol. 2, pp. 981-1146). Elsevier. doi:10.1016/B978-0-444-59428-0.00013-8

Pop, R.-A., Săplăcan, Z., Dabija, D.-C., & Alt, M.-A. (2021). The impact of social media influencers on travel decisions: The role of trust in consumer decision journey. *Current Issues in Tourism*. Advance online publication. doi:10.1080/13683500.2021.1895729

Pothipala, V., Keerasuntonpong, P., & Cordery, C. (2021). Alleviating social and economic inequality? The role of social enterprises in Thailand. *Journal of Accounting & Organizational Change*, *17*(1), 50–70. doi:10.1108/JAOC-09-2020-0127

Pritchard, A. (2014). Gender and feminist perspectives in tourism research. *The Wiley Blackwell companion to tourism*, 314-324.

Pritchard, A., & Morgan, N. (2016). Encyclopedia of Tourism. Springer Cham.

Pritchard, A., Morgan, N., Ateljevic, I., & Harris, C. (Eds.). (2007). *Tourism and gender: Embodiment, sensuality and experience.* Cabi. doi:10.1079/9781845932718.0000

Punzi, M. C., & Werner, M. (2020). Challenging the menstruation taboo one sale at a time: The role of social entrepreneurs in the period revolution. The Palgrave handbook of critical menstruation studies, 833-851.

Purcell, K. (1996). The relationship between career and job opportunities: Women's employment in the hospitality industry as a microcosm of women's employment. *Women in Management Review*, *11*(5), 17–24. doi:10.1108/09649429610122618

Putri, F. A., & Adinia, N. C. (2018). The Role of Communication in Sustainable Development Tourism: A Case Study on Community-based Tourism (Pokdarwis) in Nglanggeran Village. *Jurnal Komunikasi Indonesia*, *7*(2), 5. Advance online publication. doi:10.7454/jki.v7i2.9795

Qiu, H., Fan, D. X. F., Lyu, J., Lin, P. M. C., & Jenkins, C. L. (2019). Analyzing the Economic Sustainability of Tourism Development: Evidence from Hong Kong. *Journal of Hospitality & Tourism Research (Washington, D.C.)*, *43*(2), 226–248. doi:10.1177/1096348018777046

Quan, S., & Wang, N. (2004). Towards a structural model of the tourist experience: An illustration from food experiences in tourism. *Tourism Management*, *25*(3), 297–305. doi:10.1016/S0261-5177(03)00130-4

Rabotić, B. (2010, March). Tourist Guides in Contemporary Tourism. In *International Conference on Tourism and Environment* (pp. 353-364). Academic Press.

Radović-Marković, M. (2015). Female entrepreneurship in transition economies: An overview. *Female Entrepreneurship in Transition Economies*, 9–30. doi:10.1057/9781137444516_2

Radović-Marković, M., & Živanović, B. (2019). Fostering Green Entrepreneurship and Women's Empowerment through Education and Banks' Investments in Tourism: Evidence from Serbia. *Sustainability (Basel)*, *11*(23), 6826. Advance online publication. doi:10.3390u11236826

Ragsdale, K., & Difranceisco, W. (2006). *Where the boys are: sexual expectations and behaviour among young women on holiday. In Culture, health.* Taylor &Francis. doi:10.1080/13691050600569570

Rahman, M., Moghavvemi, S., Thirumoorthi, T., & Rahman, M. K. (2020). The impact of tourists' perceptions on halal tourism destination: A structural model analysis. *Tourism Review*, *75*(3), 575–594. doi:10.1108/TR-05-2019-0182

Ramadani, V., Rahman, M. M., Salamzadeh, A., Rahaman, M. S., & Abazi-Alili, H. (2022). Entrepreneurship education and Graduates' entrepreneurial intentions: Does gender matter? A multi-group analysis using Amos. *Technological Forecasting and Social Change*, *180*, 121693. doi:10.1016/j.techfore.2022.121693

Raman, R., Subramaniam, N., Nair, V. K., Shivdas, A., Achuthan, K., & Nedungadi, P. (2022). Women Entrepreneurship and Sustainable Development: Bibliometric analysis and emerging research trends. *Sustainability (Basel)*, *14*(15), 9160. doi:10.3390u14159160

Ramchurjee, N. (2011). *"Tourism" a vehicle for women's empowerment: Prospect and challenges.* https://www.researchgate.net/publication/256505214_Tourism_a_Vehicle_for_Women's_Empowerment_Prospect_and_Challenges

Ramchurjee, N., & Paktin, W. (2011). *Tourism a vehicle for women's empowerment: prospect and challenges.* University of Mysore.

Ramos, A. M., & Prideaux, B. (2014). Indigenous ecotourism in the Mayan rainforest of Palenque: Empowerment issues in sustainable development. *Journal of Sustainable Tourism*, *22*(3), 461–479. doi:10.1080/09669582.2013.828730

Ram, Y. (2018). Hostility or hospitality? A review on violence, bullying and sexual harassment in the tourism and hospitality industry. *Current Issues in Tourism*, *21*(7), 760–774. doi:10.1080/13683500.2015.1064364

Ranabahu, N. (2020). 'Wicked'solutions for 'wicked'problems: Responsible innovations in social enterprises for sustainable development. *Journal of Management & Organization*, *26*(6), 995–1013. doi:10.1017/jmo.2020.20

Rappaport, J., & Zimmerman, M. (1988). Citizen participation, perceived control, and psychological empowerment. *American Journal of Community Psychology*, *16*(5), 725–750. doi:10.1007/BF00930023 PMID:3218639

Rasanjali, C., Sivashankar, P., & Mahaliyanaarachchi, R. P. (2021). Women Participation in Rural Tourism: A Case of Ella, Sri Lanka. *AGRARIS: Journal of Agribusiness and Rural Development Research*, *7*(2), 256–269. doi:10.18196/agraris.v7i2.11294

Rashid, A. T. (2010). Development through social entrepreneurship: Perspectives and evidence from Bangladesh. *Canadian Journal of Development Studies*, *30*(3-4), 441–455. doi:10.1080/02255189.2010.9669311

Rasul, T. (2019). The trends, opportunities, and challenges of halal tourism: A systematic literature review. *Tourism Recreation Research*, *44*(4), 434–450. doi:10.1080/02508281.2019.1599532

Ratten, V., & Dana, L.-P. (2017). Gendered perspective of indigenous entrepreneurship. *Small Enterprise Research*, *24*(1), 62–72. doi:10.1080/13215906.2017.1289858

Ratthinan, S. P., & Selamat, N. H. (2019). Negotiating travel constraints via technology: A study of Malay Muslim women through a hierarchical constraint model perspective. *Asian Journal of Business Research*, *9*(2), 55. doi:10.14707/ajbr.190060

Ray, C. (2021). *JourneyWoman and Safe4Life Host Special Self-Defence Class for Young Women*. Retrieved from https://journeywoman.com/feature-articles/journeywoman-and-safe4life-host-special-self-defence-class-for-young-women/

Ray, C. (2022a). *Expert Solo Travel Safety Tips for Young Women*. Retrieved from https://journeywoman.com/travel-tips/solo-travel-safety-tips-for-young-women/

Ray, C. (2022b). *12+ Ways to Empower Women When You Travel*. Retrieved from https://journeywoman.com/feature-articles/12-things-you-can-do-to-empower-women-when-you-travel/

Raza, A., Shaikh, E., Tursoy, T., & Almashaqbeh, H. A. (2022). Economics and Business Perspectives of Sustainable HRM. In *Sustainable Development of Human Resources in a Globalization Period* (pp. 36–48). IGI Global. doi:10.4018/978-1-6684-4981-3.ch003

Redpointmarketingpr.com. (2023). *Social influencer marketing services*. https://redpointmarketingpr.com/services/social-influencer-marketing/

Reguera, C., & García-Izquierdo, A. L. (2021). Women as victims of court rulings: Consequences of workplace harassment in the hospitality industry in spain (2000–2016). *Sustainability (Basel)*, *13*(14), 7530. doi:10.3390u13147530

Reimer, J. K., & Walter, P. (2013). How do you know it when you see it? Community-based ecotourism in the Cardamom Mountains of southwestern Cambodia. *Tourism Management*, *34*, 122–132. doi:10.1016/j.tourman.2012.04.002

Reisinger, Y., & Mavondo, F. (2005). Travel Anxiety and Intentions to Travel Internationally: Implications of Travel Risk Perception. *Journal of Travel Research*, *43*(3), 212–225. doi:10.1177/0047287504272017

Rejeb, A., Rejeb, K., & Treiblmaier, H. (2023, April 03). And Treiblmaier, H. (2021): How augmented reality impacts retail marketing: a state-of-the-art review from a consumer perspective. *Journal of Strategic Marketing*, *31*(3), 718–748. Advance online publication. doi:10.1080/0965254X.2021.1972439

Reliance Foundation and Observer Research Foundation. (2022). *Aspirations, Access & Agency: Women transforming lives with technology*. ORF. https://www.orfonline.org/research/women-transforming-lives-with-technology/

Ribeiro, M. A., Adam, I., Kimbu, A. N., Afenyo-Agbe, E., Adeola, O., Figueroa-Domecq, C., & de Jong, A. (2021). Women entrepreneurship orientation, networks and firm performance in the tourism industry in resource-scarce contexts. *Tourism Management*, *86*(104343), 1–15. doi:10.1016/j.tourman.2021.104343

Richardson, R. B. (2021). The role of tourism in sustainable development. In Oxford Research Encyclopedia of Environmental Science. doi:10.1093/acrefore/9780199389414.013.387

Richter, L. K. (2012). Exploring the political role of gender in tourism research. In *Global tourism* (pp. 426–439). Routledge.

Riebe, M. (2012). A place of her own: The case for university-based centers for women entrepreneurs. *Journal of Education for Business*, *87*(4), 241–246. doi:10.1080/08832323.2011.591846

Rinaldi, A., & Salerno, I. (2020). The tourism gender gap and its potential impact on the development of the emerging countries. *Quality & Quantity*, *54*(5–6), 1465–1477. doi:10.100711135-019-00881-x

Ristić, D., Vukoičić, D., & Milinčić, M. (2019). Tourism and sustainable development of rural settlements in protected areas - Example NP Kopaonik (Serbia). *Land Use Policy*, *89*, 104231. doi:10.1016/j.landusepol.2019.104231

Ritchie, B. W. (2003). *Managing Educational Tourism*. Channel View Publications. doi:10.21832/9781873150528

Ritchie, B. W., & Jiang, Y. (2021). Research on tourism entrepreneurship in the aftermath of the COVID-19 pandemic: Implications and directions. *Journal of Sustainable Tourism*, *29*(6), 705–720.

Ritchie, B. W., & Jiang, Y. (2021). Risk, crisis and disaster management in hospitality and tourism: A comparative review. *International Journal of Contemporary Hospitality Management*, *33*(10), 3465–3493. doi:10.1108/IJCHM-12-2020-1480

Roberts, C., Reynolds, J., & Dolasinski, M. J. (2022). Meta-Analysis of Tourism Sustainability Research: 2019–2021. *Sustainability (Basel)*, *14*(6), 3303. doi:10.3390u14063303

Robinson, J. (1994). *Unsuitable for ladies: An anthology of women travellers*. OUP Oxford.

Robinson, M., & Novelli, M. (2007). Niche tourism: An introduction. In M. Novelli (Ed.), *Niche tourism: Contemporary issues, trends and cases* (pp. 1–11). Elsevier.

Rodrigo, P., & Turnbull, S. (2019). Halal holidays: How is value perceived by Muslim tourists? *International Journal of Tourism Research*, *21*(5), 675–692. doi:10.1002/jtr.2290

Roehl, W., & Fesenmaier, D. (1992). Risk Perceptions and Pleasure Travel: An Exploratory Analysis. *Journal of Travel Research*, *30*(4), 17–26. doi:10.1177/004728759203000403

Rogers, D. L. (2016). *The digital transformation playbook: rethink your business for the digital age*. Columbia University Press. doi:10.7312/roge17544

Rojas, L., Quiñones, D., & Rusu, C. (2020). Understanding User Needs and Customer eXperience in Tourism Area. HCII. , LNCS 12195. doi:10.1007/978-3-030-49576-3_16

Rojewski, J. W. (2005). Occupational aspirations: constructs, meaning, and application. In S. D. Brown & R. W. Lent (Eds.), *Career development and counseling: Putting theory and research to work*. John Wiley & Sons.

Roomi, M. A., & Harrison, P. (2010). Behind the veil: Women-only entrepreneurship training in Pakistan. *International Journal of Gender and Entrepreneurship*, *2*(2), 150–172. doi:10.1108/17566261011051017

Roomi, M. A., & Parrott, G. (2008). Barriers to development and progression of women entrepreneurs in Pakistan. *The Journal of Entrepreneurship*, *17*(1), 59–7. doi:10.1177/097135570701700105

Rosado, C. (2006). What Do We mean By "Managing Diversity". *Workforce Diversity*, *3*(1), 1–15.

Rosca, E., Agarwal, N., & Brem, A. (2020). Women entrepreneurs as agents of change: A comparative analysis of social entrepreneurship processes in emerging markets. *Technological Forecasting and Social Change*, *157*, 120067. doi:10.1016/j.techfore.2020.120067

Rosenfeld, K. (2022). *Travel Advisors Should Be Working With Influencers — Here's How to Start*. https://www.travelagewest.com/Industry-Insight/Business-Features/working-with-influencers

Routledge.Strzelecka, M., & Wicks, B. E. (2015). Community participation and empowerment in rural post-communist societies: Lessons from the leader approach in Pomerania, Poland. *Tourism Planning & Development*, *12*(4), 381–397. doi:10.1080/21568316.2015.1013564

Roy, S. C., & Roy, M. (2015). Tourism in Bangladesh: Present status and future prospects. *International Journal of Management Science and Business Administration*, *1*(8), 53–61. doi:10.18775/ijmsba.1849-5664-5419.2014.18.1006

Russell, R., & Faulkner, B. (1999). Movers and shakers: Chaos makers in tourism development. *Tourism Management*, *20*(4), 411–423. doi:10.1016/S0261-5177(99)00014-X

Russell, R., & Faulkner, B. (2004). Entrepreneurship, chaos and the tourism area lifecycle. *Annals of Tourism Research*, *31*(3), 556–579. doi:10.1016/j.annals.2004.01.008

Ryan, M. K., Haslam, S. A., Hersby, M. D., Kulich, C., & Atkins, C. (2007). Opting out or pushed off the edge? The glass cliff and the precariousness of women's leadership positions. *Social and Personality Psychology Compass*, *1*(1), 266–279. doi:10.1111/j.1751-9004.2007.00007.x

Rybczynski, W., & Glancy, M. (1992). Waiting for the Weekend. *Journal of Leisure Research*, *24*(3), 295–296. doi:10.1080/00222216.1992.11969895

Rymer, T. M. (1992). Growth of U.S. Ecotourism and Its Future in the 1990s. *FIU Hospitality Review*, *10*(1), 1–10.

Sabina, J. M., & Nicolae, J. C. (2013). Gender trends in tourism destination. *Procedia: Social and Behavioral Sciences*, *92*, 437–444. doi:10.1016/j.sbspro.2013.08.698

Sabuncuoğlu, A., & Gülay, G. (2014). Sosyal Medyadaki Yeni Kanaat Önderlerinin Birer Reklam Aracı Olarak Kullanımı: Twitter Fenomenleri Üzerine Bir Araştırma. *İletişim Kuram ve Araştırma Dergisi,* (38).

Saçak Gündüz, S. (2021). COVID-19 Salgını Döneminde Türkiye'de Kadın Girişimcilik Yazını [Women's Entrepreneurship Literature in Turkey During The COVID-19 Pandemic]. *Journal of Research in Entrepreneurship Innovation and Marketing*, *5*(10), 148–165.

Sadi, M. A., & Al-Ghazali, B. M. (2010). Doing business with impudence: A focus on women entrepreneurship in Saudi Arabia. *African Journal of Business Management*, *4*, 1–11.

Sagır, Z. (2018). *Cultural adaptation, mental health, and religion among Syrian women refugees* [Suriyeli kadın mültecilerde kültürel uyum, ruh sağlığı ve din] [Unpublished PhD dissertation]. Istanbul University, Istanbul, Turkey.

Şahankaya Adar, A., Dedeoğlu, S., & Kurtuluş, G. (2023). Türkiye'de Kadın Kooperatifleri. *Mevcut Durum Analizi, Kadın Güçlenmesi ve İstihdam Yaratma Potansiyeli, Çalışma ve Toplum*, *2*(77), 1171–1208.

Sahay, S. (1998a). *Women and empowerment: Approaches and strategies*. Discovery Publishing House.

Şahin, E. (2006). *Kadın Girişimcilik ve Konya İlinde Kadın Girişimcilik Profili Üzerine Bir Uygulama* [An Application on Women Entrepreneurship and the Profile of Women Entrepreneurship in Konya Province] [Master Thesis]. Selcuk University.

Sahin, G. G., & Sengün, G. (2015). The effects of social media on tourism marketing: A study among university students. *Management and Administrative Sciences Review*, *4*(5), 772–786.

Said, I., & Enslin, C. (2020). Lived experiences of females with entrepreneurship in Sudan: Networking, social expectations, and Family Support. *SAGE Open*, *10*(4). doi:10.1177/2158244020963131

Sajjad, M., Kaleem, N., Chani, M. I., & Ahmed, M. (2020). Worldwide role of Women Entrepreneurs in Economic Development. *Asia Pacific Journal of Innovation and Entrepreneurship*, *14*(2), 151–160. doi:10.1108/APJIE-06-2019-0041

Samad, S., & Alharthi, A. (2022). Untangling Factors Influencing Women Entrepreneurs' Involvement in Tourism and Its Impact on Sustainable Tourism Development. *Administrative Sciences*, *12*(2), 52. doi:10.3390/admsci12020052

Samarakoon, S., & Parinduri, R. A. (2015). Does Education Empower Women? Evidence from Indonesia. *World Development*, *66*, 428–442. doi:10.1016/j.worlddev.2014.09.002

Santos, L. D., & Varejão, J. (2007). Employment, Pay and Discrimination in the Tourism Industry. *Tourism Economics*, *13*(2), 225–240. doi:10.5367/000000007780823186

Sarfaraz, L., Faghih, N., & Majd, A. (2014). The relationship between Women Entrepreneurship and Gender Equality. *Journal of Global Entrepreneurship Research*, *2*(1), 6. doi:10.1186/2251-7316-2-6

Sarwar, S. (2022). Impact of energy intensity, green economy and blue economy to achieve sustainable economic growth in GCC countries: Does Saudi Vision 2030 matters to GCC countries. *Renewable Energy*, *191*, 30–46. doi:10.1016/j.renene.2022.03.122

Saunders, M., Lewis, P., & Thornhill, A. (2007). Research methods. Business Students, 6(3), 1-268

Saunders, M., Lewis, P., & Thornhill, A. (2007). Research methods. In Business Students (4th ed.). Pearson Education Limited.

Saunders, M. N. K., Lewis, P., Thornhill, A., & Bristow, A. (2015). *Understanding research philosophies and approaches: Research methods for business students*. Pearson.

Sayın, E. (2011). Descriptive Analysis of The Problems of Women Entrepreneurs. *Journal of Organization and Management Sciences*, *3*(1), 1309–8039.

Sayira, T., Andrews, H., & Ayesha, Q. U. A. (2021). Are Women Empowered Producers of Tourism in Pakistan? In *Women in Tourism in Asian Muslim Countries* (pp. 81–96). Springer. doi:10.1007/978-981-33-4757-1_6

Schaltegger, S., & Wagner, M. (2011). Sustainable Entrepreneurship and sustainability innovation: Categories and interactions. *Business Strategy and the Environment*, *20*(4), 222–237. doi:10.1002/bse.682

Scheepers, D., & Ellemers, N. (2019). Social identity theory. *Social psychology in action: Evidence-based interventions from theory to practice*, 129-143.

Schegg, R., & Stangl, B. (2021). The role of technology in the recovery phase of family-owned hotels. *Current Issues in Tourism*, *24*(11), 1419–1425. doi:10.1080/13683500.2020.1851388

Schein, E. H. (1990). *Career anchors: Discovering your real values* (revised edition). Pfeiffer & Company.

Schein, E. H. (1996). Career anchors revisited: Implications for career development in the 21st century. *The Academy of Management Perspectives*, *10*(4), 80–88. doi:10.5465/ame.1996.3145321

Scheyvens, R., & Hughes, E. (2019). Can tourism help to "end poverty in all its forms everywhere"? The challenge of tourism addressing SDG1. *Journal of Sustainable Tourism*.

Scheyvens, R. (1999). Ecotourism and the empowerment of local communities. *Tourism Management*, *20*(2), 245–249. doi:10.1016/S0261-5177(98)00069-7

Scheyvens, R. (2000). Promoting women's empowerment through involvement in ecotourism: Experiences from the Third World. *Journal of Sustainable Tourism*, *8*(3), 232–249. doi:10.1080/09669580008667360

Scheyvens, R., & Hughes, E. (2019). Can tourism help to "end poverty in all its forms everywhere"? The challenge of tourism addressing SDG1. *Journal of Sustainable Tourism*, *27*(7), 1061–1079. doi:10.1080/09669582.2018.1551404

Scheyvens, R., & van der Watt, H. (2021). Tourism, empowerment and sustainable development: A new framework for analysis. *Sustainability (Basel)*, *13*(22), 12606. doi:10.3390u132212606

Schintler, L., Root, A., & Button, K. (2000). Women's Travel Patterns and the Environment an Agenda for Research. *Transportation Research Record: Journal of the Transportation Research Board*, *1726*(1), 33–40. doi:10.3141/1726-05

Schlüter, R. G. (2008). Mujer y turismo. Vestimenta e interacción social en los centros turísticos de Argentina durante la década de 1930 [Women and tourism: dressing and social interaction at the tourist destinations in Argentina during the 30s decade]. *Cuadernos de Turismo*, (21), 181–199.

Schmallegger, D., & Carson, D. (2008). Blogs in tourism: Changing approaches to information exchange. *Journal of Vacation Marketing*, *14*(2), 99–110. doi:10.1177/1356766707087519

Schmitt, B. (1999). Experiential marketing. *Journal of Marketing Management*, *15*(1), 53–67. doi:10.1362/026725799784870496

Schnittker, J., & Ettl, K. (2021). Career Paths of Women Entrepreneurs in STEM–Challenges and Opportunities. In E. Laveren, C. Ben-Hafaïedh, A. Kurczewska, & Y. D. Jiang (Eds.), *Contextual Heterogeneity in Entrepreneurship Research*. Edward Elgar Publishing. doi:10.4337/9781802200201.00012

Schumpeter, J. A., & Redvers, O. P. I. E. (1934). *Theorie Der Wirtschaftlichen Entwicklung. The theory of economic development. An inquiry into profits, capital, credit, interest, and the business cycle... translated... by redvers opie.* Academic Press.

Schwab, K. (2019). *"Seriously, shit has to change." Spatial Constraint Negotiation among Solo Female Travelers.* scholarworks.umass.edu, https://scholarworks.umass.edu/ttra/2019/research_papers/96/

Schwab, K. (2019). *Because I Can, So I'm Going To: Solo Female Travelers and the Absence of Constraints.* scholarworks.umass.edu, https://scholarworks.umass.edu/ttra/2019/qual_research_papers/6/

Schwab, K. (2017). *The Fourth Industrial Revolution*. Crown Business.

Scott, L., Dolan, C., Johnstone–Louis, M., Sugden, K., & Wu, M. (2012). Enterprise and inequality: A study of avon in South Africa. *Entrepreneurship Theory and Practice*, *36*(3), 543–568. doi:10.1111/j.1540-6520.2012.00507.x

Sebora, T. C., & Theerapatvong, T. (2010). Corporate entrepreneurship: A test of external and internal influences on managers' idea generation, risk taking, and proactiveness. *The International Entrepreneurship and Management Journal*, *6*(3), 331–350. doi:10.100711365-009-0108-5

Sebova, L., Pompurova, K., & Marcekova, R. (2021). *Solo female travelers as a new trend in tourism destinations. In Culture and Tourism in a.* Springer. doi:10.1007/978-3-030-72469-6_20

Seguino, S. (2000). Gender inequality and economic growth: A cross-country analysis. *World Development*, *28*(7), 1211–1230. doi:10.1016/S0305-750X(00)00018-8

Sekaran, U., & Bougie, R. (2016). *Research methods for business: A skill building approach.* John Wiley & sons.

Selanik Ay, T., & Acar, Ş. (2016). Sınıf öğretmenlerinin girişimcilik becerisi kazandırmaya yönelik görüşleri. *Elektronik Sosyal Bilimler Dergisi*, *15*(58), 960–976. doi:10.17755/esosder.03138

Şengür, F. K., Hamutoğlu, N. B., & Lapçın, H. T. (2022). Aviator Girls Step into Entrepreneurship. *Aviation Turkey.* https://www.aviationturkey.com/en/content/aviator-girls-step-into-entrepreneurship-688

Şengür, F. K., Hamutoğlu, N. B., & Lapçın, H. T. (2022). Havacı Kızlar Girişimciliğe Adım Atıyor. *UTED*, 50. www.uteddergi.com

Şengür, F. K., Lapçın, H. T., & Hamutoğlu, N. B. (2022a). Determination of entrepreneurship tendencies of students after TUBİTAK 4004 "Aviator Girls Step into Entrepreneurship" Project. 8th International EMI Entrepreneurship & Social Sciences Congress, Aksaray, Turkey.

Şengür, F. K., Lapçın, H. T., & Hamutoğlu, N. B. (2022c). Opinions of the young aviators on entrepreneurship: TUBİTAK 4004 "Aviator Girls Step into Entrepreneurship" project. *Ege 7th International Conference on Social Sciences.*

Şengür, F. K., Lapçın, H. T., & Hamutoğlu, N. B. (2022b). Improving entrepreneurship characteristics of young female aviators: The TUBITAK 4004 Project. *3rd International Congress of Women in a Global World (WGW2022).*

Şen, Ö. (2020). Bibliyometrik Analiz Yöntemi ile Ağızdan Ağıza İletişim (Wom) Konusunun İncelenmesi. *Journal of International Social Sciences Studies*, *6*(54), 1–10. doi:10.26449ssj.1919

Seow, D., & Brown, L. (2018). *The solo female Asian tourist. In Current Issues in Tourism.* Taylor &Francis. doi:10.1080/13683500.2017.1423283

Serafimova, M., & Petrevska, B. (2018). Female entrepreneurship in tourism: A strategic management perspective in Macedonia. *Journal of Applied Economics and Business*, *6*(1), 21–32.

Serinikli, N. (2019). Kırsal Alandaki Mikro Kadın Girişimcilerin Kooperatifleşmeye Karşı Tutumları: Agro-Turizm Kadın Kooperatifleri [The Attitudes of The Micro Female Entrepreneurs in The Rural Area Towards Becoming A Cooperative: Agro-Toruism Woman Cooperative]. *Girişimcilik ve Kalkınma Dergisi*, *14*(1), 45–57.

Serinikli, N. (2019). Kırsal Alandaki Mikro Kadın Girişimcilerin Kooperatifleşmeye Karşı Tutumları: Agro-Turizm Kadın Kooperatifleri. *Girişimcilik ve Kalkınma Dergisi*, *14*(1), 45–57.

Seymen Aytemiz, O. (2005). Örgütlerde Kültürel Çeşitlilik Olgusu, Boyutları ve Etkin Yönetimi Konusunda Farklı Yaklaşımlar: Yazınsal Bir Derleme. *Yönetim Dergisi: İstanbul Üniversitesi İşletme Fakültesi İşletme İktisadı Enstitüsü*, *16*(50), 3–24.

Shaari, R., Kamarudin, D., Ju, S. O. O. N., & Zakaria, H. (2022). Effects of Leadership Types on Job Satisfaction Among Malaysian Higher Education Institutions. *Asya Öğretim Dergisi*, *10*(1), 54–70. doi:10.47215/aji.1020324

Shah, S. H. A., Jamshed, K. A., Saleem, S., Bashir, M., & Kiani, O. I. (2022). Sustainable Development in the Hospitality Industry of Europe: Role of Green HRM Practices. In Sustainable Development of Human Resources in a Globalization Period (pp. 49-67). IGI Global.

Shah, F. U. H., Fatima, U., Ahmad, M., Ahad, A., Saeed, R., Riaz, M. U., & Ghafoor, F. (2021). Factors affecting women entrepreneurship and its impact on SMEs development. *International Journal of Innovation. Creativity and Change*, *15*(7), 228–252.

Shahin, M., Ilic, O., Gonsalvez, C., & Whittle, J. (2021). The impact of a STEM-based entrepreneurship program on the entrepreneurial intention of secondary school female students. *The International Entrepreneurship and Management Journal*, *17*(4), 1867–1898. doi:10.100711365-020-00713-7

Shah, S. H. A., Cheema, S., Al-Ghazali, B. M., Ali, M., & Rafiq, N. (2021). Perceived corporate social responsibility and pro-environmental behaviors: The role of organizational identification and coworker pro-environmental advocacy. *Corporate Social Responsibility and Environmental Management*, *28*(1), 366–377. doi:10.1002/csr.2054

Shahvali, M., Shahvali, R., & Kerstetter, D. (2016). Women's Travel Constraints in a Unique Context. In M. Kozak & N. Kozak (Eds.), *Tourist Behaviour: An international perspective* (pp. 16–25). CABI. doi:10.1079/9781780648125.0016

Shaikh, E., & Tunio, M. N. (2020). Customer satisfaction and Customer loyalty: An empirical case study on the impact of benefits generated through Smartphone applications. *International Journal of Public Sector Performance Management*.

Shaikh, E., Tunio, M. N., Khoso, W. M., Brahmi, M., & Rasool, S. (2022). The COVID-19 Pandemic Overlaps Entrepreneurial Activities and Triggered New Challenges: A Review Study. *Managing Human Resources in SMEs and Start-ups: International Challenges and Solutions*, 155-182.

Shaikh, E., Azhar, H., Brahmi, M., & Zehra, N. (2022). The impact of monetary and non-monetary motivation on employees' performance: A case study of Hyderabad Electric Supply Company. *International Journal of Technology Transfer and Commercialisation*, *19*(1), 127–141. doi:10.1504/IJTTC.2022.123088

Shaikh, E., Brahmi, M., Thang, P. C., Watto, W. A., Trang, T. T. N., & Loan, N. T. (2022). Should I Stay or Should I Go? Explaining the Turnover Intentions with Corporate Social Responsibility (CSR), Organizational Identification and Organizational Commitment. *Sustainability (Basel)*, *14*(10), 6030. doi:10.3390u14106030

Shaikh, E., Khoso, I., & Chandio, F. (2019). Effects of Corporate Social Responsibility on Organizational Performance: A Conceptual and Literature Review. *Journal of Grassroot*, *53*(1).

Shaikh, E., Mishra, V., Ahmed, F., Krishnan, D., & Dagar, V. (2021). Exchange rate, stock price and trade volume in US-China trade war during COVID-19: An empirical study. *Estudios de Economía Aplicada*, *39*(8). Advance online publication. doi:10.25115/eea.v39i8.5327

Shaikh, E., Tunio, M. N., & Qureshi, F. (2021). Finance and women's entrepreneurship in DETEs: A literature review. Entrepreneurial Finance. *Innovation and Development*, 191–209.

Shaikh, E., Watto, W. A., & Tunio, M. N. (2022). Impact of Authentic Leadership on Organizational Citizenship Behavior by Using The Mediating Effect of Psychological Ownership. *ETIKONOMI*, *21*(1), 89–102. doi:10.15408/etk.v21i1.18968

Shaikh, S., Sultan, M. F., Mushtaque, T., & Tunio, M. N. (2021). Impact of COVID-19 on GDP: A serial mediation effect on international tourism and hospitality. *International Journal of Management*, *12*(84), 422–430.

Shakeela, A., Ruhanen, L., & Breakey, N. (2010). Chapter 5 Women's participation in tourism. In Tourism in the Muslim World (Bridging Tourism Theory and Practice, Vol. 2). Emerald Group Publishing Limited. doi:10.1108/S2042-1443(2010)0000002008

Shams, S. (2017). Transnational education and total quality management: A stakeholder-centred model. *Journal of Management Development*, *36*(3), 376–389. doi:10.1108/JMD-10-2015-0147

Shane, S. A. (1992). Why do some societies invent more than others? *Journal of Business Venturing*, *7*(1), 29–46. doi:10.1016/0883-9026(92)90033-N

Sharafizad, J. (2018). Informal learning of Women Small Business Owners. *Education + Training*, *60*(1), 82–103. doi:10.1108/ET-01-2017-0006

Sharma, N., Yadava, A. K., Aarif, M., Anandaram, H., Alalmai, A., & Singh, C. (2022). Business Opportunities And Challenges For Women In The Travel And Tourism Industry During Pandemics Covid-19. *Journal of Positive School Psychology*, 897–903.

Sharma, P. (2013). Women Entrepreneurship Development in India. *Global Journal of Management and Business Studies*, *3*(4), 371–376.

Sharma, P., & Varma, S. K. (2008). Empowerment of women through entrepreneurial activities of self help groups. *Indian Research Journal of Extension Education*, *8*(1), 46–51.

Sharma, R. L., & Arora, M. (2015). Communication as the mainstay of entrepreneurial leadership: A conceptual framework. *Management Dynamics*, *15*(2), 20–28. doi:10.57198/2583-4932.1081

Shaw, S. M. (1994). Gender, Leisure and Constraint: Toward a Framework for Analysis of Women's Leisure. *Journal of Leisure Research*, *26*(1), 8–22. doi:10.1080/00222216.1994.11969941

Sheffield, C. J. (1995). Sexual terrorism. In J. Freeman (Ed.), *Women: A Feminist Perspective* (5th ed., pp. 1–21). Mayfield.

Sheikhi, A. R., & Imirizaldu, I. A. (2021). Women's Footprint in Traditional Muslim Ethnic Communities: The Case of World Pottery Village in Iran. In *Women in Tourism in Asian Muslim Countries* (pp. 63–80). Springer. doi:10.1007/978-981-33-4757-1_5

Shein, A. (2015). *Communication language in Social Communication and its effect on the Touristic Attration-Facebook as a Model* [Unpublished Master Thesis]. Mohamed Khaidar University, Bascara, Algeria.

Sheng, C.-W., Shen, M.-J., & Chen, M.-C. (2008). An exploratory study of types of special interest tour preferences and preference demographic variables analysis. *International Journal of Culture, Tourism and Hospitality Research*, *2*(3), 271–284. doi:10.1108/17506180810891627

Shen, K., Yang, J., & Geng, C. (2022). How residents' attitudes to tourists and tourism affect their pro-tourism behaviours: The moderating role of Chinese traditionality. *Frontiers in Psychology*, *12*, 792324. doi:10.3389/fpsyg.2021.792324 PMID:35126243

Sheth, J. N., Newman, B. I., & Gross, B. L. (1991). Why we buy what we buy: A theory of consumption values. *Journal of Business Research*, *22*(2), 159–170. doi:10.1016/0148-2963(91)90050-8

Shin, H., Nicolau, J. L., Kang, J., Sharma, A., & Lee, H. (2022). Travel decision determinants during and after COVID-19: The role of tourist trust, travel constraints, and attitudinal factors. *Tourism Management*, *88*, 1–14. doi:10.1016/j.tourman.2021.104428 PMID:34539019

Shinnar, R. S., Giacomin, O., & Janssen, F. (2012). Entrepreneurial perceptions and intentions: The role of gender and culture. *Entrepreneurship Theory and Practice*, *36*(3), 465–493. doi:10.1111/j.1540-6520.2012.00509.x

Shivakoti, A. (2022). Women in Tourism and Hospitality-A Case Study of Bhaktapur. *Journal of Tourism & Adventure*, *5*(1), 79–96. doi:10.3126/jota.v5i1.48739

Siamak Seyfi, I. D. (2023). *The gendered effects of statecraft on women in tourism: economic sanctions, women's disempowerment and sustainability? In Gender and Tourism Sustainability*. Routledge.

Siddiqui, A.B. (2012). Problems Encountered by Women Entrepreneurs in India. *International Journal of Applied Research & Studies*, *1*(2), 1-11.

Sigala, M., & Gretzel, U. (2012). *Social media in travel, tourism and hospitality: Theory, practice and cases.* Ashgate Publishing, Ltd.

Sigala, M., Gretzel, U., & Christou, E. (2021). Technology use in the hospitality industry: Impacts on service quality and the guest experience in a post-pandemic era. *International Journal of Hospitality Management*, *94*, 102847. doi:10.1016/j.ijhm.2021.102847

Sığrı, U. (2018). *Qualitative Research Methods.* Beta Publishing.

Simão, P., & Breda, Z. (2014). Gender and leadership in the tourism sector: Th e case of the restaurants industry in the city of Aveiro, Portugal. *Revista Turismo & Desenvolvimento (Aveiro)*, *1*(21/22), 341–352.

Simionescu, L. N., Ghergina, S. C., Sheikha, H. T., & Sheikha, Z. (2021). Does board gender diversity affect firm performance? Empirical evidence from Standard & Poor's 500 Information Technology Sector. *Financial Innovation*, *7*(52), 52. doi:10.118640854-021-00265-x

Şimşek, Z., Yentur Doni, N., Gül Hilali, N., & Yildirimkaya, G. (2018). A community-based survey on Syrian refugee women's health and its predictors in Şanliurfa, Turkey. *Women & Health*, *58*(6), 617–631. doi:10.1080/03630242.2017.1321609 PMID:28430082

Sinclair, A. (2000). Women within Diversity: Risks and Possibilities. *Women in Management Review*, *15*(5/6), 237–246. doi:10.1108/09649420010372850

Singh, J. (2004). The economy of Jammu and Kashmir. Radha Krishna Anand and Co.

Singhania, O., Swain, S., & George, B. (2021). Women empowerment in a traditional masculine industry: Tourism in context. *Economics. Management and Sustainability*, *6*(2), 60–71. doi:10.14254/jems.2021.6-2.5

Sinha, C. C. (2001). *Tourism Geography: Issues. Challenges and the changing Nature of contemporary tourism.* Paper presented during the geography curriculum inservice conference.

Şit, M. (2016). Employment Contribution of Tourism Sector in Turkey. *Journal of Academic Approaches*, *7*(1), 101–117.

Skwirczy ́nska, E., Piotrowiak, M., Ostrowski, M., Wróblewski, O., Tejchman, K., Kwiatkowski, S., & Cymbaluk-Płoska, A. (2022). Welfare and Self-Assessment in Patients after Aesthetic and Reconstructive Treatments. *International Journal of Environmental Research and Public Health*, *19*(18), 11238. doi:10.3390/ijerph191811238 PMID:36141511

Slathia, P. S., Paul, N., & Nain, M. S. (2015). Socio-economic Empowerment of Rural Women Through Rural Tourism Projects in Jammu Region of J&K State in India. *Indian Journal of Extension Education, 51*(3-4), 40-43.

Small Projects Istanbul. (n.d.). *Community development.* Retrieved February 9, 2023, from https://www.smallprojectsistanbul.org/community-development/

Social Media Examiner. (2021). *Social media marketing industry report.* Retrieved from https://www.socialmediaexaminer.com/social-media-marketing-industry-report/

Sodawan, A., & Hsu, R. L. W. (2022). Halal-Friendly Attributes and Muslims' Visit Intention: Exploring the Roles of Perceived Value and Destination Trust. *Sustainability (Basel)*, *14*(19), 1200. doi:10.3390u141912002

Soleimani, S., Bruwer, J., Gross, M. J., & Lee, R. (2019). Astro-tourism conceptualisation as specialinterest tourism (SIT) field: A phenomonological approach. *Current Issues in Tourism*, *22*(18), 2299–2314. doi:10.1080/13683500.2018.1444021

Sony, K. C. (2012). Roles and challenges of women in tourism sector of western Nepal: A micro-ethnographic study. *Nepal Tourism and Development Review*, *2*(1), 32–59.

Soroushmehr, H., Kalantari, K., Fami, H., & Sarani, V. (2012). Investigation of self-help groups (SHGs) effect on rural women empowerment (Hamedan county, Iran). *Journal of Agricultural Science*, *4*, 1–11.

Sosyal Ekonomi. (n.d.). https://sosyalekonomi.org/kilickaya-kadin-girisimi-uretim-ve-isletme-kooperatifi/

Sotiriadis, M. (2017). Experiential Dimensions And Their Influence On Behavioral Intentions Within The Context Of Nature-Based Tourism. *Tourism and Hospitality Management*, *23*(1), 35–50. doi:10.20867/thm.23.1.7

Sottini, A., & Ciambotti, G. (2020). Social Entrepreneurship Toward a Sustainable Peacebuilding. *Peace, Justice and Strong Institutions*, 1-11.

Soysal, A. (2010). Characteristics of Women Entrepreneurs, the Problems They Face and Suggestions for Women who will Start a Business: A Research in Kahramanmaraş Province. *Journal of Eskisehir Osmangazi University IIBF*, *5*(1), 71–95.

Spenceley, A., & Meyer, D. (2012). Tourism and poverty reduction: Theory and practice in less economically developed countries. *Journal of Sustainable Tourism*, *20*(3), 297–317. doi:10.1080/09669582.2012.668909

Spencer, D. (2015). 'To Protect Her Honour': Child Marriage in Emergencies – the Fatal Confusion between Protecting Girls and Sexual Violence. *CARE Gender and Protection in Humanitarian Contexts*. https://insights.careinternational.org.uk/media/k2/attachments/CARE_Child-marriage-in-emergencies_2015.pdf

Spreitzer, G. M. (1995). Psychological empowerment in the workplace: Dimensions, measurement, and validation. *Academy of Management Journal, 38*(5), 1442-1465.

Sprout Social. (2020). *Sprout Social index: Edition XVII: The social media trends shaping the future*. https://sproutsocial.com/insights/data/sprout-social-index/social-media-trends/

Stainton, H. (2022). *150 types of tourism! The ultimate tourism glossary*. Retrieved from https://tourismteacher.com/types-of-tourism-glossary/

Stark, J., & Meschık, M. (2018). Women's everyday mobility: Frightening situations and their impacts on travel behaviour. *Transportation Research Part F: Traffic Psychology and Behaviour*, *54*, 311–323. doi:10.1016/j.trf.2018.02.017

Statista. (2021). *Hotel market revenue in Europe from 2016 to 2023, by country (in billion U.S. dollars)*. Retrieved from https://www.statista.com/statistics/657066/hotel-market-revenue-in-europe-by-country/

Stavroulakis, D., Karagiannis, S., Mitoula, R., & Papagrigoriou, A. (2013). Female Entrepreneurship and Agritourism Cooperatives in the Greeκ Periphery: An Overview. *American Journal of Economics and Business Administration*, *5*(4), 129–138. doi:10.3844/ajebasp.2013.129.138

Stephano, R. M. (2022). *Top 10 Medical Tourism Destinations in the World*. Global Healthcare Resources & Medical Tourism Association. Retrieved from https://www.magazine.medicaltourism.com/article/top-10-medical-tourism-destinations-world

Strachan, G., Burgess, J., & Sullivan, A. (2004). Affirmative action or managing diversity: What is the future of equal opportunity policies in organisations? *Women in Management Review*, *19*(4), 196–204. doi:10.1108/09649420410541263

Streimikiene, D., Svagzdiene, B., Jasinskas, E., & Simanavicius, A. (2021). Sustainable tourism development and competitiveness: The systematic literature review. *Sustainable Development (Bradford)*, *29*(1), 259–271. doi:10.1002d.2133

Strzelecka, M., Boley, B., & Strzelecka, C. (2017). Empowerment and resident support for tourism in rural central and Eastern Europe (CEE): The case of Pomerania, Poland. *Journal of Sustainable Tourism*, *25*(4), 554–572. doi:10.1080/09669582.2016.1224891

Su, C.P., & Wu, T.C. (2022). The dark side of solo female travel: Negative encounters with male strangers. *Innovation and Impact of Sex as Leisure in …* doi:10.4324/9781003288206-8

Sukardani, P. S. (2010). *Brand communication melalui digital influencer*. https://www.academia.edu/33955942/BRAND_COMMUNICATION_MELALUI_DIGITAL_INFLUENCER?email_work_card=reading-history

Sullivan, D. M., & Meek, W. R. (2012). Gender and entrepreneurship: A review and process model. *Journal of Managerial Psychology*, *27*(5), 428–458. doi:10.1108/02683941211235373

Sultan, M. F., Shaikh, S. K., Tunio, M. N., & Asim, M. (2022). Sustainable Leadership: A Divine Form to Lead and Succeed in a Holistic Way. In Sustainable Development of Human Resources in a Globalization Period (pp. 111-118). IGI Global. doi:10.4018/978-1-6684-4981-3.ch007

Sultan, M. F., Tunio, M. N., Aziz, A., & Shaikh, S. K. (2023). Social Innovation and Social Entrepreneurship in the Wake of COVID-19: A Perspective From the Developing Side of the World. In Frugal Innovation and Social Transitions in the Digital Era (pp. 113-118). IGI Global.

Sultana, N., & Rahman, A. (2019). Social business for sustainable development: a developing country perspective. In The Components of Sustainable Development: Engagement and Partnership (pp. 179-192). Springer Singapore. doi:10.1007/978-981-13-9209-2_12

Sultan, M. F., & Tunio, M. N. (2022). Gauging the COVID-19 on Academic Stress on Tertiary Level Female Students on the eve of COVID-19. *KASBIT Business Journal*, *15*(3), 1–10.

Sumathi, V., & Gunasundari, K. (2016). Problems faced by the women entrepreneurs in India. *International Journal of Management Research and Reviews*, *6*(3), 348.

Suriyankietkaew, S., Krittayaruangroj, K., & Iamsawan, N. (2022). Sustainable Leadership practices and competencies of SMEs for sustainability and resilience: A community-based social enterprise study. *Sustainability (Basel)*, *14*(10), 5762. doi:10.3390u14105762

Suryawati, D. A., Dewi, S. P. A. A. P., Osin, R. F., & Anggayana, I. W. A. (2021). The Role of Women in Protecting the Village and Rural Tourism in Timpag Village. *Jurnal Manajemen Pelayanan Hotel*, *5*(2), 74–79.

Suseno, Y. and Abbott, L. (2021). Women entrepreneurs' digital social innovation: Linking gender, entrepreneurship, social innovation and information systems. Information Systems Journal, 31(5), 714-744.

Suseno, Y., & Abbott, L. (2021). Women entrepreneurs' digital social innovation: Linking gender, entrepreneurship, social innovation and information systems. *Information Systems Journal*, *31*(5), 717–744. doi:10.1111/isj.12327

Sutcliffe-Braithwaite, F., & Lawrence, J. (2019). Power and its loss in the iron lady. *Biography and History in Film*, 295–317. doi:10.1007/978-3-319-89408-9_13

Su, Y., Cherian, J., Sial, M. S., Badulescu, A., Thu, P. A., Badulescu, D., & Samad, S. (2021). Does tourism affect economic growth of China? A panel granger causality approach. *Sustainability (Basel)*, *13*(3), 1349. doi:10.3390u13031349

Swain, M. B. (1993). Women producers of ethnic arts. *Annals of Tourism Research*, *20*(1), 32–51. doi:10.1016/0160-7383(93)90110-O

Swain, M. B. (1995). Gender in tourism. *Annals of Tourism Research*, *22*(2), 247–266. doi:10.1016/0160-7383(94)00095-6

Swedberg, R. (2009). *Principles of economic sociology*. Princeton University Press. doi:10.2307/j.ctvcm4g75

Syathir, S. A., Abror, A., Wardy, P. T., Muslihati, M., Sofyan, S., Sirajuddin, S., Nasri, K. M., & Zulfikar, D. A. (2021). Crisis and disaster management for halal tourism: A systematic review. *Tourism Review*, *77*(1), 129–145.

Szczurski, M. M. (2017). *Social media influencer- A Lifestyle or a profession of the XXIst century?* https://depot.ceon.pl/bitstream/handle/123456789/15166/Social%20media%20influencer.pdf?sequence=1&isAllowed=y

Tajeddini, K., Ratten, V., & Denisa, M. (2017). Female tourism entrepreneurs in Bali, Indonesia. *Journal of Hospitality and Tourism Management*, *31*, 52–58. doi:10.1016/j.jhtm.2016.10.004

Tambunan, T. (2009). Women entrepreneurship in Asian developing countries: Their development and main constraints. *Journal of Development and Agricultural Economics*, *1*(2), 27–40.

Tamer, M. (2021). *Cittaslow Hareketinde Kadın Girişimciler: Seferihisar Örneği* [Women Entrepreneurs in the Cittaslow Movement: A Case Study of Seferihisar] [Master Thesis]. Van Yüzüncü Yıl University.

Tamer, M., & Alaeddinoğlu, F. (2022). Cittaslow Hareketinde Kadın Girişimciler: Seferihisar Örneği [Women Entrepreneurs in the Cittaslow Movement: The Case of Seferihisar]. *Coğrafi Bilimler Dergisi/Turkish Journal of Geographical Sciences*, *20*(2), 500–527. doi:10.33688/aucbd.1127083

Tavakoli, R., & Mura, P. (2021). Muslim women travelers' constraints: A critical review. *Women in tourism in Asian Muslim countries*, 25-40.

Tavakoli, R., & Mura, P. (2021). Muslim women travellers' constraints: A critical review. *Women in tourism in Asian Muslim countries*, 25-40.

Tavakoli, R., & Mura, P. (2021). Muslim women travellers' constraints: A critical review. *Women in Tourism in Asian Muslim Countries*, 25-40.

Tavakoli, R., & Mura, P. (2021). Muslim women travellers' constraints: A critical review. *Women in Tourism in Asian Muslim Countries*, 25–40.

TDK. (2022). *Girişim* [Enterprise]. Available at: https://sozluk.gov.tr/

Team International Assistance for Integration-TIAFI. (n.d.). *About TIAFI*. Retrieved February 9, 2023, from https://tiafi.org/

Tekeli, H. N. (2019). Women's Employment in Tourism Sector in Turkey, Issues Faced and the Effect of Glass Ceiling Syndrome on Women Workforce in Tourism. *Social Sciences*, *14*(2), 179–191.

Tekin, Ö. A. (2017). Turizm Sektöründe Toplumsal Cinsiyet Algısı: Beş Yıldızlı Otel Çalışanları Üzerine Bir Araştırma. *Avrasya Sosyal ve Ekonomi Araştırmaları Dergisi*, *4*(12), 669–684.

Temizkan, V., Güven, E. Ö., Yılmazer, A., & Andsoy, C. (2021). Covid-19 ile Gerçekleşen Tüketici Davranışları ve Eğilimlerine Yönelik Bir Araştırma [A Research on Consumer Behaviors and Trends That Occur with Covid-19]. *Elektronik Sosyal Bilimler Dergisi*, *20*(78), 1311–1327. doi:10.17755/esosder.839996

Teng, C.-C. (2011). Commercial hospitality in restaurants and tourist accommodation: Perspectives from international consumer experience in Scotland. *International Journal of Hospitality Management*, *30*(4), 866–874. doi:10.1016/j.ijhm.2011.01.007

Tengland, P. A. (2008). Empowerment: A Conceptual Discussion. *Health Care Analysis*, *16*(2), 77–96. doi:10.100710728-007-0067-3 PMID:17985247

Teo, P., & Leong, S. (2006). A postcolonial analysis of backpacking. *Annals of Tourism Research*, *33*(1), 109–131. doi:10.1016/j.annals.2005.05.001

The Hindu Bureau. (2023, February 3). *The Hindu Bureau*. Retrieved March 24, 2023, from https://www.thehindu.com/business/tourism-to-contribute-512-bn-to-indias-gdp-by-2028-says-taai-chief/article66468359.ece

The Norwegian Refugee Council. (2019). *Left Behind: The Education Crisis for Syrian Refugee Children in Turkey*. Retrieved from https://www.nrc.no/resources/reports/left-behind-the-education-crisis-for-syrian-refugee-children-in-turkey/

The World Bank. (2018). *Women economic empowerment study*. World Bank Group. Retrieved from http://documents1.worldbank.org/curated/en/86149155111 3547855/pdf/134846-WP-PUBLIC-march-2-WB-Women-Study-EN.pdf

Thomas, K. W., & Velthouse, B. A. (1990). Cognitive elements of empowerment: An "interpretive" model of intrinsic task motivation. *Academy of Management Review*, *15*(4), 666–681.

Thomas, T. K., & Mura, P. (2019). *The 'normality of unsafety'-foreign solo female travellers in India. In Tourism Recreation Research*. Taylor &Francis. doi:10.1080/02508281.2018.1494872

Thommes, J. (2012). *Jim's astrophotography. Astrophotography by Jim Thommes*. Retrieved from http://www.jthommes. com/Astro

Thrane, C. (2008). Earnings differentiation in the tourism industry: Gender, human capital and socio-demographic effects. *Tourism Management*, *29*(3), 514–524. doi:10.1016/j.tourman.2007.05.017

Thyne, M., Lawson, R., & Todd, S. (2006). The use of conjoint analysis to assess the impact of the cross-cultural exchange between hosts and guests. *Tourism Management*, *27*(2), 201–213. doi:10.1016/j.tourman.2004.09.003

Tien, N. H., Thai, T. M., Hau, T. H., Vinh, P. T., & Long, N. V. T. (2019). Solutions for Tuyen Quang and Binh Phuoc tourism industry sustainable development. Comparative analysis. *International Journal of Research in Marketing Management and Sales*, *2*(1), 101–107. doi:10.33545/26633329.2020.v2.i1b.31

Tinker, I. (2006). Empowerment just happened: the unexpected expansion of women's organizations. *Women and gender equity in development theory and practice: Institutions, resources and mobilization*, 268-303.

TOI. (2023). Tourism News: Latest News on tourism - times of India. *The Times of India*. Retrieved March 24, 2023, from https://timesofindia.indiatimes.com/topic/tourism/news

Tokgöz, G. (2011). *Kalkınmada kadın emeği*. Varlık Yayınları.

Tomazos, K., & Butler, R. (2009). Volunteer tourism: Working on holiday or playing at work? *Tourismos*, *4*(4), 331–349.

Topçu, İ., Uslay, C., & Fuxman, L. (2021). Post-COVID-19 social media marketing in family-owned hotels: An empirical investigation in Turkey. *Journal of Hospitality and Tourism Technology*, *12*(3), 373–391. doi:10.1108/JHTT-10-2020-0137

Toptaş, A. (2020). Kültür Turizminin Kadın Girişimcilerin Yaptıkları Yeniliklere Etkisi: Boğatepe Köyü'nün Örnek Vaka Analizi ile İncelenmesi [The Impact of Cultural Tourism on the Innovations of Women Entrepreneurs: Investigation of Boğatepe Village by Case Study Analysis]. *Journal of Turkish Tourism Research*, *4*(4), 3378–3391. doi:10.26677/TR1010.2020.568

Trauer, B. (2006). Conceptualizing special interest tourism—Frameworks for analysis. *Tourism Management*, *27*(2), 183–200. doi:10.1016/j.tourman.2004.10.004

Trauer, B., & Ryan, C. (2005). Destination image, romance and place experience – an application of intimacy theory in tourism. *Tourism Management*, *26*(4), 481–491. doi:10.1016/j.tourman.2004.02.014

Trauth, E. M. (2002). Odd girl out: An individual differences perspective on women in the IT profession. *Information Technology & People*, *15*(2), 98–118. doi:10.1108/09593840210430552

Travel.State.Gov. (2023). Retrieved from https://travel.state.gov/content/travel/en/international-travel/International-Travel-Country-Information-Pages/Turkey.html

TripAdvisor. (2018). *The modern traveller journey*. https://www.tripadvisor.com/TripAdvisorInsights/w671

Tristanti, Nurhaeni, Mulyanto, & Sakuntalawati. (2022). The Role of Women in Tourism: A Systematic Literature Review. In *2021 Annual Conference of Indonesian Association for Public Administration*. KnE Social Sciences. doi:10.18502/kss.v7i5.10576

Tristanti, T., Nurhaeni, I. D. A., Mulyanto, M., & Sakuntalawati, R. D. (2022). The Role of Women in Tourism: A Systematic Literature Review. *KnE Social Sciences*, 545-554.

Tristanti, T., Nurhaeni, I. D. A., Mulyanto, M., & Sakuntalawati, R. D. (2022). The Role of Women in Tourism: A Systematic Literature Review. *KnE Social Sciences*, 545–554.

Tristanti, T., Nurhaeni, I. D. A., Mulyanto, M., &Sakuntalawati, R. D. (2022). The Role of Women in Tourism: A Systematic Literature Review. *KnE Social Sciences*, 545-554.

Tsaur, S.-H., & Teng, H.-Y. (2017). Exploring tour guiding styles: The perspective of tour leader roles. *Tourism Management*, *59*, 438–448. doi:10.1016/j.tourman.2016.09.005

Tsaur, S.-H., Tzeng, G.-H., & Wang, K.-C. (1997). Evaluating Tourist Risks from Fuzzy Perspectives. *Annals of Tourism Research*, *24*(4), 796–812. doi:10.1016/S0160-7383(97)00059-5

Tseng, M. L., Lin, C., Lin, C. W. R., Wu, K. J., & Sriphon, T. (2019). Ecotourism development in Thailand: Community participation leads to the value of attractions using linguistic preferences. *Journal of Cleaner Production*, *231*, 1319–1329. doi:10.1016/j.jclepro.2019.05.305

Tucker, H. (2007). Undoing shame: Tourism and women's work in Turkey. *Journal of Tourism and Cultural Change*, *5*(2), 87–105. doi:10.2167/jtcc089.0

Tucker, H. (2009). Recognizing emotion and its postcolonial potentialities: Discomfort and shame in a tourism encounter in Turkey. *Tourism Geographies*, *11*(4), 444–461. doi:10.1080/14616680903262612

Tucker, H., & Boonabaana, B. (2012). A critical analysis of tourism, gender and poverty reduction. *Journal of Sustainable Tourism*, *20*(3), 437–455. doi:10.1080/09669582.2011.622769

Tükeltürk, Ş. A., & Perçin, N. Ş. (2008). Turizm Sektöründe Kadın Çalışanların Karşılaştıkları Kariyer Engelleri ve Cam Tavan Sendromu: Cam Tavanı Kırmaya Yönelik Stratejiler. *The Journal of American Science*, *6*(2), 113–128.

Tunca, A. (2021). *Otel mutfaklarında cinsiyet engelleri: Kadın aşçılar üzerinde bir araştırma* [Unpublished Master Thesis]. Necmettin Erbakan Üniversitesi.

Tunio, M. N. (2020) Role of ICT in Promoting Entrepreneurial Ecosystems in Pakistan. *Journal of Business Ecosystems, 1*(2).

Tunio, M. N. (2020). Role of ICT in Promoting Entrepreneurial Ecosystems in Pakistan. *Journal of Business Ecosystems, 1*(2).

Tunio, M. N., Chaudhry, I. S., Mughal, F., & Shaikh, E. (2022). Marketing Mode and Survival of the Entrepreneurial Activities of Nascent Entrepreneurs. In Big Data Analytics (pp. 1-18). Auerbach Publications. doi:10.1201/9781003307761-1

Tunio, M. N., Soomro, A. A., & Bogenhold, D. (2017). The Study of Self-employment at SMEs Level with Reference to Poverty in Developing Countries. Business and Management Research, 6(2). doi:10.5430/bmr.v6n2p33

Tunio, M. N., Yusrini, L., & Shoukat, G. (2021). Corporate social responsibility (CSR) in Hotels in Austria, Pakistan, and Indonesia: small and medium Enterprise spillover of COVID-19. In Handbook of research on entrepreneurship, innovation, sustainability, and ICTs in the post-COVID-19 era (pp. 263-280). IGI Global.

Tunio, M. N., Yusrini, L., Shah, Z. A., Katper, N., & Jariko, M. L. (2021). How Hotel Industry Copes Up with the COVID-19: An SME Perspective. *Etikonomi, 20*(2), 213 – 224.

Tunio, M. N., Yusrni, L., Shah, Z. A., Katper, N. K., & Jariko, M. A. (2021). How hotel industry cope up with the COVID-19: An SME Perspective. *Etikonomi, 20*(2).

Tunio, M. N. (2020). Role of ICT in promoting entrepreneurial ecosystems in Pakistan. *Journal of Business Ecosystems, 1*(2), 1–21. doi:10.4018/JBE.2020070101

Tunio, M. N., Chaudhry, I. S., Shaikh, S., Jariko, M. A., & Brahmi, M. (2021). Determinants of the Sustainable Entrepreneurial Engagement of Youth in Developing Country—An Empirical Evidence from Pakistan. *Sustainability (Basel), 13*(14), 7764. doi:10.3390u13147764

Tunio, M. N., Jariko, M. A., Børsen, T., Shaikh, S., Mushtaque, T., & Brahmi, M. (2021). How Entrepreneurship Sustains Barriers in the Entrepreneurial Process—A Lesson from a Developing Nation. *Sustainability (Basel), 13*(20), 1–18. doi:10.3390u132011419

Tunio, M. N., Shah, S. M. M., Qureshi, M. A., Tunio, A. N., & Shaikh, E. (2022). Career Predilections and Options to Opt Occupation for the Youth in Pakistan. In *Developing Entrepreneurial Ecosystems in Academia* (pp. 156–170). IGI Global. doi:10.4018/978-1-7998-8505-4.ch009

Tunio, M. N., & Shaikh, E. (2020). Nascent entrepreneurs and challenges in digital market in developing countries. *International Journal of Public Sector Performance Management.*

Tunio, M. N., Shaikh, E., Niaz, S., & Katper, N. S. (2021). Multifaceted perils of the Covid-19 and implications: A Review. *Estudios de Economía Aplicada*. Advance online publication. doi:10.25115/eea.v39i2.3957

Tunio, M. N., Yusrini, L., Shah, Z. A., Katper, N., & Jariko, M. A. (2021). How Hotel Industry Cope up with the COVID-19: An SME Perspective. *Etikonomi, 20*(2), 213–224. doi:10.15408/etk.v20i2.19172

Türkay, O. (2015). Turizm İşletmelerinde Kariyer Yönetimi. In E. Pelit (Ed.), *Turizm İşletmelerinde İnsan Kaynakları Yönetimi*. Grafiker Yayınları.

Turkey, U. N. H. C. R. (2022). *UNHCR Turkey Fact Sheet - February 2022*. ReliefWeb. Retrieved February 24, 2023, from https://reliefweb.int/report/turkey/unhcr-turkey-fact-sheet-february-2022-entr

Turkiye Istatıstık Kurumu. (n.d.). https://data.tuik.gov.tr/Bulten/Index?p=Turizm-Istatistikleri-IV.Ceyrek:-Ekim-Aralik-ve-Yillik,-2022-49606#:~:text=Turizm%20geliri%202022%20y%C4%B1l%C4%B1nda%20bir,milyon%20907%20bin%20dolar%20oldu

Tur-Porcar, A., Mas-Tur, A., & Belso, J. A. (2016). Barriers to women entrepreneurship. different methods, different results? *Quality & Quantity, 51*(5), 2019–2034. doi:10.100711135-016-0343-0

TURSAB. (2022). *Sektörel Raporlar* [Sectoral reports]. Available at: https://www.tursab.org.tr/turkiye-turizm-degerlendirmesi/sektorel-raporlar

Tussyadiah, I. P., Sigala, M., & Law, R. (2021). COVID-19 and the changing landscape of social media engagement with destination and hospitality brands. *Journal of Travel Research*, *60*(1), 44–56. doi:10.1177/0047287520956997

Tuten, T. L. and Solomon, M. R. (2014). Social Media Marketing. Pearson.

Tüzünkan, D. (2019). Turizmde kadın istihdamı. In A. Vatan (Ed.), Turizmde kadın olmak içinde (pp. 67-84). Adapazarı: Değişim Yayınları.

Uatay, G., Reid, E. L., & Lee, H. Y. (2019). The impact of female travelers' travel constraints on attitude toward solo travel and travel intention. *Culinary Science &Hospitality Research*. https://www.dbpia.co.kr/Journal/articleDetail?nodeId=NODE09216304

Uduji, J. I., Okolo-Obasi, E. N., Onodugo, V. A., Nnabuko, J. O., & Adedibu, B. A. (2021). Corporate Social Responsibility and The Role of Rural Women in Strengthening Agriculture-Tourism Linkages in Nigeria's Oil Producing Communities. *Journal of Tourism and Cultural Change*, *19*(6), 754–780. doi:10.1080/14766825.2020.1826500

Ulas, S. (2020). International Journal of Marketing, Communication and New Media. *Social Media Marketing.*

Ulu, S., & Özen Kutanis, R. (2016). İşgücü piyasalarında ayrımcılığın kaynakları. *Selçuk Üniversitesi Sosyal Bilimler Enstitüsü Dergisi*, (35), 359–372.

UN (United Nations). (2015). *The General Assembly on 25 September 2015 - Transforming our world: the 2030 Agenda for Sustainable Development, A/RES/70/1*. United Nations.

UN (United Nations). (2020). *United Nations sustainable development goals*. Retrieved from https://www.un.org/sustainabledevelopment/gender-equality/

UN World Tourism Organization (UNWTO). (2019). *Global report on women in tourism*. UNWTO.

UN. (2022). *Advancing Women's Economic Empowerment in the Tourism Sector in COVID-19 Response and Recovery*. UN. https://www.unwomen.org/sites/default/files/2022-05/Advancing-womens-economic-empowerment-in-the-tourism-sector-in-COVID-19-en.pdf

UNEP. (2010). *Tourism and the Green Economy: Background Report*. United Nations Environment Programme.

UNHCR. (2005). *Handbook of self-reliance*. Retrieved from https://www.unhcr.org/44bf7b012.pdf

UNHCR. (2021a). *Figures at a Glance*. UNHCR. https://www.unhcr.org/figures-at-a-glance.html

UNHCR. (2021b). *Syria Regional Refugee Response*. Retrieved February 24, 2023, from https://data2.unhcr.org/en/situations/syria/location/113

UNHCR. (n.d.). *How volunteers help refugees*. Retrieved from https://www.unhcr.org/en-my/how-volunteers-help-refugees.html

United Nations Development Programme. (n.d.). *Sustainable Development Goals*. https://www.undp.org/content/undp/en/home/sustainable-development-goals.html

United Nations Economic and Social Council. (2002). *Commission on the Status of Women: Report on the forty-sixth session (4-15 and 25 March 2002). Official Records, 2002. Supplement No. 7 (E/2002/27-E/CN.6/2002/13)*. New York: United Nations.

United Nations Volunteers. (2018). *State of the World's Volunteerism Report 2018: The Thread that Binds*. Retrieved from https://unv-swvr2018.org/files/51692_UNV_SWVR_2018_WEB_OVERVIEW.pdf

United Nations World Tourism Organization. (2020). UNWTO World Tourism Barometer.: Vol. 18. *June 2020*. World Tourism Organization.

United Nations World Tourism Organization. (2021). *International tourism fell by 74% in 2020, the worst year in tourism history*. Retrieved from https://www.unwto.org/news/international-tourism-fell-by-almost-75-in-2020

United Nations, Global Sustainable Development Report (GSDR). (2023). https://sdgs.un.org/gsdr/gsdr2023

UNWTO Global report on women in tourism – second edition. (2010). Retrieved March 24, 2023, from https://www.e-unwto.org/doi/book/10.18111/9789284420384

UNWTO. (2011). *Global report on women in tourism*. United Nation World Tourism Organization. Online available at: http://ethics.unwto.org/content/gender-and-tourism

UNWTO. (2015). *Tourism and the Sustainable Development Goals – Journey to 2030*. United Nations World Tourism Organization.

UNWTO. (2017). *Gender and tourism*. https://www.unwto.org/gender-and-tourism

UNWTO. (2018). *Tourism Highlights* (2018 Edition). United Nations World Tourism Organization.

UNWTO. (2020). *Global Report on Women in Tourism 2020*. World Tourism Organization.

UNWTO. (2023a). *Global Report on Women in Tourism: Second Edition*. UNWTO. Retrieved from https://www.e-unwto.org/doi/pdf/10.18111/9789284420384

UNWTO. (2023b). *Women's empowerment and tourism*. Retrieved from UNWTO: https://www.unwto.org/gender-and-tourism

Ural, A., & Kılıç, İ. (2018). *Bilimsel Araştırma Süreci ve SPSS ile Veri Analizi*. Detay Yayıncılık.

Utma, S. (2019). Kadına yönelik cinsiyet ayrımcılığı ve cam tavan sendromu. *Sosyal ve Beşerî Bilimler Dergisi*, *11*(1), 44–58.

UWTO Report Women in Tourism. (2010). Retrieved from UWTO: UWTO Report Women in Tourism. 2010. Available online: https://cf.cdn.unwto.org/sites/all/files/pdf/global_report_on_women_in_tourism_2010.pdf

Valaja, E. (2018). *Solo female travellers' risk perceptions and risk reduction strategies-as expressed in online travel blog narratives*. https://lup.lub.lu.se/student-papers/record/8947243/file/8947244.pdf

Valentine, G. (1989). The geography of women's fear. *Area*, *21*, 385–390.

Valeri, M., & Katsoni, V. (2021). Female entrepreneurship in tourism. In *Gender and Tourism* (pp. 1–9). Emerald Publishing Limited. doi:10.1108/978-1-80117-322-320211001

Van der Cammen, S. (1997). Involving Maasai women. In L. France (Ed.), *The earthscan reader In sustainable tourism* (pp. 162–163). Earthscan.

Van Eerdewijk, A., Wong, F., Vaast, C., Newton, J., Tyszler, M., & Pennington, A. (2017). *A conceptual model of women's and girls' empowerment*. KIT Royal Tropical Institut. https://ec.europa.eu/commission/presscorner/detail/en/IP_12_1205

Van Nguyen, S., & Habók, A. (2021). Designing and validating the learner autonomy perception questionnaire. *Heliyon*, *7*(4), e06831. doi:10.1016/j.heliyon.2021.e06831 PMID:33981892

Van Vuuren, L. J., & Fourie, C. (2000). Career anchors and career resilience: Supplementary constructs? *SA Journal of Industrial Psychology*, *26*(3), 15–20. doi:10.4102ajip.v26i3.714

Vargas-Sánchez, A., & Moral-Moral, M. (2020). Halal tourism: Literature review and experts' view. *Journal of Islamic Marketing*, *11*(3), 549–569. doi:10.1108/JIMA-04-2017-0039

Varghese, T. (2011). Women empowerment in Oman: A study based on Women Empowerment Index. *Far East Journal of Psychology and Business*, *2*(2), 37–53.

Varinlioğlu, F. (2020). Understanding Virtual Reality Applications in Digital Heritage Through Teos. *MEGARON*, *15*(1), 161–170. doi:10.14744/megaron.2019.85619

Varkaris, E., & Neuhofer, B. (2017). The influence of social media on the consumers' hotel decision journey. *Journal of Hospitality and Tourism Technology*, *8*(1), 101–118. doi:10.1108/JHTT-09-2016-0058

Vasilić, N., Popović-Pantić, S., & Semenčenko, D. (2020). Women entrepreneurship in the time of COVID-19 pandemic. *Journal of Women's Entrepreneurship & Education*, (3-4), 23–40.

Vázquez-Maguirre, M. (2020). Building sustainable rural communities through indigenous social enterprises: A humanistic approach. *Sustainability (Basel)*, *12*(22), 9643. doi:10.3390u12229643

Ventura, A., Morillas, L., Martins-Loução, M., & Cruz, C. (2021). Women's Empowerment, Research, and Management: Their Contribution to Social Sustainability. *Sustainability (Basel)*, *13*(22), 12754. doi:10.3390u132212754

Venugopalan, T., & Kumar, D. (2017). Sustainable development through sustainable tourism in India-A case study of Kerala tourism. *Asian Journal of Research in Business Economics and Management*, *7*(12), 10–27. doi:10.5958/2249-7307.2017.00189.X

Vij, M., Upadhya, A., Agha, K., Begum, V., Kashmoola, B. W., & Senior, J. R. (2023). Female economic empowerment through tourism training in the United Arab Emirates. *Worldwide Hospitality and Tourism Themes*, *15*(1), 18–28. doi:10.1108/WHATT-08-2022-0099

Virden, R. J., & Walker, G. J. (1999). Ethnic/racial and gender variations among meanings given to, and preferences for, the natural environment. *Leisure Sciences*, *21*(3), 219–239. doi:10.1080/014904099273110

Visa. (2016). *Visa Study Estimates Households Traveling Internationally Will Surpass 280 Million By 2025.* Available at https://investor.visa.com/news/news-details/2016/Visa-StudyEstimates-Households-Traveling-Internationally-Will-Surpass-280-Million-By2025/default.aspx

Vizcaino-Suárez, L. P., & Díaz-Carrión, I. A. (2019). Gender in tourism research: Perspectives from Latin America. *Tourism Review*, *74*(5), 1091–1103. doi:10.1108/TR-02-2017-0021

Von Bergen, C. W., Soper, B., & Foster, T. (2002). Unintended Negative Effects of Diversity Management. *Public Personnel Management*, *31*(2), 239–251. doi:10.1177/009102600203100209

von der Weppen, J., & Cochrane, J. (2012). Social enterprises in tourism: An exploratory study of operational models and success factors. *Journal of Sustainable Tourism*, *20*(3), 497–511. doi:10.1080/09669582.2012.663377

Vossenberg, S. (2013). Women Entrepreneurship Promotion in Developing Countries: What explains the gender gap in entrepreneurship and how to close it. *Maastricht School of Management Working Paper Series, 8*(1), 1-27.

Vrontis, D., Makrides, A., Christofi, M., & Thrassou, A. (2021). Social media influencer marketing: A systematic review, integrative framework and future research agenda. *International Journal of Consumer Studies*, *45*(4), 617–644. doi:10.1111/ijcs.12647

Vujko, A., Tretiakova, T. N., Petrović, M. D., Radovanović, M., Gajić, T., & Vuković, D. (2019). Women's empowerment through self-employment in Tourism. *Annals of Tourism Research*, *76*, 328–330. doi:10.1016/j.annals.2018.09.004

Vukovic, D. B., Petrovic, M., Maiti, M., & Vujko, A. (2021). Tourism development, entrepreneurship and women's empowerment – Focus on Serbian countryside. *Journal of Tourism Futures*, 1-21.

Vukovic, D. B., Petrovic, M., Maiti, M., & Vujko, A. (2021). Tourism Development, Entrepreneurship and Women's Empowerment–Focus on Serbian Countryside. *Journal of Tourism Futures*, 1-21.

Vukovic, D. B., Petrovic, M., Maiti, M., & Vujko, A. (2021). Tourism development, entrepreneurship and women's empowerment–Focus on Serbian countryside. *Journal of Tourism Futures.*

Wakkee, I., Elfring, T., & Monaghan, S. (2010). Creating entrepreneurial employees in traditional service sectors. *The International Entrepreneurship and Management Journal*, *6*(1), 1–21. doi:10.100711365-008-0078-z

Wall, G., & Norris, J. (2003). Gender and tourism. *Classic reviews in tourism*, 235-264.

Wallace, G. N., & Smith, M. D. (1997). A comparison of motivations, preferred management actions, and setting preferences among Costa Rican, North American and European visitors to five protected areas in Costa Rica. *Journal of Park and Recreation Administration*, *15*, 59–82.

Wallace, T., & March, C. (1991). *Changing perceptions: Writings on gender and development*. Oxfam GB.

Walls, A. R. (2013). A cross-sectional examination of hotel consumer experience and relative effects on consumer values. *International Journal of Hospitality Management*, *32*, 179–192. doi:10.1016/j.ijhm.2012.04.009

Walter, S., & Brüggemann, M. (2020). Opportunity makes opinion leaders: Analyzing the role of first-hand information in opinion leadership in social media networks. *Information Communication and Society*, *23*(2), 267–287. doi:10.1080/1369118X.2018.1500622

Wang, Q. (2019). Gender, race/ethnicity, and entrepreneurship: Women entrepreneurs in a US south city. *International Journal of Entrepreneurial Behaviour & Research*, *25*(8), 1766–1785. doi:10.1108/IJEBR-05-2017-0156

Wang, Q. (2021). Higher education institutions and entrepreneurship in underserved communities. *Higher Education*, *81*(6), 1273–1291. doi:10.100710734-020-00611-5

Wang, X., Yuen, K. F., Wong, Y. D., & Li, K. X. (2020). How can the maritime industry meet Sustainable Development Goals? An analysis of sustainability reports from the social entrepreneurship perspective. *Transportation Research Part D, Transport and Environment*, *78*, 102173. doi:10.1016/j.trd.2019.11.002

Wang, X., & Zhang, C. (2020). Contingent effects of social norms on tourists' pro-environmental behaviours: The role of Chinese traditionality. *Journal of Sustainable Tourism*, *28*(10), 1646–1664. doi:10.1080/09669582.2020.1746795

Wan, M. M., Zhang, Y., & Li, M. (2021). Do narcissistic employees remain silent? Examining the moderating roles of supervisor narcissism and traditionality in China. *Asian Business & Management*, 1–25.

Wasilczuk, J., & Zieba, K. (2008). Female Entrepreneurship in Transitional Economies: The case of Poland. *Journal of Small Business and Entrepreneurship*, *21*(2), 153–169. doi:10.1080/08276331.2008.10593419

Waterman, A. S. (2011). Eudaimonic identity theory: Identity as self-discovery. In S. J. Schwartz, K. Luyckx, & V. L. Vignoles (Eds.), Handbook of identity theory and research (pp. 357–379). Springer Science + Business Media.

Watson, J. C. (2003). *Overcoming the Challenges of Counseling College Student Athletes*. ERIC/CASS Digest.

Waugh, J. (2017). *Solo Female Travel Tips: Why, Where, How and Safety*. Retrieved from https://solotravelerworld.com/solo-female-travel-tips/

Wearing, S. (2001). *Volunteer tourism: Experiences that make a difference*. CABI. doi:10.1079/9780851995335.0000

Wearing, S., Deville, A., & Lyons, K. (2008). The volunteer's journey through leisure into the self. In S. Wearing & N. G. McGehee (Eds.), *Journeys of discovery in volunteer tourism: International case study perspectives* (pp. 65–71). CABI. doi:10.1079/9781845933807.0063

Wearing, S., & McGehee, N. G. (2013). Volunteer tourism: A review. *Tourism Management*, *38*, 120–130. doi:10.1016/j.tourman.2013.03.002

Weaver, D. (2011). Celestial ecotourism: New horizons in nature-based tourism. *Journal of Ecotourism*, *10*(1), 38–45. doi:10.1080/14724040903576116

Weber, K., & Ladkin, A. (2009). Career anchors of convention and exhibition industry professionals in Asia. *Journal of Convention & Event Tourism*, *10*(4), 243–255. doi:10.1080/15470140903356411

Weiermair, K., & Fuchs, M. (2021). Employees in family businesses: A critical resource for creating value in tourism and hospitality. *Journal of Hospitality & Tourism Research (Washington, D.C.)*, *45*(4), 518–537. doi:10.1177/1096348020966869

Weiler, B., & Ham, S. H. (2002). Tour Guide Training: A Model for Sustainable Capacity Building in Developing Countries. *Journal of Sustainable Tourism*, *10*(1), 52–69. doi:10.1080/09669580208667152

Weiss, H. M., & Cropanzano, R. (1996). Affective events theory. *Research in Organizational Behavior, 18*(1), 74.

Wei, W., Qi, R., & Zhang, L. (2019). Effects of virtual reality on theme park visitors' experience and behaviors: A presence perspective. *Tourism Management*, *71*, 282–293. doi:10.1016/j.tourman.2018.10.024

Wei, W., Torres, E. N., & And Hua, N. (2017). The power of self-service technologies in creating transcendent service experiences, The paradox of extrinsic attributes. *International Journal of Contemporary Hospitality Management*, *29*(6), 1599–1618. doi:10.1108/IJCHM-01-2016-0029

Wei, W., Torres, E., & And Hua, N. (2016). Improving consumer commitment through the integration of self-service technologies: A transcendent consumer experience perspective. *International Journal of Hospitality Management*, *59*, 105–115. doi:10.1016/j.ijhm.2016.09.004

Welsh, D. H., Memili, E., Kaciak, E., & Al Sadoon, A. (2014). Saudi women entrepreneurs: A growing economic segment. *Journal of Business Research*, *67*(5), 758–762. doi:10.1016/j.jbusres.2013.11.040

Wen, J., & Wu, M. Y. (2020). How special is special interest tourism – and how special are special interest tourists? A perspective article in a Chinese context. *Current Issues in Tourism*, *23*(16), 1968–1972. doi:10.1080/13683500.2020.1750575

Westwood, S., Pritchard, A., & Morgan, N. J. (2000). Gender-blind marketing: Businesswomen's perceptions of airline services. *Tourism Management*, *21*(4), 353–362. doi:10.1016/S0261-5177(99)00069-2

WFTGA. (2003). *What is a tourist guide?* https://wftga.org/about-us/what-is-a-tourist-guide/

Wilkinson, P. F., & Pratiwi, W. (1995). Gender and tourism in an Indonesian village. *Annals of Tourism Research*, *22*(2), 283–299. doi:10.1016/0160-7383(94)00077-8

Williams, S. (2017). *Characteristics of Social Media Key Influencers*. https://www.the24hoursecretary.com/characteristics-of-social-media-key-influencers/

Williams, C. L. (2013). The glass escalator, revisited: Gender inequality in neoliberal times, SWS feminist lecturer. *Gender & Society*, *27*(5), 609–629. doi:10.1177/0891243213490232

Wilson, E., & Little, D. E. (2005). A "relative escape"? The impact of constraints on women who travel solo. *Tourism Review International*. https://www.ingentaconnect.com/content/cog/tri/2005/00000009/00000002/art00004

Wilson, E., & Harris, C. (2006). Meaningful travel: Women, independent travel and the search forself and meaning. *Tourism (Zagreb), 54*(2), 161–172.

Wilson, E., & Little, D. E. (2005). A 'relative escape'? The impact of constraints on women who travel solo. *Tourism Review International, 9*(2), 155–174. doi:10.3727/154427205774791672

Wilson, E., & Little, D. E. (2008). *The solo female travel experience: Exploring the 'geography of women's fear'. In Current Issues in Tourism.* Taylor &Francis., doi:10.2167/cit342.0

Wilson, R., & Lohmann, G. (2019). Airline CEOs: Who are they, and what background and skill set are most commonly chosen to run the world's largest airlines? *Transportation Research Interdisciplinary Perspectives, 2*, 100054. doi:10.1016/j.trip.2019.100054

Wingett, F., & Turnbull, S. (2017). Halal holidays: Exploring expectations of Muslim-friendly holidays. *Journal of Islamic Marketing, 8*(4), 642–655. doi:10.1108/JIMA-01-2016-0002

Winn, J. (2005). Women entrepreneurs: Can we remove the barriers? *The International Entrepreneurship and Management Journal, 1*(3), 381–397. doi:10.100711365-005-2602-8

Woman, U. N. (2010). *The world's women 2010: Trends and statistics.* Retrieved from: https://www.unwomen.org/en/docs/2010/1/worlds-women-2010

Women Inspiring Change. (2020). *Karyn Thomas.* Available at: https://orgs.law.harvard.edu/womeninspiringchange/2016-honorees/karyn-thomas/

Women, U. N. (2017). *Syrian Women in Turkey: Investing in the Untapped Potential. Economic empowerment.* Available at: https://eca.unwomen.org/en/digital-library/publications/2017/06/syrian-women-in-turkey-investing-in-the-untapped-potential

Women, U. N. (2018). *Needs assessment of Syrian women and girls under temporary protection status in Turkey.* Retrieved from: https://www2.unwomen.org/-/media/field%20office%20eca/attachments/publications/country/turkey/the%20needs%20assessmentengwebcompressed.pdf?la=en&vs=3139

Women's Entrepreneurship, Issues, and Policies. (2004). *2nd OECD conference of ministers responsible for small and medium-sized enterprises (SMEs) promoting entrepreneurship and innovative SMEs in a global economy: towards a more responsible and inclusive globalisation.* Istanbul, Turkey: OECD.

Worke, M. D., Koricha, Z. B., & Debelew, G. T. (2021). Perception and experiences of sexual harassment among women working in hospitality workplaces of Bahir Dar city, Northwest Ethiopia: A qualitative study. *BMC Public Health, 21*(1), 1–18. doi:10.118612889-021-11173-1 PMID:34116672

World Bank Group. (2020). Women, business, and the law: 2020. Author.

World Bank Group. (2022). *From Swimming in Sand to High and Sustainable Growth: A Roadmap to Reduce Distortions in the Allocation of Resources and Talent in the Pakistani Economy.* World Bank.

World Bank. (2017). *Women and Tourism: Designing for Inclusion.* World Bank Group.

World Tourism Organization (2012). *Affiliate Members Global Report, Volume 4 - Global Report on Food Tourism.* UNWTo. doi:10.18111/9789284414819

World Tourism Organization (UNWTO). (2019). *Global Report on Women in Tourism 2019.* UNWTO.

World Tourism Organization. (2011). *Global report on women in tourism 2010.* UNWTO. https://www.unwto.org/centre-stage-project-womens-empowerment-during-the-covid19-recovery

World Tourism Organization. (2019). *UNWTO Tourism Definitions*. UNWTO. doi:10.18111/9789284420858

World Tourism Organization. (2019). *UNWTO Tourism Highlights* (2019 Edition). World Tourism Organization.

World Tourism Organization. (2020). *UNWTO World Tourism Barometer and Statistical Annex, November 2020*. https://www.e-unwto.org/doi/pdf/10.18111/wtobarometereng.2020.18.1.2

World Tourism Organization. (2023, March 23). *New Guidelines Put Women's Empowerment at Heart of Tourism's Restart*. Retrieved March 24, 2023, from https://www.unwto.org/news/new-guidelines-put-women-s-empowerment-at-heart-of-tourism-s-restart

World Tourism Organization. (n.d.-a). *Gastronomy and Wine Tourism*. Retrieved from https://www.unwto.org/gastronomy-wine-tourism

World Tourism Organization. (n.d.-b). *Mountain tourism*. Retrieved from https://www.unwto.org/mountain-tourism

World Tourism Organization. (n.d.-c). *Rural tourism*. Retrieved from https://www.unwto.org/rural-tourism

World Tourism Organization. (n.d.-d). *Sports tourism*. Retrieved from https://www.unwto.org/sport-tourism

Woyo, E., & Venganai, H. (2022). Gender and tourism. In *Encyclopedia of tourism management and marketing* (pp. 387–390). Edward Elgar Publishing. doi:10.4337/9781800377486.gender.and.tourism

WTO (World Tourism Organization). (2019). *Global report on women in tourism* (2nd ed.). UNWTO. doi:10.18111/9789284420384

Wu, J., Li, Y., & Zhang, D. (2019). Identifying women's entrepreneurial barriers and empowering female entrepreneurship worldwide: A fuzzy-set QCA approach. *The International Entrepreneurship and Management Journal*, *15*(3), 905–928. doi:10.100711365-019-00570-z

Xiang, Z., Du, Q., Ma, Y., & Fan, W. (2021). COVID-19 and the economy: Impacts on and implications for hotel and travel industries. *Journal of Hospitality & Tourism Research (Washington, D.C.)*, *45*(5), 704–717. doi:10.1177/1096348020969768

Xu, X., & Pratt, S. (2018). Social media influencers as endorsers to promote travel destinations: An application of self-congruence theory to the Chinese Generation Y. *Journal of Travel & Tourism Marketing*, *35*(7), 958–972. doi:10.1080/10548408.2018.1468851

Xu, H., & Gu, H. (2018). Gender and tourism development in China. *Journal of China Tourism Research*, *14*(4), 393–404. doi:10.1080/19388160.2018.1539426

Yachin, J. M. (2018). The 'customer journey': Learning from customers in tourism experience encounters. *Tourism Management Perspectives, 28*, 201-210. https://www.sciencedirect.com/science/article/pii/S2211973618300795 doi:10.1016/j.tmp.2018.09.002

Yadav, V., & Unni, J. (2016). Women entrepreneurship: Research review and future directions. *Journal of Global Entrepreneurship Research*, *6*(1), 12. Advance online publication. doi:10.118640497-016-0055-x

Yalçın, E. C., Bayram Arlı, N., Aytaç, S., Başol, O., & Aydemir, M. (2019). Türkiye'de emek piyasalarında cinsiyete dayalı ücret ayrımcılığı: IBBS düzey 1'e göre farklılıklar. *Kocaeli Üniversitesi Sosyal Bilimler Dergisi*, *2*(38), 293–314. doi:10.35343/kosbed.524255

Yanardağ, Ö., & Avcı, M. (2012). Turizm sektöründe istihdam sorunları: Marmaris, Fethiye, Bodrum ilçeleri üzerine ampirik bir inceleme. *Ege Stratejik Araştırmalar Dergisi*, *3*(2), 39–62. doi:10.18354/esam.81733

Yanes, A., Zielinski, S., Diaz Cano, M., & Kim, S.-I. (2019). Community-Based Tourism in Developing Countries: A Framework for Policy Evaluation. *Sustainability (Basel)*, *11*(9), 2506. doi:10.3390u11092506

Yang, E. C. L., & Khoo-Lattimore, C. (2018). Constructing space and self through risk taking: A case of Asian solo female travelers. *Journal of Travel …* doi:10.1177/0047287517692447

Yang, E.C.L. (2016). Risk perception of Asian solo female travelers: An autoethnographic approach. *Women and travel: Trends, journeys and experiences …* doi:10.1201/9781315365879-21

Yang, J., Zhang, D., Liu, X., Li, Z., & Liang, Y. (2022). Reflecting the convergence or divergence of Chinese outbound solo travellers based on the stimulus-organism-response model: A gender comparison perspective. *Tourism Management Perspectives.* https://www.sciencedirect.com/science/article/pii/S2211973622000472

Yang, L. X. (1993). *An analysis of R.O.C. nationals tourism consumer behavior – the case of the residents of Taipei* [Unpublished Master's thesis]. Department of Business Administration, National Sun Yat-Sen University, Kaohsiung.

Yang, E. C. L., Khoo-Lattimore, C., & Arcodia, C. (2017). A narrative review of Asian female travellers: Looking into the future through the past. *Current Issues in Tourism*, *20*(10), 1008–1027. doi:10.1080/13683500.2016.1208741

Yang, K. (2003). Methodological and theoretical issues on psychological traditionality and modernity research in an Asian society: In response to Kwang-Kuo Hwang and beyond. *Asian Journal of Social Psychology*, *6*(3), 263–285. doi:10.1046/j.1467-839X.2003.00126.x

Yaqoob, S. (2020). The Emerging trend women entrepreneurship in Pakistan. *Journal of Arts & Social Sciences*, *7*(2), 217–230. doi:10.46662/jass-vol7-iss2-2020(217-230)

Yarnall, J. (1998). Career anchors: Results of an organisational study in the UK. *Career Development International*, *3*(2), 56–61. doi:10.1108/13620439810207536

Yavas, U. (1990). Correlates of Vacation Travel: Some Empirical Evidence. *Journal of Professional Services Marketing*, *5*(2), 3–18. doi:10.1300/J090v05n02_02

Yazıcı Ayyıldız, A. (2020). COVİD-19 Salgın Döneminde Otellerin Pazarlama Stratejileri: Kuşadası Örneği [Marketing Strategies of Hotels During the COVID-19 Pandemic Period: The Case of Kusadasi]. *BMIJ*, *8*(3), 3328–3358. doi:10.15295/bmij.v8i3.1587

Yenipınar, U. (2005). Otel işletmelerinde ücretlendirme: İzmir ili 4-5 yıldızlı otel işletmelerinde bir analiz. *Dokuz Eylül Üniversitesi Sosyal Bilimler Enstitüsü Dergisi*, *7*(3), 148–176.

Yetiş, Ş. A., & Çalışkan, N. (2020). Female Employment in Tourism Sector: An Assessment of the Current Situation. *Manisa Celal Bayar University Journal of Social Sciences*, *18*(3), 105–119.

Yıldırım, A., & Şimşek, H. (2013). *Sosyal Bilimlerde Nitel Araştırma Yöntemleri.* Seçkin Yayıncılık.

Yıldırım, A., & Şimşek, H. (2021). Sosyal Bilimlerde Nitel Araştırma Yöntemleri [Qualitative Research Methods in Social Sciences]. Ankara: Seçkin Yayıncılık.

Yıldırım, F., & Gül, H. (2021). Uluslararası kalkınma politikaları ve kadınların güçlendirilmesi üzerine bir değerlendirme. *Manas Sosyal Araştırmalar Dergisi*, *10*(1), 679–695. doi:10.33206/mjss.760005

Yıldırım, G., & Özbek, Ö. (2019). Profesyonel Turist Rehberlerine Yönelik Cinsel Taciz ile ilgili Bir Araştırma. *Turist Rehberliği Dergisi*, *2*(2), 72–87. doi:10.34090/tured.591586

Yıldırım, G., Tarınç, A., & Kılınç, C. C. (2020). Women Entrepreneurship in Tourism: The Case of Turkey. *Journal of Tourism and Gastronomy Studies*, *8*(4), 2462–2477. doi:10.21325/jotags.2020.721

Yıldız, A. N., Kaya, M., & Bilir, N. (2011). *İşyerinde şiddet*. Türkiye İşçi Sendikaları Konfederasyonu.

Yıldız, S. (2017). Toplumsal cinsiyetin şirketlere yansımasında ortaya çıkan ayrımcılık kavramlarının ardılları üzerine bir model önerisi. *ESOGÜ Sosyal Bilimler Dergisi*, *18*(1), 121–138. doi:10.17494/ogusbd.330753

Yıldız, S. Y. (2022). Instagram Influencerlarının Cinsiyetine Göre Tüketici Profillerinin Belirlenmesi. *Cumhuriyet Üniversitesi İktisadi ve İdari Bilimler Dergisi*, *23*(4), 943–953. doi:10.37880/cumuiibf.1130303

Yıldız, S., Alhas, F., Sakal, Ö., & Yıldız, H. (2016). Cam uçurum: Kadın yöneticiler cam tavanı ne zaman aşar? *Ankara Üniversitesi SBF Dergisi*, *71*(4), 1119–1146. doi:10.1501/SBFder_0000002425

Yılmaz, E., Usta, J., Akyüz, A., & Özbay, F. (2018). Livelihoods and Coping Strategies of Syrian Women Refugees in Turkey: A Gender-Sensitive Analysis. *Journal of International Women's Studies*, *19*(4), 27–44.

Yılmaz, İ., & Akay, E. (2020). Gastronomi sektöründe kadınların durumunun değerlendirilmesi. *International Journal of Social and Economic Sciences*, *10*(2), 81–86.

Yılmaz, M., Sezerel, H., & Uzuner, Y. (2020). Sharing experiences and interpretation of experiences: A phenomenological research on Instagram influencers. *Current Issues in Tourism*, *23*(24), 3034–3041. Advance online publication. doi:10.1080/13683500.2020.1763270

Yirik, Ş., & Ilgaz Yıldırım, B. (2014a). Field Study of Relations Between Personal Values of Women Entrepreneurs In Tourism Sector and Their Perception of Risk and Uncertainty Perceptions: The Case of Antalya. *Journal of Niğde University IIBF*, *7*(2), 97–111.

Yirik, Ş., & Ilgaz Yıldırım, B. (2014b). Analysis of Personal Values of Women Entrepreneurs of Tourism Sector In Terms of Their Demographic Characteristics: The Case of Antalya. *Mustafa Kemal University Journal of Graduate School of Social Sciences*, *11*(26), 361–378.

Yoopetch, C. (2021). Women empowerment, attitude toward risk-taking and entrepreneurial intention in the hospitality industry. *International Journal of Culture, Tourism and Hospitality Research*, *15*(1), 59–76. doi:10.1108/IJCTHR-01-2020-0016

Yoo, W., Choi, D. H., & Park, K. (2016). The effects of SNS communication: How expressing and receiving information predict MERS-preventive behavioral intentions in South Korea. *Computers in Human Behavior*, *62*, 34–43. doi:10.1016/j.chb.2016.03.058 PMID:32288174

Yuan, S., Shen, X., Chen, W., & Zhan, W. (2021). Service innovation and customer satisfaction in family-owned hotels: The mediating role of customer experience. *Journal of Hospitality and Tourism Management*, *48*, 83–91. doi:10.1016/j.jhtm.2021.05.004

Yüksel, İ. (2005). İşsizlik olgusunun psikolojik boyutu: Görgül bir araştırma. *Ankara Üniversitesi SBF Dergisi*, *60*(3), 255–274.

Yumuk, Y., & İli, N. D. (2022). Turist Rehberliğinde Mesleki Yatkınlık ve Kariyer Sürdürme İlişkisinin Cinsiyet Değişkeni Açısından Değerlendirilmesi. *Nevşehir Hacı Bektaş Veli Üniversitesi SBE Dergisi*, *12*(3), 1314–1333. doi:10.30783/nevsosbilen.1106689

Yung, R., & Khoo-Lattimore, C. (2019). New realities: A systematic literature review on virtual reality and augmented reality in tourism research. *Current Issues in Tourism*, *22*(17), 2056–2081. doi:10.1080/13683500.2017.1417359

Zainol, N., Esa, M. M., Muhamad, N., & Mohamad, S. S. (2022). Factors Affecting The Success of Digital Entrepreneurship among Suburban Women: Malacca. *International Journal of Academic Research in Business & Social Sciences*, *12*(12), 153–163. doi:10.6007/IJARBSS/v12-i12/15426

Zamberi Ahmad, S. (2011). Evidence of the characteristics of women entrepreneurs in the Kingdom of Saudi Arabia: An empirical investigation. *International Journal of Gender and Entrepreneurship*, *3*(2), 123–143. doi:10.1108/17566261111140206

Zapata Campos, M. J., Hall, C. M., & Backlund, S. (2018). Can MNCs promote more inclusive tourism? Apollo tour operator's sustainability work. *Tourism Geographies*, *20*(4), 630–652. doi:10.1080/14616688.2018.1457074

Zel, U. (2002). İş arenasında kadın yöneticilerin algılanması ve kraliçe arı sendromu. *Amme Idaresi Dergisi*, *35*(2), 39–48.

Zengin, B., Erkol, G., & Eker, N. (2014). Rehberlik Mesleğinde Bir Engel Olarak Cinsiyet Ayrımcılığı: Bayan Turist Rehberleri Üzerine Bir Araştırma, 15. *Ulusal Turizm Kongresi*, *1144*, 1159.

Zeyen, A., Beckmann, M., Mueller, S., Dees, J. G., Khanin, D., Krueger, N., Murphy, P. J., Santos, F., Scarlata, M., Walske, J., & Zacharakis, A. (2013). Social entrepreneurship and broader theories: Shedding new light on the 'Bigger Picture'. *Journal of Social Entrepreneurship*, *4*(1), 88–107. doi:10.1080/19420676.2012.725422

Zhang, H. Q., & Chow, I. (2004). Application of Importance-Performance Model in Tour Guides' Performance: Evidence from Mainland Chinese Outbound Visitors in Hong Kong. *Tourism Management*, *25*(1), 81–91. doi:10.1016/S0261-5177(03)00064-5

Zhang, J., & Duan, Y. (2010). The impact of different types of market orientation on product innovation performance: Evidence from Chinese manufacturers. *Management Decision*, *48*(6), 849–867. doi:10.1108/00251741011053433

Zhang, K., Chen, Y., & Li, C. (2019). Discovering the tourists' behaviors and perceptions in a tourism destination by analyzing photos' visual content with a computer deep learning model: The Case of Beijing. *Tourism Management*, *75*, 595–608. doi:10.1016/j.tourman.2019.07.002

Zhang, L., & Gao, J. (2016). Exploring the effects of international tourism on China's economic growth, energy consumption and environmental pollution: Evidence from a regional panel analysis. *Renewable & Sustainable Energy Reviews*, *53*, 225–234. doi:10.1016/j.rser.2015.08.040

Zhang, Y., Duysters, G., & Cloodt, M. (2014). The role of entrepreneurship education as a predictor of university students' entrepreneurial intention. *The International Entrepreneurship and Management Journal*, *10*(3), 623–641. doi:10.100711365-012-0246-z

Zheng, Y., Goh, E. & Wen, J. (2020). The effects of misleading media reports about COVID-19 on Chinese tourists' mental health: a perspective article. *Anatolia: An International Journal of Tourism and Hospitality Research*. . doi:10.1080/13032917.2020.1747208

Zopiatis, A., Giannikis, S., & Theocharous, A. L. (2021). Food tourism and the post-COVID era: An exploratory study of the perceptions of Cypriot family-owned restaurants. *Journal of Culinary Science & Technology*, *19*(4), 366–387. doi:10.1080/15428052.2021.1906651

About the Contributors

Gül Erkol Bayram is currently an Associate Professor in the School of tourism and hotel management, department of tour guiding, University of Sinop, Sinop, Turkey. Dr. Erkol Bayram has worked as an internal trainer and teacher in the tour guiding arena. Her doctorate is in Tourism Management from the Sakarya University, Turkey, and she completed her dissertation research on Tour Guiding in Turkey. Her core subjects are Tourism, tour guiding, tourism policy and Planning, women studies. Erkol Bayram has also worked as a professional tour guide in the tourism sector. The editor has many book chapters in the international arena and published her books as an editor in the national arena.

Syed Haider Ali Shah is Associate Professor/Research Cell Coordinator at Business Studies Department, Bahria Business School, Bahria University Islamabad Pakistan. He Holds a PhD from University of Malaya, Malaysia. Its Malaysia Top University. His area of specialization is HR Practices, Leadership and Management with special research interest in HR Practices, Green HR Practices, Talent Management, Green HR Practices, Leadership Styles and CSR Practices and Pro-environmental behavior. Dr. Haider Shah is engaged in many academic activities and professional training in both nationally and internationally. Hs has been conducting trainings through different platform like Leadership Development Centre (LDC), Bahria University and CALWASS.

Muhammad Nawaz Tunio is an Assistant Professor, at the Department of Business Administration, University of Sufism and Modern Sciences, Bhitshah, Pakistan. Dr. Tunio is Ph.D. in the Doctoral track Entrepreneurship, Innovation, and Economic Development, Alpen Adria University, Klagenfurt, Austria on the government scholarship of the Higher Education Commission of Pakistan. He was awarded a one-month young scientist research fellowship for Kent State University, Ohio. His field of research interest is entrepreneurship, careers, youth development, and self-employment. He has few publications in the above-mentioned areas. He has presented papers in international conferences and attended research workshops.

* * *

Muhammad Imran Afzal is a PhD Scholar at Northeast Normal University Changchun, Jilin, China and he is also Research Associates Officer at Center of Sustainability Research and Consultancy, Pakistan. He has interest in Developmental Cognitive Educational Neuroscience, Developmental Educational Psychology, Behavioral management Psychology and also keenly future interest in Perspective Memory, creativity, and Caregivers of Cancer Patients Psychological well-Being.

Aisha Ahmed is a BBA student in MAJU.

Zeynep Alkan completed her undergraduate education in the Department of Visual Communication Design at Near East University's Faculty of Communication in 2014. In 2018, she graduated as the top-ranking student from the Faculty of Communication. In 2020, Alkan earned her master's degree in Media and Communication Studies from the Institute of Social Sciences at Near East University, obtaining expertise in the field. She began her doctoral studies in the same year and is currently continuing her thesis research in the Department of Media and Communication Studies. Additionally, Alkan works as a part-time lecturer in the Department of Public Relations and Advertising at Near East University, Cyprus. Her research areas primarily focus on social media, influencers, and influencer marketing.

Sumitha Anantharajah is working in the School of Hospitality, Tourism and Events, Faculty of Social Sciences and Leisure Management, Taylor's University, Subang Jaya, 47500, Malaysia.

Esra Anış is a PhD student of Tourism Management at Aydın Adnan Menderes University, Turkey. She completed her undergraduate education at Yaşar University, Izmir, Turkey, Department of Tourism Management (50% scholarship, 100% English language of instruction) in 2012. Between 2010 and 2022, she worked as a front desk receptionist, Spa receptionist, service member in the F&B department, and both full-time and part-time cashier and team member in international fast food business. In 2021, she completed his master's degree at Aydın Adnan Menderes University, Institute of Social Sciences, Department of Tourism Management, by completing his thesis on " The Effect of Brand Personality on Purchasing Motivations (Motives) in Restaurant Businesses: An Application in Franchise Operation."She has been studying for a doctorate in the same university and department since 2021. Her study subjects are Food & Beverage Management, Hospitality Management, Franchising, Marketing, Brand Management, Consumer Behaviors, Experiences, Attitudes and Perceptions, Purchasing Behaviors and Motivations, Organizational Cynicism, Cittaslow, Destination Marketing, Solo Travelers, Virtual Reality (VR) and Metaverse.

Manpreet Arora is Senior Assistant Professor of Management in the School of Commerce and Management Studies, Central University of Himachal Pradesh Dharamshala, India. With around twenty one years of teaching experience, she has varied interest areas. A Gold medalist at undergraduate and distinction holder at postgraduate level, she obtained her Ph.d in International Trade from Himachal Pradesh University, Shimla, India. She is an M Phil also from the same university.She is a throughout merit holder with meritorious ranks in 10th and 12th also. Her areas of research interest include Accounting and Finance, Strategic Management, Entrepreneurship, Qualitative Research, Case Study Development, Communication Skills and Microfinance. She has been guiding research at the doctoral level and has worked in the area of Microfinance. As an active seminarian, she has attended more than hundred seminars/ conferences, and has visited several universities and colleges to deliver invited talks on Finance, Strategic Management, Qualitative Research, Business Communication, Interpersonal skills, Entrepreneurship and Skill Development. She is a motivational speaker and conducts workshops on communication and motivation. Having published more than 28 papers in various journals of national and international repute (including SCOPUS, WOS and Category Journals), she has also worked as content developer of MHRD 'e-PG Pathshala' Project and OER's for IGNOU. She has written fifty book chapters in national as well as international books/handbooks/volumes published with Routledge, CABI, Apple

Academic Press, IGI, Taylor and Francis and latest with Springer Nature etc. With four edited books in her credit she is a persistent researcher in the field of Management. She is an active social worker also and is working towards the protecting the rights of women.

Seraynur Bektas is doing her master's degree at Afyon Kocatepe University.

Rimsha Brohi is a BBA student in Mohammad Ali Jinnah University.

Nimit Chowdhary is a Professor of Tourism at the Department of Tourism and Hospitality Management, Jamia Millia Islamia, India. He researches tourism, service/ experiences, and entrepreneurship, among other focuses. He is a recipient of many awards and fellowships. In addition, he has supervised 23 PhDs, authored fifteen books, edited nine, and contributed more than 225 papers and chapters.

Dönüş Çiçek earned her bachelor's degree at Anadolu University School of Tourism and Hotel Management (1997) and master's degree at Anadolu University Tourism and Hotel Management department (2000). She had started working as a lecturer at Balıkesir University Erdek Vocational School (1998). Subsequently she has started to teach at Anadolu University Eskişehir Vocational School (2004). She got her phD degree from Eskişehir Osmangazi University at Tourism Administration department (2017). She is still working as assistant professor at Eskişehir Vocational School.

Ülker Çolakoğlu is currently an Associate Professor in the Faculty of Tourism, University of Aydın Adnan Menderes, Aydın, Turkey. Her doctorate is in Tourism Management from the University of Dokuz Eylül, İzmir, Turkey. She completed her dissertation research on "Communication strategies between managers and employers in accommodation establishments." She focuses on the management in tourism businesses. She has many publications in her fields both national and international. She has been the thesis advisor of many students at master of science and doctoral levels.

Ruth Sabina Francis, Lecturer & Programme Director, School of Hospitality, Tourism and Events, Taylor's University, has 20 years of both Industry and teaching experience. She is well known in the hospitality Industry for her expertise in Rooms Division Management. She was awarded as the Young Women Achiever in the field of Hospitality for her teaching and learning practices for the year 2016 by the centre for women empowerment in India. As a PSMBB Certified Trainer (TTT) she has conducted various training programmes for the Hospitality Industry personnel in Malaysia. Being in the Industry for seven years she has gained a lot of exposure towards sustained quality management. During her stint as Housekeeper in Chennai she has trained many staff in the area of housekeeping. Her research expertise includes areas such as Human Resources, Emotional intelligence, Organizational Citizenship Behaviour in the Hospitality field. She has published articles in well-known journals and in Scopus indexed journals. She is an active reviewer for various journals namely e-review of Tourism Research a scopus indexed body.

Seher Gülenç received the bachelor's degree with highest honor in tourism guidance from Balıkesir University, Balıkesir, Türkiye, in 2009, and the master's and PhD degrees in tourism management from Balıkesir University in 2011 and from Anadolu University, Eskişehir, Türkiye, in 2019, respectively. She currently works at Anadolu University, Faculty of Business Administration, Department of Tourism

Management as a Lecturer Dr. Her current research interests include health tourism, medical tourism, sustainable tourism and tourism management.

Hawa Imran is an ambitious student working towards achieving my dreams and goals.

Kamran Jamshed is currently a Ph.D. aspirant and holds a Master's degree in Management Studies from the Business School of Bahria University Islamabad, Pakistan. His published work is on Green Leadership, Green Training, Green Process Innovation, Ethical Leadership, Work Performance, Perceived Organizational Politics, and Political Skills. His current research interests include small businesses, family-oriented businesses, family businesses hotels, travel, and tourism industry focusing on GHRM and sustainable development, Human Resource Development, Human trafficking and sustainability in the hospitality industry, strategic management, and organizational behavior and their implications in the industry. He worked in the telecom industry of Saudi Arabia under Mobily (Etisalat) for more than 7 years on different roles and overall, he is having the industrial experience of more than 10 years.

Havilah Johnson is pursuing MCOM in CHRIST (Deemed to be University).

Naveeda Katper is Assistant Professor at IBA, University of Sindh, Pakistan. She is PhD in Finance from Malaysia. She is expert in quantitative research methods.

Asad Khan is a BBA student in MAJU.

Erdem Korkmaz received the bachelor's degree from Tourism and Hotel Management department in 2002 and master's degree in 2005 from Anadolu University, Eskişehir, Turkey. He got his phD degree from Anadolu University Tourism Management department. He is working as assistant professor at Anadolu University Eskişehir Vocational School.

Maximiliano E. Korstanje is editor in chief of International Journal of Safety and Security in Tourism (UP Argentina) and Editor in Chief Emeritus of International Journal of Cyber Warfare and Terrorism (IGI-Global US). Korstanje is Senior Researchers in the Department of Economics at University of Palermo, Argentina. In 2015 he was awarded as Visiting Research Fellow at School of Sociology and Social Policy, University of Leeds, UK and the University of La Habana Cuba. In 2017 is elected as Foreign Faculty Member of AMIT, Mexican Academy in the study of Tourism, which is the most prominent institutions dedicated to tourism research in Mexico. He had a vast experience in editorial projects working as advisory member of Elsevier, Routledge, Springer, IGI global and Cambridge Scholar publishing. Korstanje had visited and given seminars in many important universities worldwide. He has also recently been selected to take part of the 2018 Albert Nelson Marquis Lifetime Achievement Award. a great distinction given by Marquis Who´s Who in the World.

Sweety Mishra is a Junior Research Fellow with the Department of Tourism and Hospitality Management in Jamia Millia Islamia, Delhi. She is graduated with a Bachelor of Computer Application from Jaipur National University and completed her Postgraduation in Master of Travel and Tourism Management from Banaras Hindu University. She has worked in the tourism industry for one year. Her research interest areas are -Entrepreneurship, Women Entrepreneurship, Gender Studies, Women Empowerment, and Tourism Mindfulness. She has authored two books for CBSE's vocational subject on tourism.

Harun Özbey received his undergraduate education in tourism guidance at Sinop University, School of Tourism Management and Hotel Management. He is still continuing his master's degree in tourism guidance at Afyon Kocatepe University.

Muhammad Omer Rafique is an experienced researcher and trainer in the field of Islamic finance and the Halal industry. He is serving several international organizations as a Shariah Advisor. He has done his PhD in Islamic finance from University of Malaya, Kuala Lumpur.

Helena Belchior Rocha has a PhD in Social Work, is an Assistant professor at ISCTE- University Institute of Lisbon in the Department of Political Science and Public Policies and deputy director of the Transversal Skills Laboratory. Integrated researcher at CIES, Centre for Research and Studies in Sociology, linked to national and international research projects, namely 2 from Marie Curie Actions. Author of papers and communications at national and international congresses, in the areas of social work theory and methodology, environment, sustainability, community Intervention, ethics, human rights, social policies and Well-being, education and soft skills. Member of the Editorial Board of national/ international journals.

Melike Sak is completed undergraduate education in Sinop University, School of Tourism and Hotel Management, Tourist Guidance, the author is doing her master's degree in Selcuk University, Institute of Social Sciences, Travel Management and Tour Guiding Department. The author, has presented national and international papers in various tourism fields. The author has an international book chapter, national and international papers and articles. The author has various studies: sustainable tourism, health tourism, cultural heritage tourism, ecotourism, event tourism, sustainable tourism, event tourism and tourism guidance and she has participated in national and international congresses. The author has participated in national and international congresses.

Muhammad Furquan Saleem has been working an Assistant Manager Research at Research Facilitation Unit, Iqra University for the last 12 years. Previously he was affiliated with Gallup Pakistan (a leading marketing research organization). He is also doing Ph.D. with a major in Marketing from Karachi University Business School (KUBS). He completed his MPhil in Marketing from Iqra University and his MSc in Statistics from the University of Karachi.

Hatice Sari Gök is currently Assistant Professor in Yalvaç Vocational School, Departman of Travel, Tourism and Entertainment Services, Isparta University of Applied Sciences, Isparta, Turkey. She completed her doctorate in the joint program of Selçuk University and Gazi University. She completed her master's and undergraduate studies at Gazi University. She received her language education in Travel Management and Tourist Guidance Teaching program. She has also worked in the reservation department of the hotel business in the tourism sector, in the front office department. Sarı Gök has many academic publications (articles, conference proceedings, book chapters) related to tourism. His research areas include community-based tourism, destination management, tourism marketing, tourism types, gastronomy. He also serves on the referee board of several national and international journals.

Ferhan Sengur currently is a member and the Head of Department of Department of Aviation Management, Faculty of Aeronautics and Astronautics, Eskisehir Technical University, Turkey. She earned her PhD in Aviation Management and completed her Master and Bachelor degrees in Aviation Management as well. She conducted her post-doctoral research at Embry Riddle Aeronautical University, Florida, USA on a grant from the Turkish Higher Education Council in 2010. Currently she also holds the position of the Director of the Women in Science and Technology Center at Eskisehir Technical University. As a Professor at the Department of Aviation Management, Faculty of Aeronautics and Astronautics, Eskisehir Technical University, she had gained valuable experience in teaching both graduate and undergraduate levels. She had also conducted multiple international and national research projects, organized several conferences, delivered papers on national and international aviation and management symposiums, and published articles in national and international peer-reviewed journals. She also authored several chapters and books. Prof. Sengur had also consulted the Turkish State Planning Organization, Turkish Transportation Ministry and Turkish Directorate General of Civil Aviation for several National Aviation legislations and White Papers. Prof. Sengur is an expert in Airport and Airline Management and her research interests incorporate but not limited to Strategic Management, Transportation Policies, and Corporate Social responsibility. Her current research interest focuses on Airport Governance Models, Airport Public Private Partnerships, Airline Business models, and Responsible Management in Aviation.

Erum Shaikh is PhD in CSR from University of Sindh, Jamshoro, Pakistan.

Amrik Singh is working as Professor and Head in the School of Hotel Management and Tourism at Lovely Professional University, Punjab, India. He obtained his Ph.D. degree in Hotel Management from Kurukshetra University, Kurukshetra. He started his academic career at Lovely Professional University, Punjab, India in the year 2007. He has published more than 30 research papers in UGC and peer-reviewed journals and 10 publications (Scopus/Web of Science). He has published 12 patents and 03 patent has been granted in the inter-disciplinary domain and 8 copyrights granted by Registrar Govt. of India. He has authored 2 edited books and 1 textbook. Dr. Amrik Singh participated and acted as a resource person in various national and international conferences, seminars, research workshops, and industry talks. His area of research interest is Accommodation Management, Hotel Ergonomics, Hotel green practices, Work life practices, Talent Management practices, Waste management, AR VR in hospitality, Food Waste, Training and development, and Sustainability in tourism and hospitality. He is currently guiding 8 Ph.D. scholars and 2 Ph.D. scholars have been awarded Ph.D. He is also "Best Teacher Awardee" by Lovely Professional University (Twice) and Indian Hospitality Congress (Once). He has been invited to speak in seminars, conferences, and TV/Radio talk shows. He has been a part of the final Ph.D. thesis evaluation and External expert for a Ph.D. viva for other universities.

Selen Subaşı is a Ph.D. candidate at the Doctoral School of Education at the University of Szeged. Her research interests encompass lifelong learning, non-formal education, women's education, and the education of disadvantaged and vulnerable communities, including minorities and refugees. Her Master's thesis explored the educational opportunities provided to members of the Roma minority in Hungary. Currently, she is working on her thesis focusing on the education and well-being of Syrian refugees in Turkey. Selen is committed to addressing the challenges faced by marginalized populations and contributing to the advancement of inclusive and equitable education.

Khursheed Ahmad Wani is Assistant Professor in the Department of Environmental Science, Government Degree College Thidim Kreeri, J&K. Dr. Wani served at ITM University Gwalior as Assistant Professor in the Department of Environmental Science from 2011 to 2017. He has completed his M. Sc. and Ph. D. in Environmental Science from Jiwaji University, Gwalior, Madhya Pradesh. He has also qualified Jammu and Kashmir State Level Eligibility Test (JKSET). His areas of interest include Environmental management, Environmental Impact Assessment and Occupational Health. He has published research papers in various national and international journals and is the author of three books on Environmental Science. He is a member of several national and international organizations and is in the editorial board of various journals. He has organized different programmes on Environmental Awareness and has completed research project sponsored by Madhya Pradesh Council of Science and Technology, Bhopal, India. Dr. Wani has organized various national and international conferences/ seminars/ workshops as well.

Ahu Yazıcı Ayyıldız is currently Associate Professor in Faculty of Turism, University of Aydın Adnan Menderes, Aydın, Turkey. Her doctorate is in Tourism Management from the Aydın Adnan Menderes University, Turkey and she completed her dissertation research on private label usage and applications in retailing. Her core subjects are accommodation businesses and tourism marketing. She has conducted research on Marketing, Service Marketing and Customer Relationship Management within the Basic Area of Social, Humanities and Administrative Sciences. The author Dr.Ayyıldız has many book chapters and articles in national and international arena. Dr Yazıcı Ayyıldız has been invited by many universities and tourism businesses for many speeches / conferences / panel discussions and has also advised many graduate and doctoral students and continues her studies meticulously.

Serdar Yener currently works as management professor in academy. He has more than 25 years experience in public management, academic research, project management and as an air defense officer in Turkish Military. He has coordinated 3 international projects (EU, UN). He has more than 50 publications. He is working as head of dpt. And coordinates virtual courses and related activities in dpt. Serdar YENER can conduct critical thinking, project management, problem solving, soft skills researches and courses in defined domains. Also he can conduct other field practices, project management, financial management, etc.

Nuray Yıldız is a Lecturer at Cappadocia University/Mustafapaşa/Ürgüp.

Özcan Zorlu is the co-head of Tourism Guidance department at faculty of Tourism at Afyon Kocatepe University. He received his bachelor's degree inhospitality and tourism management at Balıkesir University and his master's degree in tourism management from Balıkesir University. He has received his associate professorship in tourism and written widely on specifically organizational behavior, knowledge management and alternative tourism activities in tourism.

Index

F

G

H

I

K

L

M

O

P

R

S

T

U

V

W

Recommended Reference Books

IGI Global's reference books are available in three unique pricing formats: Print Only, E-Book Only, or Print + E-Book.

Order direct through IGI Global's Online Bookstore at **www.igi-global.com** or through your preferred provider.

ISBN: 9781799887096
EISBN: 9781799887119
© 2022; 413 pp.
List Price: US$ **250**

ISBN: 9781799874157
EISBN: 9781799874164
© 2022; 334 pp.
List Price: US$ **240**

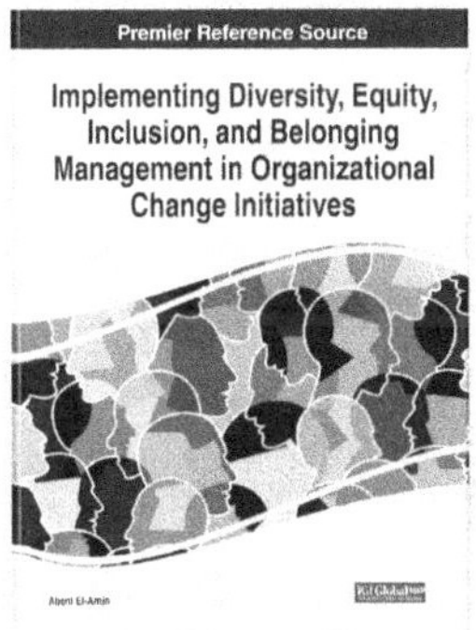

ISBN: 9781668440230
EISBN: 9781668440254
© 2022; 320 pp.
List Price: US$ **215**

ISBN: 9781799889502
EISBN: 9781799889526
© 2022; 263 pp.
List Price: US$ **240**

ISBN: 9781799885283
EISBN: 9781799885306
© 2022; 587 pp.
List Price: US$ **360**

ISBN: 9781668455906
EISBN: 9781668455913
© 2022; 2,235 pp.
List Price: US$ **1,865**

Do you want to stay current on the latest research trends, product announcements, news, and special offers?
Join IGI Global's mailing list to receive customized recommendations, exclusive discounts, and more.
Sign up at: **www.igi-global.com/newsletters.**

Publisher of Timely, Peer-Reviewed Inclusive Research Since 1988

IGI Global
PUBLISHER of TIMELY KNOWLEDGE

www.igi-global.com Sign up at www.igi-global.com/newsletters facebook.com/igiglobal twitter.com/igiglobal linkedin.com/igiglobal

www.ingramcontent.com/pod-product-compliance
Lightning Source LLC
LaVergne TN
LVHW080111011026
849160LV00013B/1072